T0185553

OXYGEN TRANSPORT
TO TISSUE X

ADVANCES IN EXPERIMENTAL MEDICINE AND BIOLOGY

OXYGEN TRANSPORT TO TISSUE X

Edited by

Masaji Mochizuki
Nishimaruyama Hospital
Sapporo, Japan

Carl R. Honig
The University of Rochester
Rochester, New York

Tomiyasu Koyama
Hokkaido University
Sapporo, Japan

Thomas K. Goldstick
Northwestern University
Evanston, Illinois

and

Duane F. Bruley
California Polytechnic State University
San Luis Obispo, California

PLENUM PRESS • NEW YORK AND LONDON

Library of Congress Cataloging in Publication Data

International Society on Oxygen Transport to Tissue. Meeting (15th: 1987: Sapporo-shi, Japan)
Oxygen transport to tissue X.

(Advances in experimental medicine and biology; v. 222)
"Proceedings of the 15th Annual Meeting of the International Society on Oxygen Transport to Tissue, held July 22-24, 1987, at Hokkaido University, Sapporo, Japan" — T.p. verso.
Includes bibliographies and indexes.
1. Tissue respiration — Congresses. 2. Oxygen transport (Physiology) — Congresses. 3. Microcirculation — Congresses. I. Mochizuki, Masaji, date. D. II. Title. III. Title: Oxygen transport to tissue 10. IV. Title: Oxygen transport to tissue ten. V. Series. [DNLM: 1. Biological Transport — congresses. 2. Oxygen — blood — congresses. 3. Oxygen Consumption — congresses. W1 AD559 v.222 / QV 312 I612 1987o]
QP177.I56   1987   574.1'2   87-32879

ISBN-13: 978-1-4615-9512-0   e-ISBN-13: 978-1-4615-9510-6
DOI: 10.1007/978-1-4615-9510-6

Proceedings of the 15th Annual Meeting of the International Society on Oxygen Transport to Tissue, held July 22-24, 1987, at Hokkaido University, Sapporo, Japan

© 1988 Plenum Press, New York
A Division of Plenum Publishing Corporation
233 Spring Street, New York, N.Y. 10013
Softcover reprint of the hardcover 1st edition 1988

INTERNATIONAL SOCIETY ON OXYGEN TRANSPORT TO TISSUE 1986-87

Officers:

President: M. Mochizuki, Sapporo, Japan
Co-President: C.R. Honig, Rochester, NY, USA
Past President: I.A. Silver, Bristol, UK
President-Elect: K. Rakusan, Ottawa, Canada
Secretary: N.S. Faithfull, Manchester, UK
Treasurer: J. Grote, Bonn, W. Germany

Executive Committee:

A. Eke, Budapest, Hungary
T.E.J. Gayeski, Rochester, NY, USA
T.K. Goldstick, Evanston, IL, USA
T. Koyama, Sapporo, Japan
J.C. LaManna, Cleveland, OH, USA
D.W. Luebbers, Dortmund, W. Germany
W. Mueller-Klieser, Mainz, W. Germany
E.M. Nemoto, Pittsburgh, PA, USA
D.F. Wilson, Philadelphia, PA, USA

SAPPORO MEETING, JULY 22-24, 1987

President: M. Mochizuki
Secretary General: T. Koyama

Local Organizing Committee:

T. Araiso
Y. Enoki
Y. Kakiuchi
A. Kamiya
Y. Kawakami
M. Kinko
T. Koyama
S. Makinoda
H. Niimi
T. Takahashi
M. Tamura
H. Tazawa
I. Yamazaki
J. Yasuda

PREFACE

The International Society on Oxygen Transport to Tissue (ISOTT) was founded in 1973 "to facilitate the exchange of scientific information among those interested in any aspect of the transport and/or utilization of oxygen in tissues". Its members span virtually all disciplines, extending from various branches of clinical medicine such as anesthesiology, ophthalmology and surgery through the basic medical sciences of physiology and biochemistry to the physical sciences and engineering.

The fifteenth annual meeting of ISOTT was held in 1987 for three days, from July 22 to 24, at Hokkaido University in Sapporo, Japan. Previously, all ISOTT meetings had been held in Europe or the USA alternatively. This time, however, the meeting was held for the first time in an Asian country. When we first started preparing for this meeting some of our members were afraid that the number of those attending would not exceed 30. Fortunately the results were quite different. We had more than 60 participants from abroad and an even greater number from Japan. In addition to three special lectures and two symposia there were a total of 88 posters presented over the three days of the meeting. These covered all aspects of physiological oxygen transport including convection, diffusion, chemical reaction, and control of oxygen demand in blood and various tissues as well as the methods, models and instrumentation for their study. The 92 papers which comprise this volume encompass all of these areas. This volume also contains a much larger contribution from Asian investigators than any previous volume.

To have this volume available quickly, the editors have opted for rapid publication consistent with having all papers of a high quality. The time constraint has made it imperative that almost all preparation and correction of the camera-ready manuscripts, although under the direction of the editors, be left to the authors themselves. We greatly appreciate their cooperation. In addition, we also wish to express our profound gratitude to Ann, Marian, and Ian A. Silver, of the United Kingdom, for their invaluable help in correcting and revising manuscripts in Sapporo. The editors are also grateful for the assistance in proofreading and typing by Charlene K. Massey and Rod D. Braun in Evanston.

For the editors

Masaji Mochizuki
Tomiyasu Koyama
Thomas K. Goldstick

September, 1987

ACKNOWLEDGEMENTS

We gratefully acknowledge the promotional assistance and financial support for the 1987 ISOTT meeting as well as for the preparation of this volume which was received from the following:

Commemorative Association for the Japan World Exposition (1970)

Tokyo Iyakuhinkogyo Kyokai
Osaka Iyakuhin Kyokai
Naito-Kinen Kagaku Shinko Zaidan
Sankyo Seimeikagaku Kenkyu Shinko Zaidan
Shimadzu Kagakugijutsu Shinko Zaidan
Akiyama-Kinen Seimeikagaku Shinko Zaidan
Hokkaido Ishi-kai
Hokkaido
City of Sapporo

CONTENTS

MATHEMATICAL MODELS

CARDIOVASCULAR SYSTEM

RESPIRATORY SYSTEM

OTHER ORGANS AND TISSUES

MATHEMATICAL
MODELS

COUPLING OF HEMODYNAMICS TO DIFFUSIONAL OXYGEN MASS TRANSPORT

K. GROEBE[1]

Dept. of Mechan. Engin., University of Rochester
Rochester, NY 14627, USA

INTRODUCTION

Maximally working skeletal muscle exhibits one of the highest O_2 consumption rates observed in any tissue. In order to understand the mechanisms which bring about these enormous oxygen fluxes, two key questions have to be addressed:

1. How can diffusional fluxes of that magnitude from the red blood cell into the muscle fiber be achieved in spite of unfavourably long diffusion paths?

2. How is it possible that red cell transit times through the capillary bed are sufficiently long to discharge enough oxygen even though blood flow values are 25-fold compared to rest and the capillary bed is significantly de-recruited due to raised tissue pressures.

The first step to take is to understand oxygen diffusion inside the muscle cell subject to certain boundary conditions on the sarcolemma (Fig. 1, left panel). Based on cryophotometrical P_{O_2} measurements in maximally working skeletal muscle by

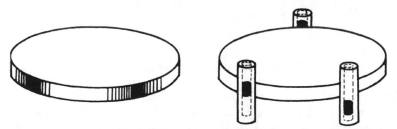

Figure 1: Schematic diagram of one-phase (left) and two-phase (right) O_2-diffusion model of the muscle fiber. In the one-phase model the boundary P_{O_2} is specified at the sarcolemma. The two-phase model includes capillaries and a diffusion layer surrounding the fiber. The boundary P_{O_2} is specified at the red cell membrane.

[1]Supported by Deutsche Forschungsgemeinschaft grant Gr 887/1–1 and by NIH grant HL37205

GAYESKI and HONIG [5–8,14], most recently FEDERSPIEL [2] demonstrated that the flat P_{O_2} profiles found in these experiments are well compatible with classical diffusion theory. One-phase models of the type used by FEDERSPIEL, however, fail to give any information on the absolute value of the P_{O_2} inside the muscle cell as that one is determined by the magnitudes of the red cell P_{O_2} inside the supplying capillary and of the P_{O_2} drop across the surrounding plasma sleeve, across the endothelium, and across the interstitial space.

Therefore, a further step is required (Fig. 1, right panel). Absolute values of the P_{O_2} at the red cell membrane can be inferred from calculations of red cell O_2 unloading under conditions of muscle capillaries. These may be employed to specify the boundary conditions in an extended, two-phase muscle fiber model which accounts for diffusion not only inside the muscle cell but also in the adjacent diffusion layer. A first approach in this direction was taken by GROEBE and THEWS who developed a 3-dimensional hybrid model of the muscle fiber and the surrounding capillaries [9]. The resulting P_{O_2} profiles in cross sections through muscle fiber and diffusion layer (Fig. 2) exhibit a steep P_{O_2} drop across the layer which, consequently, acts as a functional O_2 "diffusion barrier". Only slight variations in intracellular P_{O_2} at low levels of 0–3 $mm\,Hg$ were computed.

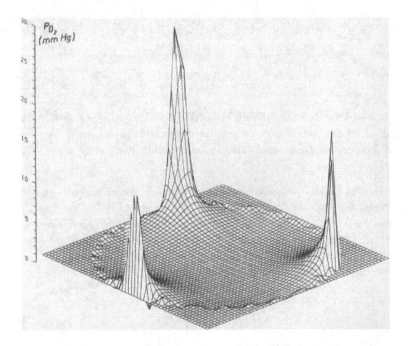

Figure 2: P_{O_2} distribution in a cross section through muscle fiber and diffusion layer which are surrounded by 3 capillaries as calculated by the two-phase diffusion model [9]. Note the steep P_{O_2} drops across the diffusion layer contrasting the merely trivial variations in fiber P_{O_2} at low levels of 0–3 $mm\,Hg$.

On the other hand and leading over to the second question addressed, mean red cell O_2 unloading times specify the time requirements which have to be met by the microcirculation to allow for sufficient oxygen delivery to the tissue. The present contribution will assess the magnitude of mean unloading time – which is considerably longer than the desaturation time of a single erythrocyte – and compare it to the mean red cell transit time through the capillary bed. The methods presented do not require any input data from microvascular in vivo observations. Therefore, they also allow estimates on transit time distributions in experimental situations where such observations are impracticable, e.g., in bulk muscles and at high performance.

METHODS

The individual steps of the analysis to follow are summarized in the flow chart of Fig. 3. We are going to deal with a number of probability distributions, the first of which will be the distribution of flow path length. From that we are going to work up our way to the distributions of blood flow, of red cell flow, and finally of red cell capillary transit time, the latter one of which will give us the parameters mean, standard deviation and skewness of transit time. By non-dimensionalizing the transit time distribution, we can apply it as a weight function to time courses of red cell O_2 unloading and find the mean O_2 desaturation time which ideally should be identical with mean transit time.

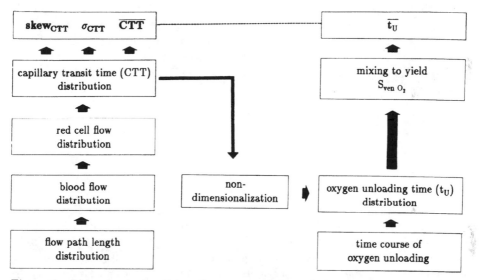

Figure 3: Flow chart visualizing the course of the calculations performed. Left: Distributions calculated to yield the distribution of capillary transit time (CTT) and its parameters mean ($\overline{CTT}$), standard deviation (σ_{CTT}), and skewness ($\mathbf{skew}_{CTT}$). Right: Application of the dimensionless transit time distribution to a time course of red cell O_2 unloading resulting in the mean unloading time $\overline{t_U}$.

In working skeletal muscle, active vasomotion is ruled out so that only passive mechanisms can cause capillary transit time heterogeneities. These are determined by variations in the lengths of the capillary pathways, leading to variable flow resistances and hence varying red cell velocities and transit times. In order to study these variations, we consider a network consisting of a number of parallel pathways which connect arteriole and venule. The lengths of the pathways have been determined to be near Γ-distributed with a standard deviation of approximately 0.4 times the mean length [10] (Fig. 4).

For each flow path, the same average inflow and outflow conditions are assumed (i.e., same pressures at origins and endings, same branching geometry at origins, same

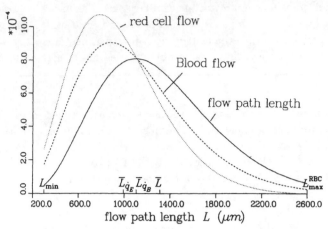

Figure 4: Densities of the distributions of flow path length (solid), of blood flow (dashed), and of red cell flow (dotted) for dog gracilis muscle at maximum performance. The respective mean values are annotated as $\overline{L}$, $\overline{L}_{\dot{q}_B}$, and $\overline{L}_{\dot{q}_E}$. L_{min} and L_{max}^{RBC} are the lengths of the longest and of the shortest red cell perfused paths, respectively. Note that the paths conducting the bulk of the red cell flow are markedly shorter than those conducting the bulk of the blood flow and these latter ones are again shorter than the average flow path. This is an effect of the inverse relation between path length and blood flow and of the nonlinear red cell distribution function.

composition of inflowing arteriolar blood). Differences between individual bifurcations are restricted to the lengths of the capillary branches. This hypothesis does not mean to say that the differences in feeding and draining conditions are negligible. Rather, it understands that it should be possible to describe the flows etc., averaged over all flow paths of a common length, in terms of these mean conditions. Capillary flow resistance is taken to be proportional to path length and to be independent of the ratio of capillary red cell flow to blood flow. This assumption is justified by measurements of resistance to flow through narrow glas capillaries which resulted in only minor differences in apparent viscosity of plasma and whole blood [3,4]. The blood flow through a single flow path is then in inverse proportion to flow path length.

Considering the distribution of red cell flow rate, we have to account for the fact that the red cells are not distributed among the capillaries in proportion to the respective blood flow values. Rather, the capillaries exhibiting higher blood flow rates obtain a disproportionately large share of red cells. SCHMID-SCHÖNBEIN et. al. [17] have studied red cell distributions at microvascular bifurcations in the rabbit's ear chamber, finding the red cell fraction Φ which enters a branch as a function of the blood flow fraction Ψ into this branch. Fig. 5 displays a typical example of their "fraction related" red cell distribution functions. The heavy line shows the type of functions employed in our calculations. Note that for very small blood flows which will occur in branches with the highest flow resistances, hence in the longest flow

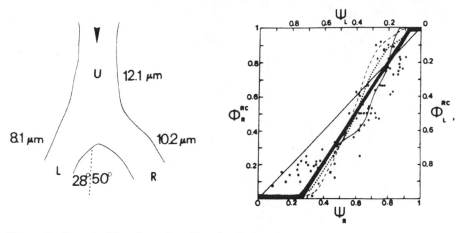

Figure 5: Sample "fraction related" red cell distribution function at a capillary bifurcation. The graph shows the red cell fraction Φ entering a branch as a function of the blood fraction Ψ into this branch. From SCHMID-SCHÖNBEIN et. al. [17]. Overlaid is an affine-linear distribution function (heavy line) as an example of the type of functions which were employed in the present calculations.

paths, no red cells at all enter the capillary branch. Fig. 6 shows non-dimensional blood flow and red cell flow through a single flow path as functions of path length. The relation of the two curves reflects the non-linear red cell distribution function. Note that red cell flow vanishes in the very longest flow paths.

The measurements by SCHMID-SCHÖNBEIN et. al. [17] indicate a fairly wide range of possible shapes of their distribution functions. On the other hand, the results depend sensitively upon the shape chosen, and there are no data available which choice might be adequate. A way out of this dilemma is to substitute the parameters of the distribution function by two other quantities. If we use the mean number of capillaries on an arteriole and the length of the longest red cell perfused flow path L_{max}^{RBC} as independent variables, resulting transit time distributions are remarkably stable for a wide range of parameter values.

In Fig. 4 the probability densities of the distributions discussed so far are plotted. The solid line shows the distribution of flow path length, dashed means blood flow and dotted red cell flow. The respective mean values are annotated as $\overline{L}$, $\overline{L}_{\dot{q}_B}$, and $\overline{L}_{\dot{q}_E}$. Note that the flow paths conducting the bulk of the red cell flow are markedly shorter than those conducting the bulk of the blood flow and these latter ones are again shorter than the average path. This is an effect of the inverse relation between path length and blood flow and of the nonlinear red cell distribution function.

A key problem in the determination of red cell transit times is how to find the ratio of red cell velocity to mean whole blood velocity, which is known to be highly variable with blood flow rate. It was first shown by FÅHRÆUS [1] that this velocity ratio is equal to the ratio of discharge hematocrit to capillary hematocrit, the latter of which is in proportion to the number of red cells per unit of flow path length. Based on this number and on measured functional capillary density, the above distributions can be used to calculate the functional dependence of a number of microrheological parameters upon flow path lengths. Fig. 7 displays scaled graphs of the discharge hematocrit (solid), of intererythrocytic gap length (dashed),   of the number of red

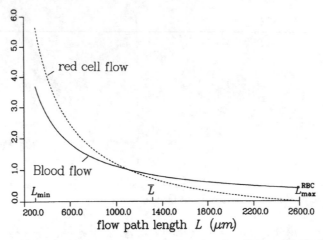

Figure 6: Non-dimensional blood flow (solid) and red cell flow (dashed) through a single flow path as functions of path length L. For L_{min}, $\overline{L}$, and L_{max}^{RBC} see Fig. 4. Note that blood flow is in inverse proportion to path length. The relation between red cell flow and blood flow reflects the non-linear red cell distribution function.

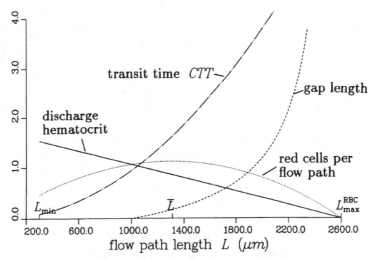

Figure 7: Normalized discharge hematocrit (solid), intererythrocytic gap length (dashed), number of red cells contained in a flow path (dotted), and red cell transit time CTT (dashdotted) as functions of path length L. For L_{min}, $\overline{L}$, and L_{max}^{RBC} see Fig. 4. $CTT(L)$ is proportional to the square of L, by this effecting a much broader transit time distribution compared to the path length distribution.

8

cells contained in a flow path (dotted), and of red cell transit time (dashdotted). As expected, the hematocrit decreases to 0 and gap length increases towards $L_{\max}^{\text{RBC}}$ with increasing path length. Transit time is in proportion to the square of path length, by this effecting a much broader transit time distribution compared to the path length distribution. Note that the number of red cells per flow path is small or 0 for the shortest and for the longest paths and exhibits a maximum near the mean path length.

From the transit time as a function of flow path length it is only a small step to the probability density of the red cell transit time distribution which is shown in Fig. 8. Compared to the distributions of flow path lengths or flows, the transit time distribution is about twice as broad and twice as much skewed. This matches well with the respective relations which have been established by direct in vivo measurements of red cell fluxes and transit times in resting and vasodilated two-dimensional skeletal muscles [16].

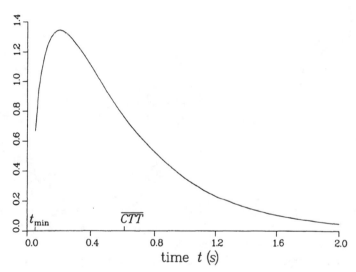

time t (s)

Figure 8: Distribution of red cell capillary transit time CTT in dog gracilis muscle at maximum performance. $t_{\min}$ represents the lower limit of transit times. The parameters of the distribution are: mean transit time $\overline{CTT}$=614 ms, standard deviation σ_{CTT}=495 ms, coefficient of variation v_{CTT}=0.65, skewness S_{CTT}=1.69. Note that variation coefficient and skewness of the path length distribution are only 0.16 and 0.79, respectively. This indicates a considerable broadening and loss of symmetry of the transit time distribution compared to the underlying length distribution which is due to the quadratic dependence of transit time upon path length. Qualitatively, the same relations have been observed in direct measurements of red cell fluxes and transit times in resting and vasodilated two-dimensional skeletal muscles [16].

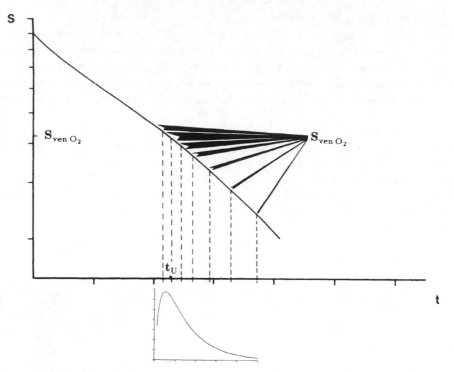

Figure 9: Schematic diagram visualizing the concept of "mean O_2 unloading time". Solid curve: Time course of red cell O_2 unloading under conditions of heavily working muscle. Abscissa: Time t. Ordinate: Mean red cell O_2 saturation S. $S_{ven O_2} =$ measured venous O_2 saturation, $t_U = O_2$ unloading time of a single red cell. The non-dimensional transit time distribution (lower panel) is employed to specify the "recipe" according to which blood at different stages of O_2 unloading is mixed (arrows) to form the venous blood.

Now we turn to the second branch of the flow chart (Fig. 3) and start dealing with red cell O_2 desaturation. To understand how the distribution of transit times relates to oxygen unloading, consider the time course of red cell O_2 desaturation under conditions of the maximally working skeletal muscle in Fig. 9 [9]. "O_2 unloading time", denoted by t_U, generally is understood to be the time an erythrocyte takes to release its oxygen, starting at some specified arterial O_2 saturation and finishing at the venous saturation $S_{ven O_2}$. In real tissue, however, the red cells pass down faster in some capillary pathways and slower in others, consequently unloading less or more oxygen, respectively. The only condition they satisfy is that, after mixing, the blood exhibits the measured venous O_2 saturation. The bearing of these heterogeneities may be quantified by forming the dimensionless counterpart of the transit time density (bottom panel in Fig. 9) and using it as a "recipe" according to which blood at different stages of O_2 unloading is to be mixed to form the venous blood (arrows of different widths in Fig. 9). Now, the mean desaturation time can be determined in such a way that the O_2 saturation of the "computer-mixed" venous blood coincides with the measured venous saturation.

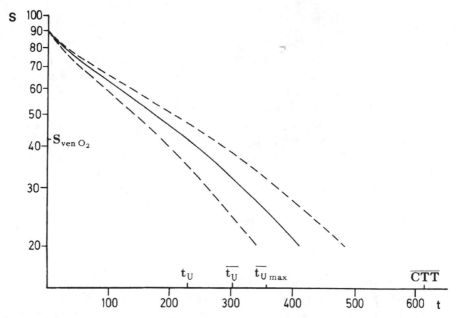

Figure 10: Time course of red cell O_2 unloading under conditions of heavily working muscle. Abscissa: Time t in ms. Ordinate: Mean red cell O_2 saturation S. Dashed lines indicate upper and lower bounds due to uncertainties in the employed data. $S_{ven\,O_2}$ = measured venous O_2 saturation, t_U = O_2 unloading time of a single red cell, $\overline{t_U}$ = mean O_2 unloading time, $\overline{t_{U\,max}}$ = upper bound for O_2 unloading time, $\overline{CTT}$ = mean capillary transit time.

RESULTS

Calculations were performed for dog gracilis muscles at maximum oxygen uptake. The data are from a paper by HONIG and ODOROFF [12] or are unpublished observations by HONIG. Functional capillary density and intracapillary red cell spacing are estimated from samples of freeze-clamped working muscles. Fig. 10 shows the time course of Fig. 9 and, furthermore, includes the summarized results of the present calculations. Dashed lines are the expected lower and upper limits of the time course which account for uncertainties in the employed data. Calculated mean O_2 unloading time $\overline{t_U}$ is 302 ms and by that 31 % larger than the desaturation time of a single erythrocyte. On the other hand, mean capillary transit time $\overline{CTT}$ calculated from the present model is 614 ms and exceeds the upper time limit $\overline{t_{U\,max}}$ expected for O_2 unloading by more than 250 ms. This may be a consequence of an underestimation of the extraerythrocytic diffusion layer in the O_2 unloading calculations. It indicates, however, that even in this most critical case for muscle O_2 supply (maximum oxygen uptake and significant capillary de-recruitment) calculated transit times are long enough to enable diffusional transport of oxygen out of the erythrocytes into the muscle fibers which corresponds to experimental obsevations.

DISCUSSION

We are aware of the fact that the present study is only a first and very crude approach to modelling red cell transit time distributions from bulk parameters, microvascular network geometry, and microrheological data available from freeze-clamped tissue samples. It appears to be an approach, however, which is worth pursuing because it is free of the principal limitations laid upon direct measurements of red cell velocity or transit time in bulk muscles at high performance by the experimental methodology.

The most severe restrictions of the model originate from our lack of quantitative information on capillary network geometry. The only quantities in this field which have been studied to a certain extent are red cell flow path length distributions [10,15] and functional capillary density [11,12,13]. In this situation, the only alternative is to adopt fixed conditions at each capillary offspring and to assume that the deviations from these conditions, present in real tissue, should cancel for a large number of branching points. As the resulting transit time distributions turn out to remain remarkably stable for a wide range of parameter values characterizing those fixed conditions, we have some confidence that our assumption of constant branching conditions should not seriously affect the results.

CONCLUSIONS

1. The model presented estimates distributions of capillary transit times from bulk parameters, functional capillary density and intracapillary red cell spacing (Fig. 11), the latter both of which can be measured in freeze-clamped muscle samples without requiring in vivo observations. It therefore is applicable to bulk muscles at high performance also.

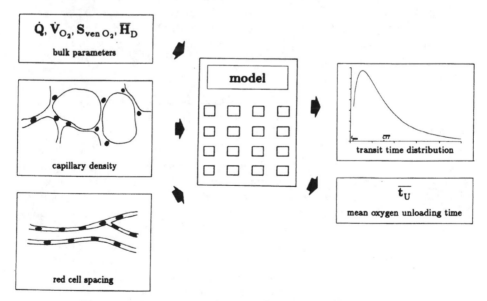

Figure 11: Input into and output from the model presented.

2. Even though blood flow through a single flow path was taken to be in inverse proportion to flow path length, calculated total blood and red cell flows are distributed similar to Gaussian. The resulting transit time distribution is by a factor of 2.0 broader and by a factor of 2.3 more skewed than the underlying path length distribution (Fig. 12). Both of these findings are in good agreement with in vivo observations of red cell fluxes and transit times in resting or vasodilated skeletal muscles.

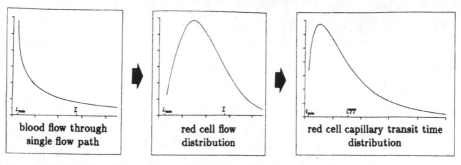

| blood flow through single flow path | red cell flow distribution | red cell capillary transit time distribution |

Figure 12: Relationship between blood flow through a single flow path and distributions of red cell flow and of transit time. Even though blood flow through a single path is in inverse proportion to path length, red cell flow is distributed similar to Gaussian, and the transit time distribution exhibits a wide scatter and a large skewness.

3. Due to transit time heterogeneities, mean O_2 unloading time (i.e. the mean value of an unloading time distribution) is 31 % longer than an individual red cell takes to desaturate from arterial to venous O_2 saturation (Fig. 10).

4. Even under the most unfavourable O_2 supply conditions of skeletal muscle at maximum performance, mean red cell transit times calculated on the basis of our transit time distributions are long enough to enable diffusional transport of oxygen out of the erythrocytes into the muscle fiber which corresponds to experimental observations (Fig. 10).

5. Mean transit times found at maximum O_2 consumption are twice as long as the expected mean times for O_2 unloading (Fig. 10). This points towards the conclusion that it is not the oxygen supply capacity which limits muscle performance.

References

[1] R. FÅHRÆUS, The suspension stability of the blood, *Physiol.Rev.* 9:241 (1929)

[2] W.J. FEDERSPIEL, A model study of intracellular oxygen gradients in a myoglobin-containing skeletal muscle fiber, *Biophys.J.* 49:857 (1986)

[3] P. GAEHTGENS, Flow of blood through narrow capillaries: Rheological mechanisms determining capillary hematocrit and apparent viscosity, *Biorheology* 17:183 (1980)

[4] P.GAEHTGENS, Mikrozirkulation, **in:** "Kreislaufphysiologie", S. 70, R. BUSSE, Hsg., Thieme, Stuttgart, 1982

[5] T.E.J. GAYESKI, C.R. HONIG, Myoglobin saturation and calculated P_{O_2} in single cells of resting gracilis muscles, *Adv.Exp.Med.Biol.* 94:77 (1978)

[6] T.E.J. GAYESKI, C.R. HONIG, Direct measurement of intracellular O_2 gradients: role of convection and myoglobin, *Adv.Exp.Med.Biol.* 159:613 (1983)

[7] T.E.J. GAYESKI, C.R. HONIG, O_2 gradients from sarcolemma to cell interior in red muscle at maximal $\dot{V}_{O_2}$, *Am.J.Physiol.* 251:H789 (1986)

[8] T.E.J. GAYESKI, C.R. HONIG, Shallow intracellular O_2 gradients and absence of perimitochondrial O_2 "wells" in heavily working red muscle, *Adv.Exp.Med.Biol.* 200:487 (1986)

[9] K. GROEBE, G. THEWS, Theoretical analysis of oxygen supply to contracted skeletal muscle, *Adv.Exp.Med.Biol.* 200:495 (1986)

[10] C.R. HONIG, M.L. FELDSTEIN, J.L. FRIERSON, Capillary lengths, anastomoses, and estimated capillary transit times in skeletal muscle, *Am.J.Physiol.* 233:H122 (1977)

[11] C.R. HONIG, C.L. ODOROFF, J.L. FRIERSON, Capillary recruitment in exercise: rate, extent, uniformity, and relation to blood flow, *Am.J.Physiol.* 238:H31 (1980)

[12] C.R. HONIG, C.L. ODOROFF, Calculated dispersion of capillary transit times: significance for oxygen exchange, *Am.J.Physiol.* 240:H196 (1982)

[13] C.R. HONIG, C.L. ODOROFF, J.L. FRIERSON, Active and passive capillary control in red muscle at rest and in exercise, *Am.J.Physiol.* 243:H196 (1982)

[14] C.R. HONIG, T.E.J. GAYESKI, W. FEDERSPIEL, A. CLARK, P. CLARK, Muscle O_2 gradients from hemoglobin to cytochrome: new concepts, new complexities, *Adv.Exp.Med.Biol.* 169:23 (1984)

[15] M.J. PLYLEY, A.C. GROOM, Geometrical distribution of capillaries in mammalian striated muscle, *Am.J.Physiol.* 228:1376 (1975)

[16] I.H. SARELIUS, Cell flow path influences transit time through striated muscle capillaries, *Am.J.Physiol.* 250:H899 (1986)

[17] G.W. SCHMID-SCHÖNBEIN, R. SKALAK, S. USAMI, S. CHIEN, Cell distribution in capillary networks, *Microvasc.Res.* 19:18 (1980)

FRACTAL ANALYSIS OF BLOOD-TISSUE EXCHANGE KINETICS

James B. Bassingthwaighte, M.D, Ph.D, Richard B. King, M.S.,
John E. Sambrook, M.S., and Brett van Steenwyk, B.S.

Center for Bioengineering WD-12, University of Washington
Seattle, Washington 98195

INTRODUCTION

Having a "fractal nature" implies that one or more features of a system or phenomenon appear to have similar characteristics when examined over a range scale. Mathematical fractals are generated by recursive expressions wherein each generation is derived from the preceding in a specific way, a precise deterministic fashion or a looser probabilistic fashion.

Several elements of the processes of delivery by flow and transmembrane transport of substrates in an organ have been observed to follow fractal kinetics. The anatomy of the vascular tree, at least the length, diameters and pressures, can be roughly described as fractals (Suwa et al, 1963, 1971). Regional flow heterogeneity in the myocardium is distinctly fractal having the same shape of distribution at increasingly fine sample volumes, and spreading with a fractal dimension of 1.2 to 1.3 (Bassingthwaighte, 1987). The flow heterogeneity may be mainly stochastic, and anatomic geometry more deterministic.

A specific application of the fractal approach is in the interpretation of indicator dilution curves for estimating membrane transport. This is done by projecting the heterogeneity of local flows to the size of the exchange unit, and using this degree of heterogeneity in the analysis of multi-capillary organ models. The fractal approach removes substantial bias in the estimates of the permeability-surface area products for the capillary and cell membranes, leaving relatively small random error.

Perhaps even more important in microcirculatory events is the fluctuating nature of the intravascular velocities and diameters. These are not strongly influenced by the pulsatile arterial pressure, but operate on a more local basis. The physiological basis for the fractal behavior needs to be worked out.

SPATIAL HETEROGENEITY OF REGIONAL MYOCARDIAL BLOOD FLOWS

Microsphere deposition has long been the method of choice for estimating regional flows, both at the organ level and regionally within organs. In the heart, the flow heterogeneity has been found to be large (Bassingthwaighte et al, 1972; Yipintsoi et al, 1973; Archie et al, 1974, King et al, 1985), but there was a nagging suspicion amongst some critics that this was due in a major way to the technique itself. King et al (1985) demonstrated that the regional flow profiles in awake baboons remained essentially constant for many hours. The stability of the distributions of microsphere depositions could not have occurred if the variance was due simply to methodologic scatter. Indeed the stability was such that any region that had a local flow greater than 150% of the mean flow for the heart at one moment, did not, on any of five other occasions over 6 to 21 hours, have an observed flow as low as the mean flow. Likewise, any region having a flow of 50% or less of the average for the heart, did not increase its flow to the average on any of five other observations times. Thus, even if microspheres were giving erroneous values, they certainly gave reproducible values.

In these hearts the total variance divided by the square of the mean flow was about 0.11. The microsphere variance was observed to be 0.005 in 200 mg pieces of about 40 g baboon hearts. King and Bassingthwaighte (unpublished) found that the temporal fluctuation was slightly more, 0.01 of the 0.11. Together methodologic noise and temporal fluctuations accounted for 0.015 of the 0.11. The major fraction of the remaining variance, 0.085 of 0.11 or 77%, was due to a stable pattern of regional flows.

Interestingly, there is no consistency in the spatial patterns from one heart to another. It appears that the high and low flow regions are disordered, but as we shall see, using fractals, they are not random.

Is the stable pattern due to reproducible bias in the microsphere technique

This could not be tested prior to the development of a new reference technique for the estimation of local flows, a "molecular microsphere" which travels dissolved in plasma and is not subject to rheologic biases at branch points in the arterial tree. The substance is iodinated desmethylimipramine (IDMI) (Little et al, 1983, 1986) which is over 99% extracted during single passage through the hearts of rabbits and sheep, and so is delivered and deposited in tissue (by binding) in proportion to local flow. Direct comparisons between microspheres and IDMI injected simultaneously were made in rabbits (Bassingthwaighte, Malone et al, 1987) and in sheep (unpublished). The distributions were very similar; in Figure 1 are shown two of the comparisons where the greatest differences were observed. There was a small systematic bias: microspheres were deposited preferentially in high flow regions and therefore were in lower concentrations than IDMI in low flow regions. This bias broadens their distributions a bit compared to IDMI, as the figure shows.

Despite the scatter and distinct bias in the microsphere-IDMI relationship, the probability density functions of microsphere depositions are not much broader than those of the IDMI distributions. The conclusion is that, although there are systematic deviations from strict accordance with local flows, these are not so marked as to cause gross overestimation of the heterogeneity of flows normally present in moderately large regions. Nevertheless, the errors are systematic and in the low flow regions are large enough that caution (and maybe corrections) should be used in interpretation. Our evidence in rabbit and sheep hearts (Bassingthwaighte, Malone et al, 1987, for rabbits) affirms what would be predicted from the theory of Fung (1973), Yen and Fung (1978), and the hydraulics experiments of Chien et al (1985), namely that one would expect a preferential deposition of 15 micron diameter microspheres into regions of higher than average flow and less deposition in low flow regions. This reinforces and explains better the results of Yipintsoi et al (1973) that suggested biasing effects of branching flow and microsphere distributions affected endo/epi gradients. The effect would be greater in larger hearts due to the larger transmural pressure gradients during systole.

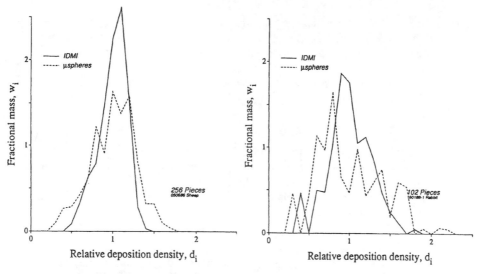

Figure 1. Probability density functions of regional left ventricular blood flows relative to the mean flow in the left ventricles of hearts of open chested rabbits and sheep. Microspheres and IDMI were injected simultaneously into the left atrium. Rabbit (right): the relative dispersions (SD/mean) were 41.1% and 24.9% for microspheres and IDMI. Sheep (left): the relative dispersions (SD/mean) were 27.6% and 17.0% for microspheres and IDMI.

Systematic Biases Due to Heterogeneity

One of the effects of flow heterogeneity is to cause dispersion of tracer that passes through the organ. The organ acts as a low pass filter, reducing the amplitude of any fluctuations in concentrations at the inflow so that they are slurred before reaching the output. The sharpness of the peak in the outflow dilution curve is reduced compared to the inflow peak. The heterogeneity, a combination of a heterogeneity of vascular velocities and a heterogeneity of vascular path lengths, can be modeled in an infinite number of fashions. None can be explicitly proven to be the correct representation unless the model represents a known, precisely described portion of a vascular network. Several simple networks were described by Bassingthwaighte and Goresky (1984), for substances of low permeability and of low intratissue diffusibility; many models will give fairly similar results because the transfer function of the organ will be dominated by the intravascular transport. There is quite likely a fractal basis for the dispersion of the intravascular indicator (Bassingthwaighte, 1987). [For highly diffusible indicators the story is different because whenever there is diffusional exchange between pathways, the transport function has a shape governed by both flow and diffusion; substances of differing diffusibility will have differing proportions of the entering material traverse different pathways. Diffusional shunting between entering and draining vessels is one mechanism producing distortion of the transport functions (Bassingthwaighte, Yipintsoi and Knopp, 1984).]

Systematic underestimates of capillary PS with heterogeneity. The traditional Crone-Renkin expression, $PS = F - ln(1-E)$, where PS is capillary permeability-surface area product, F is flow and E is maximum extraction of tracer during transorgan passage, is a good expression when flows are everywhere uniform and there is no tracer

17

reflux from tissue. But is there is intraorgan heterogeneity of flows, the mean extraction E is underestimated because the reflux from the high flow regions masks the higher extractions in the low flow regions, and PS is underestimated systematically. The error is large, 35 to 60% when the relative dispersion of flows is 30 to 50%. A method for avoiding the error is to use multicapillary models and a fractal-based estimate of the heterogeneity.

A Fractal Analysis of Spatial Heterogeneity

In estimating regional flows from the local concentration of a deposited tracer, if one chops the organ into large pieces, one gets a low estimate of the relative dispersion of flows. Cutting the tissue more finely reveals broader heterogeneity. For the heart the relationship reveals similarity of a fractal type, namely the relative dispersion is a fractal function of pieces size:

$$RD(w) = RD(w=1g) \; w^{1-D} \tag{1}$$

where $RD(w)$ is the relative dispersion (the standard deviation of the distribution divided by the mean) for a particular piece size, w, $RD(w=1g)$ is the RD at an arbitrary reference point for 1 g pieces, and D is the fractal dimension. For baboon and sheep hearts $RD(w=1g)$ is about 18% and the fractal D is about 1.2 to 1.3. The value of D would be 2.0 if the heterogeneity were random as it is for Brownian motion or molecular diffusion. The value of 1.2 suggests an ordered arrangement with continuity in flows in adjacent regions.

<u>Using the fractal D to estimate capillary PS.</u> The estimation of RD at the level of the functional microvascular unit size involves an extrapolation to a smaller size of tissue piece than one in which flow can be measured by standard techniques even using the molecular microsphere. If the fractal relationship holds down to 1 mg pieces, the unit size estimated by Bassingthwaighte, Yipintsoi and Harvey (1974), then $RD(0.001g) = 18\%(0.001)^{-0.2} = 72\%$. This large degree of heterogeneity of flows would then be used in a multicapillary model, fitting the tracer dilution curves to provide the correct overall transorgan extraction, the E in the Crone-Renkin expression.

There is some chance that this approach would overestimate the heterogeneity: it is likely that flows in neighboring microvascular units are similar and that the degree of association falls off with distance. This means that instead of the infinitely fractal relationship defined by Equation 1, the relationship flattens or curves toward a plateau at small pieces as suggested by Figure 2. Rigaut (1983) in looking at the lengths of contours of pulmonary alveolar boundaries observed a convex upward curvature toward a plateau on the plot of the logarithm of total length versus the "stride length", the length of the measuring calipers. The relationship was fractal, that is, linear on the log-log plot, at large caliper lengths, but fell below the fractal relationship at short caliper lengths, giving a concave downward curvature. This is natural, since the maximum radius of curvature of a lipid bilayer is about 30 nm, and irregularity in the contour must diminish to that of a smooth contour at caliper lengths, ε, of this order. Rigaut's empirical equation for the boundary length L, at each ε, is:

$$L(\varepsilon) = \frac{L_{max}}{1 + \alpha^c \varepsilon^c} \tag{2}$$

This expression gives curvature, plateauing at L_{max} at small ε, and curving asymptotically toward a downward sloping straight line at large ε. The scalar α and the exponent c are not readily interpretable except in the extreme case where $L_{max} \rightarrow \infty$, $\alpha^c \rightarrow \infty$, and $c \rightarrow D-1$, which is a normal fractal relationship. However, Rigaut's equation is an excellent provocation to develop expressions which express directly the fractal D and the rate of convergence to the non fractal plateau. At this point, how to handle such pseudofractal relationships in a physically or physiologically meaningful mathematical fashion is not clear.

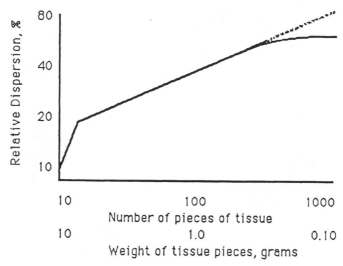

Figure 2. Deviation from a fractal relationship occurs when there is a tendency for small regions to have uniform flow.

TEMPORAL FLUCTUATIONS IN INTRAVASCULAR VELOCITIES AND FLOWS

Microscopic observations of intracapillary and intra-arteriolar velocities can now be made using modern techniques of image acquisition and analysis, as by Slaaf et al (1986, 1987), and Kislyakov et al (1987). If the vessels were rigid pipes, then the velocities would be proportional to aortic pressure, but all of such velocity observations show fluctuations which are different in form from aortic pressure. If local arteriolar flows were totally unrelated to the aortic driving pressure this would be surprising, especially for oscillations at cardiac frequency, but autoregulatory phenomena certainly inhibit flow changes at lower frequencies (Baer et al, 1984).

How to characterize fluctuations has been a problem. Autocorrelation techniques show a dependence on binning duration, i.e. on the length of intervals over which flow is averaged. Frequency transforms such as finite Fourier series representation have an appropriate general form, except that they inevitably alias the high frequency information and do not provide a good description allowing comparison of one situation to another. A fractal approach has the virtue of encompassing a wide range of frequencies, unfettered by the artifact of frequency foldover (aliasing). The key is to examine the data, not with a single sampling frequency, but with a number of binning durations, τ, extending over a wide range. The idea is expressed in Figure 3; the measurement of velocity is the average over an interval, τ. The topmost "curve" represents data binned at 0.16 second intervals. Binning over larger τ's reduces the excursions.

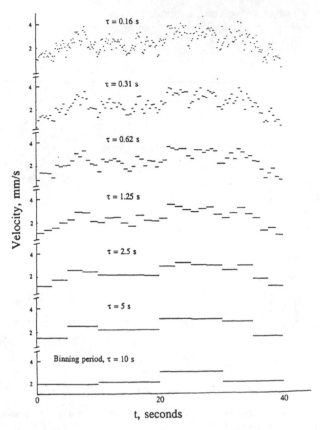

Figure 3. Diagrammatic representation of data showing temporal fluctuations. The raw "data" of the top panel are then binned over successively longer times τ for the panels below. Lower resolution detection, e.g. a filtered signal or a long time constant in the transduction system, results in underestimation of the excursions. As with the spatial fractals, projection to higher frequencies may be justified.

Taking the relative dispersions at each τ is a simple way of summarizing such data. We observe, on data taken from the published figures of Kislyakov et al (1987) that a

$$RD(\tau) = RD(\tau=1) \, \tau^{1-D_\tau}$$

where D_τ is the fractal dimension for the temporal fluctuations. The data were velocities in pial vessels of 6 or 7 microns diameter with τ in seconds. The function $RD(\tau) = 11.2 \, \tau^{-0.37}$ fitted well, the correlation coefficient being r = -0.995, over the range of τ's from 0.14 to 2.8 seconds.

Velocities cannot be infinitely fractal. High frequency fluctuations require high forces (energy, work) to produce the accelerations required to change velocity. In other words, the inertia of a moving column of blood is such that velocity changes cannot be instantaneous. The corollary is that the fractal relationship must at some small values of τ show a "concave downward" pattern. This is completely analogous to the plateauing in the spatial domain.

DISCUSSION

Mandelbrot's writings have fostered much development in the applications of fractals to describing physical features. Many interesting ideas are presented in his books (e.g. 1983). Peitgen and Richter (1986) give a nice introduction to the mathematics and some wonderful pictures. Applications to dynamic phenomena are anticipated by the efforts to describe chaos, on which Holden has edited a very useful book (Holden, 1986). Transitions from laminar to disturbed or turbulent flow may be described as mathematical predictions of instability or non-predictable behavior (Devaney, 1987).

At the membrane level, time-dependent ionic channel fluctuations have been described by Liebovitch et al (1987), in terms of a single fractal dimension replacing more cumbersome traditional descriptions. At a multicellular level, Glass et al (in Holden's book, 1986) describe chaotic cardiac rhythms, a natural evolution from the old clinical description of atrial fibrillation as an irregularly irregular rhythm.

Our approach to describing heterogeneity of regional blood flows by fractals in space and time may be novel in a way. The simplified description of "heterogeneity" by the relative dispersion or coefficient of variation is not, but the 2-parameter description of heterogeneity ($RD[w=1]$ and D_w or $RD[\tau=1]$ and D_τ) is apparently new. It is clearly useful since such a descriptor allows comparisons of data between different laboratories to be made in a meaningful way. In the past a description of RD in regional flow might have differed from that in another lab only because the sample size, w, or the binning times, τ, differed. Now we can use two parameter descriptors, which should be comparable from lab to lab and independent of the particular w's to τ's chosen by an investigator.

CONCLUSION

Fractal approaches provide another analytical tool for examining biological phenomena. In some instances they will go beyond description to allow new insight. This transition has not yet been made very often in circulatory studies. The first applications have been to structure. Now the attempt is to examine dynamic situations as well.

REFERENCES

1. Archie, J. P., D. E. Fixler, D. J. Ullyot, G. D. Buckberg, and J. I. E. Hoffman. Regional myocardial blood flow in lambs with concentric right ventricular hypertrophy. Circ. Res. 34:143-154, 1974.

2. Baer, R. W., B. D. Payne, E. D. Verrier, G. J. Vlahakes, D. Molodowitch, P. N. Uhlig, and J. I. E. Hoffman. Increased number of myocardial blood flow measurements with radionuclide-labeled microspheres. Am. J. Physiol. 246 (Heart. Circ. Physiol. 15):H418-H434, 1984.

3. Bassingthwaighte, J. B., W. A. Dobbs, and T. Yipintsoi. Heterogeneity of myocardial blood flow. In: Myocardial Blood Flow in Man: Methods and significance in coronary disease, edited by A. Maseri. Torino, Italy:Minerva Medica, 1972, p. 197-205.

4. Bassingthwaighte, J. B., T. Yipintsoi, and R. B. Harvey. Microvasculature of the dog left ventricular myocardium. Microvasc. Res. 7:229-249, 1974.

5. Bassingthwaighte, J. B., T. Yipintsoi, and T. J. Knopp. Diffusional arteriovenous shunting in the heart. Microvasc. Res. 28:233-253, 1984.

6. Bassingthwaighte, J. B., and C. A. Goresky. Modeling in the analysis of solute and water exchange in the microvasculature. In: Handbook of Physiology, Sect. 2 The Cardiovascular System, Vol IV, Microcirculation, Chapt. 13, edited by E. M. Renkin, and C. C. Michel. Bethesda, MD:American Physiological Society, 1984, p. 549-626.

7. Bassingthwaighte, J. B. Physiological heterogeneity: Fractals link determinism and randomness in structures and functions. *News in Physiol. Sci.* 2:xx-xx, 1987. (accepted)

8. Bassingthwaighte, J. B., M. A. Malone, T. C. Moffett, R. B. King, S. E. Little, J. M. Link, and K. A. Krohn. Validity of microsphere depositions for regional myocardial flows. Am. J. Physiol. 253 (Heart. Circ. Physiol. 22):H184-H193, 1987.

9. Chien, S., C. D. Tvetenstrand, M. A. F. Epstein, and G. W. Schmid-Schönbein. Model studies on distributions of blood cells at microvascular bifurcations. Am. J. Physiol. 248 (Heart. Circ. Physiol. 17):H568-H576, 1985.

10. Fung, Y. C. Stochastic flow in capillary blood vessels. Microvasc. Res. 5:34-48, 1973.

11. Glass, L., A. Shrier, and J. Bélair. Chaotic cardiac rhythms. In: *Chaos*, edited by A. V. Holden. Princeton:Princeton University Press, 1986, p. 237-256.

12. Holden, A. V., (editor). *Chaos*. Princeton: Princeton University Press, 1986.

13. King, R. B., J. B. Bassingthwaighte, J. R. S. Hales, and L. B. Rowell. Stability of heterogeneity of myocardial blood flow in normal awake baboons. Circ. Res. 57:285-295, 1985.

14. Kislyakov, Y. Y., Y. I. Levkovitch, T. E. Shuymilova, and E. A. Vershinina. Blood flow fluctuations in cerebral cortex microvessels. Int. J. Microcirc. Clin. Exp. 6:3-13, 1987.

15. Levin, M., and J. B. Bassingthwaighte. Sensitivity functions in optimizing the fitting of a transport model to observed system responses. Ann. Biomed. Eng. (in preparation)

16. Liebovitch, L. S., J. Fischbarg, J. P. Koniarek, I. Todorova, and M. Wang. Fractal model of ion-channel kinetics. *Biochim. Biophys. Acta* 896:173-180, 1987.

17. Little, S. E., and J. B. Bassingthwaighte. Plasma-soluble marker for intraorgan regional flows. Am. J. Physiol. 245 (Heart. Circ. Physiol. 14):H707-H712, 1983.

18. Little, S. E., J. M. Link, K. A. Krohn, and J. B. Bassingthwaighte. Myocardial extraction and retention of 2-iododesmethylimipramine: a novel flow marker. Am. J. Physiol. 250 (Heart. Circ. Physiol.19):H1060-H1070, 1986.

19. Mandelbrot, B. B. *The fractal geometry of nature*. San Francisco: W.H. Free-man and Co., 1983.

20. Peitgen, H. O., and P. H. Richter. *The beauty of fractals: images of complex dynamical systems*. Berlin/Heidelberg: Springer-Verlag, 1986.

21. Rigaut, J. P., P. Berggren, and B. Robertson. Resolution-dependence of stereo-logical estimations: interpretation, with a new fractal concept, of automated image analyser - obtained results on lung sections. Acta Stereol. 2(Suppl.I):121-124, 1983.

22. Slaaf, D. W., G. J. Tangelder, R. S. Reneman, and T. Arts. Methods to measure flow velocity of red blood cells *in vivo* at the microscopic level. Ann. Biomed. Eng. 14:175-186, 1986.

23. Slaaf, D. W., G. J. Tangelder, H. C. Teirlinck, and R. S. Reneman. Arteriolar vasomotion and arterial pressure reduction in rabbit tenuissimus muscle. Microvasc. Res. 33:71-80, 1987.

24. Suwa N., T. Niwa, H. Fukasawa, and Y. Sasaki. Estimation of intravascular blood pressure gradient by mathematical analysis of arterial casts. Tohoku J. Exp. Med. 79:168-198, 1963.

25. Suwa, N., and T. Takahashi. Morphological and morphometrical analysis of circulation in hypertension and ischemic kidney. Munich: Urban & Schwarzenberg, 1971.

26. Yen, R. T., and Y. C. Fung. Effect of velocity distribution on red cell distribution in capillary blood vessels. Am. J. Physiol. 235 (Heart. Circ. Physiol. 4):H251-H257, 1978.

27. Yipintsoi, T., W. A. Dobbs, Jr., P. D. Scanlon, T. J. Knopp, and J. B. Bassingthwaighte. Regional distribution of diffusible tracers and carbonized microspheres in the left ventricle of isolated dog hearts. Circ. Res. 33:573-587, 1973.

A GRAPHICAL ANALYSIS OF THE INFLUENCE OF RED CELL TRANSIT TIME, CARRIER-FREE LAYER THICKNESS, AND INTRACELLULAR PO_2 ON BLOOD-TISSUE O_2 TRANSPORT

T.E.J. Gayeski, W.J. Federspiel, and C.R. Honig

The University of Rochester, School of Medicine
601 Elmwood Avenue
Rochester, New York 14642

The Biomechanics Institute
Boston, Massachusetts 02215

This paper relates spectroscopic determinations of myoglobin (Mb) saturation and PO_2 in individual myocytes (Gayeski and Honig, 1986) to a mathematical model of O_2 transport (Federspiel, 1983). A graphical analysis of model results is used to illustrate interactions among the main determinants of O_2 transport between hemoglobin (Hb) and cytochrome. The results of the analysis are summarized in a plot called the O_2 release curve. The results are qualitatively applicable to normal and pathophysiology, and to Mb-free tissues.

METHODS

Procedures used to obtain Mb and Hb saturations in canine muscles have been described in detail (Gayeski and Honig, 1986). Briefly, a dog is anesthetized with sodium pentobarbital and intubated. For skeletal muscle, dogs breathed room air spontaneously. For myocardium, open-chested dogs are mechanically ventilated with F_IO_2 equal to 35% to maintain normal arterial blood gases. Covariates for all experiments include arterial pH, PCO_2, and PO_2, Hb concentration, Hb saturation, and systemic blood pressure.

Gracilis muscles are vascularly and neurally isolated and stimulated to achieve maximal twitch contractions. Increases in capillary density and blood flow are complete within 10 and 60 seconds respectively; the steady state for rate O_2 consumption (VO_2) is achieved within 2 minutes (Gayeski et al., 1985). Muscles were collected 3 minutes after initiation of stimulation. The muscles were rapidly frozen in situ by bringing a copper block cooled to -197°C into contact with the muscle via a pneumatically driven piston.

The left ventricular free wall was exposed through a left thoracotomy. When heart rate, cardiac output, and left ventricular diastolic pressure were in the steady state, left ventricular free wall specimens were obtained by a rapid freezing technique similar to that used for gracilis. We did not synchronize freezing with the ECG because Mb saturation is constant during the cardiac cycle (Fabel, 1968; Makino et

al., 1983). Freezing rates are about 10 μm/ms approximately 500 μm from
the surface, and are adequate to trap Mb saturation at a depth of 1 mm to
within 1% of values existing immediately prior to freezing (Clark and
Clark, 1983; Gayeski and Honig, 1986). Muscles are cleaved into blocks of
approximately 1 cm^3. The blocks are stored under liquid N_2 until ready
for cryogenic microspectrophotometry. No change in the distribution of Mb
saturations can be demonstrated after storage for as long as 5 years.

A freshly cleaved surface in either longitudinal or cross section
were prepared and the specimen was mounted on a microscope stage regulated
at -110°C. At this temperature no change in reflected spectra for either
Hb or Mb could be observed for as long as 4 hours. Measurements were
typically completed in ˜1 hour. Percent saturation was determined using a
4 wavelength method. A 5% difference in Hb saturation could be detected.
Mb saturation differences of 1.5% are distinguishable within a given cell.
Differences of 3% are distinguishable amongst a group of cells (Gayeski
and Honig, 1986).

Optical theory (Gayeski, 1981) and empirical observations (Gayeski
and Honig, 1986) indicate that the photometer receives light from
approximately 50 μm^3 when the image of the measuring diaphragm in the
optical field is 3x3 μm. The exact location of the measuring diaphragm
relative to microvessels and myocytes was visualized and noted.   To
determine radial or longitudinal gradients, the measuring diaphragm was
moved in appropriate steps, and saturation determined at each location.

PHYSIOLOGIC RESULTS AND DISCUSSION

Blood Vessel to Myocyte PO_2 Differences

The blood PO_2 of interest is that within capillaries, but this cannot
be determined with the existing cryospectrophotometer.   However, Hb
saturation can be determined in small arterioles; see Figure 1, from a

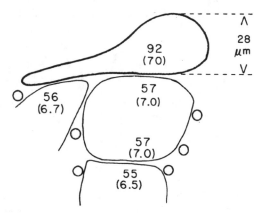

Figure 1. Percent saturation of Hb or Mb; corresponding PO_2 is shown in
parentheses. Numbers indicate location of measurements.

muscle consuming oxygen at 10 ml/100 g·min (~60% $\dot{V}O_{2max}$). Saturations are depicted in block numbers. The PO_2 estimated from the oxyhemoglobin dissociation curve is shown in parentheses. Measurements within the arteriole revealed uniform Hb saturation. The Mb saturation found in the adjacent myocytes was indistinguishable from that found within myocytes remote from arterioles. Furthermore, there was no progressive decrease in Mb saturation as the radial distance from the arteriole increased over several cell diameters. The difference between PO_2 in the arteriole and the PO_2 in equilibrium with Mb ($PmbO_2$) is about 63 torr over 5 μm, or 15 torr/μm! A short branch of the arteriole gave rise to the capillary shown at upper left. Since $PmbO_2$ in the neighboring cell was 6.7 torr it seems likely that the PO_2 gradient across that capillary was of the same order of magnitude as that across the arteriole.

Comparison of $PmbO_2$ with venous PO_2 (P_vO_2) provides further evidence that the red cell to myocyte interface is the main site of resistance to O_2 mass transport. The PO_2 of effluent blood is an estimate of mean end-capillary PO_2 because Mb eliminates diffusive shunting of O_2 (Hukman et al., 1969; Rose and Goresky, 1988). Effluent PO_2 from gracilis muscles working at 60-70% $\dot{V}O_{2max}$ averaged 20 torr (Honig et al., 1984). Coronary sinus PO_2 in dogs prepared as described averaged 27 torr (Henquell and Honig, 1976). In every muscle the **maximum** PO_2 found was much lower than P_vO_2. Thus a large ΔPO_2 between capillary blood and myocyte is probable, in accord with recent mathematical models (Federspiel, 1983; Clark et al., 1985; Federspiel and Popel, 1986; Groebe and Thews, 1987).

Variability of $PmbO_2$

Differences in Mb saturation within an individual heavily working myocyte can be more than 20% (Gayeski and Honig, 1986), but the corresponding differences in $PmbO_2$ are only ~2 torr, partly because the steep slope of the oxymyoglobin dissociation curve acts as a $PmbO_2$ buffer. The range in $PmbO_2$ in a population of heavily working myocytes is shown in Figure 2. The curves with filled symbols represent the distribution of $PmbO_2$ in 50 randomly chosen subepicardial myocytes in each of two

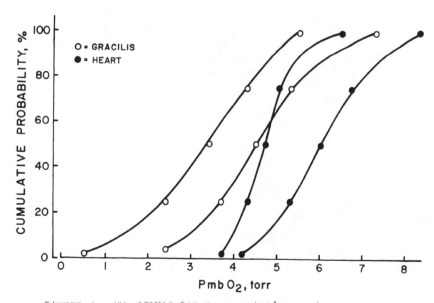

Figure 2. PO_2 ogives for representative canine muscles.

unstressed dog hearts. The open-symboled curves describe the distribution from 50 myocytes in each of two gracilis muscles working at comparable VO_2, but at a much higher percentage of VO_{2max}. The interquartile range (25th-75th percentile) is approximately 1 torr for hearts and 2 torr for gracilis.

Interaction of Diffusion Fields

$PmbO_2$ gradients between myocytes have been reported (Gayeski and Honig, 1986). Though intermyocyte $PmbO_2$ differences are small (~1 torr), the surface area between myocytes is large compared to the surface area of a capillary. For a gracilis myocyte 50 μm in diameter with 4 contiguous red cell-containing capillaries 4 μm in diameter, the ratio of the surface area of adjacent myocytes to the surface area of capillaries is ~10. Capillary hematocrit is important because about 80% of the calculated O_2 flux occurs across the capillary area immediately subjacent to red cells (Federspiel & Popel, 1986). If one takes capillary hematocrit into account, the foregoing surface area ratio would be approximately 20. Because of their large surface area, myocytes with few surrounding capillaries can receive a large fraction of their O_2 requirements at low flux density from remote capillaries via intervening myocytes. Intra- and intercellular O_2 diffusion is greatly facilitated by a flux of oxymyoglobin molecules in parallel with free O_2 (Murray, 1974; Kreuzer and Hoofd, 1987). Calculations indicate that intercellular redistribution of O_2 makes a substantial contribution to the uniformity of O_2 supply despite the small intermyocyte $PmbO_2$ difference (Clark, A. and Clark, P., personal communication).

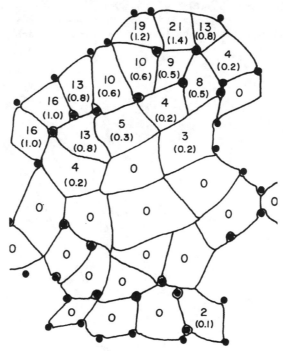

Figure 3. Map of Mb saturation and PO_2 (in parentheses) in cluster of myocytes at border of an ischemic region.

Over what distance can capillaries contribute O_2 to a myocyte via intervening myocytes? To test for the order of magnitude of this distance, we altered blood flow to part of a working gracilis and looked for the size of "border zones" between perfused and unperfused regions of muscle. Figure 3 depicts typical results when the muscle was allowed to contract at 4/s after the distal set of gracilis vessels was ligated. The distal artery accounts for about 30% of total gracilis blood flow. $PmbO_2$ fell to zero over 2 to 3 cell diameters, or 100 to 150 μm. Since capillaries 100 to 150 μm away can supply O_2 to a myocyte, the supply unit is not Krogh's single capillary and its independent diffusion field, but rather a cluster of cells and capillaries. Interaction of diffusion fields around capillaries and myocytes contributes to the uniformity of $PmbO_2$ shown in Figure 2.

Longitudinal Gradients

With the fiber axis positioned longitudinally, we used a 3x3 μm measuring diaphragm to measure the Mb saturation at sites parallel to the long axis of both cardiac and skeletal myocytes. Differences in $PmbO_2$ along a particular fiber were comparable to those seen between cells. However, the striking result is that we never observed a progressive change in $PmbO_2$ in the axis of a fiber over distances more than half total capillary length in skeletal muscle, and twice the average length of capillary network segments in cardiac muscle.

Summary of Experimental Results

1. A large ΔPO_2 between Hb and Mb is probable.

2. $PmbO_2$ is in the functional range of Mb at all locations in heart and heavily working skeletal muscle.

3. At high $\dot{V}O_2$ $PmbO_2$ varies by less than 2 torr in tissue volumes that might be served by a terminal arteriole and its sheaf of capillaries.

Significance

Spatial heterogeneities of red cell velocity, red cell spacing, capillary density, and intracapillary PO_2 are inevitable in microvascular networks (Honig et al., 1977; 1982; 1987; Sarelius and Duling, 1982; Tyml, 1987). This variability is due to anatomic constraints, random rheologic processes, and precapillary vasomotion (Fung, 1973; Schmid-Schönbein et al., 1980; Honig et al., 1982; Tyml, 1987). The resulting large heterogeneities of red cell flux are on an anatomical scale comparable to the scale of the diffusive interactions shown in Figure 3. These interactions, and functions of Mb, stabilize local $PmbO_2$. To a first approximation a red cell "sees" a spatially uniform O_2 sink in its passage through a capillary in heavily working red muscle. Its rate of O_2 unloading can therefore be modelled as though it were dropped into a Mb solution of known saturation.

MATHEMATICAL MODEL OF O_2 RELEASE FROM BLOOD

Formulation of the Model

The objective is to evaluate the rate of release of O_2. To this end we consider a one-dimensional slab model of a red cell with a characteristic half-thickness equal to 1.0 μm. The slab red cell releases O_2 into a sink maintained at fixed PO_2. The slab is covered by a layer of material 1-2 μm thick with the physical properties of plasma. The salient feature of this portion of the O_2 diffusion path is that it is devoid of a heme protein O_2 carrier. The model for transport in the red cell is identical to that described by Clark et al. (Biophysical J., 1985). This model accounts for free and facilitated transport of O_2 in the red cell. Details of the coupled transport equation that describes O_2 release from a red cell covered by a carrier-free layer can be found in Federspiel, 1983, Appendix A.

Parameter values used in the simulation are: **For the red cell:** half-thickness = 1 μm, O_2 diffusivity = 9.5×10^{-6} $cm^2 sec^{-1}$, Hb diffusivity = 1.4×10^{-7} $cm^2 sec^{-1}$, total heme concentration = 2.0×10^{-5} $mole \cdot cm^{-3}$, Hb P_{50} = 26 torr, O_2 solubility = 1.6×10^{-9} $mole \cdot cm^{-3} torr^{-1}$ and Hb-O_2 dissociation constant = 200 sec^{-1}. **For plasma:** D_{O2} and O_2 solubility = 2.5 and 0.92 times the respective values for the red cell.

The model predicts the time required to change Hb saturation from an initial value to a final value. Release of O_2 is driven by exposure of the simulated red cell to an O_2 sink. The central concept is that the PO_2 **in red cells during passage through a capillary cannot be thought of as equilibrating with the PO_2 in the subjacent myocyte** under conditions of high O_2 flux. Because of resistances to O_2 transport within red cells and in the carrier-free portion of the diffusion path, O_2 is released throughout the red cell's capillary transit time.

The O_2 Release Curve

Results of the model are described in Figure 4 as the relation between the release time for O_2 and Hb saturation. This plot, called the O_2 release curve, allows qualitative analysis of blood-tissue O_2 transport under virtually all conditions. The principal limitation to its application is uncertainty about parameters.

The lower abscissa when scaled to Hb concentration represents the O_2 content of blood; the upper abscissa indicates the PO_2 in red cells in equilibrium with Hb. O_2 saturation of blood entering capillaries is set at 90%, in accord with such data as Figure 1. Movement along the abscissae from left to right simulates extraction of O_2 by consuming tissue. The release times for O_2 shown on the ordinate are equivalent to capillary transit times for red cells in vivo. Transit times thought to exist in heavy exercise are in the range 200-500 ms (Honig and Odoroff, 1981; Groebe and Thews, 1987). This range is indicated by the horizontal lines in Figure 4. Each isopleth is calculated for a fixed distance, D_s, between Hb and the O_2 sink and a fixed PO_2 of the O_2 sink. D_s corresponds to the thickness of the carrier-free layer used in the model. In vivo this layer consists of the plasma sleeve between the red cell and endothelium, the endothelial cell, and the interstitium. O_2 release depends on functional factors as well as the anatomical thickness of the carrier-free layer. To take account of these we define an effective D_s of which the anatomical carrier-free layer is a major component. The number beside each curve denotes the PO_2 of the O_2 sink. As PO_2 in blood approaches that in the O_2 sink the driving force for diffusion is dissipated, and the release time for O_2 approaches an asymptote. Thus PO_2 in the O_2 sink, homologous to $PmbO_2$, sets the minimum Hb saturation and maximum O_2 extraction that could be attained.

The one-dimensional slab geometry for a red cell and associated carrier-free layer is useful for describing **qualitative** relationships among the parameters that influence O_2 release. Establishing quantitative relationships is more difficult, and would require a model incorporating the complex three-dimensional shape of red cells in relation to surrounding structures, the kinetics of red cell passage through the capillary, and the flux density profile through the carrier-free portion of the diffusion path. Principles can be inferred from the O_2 release curves but close numerical comparisons with experimental results are inappropriate with this version of the model.

Role of $PmbO_2$

$PmbO_2$ is a powerful determinant of O_2 extraction because the driving force for blood-tissue transport is the difference between PO_2 within red cells and PO_2 in tissue cells. To illustrate the effect of $PmbO_2$ on end-capillary Hb saturation and O_2 extraction, consider the case of a 1 μm carrier-free layer exposed to O_2 sinks at 1 torr or 5 torr. These PO_2 values represent the lower halves of the distributions shown in Figure 2. We chose a release time of 0.2 sec to represent a condition of maximum stress on the O_2 delivery system, since this value approaches the minimum transit time in heavily working muscle. Isopleths for 5 and 1 torr are almost linear over the physiological range for O_2 extraction, and their slopes are almost the same. They intersect the 0.2 sec line at 29% and 23% saturation, respectively. These saturations simulate the end-capillary saturations that would exist in vivo for the above conditions. The corresponding values for PO_2 in red cells (upper abscissa) would be 19 and 16 torr, respectively. Note that the variation in $PmbO_2$ observed in heavily working red muscle would change O_2 extraction by only ~10% even for the shortest transit times. Consequently if $PmbO_2$ for a cluster of heavily working myocytes is near Mb's P_{50}, $PmbO_2$ can be considered homogeneous from an O_2 extraction standpoint. The same conclusion holds if the effective D_S were 0 or 2 μm. The model indicates that end-capillary PO_2 in red cells would be 17 and 14 torr (upper abscissa) for $PmbO_2$ equal to 5 torr and 1 torr. The driving forces near the ends of capillaries would be approximately equal for the two $PmbO_2$ values. Thus, there is little physiological advantage in a median $PmbO_2$ below 5 torr unless maximal O_2 extraction is required. Values for median $PmbO_2$ in a large population of heart and gracilis muscles performing heavy, submaximal exercise cluster near the MbP_{50}. Though VO_{2max} can be achieved at a 10-fold lower $PmbO_2$, major metabolic compensations are required (Gayeski et al., 1987). It is of interest that only in muscles stressed to VO_2 $_{max}$ is median $PmbO_2$ near 1 torr (Gayeski et al., 1987).

The isopleth for 20 torr yields a particularly important insight into adaptation to exercise. Median $PmbO_2$ is near 20 torr at rest (Gayeski et al., 1987). Note that the isopleth for 20 torr is markedly non-linear over the entire range of O_2 extraction. If $PmbO_2$ were 20 torr and the effective D_S were 1 μm in exercise, end-capillary Hb saturation would be about 40% for red cell transit times measured (Sarelius and Duling, 1982) or calculated (Honig and Odoroff, 1981) for resting skeletal muscle. The observed venous saturation is near 25% in heavily working gracilis, and is <10% in maximally exercising dog hearts (Von Restorff et al., 1972). If $PmbO_2$ were to fall from 20 torr to 5 torr with no change in transit time almost twice as much O_2 could be extracted, solely as a result of the decrease in $PmbO_2$. Thus, $PmbO_2$ must be lowered well below resting values to achieve high VO_2, in accord with our data.

Venous saturations in resting gracilis prepared as described range from 79-91%, corresponding to P_vO_2 values of 45-60 torr. Thus a large ΔPO_2 exists between effluent blood and myocytes even at very low O_2 flux. The mechanisms responsible for this difference are poorly understood. Low capillary density and low capillary hematocrit at rest would increase O_2 flux per red cell, and necessitate a driving force greater than might be expected for such low VO_2. Also, the effective D_S could be greater at rest than in exercise. Finally, diffusive shunting between large arterioles and venules could become important at flows and VO_2 observed at rest'(Piiper, 1987). Whatever the mechanisms, $PmbO_2$ approaches venous PO_2 only in the upper quartile of its probability distribution when the muscle is at rest, and is well below P_vO_2 at all locations in exercise.

Transit Time

Transit time is the ratio of red cell path length to red cell velocity. Short arteriolar path lengths, particularly in myocardium (Bassingthwaighte et al., 1974), and high precapillary red cell velocities (Fronek and Zweifach, 1977) result in arteriolar transit times ~1/10 those in capillaries or 0.02 to 0.05 sec (Gaehtgens et al., 1970). Also, the arteriolar wall contributes to the thickness of the carrier-free layer, resulting in carrier-free layers greater than 2 μm. Notice in Figure 4 that virtually no change in saturation would be expected for such conditions (lower left), in accord with the 90% saturation of Hb shown in Figure 1. Similar conditions apply for post-capillary vessels. Capillary transit times directly measured in various muscle microcirculatory preparations at rest are on the order of 800-3000 ms (Sarelius & Duling, 1982); measurements during exercise are not available but would be of great interest. Red cell path length is about the same at rest and in exercise in rat gracilis (Honig et al., 1977), but velocity increases with volume flow. Time available for O_2 release in exercise is defended by recruitment of additional flow paths. Recruitment lowers red cell velocity by increasing aggregate cross-sectional area. Two independent

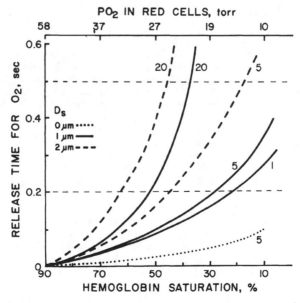

Figure 4. O_2 release curves for tissue at high $\dot{V}O_2$.

calculations predict mean transit times between 200 and 500 ms for working dog gracilis (Honig & Odoroff, 1981; Groebe & Thews, 1987).

We use the above range of transit times, $PmbO_2$ of 5 torr and effective D_S equal to 1 μm to explore the influence of transit time on O_2 extraction from a red cell. From Figure 4, end-capillary Hb saturation would be 30% and 10% for transit times of 0.2 and 0.5 sec respectively, an increase in extraction of ~33% at the longer transit time. To meet O_2 demands in the presence of short transit times, a myocyte could lower its $PmbO_2$ but this adaptation is effective only at high O_2 extraction. Alternatively, a myocyte could 1) extract a larger fraction of its O_2 from capillaries with higher hematocrit or longer transit times, 2) cause recruitment of additional capillaries, 3) derive some of its O_2 from neighboring cells served by capillaries in which O_2 supply was more favorable. Inferences about which mechanism are utilized might be obtained from probabiity distributions of end-capillary PO_2 or PO_2 in small venules.

The impact of heterogeneity in transit times will be different at rest. The isopleths for 20 torr, a value typical of $PmbO_2$ in resting muscle, become asymptotic at release times of 0.5 to 0.6 sec. The measured range of transit times is 0.8 to 3.0 sec (Sarelius and Duling, 1982). Hence, end-capillary Hb saturation and PO_2 should be nearly equal amongst capillaries. No data is yet available to test this thesis. In contrast, end-capillary PO_2 should be strongly dependent on transit times at $PmbO_2$ values characteristic of myocardium and exercising gracilis.

Influence of Carrier-Free Layer

The dotted curve in Figure 4 is calculated on the assumption that a carrier-free layer does not exist, and the PO_2 of the O_2 sink is 5 torr. If this were the case virtually all the O_2 in arterial blood could be extracted by exercising muscle if the shortest predicted transit times are about 0.2 sec. This prediction is inconsistent with measured values of PvO_2. The dramatic change induced by an effective D_S of 1 or 2 μm (solid and dashed curves labelled 5 torr) indicates that **a carrier-free layer can be a major determinant of O_2 transport**. At first it may seem surprising that a carrier-free layer approaching the thickness of a red cell can have such a large effect. The reason is, in part, that O_2 flux density through the carrier-free layer is greater than at any other point in the diffusion path from red cell membrane to cytochrome. High flux density (flux/unit area) and absence of carrier-facilitated diffusion require a steep concentration gradient for diffusion of free O_2. Because the portions of the carrier-free layer closest to the erythrocyte accommodate a greater O_2 flux density, these portions have a greater PO_2 drop than those portions further away. This effect is not addressable within our slab model, in which O_2 flux density is uniform across the carrier-free layer. However, the impact of this geometric consideration is that the anatomical component of D_S in vivo would correspond to a shorter distance than in our slab model.

Estimates of the anatomical component of the carrier-free layer are unavailable; judging from the thickness of endothelial cells, 1 μm is likely to be a lower bound. In the model D_S can be regarded as a scaling factor. For example, if $PmbO_2$ were 5 torr and transit time were 0.2 sec end-capillary saturation would be 46% if D_S were 2 μm and 29% if D_S were 1 μm. This corresponds to a 40% difference in extraction! Precise measurements of the anatomic component of the effective D_S are essential in view of its potential effect on blood-tissue O_2 transport in health, and in pathological states such as edema or diabetes.

SUMMARY OF MODELLING RESULTS

1. The O_2 release curve (the relation between the release time for O_2 and Hb saturation), is proposed as a convenient descriptor of blood-tissue O_2 transport.

2. At high $\dot{V}O_2$ $PmbO_2$ must fall well below PO_2 in red cells to permit high extraction during the capillary transit time of a red cell.

3. O_2 release is strongly influenced by the thickness of the carrier-free portion of the O_2 diffusion path, particularly at high O_2 flux.

4. In both heart and working skeletal muscle $PmbO_2$ can be regarded as a quasi-uniform O_2 sink from the standpoint of O_2 extraction, despite microvascular heterogeneities.

5. Capillary transit time, red cell PO_2, $PmbO_2$, and the effective carrier-free layer thickness interact strongly as determinants of the balance between O_2 supply and demand.

REFERENCES

Bassingthwaighte, J.B., Yipintsoi, T., and Harvey, R.B., 1974, Microvasculature of the dog left ventricular myocardium, Microvasc. Res., 7:229-249.

Clark, A., and Clark, P.A.A., 1983, Capture of spatially homogenous chemical reactions in tissue by freezing, Biophys. J., 42:25-30.

Clark, A., Jr., Federspiel, W., Clark, P.A.A., and Cokelet, G.R., 1985, Oxygen delivery from red cells, Biophys. J., 47:171-181.

Fabel, H., 1968, Normal and critical O_2-supply of the heart, in: "Oxygen Transport in Blood and Tissue," D.W. Lübbers, U.C. Luft, G. Thews, and E. Witzleb, eds., Thieme Verlag, Stuttgart.

Federspiel, W.J., 1983, Engineering analysis of two blood transport problems: Oxygen transport in a red blood cell, Ph.D. Dissertation, University of Rochester, Rochester, New York.

Federspiel, W.J., and Popel, A.S., 1986, A theoretical analysis of the effect of the particulate nature of blood on oxygen release in capillaries Microvasc. Res., 32:164-189.

Fung, Y.-C., 1973, Stochastic flow in capillary blood vessels, Microvasc. Res., 5:34-48.

Gaehtgens, P., Meiselman, H.J., and Wayland, H., 1970, Erythrocyte flow velocities in mesenteric microvessels of the cat, Microvasc. Res., 2:151-162.

Gayeski, T.E.J., 1981, A cryogenic microspectrophotometric method for measuring myoglobin saturation in subcellular volumes; Application to resting dog gracilis muscle, Ph.D. Dissertation, University of Rochester, Rochester, NY.

Gayeski, T.E.J., Connett, R.J., and Honig, C.R., 1985, O_2 transport in the rest-work transition illustrates new functions for myoglobin, Am. J. Physiol., 248:H914-921.

Gayeski, T.E.J., and Honig, C.R., 1986, O_2 gradients from sarcolemma to cell interior in a red muscle at maximal $\dot{V}O_2$, Am. J. Physiol., 251:H789-H799.

Gayeski, T.E.J., Connett, R.J., and Honig, C.R., 1987, The minimum intracellular PO_2 for maximum cytochrome turnover in red muscle in situ, Am. J. Physiol., 252:H906-H915.

Groebe, K., and Thews, G., 1987, Theoretical analysis of oxygen supply to contracted skeletal muscle, Adv. Exp. Med. Biol., in press.

Henquell, L., and Honig, C.R., 1976, O_2 extraction of right and left ventricles, Proc. Soc. Exp. Biol. Med., 152:52-53.

Honig, C.R., and Odoroff, C.L., 1981, Calculated dispersion of capillary transit times: significance for oxygen exchange, Am. J. Physiol., 240:H199.

Honig, C.R., Feldstein, M.L., and Frierson, J., 1977, Capillary lengths, anastomoses, and estimated capillary transit times in skeletal muscle, Am. J. Physiol, 233:H122-H129.

Honig, C.R., Odoroff, C.L., and Frierson, J.L., 1982, Active and passive capillary control in red muscle at rest and in exercise, Am. J. Physiol., 243:H196-H206.

Honig, C.R., Frierson, J.L., and Gayeski, T.E.J., 1988, Anatomic determinants of O_2 flux density at coronary capillaries, Am. J. Physiol., in press.

Honig, C.R., Gayeski, T.E.J., Federspiel, W., Clark, A., Jr., and Clark, P., 1984, Muscle O_2 gradients from hemoglobin to cytochrome; new concepts, new complexities, Adv. Exp. Med. Biol., 169:23-38.

Hukmann, W., Niesel, W., and Grote, J. 1969, Untersuchungen über die Bedingungen für die Sauerstoffversorgung des Myokards an perfoundierten Rattenherzen, Pflügers Arch., 294:250-255.

Kreuzer, F., and Hoofd, L., 1987, Facilitated diffusion of oxygen and carbon dioxide, in: "Handbook of Physiology, Section 3: The Respiratory System, Volume IV: Gas Exchange," L.E. Farhi, and S.M. Tenney, eds., American Physiological Society, Bethesda, MD.

Makino, N., Kanaide, H., Yoshimura, R., and Nakimura, M., 1983, Myoglobin oxygenation remains constant during the cardiac cycle, Am. J. Physiol., 245:H237-H243.

Murray, J.D., 1974, On the role of oxymyoglobin in muscle respiration, J. Theor. Biol., 47:115-126.

Piiper, J., 1988, Role of diffusion shunt in transfer of inert gases and O_2 in muscle, Adv. Exper. Med. Biol., In Press.

Rose, C., and Goresky, C.A., 1988, In vivo comparison of non-gaseous metabolite and oxygen transport in the heart, Adv. Exper. Med. Biol., In Press.

Schmid-Schönbein, G.W., Skalak, R., Usami, S., and Chien, S., 1980, Cell distributions in capillary networks, Microvasc. Res., 19:18-44.

Sarelius, I.H., and Duling, B.R., 1982, Direct measurement of microvessel hematocrit, red cell flux, velocity, and transit time, Am. J. Physiol., 243:H1018-H1026.

Tyml, K., 1987, Red cell perfusion in skeletal muscle at rest and after mild and severe contractions, Am. J. Physiol., 252:H485-H493.

von Restorff, W., Holtz, J., and Bassenge, E., 1977, Exercise induced augmentation of myocardial oxygen extraction in spite of normal coronary dilatory capacity in dogs, Pflügers Arch., 372:181-189.

ACKNOWLEDGEMENT

We are indebted to A. Clark and P.A.A. Clark for estimates of intercellular O_2 fluxes, and for stimulating discussions of all aspects of O_2 transport over the years. Our research is supported by grants HL03290 and HL37106, from the United States Public Health Service.

EFFECTS OF PHYSIOLOGICAL FACTORS ON OXYGEN

TRANSPORT IN AN IN VITRO CAPILLARY SYSTEM

D. D. Lemon, E. J. Boland, P. K. Nair,
J. S. Olson, and J. D. Hellums

Biomedical Engineering Laboratory and
Department of Biochemistry
Rice University
Houston, TX 77251

INTRODUCTION

Mathematical simulation of oxygen transport in the microcirculation normally requires the introduction of a number of simplifying assumptions to reduce the very complicated physical situation to a form that is tractable. Unfortunately, there are difficulties in designing and executing experiments that will serve to test these assumptions critically. As a result, various workers use different sets of inadequately tested simplifying assumptions. Somewhat analogous difficulties exist in measurement of detailed oxygen concentrations and transport rates in vivo. Considerable progress has been made in these important measurements, but there are experimental difficulties in exact determination of several important parameters, including the capillary wall boundary conditions and the capillary dimensions.

These difficulties have constituted the incentive for our continuing development of an in vitro system in which the capillary dimensions are determined precisely by light and electron microscopy; the flow rate is carefully regulated; the inlet concentration of red cells or hemoglobin is controlled independently; the fractional saturation of hemoglobin is measured spectrophotometrically; and the boundary conditions in the silicone rubber capillary bed can be computed by established mathematical techniques. Accurate measurement of these variables is proving to be useful in testing and improving mathematical methods for simulation of microcirculatory oxygen transport.

Here we will review some of the results on transport of oxygen to and from hemoglobin solutions and red cell suspensions flowing in the artificial capillary system--with emphasis on effects of physiological factors. In addition we will review the mathematical simulation of oxygen transport in the experimental system. Other results and more details have been presented elsewhere (Boland et al., 1984; 1987; Stathopoulas et al., 1986; Lemon et al., 1987).

METHODS

The experimental system has been described earlier (Boland et al., 1987), so only a brief outline will be given here. A capillary embedded in a silicone rubber film of rectangular cross section is perfused with a hemoglobin solution or a red cell suspension at a known, controlled flow rate and with known, controlled inlet oxygen saturation. The external (planar) surfaces of the film are suffused with air or oxygen (in oxygen uptake experiments), or with nitrogen (in oxygen release experiments). Thus, the hemoglobin solution is oxygenated (in the uptake experiments) or deoxygenated (in the release experiments) as it flows through the capillary. In a release experiment, for example, oxygen from the lumen of the capillary diffuses through the capillary wall into the silicone rubber film, then through the silicone rubber into the gas space surrounding the surfaces of the film. The oxygen saturation of the hemoglobin diminishes as the solution flows through the capillary.

In the experiments to be discussed here the capillaries were 27 to 28 μm in diameter and 0.5 cm in length, embedded in a silicone rubber film of 170 μm thickness. At various axial positions along the capillary a microspectro-photometer is used to determine oxygen saturation of hemoglobin. The measurement yields a space-averaged oxygen saturation over a square field of view 27 μm X 27 μm.

The experiments were simulated mathematically taking into account the hemoglobin-oxygen chemical reaction kinetics, axial convection, and radial diffusion. The rate of the reversible reaction per unit volume, F, expressed as an oxygen association rate, is given by

$$F = k' C_1 C_3 - k C_2 \qquad (1)$$

where C_1, C_2, and C_3 represent the concentrations of free oxygen, oxyhemoglobin, and hemoglobin, respectively. The sum of C_2 and C_3, the total hemoglobin concentration, is assumed to be invariant with spatial position. Also k' is the bimolecular association rate coefficient, assumed constant in this work; and k is the bimolecular disociation rate coefficient, varied as a function of hemoglobin oxygen saturation by the method of Moll (1969) as modified by Vandegriff and Olson (1984). This variation of k forces the kinetic expression to be compatible with the Hill equation for equilibrium.

The partial differential equations of oxygen and oxyhemoglobin transport are given below.

$$U \frac{\partial C_1}{\partial z} = D_1 \frac{\partial^2 C_1}{\partial r^2} + \frac{1}{r} \frac{\partial C_1}{\partial r} - F \qquad (2a)$$

$$U \frac{\partial C_2}{\partial z} = D_2 \frac{\partial^2 C_2}{\partial r^2} + \frac{1}{r} \frac{\partial C_2}{\partial r} + F \qquad (2b)$$

where z, r are axial and radial space coordinates, and D_1, and D_2 are the diffusion coefficients for free oxygen and oxyhemoglobin, respectively. U denotes the axial velocity for laminar, fully developed flow in a cylindrical conduit given by

$$U = \frac{2Q}{\Pi R^2} \left[1 - \left({r}/{R} \right)^2 \right] \qquad (3)$$

where Q is the volumetric flow rate, and R is the radius of the capillary.

Determination of the boundary conditions on equations (2) requires consideration of the diffusion of oxygen in the silicone rubber film. Solution of this diffusion problem also makes it possible to analyze and interpret the experimental data in a way that is device-independent. In this work we are focusing attention on the oxygen transport in the blood, and we require knowledge of the capillary wall oxygen tension. The silicone rubber film which contains the capillary is rectangular in cross section. Thus, determination of the boundary concentration requires solving the diffusion equation in the rectangular (in cross section) film of silicone rubber which surrounds the capillary. The problem is treated as a two-dimensional Dirichlet problem. In other words, Laplace's equation is solved in the rubber film treating the capillary wall as at a uniform concentration at each axial position.

Solution to Laplace's equation for this configuration have been developed by Balcerzak and Raynor (1961). From their work it is clear that the asymptotic solution for an infinitely wide strip is applicable. This solution can be expressed in a simple closed form:

$$D_1 \frac{\partial C_1}{\partial r} = \frac{K \; \Delta p}{R \; \ln(2a/\Pi R)} \tag{4}$$

Where K denotes the oxygen permeability of silicone rubber, Δp denotes the oxygen tension difference from the capillary wall to the external face of the silicone rubber film, and "a" denotes the film thickness. The left-hand side of equation (4) corresponds to the capillary wall oxygen flux. Equation (4) is used as a boundary condition on equations (2), as well as for analysis of the experimental results.

RESULTS AND DISCUSSION

Comparison with Mathematical Model

Example results from the experiments on oxygen transport are given in Figs. 1 and 2 in comparison with calculated results from equations (1) - (4). All the curves for hemoglobin solutions in the figures are from the mathematical model. In all cases the chemical reaction, diffusion, and permeability constants were obtained from the literature or independent measurements, and the geometrical and flow parameters were measured directly. Thus, the correspondence between the observed and calculated saturation curves in Figs. 1 and 2 serves to validate both the mathematical model and the experimental system.

Dependence on Oxygen Tension

Detailed calculations have been made of the oxygen tension distribution throughout both the capillary and the surrounding silicone rubber film by means of equations (1) - (4), and the more detailed solution of Balcerzak and Raynor (1961). These results make it possible to compare the intracapillary resistance to oxygen transport to that of the silicone rubber film. It was found that the intracapillary resistance constituted 60 to 80% of the overall resistance. This finding on resistances is important in that if the resistance in the silicone rubber dominated, it would not be possible to determine intracapillary resistance accurately.

The overall oxygen tension gradient in the uptake experiments is of the order of 150 mmHg when air is used external to the capillary--corresponding approximately to the oxygen tension in the gas space surrounding the capillary. The overall gradient in the release experiments is an order of magnitude lower since it is limited to the dissolved oxygen tension (roughly of the order of the P_{50}, which ranged from 10 to 30 mmHg).

Oxygen uptake and release experiments were carried out on hemoglobin solutions and on red blood cell suspensions of the same hemoglobin content

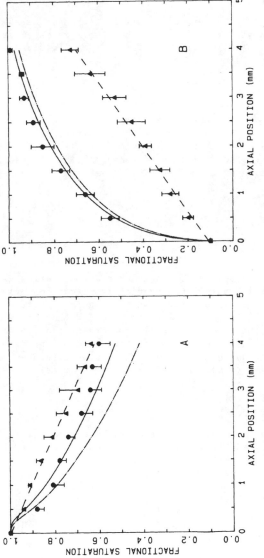

Figure 1: Comparison of O_2 transport results for hemoglobin solutions with red cell suspension of the same hemoglobin content flowing in a 27 μm artificial capillary. The experiments were carried out at a flow rate of 12 μℓ/hr (A), and 23 μℓ/hr (B). **Data points**, mean ± SD for 8 replicate experiments at 37°C, pH 7.4, with a 1.0 mM hemoglobin content. **Circles**, experimental results for hemoglobin solutions O_2 half-saturation pressure of hemoglobin; $(P_{50}) = 13$ **mmHg** ; **solid curves** theoretical simulation curves for same conditions. **Triangles and broken lines**, experimental findings on red cell suspensions $(P_{50} = 20$ mmHg). **Dashed-dotted line**, plot of theoretical curve for a hemoglobin solution with a P_{50} of 20 mmHg.

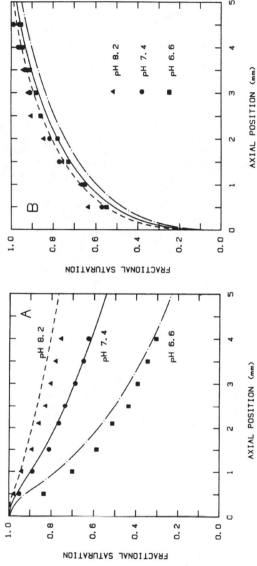

Figure 2: Effect of pH on O_2 exchange. **Triangles,** observed data taken using hemoglobin at pH 8.2. **Circles,** data for pH 7.4. **Squares,** data for pH 6.6. In all cases, heme concentration was 2 mM. **Dashed, solid,** and **dashed-dotted lines,** theoretical curves calculated for pH 8.2, 7.4 and 6.6, respectively. A: observed (avg SD = 0.03) and calculated (root-mean-square difference = 0.04) profiles of O_2 release using capillary 24. Samples were equilibrated with 1 atm of O_2 before their entry into capillary. Flow rate was always 10 µℓ/h. B: observed (avg SD = 0.02) and theoretical (avg SD = 0.04) profiles of O_2 uptake in capillary 37. Air (21% O_2) was used in capillary gas space. Flow rate in each case was 42 µℓ/h.

(Figs. 1 and 2). Because of the large difference in overall gradients, oxygenation occurred much more rapidly than deoxygenation. The key parameter is the capillary residence time of the solution element being examined. This time is determined by the bulk axial flow rate and the distance of the element from the entrance to the capillary. For example, in the uptake experiment shown in Fig. 1, the residence time required for 50% saturation of the hemoglobin solution was ~0.050 s, which corresponded to an axial position ~0.5 mm downstream from the entrance at a linear flow rate of 11.2mm/s. In the release experiment the residence time for 50% deoxygenation was ~1 s.

Comparison of Red Cell Suspensions with Homogeneous Solutions

In the case of uptake, the rate of oxygenation of the red blood cell sample was approximately five times smaller than that of an equivalent concentration of free hemoglobin flowing through the capillary at the same velocity (Fig. 1B).

In the O_2 release experiments (Fig. 1A) the difference between the transport curve for red cells and that for free hemoglobin is much smaller than the difference observed between the corresponding curves for uptake (Fig. 1B). However, in the case of release experiments, direct comparison between the data for the cell suspension and free hemoglobin is difficult due to affinity differences. The cell suspension exhibited a P_{50} of 20 mmHg compared with 13 mmHg for the hemoglobin solution. The rate of oxygen release in the capillary is roughly proportional to the P_{50} of the hemoglobin sample. Consequently, the higher P_{50} of the erythrocyte suspension partly compensates for the extra resistance caused by packaging the hemoglobin in discrete cellular units. The dashed-dotted line represents a theoretical curve for a free hemoglobin sample with a P_{50} equal to that of the cell suspension. Thus, for release experiments, one should compare the experimental results for red cells (triangle and broken line in Fig. 1A) with the bottom theoretical curve (dashed-dotted line of Fig. 1A) for a hemoglobin solution exhibiting the same P_{50} value. A similar theoretical calculation was carried out for the uptake experiments, and as shown in Fig. 1B (dashed-dotted line), the P_{50} value of the sample has little effect on the uptake curve. Thus measurements of O_2 uptake allow a more direct experimental evaluation of the extra resistance to transport caused by using cell suspensions. The results in Fig. 1 support the idea developed by Hellums (1977), and Baxley and Hellums (1983) in previous theoretical work; that the resistance to O_2 transport in the blood itself is a significant fraction of the total resistance in the microcirculation and is much higher than that suggested by a number of investigators.

Physiological Effects

The equilibrium parameters exert a marked influence on the O_2 release process (Fig. 2A). In deoxygenation experiments, the net flux from the capillary to the external gas space is governed by the O_2 tension in the capillary, which in turn is determined by the equilibrium binding parameters of the hemoglobin molecules. For example, at 50% saturation, the average O_2 tension in the capillary is approximately equal to the O_2 half-saturation pressure of hemoglobin (P_{50}) of the protein sample, and as a result, the rate of efflux is approximately proportional to this value. If the affinity of hemoglobin is very high, the O_2 tension must be lowered to a very small value before net release from the protein occurs.

In contrast to the situation for release, the rate of oxygenation is roughly independent of the equilibrium parameters for O_2 binding. In this case, the O_2 flux is determined primarily by the external O_2 tension, which is always 10-20 times greater than the tension in the capillary fluid. As a result, the exact value of P_{50} for the sample exerts little influence on the observed saturation distributions along the capillary.

42

The equilibrium parameters for O_2 binding to hemoglobin are regulated in vivo by pH and organic phosphate concentration. The effects of pH on the speed of O_2 uptake and release in the capillary system are shown in Fig. 2. The P_{50} values for these solutions were ~2, 5, and 12 mmHg for pH 8.2, 7.4, and 6.6, respectively. As expected, the oxygenation curves exhibited little dependence on hydrogen ion concentrations, whereas decreasing the pH caused a marked increase in extent of deoxygenation at each axial position. The latter effect correlates directly with the larger P_{50} at the lower pH. Under acid conditions, the O_2 tension in the capillary required for net release from hemoglobin was much higher, larger diffusion gradients occurred, and the rate of transport was higher.

Similar results (Lemon et al., 1987) were obtained when inositol hexaphosphate was added to the hemoglobin solutions at pH 7.4. This organic phosphate exerts a larger effect than 2,3-diphosphoglycerate on the affinity of hemoglobin for O_2 and causes the P_{50} to rise from 5-7 to 30-40 mmHg. The addition of inositol hexaphosphate exerted only small changes in the saturation distribution for O_2 uptake, whereas a dramatic increase in the speed of deoxygenation was observed.

The effects of temperature were also studied (Lemon et al., 1987). The largest effect is a roughly threefold increase in the P_{50} for O_2 binding to hemoglobin in going from 25° (~6 mmHg) to 37° (~15 mmHg). An increase in the rate of deoxygenation was observed consistent with the change in P_{50}. The O_2 uptake curves changed very little with increasing temperature.

Although many of the results presented in this paper were obtained with hemoglobin solutions flowing through a vessel about the size of an arteriole, the basic conclusions appear to apply to the more physiological situation with red cells and smaller capillaries. First, since the absolute gradients for uptake are roughly 10 times greater than those observed for release, much longer transit times are required for complete deoxygenation, and this explains the need for longer capillaries in aerobic muscle tissues (Honig et al., 1977). Second, the decrease in blood pH and elevation in temperature produced by heavy exercise enhances both the rate and extent of deoxygenation in the microcirculation, whereas these changes in pH and temperature exert little or no effect on the rate of O_2 uptake in alveolar capillaries. As a result, the overall efficiency of O_2 transport from the lungs to respiring tissues is enhanced. Third, long-range adaptation to high altitudes or continuous exercise by elevation of intracellular 2,3-diphosphoglycerate levels produces similar effects. The rate of deoxygenation is enhanced significantly with little or no effect on the already rapid rate of O_2 uptake.

REFERENCES

Balcerzak, M. J., and Raynor, S., 1961, Steady state temperature distribution and heat flow in prismatic bars with isothermal boundary conditions, Int. J. Heat Mass Transfer 3:113-125.

Baxley, P. T., and Hellums, J. D., 1983, A simple method of simulation of oxygen transport in the microcirculation, Ann. Biomed. Eng., 11:401-416.

Bird, R. B., Stewart, W. E., and Lightfoot, E. N., 1960, Transport Phenomena, 42-47.

Boland, E. J., Nair, P. K., Lemon, D. D., Olson, J. S., and Hellums, J. D., 1987, An in vitro capillary system for studies on microcirculatory O_2 transport, J. Appl. Physiol., 62:791-797.

Hellums, J. D., 1977, The resistance to oxygen transport in the capillaries relative to that in the surrounding tissue, Microvascular Res., 13:131.

Honig, C. R., Feldstein, M. L., and Frierson, J. L., 1977, Capillary lengths, anastomoses, and estimated capillary transit times in skeletal muscle, Am. J. Physiol., 233(Heart Circ. Physiol. 2):H122-H129.

Lemon, D. D., Nair, P. K., Boland, E. J., Olson, J. S., and Hellums, J. D., 1987, Physiological factors affecting O_2 transport by Hemoglobin in an in vitro capillary system, J. Appl. Physiol., 62:798.

Moll, W., 1969, The influence of hemoglobin diffusion on oxygen uptake and release by red cells, Respir. Physiol., 6:1-15.

Stathopoulos, N. A., and Hellums, J. D., 1986, Oxygen transport studies of normal and sickle erythrocyte suspensions in artificial capillaries, in "Oxygen Transport to Tissue VIII," I. S. Longmuir, ed., Plenum Publishing Corp., New York.

Vandegriff, K. D., and Olson, J. S., 1984, The kinetics of O_2 release by human red blood cells in the presence of external sodium dithionite, J. Biol. Chem., 259:12609-12618.

Vandegriff, K. D., and Olson, J. S., 1984, Morphological and physiological factors affecting oxygen uptake and release by red blood cells, J. Biol. Chem., 259:12619-12627.

IN VIVO COMPARISON OF NON-GASEOUS METABOLITE AND

OXYGEN TRANSPORT IN THE HEART

C. P. Rose, C. A. Goresky, and G.G. Bach
McGill University

J. B. Bassingthwaighte and S. Little
University of Washington

INTRODUCTION

Oxygen transport has traditionally been approached as a specialized subject with little connection to the large amount of data on transport of other substances, equally essential for steady-state metabolism. Heuristically, there is no reason to expect a major difference but measurements of tissue PO_2 with oxygen electrodes in organs with high oxygen consumptions have yielded data which are incompatible with the classical Krogh-cylinder model of capillary-tissue oxygen transport. A number of alternative models, including diffusional shunting and flow heterogeneity, have been developed on the assumption that oxygen transport is a special case, with little or no consideration of the overall nature of organ transport as reflected in the transport of other substances equally essential for metabolism. As we shall show, when examined in this light, oxygen transport is not essentially different from that of other substances. With the understanding afforded by this approach and recent developments based on it, future investigational effort can now be profitably directed at more complex problems, such as the role of impaired oxygen transport in certain pathological states of vital organs.

FREE FATTY ACIDS

In the case of the heart there is, we think, a direct analogy between the transport of free fatty acids (FFA) and oxygen. After fasting and during aerobic exercise, FFA are the major substrate for energy production in the heart. Like oxygen, FFA are highly extracted (up to 60% when circulating lactate is low) and like oxygen they are bound to a carrier in the blood (albumin, in the case of FFA, and hemoglobin, in the case of oxygen) Also, like oxygen it had been assumed that, being lipid soluble, FFA would diffuse so rapidly into the myocyte that venous outflow concentration would reflect the tissue concentration. Utilizing the multiple indicator dilution technique with labeled palmitate as an example of an FFA, Rose and Goresky (1976) refuted these assumptions. Labeled albumin was used as an intravascular reference and labeled sucrose as an inert marker which permeates the capillaries via the aqueous pores and distributes in the interstitial space. All three tracer were injected simultaneously in the coronary artery of an intact, working heart and venous samples were collected rapidly from the coronary sinus. The activity of each tracer was measured in each sample and divided by the respective activity in the injection mixture, to give the normalized outflow concentration. Figure 1 shows two sets of data before and after intravenous lactate infusion. The form of the albumin outflow is determined solely by the distribution of large and small vessel transit times, of the sucrose, additionally, by the volume of the interstitial space, and of the palmitate, also by permeation into and metabolism within the myocyte. When intracellular sequestration is inhibited by providing an alternative substrate (lactate), more of the labeled FFA that entered the myocyte returns to the capillary. Thus, outflow of the labeled FFA is composed of two components, an early exiting portion which has never left the capillary and a later or returning component which is material which has left and which returns later in time.

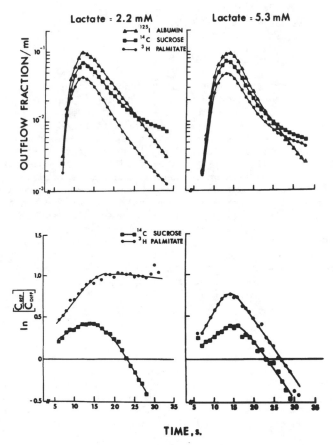

Figure 1. Normalized coronary sinus outflow dilution curves after simultaneous injection of labeled albumin, sucrose and palmitate into the coronary artery of an anesthetized dog before and after the infusion of lactate. The lower panels are the corresponding log ratio curves where C_{Ref} represents albumin and C_{Diff} represents sucrose or palmitate.
Reproduced from Rose, C.P., Goresky, C.A., 1977, Constraints on the uptake of labeled palmitate by the heart: the barriers at the capillary and sarcolemmal surfaces and the control of intracellular sequestration, <u>Circ. Res.</u>, 41: 534-545 by permission of the American Heart Association, Inc.

On inspection of this data a few features are immediately apparent. First, the rate constant for capillary permeation of the FFA (which is roughly indicated by the ratio of albumin to palmitate at the peak of the curves) is only slightly greater than that for sucrose, a substance which permeates the capillary endothelium only via the aqueous pores. Second, most of the FFA remaining in the coronary sinus has not even left the capillary in its passage. Thirdly, there is no change in the throughput component when the net transcapillary flux of FFA was reduced by lactate competition.

Modeling of the processes involved in the exchange process revealed that, in the steady state, calculated intracellular concentrations of FFA were only one-twentieth of the plasma concentration. This prediction was verified a few years later by van der Vusse et al (1982) who measured tissue FFA concentrations in rapidly frozen samples of myocardium. The explanation is that, even for a substrate as important as FFA, the tissue concentration will be lower than the venous concentration when there is sequestration behind a relative barrier to diffusion (Goresky et al, 1983).

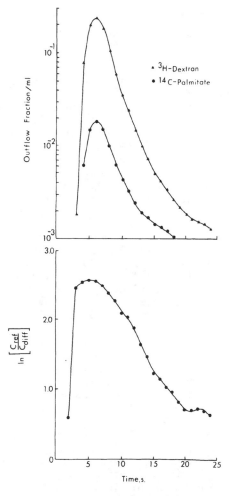

Figure 2. Normalized coronary sinus outflow dilution curves after simultaneous injection of labeled dextran and palmitate into the aortic root of an isolated rabbit heart preparation. Compare with Figure 1. The absence of albumin in the perfusion fluid caused a large increase in the extraction of palmitate. Unpublished data of Bassingthwaighte and Little.

The nature of this barrier is revealed when tracer amounts of palmitate are dissolved in water and dextran is used as a reference, in order to eliminate albumin from the perfusion system in an isolated heart preparation (Little and Bassingthwaighte, 1987). Figure 2 shows that when albumin is removed the extraction of FFA increases to 95% and the estimated rate constant for FFA diffusion from the capillary increases by an order of magnitude. It appears that, within physiologically attainable plasma FFA concentrations, the capacity to sequester FFA is limited only by the diffusion barrier. If we can assume that the permeability of the capillary endothelium was no different in the two situations, then the increase in the transport rate constant was due only to the elimination of albumin binding sites in the plasma within the capillary. The true permeability may be even higher than estimated from this data because it is impossible to dissolve even tracer amounts of FFA in water without formation of micelles, which would tend to limit diffusion. The appendix shows mathematically why an increase in the apparent vascular volume of distribution of a permeating substance will reduce the rate constant for permeation. This is a general phenomenon and and applies to any substance with a large relative volume of distribution or binding capacity in the capillary. Of course, the best example of such a substance is oxygen.

OXYGEN

August Krogh (1919) imagined a capillary surrounded by a large cylindrical tissue space homogeneously consuming oxygen. While the model was useful in proving that physical diffusion of oxygen was sufficient to explain the delivery of oxygen to tissue at physiologic rates of oxygen consumption in resting skeletal muscle, the model did not include blood flow rate, capillary length or potential inhomogeneities in oxygen diffusion associated with membranes. Motivation for more complex modeling came from the accumulation of tissue PO_2 histograms from various organs under various conditions. As a general rule, the smaller the polarographic electrode the more likely the PO_2 histogram will represent the true intracellular PO_2. Whalen and co-workers (Whalen, 1973; Schubert et al, 1978) have been the only investigators to consistently use electrodes with tip diameters on the order of $1\mu m$. Electrodes larger than this are likely to interfere with the microcirculation or give a mixture of capillary and tissue rather than a true extravascular value. Neo-Krogh models with varying capillary geometries and diffusional interaction have been developed (Metzger, 1969; Grunewald and Sowa, 1978; Schubert et al., 1978) and predicted random PO_2 histograms have been compared with the data. There is agreement in cases of low oxygen consumption but not in the more interesting situations.

In the liver, an organ with a relatively large vascular space and discontinuous capillaries, tissue PO_2 measured with surface electrodes is between arterial and venous values (Matsumura et al, 1986). The earlier measurements of Kessler (1967) using penetrating $50\mu m$ electrodes probably were too low because the electrode interfered with local flow in the sponge-like liver. In the case of the heart, there is a major discrepancy between the predictions of neo-Krogh modeling and tissue PO_2 histograms. The problem is was succinctly described by Lubbers in 1982:

> "We are able to show that...an oxygen pressure field can be characterized...by a PO_2 frequency histogram obtained from 100-200 local PO_2 measurements using classes of 5 mmHg...For example, in a dog with a PO_2 in the sinus coronarius of 18 mmHg, the maximum of the histogram is in the class of 10-20 mmHg (32%); 11% of values are below 10 mmHg, i.e. about 43% of PO_2 values are equal to or below the venous PO_2. Since, due to the small capillary distances in the myocardium, [predicted] PO_2 [difference]

between capillary and tissue is only in the range of 2-3 mmHg, this cannot be explained by a simple capillary model."
Data obtained using smaller electrodes are even more anomalous (Whalen, 1973). The discrepancy between data and model is not resolved by inserting heterogeneity of capillary flows or capillary diffusional interaction into the standard model (Grunewald and Sowa, 1978).

Another set of observations also emphasizes the large gradient between venous and tissue PO_2. In situ, the myocardium begins to fail mechanically and produce lactate when blood flow is reduced such that the PO_2 in the coronary sinus is 9-11 torr (Case, 1966). Isolated cardiac myocytes, however, maintain oxygen consumption down to a PO_2 of 0.2 torr (Wittenberg and Robinson, 1981). There are two orders of magnitude between the two definitions of hypoxia!

There are two and only two explanations for these anomalies; both necessitate a rejection of the classical Krogh model. Either there is a diffusional shunt at the level of the arterioles and venules which would lower arteriolar PO_2 and raise venous PO_2 relative to tissue PO_2, as proposed by Schubert et al (1978), or the diffusion rate of oxygen from blood to tissue is much less than that measured across a slab of muscle tissue.

There is a relatively easy way to discover which explanation is the correct one. If there were a large diffusional shunt, coronary sinus blood samples collected sequentially after a bolus of blood containing tracers for erythrocytes and oxygen was injected into a coronary artery would show tracer oxygen appearing before the erythrocytes, which would necessarily have to take the longer pathway though the tissue. On the other hand, if there were a resistance to diffusion the outflow pattern would resemble that of other substances such as FFA undergoing barrier-limited transport (Rose and Goresky, 1976). Using $18O_2$, a heavy isotope of oxygen as the oxygen tracer we performed such an experiment (Rose and Goresky, 1985). The results are shown in Figure 3 and were unequivocal; there was no precession of oxygen before the erythrocytes but there was evidence for a barrier between erythrocytes and mitochondria, probably at the level of the endothelial membrane. Previously, the permeability of continuous capillaries to oxygen had been measured in only one organ, the rete mirabile of the eel, which consists only of capillaries. During counter-current perfusion of this organ with erythrocyte-free, oxygenated buffer the endothelial barrier reduced the apparent diffusion coefficient of oxygen to 0.5% of its free diffusion coefficient in water (Rasio and Goresky, 1983); the permeability was about twice that for tracer water in the same system. In the heart, on the other hand, the relatively high but finite permeability of the endothelial cell membrane to oxygen is decreased by two orders of magnitude by the hemoglobin binding of tracer oxygen to yield an apparent rate constant for oxygen transfer no greater than that for the sodium ion, which diffuses only via aqueous pores in the capillary. The analogous experiment to the albumin-free FFA experiment would be a tracer oxygen experiment with hemoglobin free perfusate. Unfortunately, our quadrupole mass spectrometer is not sensitive enough to detect such small amounts of heavy oxygen as would dissolve in water against a background of the much more abundant, dissolved carbon dioxide. We would predict that when such data becomes available it will look something like that in Figure 2 if tracer oxygen is substituted for tracer FFA. The factor by which hemoglobin binding reduces the rate constant for oxygen transport can be calculated from knowledge of the ratio of the oxygen content of plasma to that of red cells in fully oxygenated blood, 0.3:20.0 = 0.015. Now, since the ratio of sodium to oxygen rate constants is about 1.4 (Rose and Goresky, 1985), the ratio of oxygen to sodium capillary permeabilities is about 94. Thus, the large intravascular hemoglobin binding capacity reduces the rate constant for oxygen exchange by two orders of magnitude and the diffusional limitation offered by the capillary endothelial membrane is effectively magnified by this factor. This phenomenon was not envisioned by Krogh and has not been included in any of the previous attempts to model in vivo oxygen transport. However, it could have been predicted from the FFA data which has been available for more than ten years.

Modeling of the transport process also gives an estimate of the ratio of extravascular to intravascular oxygen concentrations. We estimate that the extravascular concentration is at least a factor of three lower than the blood concentration at any point along the capillary. This estimate is dependent on the details of the modeling which omits some aspects of oxygen

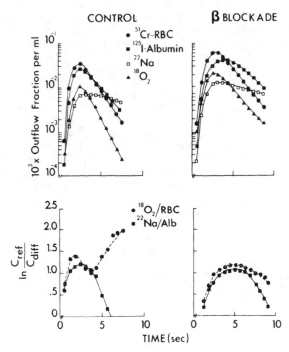

Figure 3. Normalized coronary sinus outflow curves after simultaneous injection of labeled red cells, albumin, sodium ion and oxygen into the coronary artery of an anesthetized dog. Note that the peak extraction of oxygen is no greater than that of sodium ion and that reduction of oxygen consumption after β-blockade causes a change only in the returning component. Reproduced from Rose, C.P., Goresky, C.A., 1985, Limitations of tracer oxygen uptake in the canine coronary circulation, Circ. Res., 56: 57-71 by permission of the American Heart Association, Inc.

transport such as nonlinear hemoglobin-oxygen binding and potentially zero-order sequestration. Data from hemoglobin-free, isolated hearts, in which myoglobin saturation is used to estimate intracellular PO_2, suggest that the transcapillary gradient may be as much as two orders of magnitude (Araki et al, 1983). This preparation is, however, characterized by huge perfusion rates and interstitial edema and certainly does not accurately reflect the physiological situation. The data indicate that even if the transcapillary gradient is only one order of magnitude, the normal heart functions at an intracellular PO_2 no more than an order of magnitude above the critical PO_2 for isolated myocytes, which is estimated at about 0.2 mmHg (Wittenberg and Robinson, 1981). While the oxygen transport system is adequate for normal heart function, it is not unlikely that its limitations become critical when oxygen consumption is chronically increased, for example in valvular heart disease. In this situation the eventual deterioration in heart function which results in the symptoms of heart failure could be caused by some dysfunction related to chronic tissue hypoxia.

The large capillary-tissue oxygen concentration gradient predicted from our data in the heart was supported by data from exercising skeletal muscle obtained by Gayeski and Honig (1986) who measured oxygen saturation of myoglobin microcryophotometrically in cross sections of muscle fibers, freeze-clamped during twitch contraction. They found very low PO_2's of about 2-3 mm Hg, much lower than the venous PO_2, and exhibiting little intracellular gradient. If we can assume by analogy with the heart that there is no large vessel diffusional shunt then this data can be explained by a relative diffusion barrier at the level of the endothelium. An alternative explanation has been proposed by Groebe and Thews (1986). They have proposed that the steep capillary-tissue gradient is the result of high flux densities through a "resistance layer" composed of endothelium and interstitial space which do not contain hemoglobin or myoglobin. If this formulation is correct then any apparent barrier should disappear at low rates of oxygen consumption and outflow dilution curves should conform to the flow-limited or delayed-wave model of exchange (Goresky et al, 1970). We believe that this is unlikely because the data from the heart suggest that when net transcapillary flux is lowered in the case of both FFA and oxygen there is no change in the apparent throughput component; the barrier appears to present regardless of the rate of transport. A definitive answer must await outflow dilution data from resting skeletal muscle or unloaded heart.

APPENDIX

Mathematical description of the effect of intravascular binding on uptake rate constant and transit time

The conservation equation for a single capillary and an associated extravascular space in which oxygen is bound reversibly to moving red cells is

$$\frac{\partial u}{\partial t} + W \frac{\partial u}{\partial t} + \gamma \frac{\partial v}{\partial t} + R_c \, \beta \left[\frac{\partial y}{\partial t} + W \frac{\partial y}{\partial x} \right] + R_m \, \theta \, \frac{\partial z}{\partial t} + R_m \, \theta \, k_3 z = 0,$$

where   u is the plasma concentration in the plugs of plasma between the red cells,

             v is the plasma concentration in the small annulus of plasma around the red cells, y is the concentration in the red cells,

z is the concentration in the extravascular space,

W is the velocity of the red cells,

β is the ratio of red cell volume to moving plasma volume such that hematocrit = $\beta/(1+\beta)$,

γ describes the ratio of the volume of plasma in excess of that moving plug-like with the red cells,

θ is the ratio of extravascular volume to plasma volume for oxygen,

R_c is the ratio of red cell to plasma oxygen concentration,

R_m is the ratio of extravascular to plasma oxygen concentration in an equilibrium situation, and,

k_3 is the first-order rate constant for irreversible sequestration within the extravascular space.

Note that oxygen uptake is close to zero-order in the physiological situation and that the sigmoidal nature of the hemoglobin-oxygen dissociation curve has not been accounted for. Since this simplified model appears to explain the available data to a first approximation, these refinements have not been included. As more precise data becomes available, their inclusion may become necessary.

The mass balance equation for the extravascular space is

$$\frac{\partial z}{\partial t} - k_1 u + k_2 z + k_3 z = 0,$$

where k_1 is the rate constant for outward transport from the capillary and,

k_2 is the rate constant for inward transport to the capillary from the extravascular space.

The details of the simultaneous solution of these two partial differential equations will not be given here because of space limitations. The final result is,

$$u(x,t) = \frac{q_0}{F_c} \cdot e^{-\frac{k_1 R_m \theta}{(1 + R_c\beta)}\frac{x}{W}} \cdot \delta\left[t - \frac{(1 + \gamma + R_c\beta)}{(1 + R_c\beta)}\frac{x}{W}\right]$$

$$+ \frac{q_0}{F_c} \cdot e^{-\frac{k_1 R_m \theta}{(1 + R_c\beta)}\frac{x}{W}} \cdot e^{-(k_2 + k_3)\left[t - \frac{(1 + \gamma + R_c\beta)}{(1 + R_c\beta)}\right]}$$

$$\cdot \sum_{n=1}^{\infty} \frac{\left[\frac{k_1 k_2 R_m \theta}{(1 + R_c\beta)}\frac{x}{W}\right]^n \left[t - \frac{(1 + \gamma + R_c\beta)}{(1 + R_c\beta)}\right]^{n-1}}{n ! \, (n-1)!}$$

$$\cdot S\left[t - \frac{(1 + \gamma + R_c\beta)}{(1 + R_c\beta)}\right],$$

where q_0 is the amount of tracer injected and F_c is the flow rate in the capillary.

The first term in the above expression represents that part of the inflow that does not leave the capillary and the second term is the part which has left and returns later in time. The important point here is that the rate constant for capillary exchange, k_1, is modified by a factor,

$R_m \theta / (1 + R_c \beta)$, which, for oxygen, is roughly the ratio of plasma to red cell oxygen contents.

In the case of the FFA all of the tracer is confined to the plasma phase so that $\gamma = 0$. Also, the effect of binding to albumin can be modeled by substituting

$$\frac{[\text{free FFA}]}{[\text{total FFA}]} = \frac{1}{1 + R_c \beta} .$$

The FFA also bind to albumin in the interstitial space and consequently encounter a resistance at the sarcolemmal membrane which is not included in this simplified model.

REFERENCES

Araki, R., Tamura, M., Yamazaki, I., 1983, The effect of intracellular oxygen concentration on lactate release, pyridine nucleoside reduction, and respiration rate in rat cardiac tissue, Circ. Res., 53: 448-455.

Case, R.B., 1966, Effect of low PO2 on left ventricular function, in Proceedings Int. Sym. Cardiovasc. Respir., Effects of Hypoxia, pp. 191-207, Karger, Basal/New York.

Gayeski, T.E.J., Honig, C.R., 1986, O_2 gradients from sarcolemma to cell interior in red muscle at maximal VO_2, Am. J. Physiol., 251: H789-H799.

Goresky, C. A., Ziegler, W.H., Bach, G.G., 1970, Capillary exchange modeling: barrier-limited and flow-limited distribution, Circ. Res., 27: 739-634.

Goresky, C.A., Bach, G.G., Rose, C.P., 1983, Effects of saturating metabolic uptake on space profiles and tracer kinetics, Am. J. Physiol., 244: G215-G232.

Groebe, K., Thews, G., 1986, Theoretical analysis of oxygen supply to contracted skeletal muscle, in Adv. Exp. Med. Biol., 200, Oxygen Transport to Tissue VIII, ed. I.S. Longmuir, 495-514.

Grunewald, W.A., Sowa, W., 1978, Distribution of the myocardial tissue PO_2 in the rat and the inhomogeneity of the coronary bed, Pflugers Arch., 374: 57-66.

Kessler, M., 1967, Normale und kritische Sauerstoffversorgung der Leber bei Normo- und Hypothermie., Habil.-Schrift, Marburg/Lahn.

Krogh, A., 1919, The number and distribution of capillaries in muscles with calculations of the oxygen pressure head necessary for supplying tissue, J. Physiol., (London), 82: 490-415.

Little, S., Bassingthwaighte, J.B., 1987, unpublished data

Lubbers, D.W., 1982, Oxygen supply to the myocardium, in Microcirculation of the Heart, H. Tillmans, W. Kubler, H. Zebe, eds., Springer-Verlag, Berlin, p 119.

Matsumura, T., Kauffman, F.C., Meren, H., Thurman, R.G., 1986, O uptake in periportal and pericentral regions of liver lobule in perfused liver, Am. J. Physiol., 250: G800-G805.

Metzger, H., 1969, Distribution of oxygen partial pressure in a two-dimensional tissue supplied by capillary meshes and concurrent and countercurrent systems, Math. Biosciences., 5: 143-154.

Rose, C.P., Goresky, C.A., 1977, Constraints on the uptake of labeled palmitate by the heart: the barriers at the capillary and sarcolemmal surfaces and the control of intracellular sequestration, Circ. Res.,41: 534-545.

Rose, C.P., Goresky, C.A., 1985, Limitations of tracer oxygen uptake in the canine coronary circulation, Circ. Res., 56: 57-71.

Rasio, E.A., Goresky, C.A., 1979, Capillary limitation of oxygen distribution in the rete mirabile of the eel (<u>Anguilla anguilla</u>), <u>Circ. Res.</u>, 44: 498-504.

Schubert, R.W., Whalen, W.J., Nair, P., 1978, Myocardial PO_2 distribution: Relationship to coronary autoregulation, <u>Amer. J. Physiol.</u>, 234: H361-H370.

van der Vusse, G.J., Roeman, Th.H.M., Prinzen, F.W., Coumans, W.A., Reneman, R.S., 1982, Uptake and tissue content of fatty acids in dog myocardium under normal and ischemic conditions, <u>Circ. Res.</u> 50: 538-546.

Whalen, W.J., Intracellular PO_2 in heart and skeletal muscle, <u>Physiologist</u> 14: 69-82, 1971.

Wittenberg, B.A., Robinson, T.F., 1981, Oxygen requirements, morphology, cell coat, and membrane permeability of calcium tolerant myocytes from hearts of adult rats, <u>Cell Tissue Res.</u>, 216: 231-251.

ROLE OF DIFFUSION SHUNT IN TRANSFER OF INERT GASES AND O_2 IN MUSCLE

Johannes Piiper

Abteilung Physiologie, Max-Planck-Institut für
experimentelle Medizin, D-3400 Göttingen, F.R.G.

INTRODUCTION

Diffusion shunt designates diffusive exchange between precapillary (arterial) and postcapillary (venous) vessels, leading to functional by-passing of the capillary bed. Diffusion shunt is expected to occur where arterial and venous vessels lie sufficiently close to each other. They may belong to the same capillary area or to different capillary areas (Fig. 1).

In general, exchange of gases between vessels of any size is possible. It occurs according to the diffusive conductance or diffusing capacity, which is determined by the pertinent physical constants (diffusivity and solubility) and geometry (surface area and distance), and by the effective partial pressure differences. Thus also the extent of extracapillary gas transfer and diffusive shunting is determined by these parameters. The major part of gas exchange takes place in capillaries (between capillary blood and tissue) due to the large surface area and the short diffusion distances. But the differences between capillary and extracapillary

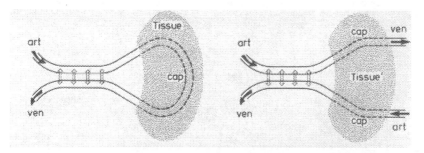

Fig. 1. Schema of diffusion shunt by gas exchange (shown by open double arrows) between arterial (art) and venous (ven) vessels. Capillaries (cap) are represented by broken lines and tissue supplied by capillaries, by shading. Exchange between artery and vein perfused by same blood (left) or of differing origin (right).

vessels with respect to these parameters are quantitative, not qualitative. Thus extracapillary gas exchange including arterio-venous shunting
is expected to occur, only its extent and significance are open to
question.

In this report, experimental evidence for diffusion shunt obtained by
inert gas washout from muscle is reviewed, and simple models allowing some
predictions also for diffusive shunting of O_2 and CO_2 in steady state are
investigated.

EXPERIMENTAL EVIDENCE FROM INERT GAS WASHOUT

In experiments on isolated blood-perfused gastrocnemius muscles of
the dog, the washout of various inert test gases was studied, by gas-
chromatographic analysis of venous blood, after equilibration of the
muscle by blood containing the test gases administered via inspired gas
(Piiper and Meyer, 1984). The washout kinetics of all test gases was non-
monoexponential and was probably due to unequal distribution of blood flow
to tissue volume. This has been confirmed in the same preparation by use
of intraarterially injected, embolizing microspheres (Piiper et al.,
1985). The washout could be approximated by a sum of three exponential
components characterized by their rate constants, k (dimension, time^{-1}).
In a number of experiments on both resting and stimulated muscles the
simultaneously measured washout kinetics of methane (CH_4) and sulfur hexa-
fluoride (SF_6) were compared. It was found that on the average the ratio
$k(CH_4)/k(SF_6)$ was 1.0.

If the washout were limited by perfusion only, the k ratio is expected
to be determined by the ratio of the respective blood/tissue partition
coefficients, and the predicted $k(CH_4)/k(SF_6)$ ratio would be 2.0. Diffusion limitation in blood/tissue equilibration would increase the
$k(CH_4)/k(SF_6)$ ratio from the value predicted for pure perfusion limitation, 2.0, to a higher value. The maximum value, reached at exclusive
diffusion limitation, would amount to 7.3 (= CH_4/SF_6 ratio of diffusion
coefficients).

The only reasonable model able to explain the experimental findings
is diffusion shunt by veno-arterial back diffusion. By this process the
better diffusible gas would more extensively diffuse from venous to
arterial blood and thus would become more delayed in washout compared to
the better diffusible gas. This means that the $k(CH_4)/k(SF_6)$ ratio would
be decreased. Model calculations show that the limiting value, attained in
the case that diffusion shunt is the process predominantly limiting the
washout rate, is 0.55. Thus, indeed, the experimental value, 1.0, is in
the range explainable by veno-arterial back diffusion.

In other experiments on the same dog gastrocnemius preparation, the
clearance of locally injected radioactive xenon was used to calculate
local tissue perfusion according to Lassen et al. (1964). The tissue
perfusion thus determined was on the average 60% of the muscle blood flow
measured by venous outflow (Cerretelli et al., 1984). The shunting of
considerable extent may well be in part due to diffusion shunt, in particular since there is no evidence for anatomical shunt vessels (arterio-
venous anastomoses) in skeletal muscle.

THEORETICAL MODELING AND PHYSIOLOGICAL CONDITIONS

Washout

In modeling of inert gas washout from tissue use was made of the simplest model comprising blood flow ($\dot{Q}$), tissue-blood diffusive conductance (D), and diffusive conductance for veno-arterial back diffusion or diffusion shunt (D´) (Piiper et al., 1984 a, b; Scheid and Piiper, 1986). Based on anatomy, the counter-current arrangement of flow in adjacent artery and vein was assumed (Fig. 2). For test gas partial pressure in both tisse (P_T) and effluent venous blood (P_v), the washout is governed by the same rate constant, k:

$$ k = \frac{-dP_v/dt}{P_v} = \frac{-dP_T/dt}{P_T} \tag{1} $$

The constant k can be viewed as determined by three parameters (A, B and C) each reflecting a mechanism of gas transport:

$$ k = \frac{A}{B + C} \tag{2} $$

The parameters A, B and C are determined by $\dot{Q}$, D, D´, tissue volume (V), the solubility of inert gas in blood (β) and in tissue (α):

$$ A = \frac{\dot{Q} \cdot \beta}{V} \tag{3} $$

$$ B = \frac{1}{1 - \exp[\,-D/(\dot{Q}\beta)\,]} \tag{4} $$

$$ C = \frac{D´}{\dot{Q} \cdot \beta} \tag{5} $$

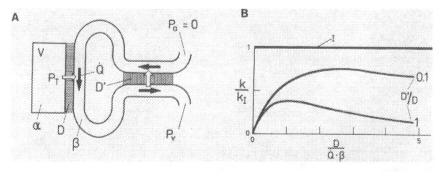

Fig. 2. Modeling of gas washout limited by tissue/blood diffusion and diffusion shunt (veno-arterial back diffusion). A. Model. For explanation of symbols, see text. B. Washout rate constant, k, standardized to rate constant for perfusion-limited washout, k_I, plotted against $D/(\dot{Q}\beta)$. The cases $D´/D = 0.1$ and 1 represent different conditions for veno-arterial back diffusion. After Piiper et al. (1984a).

I. In the case of pure perfusion limitation $D = \infty$ and $D' = 0$, and thus $B = 1$ and $C = 0$, yielding: $k = A = (\dot{Q} \cdot \beta)/(V \cdot \alpha)$, i.e. washout rate is given by the product of specific blood flow ($\dot{Q}/V$) and the blood/tissue partition coefficient (β/α). This case is identical with the classical model introduced by Kety (1951).

II. When there is, in addition to flow limitation, tissue/blood diffusion limitation, but no veno-arterial back diffusion, $B > 1$ and $C = 0$. Thus k (= A/B) is smaller than in the case of pure diffusion limitation.

III. With perfusion-limited washout delayed by diffusion shunt, in the absence of tisse/blood diffusion limitation, $B = 1$ and $C > 0$. Increase of D' decreases k due to veno-arterial back diffusion.

IV. Of particular interest is the comprehensive case in which gas transfer is limited by all three processes, blood flow, tissue-blood diffusion and veno-arterial back diffusion. In this case, $B > 1$, $C > 0$. Decreasing blood flow ($\dot{Q}$) or increasing diffusivity (leading to increasing D and D') lead to diminished tissue-blood limitation, increasing k/A, but concomitantly produce more veno-arterial diffusion shunt, thereby decreasing k/A. The overall result is a relative independence of k/A (= k/k_I) of changes in blood flow diffusivity (Fig. 2, B).

Steady state vs. washout

The models are applicable not only to washout, but also to transfer of O_2 and CO_2 in steady state. For this case the term equilibration efficiency, E, may be used. E designates the degree of tissue/blood equilibration reached or the ratio of actual transport rate, $\dot{Q} \cdot \beta \cdot (P_a - P_v)$, to ideally possible transport rate (i.e. in absence of both tissue-blood diffusion limitation and diffusion shunt), $\dot{Q} \cdot \beta \cdot (P_a - P_T)$:

$$E = \frac{P_a - P_v}{P_a - P_T} \qquad (6)$$

(the differences $P_a - P_v$ and $P_a - P_T$ are positive for O_2, negative for CO_2 and for inert gases during washout).

The relationship between E and k is:

$$E = k/A \qquad (7)$$

and it follows that

$$E = \frac{1}{B + C} \qquad (8)$$

Tissue-blood limitation ($B > 1$) and diffusion shunt ($C > 0$) thus reduce steady state gas exchange in a manner analogous to their delaying effect on inert gas washout.

Respiratory gases O_2 and CO_2

A particular characteristic of O_2 and CO_2 is their chemical binding in blood. This leads to high values for the effective blood/tissue partition coefficient, β/α, whereas for inert gases this ratio is not far from

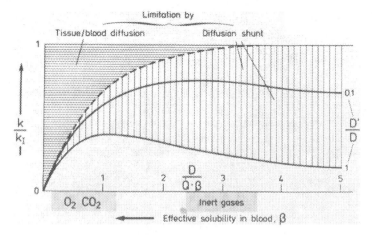

Fig. 3. O_2 and CO_2 vs. inert gases in tissue washout model. Same plot as in Fig. 2.

unity. Moreover, since the O_2 and CO_2 dissociation curves of blood are curvilinear, the β/α ratio depends on the partial pressure range.

The expected difference between respiratory and inert gases is illustrated in Fig. 3. Whereas inert gases appear to be largely limited in washout by diffusion shunt, for O_2 and CO_2 the effect should be smaller, and that of tissue-blood diffusion limitation much more prominent.

Due to increasing slope of the O_2 dissociation curve of blood, diffusion shunt for O_2 is expected to be reduced in hypoxia, but may play a considerable role in hyperoxia. Thus concerning O_2 supply to muscle, whenever O_2 demand is increased and muscle Po_2 drops, the waste of O_2 by diffusion shunt appears to be automatically reduced. The same mechanism would apply when local Po_2 is reduced by low blood flow.

Myoglobin in the myocytes appears to play a considerable role in O_2 transport by facilitated diffusion (Kreuzer, 1970). In terms of our model, this means increased D and D´ and thus decreased blood/tissue diffusion limitation, but increased diffusion shunt. But the significance of such effects cannot be evaluated.

PHYSIOLOGICAL SIGNIFICANCE

The delaying effect of diffusion shunt on inert gas washout and washin is expected to play a role in determining the kinetics of anesthesia gases and the behavior of inert gases in hyperbaria. The obvious effect of diffusion shunt in O_2 transport is to reduce the availability of O_2 by diminishing Po_2 in blood entering the capillaries. This means that, other conditions remaining the same, Po_2 in tissue is reduced as compared to a model without diffusion shunt (Fig. 4).

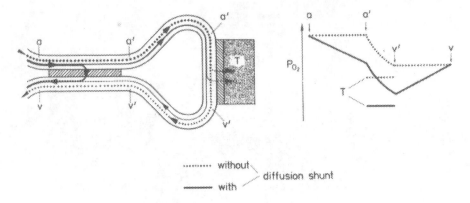

Fig. 4. Diffusion shunt and steady state O_2 transfer. See text for explanations. After Piiper et al. (1984a).

To what degree tissue P_{O_2} and P_{CO_2} are affected by diffusion shunt is open to investigation. Certainly differences are expected depending on the anatomical structure of the vascular systems in different organs. In lungs, the anatomical separation of pulmonary arterial and venous vessels is believed to play a role in preventing loss of O_2 from arterialized blood into mixed venous blood.

In the kidney, the significance of counter–current arrangements in renal tubules and blood vessels is well known. The low P_{O_2} in urine appears to be in part due to diffusion shunt of O_2 in the vascular loops supplying the renal medulla (Rennie et al., 1958). A decisive role is played by counter–current diffusion of gases in the secretion of gases into the swim bladder (cf. Fänge, 1983).

In skeletal muscle, and in many other organs, the arterial branches are accompanied by venous branches providing a substrate for diffusion shunt. Enhanced O_2 shunting is expected at low blood flow, at high P_{O_2} and high arterio–venous P_{O_2} difference. Morphometric data and physiological measurements are needed for a quantitative estimation of the diffusion shunt of O_2 in various tissues and physiological conditions.

SUMMARY

Diffusion shunt is diffusive gas exchange between arterial and venous vessels. Evidence for diffusion shunt had been obtained in washout studies in the gastrocnemius muscle of the dog. According to models, diffusion shunt is expected to be enhanced at low blood flow, and for gases of high diffusivity. Shunting of O_2 should be reduced in comparison to inert gases because of chemical binding in blood.

REFERENCES

Fänge, R., 1983, Gas exchange in fish swim bladder, Rev. Physiol. Biochem. Pharmacol. 17: 112-158.

Kety, S., 1951, The theory and applications of the exchange of inert gas at the lung and tissues, Pharmacol. Rev. 3: 1-41.

Kreuzer, F., 1970, Facilitated diffusion of oxygen and its possible significance: a review, Respir. Physiol. 9: 1-30.

Lassen, N.A., Lindbjerg, I., and Munck, O., 1964, Measurement of blood flow through skeletal muscle by intramuscular injection of xenon[133], Lancet 1: 686-689.

Piiper, J., and Meyer, M., 1984, Diffusion-perfusion relationships in skeletal muscle: models and experimental evidence from inert gas washout, in: "Oxygen Transport to Tissue V", D.W. Lübbers, H. Acker, E. Leniger-Follert and T.K. Goldstick, eds., Plenum Press, New York and London, pp. 457-466.

Piiper, J., Meyer, M., and Scheid, P., 1984a, Dual role of diffusion in tissue gas exchange: blood-tissue equilibration and diffusion shunt, in: "Oxygen Transport to Tissue VI", D. Bruley, H.I. Bicher and D. Reneau, eds., Plenum Press, New York and London, pp. 85-94.

Piiper, J., Meyer, M., and Scheid, P., 1984b, Dual role of diffusion in tissue gas exchange: blood-tissue equilibration and diffusion shunt, Respir. Physiol. 56: 131-144.

Piiper, J., Pendergast, D.R., Marconi, C., Meyer, M., Heisler, N., and Cerretelli, P., 1985, Blood flow distribution in dog gastrocnemius muscle at rest and during stimulation, J. Appl. Physiol. 58: 2068-2074.

Rennie, D.W., Reeves, R.B., and Pappenheimer, J.R., 1958, The oxygen tension of urine and its relation to intra-renal blood flow, Am. J. Physiol.. 195: 120-132.

Scheid, P., and Piiper, J., 1986, Inert gas wash-out from tissue: model analysis, Respir. Physiol. 63: 1-18.

DIFFUSION PATHWAY OF OXYGEN IN OX LUNG

T. Koyama and T. Araiso

Section of Physiology
Research Institute of Applied Electricity
Hokkaido University
060 Sapporo, Japan

INTRODUCTION

Since hemoglobin in erythrocytes in the lung capillary is separated from alveolar air by pneumocytes, vascular endothelium, the blood plasma layer and the erythrocyte membrane, inhaled oxygen molecules diffuse from alveolar air through cell membranes and cytosol to reach the hemoglobin molecules. The membrane viscosity of the phospholipid bilayer of isolated pneumocytes, endothelial cells and erythrocytes was estimated by anisotropy decay measurements with a nanosecond fluorometer (Araiso et al. 1986) in a previous study (Koyama et Araiso, 1987). It ranged from 47 to 132 mPa.sec. This range of viscosity seemed to be higher than that of the cytosol and blood plasma. It is, therefore, probable that the cell membranes could be a barrier to oxygen diffusion. Since the diffusion coefficient is inversely proportional to the membrane viscosity, the distribution of the conductance for oxygen diffusion in the barrier can be calculated. Hence the influence of the microstructure of the pathway of oxygen in the lung on the oxygenation rate of hemoglobin in erythrocytes can be estimated. In the present study a preliminary trial was made to analyse the effect of the cell membranes on the diffusion of oxygen.

MEASUREMENT OF THE VISCOSITY OF BLOOD PLASMA AND HEMOGLOBIN SOLUTION IN ERYTHROCYTES

Blood samples were obtained from slaughtered oxen and centrifuged to separate erythrocytes from blood plasma. Erythrocytes were washed twice with saline. The column of erythrocytes was sonicated with an ultrasonic sonicator (Sonifier, Branson Co., West Germany). The viscosity of the sonicated erythrocytes, as well as that of the blood plasma, was found to be 9 and 1.7 mPa.sec, measured with a rotating cone plate viscosity meter (Biorheolizer, Tokyoseiki Co., Japan) at the shear rate of 79 sec^{-1}. The viscosity of the sonicated erythrocytes was assumed to represent that of the hemoglobin solution in erythrocytes.

MODELING OF THE PATHWAY OF OXYGEN

The thickness of one phospholipid bilayer is 5 nm as shown electron-

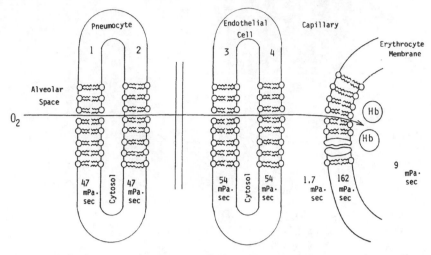

Fig. 1. Schematic illustration of the microstructures on the pathway
of oxygen in the lung. Their width and viscosity are shown.

microscopically. Oxygen molecules pass through four phospholipid bilayers
to reach the lumen of the capillary as schematically shown in Fig.1. First
the molecules traverse the pneumocyte membrane (1) lining the alveolar
space, next they pass through the pneumocyte membrane (2) and the endothelial membrane (3) adjacent to the basement membrane, and finally
through the endothelial membrane (4) facing the capillary lumen. Fig. 1
also shows the values for the viscosities of pneumocytes and endothelial
cells obtained in the previous study. For the ease of calculation these
four phospholipid bilayers are assumed to be unified in one layer 20 nm
thick. Both pneumocytes and endothelial cells contain cytosol between their
two phospholipid bilayers and a thin stagnant layer of blood plasma remains
on the interior surface of the capillary vessel. Let assume that the total
thickness of the cytosol and the stagnant layer is 520 nm. The simplified
model of the pathway for oxygen molecules consists of four steps as
schematically represented in Fig. 2.
The diffusion coefficient, D, of each step of the simplified pathway model
was estimated by the equation, $D = kT/2\eta r$, where k, T, η and r represent
Boltzman's constant, absolute temperature, the viscosity and the radius of
the oxygen molecule, respectively.

CALCULATION

The time course of the distribution of oxygen partial pressure was calculated in a case of sudden change in the oxygen partial pressure at the alveolar surface. A thin hemoglobin film of 5 nm in thickness was assumed adjacent to the phospholipid bilayer of the erythrocyte to shorten the calculation time.

The following established conditions (Bruley et Knisely, 1970) were used in
the calculation for the transient behavior of oxygen partial pressure:

1. Oxygen diffusion in the tissue results from an axial gradient in the x direction of oxygen partial pressure.
2. The diffusion coefficient is different in each step of the simplified model as noted in Fig. 2.
3. The solubility of oxygen is different in each step.
4. Differences in oxygen partial pressure at the interfaces between neighboring steps are negligible.
5. Oxygen uptake by hemoglobin depends on the partial pressure of the free oxygen physically dissolved in the hemoglobin solution and on the pH of the solution.
6. The combination velocity is sufficiently high.

Then, the resulting equations describing the unsteady state behavior of the model are as follows:

Phospholipid bilayer of pneumocytes and endothelial cells: $\dfrac{\delta P}{\delta t} = \alpha_{PL} D_{PL} \dfrac{\delta^2 P}{\delta x^2}$

where t=time in sec, x=variable length in cm and α_{PL} = oxygen diffusion coefficient for phospholipid bilayers of pneumocytes and endothelial cells (PL).

Interface between PL and pl: $P_{i,j\ PL} = P_{i+1,j\ pl}$

Cytosol + blood plasma (pl): $\dfrac{\delta P}{\delta t} = \alpha_{pl} D_{pl} \dfrac{\delta^2 P}{\delta x^2}$

Interface between pl and erythrocyte membrane (ErPL): $P_{i,j\ pl} = P_{i+1,j\ ErPL}$

Erythrocyte membrane: $\dfrac{\delta P}{\delta t} = \alpha_{ErPL} D_{ErPL} \dfrac{\delta^2 P}{\delta x^2}$

Interface between the phospholipid bilayer of the erythrocyte and hemoglobin solution: $P_{i,j\ ErPL} = P_{i+1,J\ Hb}$

Hemoglobin solution:

$$\frac{\delta P}{\delta t} = \frac{D_{Hb}}{\left[1 + \dfrac{Nkn P^{n-1}}{c + (1+kP^n)} \right]} \times \frac{\delta^2 P}{\delta x^2}$$

where c=oxygen solubility coefficient for blood, 3.42×10^{-5} ml O_2/ml

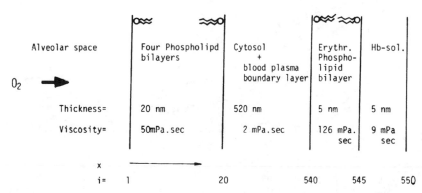

Fig. 2. A simplified model of the pathway of oxygen molecules.

blood.mmHg^{-1}, k=constant which depends on blood pH, 1×10^{-3}, N=oxygen capacity of erythrocytes, 0.4 mlO$_2$, n=constant which depends on blood pH, 2.20.

The initial condition: P=0 at 0<x<550 nm and t<0.
The boundary conditions: P=1 at x=0 and t>0,
dP/dx =0 at x=550 nm.

The partial derivative equation was first-order differentiated with respect to x and t to allow numerical integration. The resulting expression is:

$$\frac{P_{i,j+1} - P_{i,j}}{\Delta t} = \alpha D \frac{P_{i+1,j} - 2P_{i,j} + P_{i-1,j}}{\Delta x^2}$$

where i and j represent axial increment number and time increment number, respectively. Since the value of $\alpha D \Delta t / \Delta x^2$ must be smaller than 0.5, the values for Δx and Δt were selected to be 10^{-7}cm (=1 nm) and 2.5×10^{-8}sec (25 nsec), respectively.

RESULTS

The following three conditions were selected in the calculation. #1; the value of D and solubility are equal to those in cytosol and blood plasma for all steps in the pathway model. #2; the value of D is different as shown by the viscosity values in Fig. 2 and both α_{PL}/α_{pl} and $\alpha_{ErPL}/\alpha_{pl}$ are 5. #3; the value of D is different and the two solubility ratios are 1. The time course of oxygen partial pressure is shown in Fig. 3. The oxygen partial pressure in the hemoglobin solution is the highest in condition #1. That in #2 is 11 % lower than that in #1 at five msec after the onset of oxygenation, that in #3 is 50 % lower than that in #1.

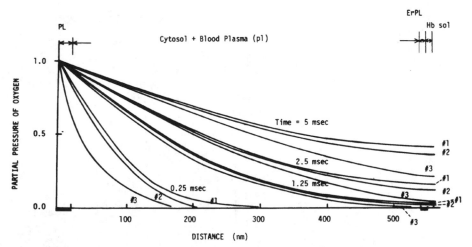

Fig. 3. The time course of the distribution of oxygen partial pressure at three values of solubility of oxygen to phospholipid bilayer. The abbreviations and thin arrows above indicate the microstructures on the diffusion pathway and their assumed distances.

DISCUSSION

The high viscosity of phospholipid bilayers causes a reduction in the rate of oxygenation of the tissue as seen in curve #3. The high solubility of oxygen in lipid makes the difference in oxygenation of hemoglobin solution smaller as can be seen in curve #2. Since the solubility of oxygen in erythrocyte membranes is higher than in blood plasma (McCabe, 1986), condition #2 seems to be more realistic for the oxygen pathway in the lung. The small thickness, and the high oxygen solubility, of the phospholipid bilayers minimize the effect of their high viscosity on oxygen diffusion. A difference of 11 % in the oxygen partial pressure at 5 msec still remains between curve #2 and #1. This result suggests that the phospholipid bilayers of cell membranes may offer a small resistance to oxygen diffusion.

The anisotropy decay of the fluorophore, DPH, in cell membranes indicates the friction which is caused by the collision of wobbling phospholipid molecules (Araiso and Koyama, unpublished data). The value of membrane viscosity which is estimated by the anisotropy decay measurement is comparable to the value of the viscosity measured with photo-bleaching method (Koppel et al., 1980) which is based on the translational diffusion of small molecules in cell membranes (Araiso and Koyama, 1987). The membrane viscosity, therefore, has a physical expression affecting the rate of diffusion of small molecules in the phospholipid layers. There remains still the question whether oxygen molecules flowing between ordered phospholipid molecules encountered the same degree of viscosity as under the wobbling conditions involved in the anisotropy measurement (Koyama et Araiso, 1986).

SUMMARY

The diffusion coefficients of cell membranes of pneumocytes plus endothelial cells, cytosol plus blood plasma, erythrocyte membranes, and hemoglobin solution in erythrocytes were estimated from the fluorometrically measured membrane viscosity. The time course of oxygen partial pressure distribution was numerically calculated in a model for the pathway of oxygen in the lung. The high viscosity of the phospholipid bilayers seems to cause a reduction in the rate of oxygenation of the hemoglobin solution.

REFERENCES

T. Araiso, Y. Shindo, T. Arai, J. Nitta, Y. Kikuchi, Y. Kakiuchi and T. Koyama, 1986, Viscosity and order in erythrocyte membranes studied with nanosecond fluorometery, Biorheology, 23: 435-552.

T. Araiso and T. Koyama, 1987, Fluidity and lipid dynamics in biomembrane from a fluorescence depolarization study, J. Physiol. Soc. Jap. 49: 1-11, (in Japanese).

D. F. Bruley and M. H. Knisely, 1970, Hybridsimulation - Oxygen transport in the microcirculation, Chem. Engin. Progr. Symp. Ser. 66: 22-32.

D. E. Koppel, M. P. Sheetz, and M. Schindler, 1980, Lateral diffusion in biological membranes: a normal-mode analysis of diffusion on a spherical surface, Biophys. J. 30: 187-192.

T. Koyama, and T. Araiso, 1987, Dynamic structure of phospholipid bilayers on the path for oxygen diffusion in the ox lung, Oxygen transport to tissue IX, Eds I. A. Silver and A. Silver, Plenum Press, New York and London, (in press).

M. McCabe, 1986, The solubility of oxygen in erythrocyte ghosts and the flux of oxygen across the red cell membrane, Oxygen transport to tissue vol. VIII, Ed. I. S. Longmuir, Plenum Press, New York and London, pp. 13-20.

OXYGEN DELIVERY TO TISSUE: CALCULATION OF

OXYGEN GRADIENTS IN THE CARDIAC CELL

E.M. Chance[1] and Britton Chance[2]

[1]Biochemistry Dept., University College London, London WC1E
6BT U.K.

[2]Biochemistry/Biophysics Department, University of
Pennsylvania and Institute for Structural and Functional
Studies, University City Science Center, Phila., PA 19104

In this study, a two dimensional mathematical model of a cross
section of a single heart muscle cell 30.0 μ in diameter, was constructed
from which one would be able to calculate the rate of oxygen delivery
from the capillary blood supply to the mitochondria by passive oxygen
diffusion, myoglobin facilitation of oxygen delivery, and subsequent
utilization of oxygen by cytochrome oxidase. From such calculations, the
conditions under which the two hemoprotein indicators, myoglobin and
cytochrome oxidase, behave in a coherent manner, as observed by Tamura
and his colleagues (1978), could be established and the loci of oxygen
gradients identified.

The network, based on mass action chemistry and Fick's Law diffusion
was programmed in two dimensions and the resultant system of ordinary
differential equations solved on the Amdahl 5890/300 computer at the
University of London Computer Centre using the FACSIMILE program (Chance,
E.M. et al., 1977). The use of the FACSIMILE program for the solution of
such problems is fully explained in Curtis (1983). Discretisation
(Curtis, 1983) was carried out in a similar manner using two dimensional
rectangular co-ordinates rather than the system of one dimensional polar
co-ordinates used for the analysis of chemical wave formation in cell
free yeast suspensions (Hess et al., 1981), using chemistry coupled with
diffusion as a basis for the calculation. The mesh consists of 26 mesh
points and 25 elements in the x or lateral direction parallel to the
plasma membrane with an increment of 0.25 μ resulting in a total distance
of 6.25 μ from the capillary oxygen point source (the origin) to the end
of the mesh. By symmetry another capillary is located an additional 6.25
μ from the end giving a boundary condition of constant and equal
capillary oxygen tension (PO_2) at either boundary 12.5 μ apart. Oxygen
diffuses laterally in the tissue and cytosol, a process which is
facilitated by myoglobin for those oxygen molecules which are available
to react with it.

_____
Supported by National Institutes of Health Grants HL-18708 and
NS-22881.

The geometry of the mesh is a plane passing vertically through an idealized cylindrical shaped cardiac cell from the center to the plasma membrane and symmetrical in all directions as represented in Figure 1.

In these calculations, the mitochondria are lumped together as a single volume and act as the ultimate sink for oxygen. No oxygen molecules diffuse further into the myofibril space nor is any attempt made to represent ATP synthesis or cardiac muscle contraction at present.

Tissue oxygen (TiO_2) passively diffuses across the cell membrane into the cytoplasm becoming cytoplasmic oxygen (CyO_2). Myoglobin equilibrates with tissue and cytoplasmic oxygen in a cytoplasmic myoglobin compartment whose inward boundary is 0.25 μ from the plasma membrane, but not with oxygen in the mitochondrial compartment (MwO_2). Cytoplasmic oxygen diffuses from the plasma 0.5 μ in the inward y-direction to the mitochondrial membrane where an additional diffusion barrier is imposed as it diffuses into the mitochondria and is concentrated at least ten fold, depending on the metabolic state of the cell. At the mesh points, the following chemical and transport reactions take place:

Reaction	Oxygen Transport
$TiO_2 + Mb \xrightleftharpoons[k_2^{Mb}]{k_1^{Mb}} Mb.O_2$	Lateral and Inward (x,y)
$CyO_2 + Mb \xrightleftharpoons[k_2^{Mb}]{k_1^{Mb}} Mb.O_2$	Lateral and Inward (x,y)
$TiO_2 \xrightleftharpoons[TrTiO_2]{TrTiO_2} CyO_2$	Lateral and Inward (x,y)
$CyO_2 \xrightleftharpoons[TrCyO_2]{TrCyO_2} MwO_2$	Inward (y)
$MwO_2 + aa_3^{2+} \xrightleftharpoons[k_2^{a}]{k_1^{a3}} aa_3^{3+}$	None
$aa_3^{3+} \xrightarrow{k_t^{a3}} aa_3^{2+}$	None

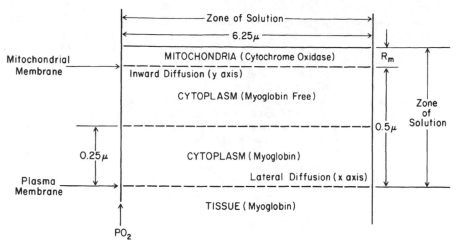

Fig. 1. A schematic diagram of the geometry of the two dimensional diffusion calculation. The calculations of concentration refer to these specific compartments, cytoplasm (myoglobin) and mitochonria (cytochrome oxidase). The capillary delivers oxygen perpendicular to the plane of the diagram and at a point source in the lower left-hand corner.

The following constraints are applied to the calculation:

a) Mb is located in the tissue and at the plasma membrane;
b) Mb reacts with TiO_2 and CyO_2 within 0.25 μ from the plasma membrane;
c) The tissue volume and the cytoplasmic volume (V_c) are equal;
d) The rate of diffusion of oxygen across the plasma and mitochondrial membrane is ten-fold slower than in solution;
e) At low coherence (<50%) and non coherence, V_c/V_m = 0.1;
f) At high coherence (>50%), the oxygen flux is high and constant V_c/V_m calculated by non-linear least squares analysis;
g) The mitochondrial volume (V_m) $\leq V_c/10$.

We have used the term coherence to indicate equal fractional responses of MbO_2 to deoxygenation and cytochrome aa_3 to reduction as the oxygen concentration is decreased (Fig. 2 below). In the mathematical sense, coherence is defined by a set of initial values whose solution in the steady state is an equal percent myoglobin oxygenation and cytochrome oxidation. The ratio of myoglobin molecules to cytochrome oxidase molecules was constant at ten and the initial Mb concentration at 1 mM. The values of k_1^{Mb} and k_1^{a3} were set equal to 1 x 10^7 $M^{-1}s^{-1}$, and k_2^{Mb} and k_2^{a3} were 50 s^{-1} and 5 s^{-1}, giving binding constants of 1 and 0.02μM, respectively. The value used for the diffusion constant for oxygen was $1920μ^2$ s^{-1} and the transport coefficients calculated by dividing the appropriate diffusion constant by the square of the spatial increment 0.5 μ and corrected for the volume of the compartment. The values of these preceding parameters are assumed to be known and were not varied in any of the calculations reported in this paper. By constraining the steady state percentage redox level of the hemoproteins to be equal, the unknown parameters PO_2 and k_t^{a3} are estimated by non-linear least squares analysis and the spatial relationships between the metabolic concentrations calculated and the loci of oxygen gradients identified by contour mapping. From the non-linear least squares analysis, the parameters which are functions of the initial values including the K_m for cytochrome oxidase, the overall oxygen uptake (VO_2) and the initial concentration of cytochrome oxidase (aa_{30}) can be calculated.

Two indicator coherence can be calculated with high accuracy from the two dimensional model described above. PO_2 values in the range of 0.4 to 77 torr, VO values chosen to range from 19 to 250 ml O_2/100g/min in order to cover the maximum that could be expected from cells and tissues and hemoprotein redox levels from 5 to 95% are obtained. Oxygen gradients between myoglobin and cytochrome oxidase as high as 12 torr/μ were observed depending on the length of the cytoplasmic diffusion vector and the magnitude of the diffusion barriers imposed. These results are summarized in Table 1. Values marked with * were constrained during the calculations. The K_m values refer to (MbO_2); cytochrome aa_3 are flux dependent.

Table 1. Coherence Calculation

Coherence	K_m	PO_2	VO_2 (flux)	Gradient	V_M/V_C	aa_{30}
(%)	(torr)	(torr)	(ml O_2/ 100g/min)	(torr/μ)		mM
5.0	2.90	0.4	18.9	0.6	0.100*	1.00*
10.0	2.93	0.9	38.4	1.3	0.100*	1.00*
20.0	3.00	2.0	79.4	2.7	0.100*	1.00*
30.0	3.08	3.4	122.9	4.4	0.100*	1.00*
40.0	3.15	5.1	169.2	6.2	0.100*	1.00*
50.0	3.23	7.4	218.5	8.2	0.100*	1.00*
60.0	3.36	9.9	250.0*	9.7	0.0760	1.32
70.0	3.50	13.2	250.0*	10.0	0.0512	1.95
80.0	3.61	19.7	250.0*	10.6	0.0410	2.44
90.0	3.70	38.7	250.0*	11.5	0.0354	2.82
95.0	3.76	76.7	250.0*	12.2	0.0335	2.99

A crucial test of the model is its response to low values of flux where non-coherence is observed. Table 2 gives the solution for a constrained value of VO_2 of 2.5 ml O_2/100g/min. It is observed that the response is non coherent.

Table 2. Non Coherence Calculation

$aa_3{}^{3+}$	$Mb.O_2$	K_m	PO_2	VO_2 (flux)	Gradient	aa_{30}
(%)	(%)	(torr)	(torr)	(ml O_2/μ 100g/min)	(torr/μ)	mM
5.0	1.6	1.04	0.09	2.50*	0.074	1.00*
10.0	2.7	0.90	0.14	2.50*	0.071	1.00*
20.0	5.3	0.83	0.25	2.50*	0.069	1.00*
30.0	8.5	0.81	0.39	2.50*	0.069	1.00*
40.0	12.4	0.80	0.57	2.50*	0.069	1.00*
50.0	17.3	0.79	0.83	2.50*	0.070	1.00*
60.0	23.7	0.78	1.22	2.50*	0.072	1.00*
70.0	32.4	0.78	1.86	2.50*	0.075	1.00*
80.0	45.1	0.78	3.15	2.50*	0.080	1.00*
90.0	64.7	0.78	7.01	2.50*	0.090	1.00*
95.0	79.5	0.78	14.7	2.50*	0.102	1.00*

The results of the coherence and non coherence calculations are presented graphically in Figure 2 and closely resemble the experimental data (Tamura et al., 1978; Chance, B. et al., 1987).

The oxygen gradients formed in these calculations are clearly shown in the contour maps in Figures 3 and 4. At 5% coherence (Figure 3) there are step gradients across the mitochondrial membrane and near the capillary.

Figure 4 shows contours of oxygen at 95% coherence. At high oxygen flux and blood flow, gradients in the cytoplasm become significant. In order to achieve these conditions in the calculations, the relative volumes of the myoglobin and cytochrome oxidase compartments must change 3.35-fold in order to achieve coherence at high flux.

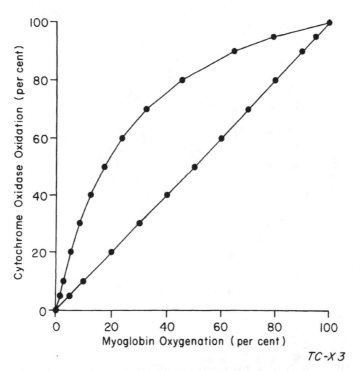

TC-X 3

Fig. 2. The calculated coherence diagram for myoglobin and cytochrome oxidase. The straight line represents coherence under the conditions given in Table 1. The upper curve shows a non-coherent response at constant low flux (Table 2).

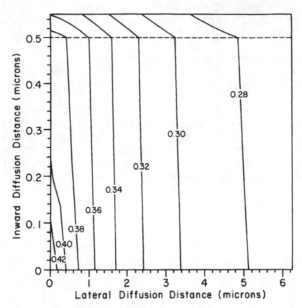

Fig. 3. Contour map of coherence calculation at 5% oxidation. The plasma membrane is the x-axis and the horizontal line at 0.5 locates the mitochondrial membrane and the maximum value of y=0.551μ. PO_2 is held constant at the origin at 0.436 torr and the contour gradient 0.0217 torr, R_m=0.0509μ and V_m/V_c = 0.1, VO_2=18.9 ml/100g/min. The contour values are torr.

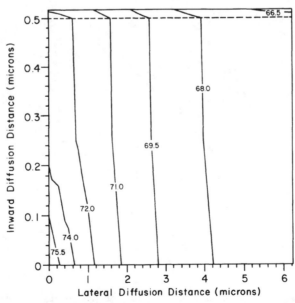

Fig. 4. Contour map similar to Fig. 3 at 95% coherence is shown. PO_2=76.9 torr and the gradient between contours 1.5 torr. The maximum value of y decreased from 0.551 to 0.517μ, R_m=0.017μ, V_m/V_c= 0.0335, VO_2=250 ml O_2/100g/min.

These calculations define sets of conditions necessary to observe oxygen gradients together with coherent and non-coherent responses of the percentage oxygenation and oxidation of the two hemoprotein indicators, myoglobin and cytochrome oxidase. In all cases, the maximum oxygen gradient between the tissue and the oxygen available to react with cytochrome oxidase occurred along the y-axis. The magnitude of the gradient may be estimated graphically from the contour map in Figure 4. The oxygen concentration at the origin is 77 torr and the 72 torr contour intersects the y axis approximately $0.5\ \mu$ from the plasma membrane, yielding a rough estimate of the gradient to be 10 torr/μ. These gradients come about because myoglobin does not equilibriate directly with oxygen in the mitochondria because the mitochondrial membrane is impermeable to it, and oxygen must passively diffuse through a myoglobin free volume and cross two membranes before it is available to react with cytochrome oxidase. The myoglobin free space reported here has a radius of $0.5\ \mu$ but similar results may be obtained with smaller values thus making a better approximation of the geometry of the inner and outer mitochondrial membranes.

Coherence can be achieved at a constant, high flux by varying the relative volumes of the myoglobin and cytochrome oxidase compartments, which increase the total cytochrome oxidase concentration relative to total myoglobin which was constant at 1 mM in all the calculations reported in this study. As can be seen in Table 1, at a value of VO_2 of 250 ml O_2/100g/min and at 60% coherence, the total cytochrome oxidase is 1.32 mM as calculated for the mitochondrial matrix space. At 95% coherence, it has risen to three times that of myoglobin. For values of coherence between 5% and 40%, myoglobin and cytochrome oxidase are equimolar.

Non coherence occurs at low turnover numbers such that the K is equivalent to the binding of oxygen to cytochrome oxidase. The resultant state of myoglobin oxygenation depends on the PO_2 required to oxidize cytochrome oxidase to the required level and does not depend on flux to any measurable degree.

REFERENCES

Chance, B., Chance, E.M. et al., 1987, On the Location of Steep Oxygen Gradients in Rapidly Respiring Tissues, Proc. Natl. Acad. Sci. USA in preparation.

Chance, E.M., Curtis, A.R., Jones, I.P. and Kirby, C.R., 1977, FACSIMILE: A Computer Program for Flow and Chemistry Simulation, and General Initial Value Problems, A.E.R.E. Report No. R8775, H.M. Stationary Office, London.

Curtis, A.R. (1983) An Introduction to the FACSIMILIE Program, AERE Rpt. R-3134, H.M. Stationary Office, London.

Hess, B., Chance, E.M., Curtis, A.R. and Boiteux, A., 1981, in "Complex Dynamic Structures, Snergetics," Vial, C. and Pacault, A., Eds. Springer Verlag, Berlin, 172-179.

Tamura, M., Oshino, N., Chance, B and Silver, I.A., 1978, Optical Measurements of Intracellular Oxygen Concentration of Rat Heart in vitro. Arch. Biochem. Biophys. 191, 8-82.

THREE DIMENSIONAL RECONSTRUCTION OF BRANCHED TREE STRUCTURES FROM

SERIAL SECTIONS

Nathan A. Busch and Ian A. Silver

Department of Pathology, University of Bristol, Bristol

United Kingdom

INTRODUCTION

The quantative analysis of oxygen transport to tissue requires knowledge of the spatial configuration of the oxygen sources and sinks, mathematical descriptions of the transport phenomena, and experimentally measured tension. The techniques for measuring the oxygen tension in tissue are readily available. Acceptable mathematical description of the transport phenomena is precluded by defining acceptable spatial configuration of the microcirculation of the tissue. The current mathematical descriptions of the transport rely primarily on deterministic models of the phenomena and an assumed simplistic geometry for the system. To permit the development of usable mathematical models for the transport then requires that a technique be available to construct from cross-section images of tissue the micro-structure of the system. The micro-structure includes both the micro-circulatory system as well as the location of the sinks for oxygen. To this end an automatic three dimensional reconstruction process has been developed which uses serial cross-section images of the tissue. The possibility of automating the reconstruction process has received considerable attention recently by Adair, et al., (1981), Batnitzky, et al., (1981), Edelstein, et al., (1981), Moseley, (1982), Bajcsy, et al., (1983), and Hull, et al., (1984).

Statement of the Problem

Consider an object which contains a bifurcation along its principle axis. Slice the object immediately below and above the bifurcation with parallel planes which are separated by a known distance. The projection of the object onto the plane slicing it immediately below the bifurcation will be a closed planar curve. The projection onto the plane slicing the object immediately above the bifurcation will be two closed planar curves, which may or may not be intersecting. The problem is to reconstruct the object which is between the two parallel slicing planes based only upon information provided by the projection of the object onto the slicing planes, and knowledge of the distance between the two planes.

The procedure to reconstruct the object which is between the two planes is:

1) Identify the relevant areas in the cross section. This involves identifying the conic sections on each parallel slicing plane.

2) Determine the interconnections between areas in adjacent slices, selecting only the best possible paths between objects on adjacent slices such that the original structure is obtained.

3) Generate the set of triangular tile facets representing the structure surface between the two adjacent slices such that the surface area is minimal.

Object of Identification

The cross-sections shown in Figure 1a represent the objects found on a serial section. The areas consist of a boundary of varying thickness defining the outline of the structure of interest. Inside this boundary is an area of pixels with opposite value. In some cases this area contains clusters of pixels with the same value as the boundary. In an extreme case these clusters could contain other areas of the opposite pixel value and so on. The convention used in this paper is that pixels shown as dark in the figures are considered to have a value of one, while those shown as white are of value zero. Clusters of pixels with a value of one are known as 'objects' and those with a value of zero are known as 'holes'. The "background" pixels have values of zero and are a special case of a hole.

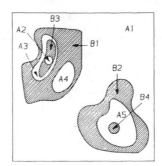

a) Section containing objects and holes

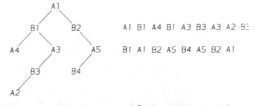

b) Tree representation   c) String representation

Figure 1. Representation of a Cross-Section Image

The algorithm used for the identification of objects of interest in the image is that presented by Buneman (1969). This algorithm represents the topology of a binary image as a tree. The tree description of the objects and holes of the image in Figure 1a are given in Figure 1b. Each node refers to a connected component in the image; 'a' for white, and 'b' for black. The topmost node represents the background and the downward branches represent a succession of objects and holes. This tree may be represented by a string of its nodes in the manner shown in Figure 1c. The algorithm described consists of a set of rules which are implemented during a single pass over the image and which produces the topological string representation of the objects and holes in the serial section image. Each node on the tree is assigned a level whose numerical value depends on the number of nodes between that node and the root of the tree. (As is shown in Figure 1b.) The level of the objects in the string which describes the tree can be easily determined in a single pass through the string.

This procedure thus gives a full topological description of the tree in terms of the presence of objects and holes. If the objects are to be related to objects in other slices and connecting surfaces constructed, it will be necessary to determine the positions and shapes of the objects and holes with respect to a global coordinate system. To accomplish this, use is made of the fact that one of two possible rules is always applied when, and only when, the last pixel of an object or hole is scanned. When either of these rules has been applied, the coordinates of the pixel together with other known details of the object (or hole) are written out to a file. This file is then read by the next part of the program which determines the geometric properties of the holes or objects of interest.

Determination of Interconnections

Once the topology of the image has been determined the next step is to define the relationships between objects in adjacent slices. This is performed in two steps: The first is to calculate a set of parameters which describe the object and the second is to take pairs of objects, one from each of two adjacent slices and to use these parameters to evaluate a weighting function. This function would ideally return one value if object "i" in slice "n" and object "j" in slice "n+1" are connected, but a second value if there is no connection.

In practice such a perfect function does not exist. It is therefore necessary to select a function which is related to the probability of the two objects being connected. The function described in the previous paragraph, (for convenience denoted $f(n,i,j)$) is then evaluated for all "j" and the value of "j" found for which $f(n,i,j)$ has the greatest value. This is the object in the slice n+1 which is considered to be the most likely to be connected to object "i" in slice "n".

It is necessary to consider three different cases.

1)   Object "i" is not connected to any objects in slice n+1.

2)   Object "i" is connected to one object in slice n+1.

3)   Object "i" is connected to two or more objects in slice n+1.

The first case can be dealt with by setting a threshold value for the function $f(n,i,j)$, such that an interconnection is only considered to exist if $f(n,i,j)$ is greater than the threshold value.

The second case is the simplest situation and is handled as outlined above. An example of the third case is shown in Figure 2. The structure of branches between the two sections such that object "1" in section 1 is connected to objects "1" and "2" in section 2. The figure also shows that objects "2" and "3" in section 1 are both connected to object "3" in section 2. This latter event is covered by the second case. Since there is nothing unique about scanning the slices in the order one to n, as opposed to the order n to one, it is possible to scan two sections in either direction. The changes that the scanning direction makes is shown in Figure 2b, which is simply an inversion of Figure 2a. The result is that what was a branch in the structure becomes a junction and vice versa. Branches are consequently dealt with simply by scanning the sections in both directions and combining the results of two scans. Thus if a satisfactory function f(n,i,j) can be found, it should be possible to determine a set of interconnections that are a good approximation to the original structure.

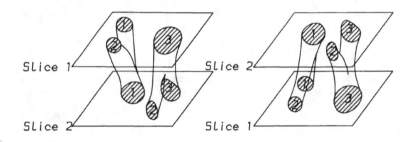

a) Original orientation   b) Inverted

Figure 2. Bifurcation and Slice Inversion.

Selection of a Suitable Function

The interconnection process assumes that the path followed by any one branch of the structure has a limited curvature. The exact limit is not defined but, as will be discussed later, it is related to the size of the object being considered and the distance between the sections. The method used to determine the chosen weighting function is a follows;

1)   Determine the current path of the branch.

2)   Extrapolate the path into the next slice and determine the coordinates of intersection (x,y).

3)   Determine the proximity of (x,y) to objects in the next slice.

4)   Select closest object.

Path Definition and Extrapolation

The path of a branch is defined in terms of the coordinates of the centroids of the objects through which it passes. The method to determine the current path of the branch depends on the position of the slice that is being considered at the time. When the first section is being examined, there is no previous history that can be used to establish a path for the branch. Therefore the policy adopted is to extrapolate vertically into the next section. Once the interconnections between the first two sections have been determined, a linear path is assumed and this line is extrapolated into the next section. Having then established the interconnections between objects in the first three layers, a second order polynomial is used to define the path. It would be possible to continue this process using polynomials of increasing order as more slices are added. However, high order polynomial curve fits can oscillate wildly between sample points and extrapolation yields spurious results. For this reason, second order polynomials are used to define the paths of objects passing through third and higher sections. This extrapolation yields a value for coordinate pair (x,y) mentioned in step 2 in the preceeding paragraph. The next two steps are considered below.

Proximity Check

The proximity of the extrapolated path of a branch to an object could be defined in a number of ways. The simplest might be to calculate the distance (x,y) and the coordinates of the centroid of the object. Since a path is defined in terms of the centroids of objects, the coordinates of the centroids must be calculated at any rate. This does not give any consideration to the size of an object. The ideal method might be to determine whether (x,y) lies within the boundary of an object; unfortunately this is not a trivial problem. The solution adopted was to make use of some of the known characteristics of the objects being considered. Since the software was originally developed to reconstruct networks of blood vessels, the structural properties of these were used as a basis for interpretation. Cross sections of blood bessels may be classified according to the degree of re-entry of their projected shape onto the plane of the cross-section. The degree of re-entry of the shapes and could perhaps be defined mathematically by stating that a shape belongs to one class if its centroid lies within the shape and to another if it does not. Such shapes can be simply characterized by a particular dimension, an aspect ratio and an orientation. These were defined indirectly in terms of geometric properties that are easily calculated at the same time as the coordinates of the centroid. The actual properties used are the second moments of area, the orientation of the principal axes, and the area.

The parameters calculated for each object are:

1) Coordinates of the centroid.

2) Area.

3) Second moments of area.

4) Orientation of the principal axes relative to the image axes.

These values are initially calculated in the image axes and then recalculated to give values in a coordinate system parallel to the image axes but with the origin at the centroid of the object. The second moments of area will vary with the orientation of the axes. The orientation of the axes is given as the angle of rotation about the positive x axis such that the product moments are identically zero. The set of axes in which the product moments are zero and which has its origin at the centroid of the object is called the principle axis of the object.

The weighting function used in determining the interconnections is calculated using one of two techniques. The ratio of the second moments of area (x) and (y) is calculated. If the ratio has a value greater than 0.5, the object is considered to be a reasonable approximation to a circle. The expected radius of the object on the adjacent slice is computed as the square root of the area of the current object divided by pi. Let (a) be the distance between the expected center of gravity on the next slice and the center of gravity of the object currently under investigation as the target object in the interconnection path. The value of the weighting function is then;

$$w = \begin{cases} 100(1-a/r) & \text{if } a < r, \\ 0 & \text{if } a > r, \end{cases}$$

If the ratio of the second moments of area (x), and (y) is less than 0.5, then the dimensions of a rectangle b, and h are calculated. The value of the dimension b is the square root of the second moment in y divided by twelve times the current object area. The value of the dimension h is the square root of the second moment in x also divided by twelve times the current object area. The area of the target object on the next slice is A = bh.

The expected center of gravity on the next slice (x,y) is transformed to the principal coordinate set of the object, giving coordinates (x',y'). Absolute values are then determined to move the point into the first quadrant. If the point lies within the rectangle, then the weighting function is assigned a value of 100. If the point lies outside of the rectangle, the program determines in which of the three zones the point lies.

As demonstrated in Figure 3, for each of the three zones there is a characteristic length, (d) and a reference line or point. The distance of (x',y') from the reference point or line is (a). The weighting function is then determined from:

$$w = \begin{cases} 100 & \text{, if } a < d, \\ 100(a-d)/d & \text{, if } d < a < 2d, \\ 0 & \text{, if } 2d < a. \end{cases}$$

For all objects on the current slice the expected center of gravity on the next slice is computed. Then all the objects on the next slice (even those that have been identified as already being connected to objects on the current slice) are investigated to determine possible connection to the current object. The weight function values for all of the objects on the next slice are sorted according to magnitude, and the object with the highest weight function value is selected as being connected. If no object on the next slice has a weight function value greater than zero, then there are no identifiable connections between the current object and objects on the next slice. The structural interconnection path is terminated at the current object.

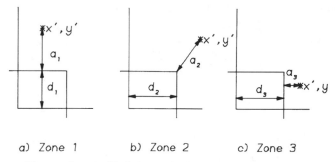

a) Zone 1   b) Zone 2   c) Zone 3

Figure 3. Definition of three possible zones.

Surface Generation

Having determined the shapes of objects and their interconnections, all the necessary information is available for generating the surfaces of the structure being analyzed. The first stage of this process was to examine each object of interest and reduce its outline definition to no more than twenty points. In cases where the outline initially involved more than twenty pixels, twenty pixels coordinates were chosen from the list at approximately equal intervals. The surfaces between the two interconnected objects was formed by a series of triangular tiles.

Figure 4 shows the outlines of two objects. The smaller of the two should be considered to be above the page and therefore also above the other object. The aim of the algorithm is to join each point on the small object to at least one point on the larger object and vice versa to form triangular facets. The surface represented by the set of triangular tiles is required to be of minimum area.

The numbering of the points on the objects is in increasing order around the object. The direction is anti-clockwise for objects, and clockwise for holes. The first point on each contour is the first pixel of the contour pixel list as determined in the object identification and proximity determination phase of the reconstruction process. By definition it is the leftmost pixel on the last scan-line contained within the object. The two closed contour lines are disconnected between the last and first pixel and "unwrapped". When adjacent, the edges of the triangular tiles between the two lines may be formed.

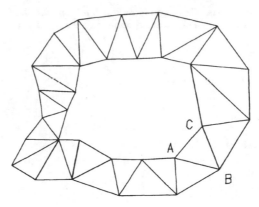

Figure 4. Two object contours and a possible connection.

First the contour with the fewest points is chosen. The initial
point of this contour is selected (labeled A). The next point on the
contour is then found (labeled C). All the points on the other contour
are then scanned and the point B selected that has the minimum value
for a weighting function (as described below). The next point on the
shorter contour is then considered; Point B and the following points on
the longer contour are examined in turn to find the point that gives
the minimum value for the weighting function (w) subject to the
condition that no search goes around the contour past point B. This
process is repeated for the remaining points on the shorter contour.
The result of this is a list of numbers such that points i and i+1 on
the shorter contour are both joined to the point j on the longer.

Consider the three points A, B, and C for which the weighting
function is to be determined. Points A and C are on the shorter of the
contours and point B is on the other contour. Points A and C will have
the same z coordinate and point B will have a different z coordinate.
However all the points with which B is to be compared have the same z
coordinate as B. Two weighting functions (w(1) and w(a)) are used and
both are based upon two physical attributes of the triangle with
verticies (A,B,C). The first is the sum of the lengths of the line ABC
(AB+BC). The second is the angle of CAB.

The weighting function w(1) and w(a) become smaller as the
length of the line ABC is reduced and as the angle CAB approaches 90
degrees respectively. The relative significance of these two factors
is adjusted by a predetermined factor labeled "FACTR". For FACTR = 0,
variation in the angle CAB does not affect the value of the weighting
function. For increasing values of FACTR, the length of ABC has a
decreasing influence on the value of the weighting function.

The difference between the two routines lies in the way in which
the angle CAB is computed. Function w(1) computes the angle CAB as the
difference in the angle between lines CA and AB. This produces a value
in the range −180 to +180 degrees. The most desirable angle depends on

whether the contours being examined go clockwise around the object or counter clockwise. If clockwise, and angle of 90 degrees is considered optimal while for the counter clockwise, -90 is used. When the contours have been produces by the program used here, the points will be listed in a clockwise direction for contours at even levels and counter clockwise for objects at odd levels.

Function w(a) calculates the magnitude of the angle CAB using the cosine formula. Thus the value returned is always positive. In fact the angle itself was never actually determined since the weighting function was calculated from the cosine of the angle.

The policy adopted is as follows. For matching the first point of a contour w(1) is used with FACTR set to 1.0. For this value the weighting function is most influenced by the angle. Thus the point chosen will always lie on the correct side of the line AB. For subsequent points function w(a) is used with FACTR having a value of 0.3. To gain some impression of the values used for FACTR, consider two cases. First, a small triangle is to be matched to a 20 point contour of considerably larger area. If only the length is used for the matching criteria, then the variation in the weighting function around the larger contour is small and a point on the wrong side of the line could be easily chosen. If FACTR is set to a large value, then the angle will dominate. For function w(1) this will ensure a reasonable starting point but for function w(a), there are two points which will return angle CAB of 90 degrees.

The reason for using function w(a) is that it requires less time to evaluate than w(1). For total reliability function w(1) could be used for all matching. However, the system used above has been found to be reliable for structures examined during program testing.

This technique can fail if the object in one layer is not directly above the object to which it is connected. This could occur for example if the objects are part of a cylinder that is not normal to the plane of slicing. This is avoided by translating all coordinates in the shorter contour so that its centroid is vertically above or below that of the other object. After matching, the reverse translation is applied to restore the original coordinates.

Once the points on the outlines have been connected, the triangular tiles can be defined by three sets of x, y, and z coordinates. At the same time the normal to the tile surface is calculated using an algorithm given by Sutherland, et al. (1974) and the normal is then represented by a fourth coordinate set. These four coordinate sets are then written to a file for use by the surface graphics algorithm.

DISCUSSION AND CONCLUSION

The techniques described above have been implemented and have proved capable of reconstructing tree-like structures. The emphasis of the technique is on the minimization of computational effort rather on the ultimate in precision. For the problem considered there is no exact solution with which the methods can be tested, however in practice all the test cases were successfully reconstructed and displayed.

The importance of this work to the field of oxygen transport to tissue is that an automatic process is now available for the three dimensional reconstruction of micro-structures of tissue from serial cross-section images. The technique produces a list of triangular tiles defining the surface of the structures. The set of triangular tiles is immediately amenable to the definition of surface structures used in finite element analysis of oxygen transport. The set is also precisely the surface definition used in the stochastic analysis techniques.

SUMMARY

A process is described for the automatic registration and reconstruction of anatomical tree structures such as capillary networks in the microstructure of tissue. The technique uses images of, or images representing, a set of closely spaced parallel slices called "serial sections". The source of the serial sections may be MRI, multi-planar X-ray, multi-planar infra-red scans, or simple histological sections. A tree structure is defined as a structure consisting of a network of nested ducts, vessels or solid cores which branch and join with another such that one structure may appear as more than one distinct area in a single section, or in a series of sections. The reconstruction of such an object poses many problems which do not occur when restructuring objects which are so shaped that they can give rise to only one area in each section.

REFERENCES

Adair, T., Karp, P., Stein, A., Bajcsy, R., and Reivich, M., 1981, Computer Assisted Analysis of Tomographic Images of the Brain, J. Computer Assisted Tomography, Vol. 5, 929-932.

Bajcsy, R., Liberson, R., and Reivich, M., 1983, A computerized System for the Elastic Matching of Deformed Radiographic Images to Idealized Atlas Images, J. Computer Assisted Tomography, Vol. 7, 618-625.

Batnitzky, S., Price, H.I., Cook, L.T., and Dwyer, S.J., 1981, Three-Dimensional Computer Reconstruction from Surface Contours for Head CT Examination, J. Computer Assisted Tomography, Vol. 5, 60-67.

Buneman, O. P., 1969, A Grammar for the Topological Analysis of Plane Figures, Machine Intelligence, Vol. 5, Edinburgh University Press.

Edelstein, W. A., Hutchinson, J.M.S., Mallard, F.W., Johnson, J.R., Redpath, T.W., 1981, Human Whole Body NMR Tomographic Imaging of Normal Sections, Br. J. Radiology, Vol. 54, 149-151.

Hull, R. G., Rennie, J.A.N., Eastmond, C.J., Hutchinson, J.M.S., and Smith, F.W., 1984, Nuclear Magnetic Resonance (NMR) Tomographic Imaging for Popliteal Cysts in Rheumatoid Arthritis, Annals of Rheumatoid Disease, Vol. 43, 56-59.

Moseley, I., 1982, Recent Developments in Imaging Techniques, Br. Med. J., Vol. 284, 1141-1144.

Sutherland, I. E., Sproull, R.F., and Schumaker, R.A., 1974, A Characterization of Ten Hidden Surface Algorithms, Computing Surveys, Vol. 6, No.1, 15-30.

MATHEMATICAL ANALYSIS OF NETWORK TOPOLOGY IN THE CEREBROCORTICAL MICROVASCULATURE[1]

Antal G. Hudetz[1], Karl A. Conger[2], Miklos Pal[3], and Charles R. Horton[1]

[1]Center for Rehabilitation Science and Biomedical
 Engineering, Louisiana Tech University, Ruston, LA, USA
[2]Department of Pathology, University of Alabama at
 Birmingham, Birmingham, AL, USA
[3]Experimental Research Department, Semmelweis Medical
 University, Budapest, Hungary

INTRODUCTION

Pathological conditions such as shock, ischemia and traumatic injury often result in a decreasing number of physiologically perfused capillaries (Dintenfass, 1971; Crowell and Olsson, 1972; Fischer, 1973; Little et al., 1976). Blood flow is diverted to various bypass routes in the microvasculature resulting in deficient transport of oxygen and other nutrients to the tissue. Microvascular conductance may quickly be reduced to zero even if several capillaries remain patent (Hudetz et al., 1984; Hudetz and Werin, 1985). The extent to which these events happen depends on the number and availability of capillary bypass routes – characteristics described by network topology. Also, the mathematical modelling of microvascular blood flow and oxygen transport to tissue requires the precise knowledge of the network topology and geometry.

Aside from the qualitative description of microvascular geometry, data on cerebral capillary morphology are scarce and mostly limited to the upper cortical layer. Both SEM morphology (Motti et al., 1986) and in vivo microscopy (Ma et al., 1974, Pawlik et al., 1981) provide information from the superficial layers of tissue, however, the organization of the capillary bed is significantly different in various layers of the cerebral cortex (Duvernoy et al., 1981). On the other hand, the reconstruction of three-dimensional architecture of cerebrocortical capillary networks from serial sections has been a technically difficult task (Wiederhold et al., 1976).

This paper represents a preliminary report on our attempt to reconstruct the three-dimensional topology of the deep cerebrocortical capillary network. The implications of network topology on capillary hemodynamics is estimated by simulating the distribution of blood flow in

[1]This work was supported in part by the USPHS grant No. NS-08802, by the Ministry of Health, Hungary (3.01.3) and by the State of Louisiana, Division of Rehabilitation Services.

a mathematical model of the reconstructed network. The characteristics of transit of blood cells through the network is calculated by a probabilistic simulation.

METHODS

Reconstruction of microvascular topology

The three-dimensional branching pattern of cortical capillary networks was reconstructed from thick histological sections by the method of optical sectioning. The capillaries were visualized in rats by the injection of india ink- gelatine mixture following perfusion fixation at normal blood pressure. Approximately 400 um thick coronal sections were cut and cleared in glycerol to obtain an adequate depth of view.

A grid with 1 mm cell size was overlaid the section for orientation. At 50X magnification the image covered approximately one grid cell. The image picked up by a CCTV camera (Panasonic, vidicon) was observed on a 19 inch video monitor. The coordinates of capillary branch points were determined using an X-Y digitizer (Radio Shack) placed over the monitor. A schematic of the image analysis system is displayed in Figure 1. The X and Y coordinates, together with the vertical position of the microscope stage, were input to an APPLE II+ microcomputer. Data analysis and mathematical modelling were performed using an IBM AT computer.

In addition to the capillary branch points, the arteriolar inputs and venular outputs of the capillary system within a grid cell were recorded. Arteries and veins could be identified most easily based on the characteristics of their arborization and by tracing them from the brain surface at low magnification (Duvernoy et al., 1981).

Computer simulation of capillary blood flow

A wire frame model of the capillary network was constructed substituting each capillary segment by a straight line between connected branch points. The distribution of blood flow in the model was calculated

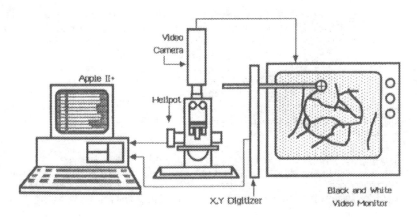

Fig. 1. Schematic of the image analysis system used for 3-D capillary reconstruction. The image was displayed on a large screen video monitor. The location of capillary branch points were digitized and the data were fed into the computer.

based on the analogy with resistor networks. The hydraulic resistance of each capillary was assumed to be proportional to its length. Equations for flow were written for each capillary segment and at each branch point the flow balance was written as a second equation. The resulting system of equations was solved by standard computer methods using matrix inversion.

In addition to the distribution of blood flow, the path lengths for transit of blood cells was estimated using a probabilistic approach. Individual transits from arteriole to venule were simulated by Monte Carlo method. At each bifurcation the probability of selecting one of the channels by the cell was determined according to the rule proposed by Yen and Fung (1978) for erythrocytes:

$$H1/H2 - 1 = a (V1/V2 - 1)$$

where H1 and H2 are the fractional volume flows of red cells and V1 and V2 are the mean flow velocities in two branches of equal cross section, respectively. The dimensionless constant "a" is, in general, a function of the ratio of cell diameter to tube diameter, the shape and rigidity of erythrocytes, and the feed hematocrit. At normal blood flow, the mean tissue hematocrit is between 30 and 32 in the rat brain cortex (Cremer and Seville, 1983). At this hematocrit, and for capillaries whose diameter is approximately equal to that of the red cell, a=1, approximately. In our simulation technique, random numbers with uniform distribution between 0 and 1 were generated. A path along branch 1 of a bifurcation was selected whenever the random number was less than V1/(V1+V2), and a path along branch 2 was selected otherwise. The total path length was given as the sum of selected segment lengths. 1000 transits per network were simulated.

RESULTS

Capillary topology

Capillary pathways from arteriole to venule were examined in the displayed image of the cortical vascular network at low magnification (25X). The topological length of A-V routes, that is, the number of nodes along these routes were determined. The number of nodes along three selected A-V routes in a particular field were: 6.5+7, 6.4+1.3 and 6.7+0.8. The number of nodes forming closed capillary loops was also noted. These varied between 4 and 7 in the given sample. Because the direction of blood flow was not known in the vessels, the observed anatomical paths may or may not correspond to physiological paths of blood flow.

Subsequently, the 3-dimensional topology of the capillary network was reconstructed by computer. In order to allow for differentiation between different depths in the tissue, the Z coordinate was represented in pseudocolor. Figures 2 and 3 display examples of 3-D capillary reconstruction. It is apparent that the maximum depths were in most cases smaller than the thickness to which the tissues were sectioned, probably due to some uncontrolled shrinkage of the samples during the clearing process.

Capillary hemodynamics

In order to obtain an estimate of physiological capillary path lengths, a technique for computer simulation of transit was established. Figure 3 displays the result of calculated capillary blood flow

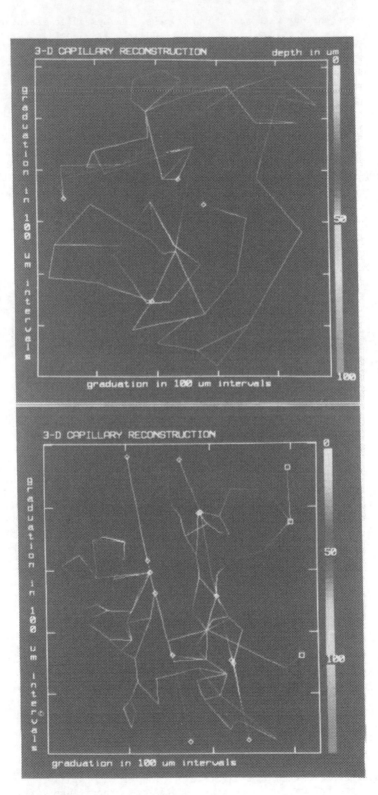

Fig. 2. (facing page) representation of reconstructed cortical
capillary networks (top: sample #6, bottom: sample #7).
Diamonds and squares indicate the location of a connection to an
arteriole or venule, respectively. No venule was seen in the
displayed field in sample #6.

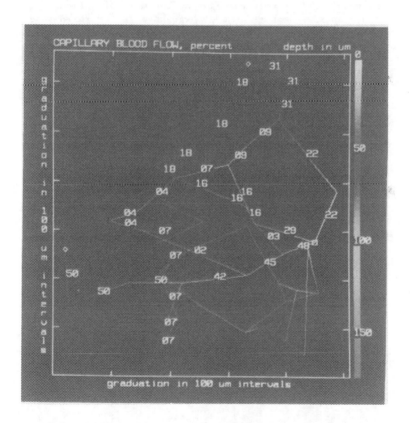

Fig. 3. Distribution of capillary blood flow calculated in a
reconstructed cortical capillary network (sample #4). Depth is
coded in pseudocolor. Blood flow values are displayed at the
corresponding capillary segments in percent of the total flow
through the system. A deep network of capillaries was apparently
not connected to the root of the system within the dimensions of
the observed tissue.

distribution in one of the reconstructed capillary networks (sample #4). The blood flows are given in percent of the total flow through the system. There were two arterioles and one venule in this network. Equal volume of blood flow was assumed to enter the two arterioles.

The transit of blood cells was simulated for the same network (sample #4). Fifty percent of the transits were simulated with one arteriole as input and 50 percent was done with the other as input. As a result of simulation, four distinct paths of different lengths were obtained: 233, 273, 285 and 308 micrometers. Note that these data do not reflect actual capillary lengths because the curvature of vessels was disregarded. Figure 4 illustrates the frequency of occurrence of these paths. A Gaussian distribution was fitted to the data points with a standard deviation of about 60 micrometers.

DISCUSSION

A fundamental characteristic of the cerebrocortical capillaries is their irregular pattern (Wiederhold et al., 1976; Duvernoy, 1981; Pawlik et al., 1981). In man, significant differences exist between the vascularization of various cortical layers. The multidirectional orientation of capillary meshes is most expressed in the third vascular layer (Duvernoy et al., 1981). In the present study we have seen similar differences in various layers of the rat brain cortex. In the region of interest of this study (third and fourth vascular layers) a highly irregular arrangement of capillaries was observed. The estimated blood flow velocities suggested hemodynamic heterogeneity too, although the

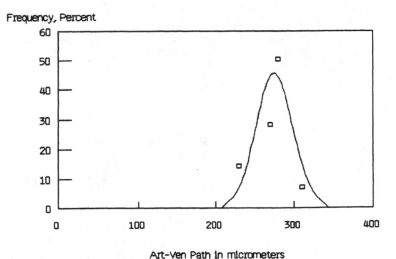

Fig. 4. Frequency distribution plot of transit path lengths (squares) calculated in sample #4. The continuous line represents Gaussian distribution fitted to the calculated data.

lack of data on actual capillary lengths did not allow a definite
conclusion. Regional differences may explain why in certain studies
(Motti et al., 1981) the capillary network seems to be more regular than
in most others.

In this study the distribution of blood flow was calculated assuming
uniform and Newtonian blood viscosity. Recent experimental findings
indicate that red cells are distributed uniformly across capillary
networks of the hamster cheek pouch and cremaster muscle (Desjardins and
Duling, 1987). As predicted previously by mathematical models (Papenfuss
and Gross, 1981), this is probably due to the self adjustment of tube
hematocrit which controls the resistance to flow. In the brain cortex,
capillary hematocrit is almost constant if the red cell flow velocity is
more than 2.0 mm/sec (Yamakawa et al., 1986). Thus, the apparent
viscosity is expected to be uniform in the network. For channels with
very low velocity, the capillary resistance will have to be adjusted to
account for the increased viscosity. This remains to be done until an
appropriate mathematical solution to the problem is obtained (Secomb et
al., 1985).

It had been expected that A-V path lengths were distributed similar
to a gamma distribution (Pawlik et al., 1981). Due to the lack of short
A-V paths the Gaussian distribution seemed to be more appropriate for a
fit in this study. It should be noted, however, that the path lengths
obtained here are, at best, illustrative for the methodology applied in
this study. Capillaries of the cerebral cortex may run a remarkably
irregular course in deep cortical layers (Duvernoy et al., 1981). It will
be essential to acquire data on 3-dimensional capillary tortuosity
(Pawlik et al., 1981) and apply those to the simulation. In spite of
these difficulties, the qualitative examination of the networks suggests
that the cerebrocortical microvasculature is essentially different from
the parallel capillary model, and that there is abundant structural
possibility for local bypass flow formation in the brain cortex.

SUMMARY

The three-dimensional branching pattern of deep cerebrocortical
capillary networks was reconstructed from histological sections. The
distribution of blood flow in a mathematical model of the reconstructed
network was calculated. The transit of red blood cells through the
network was simulated by computer, and the total path length traveled by
the cells was estimated. The results support both anatomical and
hemodynamic heterogeneity of the cerebrocortical microvascular system.

ACKNOWLEDGEMENT

The authors express their gratitude to Mr. Paul Copland and Ms.
Cynthia Pinell of the Technology Transfer Office for their support and
preparation of all illustrations.

REFERENCES

Cremer, J.E., and Seville, M.P., 1983, Regional brain blood flow, blood
    volume, and haematocrit values in the adult rat, J. Cereb. Blood
    Flow Metabol., 3:254.

Crowell, R. M., and Olsson, Y., 1972, Impaired microvascular filling
    after focal cerebral ischemia in the monkey, Neurology, 22:500.

Desjardins, C., and Duling, B. R., 1987, Microvessel hematocrit: measurement and implication for capillary oxygen transport, Am. J. Physiol., 252:H494.

Dintenfass, L., 1971, "Blood Microrheology - Viscosity Factors in Blood Flow, Ischemia, and Thrombosis", Appleton-Century-Crafts, Inc., New York.

Duvernoy, H. M., Delon, S., and Vannson, J. L., 1981, Cortical blood vessels of the human brain, Brain Res. Bull., 7:519.

Fischer, E. G., 1973, Impaired perfusion following cerebrovascular stasis, Arch. Neurol., 29:361.

Hudetz, A. G., Conger, K. A., Kovach, A. G. B., Halsey, J. H., and Hino, K., 1984, Distribution of blood flow in partially closed cerebral capillary networks, in: "Oxygen Transport to Tissue VI," D. F. Bruley, H. I. Bicher, D. D. Reneau, eds., Plenum Press, New York.

Hudetz, A. G. and Werin, S., 1986, Percolation and transit in microvascular networks, in: "Oxygen Transport to Tissue VIII," Ian S. Longmuir, ed., Plenum Press, New York.

Little, J. R., Kerr, F. W. L., and Sundt, T. M., 1976, Microcirculatory obstruction in focal cerebral ischemia: An electron microscopic investigation in monkeys, Stroke, 7:25.

Ma, Y.P., Koo, A., Kwan, H.C., and Cheng, K.K., 1974, On-line measurement of the dynamic velocity of erythrocytes in the cerebral microvessels in the rat, Microvasc. Res., 8:1.

Motti, E. D. F., Imhof, H.-G., and Yasargil, M. G., 1986, The terminal vascular bed in the superficial cortex of the rat, J. Neurosurg., 65:834.

Papenfuss H.-D., and Gross, J. F., 1981, Microhemodynamics of capillary networks, Biorheology, 18:673.

Pawlik, G., Rakl, A., and Bing, R.J., 1981, Quantitative capillary topography and blood flow in the cerebral cortex of cats: an in vivo microscopic study, Brain Res., 208:35.

Secomb, T. W., Skalak, R., Ozkaya, N., and Gross, J.F., 1985, Flow of axisymmetric red blood cells in narrow capillaries, J. Fluid. Mech., 163:405.

Wiederhold, K.-H., Bielser, W., Schultz, U., Jr., Veteau, M.-J., and Hunziker, O., 1976, Three dimensional reconstruction of brain capillaries from frozen serial sections, Microvasc. Res., 11:175.

Yamakawa, T., Niimi, H., Sugiayama, I., and Yamaguchi, S., 1986, Red blood cell flow distribution and capillary hematocrit in the cerebral cortex microcirculation of cat: intravital microscopic study, Proc. IUPS., 16:222.

Yen, R. T., and Fung, Y. C., 1978, Effect of velocity distribution on red cell distribution in capillary blood vessels, Am. J. Physiol., 235:H251

ANALYSIS OF OXYGEN DELIVERY TO TISSUE

BY MICROVASCULAR NETWORKS

T.W. Secomb and R. Hsu

Department of Physiology
University of Arizona
Tucson AZ 85724 U.S.A.

INTRODUCTION

Theoretical modeling of oxygen delivery to tissue provides a means to obtain information about oxygen concentration fields in tissue at the scale of individual capillaries. Such information is difficult or impossible to obtain with available experimental techniques. Theoretical models can provide insights into the relationship between perfusion and metabolic demand in tissue, and also contribute to the understanding of metabolic mechanisms of blood flow regulation. The value of theoretical analyses was recognized in the classic work of Krogh (1918) on oxygen diffusion from parallel arrays of capillaries, and since then most models of oxygen delivery to tissue have been modifications and extensions of the Krogh model (Middleman, 1972). Generally, these models have retained the assumption that each point in the tissue receives oxygen only from the nearest capillary. However, this assumption was relaxed, for the case of multiple parallel capillaries, in studies by Popel (1978, 1980), Salathe (1982) and Klitzman et al (1983).

Duling and Berne (1970) demonstrated significant drops in partial pressure of oxygen (PO_2) in precapillary vessels of the hamster cheek pouch, suggesting that arterioles are able to supply a significant amount of oxygen to tissue. This has been confirmed experimentally by Pittman et al (1985) and theoretically by Popel and Gross (1979). Also, Duling and Berne (1970) pointed out that precapillary oxygen losses would increase with increasing metabolism or decreasing flow rate. This would allow changes in the relative contributions of arterioles and capillaries to oxygen delivery. These findings have important implications for theoretical modeling of oxygen transport to tissue. Firstly, they show that arteriolar contributions to oxygen delivery cannot be neglected. Secondly, they indicate that each point in the tissue does not necessarily receive oxygen only from the nearest vessel segment, and that the vessel segment responsible for oxygen delivery to a particular point in the tissue may vary with blood flow and metabolic rate. In particular, a relatively sparse network of vessels would suffice to supply the tissue at very low metabolic rates, and the arterioles could provide a large fraction of the requirement. As metabolic rate increases, an increasingly dense vascular supply must be called upon, and one would expect the contribution of the capillaries to become more important.

In this paper we describe a model for oxygen delivery from a network of microvessels of arbitrary geometry to a finite region of tissue, in which no a priori assumptions are made concerning the tissue region supplied by each vessel segment. This permits quantitative investigation of the effects described in the previous paragraph. It is based on the classical Green's function method (Green, 1828) for solving mathematical problems in potential theory. The essential idea is that the spatial field resulting from a distribution of sources or sinks can be expressed as a sum or integral of fields resulting from individual point sources. Thus, the field is determined by computing the strengths of the sources.

ASSUMPTIONS OF THE MODEL

The aim of the model is to simulate oxygen delivery by a network of microvessels to a finite region of tissue, without restricting the network geometry or prescribing the vessel segment which supplies oxygen to each particular location in the tissue. Consequently, the PO_2 at each point in the tissue depends on the oxygen delivery from all vessel segments within the tissue region, and the equations governing extravascular oxygen diffusion must be solved simultaneously with the equations for intravascular oxygen transport throughout the entire region. Some geometrical restrictions are required for technical reasons associated with the method of solution: the radius of each vessel segment must be much smaller than its length, its distance from the boundary and from other non-connected segments, and its radius of curvature.

The metabolic rate of oxygen consumption is assumed constant and uniform throughout the tissue region, which is assumed to be cuboidal. This is the most convenient shape from a mathematical point of view, and allows for regularly repeating structures consisting of multiple cuboidal regions. Such arrangements are consistent with anatomical studies of skeletal muscle (Myrhage and Eriksson, 1980). The diffusive oxygen flux is assumed to be zero on all boundary surfaces. This will be a valid approximation if the tissue region is large enough that its net oxygen supply is large compared to the flux across its boundaries. Within the tissue space, oxygen is assumed to have a uniform diffusivity, and a linear relationship between oxygen concentration and PO_2 in tissue is assumed, with a uniform solubility. Myoglobin binding of oxygen in the tissue is thus neglected, an approximation which is justified if the PO_2 is sufficiently high to ensure that myoglobin is nearly saturated (above 5-10 mmHg).

Oxygen transport along blood vessels is assumed to occur purely by convection. The blood is assumed to be well-mixed, in the sense that PO_2 is uniform across the vessel cross-section. In smaller vessels, this condition may not be satisfied if diffusive resistance within the vessel is significant (Hellums, 1977). It is possible to modify the model to include such effects but we omit the details here. The nonlinear relationship between oxygen content of blood and PO_2 is modeled using Hill's equation, and transient effects associated with the kinetics of oxyhemoglobin dissociation are neglected.

MATHEMATICAL FORMULATION

Figure 1 shows a geometric configuration of the type to be considered. A cartesian coordinate system (x_1, x_2, x_3) is used, and the oxygen concentration at each point in the tissue is denoted by $C(x_1, x_2, x_3)$. In the tissue, C satisfies

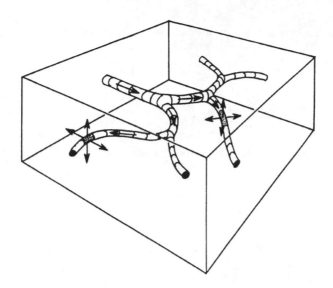

Figure 1. Example of configuration of tissues and vessels to be modeled. The division of each segment into discrete subregions in the computation is indicated. Arrows indicate convective and diffusive pathways taken by oxygen reaching tissue through the two segments shown shaded.

$$D\nabla^2 C = M \tag{1}$$

where M is the metabolic rate of oxygen consumption, D is the oxygen diffusivity, and ∇^2 denotes the Laplacian, $\partial^2/\partial x_1^2 + \partial^2/\partial x_2^2 + \partial^2/\partial x_3^2$. At each point of the tissue boundary, zero oxygen flux is assumed, and hence

$$\frac{\partial C}{\partial n} = 0 \tag{2}$$

where n denotes distance normal to the boundary. In the tissue, PO_2 and oxygen concentration are linearly related

$$C = \alpha P \tag{3}$$

where α is the solubility. We assume $D = 1.5 \times 10^{-5}$ cm^2/s and $\alpha = 3 \times 10^{-5}$ cm^{3}O$_2$/cm^3/mmHg (Popel and Gross, 1979).

Each vessel segment is assumed to have a uniform circular cross-section of radius r_0, and constant volume flow rate Q. A cylindrical coordinate system (r, θ, z) is set up in the segment. (If the segment is curved, z denotes arc length along the segment.) The oxygen concentration within the vessel is defined by $C_V(z)$. The corresponding PO_2 is $P_V(z)$, and is related to C_V according to Hill's equation

$$C_V = C_S \, f(P_V/P_{50}) \quad \text{where} \quad f(x) = x^r/(1 + x^r) \tag{4}$$

C_S is the concentration of oxygen in fully saturated blood, here taken as 0.1 cm^{3}O$_2$/cm^3, $P_{50} = 26$ mmHg is the PO_2 at which 50% saturation occurs, and r = 2.55. The rate of oxygen efflux per unit length of the segment is denoted by q(z). Conservation of mass then implies that

$$Q \frac{dC_V}{dz} = -q(z) \tag{5}$$

The oxygen concentration in the tissue surrounding a given segment is described by $C(r, \theta, z)$. For matching of the extravascular and intravascular oxygen fields, oxygen flux and PO_2 must be continuous at the blood-tissue boundary. Continuity of oxygen flux implies that

$$q(z) = -Dr_0 \int_0^{2\pi} \frac{\partial C(r, \theta, z)}{\partial r}\Big|_{r=r_0} d\theta \qquad (6)$$

The local oxygen concentration in the tissue near the vessel is approximated as

$$\overline{C}(z) = \frac{1}{2\pi} \int_0^{2\pi} C(r_0, \theta, z) \, d\theta \qquad (7)$$

The local PO_2 is then $\overline{P} = \overline{C}/\alpha$ and continuity of PO_2 requires that $\overline{P} = P_v$. Then q and $\overline{C}$ are related by

$$Q \, C_s \frac{d}{dz} f(\overline{C}/\alpha P_{50}) = -q(z) \qquad (8)$$

THE GREEN'S FUNCTION METHOD

To model oxygen delivery from microvascular networks of arbitrary and possibly complex geometry, an efficient computational method is essential. Finite difference and finite element methods would lead to unacceptably large numbers of unknown parameters, because of the three-dimensionality of the problem and the need to resolve individual vessels with the grid or mesh chosen. The Green's function method used here takes advantage of the linearity of equation (1), to reduce the number of unknowns to the number of nodes required to represent the source strength distributions of the vessel segments (cf. Fleischman et al, 1986a,b).

We define a modified Green's function $G(\underline{x}; \underline{x}^*)$ as the solution to

$$\nabla^2 G = -\delta(\underline{x}, \underline{x}^*) + 1/V \qquad (9)$$

together with the boundary condition (2), where $\underline{x} = (x_1, x_2, x_3)$ denotes a point in the tissue space, $\underline{x}^*$ denotes a point on the blood-tissue boundary, V is the volume of the tissue and δ denotes the Dirac delta function. Note that G contains an arbitrary additive constant. We decompose G into three parts:

$$G = G_1 + G_2 + G_3 \qquad (10)$$

where G_1 represents the singular part and corresponds to a point source in an infinite medium:

$$G_1 = \frac{1}{4\pi |\underline{x} - \underline{x}^*|} \qquad (11)$$

The second part G_2 is a second order polynomial in x_1, x_2 and x_3, chosen to satisfy

$$\nabla^2 G_2 = 1/V \qquad (12)$$

and to make the total net flux across each of the six surfaces due to $G_1 + G_2$ equal zero. The third part, G_3, is a smooth function which satisfies $\nabla^2 G_3 = 0$ and cancels the local fluxes due to $G_1 + G_2$ on each of the six planes of the tissue boundary. To obtain a compact representation of G_3, we take advantage of the fact that it satisfies Laplace's equation, and write it as the sum of three double Fourier series with undetermined coefficients:

$$G_3 = \sum_{m=0}^{\infty} \sum_{n=0}^{\infty} \left[[A_{m,n} \cosh k_{m,n}(x_1-\ell_1) + B_{m,n} \cosh k_{m,n}x_1] \cos \frac{m\pi x_2}{\ell_2} \cos \frac{n\pi x_3}{\ell_3} \right.$$

$$+ [A'_{m,n} \cosh k'_{m,n}(x_2-\ell_2) + B'_{m,n} \cosh k'_{m,n}x_2] \cos \frac{m\pi x_3}{\ell_3} \cos \frac{n\pi x_1}{\ell_1}$$

$$\left. + [A''_{m,n} \cosh k''_{m,n}(x_3-\ell_3) + B''_{m,n} \cosh k''_{m,n}x_3] \cos \frac{m\pi x_1}{\ell_1} \cos \frac{n\pi x_2}{\ell_2} \right] \quad (13)$$

where ℓ_1, ℓ_2 and Ω_3 are the edge lengths of the region and

$$k_{m,n} = \pi \left(\frac{m^2}{\ell_2^2} + \frac{n^2}{\ell_3^2} \right)^{1/2}, \ k'_{m,n} = \pi \left(\frac{m^2}{\ell_3^2} + \frac{n^2}{\ell_1^2} \right)^{1/2}, \ k''_{m,n} = \pi \left(\frac{m^2}{\ell_1^2} + \frac{n^2}{\ell_2^2} \right)^{1/2}$$

The Fourier coefficients are determined by applying the no-flux boundary conditions, yielding

$$A_{m,n} = \frac{4}{\ell_2 \ell_3 \sinh k_{m,n}\ell_1} \int_0^{\ell_3} \int_0^{\ell_2} \frac{\partial}{\partial x_1}(G_1 + G_2)\Big|_{x_1=0} \cos\frac{m\pi x_2}{\ell_2} \cos\frac{n\pi x_3}{\ell_3} \, dx_2 dx_3 \quad (14)$$

The coefficients $B_{m,n}$, $A'_{m,n}$, $B'_{m,n}$, $A''_{m,n}$ and $B''_{m,n}$ are determined similarly. The infinite series (13) may be truncated when remaining terms are negligible.

The oxygen concentration in the tissue can be expressed in terms of the Green's function corresponding to a point source:

$$C(\underline{x}) = \int \int_S G(\underline{x}; \underline{x}^*) \, q_0(\underline{x}^*) \, dS + G_0 \quad (15)$$

where $q_0(\underline{x}^*)$ denotes a distribution of sources on the blood-tissue boundaries of all the vessel segments, and the integral is taken over the complete boundary surface S. The unknown constant G_0 results from the arbitrary additive constant in the Green's function. For the moment, we consider the case of a single segment of length ℓ. The generalization to multiple segments is readily made. Because of the geometrical assumptions, the oxygen source strength is approximately independent of θ, and can be written in local cylindrical coordinates as

$$q_0(r_0, \theta^*, z^*) \approx q(z^*)/2\pi r_0 \quad (16)$$

The mean tissue concentration $\overline{C}$ near the vessel may then be expressed in terms of $q(z^*)$:

$$\overline{C}(z) = \int_0^{\ell} \overline{G}(z, z^*) \, q(z^*) \, dz^* + G_0 \quad (17)$$

where $\overline{G}$ is obtained by averaging G with respect to θ and θ^*, the angular positions of the field and source points with respect to the vessel axis. The averaged Green's function G is written as the sum of three parts:

$$\overline{G} = \overline{G}_1 + \overline{G}_2 + \overline{G}_3, \tag{18}$$

where

$$\overline{G}_m(z,z^*) = \frac{1}{4\pi^2} \int_0^{2\pi} \int_0^{2\pi} G_m(r_0,\theta,z;r_0,\theta^*,z^*) \, d\theta \, d\theta^* \tag{19}$$

for m=1,2,3. It can be shown that

$$\overline{G}_1 = \frac{k \, K(k)}{2\pi r_0} \tag{20}$$

where $K(k)$ is the complete elliptical integral of the first kind

$$K(k) = \int_0^{\pi/2} \frac{d\theta}{(1 - k^2\sin^2\theta)^{1/2}} \quad \text{and} \quad k = (1 + [(z - z^*)/2r_0]^2)^{-1/2} \tag{21}$$

Since the vessel radius is small and the functions G_2 and G_3 are both smooth, we approximate their values in (19) by their center-line values. This completes the calculation of the Green's function $\overline{G}$. The numerical evaluation of the integrals (14) is the most time-consuming part of the computation. However, it depends only on the geometrical parameters and only needs to be done once for a given geometrical configuration.

The solution of the problem for particular blood flows and metabolic rate requires simultaneous solution of (8) and (17) for the two unknown functions $\overline{C}(z)$ and $q(z)$. First, (8) is integrated to give

$$f(\overline{C}(z)/\alpha P_{50}) = f(\overline{C}(0)/\alpha P_{50}) - (Q \, C_S)^{-1} \int_0^z q(z) \, dz \tag{22}$$

Then $\overline{C}(z)$ is eliminated from (17) and (22), yielding

$$\int_0^\ell \overline{G}(z,z^*) \, q(z^*) \, dz^* + G_0$$

$$= \alpha P_{50} \, f^{-1}[\, f(\overline{C}(0)/\alpha P_{50}) - (Q \, C_S)^{-1} \int_0^z q(z) \, dz \,] \tag{23}$$

The numerical procedure for solving the integral equation (23) may be summarized as follows. Each vessel segment is divided into N small pieces of length Δz_i (i=1,2,...,N), with uniform source strength q_i on each piece. The Green's function $\overline{G}$ is integrated numerically along each piece, so that the left hand side of equation (23) becomes a system of linear expressions in the unknowns q_i. Because of the nonlinearity of the right hand side, an iterative approach is used. At each iteration the right hand side is linearized about the values of q_i obtained from the previous iteration. A system of linear equations is then solved to obtain improved estimates for q_i. This procedure is repeated until convergence is achieved, usually within a few iterations. The metabolic rate in the tissue is obtained by summing the oxygen fluxes from the pieces of the segments, and dividing the sum by the volume of the region. Adjustment of the metabolic rate is achieved by varying the parameter G_0.

100

RESULTS

For a simple example, we consider first a single straight vessel segment lying in the center of the tissue region, with dimensions as shown in figure 2. This configuration is analogous to that assumed in

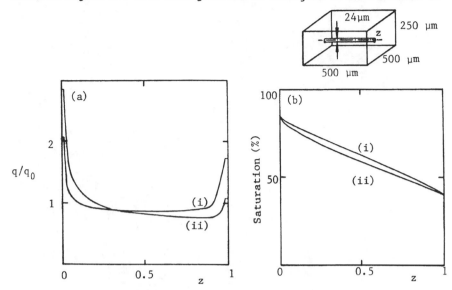

Figure 2. Variation of (a) source strength q (relative to mean source strength q_0) and (b) blood oxygen saturation, with distance z along a vessel segment in a cuboidal tissue region. Curves (i) are computed for a flow rate of 5×10^{-8} cm^3/s and a metabolic rate of 3.5×10^{-5} cm^{3}O$_2$/cm^3/s. Curves (ii) are computed assuming that both flow rate and metabolic rate are reduced by a factor of 10 relative to case (i).

Krogh-type models. The following parameter values are assumed: P(0) (inflow PO$_2$) = $2P_{50}$ = 52 mmHg, r_0 = 12 μm. Profiles of source strength and intravascular oxyhemoglobin saturation are shown for two different metabolic rates and volume flow rates, chosen so that the outflow oxygen saturation is the same in each case. Figure 2 shows a non-uniform variation of source strength along the vessel segment. The swings in q(z) near each end of the segment result from the finite length of the segment, and are not important from a physiological point of view. The asymmetry in the profile reflects the fall in saturation along the segment. Since metabolic uptake is the same in the upstream and downstream halves of the tissue domain, there is a net diffusion of oxygen in the tissue towards the downstream end of the domain. This phenomenon is excluded in the classical Krogh cylinder model, but is also predicted in modifications of the Krogh model which allow for axial diffusion.

In figure 3, results are presented for a network consisting of a central straight arteriole and four perpendicular capillary branches. This idealized geometry was chosen for simplicity, although the method is also applicable to more complex and realistic geometries. When flow rate and metabolic rate are varied, changes in the relative contributions of the network segments to total oxygen delivery are clearly demonstrated. At higher flow rate and metabolic rate, the contribution of the distal branches to total delivery increases. Comparison of the results for a single vessel and a network shows that this shift becomes more marked as number of segments in the network increases. Figure 3 also shows effects of interactions between adjacent vessel segments. For instance, the

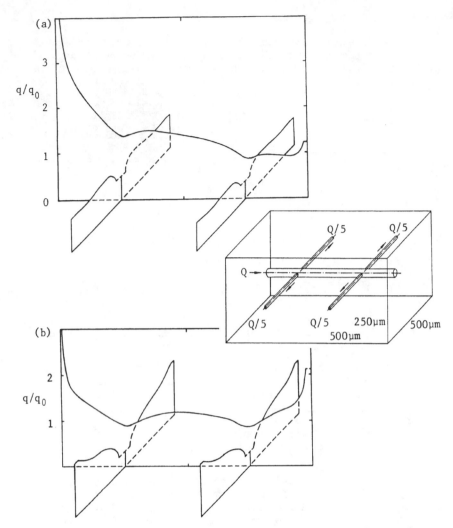

Figure 3. Variation of oxygen source strength (vertical axis) with position in an idealized microvascular network. (a) Flow rate in feeding vessel = 5 x 10^{-9} cm^3/s, metabolic rate = 3.5 x 10^{-6} $cm^3 O_2$/cm^3/s. (b) Flow rate in feeding vessel = 5 x 10^{-8} cm^3/s, metabolic rate = 3.5 x 10^{-5} $cm^3 O_2$/cm^3/s.

source strength of the parent vessel has a local minimum at the point where a daughter vessel branches off. In this region, the daughter vessel produces a local increase in the PO_2 near the parent vessel.

CONCLUSIONS

The Green's function approach permits modeling of oxygen delivery to tissue by networks of vessel segments, at a reasonable computational cost. It allows modeling under less restrictive assumptions than are involved in the Krogh model and its variants. In particular, it is not necessary to make a priori assumptions concerning the location of the vessel segment which provides oxygen to a given point in the tissue. The model thus permits more realistic predictions of spatial oxygen fields than has previously been possible.

Preliminary results have been obtained for a single vessel segment
and for an idealized seven-segment network. The results shown in figures
2 and 3 confirm the occurrence of shifts in the distribution of source
strength with changes in metabolic rate. At very low metabolic rates,
the proximal part of the network can supply a large part of the total
demand, and extravascular and intravascular PO_2's are nearly equilibrated
in the most distal parts. As metabolic rate increases, the distal parts
of the network supply an increasingly large proportion of the total
requirement. These findings are particularly relevant to the understanding of oxygen transport to skeletal muscle, in which very large changes
occur in flow rate and metabolic demand between resting and exercising
states.

This work was supported by NIH Grant HL17421.

REFERENCES

Duling, B.R. and Berne, R.M. (1970) Longitudinal gradients in periarter-
    iolar oxygen tension. Circ. Res. 27:669-678.
Fleischman, G.J., Secomb, T.W. and Gross, J.F. (1986a) Effect of extra-
    vascular pressure gradients on capillary fluid exchange. Math.
    Biosci. 81:145-164.
Fleischman, G.J., Secomb, T.W. and Gross, J.F. (1986b) The interaction
    of extravascular pressure fields and fluid exchange in capillary
    networks. Math. Biosci. 82:141-151.
Green, G. (1828) Essay on the Application of Mathematical Analysis to
    the Theory of Electricity and Magnetism. Nottingham.
Hellums, J.D. (1977) The resistance to oxygen transport in the capil-
    laries relative to that in the surrounding tissue. Microvasc. Res.
    13:131-136.
Krogh, A. (1918) The number and the distribution of capillaries in
    muscle with the calculation of the oxygen pressure necessary for
    supplying the tissue. J. Physiol. 52:409-515.
Klitzman, B., Popel, A.S. and Duling, B.R. Oxygen transport in resting
    and contracting hamster cremaster muscles: Experimental and theo-
    retical microvascular studies. Microvasc. Res. 25:108-131.
Middleman, S. (1972) Transport phenomena in the cardiovascular system.
    Wiley-Interscience, New York.
Myrhage, R. and Eriksson, E. (1980) Vascular arrangements in hind limb
    muscles of the cat. J. Anat. 131:1-17.
Pittman, R.N., Ellsworth, M.L. and Swain, D.P. (1985) Measurements of
    oxygen transport in skeletal muscle microcirculation. Proc. 7th
    Ann. Conf. IEEE Eng. in Med. Biol. Soc., pp. 536-540.
Popel, A.S. (1978) Analysis of capillary-tissue diffusion in multicapil-
    lary systems. Math. Biosci. 39:187-211.
Popel, A.S. (1980) Oxygen diffusion from capillary layers with concur-
    rent flow. Math. Biosci. 50:171-193.
Popel, A.S. and Gross, J.F. (1979) Analysis of oxygen diffusion from
    arteriolar networks. Amer. J. Physiol. 237:H681-689.
Salathe, E.P. (1982) Mathematical modeling of O2 transport in skeletal
    muscle. Math. Biosci. 58:171-184.

OXYGEN TRANSPORT TO TISSUE DURING RECURRENT BLOOD FLOW SUPPLY BY GROUPED CAPILLARIES IN SKELETAL MUSCLE WITH OR WITHOUT FACILITATED DIFFUSION

Ikuo Ohta, Aki Ohta, Masahiro Shibata and Akira Kamiya

Research Institute for Applied Electricity
Hokkaido University
060, Sapporo, Japan

INTRODUCTION

Intravital observations of the microcirculation in various skeletal muscles have revealed that the capillary red cell velocity in the resting state is neither steady nor uniform (Wiedeman, 1984). It often fluctuates periodically and capillaries seem to repeat intermittently open and closed phases of flow with a frequency of 0.05 Hz to 0.2 Hz (Lindbom et al., 1980; Shibata et al., 1983). In the previous study (Fukuoka et al., 1983), the effects of such an intermittent capillary flow on oxygen (O_2) transport to tissue was analyzed with a dynamic computor simulation. The result demonstrated that when a tissue region is supplied with O_2 recurrently from one of the surrounding capillaries, the lowest level of O_2 tension (PO_2) in the tissue becomes substantially higher than that perfused continuously by a certain fixed capillary with the same blood flow. However, further detailed observation of the skeletal muscle microcirculation has shown that tissue is supplied with O_2 by several grouped capillaries, which originate from the same arteriole and repeat the open and closed phases simultaneously in each group but out of phase to other groups (Shibata et al., 1985). In this study we tried to evaluate the effect of the intermittent capillary flow on oxygen transport to tissue by employing a more realistic model of recurrent blood flow supply by grouped capillaries. The influence of the facilitated diffusion by myoglobin (Mb) on the transport (Kreuzer, 1970) was also analysed in this model.

MATHEMATICAL MODEL AND COMPUTATIONAL PROCEDURE

As shown in Fig.1, it is well known that most capillaries are uniformly distributed in the mammalian skeletal muscle, running parallel to the straight muscle fibers and that only 1/4 or 1/3 of the total capillaries are open for blood flow supply in the resting state (Folkow and Neil, 1971). This allows us to assume that capillaries repeat the open and close phases approximately in the ratio 1 : 3. In the previous study, we simulated the O_2 tension (PO_2) distribution in the tissue under such an intermittent and recurrent blood flow supply as shown in Fig.2,(a) (single capillary mode) in which some one of the neighboring four capillaries is always open but the open channel moves to the next capillary at a certain interval. In this study, however, the recurrence mode of the open capillary channel has been improved in a more realistic one as shown in Fig.2,(b) (grouped capillary mode) in which 4 neighboring capillaries

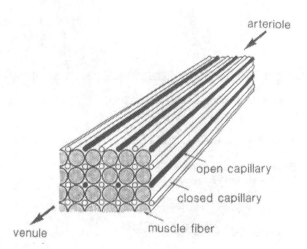

Fig.1 Schematic illustration of the capillary distribution in the resting skeletal muscle.

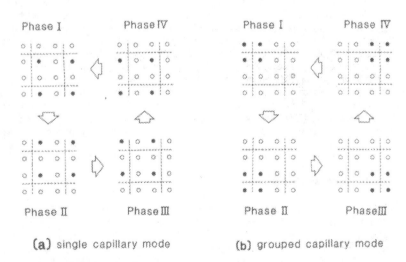

Fig.2 Schematic illustrations of the models of recurrent blood flow supply to tissue; the single capillary mode (a) and the grouped capillary mode (b).

originated from a single arteriole are open to supply O_2 to a certain tissue region while capillaries in other 3 groups surrounding it are closed but the open group moves to the next at a certain interval. As is seen in Fig.2, this mode of blood flow supply also brings about 4 phases (phase I, II, III, and IV) in accord with the group of open capillaries as the case of the single capillary mode.

The symmetrical geometry of the open and closed capillaries in Fig.2, (b) demonstrates that if the interested tissue region is sufficiently far from the tissue edge, there exist 4 planes of symmetry in the tissue as indicated by broken lines in the figure, at any moment of 4 phases. In the computer simulation of the O_2 transport, these planes can be treated as nonpermeable wall. Thus, the tissue region to be calculated is limited to a rectangular space including 4 capillaries inside and surrounded with the complete diffusion barriers to O_2 as shown in Fig.3, (a).

The mathematical formulations for the dynamic simulation in the above rectangular space are expressed in the orthogonal coordinates x, y and z as shown in Fig.3,(a). Then, O_2 diffusion across the capillary wall and in the tissue space as well as O_2 convection by capillary blood flow are formulated by the following differential equations in terms of PO_2 changes in the open capillary (Pc) and in the tissue (Pt) with time t respectively;

For the open capillary channel,

$$\partial Pc/\partial t = -Vc(\partial Pc/\partial z) + D(\alpha t/\alpha c)(2/Rc)(grad*P) + D\partial^2 Pc/\partial z^2 \qquad (1)$$

and for the tissue space

$$\partial Pt/\partial t = D'[\partial^2 Pt/\partial x^2 + \partial^2 Pt/\partial y^2 + \partial^2 Pt/\partial z^2] - \dot{Q}t/\alpha t \qquad (2)$$

where
Vc : Capillary red cell velocity (200–1500 μm/s)
Rc : Capillary radius (2 μm)
grad*P : PO_2 gradient across capillary wall
D : O_2 free diffusion coefficient (1500 μm^2/s)
αt : O_2 solubulity in plasma or interstitial fluid (3×10^{-5}/mmHg)
αc : O_2 solubility in blood (3.8×10^{-3}/mmHg for Pc < 47mmHg, 5.8×10^{-4}/mmHg for Pc > 47mmHg)
D' : O_2 diffusivity in the tissue
$\dot{Q}o_2$: O_2 consumption rate of the tissue (0.6–1.5×10^{-4} /s)

The values in the parentheses are data used in this study. With respect to the O_2 solubility in blood, the slopes of the O_2 saturation curve of blood simulated with a broken line as shown in Fig.3,(b) are employed. In order to evaluate the effect of the facilitated diffusion by Mb, O_2 diffusivity in tissue (D') is determined as,

$$D' = D \qquad \text{for Pt > 10mmHg} \qquad (1500 \ \mu m^2/s)$$
$$D' = DpCp\alpha m/\alpha t + D \qquad \text{for Pt < 10mmHg} \qquad (3900 \ \mu m^2/s)$$

where
Dp : Mb diffusivity in the tissue (93 μm^2/s)
Cp : Mb concentration in the tissue (0.45 mmol/kg = 0.0077)
αm : O_2 solubility in Mb solution. (0.1/mmHg for Pt < 10mHg, 0 for Pt > 10mmHg)

With respect to the O_2 solubility in Mb solution (αm), the slopes of O_2 dissociation curve for Mb simulated with a broken line as shown in Fig.3,(c) are employed.

The boundary conditions for the above equations in this model are;

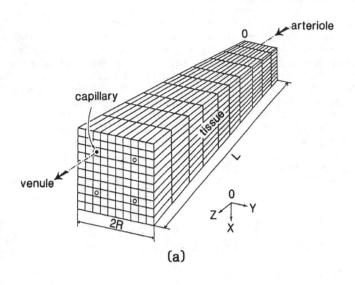

(a)

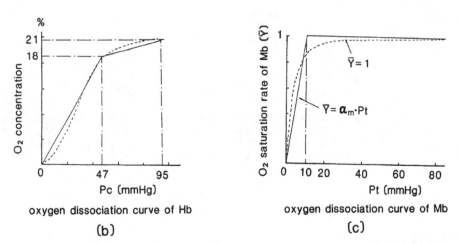

oxygen dissociation curve of Hb

(b)

oxygen dissociation curve of Mb

(c)

Fig.3 A rectangular model of the capillary-tissue system employed in the present dynamic computer simulation (a), O_2 dissociation curves of Hb (blood)(b) and of Mb (c) fitted with broken lines.

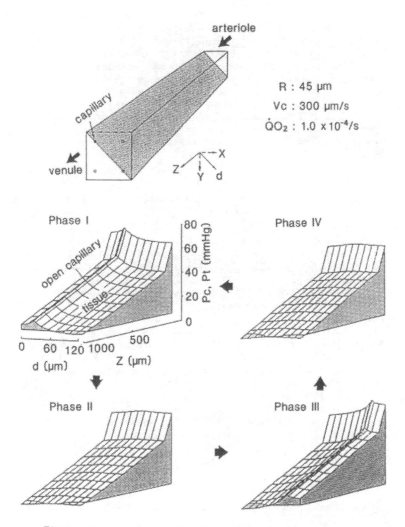

Fig.4 An example of the simulated PO$_2$ distribution.

$$\left.\partial Pt/\partial x\right|_{x=0} = \left.\partial Pt/\partial x\right|_{x=2R} = 0, \quad \left.\partial Pt/\partial y\right|_{y=0} = \left.\partial Pt/\partial y\right|_{y=2R} = 0,$$

$$\left.\partial Pt/\partial z\right|_{z=0} = \left.\partial Pt/\partial z\right|_{z=L} = 0.$$

where
   R : Intercapillary distance (30-75 µm)
   L : Capillary length (1000 µm)

To carry out practically the digital computation, the tissue space was divided into uniform rectangular compartments. Usually x, y and z axes were divided into 10 sections respectively. By employing the iterative method, PO_2 change in every compartment for a minute time interval (0.01 s) was calculated consecutively. The initial values for tissue and capillary PO_2 distribution were given from the steady state solution of the Krogh-Erlang equation. When the calculation of the phase I for duration of 5 to 10 s was finished, that of the phase II was followed. The computation was continued recurrently in the order of phase I, II, III, IV, and again I,..., until numerical values of PO_2 distribution were converged into a certain cyclic manner. Even when the value of PO_2 in a tissue compartment became negative, the computation was continued with no special modification.

RESULTS

Fig.4 shows an example of calculated results obtained in the model of the grouped capillary mode. The PO_2 distribution on the hatched plane in the upper panel at the end of the each phase are depicted. Cyclic changes in tissue PO_2 in accord with the phase shift are demonstrated. The values of intercapillary distance (R), capillary flow velocity (Vc) and tissue O_2 consumption rate ($\dot{Q}O_2$) are also shown in the figure. The effect of facilitated diffusion by Mb is not included.

Fig.5 illustrates the effects of the recurrent blood flow supply by grouped capillaries on the time course of the tissue PO_2 with and without the ffect of facilitated diffusion. The data obtained at the points designated as K = 5 and K = 10 in the upper panel are shown. The broken lines in the lower pannels indicate the level of steady state flow supply when the open capillaries are fixed at the state of phase IV. It is evident that PO_2 by the recurrent blood flow supply is always higher than the level by steady flow supply. This implies that the lowest tissue PO_2 level can be enhanced with the recurrent blood flow supply by recruitment and release of the different groups of capillaries. It is also obvious from this figure that the facilitated diffusion is quite effective to increase the lowest tissue PO_2. Although the time courses of tissue PO_2 near the midcapillary (K=5) show no difference with and without facilitate diffusion, it increases the tissue PO_2 near the venous end of capillary (K=10) by approximately 7 to 8 mmHg. The effect of the recurrent flow supply remains unchanged with the facilitate diffusion.

DISCUSSION AND CONCLUSION

In this study, we evaluated the effect of the recurrent blood flow supply by grouped capillaries on tissue PO_2 distribution by a dynamic computer simulation. When the results obtained in this study is compared with those in the previous study obtained in the single capillary model (Fukuoka et al., 1983), the difference between two models is obvious as shown in Fig.6. In this figure, the abscissa R indicates the intercapillary difference and the ordinates ΔPt represents the mean tissue PO_2 difference from the steady flow model at the lowest PO_2 region. It is evident that as R increases, ΔPt increases in both models but the enhancing effect of the grouped capillaries becomes larger than that of the

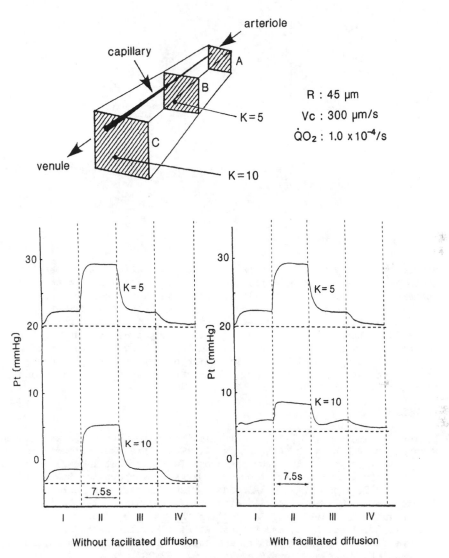

Fig.5 A typical results of the computer simulation showing the effects of recurrent blood flow supply by grouped capillaries and the facilitated diffusion on tissue PO_2.

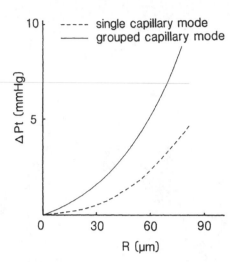

Fig.6 Difference in the effects of single and grouped capillary modes on the lowest tissue PO_2.

single capillary model. The magnitude of this effect is comparable with that of the facilitated diffusion by Mb and the above tendency remains unchanged with or without facilitated diffusion as seen in Fig.5.

As the summary of these results, it may be concluded that the recurrent blood flow supply to tissue by grouped capillaries, which is often observed in the microcirculation of the resting skeletal muscle, plays an important physiological role in increasing the lowest level of tissue PO_2 and in relieving the tissue portion which otherwise may fall in the state of oxygen debt and dysfunction. This effect seems to be as potential as the facilitate diffusion.

REFERENCES

Folkow, B. and Neil, E., 1971, "Circulation" Oxford Univ. Press, London.
Fukuoka, M., Shibata, M. and Kamiya, A. 1985, Effect of intermittent capillary flow on oxygen transport in the skeletal muscle studied by dynamic computer simulation. In "Oxygen Transport to Tissue VIII", pp 323-332, Kreuzer, F. et al. eds., Plenum Press, New York.
Kreuzer, F. 1970, Facilitated diffusion of oxygen and its possible significant; A Review. Respiration Physiol., 9, 1-30.
Lindbom, L., Tuma, R. E. and Arfors, K. E. 1980, Influence of oxygen on perfused capillary density and capillary red cell velocity in rabbit skeletal muscle. Microvasc. Res. 19, 197-208.
Shibata, M. and Kamiya, A. 1983, Local and neural regulation of microcirculation in the rabbit tenuissimus muscle. In "Intravital Observation of Organ Microcirculation" pp 52-61, Tsuchiya, M. et al., eds., Excerpta Medica, Amsterdam.
Shibata, M. and Kamiya, A. 1985, Microcirculatory responses to carotid sinus nerve stimulation at various ambient O_2 tension in the rabbit tenuissimus muscle. Microvasc. Res. 30, 333-345.
Wiedeman, M. 1984, Archtecture. In "Handbook of Physiology" Sect.II, Vol. IV (Microcirculation) pp11-40, Renkin, E. and Mitchel, C. C. eds., American Physiological Society, Bethesda.

112

A MODEL FOR TRANSMEMBRANE OXYGEN FLUX BY DIPOLE OSCILLATION AND FOR

SUPEROXIDE EXTRUSION BY PHAGOCYTIC CELLS

Michael McCabe

Department of Chemistry and Biochemistry
James Cook University of North Queensland
Townsville Q4811, Australia

INTRODUCTION

Classical theories for the flux of low molecular weight substances across cell membranes conceive of migration as being either dependent on a favourable partition coefficient between aqueous and lipid phases, or as a consequence of diffusion through appropriate holes within the membrane or occurring as a consequence of a reversible binding to a carrier molecule situated within the membrane. It has been suggested from time to time that the transport via a carrier might be accomplished by a rotation of the carrier protein, but there seems to have been little detailed discussion of the possible details and consequences of such a mechanism.

Brownian motion will generate a spontaneous tendency for a spheroidal and partly hydrophobic carrier protein (located within a membrane) to rotate or to oscillate within the essentially lipid phase of the membrane. Inter-molecular hydrophobic bonds would act as an axis for free rotation within the plane of the membrane. Additionally, since the carrier protein will almost certainly possess a significant dipole, its rotational diffusion will be modified by the proximity of the resting potential of the cell membrane. The consequence of this would be a tendency to convert rotation into a sustained harmonic oscillation. Provided that the amplitude of this oscillation reaches π radians, then sites on the carrier must inevitably oscillate from one side of the membrane to the other, and would moreover make finite stops on each side as the direction of the oscillation reverses.

Finally, as the protein oscillates (or rotates) a surge of charge would be induced within the protein, and this oscillation of charge could also be available to the carrier to facilitate loading and unloading in a vectorial manner, i.e. the tendency of the surge of charge would be to promote a loading up on one side and an unloading on the other. Any charge change of the carrier protein associated with substration and/or product liberation would also be utilised to enhance the oscillation, and to extend the time period which the (substrate bound) site subsequently spends on the alternate side of the membrane, thus facilitating unloading. The concept thus provides a mechanistic model for the phenomenon of co- and anti-transport of charged ions (such as sodium). Furthermore the concept permits a simple model for the vectorial generation of product by a membrane associated enzyme (i.e. one which loads substrate on one side of the membrane while liberating products onto the other side). Finally the proposed mechanism provides an additional

113

reason for the existence of the cell membrane resting potential, as well as a basis for the known relationship between transmembrane flux and potential.

NECESSARY DIMENSIONS FOR A SUITABLE OSCILLATING CARRIER

It seems generally accepted that the Danielli and Davson model of the cell membrane is correct with the proviso that occasional hydrophobic proteins exist which essentially span the lipid part of the membrane. Most cell membranes seem to have a central lipid layer of approximately 30 Å thickness with a total membrane thickness of around 100 Å. Numerous physical measurements have confirmed the general validity of this picture of the membrane. Additionally it seems likely from experimental evidence that many of the proteins which actually penetrate and span the membrane (rather than simply laying along its plane), have molecular weights ranging from 30,000 to several hundred thousand (Singer and Nicolson, 1972; Poo and Cone, 1974). For a protein molecule to span the membrane and be capable of rotational type oscillations, it would therefore need to have a minimum radius of 15 Å (if the aqueous part of the bilayer is absent in that region) up to a maximum requirement of 50 Å radius. Moreover it must be spheroidal, i.e. a sphere or oblate ellipsoid with ahydrophobic axis which would align itself into the plane of the membrane.

For the simplest case of a sphere, the molecular weight is given by

$$M = 4 \pi r^3 N/3\bar{v}$$

where r is the effective radius (and equal to half the thickness of the membrane), N is the Avogadro number, and $\bar{v}$ is the partial specific volume (normally between 0.7 and 0.75). Thus the limits of required molecular weights for this shape protein can be found as 11,500 and 450,000. More generally for a suitable ellipsoid, the molecular weight is given by

$$M = 4 \pi r^2 a/3\bar{v}$$

(Springall, 1954) where a is the axis laying in the plane of the membrane. A minimal value for a would generate a wheel like structure whose thickness would be limited by the dimensions of the amino acid constituents and their secondary structure. If we assume that the maximum ratio of major to minor axes is 3, then the minimum molecular weight to completely span the full thickness membrane is reduced from 450,000 to 150,000.

ROTATION OF CARRIER PROTEINS IN THE ABSENCE OF A CELL MEMBRANE POTENTIAL

Under Brownian impact from solvent and other surrounding molecules, carrier spheroidal protein molecules will exhibit rotational diffusion characterised by a rotational diffusion constant which is defined by a partial differential equation analogous to Ficks 2nd law, i.e.

$$\partial f(\theta)/\partial t = H \partial^2 f(\theta)/\partial\theta^2$$

where θ is the angle through which a solute particle has rotated from the field axis, and H is the rotational diffusion constant. For a spherical particle of radius r, and in a medium of viscosity η, then

$$H = kT/8\pi\eta r^3$$

(Perrin, 1909) while for an oblate ellipsoid when the ratio of major to minor axes is large, i.e. a/b << 1, and the structure resembles a disc, then

$$H_b = H_a = 3kT/32\eta b^3$$

(Perrin, 1943) where k is Boltzmans constant, T is the absolute temperature, b becomes the half thickness of the membrane, and η is the microscopic viscosity of the cell membrane.

The microscopic viscosity of the cell membrane.has been the subject of some experimentation since the rotational diffusion of naturally occurring steroid lipid components of the membrane have been measured. Additionally rotational diffusion constants have been measured for a variety of artificial lipid soluble labels capable of detection by ESR or fluorescence techniques. All of these measurements suggest that a rapid rotation of such lipid components occurs with values of between 10^7 and 10^8 revs/sec. These measurements allow estimates to be made for an appropriate microscopic viscosity of the lipid phase of the membrane. Edidin (1974) has calculated values for such viscosities and finds them to fall between 1 and 10 poise. The microscopic viscosity of the aqueous parts of the membrane sandwich will more likely resemble the viscosity of water since Laurent (1972) has shown that the rotation of proteins is essentially unchanged even when the measurements are made in very concentrated polymers (which express a large macroscopic viscosity), i.e. the protein will rotate freely within a pure solvent compartment which is effectively being constantly swept and preserved as a cavity. Thus the viscosity of the lipid part of the membrane will be dominant.

Substituting the lipid values as the appropriate viscosities will thus enable a prediction for the rotational diffusion coefficients for carrier proteins with the required dimensions. The results are shown in Table 1. It can be seen that even the largest carriers in the most viscous membranes seem capable of generating an adequate flux by this mechanism.

DIPOLE MOMENTS OF PROTEINS

These have been investigated by Debye, and later Wyman (1936) who showed that the magnitudes of the dipoles for proteins are generally significant (since the number of ionic groups is large and the molecular dimensions are also large). However generally the ionic groups of proteins are rather evenly distributed throughout the molecule, so that values tend often to be several hundred Debye units rather than the thousands which are theoretically possible (if all opposed charges were to be located at opposite ends of the particle), i.e. $\mu_{obs}/\mu_{max} \simeq 0.02$. Nevertheless values of several hundred D bye are already substantial when compared with the values for a normal covalent molecule (0 to 5D) or a small dipolar amino acid (20D). Thus the

Table 1. Calculated rotational diffusion coefficients at 37° for carrier proteins as a function of shape and dimension for limiting predicted membrane microscopic viscosities

Shape	Mol. wt.	Particle Dimensions r a($\overset{\circ}{A}$)	Microscopic Viscosity (poise)	Approx. Rotational Diffusion Constant (sec^{-1})
Minimal sphere	11,500	15 15	1	5.0×10^5
	11,500	15 15	10	5.0×10^4
Oblate spheroid (wheel)	150,000	50 16	1	3.1×10^5
	150,000	50 16	10	3.1×10^4
Maximal sphere	450,000	50 50	1	1.36×10^4
	450,000	50 50	10	1.36×10^3

carrier proteins will be expected to carry a significant dipole which would be the vectorial sum of a permanent dipole and an induced dipole (produced by the induced displacement of charge consequent on the orientation of the carrier in the membranes electric field). The permanent component of the dipole will tend to produce a preferred orientation relative to the membrane, while the induced dipole will oscillate as the molecule oscillates in the electric field, and tend to cause a damping of the oscillation. The oscillation of dipole as the molecule rotates (or vibrates about an equilibrium position determined by the permanent dipole, should it be strong enough) will have the effect of periodically modulating the strength of covalent or hydrogen type bonding of substrate (or product should the carrier protein also be an enzyme). Moreover this periodic change should generate a maximal bond strength on one side of the membrane which becomes minimal after the rotation to the other side of the membrane. Thus the carrier molecule will be oscillating or rotating under the influence of two opposed forces, a Brownian generated force promoting rotation, and a restoring force consequent of the dipole and the electric field strength generated by the membrane resting potential, and proportional to the angle of twist from the equilibrium (resting) position. For a spherical protein acting as a rigid body, the frequency of the vibration is given by

$$\nu = \frac{1}{2\pi} \sqrt{\frac{\text{moment of torsion}}{\text{moment of inertia}}}$$

Thus μ the frequency can be evaluated for any particular conformation and dipole of the carrier.

APPLICATION TO THE VECTORIAL GENERATION OF PRODUCT

It is thus possible to conceive of an essentially spheroidal enzyme complex, which is lodged in the plasma membrane and which is capable of loading up substrate on one side of the plasma membrane (the loading thus generating a charge change associated with the substration). The substrate induced dipole then generates an added momentum to the (Brownian) tendency to rotate. The rotation then induces an oscillation of the (inducible) dipole, which in turn then promotes the dissociation of the product from the enzyme (but now on the other side of the plasma membrane).

The identity of the enzyme responsible for the respiratory burst in phagocytic cells has been a controversial issue for several years, however there now seems to be a consensus, that the responsible enzyme is a plasma membrane associated NADPH oxidase that donates electrons in a "b" type cytochrome (cytochrome b_{245}) en route to the reduction of oxygen (Cross et al., 1985). The phagocytic cell has a problem since the free radical oxygen which it is generating when this enzyme is activated, is extremely toxic and if liberated inside the phagocyte would undoubtedly damage that cell rather than the foreign target cell immediately adjacent to it. Thus there are substantial grounds for believing that the superoxide product must be generated by addition of substrate from the inside of the cell (since the NADPH is generated via the hexose monophosphate shunt system which exists within the cytoplasm) with the formation/liberation of superoxide taking place on the outside of the cell plasma membrane. At present any proposed mechanism for the generation of superoxide at the plasma membrane is speculative since little is known of the structure of the plasma membrane NADPH oxidase responsible for the respiratory burst. A part of the reason must be that this enzyme system is notoriously unstable following its extraction from the membrane ($t_{\frac{1}{2}} \simeq 2$ to 4 min)(Cross et al., 1984; Light et al., 1981). However there is a significant electron transfer within the complex during superoxide synthesis as well as a large charge change associated with the liberation of product – all of which would generate a torque on the enzyme and cause it to rotate unless it were firmly anchored into the membrane.

REFERENCES

Cross, A., Parkinson, J., and Jones, O., 1984, The superoxide generating
     oxidase of leucocytes, Biochem. J., 223:337-344.
Cross, A., Parkinson, J., and Jones, O., 1985, Mechanism of the superoxide-
     producing oxidase of neutrophils, Biochem. J., 226:881-884.
Edidin, M., 1974, in: "Transport at the Cellular Level," Symp. Soc. Exp.
     Biol. XXVIII, Cambridge University Press, 1-14; Ann. Rev. Biophys.
     Bioeng, 3:179-201.
Laurent, T.C., and Obrink, B., 1972, On the restriction of the rotational
     diffusion of proteins in polymer networks, Eur. J. Biochem., 28:
     94-101.
Light, D., Walsh, C., O'Callaghan, A., Goetzl, E., and Trauber, A., 1981,
     Characteristics of the cofactor requirements for the superoxide-
     generating NADPH oxidase of human polymorphonuclear leukocytes,
     Biochemistry, 20:1468-1476.
Perrin, F., 1934, Brownian movement of an ellipsoid I-dielectric dispersion
     of an ellipsoid molecule, J. Phys. Radium, (7) 5:497-511.
Perrin, J., 1909, The Brownian movement of rotation, C.R. Acad. Sci., Paris,
     149:549-551.
Poo, Mu Ming, and Cone, R.A., 1974, Laterial diffusion of rhodopsin in the
     photoreceptor membrane, Nature, London, 247:438-441.
Poo, Mu Ming and Cone, R.A., 1973, Lateral diffusion of rhodopsin in
     Necturus rods, Exp. Eye Res., 17:503-510.
Singer, S.J., and Nicolson, G.L., 1972, Structure and chemistry of mammal-
     ian cell membranes, Amer. J. Pathol., 65:427-438.
Singer, S.J., and Nicolson, G.L., 1971, Fluid mosaic model of the structure
     of cell membranes, Science, 175:720-731.
Springall, H., 1954, in: "The Structural Chemistry of Proteins," Butter-
     worth Scientific Publications, London, 168- and seq.
Wyman, J., 1936, The dielectric constant of solutions of dipolar ions, Chem.
     Rev., 19:214-239.

MOLECULAR
MODELS

FACTORS MODULATING THE OXYGEN DEPENDENCE OF MITOCHONDRIAL OXIDATIVE

PHOSPHORYLATION[*]

David F. Wilson and William L. Rumsey

Department of Biochemistry and Biophysics, Medical School
University of Pennsylvania, Philadelphia, PA 19104

The oxygen consumed by mitochondrial oxidative phosphorylation is reduced to water by the enzyme cytochrome c oxidase. This reaction has the overall stoichiometry:

$$4H^+ + O_2 + 4c^{2+} + 2ADP + 2Pi \longrightarrow 4c^{3+} + H_2O + 2ATP \qquad (1)$$

Electrons from cytochrome c and hydrogen ions from the aqueous phase are required for reduction of dioxygen to water. In order to understand the role of cytochrome c oxidase in the regulation of mitochondrial respiration, one must keep in mind that the reduction of dioxygen to water is irreversible under all metabolic conditions. This means that in any steady state the rate of respiration is equal to the rate of electron transfer through cytochrome c oxidase. In order for the mitochondrial respiratory rate to change it is necessary for the rate of oxygen reduction by cytochrome c oxidase to change by the same amount and in the same direction. Metabolic effectors of the rate of mitochondrial respiration in vivo and in vitro ultimately do so by modulating the rate of electron transfer through cytochrome c oxidase. Part of the energy available in the cytochrome c oxidase reaction is released as heat and part is conserved by the synthesis of ATP. The latter occurs through a coupling mechanism which involves reaction intermediates, i.e., there is no direct interaction of ADP, Pi or ATP with the enzyme. Energy coupling does, however, strongly modulate the chemical and kinetic properties of cytochrome c oxidase. Among the [ATP]/[ADP][Pi] dependent properties of[1] the enzyme are the absorption spectra of the oxidized and reduced enzyme and the rate constants for electron transfer between redox components in the enzyme[2,3]. It is the purpose of this paper to review our current knowledge concerning the metabolic factors which influence the rate of oxygen reduction by cytochrome c oxidase, with particular attention to those which modify the dependence of this rate on oxygen pressure.

General considerations concerning factors which influence the rate of the cytochrome c oxidase reaction.

Cytochrome c oxidase appears to be regulated exclusively by its substrates and products. There is no evidence for the existence of

[*] Supported by grant GM-21524 from the National Institutes of Health.

121

allosteric activators or inhibitors of the type common to other metabolic
pathways such as glycolysis. The source of reducing equivalents for cytochrome c oxidase is cytochrome c, a heme protein present at a stoichiometric ratio to cytochrome c oxidase (normally 2:1). In short term experiments, completed in a few minutes, the total amount of these cytochromes
remains constant and the rate is dependent on the fraction of the cytochrome
c reduced. As noted earlier, ATP, ADP and Pi interact with cytochrome c
oxidase energetically through the coupling mechanism (as [ATP]/[ADP][Pi])
and can change the rate of electron transfer by large amounts[2,3]. The
H^+ ions are taken up from the intramitochondrial space[4] and have a large
effect on the kinetics of the cytochrome c oxidase reaction[2]. The other
reactant is dioxygen itself. Thus, the respiratory rate can be expressed as
a function of four independent variables:

    1. The level of reduction of cytochrome c.

    2. The free energy of synthesis of ATP (usually expressed as the
cytosolic [ATP]/[ADP][Pi]).

    3. The H^+ concentration.

    4. The dioxygen pressure.

In general, data analysis is facilitated if the experimental conditions are
chosen to have no more than two independent variables.

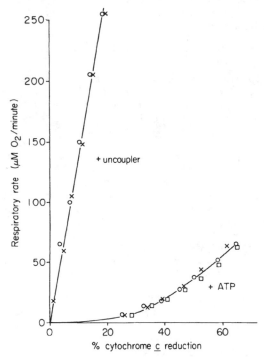

Figure 1. Dependence of the rate of mitochondrial respiration on the level
of reduction of cytochrome c. Rat liver mitochondria were suspended
in a mannitol-sucrose assay medium, final pH 7.8, to a concentration of
approximately 1.2 uM cytochrome c. Ascorbate (8 mM) was added and
various concentrations of N,N,N',N'-tetramethylparaphenylenediamine
added as a reductant for cytochrome c. The respiratory rate and the
level of reduction of cytochrome c were measured for two different
metabolic conditions and plotted on the abscissa and ordinate. The
mitochondria were treated with either 1.5 mM ATP or with uncoupler as
indicated on the figure and the different symbols indicate separate
experiments.

Regulation of mitochondrial respiration at constant pH and high oxygen concentrations

When the pH is held constant and the oxygen concentration is high, the independent variables are the state of reduction of cytochrome c and the cytosolic $[ATP]/[ADP][P_i]$. Measurement of respiratory rate as a function of cytochrome c reduction provided the relationship shown in Figure 1. At low $[ATP]/[ADP][Pi]$ values the respiratory rate increased linearly with further reduction of cytochrome c. As $[ATP]/[ADP][Pi]$ increased, the respiratory rate at each level of cytochrome c reduction decreased and the dependence on cytochrome c reduction became strikingly hyperbolic. A "respiratory control ratio" may be calculated for each level of reduction of cytochrome c. At 15% reduction, a value within the in vivo range, the respiratory rate at low $[ATP]/[ADP][Pi]$ divided by that at high $[ATP]/[ADP][Pi]$ is greater than 100 (for more discussion, see ref. 2,3).

The state of reduction of cytochrome c can be related to the intramitochondrial $[NAD^+]/[NADH]$ ratio through the near equilibrium reaction:

$$NADH_m + 2c^{3+}_c + 2ADP_c + 2Pi_c \longleftrightarrow NAD^+_m + 2c^{2+} + 2ATP_c \qquad (2)$$

where the subscripts m and c refer to the intramitochondrial and cytosolic compartments, respectively. An indirect approach based on equation 2 is often applied more readily to intact cells and tissues than direct measurement, particularly where optical measurements are difficult. Enzymatic analysis of rapidly quenched samples can be used to measure the metabolites appropriate for calculation of the intramitochondrial $[NAD^+]/[NADH]$ ratio and the cytoplasmic $[ATP]/[ADP][Pi]$. The level of cytochrome c reduction can then be calculated from equation 3:

$$K_{eq} = \left(\frac{[c^{2+}]}{[c^{3+}]}\right)^2 \frac{[NAD^+_m]}{[NADH_m]} \left(\frac{[ATP_c]}{[ADP_c][Pi_c]}\right)^2 \qquad (3)$$

where K_{eq} is the equilibrium constant for equation 2. The dependence of the cytochrome c oxidase reaction of isolated mitochondria on cytochrome c reduction, $[ATP]/[ADP][Pi]$, pH and oxygen concentration has been fitted to a model for the reaction mechanism of cytochrome c oxidase. This model has been tested for its ability to describe the interrelationships of the first three of these variables in intact cells and tissues at high oxygen pressures. The cytochrome c terms in the rate equations were replaced with the appropriate expression in $[NAD^+]/[NADH]$ and the resulting equations express the rate of cytochrome c oxidase as a function of $[NAD^+]/[NADH]$ and $[ATP]/[ADP][Pi]$[2,3]. (It should be noted that although these rate equations give a good qualitative fit to the behavior of mitochondria both in vitro and in vivo, additional data are required for complete quantitative analysis.) At high oxygen pressures and constant pH there are two independent variables, making it useful to give a "three dimensional" representation of the data. This was achieved by graphing the predicted respiratory rate at constant $[NAD^+]/[NADH]$ values as a function of log $[ATP]/[ADP][Pi]$. The cytoplasmic $[ATP]/[ADP][Pi]$, intramitochondrial $[NAD^+]/NADH]$, and total cytochrome c were measured for each cell suspension or tissue and the measured respiratory rate (y axis) was graphed against the log $[ATP]/[ADP][Pi]$ (x axis) and the log $[NAD^+]/[NADH]$ (z axis). Two points are used to plot each set of experimental data (y,x and y,z). The smaller the separation of the two points, the more closely the cellular data fit the equation describing the cytochrome c oxidase activity of isolated mitochondria. In general, the fit is good, despite the fact that the $[ATP]/[ADP][Pi]$ values for the cells ranged from 800 M^{-1} to more than 10^4 M^{-1} and the $[NAD^+]/[NADH]$ ratio ranged from 1 to 10^3.

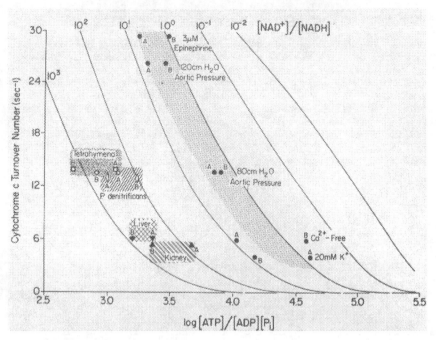

Figure 2. The dependence of cellular respiration on the cytoplasmic
[ATP]/[ADP][Pi] and intramitochondrial [NAD$^+$]/[NADH]. The measured
parameters for the indicated cells and tissues are plotted on a "three
dimensional" axis. The third axis ([NAD$^+$]/[NADH]) was generated by
fitting the data for suspensions of isolated mitochondria to a rate
expression for high oxygen concentrations[2] and plotting the predicted
respiration of mitochondria as a function of [ATP]/[ADP][Pi] at the
indicated [NAD$^+$]/[NADH] values. The respiratory rates were
normalized to constant cytochrome content by expressing them as the
turnover number for cytochrome c. The values for the cytoplasmic
[ATP]/[ADP][Pi] of a Langendorff preparation of perfused rat heart at
various work loads were calculated from the creatine/creatine-phosphate
ratio using a creatine phosphokinase equilibrium constant of 1.51 x
10^8 M^{-1} and a pH of 7.4. If the calculations had been made for a
different cytoplasmic pH or equilibrium constant the [ATP]/[ADP][Pi]
values would be correspondingly different.

Two important aspects of the relationship are readily apparent:
    1. At constant [NAD$^+$]/[NADH] the respiratory rate is strongly
dependent on [ATP]/[ADP][Pi].
    2. At constant [ATP]/[ADP][Pi] the respiratory rate is strongly
dependent on [NAD$^+$]/[NADH].
    It is noteworthy that the [NAD$^+$]/[NADH] ratio is very different among
various cells and tissues. In general, the smaller the [NAD$^+$]/[NADH]
ratio (the more reduced the mitochondrial NAD couple), the higher the
cytoplasmic [ATP]/[ADP][Pi]. For example, the turnover numbers for cytochrome c of hepatocytes and of myocytes in arrested heart are essentially
the same although the [ATP]/[ADP][Pi] values differ by 10 fold. This is
possible because the intramitochondrial [NAD$^+$]/[NADH] of hepatocytes is
100 fold greater than that of the myocytes.

    In isolated perfused rat heart (Langendorff preparation), increasing
the perfusion pressure results in increased respiration, an increase which
occurs with an essentially unchanged [NAD$^+$]/[NADH][5]. In this case,
almost all of the increase in respiration results from the decrease in

[ATP]/[ADP][Pi]. On the other hand, increasing the beat rate of dog heart causes increased respiration with little or no change in [ATP]/[ADP][Pi][6]. A decrease in [NAD$^+$]/[NADH], presumably due to increased intracellular calcium concentrations, accounts for the increased respiratory rate.

The oxygen pressure dependence of mitochondrial oxidative phosphorylation at constant [ATP]/[ADP][Pi]

Attempts to determine the oxygen dependence of mitochondrial function have been complicated by the absence of rapid and quantitative methods for measuring oxygen pressures below about 5 torr. This limitation has recently been removed by development of a new method for measuring oxygen pressure based on its effect on the phosphorescence lifetime of selected phosphors[7,8]. This method has a response time of milliseconds and is accurate for oxygen pressures from above air saturation down to 10^{-8} torr.

At constant [H$^+$] and [ATP]/[ADP][Pi] the rate of mitochondrial respiration is dependent on the state of reduction of cytochrome c and the oxygen pressure. Since oxygen is being rapidly consumed, its pressure cannot be held constant unless it is continuously added to the medium. Such an addition induces diffusion gradients within the sample volume and the effect of these diffusion gradients is difficult to quantitate. On the other hand, the total amount of cytochrome c is constant, small relative to the oxygen concentration, and responds to changes in cytochrome c oxidase in milliseconds. Thus, as oxygen is depleted from the medium, the level of cytochrome c reduction is effectively in a steady state with cytochrome c oxidase at each oxygen pressure.

Experimentally, oxygen concentration and cytochrome c reduction can be measured simultaneously, allowing the respiratory rate to be calculated as a function of oxygen pressure and the overall dependence of mitochondrial oxidative phosphorylation to be inferred. A typical experiment for suspen-

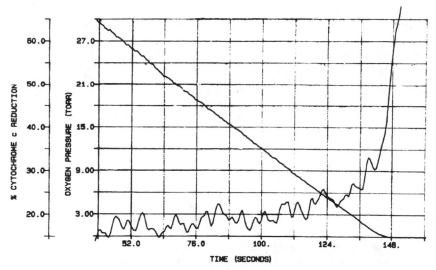

Figure 3. The oxygen dependence of cytochrome c reduction and oxygen concentration in suspensions of respiring mitochondria. The reduction of cytochrome c was measured as the absorbance at 550 nm minus that at 540 nm and oxygen concentration by quenching of phosphorescence. The mitochondria were respiring in the presence of 1.5 mM ATP with glutamate (6 mM and malate (6 mM) as oxidizable substrates.

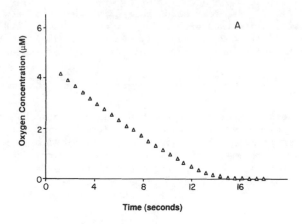

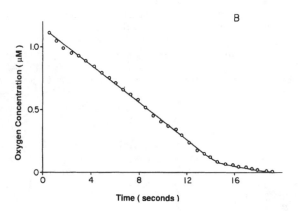

Figure 4A,B. The oxygen dependence of the respiratory rate of suspensions
of rat liver mitochondria. A. The mitochondria were well coupled and
respiring in the presence of 0.87 mM ATP. The oxidizable substrate was
glutamate (8 mM), succinate (8 mM) and malate (1 mM) and the pH was
7.4. Only the data for oxygen concentrations less than 6 uM are shown.
B. The experimental conditions were the same as for A except that an
uncoupler was added to completely uncouple oxidative phosphorylation.
In addition the mitochondria were present at a concentration approximately 20 times less than for A to allow for the nearly 10 fold
increase in respiratory rate which occurred when the mitochondria were
uncoupled.

sions of mitochondria at pH 7.0 and high [ATP]/[ADP][Pi] is presented in
Figure 3. Cytochrome c reduction is observed to increase as the oxygen
pressure falls below approximately 20 torr and (progressively) increases
further as the oxygen pressure decreases. In contrast, when [ATP]/[ADP][Pi]
is low or uncoupler is added, the reduction of cytochrome c does not begin
until the oxygen pressure falls below the limits of measurement, in this
case approximately 1 torr$^{10-12}$. Thus, the oxygen pressure dependence of
cytochrome c oxidase is much greater at high [ATP]/[ADP][Pi] than at low
[ATP]/[ADP][Pi]. Under physiological conditions (high [ATP]/[ADP][Pi]) this
oxygen dependence extends through most of the range of oxygen pressure
expected to exist in the cytoplasm of cells in tissue (to > 30 torr). This
oxygen dependence of the rate of respiration can be seen in Figure 4 (data
taken from ref. 10). At high [ATP]/[ADP][Pi], the rate of oxygen

consumption begins to decrease when the oxygen concentration falls below
approximately 5 torr. At low [ATP]/[ADP][Pi], the rate is unchanged until
the oxygen pressure falls below 0.1 torr. The apparent P_{50} values
estimated by fit to a Michaelis-Menten relationship were 0.5-0.6 torr for
high [ATP]/[ADP][Pi] and less than 0.06 torr for low [ATP]/[ADP][Pi].
The oxygen dependence of mitochondrial oxidative phosphorylation in cells.

   In suspensions of isolated cells it is not possible to hold the cytoplasmic [ATP]/[ADP][Pi] completely constant. With glucose as oxidizable
substrate, the decrease with decreasing oxygen concentration is blunted by
glycolytic ATP production but is still significant (see for examples refs.
13-15). When the aerobic/anaerobic transition occurs rapidly (2-4 minutes),
the changes are minimized. Measurements of oxygen concentration and cytochrome c reduction that were made while oxygen was being depleted indicate
cytochrome c reduction is significantly increased by the time the oxygen
pressure falls to about 30 torr. This reduction increases progressively as
the oxygen pressure is decreased further[13]. When the [ATP]/[ADP][Pi] is
lowered by adding uncouplers of oxidative phosphorylation the cytochrome c
becomes more oxidized at all oxygen concentrations. More importantly, no
oxygen pressure dependent increase in cytochrome c reduction is observed
until the oxygen pressure is less than about 3 torr (unpublished results).

   When the rate of cellular respiration is plotted against oxygen
pressure, the respiratory rate is constant at oxygen pressures above about
20 torr. Below this oxygen pressure the observed oxygen dependence is
different for different cell lines (see for examples 13,15). Data that was
obtained from a human neuroblastoma cell line (CHP-404) are presented in
Figure 5. When the cells were suspended in a Krebs-Henseleit medium
containing 10 mM glucose as substrate, the respiratory rate above 20 torr
was essentially constant and the data are not included in the figure. Below
20 torr the respiratory rate decreased with decreasing oxygen pressure, the
dependence giving a good fit to a Michaelis-Menten equation with a P_{50} of
1.1 torr and a V_{max} of 0.17 torr/sec. Decreasing the intracellular
[ATP]/[ADP][Pi] by addition of uncoupler resulted in the V_{max} increasing
to 0.37 torr/sec and the P_{50} decreasing to less than 0.4 torr.

What is the magnitude of the diffusion gradient for oxygen between the
extracellular medium and the mitochondria?

   One of the most difficult aspects of understanding oxygen delivery to
tissue has been to evaluate the diffusion barriers as oxygen moves from the
red cell to the mitochondria. The oxygen must diffuse out of the red cell,
and then through the plasma and vessel wall, the extracellular space, the
plasma membrane and part of the cytoplasm of the cell before it can be used
by the mitochondria. The results presented in this paper are relevant to
diffusion of oxygen from the extracellular space to the mitochondria. There
appears to be no diffusion limitation for oxygen utilization by suspensions
of isolated mitochondria, as indicated by the fact that when uncoupler is
added the respiratory rate increased by 10 fold and the P_{50} for
respiration decreased from 0.6 torr to less than 0.06 torr. This is not
surprising because the respiratory chain components are in the inner
mitochondrial membrane, separated from a well stirred suspending medium only
by the highly porous outer membrane and a short distance. The maximum
possible diffusion gradient is set by the observation that uncoupled
mitochondria respire at maximal rates to oxygen pressures in the suspending
medium of less than 0.06 torr. There is a greater possibility for a
diffusion limitation in cells, where the cell membrane and cytoplasm may
contribute to the diffusional resistance. The measured P_{50} for cellular
respiration is greater in normal cells than for cells in which the
mitochondria are uncoupled (1.1 torr vs < 0.4 torr). This occurs despite the
increase of more than two fold in respiratory rate of the latter. Thus the

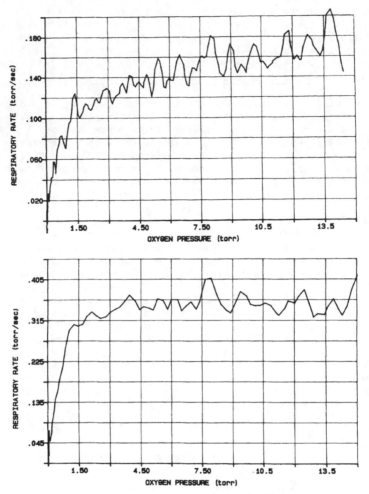

Figure 5. The oxygen pressure dependence of the respiratory rate of human
neuroblastoma cells. Cultured human neuroblastoma cells (CHP-404) were
suspended in Krebs-Henseleit medium containing 1.3 mM Ca^{2+} and 10 mM
glucose. The suspension was placed in a sealed glass chamber and the
oxygen pressure measured by the phosphorescence lifetime of
Pd-coproporphyrin. Two measurements were made per second and then the
data arrays were plotted using the Asystant software of MacMillan
Software Co. Upper curve: Normal cells. Lower curve: The same cell
suspension after treatment with an uncoupler of mitochondrial oxidative
phosphorylation.

oxygen availability to mitochondria is not usually diffusion limited in
suspensions of neuroblastoma cells. The oxygen gradient which may exist is
set by the uncoupled respiration as less than 0.4 torr.

There has been substantial debate in the literature concerning the
oxygen distribution in cells. Direct measurements indicate that the oxygen
solubility and diffusion constant of oxygen in the membranes of mammalian
cells is high, such that the membrane provides less resistance than a comparable thickness of water[16]. Similarly the oxygen diffusion constant in
media simulating the cytoplasm, such as concentrated hemoglobin solutions,
have diffusion constants similar to those of saline solutions[17]. This
indicates that oxygen diffusion from the extracellular medium to the

mitochondria is similar to diffusion through the same thickness of unstirred
saline. Both theoretical calculations[18] based on oxygen diffusion to an
oxygen consuming locus (mitochondria) and comparisons of mitochondrial[19]
oxygen consumption rates to those of similar sized oxygen electrodes
indicate that the oxygen gradients from the extracellular medium to the
mitochondria should be shallow. Measurements of the oxygen distribution in
myocytes of working muscle[20-22] show only very slight oxygen gradients (< 0.1
torr) within the cells. The current measurements of the oxygen
dependence of cellular respiration provide evidence that in neuronal cells
the oxygen gradient from the extracellular medium to the mitochondria is
less than 0.4 torr even when the mitochondria are uncoupled.

The data summarized above is in contrast to some interpretations of
measurement of the oxygen dependence of mitochondrial and cellular
respiration using "steady state" methods (see for examples ref. 23-25). In
this method, oxygen is continuously added to the medium to hold the measured
oxygen pressure either constant or slowly changing. Oxygen addition was
usually accomplished by having a gas phase with a variable oxygen pressure
above the the stirred reaction medium. When the oxygen pressure in the
liquid is constant, the rate of oxygen consumption is equal to the rate of
oxygen addition at the gas-liquid interface. Since the oxygen is being
added only to a small volume of liquid at the interface, this oxygen must be
continuously stirred into the medium. For most of the experimental conditions used, the time required to stir in a reagent added to the liquid
interface was a few seconds while the oxygen consumption rate was 0.5 to 1.5
torr/sec. This means that at oxygen pressures less than about 10-15 torr,
strong oxygen gradients with the attendant heterogeneous oxygen pressures
exist in the stirred liquid. The reported increases in P_{50} for oxygen
associated with increasing respiratory rate in mitochondria and cells (see
however ref. 26) appear to reflect the resulting increases in steepness of
the oxygen gradients from the gas-liquid interface to the oxygen sensor.
Jones and coworkers[24,25] have suggested there is a large difference in
oxygen pressure between the extracellular medium and the mitochondria (> 2.5
torr in suspensions of isolated hepatocytes) and that a region of very low
oxygen permeability surrounds the mitochondria. These suggestions are,
however, based on the use of erroneously high oxygen affinities for coupled
mitochondria and the assignment of the above mentioned oxygen pressure
gradients between the gas-liquid interface and the oxygen sensor to
diffusion from the extracellular medium to the mitochondria.

In summary. The rate of mitochondrial respiration at a constant cytochrome concentration is a function of four independent variables: 1. The
level of reduction of the intramitochondrial NAD couple; 2. The cytoplasmic
energy state ([ATP]/[ADP][Pi]); 3. The intramitochondrial pH; 4. The
pressure of dioxygen. A change in any one of these variables may result in
a change in the respiratory rate or it may be partially or entirely compensated for by an opposing change in one of the other independent variables.
Under physiological conditions the rate of ATP utilization usually
determines the respiratory rate and the balance of the four independent
variables is set by regulatory mechanisms of the cells and tissues.

The oxygen dependence of mitochondrial oxidative phosphorylation is due
to the oxygen dependence of the cytochrome c oxidase reaction. This dependence is strongly affected by pH and energy state. At high [ATP]/[ADP][Pi] it
extends to at least 20 torr at pH 7.0 and at to least 100 torr at pH 7.8.
When the energy state is low, the oxygen dependence is independent of pH and
does not extend above 2 torr. The respiratory rates of isolated mitochondria and cells in physiological metabolic states are not limited by intracellular oxygen diffusion. The oxygen diffusion gradients from the extracellular medium to the mitochondria of neuroblastoma cells are less than 0.4
torr.

References

1. D.F. Wilson, M. Erecinska and P. Nicholls, An energy dependent transformation of a ferricytochrome of the mitochondrial respiratory chain, FEBS Letters, 20:61-65 (1972).
2. D.F. Wilson, C.S. Owen and A. Holian, Control of mitochondrial respiration: A quantitative evaluation of the roles of cytochrome c and oxygen, Arch. Biochem. Biophys., 182:749-762 (1977).
3. D.F. Wilson, C.S. Owen and M. Erecinska, Quantitative dependence of mitochondrial oxidative phosphorylation on oxygen concentration: a mathematical model, Arch. Biochem. Biophys., 195:494-504 (1979).
4. N.L. Greenbaum and D.F. Wilson, Dependence of the 3-OH-butyrate dehydrogenase and cytochrome c oxidase reactions on intramitochondrial pH, in "Membrane Biochemistry and Bioenergetics" , in press.
5. M. Erecinska, D.F. Wilson and K. Nishiki, Homeostatic regulation of cellular energy metabolism: experimental characterization in vivo and fit to a model. Am. J. Physiol. 234(3):C82-C89 (1978).
6. R.S. Balaban, H.L. Kantor, L.A. Katz and R.W. Briggs, Relation between work and phosphate metabolite in the in vivo paced mammalian heart, Science, 232:1121-1123 (1986).
7. J.M. Vanderkooi and D.F. Wilson, A new method for measuring oxygen in biological systems, Adv. Exptl. Med. Biol., 200:189-193 (1986).
8. D.F. Wilson, J.M. Vanderkooi, T.J. Green, G. Maniara, S.P. DeFeo and D.C. Bloomgarden, A versitile and sensitive method for measuring oxygen, Adv. Exptl. Med. Biol. in press (1987).
9. J.M. Vanderkooi, G. Maniara, T.J. Green and D.F. Wilson, An optical method for measurement of dioxygen concentration based upon quenching of phosphorescence, J. Biol. Chem., 262:5476-5482 (1987).
10. D.F. Wilson, W.L. Rumsey, T.J. Green and J.M. Vanderkooi, The oxygen dependence of mitochondrial oxidative phosphorylation measured by a new optical method for measuring oxygen concentration, Submitted for publication.
11. D.F. Wilson and M. Erecinska, Effect of oxygen concentration on cellular metabolism, Chest, 88S:229s-232s (1985).
12. D.F. Wilson and M. Erecinska, The oxygen dependence of cellular energy metabolism, Adv. Exptl. Med. Biol., 194:229-239 (1986).
13. D.F. Wilson, M. Erecinska, C. Drown and I.A. Silver, The oxygen dependence of cellular energy metabolism, Arch. Biochem. Biophys., 195:485-493 (1979).
14. D.F. Wilson, M. Erecinska, I.A. Silver, C.S. Drown and K. Nishiki, Metabolic sensing of cellular oxygen tension, Adv. Physiol. Sci., 10:391-398 (1981).
15. T. Kashiwagura, D.F. Wilson and M. Erecinska, Oxygen dependence of cellular metabolism: the effect of O_2 on gluconeogenesis and urea synthesis in isolated hepatocytes. J. Cell. Physiol., 120:13-18 (1984).
16. S. Fischkoff and J.M. Vanderkooi, Oxygen diffusion in biological and artificial membranes determined by the fluorochrome pyrene, J. Gen. Physiol., 65:663-676 (1975).
17. A. Klug, F. Kreuzer and J.W. Roughton, The diffusion of oxygen in concentrated hemoglobin solutions, Helv. Physiol. Pharmacol. Acta, 14:121 (1956).
18. A. Clark, Jr., P.A.A. Clark, R.J. Connett, T.E.J. Gayeski and C.R. Honig, How large is the drop in PO_2 between cytosol and mitochondrion, Am. J. Physiol., 252:C583-C587 (1987).
19. D.F. Wilson, Regulation of in vivo mitochondrial oxidative phosphorylation, in "Membranes and Transport", A.N. Martonosi, ed., pp. 349-355, Plenum Press, New York (1982).
20. T.E. Gayeski and C.R. Honig, O_2 gradients from sarcolemma to cell interior in red muscle at maximal V_{O_2}, Am. J. Physiol., 251:H789-H799 (1986).

21. T.E. Gayeski and C.R. Honig, Shallow intracellular O_2 gradients and the absence of perimitochondrial "wells" in heavily working red muscle, Adv. Exptl. Med. Biol. 200:487-494 (1986).

22. B.A. Wittenberg and J.B. Wittenberg, Oxygen pressure gradients in isolated cardiac myocytes, J. Biol. Chem., 260:6548-6554 (1985).

23. N. Oshino, T. Sugano, R. Oshino and B. Chance, Mitochondrial function under hypoxic conditions: the steady states of cytochromes $a+a_3$ and their relation to mitochondrial energy states, Biochim. Biophys. Acta, 368:298-310 (1974).

24. D.P. Jones and F.G. Kennedy, Analysis of intracellular oxygenation of isolated adult cardiac myocytes, Am. J. Physiol., 250:C384-C390 (1986).

25. D.P. Jones, Intracellular diffusion gradients of O_2 and ATP, Am. J. Physiol., 250:C663-C675 (1986).

26. H. Degn and H. Wohlrab, Measurement of steady-state values of respiration rate and oxidation levels of respiratory pigments at low oxygen tensions: a new technique, Biochim. Biophys. Acta, 245:347-355 (1971).

THE CYTOSOLIC REDOX IS COUPLED TO VO_2. A WORKING HYPOTHESIS

Richard J. Connett

Department of Physiology
University of Rochester
Rochester, New York

In aerobic tissues the match between energy demand (ATPase activity) and energy supply (ATP production) requires close integration of the glycolytic and mitochondrial metabolic controls. Since most ATPase activity occurs in the cytosol this integration must be mediated by cytosolic signals. While there is uncertainty about the mechanism, there is general agreement that recruitment of mitochondrial ATP production is strongly coupled to changes in cytosolic phosphorylation state. Although the role of mitochondrial redox has been identified, cytosolic redox has been left out of models of mitochondrial control. Results from three different kinds of studies suggest that in the cell there is a strong interaction between the mitochondrial and cytosolic redox states and mitochondrial metabolism. These results include studies on: 1) intact tissues such as liver (Berry et al, 1980), heart (Kauppinen et al, 1983) and red skeletal muscle (Connett et al, 1986); 2) reconstituted glycolytic and mitochondrial systems (Jong & Davis, 1983) and 3) metabolic models of heart metabolism that include kinetic submodels of most of the individual enzymes (Kohn, 1983). The objective of this paper is to describe a distributed equilibrium hypothesis that accounts for a cooperative interaction between the glycolytic and oxidative metabolic systems under aerobic conditions.

Studies on Intact Tissue

In working skeletal muscle the cytosolic lactate/ pyruvate ratio and hence the cytosolic redox rises as the rate of work output and oxygen consumption (VO_2) rises (Connett et al, 1986). Direct measurements of intracellular PO_2 demonstrate that the cytosol goes reduced in the presence of sufficient oxygen in the cells to support maximal VO_2 at all mitochondria (Gayeski et al, 1987; Connett et al, 1986). These observations led to the suggestion that glycolysis may be necessary for efficient recruitment of mitochondrial oxidative phosphorylation (Connett et al, 1986; Connett et al, 1985).

Studies using isolated hepatocytes demonstrated that the lactate/pyruvate ratio is proportional to the ATP demand in the face of various lactate loads. Graded uncoupling of the mitochondria led to graded increases in the lactate/pyruvate ratio. These results were interpreted as demonstrating control of the cytosolic redox by the energy state of the mitochondria (Berry et al, 1980).

Studies on Reconstituted Systems

Jong and Davis (Jong & Davis, 1983) used isolated liver mitochondria and a reconstituted muscle glycolytic system that included the mitochondrial redox shuttle enzymes. They varied the demand by varying the amount of added ATPase activity. When mitochondrial oxidative phosphorylation was operating the lactate/pyruvate ratio in this system was a function of the ATPase rate even with a net glycolytic flux. When the mitochondrial electron transport was blocked and glycolysis was the sole source of ATP, the glycolytic rate was higher and the lactate/pyruvate ratio increased continuously. They concluded that in the normal, coupled situation the mitochondria determine the cytosolic redox in a manner related to the rate of oxidative phosphorylation. The cytosolic redox in turn was thought to directly or indirectly regulate glycolysis.

Studies Using Models

There are two types of approaches to modelling of metabolic processes: 1) analytic modelling, which attempts to generate an abstract description of a particular set of data and 2) synthetic modelling, which attempts to build a complete description of the system based on the known enzymatic structure. Both approaches have led to the conclusion that the transfer of reducing equivalents between the cytosolic and mitochondrial pools is a critical parameter in pathway recruitment and coupling of supply to demand (James, 1982; James, 1980; Kohn, 1983). Two examples are given below.

Analytic Model. James (1980) analyzed data from studies of the interaction between glucose synthesis and lactate consumption in isolated hepatocytes. A simple model related the rate of gluconeogenesis to the variable log ([lactate]/[pyruvate]), i.e. the rate of energy turnover was related to the cytosolic redox potential. This model was later extended to include the rate of shuttling of reducing equivalents between the cytosolic and mitochondrial pools as a critical coupling parameter (James, 1982).

Synthetic Model. A metabolic model of the working perfused heart which includes most of the enzymatic processes and accurately predicts all measured metabolite concentration changes has been developed (Kohn & Garfinkel, 1983). Sensitivity analysis of the model indicated that the major variable accounting for lactate production and glycolytic flux was the transport of reducing equivalents between the mitochondrial matrix and the cytosolic pools (Kohn, 1983).

Development of the Working Hypothesis

In order to generate a description of the integration of cytosolic redox with mitochondrial metabolism we have to take into account a number of regulatory elements. These include: 1. the coupling of energy demand to mitochondrial metabolism, 2 the mechanism for transport of reducing equivalents between mitochondrial and cytosolic pools, and 3. How the combination of demand and cytosolic redox can regulate glycolysis in a manner that permits coupling of cytosolic redox to mitochondrial function.

Coupling to Demand

With a few exceptions (e.g. urea synthesis in liver) the major effect of an increase in physiological demand is the activation of ATPase activity in the cytosol. This leads an increase in the concentration of phosphate and ADP and, unless buffered by creatine phosphate, a decrease in [ATP]. Various hypotheses and mathematical descriptions have been put forward to relate these changes in cytosolic phosphorylation state to the rate of oxidative phosphorylation and $\dot{V}O_2$. For the purposes of this working hypothesis we will use a thermodynamic formulation in terms of the redox and phosphorylation driving forces:

$$\dot{V}O_2 = A_1 \cdot F_{redox} + A_2 \cdot F_{phosphorylation}$$

Where A1 and A2 are the coupling coefficients and the F's are the driving forces.

The driving forces can be written in terms of the free energy changes and interpreted as either a linear nonequilibrium description or an equilibrium description depending on the quantitative fit to data (van der Meer et al, 1978; Nishiki et al, 1978). If the standard free energy changes for both the redox and phosphorylation potentials are grouped and it is assumed that the oxygen supply is saturating, as has been found in working red muscle (Gayeski et al, 1987), equation 1 can be approximated by:

$$\dot{V}O_2 = A_o + A_1' \cdot \log\{\frac{[NADH][H^+]}{[NAD]}\}_m + A_2' \cdot \log\{\frac{[ATP]}{[ADP][Pi]}\} \tag{1}$$

Where: Ao includes the standard free energy changes and the oxygen term and A_1' & A_2' are modified coupling coefficients.

From equation 2 we see that $\dot{V}O_2$ is the result of the balance between: 1) the mitochondrial redox potential which tends to drive electron transport and generate electrical and pH gradients, and 2) the coupled phosphorylation sites which tend to generate a phosphorylation potential at the expense of the electrochemical gradients. This formulation has generally been applied using mitochondrial redox potential and cytosolic phosphorylation potential. Thus as cytosolic ATPases are activated in response to a physiological demand the cytosolic phosphorylation potential falls and $\dot{V}O_2$ increases in response to the change in driving force. This is associated with a decrease in the mitochondrial membrane electrical and/or pH gradient (Kauppinen, 1983; Davis & Lumeng, 1975). If the mitochondrial redox potential is unchanged the $\dot{V}O_2$ will be linearly related to the change in phosphorylation potential and electrical and/or pH gradients will fall in proportion to the rate of oxidative phosphorylation.

Cytosolic-Mitochondrial Redox Shuttle

Mammalian cells appear to transport reducing equivalents between the cytosolic and mitochondrial pools primarily via a complex series of steps known as the aspartate-malate redox shuttle system (Williamson et al, 1973). The steps involved are illustrated in figure 1 below. Malate dehydrogenase and aspartate-2-oxoglutarate transaminase occur in both the cytosolic and mitochondrial spaces. These are thought to operate very close to equilibrium with their local substrate and product pools. Two translocase systems are primarily responsible for transport across the mitochondrial membrane: The electroneutral 2-oxoglutarate-malate exchanger and the electrogenic transporter which catalyzes the exchange of the aspartate anion with a proton neutralized glutamate anion (LaNoue & Schoolwerth, 1979). Because of the charge imbalance and the proton requirement the distribution of aspartate and glutamate between the mitochondrial matrix and the cytosol will depend on the membrane potential and proton gradient across the mitochondrial membrane (Kauppinen et al, 1983; Davis et al, 1980). A number of studies have demonstrated that the redox potential gradient across the mitochondrial membrane is linearly related to the membrane potential and/or the proton motive force across the membrane (e.g. Kauppinen et al, 1983; Davis et al, 1980). Quantitative analyses suggest that this may reflect an equilibrium of the redox shuttle. If the equilibrium constants for the transaminase and malate dehydrogenase are identical in both the cytosol and the mitochondrial matrix then the distribution of reactants between the mitochondrial matrix and cytosolic pools will depend only on the equilibrium position of the two transport systems. The effect of this equilibrium is to generate a redox <u>gradient</u> across the mitochondrial membrane which will be a function of the energetic state of the mitochondrion. This can be described as shown in equation 2 below:

$$\left(\frac{[NAD]}{[NADH][H^+]}\right)_c \cdot \left(\frac{[NADH][H^+]}{[NAD]}\right)_m = f(V_m, pH_m-pH_c, K_{eq}) \qquad (2)$$

or

$$\log\left(\frac{[NADH][H^+]}{[NAD]}\right)_m - \log\left(\frac{[NADH][H^+]}{[NAD]}\right)_c = g(V_m, pH_m-pH_c, K_{eq})$$

Where: the subscripts m and c refer to the mitochondrial matrix and the cytosol respectively and V_m is the mitochondrial membrane potential. Keq includes the distribution ratios of both translocases.

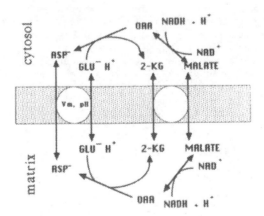

Fig. 1. Redox Shuttle

The shaded area represents the mitochondrial membrane. The reactions include: glutamate-oxaloacetate transminase and malate dehydrogenase reactions in both the cytosolic and the mitochondrial matrix compartments and the 2-ketoglutarate-malate and aspartate-glutamate transmembrane exchange reactions.
Abbreviations: OAA = oxaloacetate, MAL = malate, ASP = aspartate, Glu = glutamate and 2-KG = 2-ketoglutarate.

An equilibrium of this sort can account for the observations that the mitochondrial matrix is more reduced than the cytosol and accounts for the mitochondrial regulation of the cytosolic redox. The equilibrium requires that the redox gradient be constant under conditions of constant pH and electrical gradients. However, as discussed in the previous section, when energy demand changes the electrical gradient across the mitochondrial membrane changes. If the electrical gradient changes the redox gradient must follow this change. The result can be a change in mitochondrial redox, a change in cytosolic redox or some combination of the two.

Studies on the regulation of the tricarboxylic acid cycle (TCA cycle) have indicated that the flux is very closely regulated by redox state in the mitochondrial matrix (LaNoue et al, 1970; Williamson et al, 1976). Thus under conditions of adequate substrate supply, the TCA cycle control will operate to keep the mitochondrial redox very close to a constant under conditions of steady-state energy turnover. A negligible change in mitochondrial redox with changes in steady-state energy turnover has been observed in both heart and skeletal muscle (Nishiki et al, 1978; Kauppinen et al, 1983; Olgin et al, 1986). This stability of the mitochondrial redox can then account for the observations of lactate/pyruvate ratios that increase with increases in VO_2 or ATP turnover rates. If on the other hand, the cytosolic redox is fixed or the mitochondrial redox cannot be maintained, then the mitochondrial redox should shift in the oxidized direction. This has been observed in liver when the lactate/pyruvate ratio was fixed by perfusion (van der Meer et al, 1978). The mitochondrial matrix should become oxidized both in vivo, when inadequate sources of pyruvate and lactate are present, and in studies with isolated mitochondria in the absence of the cytosolic components of the shuttle. These changes have been observed in many studies.

Role of the Cytosolic Glycolytic System

In the absence of metabolic forces maintaining the mitochondrial redox, the change in phosphorylation potential required to generate a given VO_2 and ATP supply will be increased. This would lead to a larger decrease in the membrane electrical potential gradient and hence a larger decrease in the redox gradient. Unless stabilized in some manner this will behave as a positive feedback system. Thus it is very important for maintenance of efficient energy turnover to have several controls operating to protect the mitochondrial redox potential. We have already mentioned the regulation of the TCA cycle by mitochondrial redox. If glycolysis can be recruited with the onset of ATPase activity, the cytosol will be reduced in parallel with the decrease in the membrane redox gradient. This helps stabilize the redox potential in the mitochondrial matrix and generates substrate in support of the TCA cycle. Thus while the mitochondrial metabolism regulates the ultimate steady-state position of the cytosolic redox, the recruitment of glycolysis helps generate cytosolic reducing equivalents and buffers the mitochondrial redox. A close coupling of the rate of oxidative phosphorylation with the cytosolic phosphorylation potential and ATP turnover can then occur with the mitochondrial redox term being relatively stable.

How Can a Redox Potential Regulate Glycolysis?

An outline of the glycolytic pathway is shown in figure 2. The pathway can be divided into two sections: the upper portion, which involves the allosterically regulated enzymes and the lower section between the trioses and lactate, which is predominantly regulated by mass action considerations.

The primary site of control of the rate of glycolysis is at phosphofructokinase (PFK) in the upper portion of the pathway. This much studied enzyme is sensitive to a number of allosteric regulators. Analysis of the recruitment of glycolysis in working heart (Kohn, 1983) and during the early phase of a rest-work transition in red skeletal muscle (Connett, 1987) indicate that the major factors accounting for the flux through PFK appear to be pH and the relative concentrations of the various forms of the adenine nucleotides. The effect of the latter also includes pH since the inhibitory form of ATP appears to be $ATPH^{3-}$ (Waser et al, 1983). Other known regulators include hexosebisphosphates and compounds derived from mitochondrial metabolism such as ammonium ion and citrate. Recent studies on hexosebisphosphate levels suggest that while these may be important in transitions between gluconeogenesis and glycolysis in the liver they are not involved during changes in the rate of energy turnover in skeletal muscle (Minatogawa & Hue, 1984). In aerobically metabolizing red muscle and heart, citrate and ammonium also appear to play a minor role (Kohn, 1983; Connett, 1987).

The lower portion of the glycolytic pathway includes as major components: 1. the steps between the triose phosphates and 3-phosphoglycerate (3PG) which are near equilibrium. 2. phosphoglycerate mutase (PGM) which has a high Km for 3PG (Berry et al, 1980). This high Km makes flux in the lower portion of the pathway very sensitive to concentration of 3PG as well as [ADP]. 3. enolase and pyruvate kinase which respond to the supply of carbon substrate and ADP.

The question of how a redox signal interacts with these adenine nucleotide and substrate concentration controls remains. Krebs and coworkers suggested some years ago, that the cytosolic phosphorylation state and redox interact via the glyceraldehydephosphate dehydrogenase (GAPDH) and phosphoglycerate kinase equilibria. Lactate dehydrogenase (LDH) will also play an important role in this interaction because of the

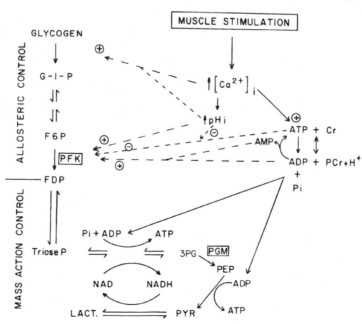

Fig. 2. Glycolytic Pathway with regulatory interactions discussed in the text.

ability to specifically transfer NADH in coupled fashion between GAPDH and LDH (Srivastava & Bernhard, 1986). Recent studies suggest that all of these enzymes are physically associated in vivo so that the combination could stay very close to equilibrium under all conditions (Clarke et al, 1980). Equations 3 & 4 below describe these combined equilibria:

$$\frac{[MgATP]}{[MgADP][HPO_4^{2-}]} = \frac{[NAD^+][DHAP]}{[NADH][H^+][3PG]} \cdot \frac{K_{GAPDH} \cdot K_{PGK}}{K_{TPI}} \tag{3}$$

$$\frac{[MgATP]}{[MgADP][HPO_4^{2-}]} = \frac{[Pyr][DHAP]}{[Lact][3PG]} \cdot \frac{K_{GAPDH} \cdot K_{PGK}}{K_{LDH} \cdot K_{TPI}} \tag{4}$$

Where equation 3 shows the effect of triosephosphate isomerase (TPI), glyceraldehyde phosphate dehydrogenase (GAPDH) and phosphoglycerate kinase (PGK) and equation 4 includes in addition lactate dehydrogenase (LDH).

A change in the phosphorylation ratio will directly affect the state of the equilibrium, giving rise to a change in the cytosolic redox (equation 3) which will be reflected in the lactate/pyruvate ratio (equation 4). Similarly if there is a change in the cytosolic redox both the phosphorylation ratio and the glycolytic ratio (DHAP/3PG) will be changed. Thus both the adenine nucleotide signal operating on PFK and the substrate concentration (Berry et al, 1980PG) in the lower portion of the pathway depend on the redox state.

Is This Distributed Equilibrium Consistent With the Kinetics in the Integrated System?

The best clue comes from a sensitivity analysis of a complex kinetic model of heart metabolism developed by Kohn and Garfinkel (Kohn & Garfinkel, 1983). This model includes fatty acid metabolism, the TCA cycle, the redox shuttle and the glycolytic system in working heart. It has 13 free parameters and computes the steady-state concentrations of 481 metabolic intermediates via 641 chemical reactions. The computed concentrations and rates agree with those measured in the perfused working heart (~50 variables). The analysis indicates that the parameter having the strongest influence on the rate of lactate production is the rate of the redox shuttle. When the factors accounting for the regulation of phosphofructokinase are examined, it is found that this control is dominated by the cytosolic pH and the distribution of the adenine nucleotides between the various charged forms that serve as activators and inhibitors of PFK. The control of the cytosolic adenine nucleotides is, in turn, distributed between creatine kinase, the glycolytic system and the mitochondrial enzymes. The overall picture is of a set of near-equilibrium steps which control the concentrations of metabolic intermediates and these concentrations in turn regulate the rates of flux through "unidirectional" enzymes controlling the pathway fluxes. The result is that pathway fluxes are extremely sensitive to the state of near-equilibrium steps and in particular the redox interactions between the mitochondria and cytosol. The net effect of this complex model is a picture similar to the less complex distributed equilibrium hypothesis developed here.

Summary of Hypothesis

Figure 3 is a diagrammatic summary of the working hypothesis, Three major components are involved:

1. The combination of the cytosolic phosphorylation and mitochondrial redox potentials define the energy state of the mitochondrion. This in turn is reflected in both the turnover rate of oxidative phosphorylation and the mitochondrial membrane potential gradient.

2. An equilibrium gradient of redox exists between the mitochondrial matrix and the cytosol. This equilibrium includes the energy state of the mitochondrion as reflected in the membrane potential and the pH gradient. If the mitochondrial redox term can be maintained constant, a linear sensitivity to the ATP demand will occur and the variation in the redox gradient will be absorbed by the cytosol where it is observed as an increase in the lactate/pyruvate ratio.

3. The interaction of cytosolic redox with the cytosolic phosphorylation state in the lower half of the glycolytic pathway and the effect of mitochondrial metabolism on cytosolic adenine nucleotide concentrations in turn regulate PFK and glycolysis to maintain the cytosolic redox as set by the mitochondrion.

The combined action of these three components permits the state of the integrated system to be determined by mitochondrial metabolism while many of the effects appear in the cytosol as changes in redox as well as [ATP],[ADP],etc. Under any transient situation where the ATP supply is not adequately matched to the demand, the phosphorylation potential will fall. The effect of this will be to both activate PFK and drive the cytosol toward a more reduced state with an increase in the level of 3PG. This in turn will promote pyruvate production from 3PG. The recruitment of glycolysis thus leads to both an elevation of the size of the lactate + pyruvate pool and a reduction of the cytosol. This same signal at the mitochondrion drives oxidative phosphorylation and decreases the

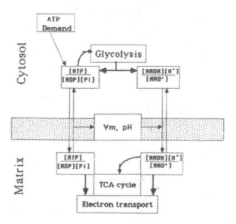

Fig. 3. Summary of interactions of metabolic pathways and the cytosolic and mitochondrial phosphrylation and redox potentials.

mitochondrial membrane potential leading to a decrease in the transmitochondrial membrane redox gradient. The combination of the effect of the cytosolic reduction and the recruitment of TCA cycle metabolism will be to restore mitochondrial redox and leave the cytosol more reduced. With time recruitment of both glycolytic and mitochondrial ATP production will lower the activation of PFK and slow the rate of glycolysis. The more reduced cytosol in combination with the partially restored phosphorylation ratio will, via the equilibrium in equation 4, stabilize to a condition with relatively more DHAP and less 3PG. This will inturn slow the flux from the trioses to pyruvate. At steady-state, as long as the mitochondrial system is maintaining the ATP supply and balancing the cytosolic redox, glycolysis is only recruited to maintain a stable pyruvate level and lactate appropriate to the mitochondrially set redox potential. That is, the glycolytic flux serves only to replenish lactate lost to the circulation and pyruvate and reducing equivalents consumed by the mitochondrion. Thus tissue lactate can be quite high while the steady-state glycolytic flux can be very small (Connett et al, 1986). The consequence of this operation is a strong bias in the system toward efficient mitochondrial ATP production that is supported and facilitated by the glycolytic system. This aerobic function appears to be the principal contribution of glycolysis to the normal operation of cardiac and red skeletal muscle.

In conclusion, we have outlined a simple working model accounting for the interaction of cytosolic and mitochondrial pathways. It is based on a distributed set of equilibria but is consistent with a more detailed model based on enzyme kinetics. It is also consistent with much of the data obtained from studies of metabolic regulation in both heart and red skeletal muscle.

Acknowlegements

This work was supported by USPHS grants AR36154 and HL03290.

References

Berry, M.N., A.R. Grivell, and P.G. Wallace, 1980, Energy-dependent regulation of the steady-state concentrations of the componenets of the lactate dehydrogenase reaction in liver, FEBS Lett., 119:317-322.

Clarke, F.M., F.D. Shaw, and D.J. Morton, 1980, Effect of electrical stimulation postmortem of bovine muscle on the binding of glycolytic enzymes. Functional and structural implications, Biochem. J., 186:105-109.

Connett, R.J., 1987, Glycolytic regulation during an aerobic rest-work transition in dog gracilis muscle, J. Appl. Physiol., in press 105-109.

Connett, R.J., T.E.J. Gayeski, and C.R. Honig, 1985, Energy sources in fully aerobic rest-work transitions: A new role for glycolysis, Am. J. Physiol., 248:H922-H929.

Connett, R.J., T.E.J. Gayeski, and C.R. Honig, 1986, Lactate efflux is unrelated to intracellular PO2 in a working red muscle in situ, J. Appl. Physiol., 61:402-408.

Davis, E.J., J. Bremer, and K.E. Akerman, 1980, Thermodynamic Aspects of Translocation of reducing equivalents by mitochondria, J. Bio. Chem., 255:2277-2283.

Davis, E.J., and L. Lumeng, 1975, Relationships between the phosphorylation potentials generated by liver mitochondria and the respiratory state under conditions of adenosine diphosphate control, J. Biol. Chem., 250:2275-2282.

Gayeski, T.E.J., R.J. Connett, and C.R. Honig, 1987, Minimum intracellular PO2 for maximum cytochrome turnover in red muscle in situ, Am. J. Physiol., 252:H906-H915.

James, A.T., 1980, Liver redox resistance: A dynamic model of gluconeogenic lactate metabolism, J. Theor. Biol., 83:623-646.

James, A.T., 1982, A note on the linear relation between lactate redox potential and the hydrogen shuttle flux, J. Theor. Biol., 94:129-133.

Jong, Y.A. and E.J. Davis, 1983, Reconstruction of steady-state in cell-free systems. Interactions between glycolysis and mitochondrial metabolism: Regulation of the redox and phosphorylation states, Arch. Biochem. Biophys., 222:179-191.

Kauppinen, R., 1983, Proton electrochemical potential of the inner mitochondrial membrane in isolated perfused rat hearts, as measured by exogenous probes, Biochim. Biophys. Acta, 725:131-137.

Kauppinen, R.A., J.K. Hiltunen, and I.E. Hassinen, 1983, Mitochondrial membrane potential, transmembrane difference in the NAD+ redox potential and the equilibrium of the glutamate-aspartate translocase in the isolated perfused rat heart, Biochim. Biophys. Acta, 725:425-433.

Kohn, M.C., 1983, Computer simulation of metabolism in palmitate-perfused rat heart. III. Sensitivity analysis, Ann. Biomed. Eng., 11:533-549.

Kohn, M.C., and D. Garfinkel, 1983, Computer simulation of metabolism in palmitate-perfused rat heart. II. Behaviour of complete model, Ann. Biomed. Eng., 11:511-531.

LaNoue, K.F., and A.C. Schoolwerth, 1979, Metabolite transport in mitochondria, Ann. Rev. Biochem., 48:871-922.

LaNoue, K., W.J. Nicklas, and J.R. Williamson, 1970, Control of citric acid cycle activity in rat heart mitochondria, J. Biol. Chem., 245:102-111.

Minatogawa, Y., and L. Hue, 1984, Fructoses-2,6-bisphosphate in rat skeletal muscle during contraction, Biochem. J., 223:73-79.

Nishiki, K., M. Erecinska, and D.F. Wilson, 1978, Energy relationships between cytosolic metabolism and mitochondrial respiration in rat heart, Am. J. Physiol., 234:C73-C81.

Olgin, J., R.J. Connett, and B. Chance, 1986, Mitochondrial redox changes during rest-work transition in dog gracilis, Adv. Exp. Med. Biol., 200:545-554.

Srivastava, D.K., and S.A. Bernhard, 1986, Metabolite transfer via enzyme,enzyme complexes, Science, 234:1081-1086.

van der Meer, R., T.P.M. Akerboom, A.K. Groen, and J.M. Tager, 1978, Relationship between oxygen uptake of perifused rat-liver cells and the cytosolic phosphorylation state calculated from indicator metabolites and a redetermined equilibrium constant, Eur. J. Biochem., 84:421-428.

Waser, M.R., L. Garfinkel, M.C. Kohn, and D. Garfinkel, 1983, Computer modelling of muscle phosphofructokinase kinetics, J. Theor. Biol., 103:295-312.

Williamson, J.R., C. Ford, J. Illingworth, and B. Safer, 1976, Coordination of citric acid cycle activity with electron transport flux, Circ. Res. Suppl 1, 38:I-39-I-51.

Williamson, J.R., B. Safer, K.F. LaNoue, C.M. Smith, and E. Walajtys, 1973, Mitochondrial-cytosolic interactions in cardiac tissue: Role of the malate-aspartate cycle in the removal of glycolytic NADH from the cytosol, in: Symposia of Society for Experimental Biology XXVII Rate control of Biological processes, Davies, D.D. ,ed., Cambridge Univ. Press, Cambridge. pp. 241-281.

PATTERNS OF O_2-DEPENDENCE OF METABOLISM

P.W. Hochachka

Department of Zoology & Sports Medicine Division
University of British Columbia
Vancouver, B.C., Canada V6T 2A9

SUMMARY

1.   In O_2 regulating systems, mitochondrial O_2 uptake is stabilized as O_2 availability declines by means of metabolite signals that simultaneously activate glycolysis; the consequent Pasteur effect is an attempt to make up the energy deficit arising from O_2 limitation.

2.   In O_2 conforming systems, the regulatory link between the ETS and glycolysis is seemingly lost. The advantage of O_2 conformity is that it avoids the Pasteur effect; the cost is an exaggerated dependence of mitochondrial respiration on O_2 availability.

3.   The $\dot{V}O_{2(max)}$ of man and other low-altitude adapted animals follows the O_2 conforming pattern; at altitudes equivalent to the peak of Everest, the $\dot{V}O_{2(max)}$ is only slightly greater than RMR. Again, key regulatory interactions between the ETS and glycolysis seem to be missing, so the energy deficit is tolerated (lactate production during exercise to exhaustion is less in hypobaric hypoxia than in normoxia).

4.   The O_2 conformity of $\dot{V}O_{2(max)}$ in mammals may be explained by inherently inefficient O_2 delivery systems in which low K_m and low k_{cat} cytochrome oxidase function would be selected. O_2-limited maximum mitochondrial respiration helps to explain what would otherwise be a perplexing observation: why over a 10^4 range of mass-specific muscle metabolic rates, the peak O_2 uptake rates per unit mitochondrial volume are always the same at $\dot{V}O_{2(max)}$.

5.   The concept of O_2-limited mitochondrial respiration predicts that more efficient O_2 delivery systems, such as tracheoles found in insect flight muscles, should support much higher in vivo cytochrome oxidase turnover rates. As far as can be currently evaluated, this prediction is realized. Thus, while other control parameters are almost certainly involved, an important regulatory (limiting) role for O_2 at maximum rates of mitochondrial respiration may be the rule in homeotherms, not the exception.

INTRODUCTION

In considering the control of O_2 demand by the electron transfer system (ETS), it may be useful to recall that most theories in this area assume three possible classes of regulatory metabolites: O_2 per se, adenylates (ATP, ADP, AMP, or related intermediates such as P_i), or metabolites which influence the redox poise of the ETS such as the NADH/NAD^+ couple.

Although these are obviously linked (in metabolic and thermodynamic terms), only O_2 interfaces mitochondrial function with the external environment, which is why its roles (i) as a substrate per se and (ii) as a regulatory metabolite potentially capable of setting ATP turnover rates are of particular interest to comparative biochemists and physiologists. This analysis attempts to evaluate the regulatory significance of O_2 under conditions of basal or resting metabolic rate (RMR) and of maximum sustainable rate, $\dot{V}O_{2(max)}$, with minimal focus on the above second and third alternatives.

Metabolic biochemists accustomed to working with isolated mitochondria assume that the K_m for O_2 is a fraction of a torr (0.5 uM or less) and that plots of respiration rates versus O_2 availability are far left-shifted on the in vivo O_2 axis. In contrast, most comparative studies indicate two fundamental responses to varying O_2. In O_2 regulators, RMR is regulated at a fairly stable rate down to low O_2 availability, and this pattern is observable in organisms, organs, tissues, and even cells. It is periodically, if rarely, observed in these kinds of systems (Gayeski et al., 1987; Fisher and Dodia, 1981) that the in vivo O_2 saturation curves display apparent K_m values similar to mitochondria; usually the K_m values are much higher (see Maren et al., 1986, for example).

In a second kind of response, termed O_2 conformity, the mismatch between mitochondrial and in vivo apparent K_m values is even greater, and O_2 uptake rates fall steadily with O_2 availability; in some cases, plots of aerobic metabolic rate versus O_2 availability pass through the origin, in which case complete arrest of ETS function presumably coincides with total O_2 lack. In other cases, O_2 uptake is blocked before O_2 availability falls to zero (Edelstone et al., 1984; Burggren and Randall, 1978; Mangum and Van Winkle, 1973). Phylogenetic surveys of the two kinds of responses may be found in Prosser (1986) or in comparative textbooks (for example, Schmidt-Nielsen, 1979). Traditionally, most of these kinds of studies are with lower animals, but some are also available for the whole organism level in mammals. In several small mammals, for example, $\dot{V}O_2$ increases as PaO_2 increases up to some critical (species-specific) value, above which $\dot{V}O_2$ remains essentially constant no matter how much more O_2 availability increases (Rosenmann and Morrison, 1974). An O_2 conforming pattern also is typically seen in patients suffering acute respiratory distress syndrome (Gilston, 1985) and in the mammalian fetus (see, for example, Edelstone, 1984).

At the organ, tissue, and cell level, different organs and tissues express differing metabolic responses to varying O_2 availability. The metabolic response of the mammalian brain represents an extreme (O_2 regulatory) end of a spectrum of responses found in mammalian tissues (Jones and Traystman, 1984). Liver in some species shows an intermediate pattern: As O_2 delivery (the sum of both hepatic arterial and portal flows and O_2 contents) declines, liver $\dot{V}O_2$ initially remains constant but even at fairly high O_2 delivery rates, $\dot{V}O_2$ begins to decline, and may fall to zero before O_2 availability falls to zero (Edelstone et al., 1984). For other species, plots of liver $\dot{V}O_2$ versus $[O_2]$ go through the origin (Maren et al., 1986). This complex O_2 conforming pattern (of $\dot{V}O_2$ varying with O_2

availability up to some critical value) is probably the most common pattern for mammalian cells and tissues. However, in skeletal muscles of the cat no plateau in RMR is reached even at very high O_2 tensions or very high O_2 delivery rates (Whalen et al., 1973), while $\dot{V}O_2$ in dog gracilis is seemingly independent of O_2 down to very low values (Gayeski et al., 1987).

From such analyses we can conclude that in all animal tissues ETS function declines when O_2 availability is adequately depressed, but there are major variations in the O_2 level at which this occurs. O_2 regulating systems are in effect "left shifted" on the $[O_2]$ axis relative to O_2 conformers. By analysing the kinds of regulatory mechanisms operating in these two kinds of systems, it may be possible to better appreciate the role of O_2 in metabolic regulation.

SUSTAINING STABLE RESPIRATION AS O_2 AVAILABILITY VARIES

Of the two kinds of responses, O_2 regulating systems are the better studied, and mechanisms have been proposed to explain the response. Most of these are consistent with the assumptions of Michaelis-Menten kinetics and are basically substrate concentration-dependent models. Wilson and his colleagues (1979), for example, suggest that compensatory regulation of the phosphate potential (or of some related signal), of the redox potential, or specifically of the reduction state of cytochromoe c are possible mechanisms for sustaining stable respiration rates in the face of varying O_2 concentration. Other workers, such as Kadenbach (1986) suggest more complex allosteric control models focussing upon cytochrome oxidases; in animal tissues this enzyme possesses many regulatory subunits (as many as 10) some of which may serve to develop tissue-specific regulatory properties and respiratory capacities. In isosteric and allosteric models, however, the same effector metabolites (especially ATP, ADP, AMP, and P_i, separately or in combination) are assumed to play key roles in the O_2 regulatory response. An under-emphasized fall-out of all such models is that the same metabolites serve as regulatory links between stabilizing ETS function and activating glycolysis during O2 limitation. This problem is reviewed elsewhere, so suffice to mention here that adenylate concentration changes involved in stabilizing ETS function at varying $[O_2]$ values also lead to glycolytic activation via effects on key enzymes in the pathway (see Storey, 1985, 1987). Because of these controlling links, O_2 regulators typically display large Pasteur effects (5-15 fold increases in glucose consumption rates) and come within about 50% of making up the energy deficit (Hochachka, 1985, 1987).

LOSS OF ETS-GLYCOLYSIS REGULATORY LINKS IN O_2 CONFORMERS

While a critical property of the O_2 regulatory response is that the metabolite signals mediating stabilized respiration also automatically mediate the Pasteur effect, perhaps the most distinguishing feature of O_2 conformers is the loss of these key regulatory interactions between the ETS and glycolysis. The way in which this loss is expressed varies, and at least three patterns can be distinguished from current information.

In one kind of response, expressed by turtle brain, initial metabolite changes are similar to those observed in O_2 regulating systems. For example, phosphocreatine (PCr) and ATP levels begin to fall, P_i levels begin to rise, and lactate production rates are initially fairly high. As the hypoxic episode continues, however, the similarities with O_2 regulators end: Cytochrome aa_3 becomes more reduced even while respiration rates are falling, adenylate concentrations return towards normoxic levels, and lactate production rates fall to less than 1/5 the rates observed in initial phases of O_2 lack (Lutz et al., 1984). The energy deficit obviously is not made up by anaerobic mechanisms. A slow-motion version of this pattern occurs in the hypoxic lungless salamander (Gatz and Piiper, 1979).

Ischemic mammalian muscle illustrates a second type of O_2 conforming pattern. In this case, as the tissue is O_2 depleted, PCr and P_i concentrations gradually change, but the adenylates are initially largely unperturbed. Lactate production rates are low and remain fairly constant throughout the ischemic period, representing maximally about 1/3 the ATP turnover rate found in normoxic resting muscle (Harris et al., 1986). As in pattern one, the energy deficit is tolerated.

A third type of pattern is evident in epaxial muscle of lungfish submerged for 12 hours in a state so hypoxic that the PaO_2 is 5 torr or less and the organism as a whole is relying upon anaerobic glycolysis for 50% or more of its depressed ATP turnover rates. Interestingly, under these conditions of hypoxia, the epaxial muscle displays no change in PCr, ATP, ADP, AMP, or P_i levels; that is presumably why there is no compensatory activation of glycolysis *at all* (no measurable increase in muscle lactate over the 12 hr hypoxic episode) (Dunn et al., 1983). This tissue illustrates metabolic arrest as a strategy for surviving periods of O_2 limitations particularly well (Hochachka, 1986a,b; Guppy et al., 1986). When O_2 availability falls in this kind of system, there is a total absence of the usual regulatory links between the ETS and glycolysis. O_2 uptake rates thus necessarily fall without a Pasteur effect.

POTENTIAL SIGNALS CONTROLLING O_2 CONFORMITY

In addressing this problem in O_2 regulators, most workers emphasize roles for various metabolites, especially the adenylates and reduced ETS substrates per se. In O_2 conformers, the adenylates can be ruled out as playing key controlling functions at least in extreme cases (such as the lungfish muscle and turtle brain examples above); this is because their concentrations simply do not change enough during O_2 lack. Changes in reduction state of ETS components, while possible regulatory signals, should compensate for declining O_2 availability and stabilize respiration, but the reverse is observed. At least tentatively therefore we rule out such redox changes as having anything to do with mediating O_2 conformity. PCr and P_i concentrations changes are potential regulatory signals, but the direction of change should favour increasing, not decreasing, respiration. That is why these too can be ruled out; they are necessarily ruled out in cases such as lungfish muscle in which their concentration does not change during hypoxia. Lactate and H^+ concentrations could have a role in O_2 conformity as they clearly increase in most cases. Unfortunately, they increase even more in O_2 regulators under hypoxia, which show stabilized, not declining, respiration rates; thus they too can be ruled out at least tentatively as regulatory metabolites.

Our analysis in fact can rule out every metabolite thus far examined in O_2 conformers during hypoxia except O2 itself. Because we cannot find another plausible candidate, we propose that the main metabolite signal to which respiration of O_2 conformers is responding is O_2 per se. It appears that as O_2 concentrations drop, cellular ATPase (cell work) rates also proportionately drop so as to effectively "clamp" the phosphate and redox potentials at steady state values where energy supply and energy demand are in balance. Not only does this interpretation explain why these regulatory signals are now not available for stabilizing respiration and activating glycolysis in the hypoxic zone; it also explains why the O_2 dependence of O_2 uptake is so pronounced. That is, with phosphate and redox potentials "clamped", O_2 seems to serve both as a substrate _per se_ and as a regulatory parameter seemingly dominant in setting the respiration rates of mitochondria.

We would be less hasty in arriving at this conclusion if it were not consistent with several observations already in the literature. For example, respiration in the lungless salamander is based entirely upon diffusion of O_2 across external barriers and except perhaps at very high ambient tensions, O_2 uptake rates of mitochondria of this organism are assuredly diffusion determined. In this event, we would predict an O_2 conforming response as ambient O_2 tensions decline, as indeed is observed (see Gatz and Piiper, 1979). Secondly, in numerous O_2 conformers, VO_2 is directly proportional to O_2 tensions in the arterial blood and drops to zero before arterial O_2 tensions do. This result is predictable if VO_2 is tracking O_2 tensions; presumably mitochondrial respiration is simply responding to available intracellular O_2 (Burggren and Randall, 1978).

A third line of _in vivo_ evidence suggesting that O_2 _per se_ is the metabolite signal for O_2 conformity comes from studies of organisms which under some circumstances behave as O_2 regulators, while under others behave as O_2 conformers. Pseudemys scripta is an air-breathing aquatic turtle that behaves as an O_2 regulator when O_2 tensions in air decline. When in water, on the other hand, gas exchange across the lung is impossible as is the O_2 regulatory response. A residual O_2 consumption is dependent entirely upon diffusion across external barriers (the skin and cloaca). We would predict - and it is well known - that in water this aquatic turtle is a classical O_2 conformer with VO_2 behaving as if it were moving down an O_2 saturation curve (see Jackson, 1968; Hochachka and Guppy, 1987).

O_2 REGULATORS AT $\dot{V}O_{2(MAX)}$ BEHAVE LIKE O_2 CONFORMERS

From the analysis thus far it appears that the advantage of O_2 conformity is that it avoids the Pasteur effect; its cost is an exaggerated ETS dependence on O_2 availability (because many, and in the extreme, all of the relevant metabolite signals needed for stabilizing respiration with respect to O_2 are missing). This conclusion, which implies that the differences between the two kinds of responses are more quantitative than qualitative, raises interesting insights into the role of O_2 limitation during $\dot{V}O_{2(max)}$ demanding performance.

It is widely appreciated that on transition from rest to maximum sustainable work, the aerobic metabolic rates of mammals increase by about 10 fold, the so-called scope for aerobic activity (Taylor et al., 1981). When inspired O_2 is reduced, as at high altitude, the aerobic scope in man is also substantially reduced (West, 1983). During sustainable muscle work, changes in adenylate concentrations are fairly modest because these are strongly buffered by creatine phosphokinase (Kushmerick, 1985; Dudley and Terjung, 1985); since absolute work rates are lower, this stability is assumed to be exaggerated at high altitude. Presumably for this reason, despite falling aerobic metabolic rates, anaerobic contributions to ATP production are not activated and lactate formation is minimized during exercise at extreme altitudes (West, 1986). At altitudes equivalent to the peak of Everest, the scopes for activity of animals such as man are reduced to near zero. At these altitudes, the concentration of O_2 in the plasma arriving at the working muscles should still exceed 10 uM; although this is higher than mitochondrial K_m values, these muscles are behaving as if O_2 were distinctly limiting, yet they do not activate glycolysis. Thus during maximally activated aerobic metabolism, O_2 regulators behave like O_2 conformers, presumably for the same reason as before: because metabolite signals mediating stabilized respiration while concomitantly activating glycolysis are minimized, over-ridden, or reversed. The advantage of this arrangement is not that ETS function declines with declining O_2 but that it avoids the Pasteur effect. During maximal muscle work, this may be far more critical than during RMR, for if the ATP turnover rates required were sustained glycolytically, the amounts of lactate and protons generated would be enormous - perhaps 50 umol per g muscle per min, which in 10 min could lead to 0.5 M (!) lactate concentrations. O_2 conformity under these conditions is easily understandable.

Additional evidence pointing to a regulatory role for O_2 during aerobic work comes from studies of the effects of size on the scope for aerobic activity in mammals. The equation relating body mass and metabolism can be given as:

$$\dot{V}O_2/m_b = 0.676\ m_b^{-0.25}$$

where $\dot{V}O_2/m_b$ is the mass specific rate of basal metabolism in ml O_2 $h^{-1}g^{-1}$ and m_b is body mass in grams. If replotted as log $\dot{V}O_2$ versus log body mass, the allometric exponent, which is the slope of the line, is about 0.75. In both kinds of plots, the allometric exponents are the same for basal or maximum aerobic metabolism (Schmidt-Nielsen, 1984; Taylor et al., 1981), which means that the scope for activity is similar in big and small animals.

It may surprise some readers that the differences between resting and maximum rates within a species are modest compared to the magnitude of the scaling impact on O_2 consumption of organisms. Compared to a 10 fold aerobic scope for activity in mammals, the mass specific RMR values in ml O_2 $g^{-1}h^{-1}$ for shrews, men, and elephants differ by a full 100 fold (7.4, 0.21, and 0.07, respectively). Extrapolating these numbers to whales of 10^8 g mass yields $\dot{V}O_2/m_b$ of only 0.002 ml O_2 $g^{-1}h^{-1}$ (Schmidt-Nielsen, 1984). This means that $\dot{V}O_2/m_b$ values for shrews are some 4000 times greater than for whales, while for hovering hummingbirds (which display the highest working metabolic rates known amongst homeotherms) these values are some 10^5 times greater than the predicted RMR for whales (Suarez et al., 1986). The effects of size thus greatly outweigh the effects of exercise and even surpass the effects of temperature and lungs, for the predicted $\dot{V}O_2/m_b$ values for whales are only 1/5 the rates in bathypelagic fishes or in lungless salamanders, at $5°$ and $13°C$, respectively (Torres et al., 1979; Gatz and Piiper, 1979).

At the level of working muscles, this incredible range of metabolic intensity can be quantitatively accounted for by the total volume of mitochondria per muscle decreasing systematically along the mouse-to-elephant curve (Hoppeler et al., 1987). Nevertheless, as we have emphasized elsewhere (Hochachka et al., 1987), at all size ranges the control mechanisms primarily responsible for the transition from RMR to $\dot{V}O_{2(max)}$ must be the same because the allometric exponents for RMR and $\dot{V}O_{2(max)}$ are the same. What is more, the dominant limiting mechanism must be the same because at $\dot{V}O_{2(max)}$ in mammals, irrespective of animal size, the O_2 uptake per ml of mitochondria is always the same, about 200 umol O_2 per ml mitochondria per minute (Hoppeler et al., 1987; Taylor, 1987). In fact, the same value is observed for hummingbird flight muscle during hovering (Suarez et al., 1986).

This is a remarkable observation. It means that over a range not only of 10 fold (rest to work transition), but over 10^4 fold (assuming constant scaling from hummingbird to whale), mitochondrial respiration rates seem to peak out at about the same value. What is more, this peak value is only 1/10th or so of the catalytic potential of cytochrome oxidase (Table 1). Why should this be so? We propose that the answer is a limitation of O_2 delivery systems in homeotherms. Given the dependence of O_2 delivery on pipes and pumps (gas exchange plus cardiovascular systems), evolutionary pressure may well have selected for low K_m and low k_{cat} cytochrome oxidase function. The operational advantage of low K_m and low k_{cat} values for all enzymes is high competitiveness for low substrate concentrations (O_2 in this case), which is indeed a hallmark characteristic of cytochrome oxidases in mammalian tissues (Kadenbach, 1986). This conceptualization also explains why in homeotherms cytochrome oxidase functions well below its catalytic potential, where its high affinity for O_2 has its greatest effect and thus pays its greatest dividend (Fersht, 1986).

Table 1. Maximum O_2 fluxes in mitochondria from muscles of mammals and birds compared to flight muscles of insects, as a measure of effectiveness of capillary versus tracheole O_2 delivery systems.

Metabolic System	Maximum O_2 Fluxes in umol O_2 Per ml Mitochondria Per Min	
	Observed	Theoretical Maximum[#]
Mammalian or bird skeletal muscles at maximum aerobic work rate*	200	2400
Insect flight muscle**	1800	2400

[#]These estimates assume (i) that the turnover number for cytochrome oxidase is 500 umol cytochrome *c* oxidized (or 125 umol O_2 consumed) per umol cytochrome oxidase per sec, (ii) that 1 mg mitochondrial protein is equivalent to 2 ul mitochondrial volume, and (iii) that the concentration of cytochrome oxidase is 0.65 nmol per mg mitochondrial protein (Tzagoloff, 1983).
*From Hoppeler et al. (1987).
**From Sacktor (1976), assuming insect flight muscle is 40% mitochondrial by volume, a value entirely comparable to that found in muscles of the smallest homeotherms (Suarez et al., 1986).

If this interpretation is correct, it would predict for systems which are *not* similarly selected to function under limiting O_2 delivery rates, (i) that cytochrome oxidase turnover numbers in vivo should be much higher than in mammals and birds, and (ii) in vivo turnover numbers should more closely approach the known turnover number of cytochrome oxidase. Fortunately, nature supplies many "experimental tests" of our prediction in the flight muscles of insects. While the ultrastructure and biochemistry of these muscles are specialized (Sacktor, 1976), they nevertheless display major similarities with the muscles of the smallest homeotherms. There is, however, a most significant difference in the mode of O_2 delivery to the working mitochondria: Trachea and tracheoles delivering gaseous O2 almost directly to cytochrome oxidase versus capillaries delivering O2-loaded red blood cells to the vicinity of the sarcolemma. Comparative biochemists have known for decades that the former is the more effective O_2 delivery system. Because of its efficiency, insect flight muscle can sustain glucose-based or fat-based metabolic rates equivalent to about 700 umol O_2 per g muscle per min (Armstrong and Mordue, 1985; Sacktor, 1976), or 1800 umol O_2 per ml mitochondria per minute. This value is about an order of magnitude higher than in homeotherms and indicates an in vivo cytochrome oxidase turnover number impressively close to the k_{cat} maximum (Table 1). Both results are in satisfactory agreement with our basic premise that O_2

(delivery or diffusional) limitation may be the rule rather than the
exception in the animal kingdom.

ACKNOWLEDGMENTS

This work was supported by NSERC (Canada)

REFERENCES

Armstrong, G., and Mordue, W., 1985, Oxygen consumption of flying locusts.
    Physiol. Entomology 10, 353-358.
Burggren, W.W., and Randall, D.J., 1978, Oxygen uptake and transport during
    hypoxic exposure in the sturgeon Acipenser transmontanus, Resp.
    Physiol., 34, 171-183.
Dudley, G.A., and Terjung, R.L., 1985, Influence of aerobic metabolism on
    IMP accumulation in fast-twitch muscle, Am. J. Physiol., 248, C37-
    C42.
Dunn, J.F., Hochachka, P.W., Davison, W., and Guppy, M., 1983, Metabolic
    adjustments to diving and recovery in the African lungfish, Am. J.
    Physiol., 245, R651-R657.
Edelstone, D.I., 1984, Fetal compensatory responses to reduced oxygen
    delivery, Seminars in Perinatology, 8, 184-191.
Edelstone, D.I., Paulone, M.E., and Holzman, I.R., 1984, Hepatic
    oxygenation during arterial hypoxemia in neonatal lambs, Am. J.
    Obstet. Gynecol., 150, 513-518.
Fersht, A., 1985, "Enzyme Structure and Mechanisms," W.H. Freeman & Co.,
    N.Y., pp. 1-475.
Fisher, A.B., and Dodia, C., 1981, The lung as a model for evaluation of
    critical intracellular PO_2 and PCO_2, Am. J. Physiol., 241, E47-E50.
Gatz, R.N., and Piiper, J., 1979, Anaerobic energy metabolism during severe
    hypoxia in the lungless salamander Desmognathus fuscus
    (Plethodontidae), Resp. Physiol., 38, 377-384.
Gayeski, T.E.J., Connett, R.J., and Honig, C.R., 1987, Minimum
    intracellular PO_2 for maximum cytochrome turnover in red muscle in
    situ. Am. J. Physiol., H906-H915.
Gilston, A., 1985, ARDS: Another approach, Int. Crit. Care Digest, 4, 1-2.
Guppy, M., Hill, R.D., Schneider, R.C., Qvist, J., Liggins, G.C., Zapol,
    W.M., and Hochachka, P.W., 1986, Micro-computer assisted metabolic
    studies of voluntary diving of Weddell seals, Am. J. Physiol., 250,
    R175-R187.
Harris, K., Walker, P.M., Mickle, D.A.G., Harding, R., Gatley, R., Wilson,
    G.J., Kuzon, B., McKee, N., and Romaschin, A.D., 1986, Metabolic
    response of skeletal muscle to ischemia, Am. J. Physiol., 250, H213-
    H220.
Hochachka, P.W., 1985, Assessing metabolic strategies for surviving O_2
    lack: role of metabolic arrest coupled with channel arrest, Mol.
    Physiol., 8, 331-350.
Hochachka, P.W., 1986a, Defense strategies against hypoxia and hypothermia,
    Science, 231, 234-241.
Hochachka, P.W., 1986b, Metabolic arrest, Intensive Care Med., 12, 127-133.
Hochachka, P.W., 1987, Metabolic suppression and oxygen availability, Can.
    J. Zool., in press.
Hochachka, P.W., and Guppy, M., 1987, "Metabolic Arrest and the Control of
    Biological Time," Harvard University Press, Cambridge, Mass., pp.
    1-227.
Hochachka, P.W., Emmett, B., and Suarez, R.K., 1987, Limits and constraints
    in the scaling of oxidative and glycolytic enzymes in homeotherms,
    Can. J. Zool., in press.

Hoppeler, H., Kayar, S.R., Claassen, H., Uhlmann, E., and Karas, R.H., 1987, Adaptive variation in the mammalian respiratory system in relation to energetic demand: III. Skeletal muscles: Setting the demand for oxygen, Resp. Physiol., 69, 27-46.

Jackson, D.C., 1968, Metabolic depression and oxygen depletion in the diving turtle, J. Appl. Physiol., 24, 503-509.

Jones, M.D., Jr., and Traystman, R.J., 1984, Cerebral oxygenation of the fetus, newborn and adult, Seminars in Perinatology, 8, 205-216.

Kadenbach, B., 1986, Mini Review: Regulation of respiration and ATP synthesis in higher organisms: hypothesis, Bioenergetics & Biomembr., 18, 39-54.

Kushmerick, M., 1985, Patterns in mammalian muscle energetics, J. Exp. Biol., 115, 165-177.

Lutz, P.L., McMahon, P., Rosenthal, M., and Sick, T.J., 1984, Relationships between aerobic and anaerobic energy production in turtle brain in situ, Am. J. Physiol., 247, R740-R744.

Mangum, C.P., and Van Winkle, W., 1973, Responses of aquatic invertebrates to declining oxygen conditions, Am. Zool., 13, 529-541.

Meren, H., Matsumura, T., Kaufman, F.C., and Thurman, R.G., 1986, Relationship between oxygen tension and oxygen uptake in the perfused rat liver, in: "O_2 Transport to Tissue," I.S. Longmuir, ed., Vol. 8, pp. 467-476.

Prosser, C.L., 1986, "Adaptational Biology: Molecules to Organisms", Wiley & Sons, New York, pp. 1-784.

Rosenmann, M., and Morrison, P., 1974, Physiological responses to hypoxia in the trundra vole, Am. J. Physiol., 227, 734-739.

Sacktor, B., 1976, Biochemical adaptations for flight in the insect, Biochem. Soc. Symp., 41, 111-131.

Schmidt-Nielsen, K., 1979, "Animal Physiology: Adaptation and Environment," Cambridge University Press, Cambridge, England, pp. 1-560.

Schmidt-Nielsen, K., 1984, "Scaling. Why Is Size So Important?" Cambridge University Press, Cambridge, pp. 1-241.

Storey, K.B., 1985, A re-evaluation of the Pasteur effect: new mechanisms in anaerobic metabolism, Mol. Physiol., 8, 439-461.

Storey, K.B., 1987, Suspended animation, Can. J. Zool., in press.

Suarez, R.K., Brown, G.S., and Hochachka, P.W., 1986, Metabolic sources of energy for hummingbird flight, Ann. Rev. Physiol., 251, R537-R542.

Taylor, C.R., 1987, Structural and functional limits to oxidative metabolism: insights from scaling, Ann. Rev. Physiol., 49, 135-146.

Taylor, C.R., Maloiy, G.M.O., Weibel, E.R., Langman, V.A., Kamau, J.M.Z., Seeherman, H.J., and Heglund, N.C., 1981, Design of the mammalian respiratory system. III. Scaling maximum aerobic capacity to body mass: wild and domestic mammals, Resp. Physiol., 44, 25-37.

Torres, J.J., Belman, B.W., and Childress, J.J., 1979, Oxygen consumption rates of midwater fishes off California, Deep-Sea Res., 26A, 185-197.

Tzagoloff, A., 1983, "Mitochondria", Plenum Press, New York, pp. 1-342.

West, J.B., 1983, Climbing Mt. Everest without oxygen: An analysis of maximal exercise during extreme hypoxia, Resp. Physiol., 52, 265-279.

West, J.B., 1986, Lactate during exercise at extreme altitude, Fed. Proc., 45, 2953-2957.

Whalen, W.J., Buerk, D., and Thuning, C.A., 1973, Blood flow-limited oxygen consumption in resting cat skeletal muscle, Am. J. Physiol., 224, 763-768.

Wilson, D.F., Owen, C.F., and Erecinska, M., 1979, Quantitative dependence of mitochondrial oxidative phosphorylation on O_2 concentration: A mathematical model, Arch. Biochem. Biophys., 195, 494-504.

TRANSDUCING CHEMICAL ENERGY INTO MECHANICAL FUNCTION:

A COMPARATIVE VIEW

C.L. Gibbs and I.R. Wendt

Department of Physiology
Monash University
Clayton. Vic.

There is little doubt that ATP hydrolysis is the driving reaction for the energy expended during contractions of skeletal, cardiac and smooth muscle. In particular it would appear that we can account for most of the initial metabolism of all muscle types in terms of four major classes of ATPases: i) the Na^+-K^+ ATPase ii) the Ca^{++} ATPases of the sarcoreticular and sarcolemmal Ca^{++} pumps iii) the actin activated myosin ATPase of the contractile proteins and iv) the ATPases involved in protein phosphorylation (Gibbs & Chapman, 1979; Homsher, 1987). The ultimate energy source for contraction comes from the reactions of intermediary metabolism and oxidative phosphorylation. In most, but not all, muscles the resynthesis of ATP consumed in the contractile event is achieved by mitochondrial oxidative phosphorylation: there is some ATP production in the cytoplasmic glycolytic pathway and in smooth muscles in particular aerobic glycolysis can underwrite a sizeable fraction of total ATP production. The combined activity of these ATP regenerating reactions is called the recovery metabolism.

In Table 1 we list some data obtained in various muscles of the rat over the last 15 years in our laboratory. This table emphasizes the large changes in energy flux that can take place across the different muscle types and the data illustrate some generalities namely that smooth muscles have the lowest ratios of active to basal energy flux; values in the range of 2 - 4 being usual in the literature, cardiac and slow twitch muscles have ratios in the 10 to 20 range, and fast striated muscles can achieve ratios greater than 100. The differences in these ratios seem to allow the three muscle types to structure their short term energy reserves (ATP + PC) differently with resultant differences in the absolute high energy phosphate levels and PC/ATP ratios. Interestingly, however, there is little difference in the cytoplasmic phosphorylation potential which means that ΔG_{ATP} is similar in all muscle tissues ($\sim$ -60 kJ/mole).

Isometric twitch contractions

The isometric mechanical (force) output of a muscle will be proportional to the average number of force-generating cross bridges in existence multiplied by their tension-time integral and the energy output will be determined by the number of cross bridges (CB) turned

Table 1

Rat myothermic data (27°C)

	Basal (mW.g-1)	Active (mW.g-1)	Recovery heat time constant(s)	Mitochondrial Volume (%)
Skeletal				
EDL	1.7_T	136_I (tetanus)	30	7*
SOL	1.3_T	21_I (tetanus)	34	10*
Cardiac				
ventricle	6_T	21_T (3 twitches/s) 50_T (6 twitches/s)	10-20	~30
Smooth				
anococcygeus	2.8_T	7_T (tetanus)	30	4*

I = initial   T = total   *Data from Davey and Wong (1980)

over. It is possible to alter the force by either changing the number of cycling cross bridges or their cycle rate but in either of these two situations there should be a proportional change in energy flux if one ATP is split per crossbridge cycle. However if a muscle can alter the tension time integral of each CB cycle (the so called 'on' time) then a change in the mechanical output relative to the amount of ATP split could be achieved and we might expect to pick up a change in the energy cost of stress (force/cross sectional area) development.

If one looks at the skeletal muscle data in the energetic literature and subtracts the activation heat (i.e. heat due to activation processes, largely Ca^{++} translocation) from the twitch heat $(mJ.g^{-1})$ and divides the residue by the developed stress $(mN.mm^{-2})$ it is possible to get some estimate of the isometric transduction efficiency. For frog (R. pipiens) sartorius and semitendinosus muscles at 0°C Rall (1979) reports mean heat:stress slopes of 0.043 and 0.048 mJ/g per mN/mm^2 respectively. At 21°C values of 0.087 and 0.050 have been reported for the chicken posterior latissimus dorsi, a phasic muscle, and the anterior latissimus dorsi, a tonic muscle, respectively. At first sight these values seem quite dissimilar and suggest large differences in transduction efficiency between these two muscles. However, as Rall has pointed out, in PLD about 1/3 of the twitch heat is degraded internal work whereas in ALD only about 1/7 of the heat can be accounted for in this manner and when corrections for internal work are made the difference in heat:stress slopes is quite small. Our own laboratory data (Gibbs & Chapman, 1974) suggest that in sartorii from the frog (H. aurea) the ratio is 0.076 whereas in toad sartorii (B. marinus) the ratio is 0.053. This result implies a greater transduction efficiency in toad muscles.

It should also be noted that temperature has little effect on the heat:stress ratio. Rall (1979) has shown in amphibian skeletal muscle that over a temperature range from 0 to 20°C the activation heat does not alter in magnitude and there is little change in the slope of the heat stress relationship. This result is remarkable in one respect in that an increase in temperature is well known to increase myosin ATPase activity and hence the CB cycle rate and shortening velocity. The simplest biochemical interpretation surely must be that approximately the same number of CB's are used to reach peak force (obviously this happens much more rapidly as the temperature is raised). If however the experimental data is rearranged so that heat is plotted against the stress-time integral then the increased cost of a unit of tension-time (at a high temperature) is immediately apparent.

With regard to cardiac muscle there have been several myothermic studies on a range of mammalian and amphibian hearts. In a comparative energetic study by Loiselle & Gibbs (1979) there was little difference in the slope of the heat:stress relation between guinea pig, cat and rat cardiac muscles (see their Fig. 6). It is important to realise that in studies of cardiac muscle with our experimental protocol recovery heat is usually present so that the slope of the heat:stress relationship has to be divided by 2 to allow comparison with most skeletal muscle data. This results in initial heat:stress ratios in the range 0.05 to 0.07 for rat, guinea pig and cat papillary muscles. In a recent paper using rabbit papillary muscles (Gibbs, Loiselle & Wendt, 1987) the mean slope (corrected for recovery heat) ranged between 0.07 and 0.08 irrespective of changes in the extracellular Ca^{++} level from 0.625 to 5.0 $mmol.l^{-1}$. These slope values are higher than most of the frog and toad skeletal muscle data but as the measured series elasticity is considerably greater in cardiac muscle, about 6.9% compared to 2-3% l_o, the transduction efficiency will not be vastly different. In cardiac muscle as in skeletal muscle this slope is not affected by temperature, and nearly all the rapidly acting inotropic agents have no statistically significant effect on the slope (Gibbs, 1982). The temperature and drug results are somewhat startling as it is quite clear from recent experiments upon rat papillary muscles (Rossmanith et al., 1986) that agents such as the catecholamines greatly increase the cycle rate and yet in the presence of such agents the heat:total stress lines remain parallel suggesting a constant transduction efficiency. A similar inference can be drawn from the recent whole heart oxygen consumption studies of Suga and colleagues (Suga et al., 1983). We have considered the implications of their data elsewhere (Chapman & Gibbs, 1985).

It does appear however that it is possible to produce changes in the slope of the cardiac heat:stress relationship by using procedures which alter the distribution of myosin between its three isoenzyme forms V_1, V_2 and V_3. In the hyperthyroid state the myosin pattern shifts towards domination by V_1 whereas in the hypothyroid state V_3 is the most abundant form. This means that in the hyperthyroid state the actin activated myosin ATPase is high and in the hypothyroid state it is low relative to controls. It does appear that in these conditions the slope of the heat:stress relationship alters being steeper in the hyperthyroid case and less steep in the hypothyroid (Alpert, Mulieri & Litten, 1979; Loiselle, Wendt & Hoh, 1982). Alpert and colleagues have shown that pressure overload hypertrophy in rabbits also lowers myosin ATPase activity and by implication reduces the slope of the heat:stress relationship. They attribute the increase in the economy of force generation in the latter state to a decrease in cycle rate and to a somewhat longer 'on' time i.e. increased unit of tension time per CB. The converse argument is used to explain their hyperthyroid energetic data. There can be dangers in this argument however as can be seen from Figure 1 where we plot heat:stress relations from rabbit and toad cardiac tissue and find very similar slopes. If the same heat data is plotted against the stress time integral then the toad muscle appears to have a much higher economy but from a physiological point of view it may be energetically irrelevant in terms of a single contraction.

In the smooth muscle literature there is hardly any data which relates to twitch type responses because the mechanical response to a single stimulus is so small. In the rabbit rectococcygeus the twitch:tetanus ratio is normally about 0.2 but can be increased to about 0.6 by using a summated response to generate the "twitch" (2 stimuli at

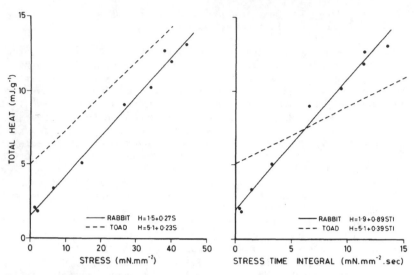

Figure 1. Heat:stress relation (left) and heat:stress time integral relation (right) for rabbit (——) and toad (---) ventricular muscle at 27°C.

5.5 Hz, Gibbs, 1984). Under such conditions the slope of the heat:stress ratio is about 0.12 after allowing for recovery heat and taking the activation heat contribution at 30% of the total energy output. This value is high relative to the skeletal and cardiac data but if allowance is made for the high series elasticity of smooth muscle the ratio would be about 0.1.

Isotonic twitch contractions

The energy expenditure of working skeletal muscle (twitches or very brief tetani) was considered in the classical paper of Fenn (1923). He established that at 0 to 8.5°C the energy output of a muscle does not depend solely upon its initial length but is modulated by the load lifted. Fenn also reported that although heat production was fairly constant at any load level, total enthalpy (heat + work) altered because work output varied with the different afterloads. This result, led Fenn to conclude that there was less energy liberated in an isometric contraction than in any contraction in which a muscle is allowed to shorten. This latter conclusion although approximately true for twitches or brief tetani at 8°C or below does not hold at higher temperatures as was clearly shown in subsequent studies (for a review see Woledge, 1971). In Fig. 2 we reproduce some twitch and summated contraction results taken from frog (Hyla aurea) sartorius experiments at 10 and 20°C (Gibbs & Chapman, 1974) which emphasize the effects that temperature and the activation level have on the shape of the enthalpy: load curve. Increasing the initial muscle length also altered the enthalpy:load relationship increasing the enthalpy output at all loads less than P_o.

As mentioned above it is well established that in skeletal muscle the activation heat component is not much altered by temperature nor is the relationship between isometric heat production and stress development. This must mean that the shape of the enthalpy curve is being dominated by the extra energy associated with muscle shortening and consequently by the work output and shortening heat terms. We have argued on many occasions that the thermodynamic distinction between a shortening heat component and a stress-dependent heat component is quite unclear at the level of a crossbridge where the free energy of ATP is either transduced into work or appears as degraded heat together with

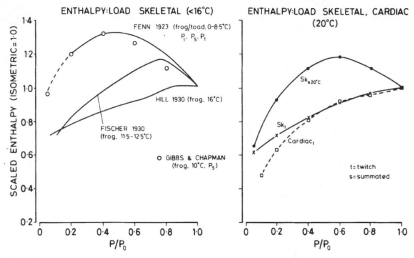

Figure 2. Enthalpy:load relations obtained in frog sartorii. Left panel, data obtained at low temperature (<16°C). Fenn's 1923 data (solid line) is compared with summated 10°C data (0) of Gibbs and Chapman (1974). Right panel, same relationships at 20°C - twitch and summated data compared with cardiac.

the entropic heat (Gibbs & Chapman, 1979). Nonetheless, no one seriously questions the concept that when muscle shortening takes place the crossbridge turnover rate must increase, whether this is always accompanied by a 1:1 increase in ATP hydrolysis is perhaps less clear. The values of the shortening heat coefficients measured in different laboratories in the same muscles seem in good agreement (see Homsher, 1987; Ford & Gilbert, 1987). However, there are still problems in terms of the coefficient's dependence upon load, myofilament overlap, shortening velocity and distance shortened. In experiments with the rectus femoris muscle of the tortoise Woledge (1968) reported a shortening heat coefficient that was only 1/6 as large as that reported for frog sartorius muscles. Several authors have considered the source of shortening heat (see Woledge, 1971; Homsher, 1987; Ford & Gilbert, 1987). In particular most speculation centres on the complexity of the various steps in the CB cycle; some of the reactions are exothermic some endothermic, and the suggestion is that with shortening there can be redistributions of the amount of reactants at different stages in the CB sequence.

Although we have not specifically tried to measure shortening heat in studies with mammalian skeletal muscles (rat EDL and soleus at 27°C) it was quite clear to us that in lightly loaded tetani the initial heat was only marginally in excess of that measured at the commencement of an isometric tetanus, see Fig. 3a - most of the extra energy appearing as work (Gibbs & Gibson, 1972; Wendt & Gibbs, 1973). In 1970 we reported data that casts doubt about the importance of shortening heat in the cardiac energy balance sheet (Gibbs & Gibson, 1970). When twice the external work was subtracted from the twitch enthalpy there was only evidence of an energy discrepancy at light loads and superimposition of heat traces from isometric and isotonic twitches gave no evidence of an increased heat rate during the phase of cardiac muscle shortening. We have recently repeated such measurements at 27°C with a more slowly contracting preparation (ventricular strip from the heart of Bufo marinus) and faster vacuum deposition thermopiles. The result is the same as that reported previously and is shown in Fig. 3b. Cardiac muscle does however have a high series elasticity compared to

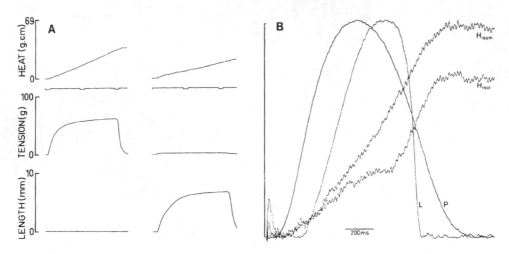

Figure 3a. Left panel, soleus muscle isometric force and heat production, 2s tetanus at 27^0C; right panel, isotonic shortening and heat production, 5g load. Muscle was 21 mm long, blotted wt 63 mg. **Fig. 3b.** Heat production in an isotonic and isometric twitch contraction of toad ventricle. Vertical calibration, $H = 8.8$ mJ.g^{-1}, $L = 2.3$ mm, $S = 52$ mN.mm^{-2}. Muscle weighed 8.5 mg, length 13.2 mm.

skeletal muscle and the possibility must exist that internal shortening of the contractile element is so great under isometric conditions that the initial isotonic and isometric heat rates are similar. We doubt this interpretation but the question will only be satisfactorily resolved when a cardiac preparation can be sarcomere length clamped on a thermopile.

    Efficiency. It has been of interest to muscle physiologists to look at the mechanical efficiency, the ratio of work to total enthalpy, of many muscles. Unfortunately there are considerable problems with this concept particularly when applied to the initial energy cycle, since some of the measured heat output relates to the entropy change that takes place. Over the total (initial + recovery) cycle of a contraction that is being fueled by carbohydrate oxidation there is some evidence that $\Delta G \simeq \Delta H$ and therefore that the entropic contribution is close to zero, however over the initial cycle there is no evidence that this state of affairs exists. Indeed ΔH_{ATP} is usually taken to be -48 kJ/mol whereas according to recent information about the levels of free ADP and P_i in the cytoplasm ΔG_{ATP} _in vivo_ is probably in the range -55 to -65 kJ/mol. ΔG may fall towards the mid 40's as contractile activity continues (see Gibbs, 1985; Kentish & Allen, 1986 for reviews).

    Ideally it is more meaningful to measure the thermodynamic efficiency of a contraction. In this measurement we are asking the question as to what fraction of the free energy made available when ATP is hydrolysed is transduced into work. Even with this definition the answer as applied to a muscle contraction can be misleading because some of the ATP used during a contraction will be used to do internal mechanical work and ion translocation work against concentration and/or electrical gradients. When allowance is made for these processes it is quite obvious that the crossbridge thermodynamic efficiency is extremely high perhaps greater than 70% (Gibbs & Chapman, 1985). Such a value is in excess of the efficiency of substrate oxidation which biochemists estimate to be about 65%.

Because of the uncertainty of many of the allowances that need to be made for the totality of processes that contribute to the energy balance sheet scientists will continue to make mechanical efficiency measurements. Under the best conditions (ergometer shortening at optimal velocity) mechanical efficiency values of about 0.4 are achieved over the initial cycle in single twitches of frog sartorius muscles (Hill, 1964). In contractions using conventional lever systems much lower values are found. The values decrease as temperature is raised and the twitch to tetanus ratio falls: mechanical efficiency can be restored to high values by optimizing the stimulus parameters at higher temperatures (Gibbs & Chapman, 1974). We found, as Hill has reported, that toad sartorii are somewhat more efficient than frog sartorii but neither approach the efficiency of the rectus femoris of the tortoise where an initial mechanical efficiency of 0.77 (tetanic contractions) has been found (Woledge, 1968). It is not too clear whether this very high efficiency results from the low shortening heat coefficient (mentioned above) or from this muscle's unexpectedly low tetanic maintenance heat rate.

The mechanical efficiency of cardiac muscle is altered by temperature, extracellular Ca^{++} concentration and initial length (Gibbs & Burke, unpublished). It reaches a maximum of about 25% at low loads (0.2-0.4 P_o). This value is calculated using the total heat, which means that over the initial cycle cardiac muscle is at least as efficient as skeletal muscle.

The energetics of maintained force production

It has been established for over sixty years that the heat production (H) during isometric contractions of frog sartorius muscles could be fitted by an equation of the type

$$H = A + bt$$

where A and b were constants and t was the stimulus duration. It was found subsequently that the heat-tension ratio ($H/P_o l_o$) plotted against stimuli duration (t) produced a similar linear relationship in most types of muscle. Several authors have noted that the term A, which relates to the energy cost of tension development is roughly the same for all muscles but that the constant 'b' which relates to the energy cost of tension maintenance varies widely (by greater than three orders of magnitude) between different muscles. Although we believe that in twitches peak stress is the preferred index of energy cost, as soon as force plateaus it is obvious that we need an index that involves time and for that reason many authors in tetanic studies plot energy production against force (or stress)-time integral.

In 1967 Bárány showed that there was a linear relationship between shortening velocity and actin activated myosin ATPase activity. Subsequently many authors have noted a relationship between the heat rate during a tetanus plateau (maintenance heat) and a muscles' myosin ATPase activity. In skeletal muscle there is general agreement that 65 to 75% of the maintenance heat rate relates to CB activity with most of the balance being attributed to the cost of Ca^{++} pumping by the sarcoplasmic reticulum. There are very clear species differences (in general the smaller the animal type the higher the myosin ATPase activity) and we and others (Gibbs & Gibson, 1972; Wendt & Gibbs, 1973; Crow and Kushmerick, 1982) have shown large differences between the cost of force maintenance in fast and slow muscles (EDL versus soleus). These energetic differences follow the known pattern of myosin isoenzyme development and cross-innervation of muscles produces the expected changes in their energetics.

The maintenance of force at extremely low energy cost reaches its economic peak in smooth muscle, particularly in those smooth muscles where force or tone has to be maintained for long periods of time (hours rather than seconds). The energetic literature has been reviewed recently (Gibbs, 1984). The first reliable paper on smooth muscle energetics was that of Bozler in 1930 using the snail retractor pharynx muscle and we have made extensive investigations with a mammalian smooth muscle. The rabbit rectococcygeus preparation is relatively fast, its V_{max} being about 0.6 muscle lengths/s at $27^{o}C$, and yet it can maintain a tetanic stress similar to that of frog sartorius muscle for about 1/50 of the energy cost. It is apparent from oxygen consumption and chemical studies that vascular and 'catch' smooth muscles are at least an order of magnitude more economic. Although there is good correlation between heat rate and myosin ATPase activity in smooth muscles it must be said that in maintained tetanic contractions the muscles are often 4 to 10 times more economic than their CB turnover rate would predict, see Paul, Gluck & Ruegg (1976).

Factors that can regulate the economy of force maintenance

Bozler (1930) reported that with repetitive activity he could find a six to tenfold increase in the economy with which the retractor pharynx preparation could maintain tension: there was no sign of any mechanical deterioration except for a progressive prolongation of relaxation. In 1981 Dillon, Aksoy, Driska and Murphy showed that in arterial smooth muscle there was a decrease in the extent of myosin light chain phosphorylation throughout a maintained contraction and they reported that this decline in phosphorylation correlated with a decrease in V_{max} as measured by quick releases at various times. Since in smooth muscle myosin must be phosphorylated for actin activated myosin ATPase activity to be turned on it was suggested that force maintenance occurred through a special crossbridge state ("latch bridge") that resulted from myosin dephosphorylation where myosin could react with actin to produce force but either would not cycle or would cycle only slowly thus constituting an internal load. Recently Butler, Siegman & Mooers (1986) presented evidence from experiments with rabbit taenia coli that argued against the latch bridge internal load hypothesis. They suggest that, as appears to be the case in mouse soleus muscle, cross-bridges progressively cycle more slowly with time. At present it is not clear what causes this down regulation of CB activity. We have looked at this phenomenon of increasing smooth muscle economy over time in our own laboratory in the rat anococcygeus muscle. Over a 10 minute period of force maintenance the isometric heat rate, which is initially quite high for a smooth muscle (~ 9 $mW.g^{-1}$) falls 50% or perhaps more and V_{max} as estimated by the slack test technique falls in a similar fashion.

In chemical studies on mouse soleus and EDL muscles Crow and Kushmerick (1982) found that EDL muscles used energy at about 3.4 times the rate of soleus at the commencement of a tetanus but that after about 3s the energy liberation in EDL dropped back to be comparable with that of soleus. In spite of much recent investigation the reason for this decrease in cycle rate is not yet known.

In recent years it has been shown that there is considerable complexity in the cyclic interactions of actin, myosin and the nucleotides and there seems to be considerable experimental evidence

160

that it is possible to uncouple the isometric and isotonic sensors of cross bridge kinetics. It is clear that CB activity can be altered by several of the products of mechanical activity and in particular the rising levels of inorganic phosphate that accompany severe exercise on hypoxia can account for many of the mechanical effects seen in fatigue. In cardiac muscle Kentish & Allen (1986) have argued that acute contractile failure is explicable in terms of changes in intracellular pH and P_i accumulation. Many of these CB effects can be expected to considerably change muscle energy flux both by altering the rate of CB turnover and by modulating the free energy available both for the CB power stroke and perhaps for the ion pumps (Gibbs, 1985) but present experiments suggest that in cardiac muscle, at least, the fall in the free energy of ATP hydrolysis is not the direct cause of reduced force production in hypoxia or ischemia.

Summary

The energetics of muscle contraction can be understood in terms of the major cellular ATPases. The twitch isometric transduction efficiency is relatively constant across muscle types and species. Although many of the factors that alter the shape of the enthalpy:load relation in isotonic twitch contractions have been identified our molecular understanding is unsatisfactory and more studies are needed of mammalian muscles working closer to $37^{o}C$. The thermodynamic efficiency of CB activity seems quite high, probably in excess of 70%. During maintained (tetanic) force there can be greater than a 1000 fold difference in energy usage across muscle types and there are factors that can down regulate CB activity: these factors remain to be fully identified in both skeletal and smooth muscles. The very diversity of muscle types and the different biochemical solutions that have emerged to match energy supply and demand should lead to important insights into the contractile mechanism. The corollary however also applies, it may be dangerous to take results obtained in one muscle type under a particular set of conditions, and extrapolate those findings to muscles in general.

Acknowledgement. The authors are indebted to their colleagues Dr. J.B. Chapman, Dr. W.R. Gibson and Dr. D.S. Loiselle for stimulating collaboration over many years. We are grateful to Ms. L. Hepburn and Mrs. J. Poynton for help in preparing the manuscript. This work is supported by a project grant from the National Health and Medical Research Council of Australia.

References

Alpert, N. R., Mulieri, L.A., and Litten, R. Z., 1979, Functional significance of altered myosin ATPase activity in enlarged hearts, Am. J. Cardiol., 44:947.

Bárány, M., 1967, ATPase activity of myosin correlated with speed of muscle shortening, J. Gen. Physiol., 50:197.

Bozler, E., 1930, The heat production of smooth muscle, J. Physiol., 69:442.

Butler, T. M., Siegman, M. J., and Mooers, S. U., 1986, Slowing of cross-bridge cycling in smooth muscle without evidence of an internal load, Am. J. Physiol., 251: C945.

Crow, M. T., and Kushmerick, M. J., 1982, Chemical energetics of slow - and fast-twitch muscles of the mouse, J. Gen. Physiol., 79:147.

Dillon, P. F., Aksoy, M. O., Driska, S. P., and Murphy, R. A., 1981, Myosin phosphorylation and the crossbridge cycle in arterial smooth muscle. Science 211:495

Fenn, W. O., 1923, A quantitative comparison between the energy liberated and work performed by the isolated sartorius of the frog, J. Physiol., 58:175.

Ford, L. E., and Gilbert, S. H., 1987, The kinetics of heat production in response to active shortening in frog skeletal muscle, J. Physiol., 385:449.

Gibbs, C. L., 1982, Modification of the physiological determinants of cardiac energy expenditure by pharmacological agents, Pharm. & Therap., 18:133.

Gibbs, C. L., 1984, Smooth muscle heat production, in: "Biochemistry of Smooth Muscle", Stephens N. L., ed, CRC Press, Boca Raton, Florida.

Gibbs, C. L., 1985, The cytoplasmic phosphorylation potential. Its possible role in the control of myocardial respiration and cardiac contractility, J. Mol. Cell. Cardiol., 17:727.

Gibbs, C. L., and Chapman, J. B., 1974, The effect of stimulus conditions and temperature upon the energy output of frog and toad sartorii, Am. J. Physiol., 227:964.

Gibbs, C. L., and Chapman, J. B., 1979, Cardiac energetics: Handbook of Physiology, Circulation, Washington D.C., Am. Physiol. Soc, Sect 2, Vol. 1, Chapter 22, 775.

Gibbs, C. L., and Chapman, J. B., 1985, Cardiac mechanics and energetics: chemomechanical transduction in cardiac muscle, Am. J. Physiol., H199.

Gibbs, C. L., and Gibson, W. R., 1970, Energy production in cardiac isotonic contractions, J. Gen. Physiol., 56:732.

Gibbs, C. L., and Gibson, W. R., 1972, The energetics of rat soleus muscle, Am. J. Physiol., 223:864.

Gibbs, C. L., Loiselle, D. S., and Wendt, I. R., 1987, Activation heat in cardiac muscle, J. Physiol., In press.

Hill, A. V., 1964, The variations of total heat production in a twitch with velocity of shortening, Proc. Roy. Soc. B, 159:596.

Homsher, E., 1987, Muscle enthalpy production and its relationship to actomyosin ATPase, Ann. Rev. Physiol., 49:673.

Kentish, J. C., and Allen, D. G., 1986, Is force production in the myocardium directly dependent upon the free energy change of ATP hydrolysis, J. Mol. Cell. Cardiol., 18:879.

Loiselle, D., and Gibbs, C. L., 1979, Species differences in cardiac energetics, Am. J. Physiol., 237:H90.

Loiselle, D. S., Wendt, I. R., and Hoh, J. F. Y., 1982, Energetic consequences of thyroid modulated shifts in ventricular isomyosin distribtuion in the rat, J. Mus. Res. Cell. Motil., 3:5.

Paul, R. J., Gluck, E., and Ruegg, J. C., 1976, Cross bridge ATP utilization in arterial smooth muscle, Pflugers Archiv., 361:297.

Rall, J. A., 1979, Effects of temperature on tension-dependent heat, and activation heat in twitches of frog skeletal muscle, J. Physiol., 291:265.

Rossmanith, G. H., Hoh, J. F. Y., Kirman, A., and Kwan, L. J., 1986, Influence of V_1 and V_3 isomyosins on the mechanical behaviour of rat papillary muscle as studied by pseudo-random binary noise modulated length perturbations, J. Mus. Res. Cell. Motil., 7:307.

Suga, H., Hisano, R., Goto, Y., Yamada, O., and Igashari, Y., 1983, Effect of positive inotropic agents on the relation between oxygen consumption and systolic pressure-volume area as predictor of cardiac oxygen consumption, Am. J. Physiol., 240:H39.

Wendt, I. R., and Gibbs, C. L., 1973, The energetics of rat extensor digitorum longus muscle, Am. J. Physiol., 224:1081.

Woledge, R. C., 1968, The energetics of tortoise muscle, J. Physiol., 197:685.

Woledge, R. C., 1971, Heat production and chemical change in muscle, Proc. Biophys. Molec. Biol., 22:39.

LIVER ORGANELLE CHANGES ON EXPOSURE TO HYPOXIA

I. S. Longmuir, W. F. Betts, and M. Clayton

Department of Biochemistry
North Carolina State University
Raleigh, North Carolina 27695-7622

INTRODUCTION

The effects of severe hypoxia on mammals are well documented. If hypoxia occurs abruptly, death may follow. However, if the arterial oxygen tension falls gradually, the organism may survive. It has become clear that many changes in the tissue occur during such hypoxia. Tissue changes which take place under these conditions may perform an acclimatizing role to assist in the organism's survival. The purpose of this paper is to study changes in hypoxic tissue at the subcellular level that may constitute the mechanism of acclimation.

The nature of oxygen transport has been investigated by many workers. Longmuir and Bourke (1959) have shown that the uptake of oxygen by the liver does not conform to the Warburg model. To resolve this dilemma, Longmuir (1970) hypothesized a tissue oxygen carrier perhaps located in the endoplasmic reticulum (e.r.). In work done by Gold (1969) it was noted that the carrier was more likely to be a fixed site carrier rather than a mobile one. Both these ideas follow a reasonable assumption that the e.r. is continuous with the plasma and mitochondrial membranes and that the fixed site carrier could be the cytochrome P-450 on the e.r. (Longmuir, 1970).

Since previous studies have shown that hypoxic liver tissue shows two distinct sets of changes, an increase in cytochrome P-450 (Longmuir and Pashko, 1976) and various alterations in the e.r. (Betts and Longmuir, 1986), the question arises as to how the changes may be measured accurately and how they may benefit a hypoxic animal. It is the purpose of this paper to explore and evaluate these changes.

MATERIALS AND METHODS

Hypoxic Exposure

ICR Dublin mice aged 7-9 weeks were placed in a warm desiccator evacuated to give a P_{IO2} of 100 mm Hg. A leak was introduced into the chamber in order to maintain the pressure constant and an air flow of at least 1 liter per minute to remove carbon dioxide. The temperature was checked several times during the exposure to insure that the animal did not become hypothermic. The animals were maintained in this hypoxic environment for 4 hours.

Tissue Preparation, Cytochrome P-450 Assay, and Measurement of Endoplasmic
Reticulum

Control and hypoxic mice were sacrificed by cervical dislocation. The
livers were exposed by abdominal incisions and were perfused with warm
normal saline via the portal vein. The livers were then excised and divided,
separating one part for electron microscopic (EM) preparation and the other
for cytochrome P-450 determination.

Livers separated for EM were then placed in 3% glutaraldehyde (GTA) at
4°C for one hour. After the initial fixation the liver tissue was rinsed
several times with buffer and post-fixed with 1% osmium tetroxide. Dehydration was carried out with ethanol at 4°C, and the tissue was then infiltrated
with 1:1 propylene oxide:Medcast epoxy resin solution. Embedding and polym-
erization were carried out utilizing Luft's procedure (1961) in the epoxy
resin mentioned before. After embedding, the tissue was thin sectioned,
placed on grids, and stained. Staining was accomplished by a double staining
method utilizing uranyl acetate in 70% ethanol and lead hydroxide chelated
with citrate. After staining, the grids were rinsed with distilled water,
dried, and allowed to stand overnight before viewing.

It was noted in this study that the centrilobular cells appear to have
a different amount of e.r. than the perilobular cells. Therefore a comparison was carried out of each of these areas separately, between normoxic and
hypoxic liver tissue. Analysis of EM micrographs were performed using
Weibel's method of curved line stereology (1967), studying the rough e.r.
(r.e.r.) to remove uncertainty of identification. The number of intersections between evenly spaced grid lines and r.e.r. were counted and used as
an indicator of e.r. length in each sample. The total number of e.r.
elements were also counted.

A spectrophotometric determination of the concentration of cytochrome
P-450 as described by Pashko (1977) was made on the remaining tissue. The
liver tissue separated for this determination was blotted dry, weighed, and
homogenized with a Potter-Elvehjem homogenizer. The concentration of cytochrome P-450 was determined by taking the difference in spectra between a
reference sodium dithionite ($Na_2S_2O_4$) reduced homogenate and a sample,
untreated homogenate. Both samples had been previously treated with carbon
monoxide (CO) so as to bring hemoglobin and cytochrome a_3 to a reduced
state. Thus, only in the cuvette with sodium dithionite will the cytochrome
P-450 be reduced and react with CO. The absorbance difference between the
sample and the reference cuvette was then measured at 450 and 490 nm.
Utilizing a millimolar extinction coefficient for cytochrome P-450 of 91,
the concentration of P-450 was calculated.

Cell and organelle areas were measured by planimetry. Liver glycogen
was measured by Stu's method (1970).

RESULTS

Elevation of the concentration of cytochrome P-450 followed hypoxia
in accordance with the findings of Longmuir and Pashko (1976).

Table 1 shows the means of the ratios between the number of intersections to the number of e.r. elements. The figures indicate a difference
between perilobular and centrilobular areas in the cell. These numbers
are proportional to the length of the e.r. Although in most samples the
centrilobular area of the cell had more elements and more intersections
than the perilobular area, these ratios were the same. Most importantly,
shown in Table 1 is a comparison of these regions under normoxic and hypoxic

Table 1. Mean Ratios of Intersections to e.r. Elements
(i.e. proportional to the length of each element)[a]

Normoxic Tissue	Hypoxic Tissue
Perilobular	
1.8577 (8)	1.1989 (8)[b]
Centrilobular	
1.8669 (8)	1.3169 (8)[c]

[a]The perilobular were found to be highly significant
($P < 0.01$) between hypoxic and normoxic tissue and significantly different ($P < 0.05$) for the centrilobular
samples. All data were compared via the Student's t
test.
[b]Significantly different than 99% level.
[c]Significantly different than 95% level.

conditions. The perilobular normoxic and hypoxic ratios were found to be
different ($P < 0.01$) utilizing the Student's t test. Analogous ratios for
the centrilobular regions were also found to be significantly different
($P < 0.05$). In both cases the ratio had decreased under hypoxic conditions.

An additional analysis of the number of polyribosomes per 10-millimeter
length (in the EM photograph) of e.r. element were counted. The data concerning these mice appear in Table 2 and contain data comparing the mean
values of these measurements. Again a difference ($P < 0.05$ was seen between
normoxic and hypoxic tissue, with the hypoxic tissue showing a greater number of polyribosomes.

There were no significant changes in cell or organelle areas nor in
the glycogen concentration or the liver weight.

Table 2. Mean of the Number of Polyribosomes Per 10 mm
Length of e.r. in Photographic Plate (length
equal to 0.54 micron of cell)

Normoxic Tissue	Hypoxic Tissue
Set (1)	
7.666 (12)	10.633 (30)[a]
Set (2)	
7.333 (12)	10.407 (30)[a]

[a]Significantly different than 95% level.

DISCUSSION

Kagawa et al. (1981) presented a theoretical model of intracellular oxygen transport based on heterogeneous oxygen diffusivity, in which oxygen moved more rapidly along channels to the site of utilization. This theoretical model, which assumed the channels amounted to 10% of the cell volume, yielded a relationship which fitted approximately experimental data of Longmuir (1975).

To date there has been no complete three-dimensional reconstruction of a high magnification image of the whole cell. Each e.r. picture consists at most of about 0.5% of a cell in our study. Thus it is only possible to arrive at a picture of the whole cell by induction from a series of EM pictures. Those that we have examined in this study are consistent with a model of the e.r. double membrane continuous with both the plasma and mitochondrial membranes.

Since the e.r. constitutes about 4% of the volume of the cell, and oxygen diffuses about six times as fast along membranes as through water (Vanderkooi et al., 1976) and may diffuse through cytosol about half as fast as through water (Ho et al., 1986), a significant proportion of oxygen will travel along the e.r. This proportion may be further increased by the anatomical relations described earlier.

Our results show that acclimation to hypoxia reduces the length of each e.r. element (i.e. the ratio of intersections to total number). Thus the length of a major oxygen pathway is reduced by about the same amount as the capillary P_{O_2}. Thus oxygen transport by this route will be restored to its normoxic level.

The Kagawa model did not contain the hypothesis that all the sites of oxygen utilization were within the fast channels, and when this is factored in, there may be a better fit with the kinetic data.

SUMMARY

Exposure of mice to hypoxia reduces the length of each element of endoplasmic reticulum. This may shorten the path length of oxygen within the cell and have an adaptive advantage.

There were no significant changes in the other organelles.

REFERENCES

Betts, W. F., and Longmuir, I. S., 1986, Effect of hypoxia on the endoplasmic reticulum in mouse liver cells, in: "Advances in Experimental Medicine and Biology," Vol. 200, "Oxygen Transport to Tissue VIII," I. S. Longmuir, ed., Plenum Press, New York.

Gold, H., 1969, Kinetics of facilitated diffusion of oxygen in tissue slices, J. Theoret. Biol., 23:455.

Ho, C. S., Ju, L., and Ho, C. T., 1986, Measuring oxygen diffusion coefficients with polarographic oxygen electrodes. II. Fermentation media, Biotech. Bioeng., 28:1086.

Kagawa, T., Mochizuki, M., Longmuir, I. S., and Koyama, T., 1981, Effect of diffusion heterogeneity on oxygen tension in tissue, in: "Advances in Physiological Sciences," Vol. 25, "Oxygen Transport to Tissue," A. G. B. Kovách, E. Dóra, M. Kessler, and I. A. Silver, eds., Pergamon Press, Akadémiai.

Longmuir, I. S., 1976, The measurement of the fraction of oxygen carried by facilitated diffusion, in: "Oxygen Transport to Tissue II," J. Grote, D. Reneau, and G. Thews, eds., Plenum Press, New York.

Longmuir, I. S., and Bourke, A., 1959, Application of Warburg's equation to tissue slices, Nature, 184:635.

Longmuir, I. S., and Sun, S., 1970, A hypothetical tissue oxygen carrier, Microvascular Res., 2:287.

Longmuir, I. S., and Pashko, L., 1976, The induction of cytochrome P-450 by hypoxia, in: "Oxygen Transport to Tissue II," J. Grote, D. Reneau, and G. Thews, eds., Plenum Press, New York.

Luft, J. H., 1961, Improvements in epoxy resin embedding methods, J. Biophys. Biochem. Cytol., 9:409.

Pashko, L., 1977, Cyclic variations in cytochrome P-450 and associated physiological changes, Ph.D. thesis, North Carolina State University, Raleigh.

Stu, L., Russell, J. C., and Taylor, A. W., 1970, Determination of glycogen in small tissue samples, J. Appl. Physiol., 28:234.

Vanderkooi, J. M., and Callis, J. B., 1976, Pyrene as a probe of lateral diffusion of oxygen in membranes, Biochemistry, 13:4000.

Weibel, E. R., 1976, "Quantitative Methods on Morphology," Springer-Verlag, New York.

CONTROL OF NON-RESPIRATORY METABOLISM BY TISSUE OXYGEN

I. S. Longmuir

Department of Biochemistry
North Carolina State University
Raleigh, North Carolina 27695-7622

For a number of years the experimental observation that extraction of many intracellular enzymes resulted in their inactivation was considered a pure artifact. The mechanism of inactivation was shown to be the S-thiolation of thiol groups necessary for enzymic activity. How then could these enzymes remain active *in situ*? In 1970 Brian Hartley suggested that cells may be totally anoxic. On closer examination, this apparently heretical view had real merit. The major oxygen-consuming enzyme, cytochrome a-a3, can use other electron acceptors, so oxygen might not be the physiological one. Early attempts to measure intracellular oxygen with very small polarographic electrodes appeared to show complete anoxia (Steele, 1960). The findings with larger electrodes could perhaps be criticized on the grounds that they had torn the cell wall, permitting oxygen to leak in. In general, studies with leuco-dyes could also have alternate explanations in that the color changes could be brought about by other electron acceptors. However, the introduction of dyes whose fluorescence is specifically quenched by oxygen in biological systems showed that the anoxia theory was untenable (Longmuir and Knopp, 1972). How then do thiol enzymes retain their activity? In 1976 Chance showed that perfusion of liver with solutions having a very high partial pressure of oxygen resulted in an efflux of oxidized glutathione (GSSG). This suggested a protective role for glutathione (GSH), since this molecule can reactivate dithiol enzymes previously inactivated by oxidation. However, it seemed unlikely that this function would occur continuously, since it violated an important dictum by Chance (1955) on cellular molecular economy.

Recently Ziegler (1985) has proposed the novel hypothesis that the reduction of oxidized enzymes is not a "repair" but a regulatory mechanism. The relation of this mechanism to tissue oxygen is as follows:

$$E\begin{smallmatrix} SH \\ SH \end{smallmatrix} \quad GSSG \quad NADPH$$

$$E\begin{smallmatrix} S \\ I \\ S \end{smallmatrix} \quad 2GSH \quad NADP^{+}$$

where

$$E\begin{smallmatrix} SH \\ SH \end{smallmatrix} \quad \text{is dithiol enzyme}$$

$$E I \begin{smallmatrix} S \\ \\ S \end{smallmatrix} \text{ is S-thiolated enzyme}$$

GSSG is oxidized glutathione

GSH is glutathione

NADPH and $NADP^+$ are reduced and oxidized nicotinamide dinucleotide phosphate, respectively.

The $NADPH/NADP^+$ ratio is related to local P_{O_2}. Thus P_{O_2} will determine the percentage of S-thiolation of various enzymes. (S-Thiolation is the correct chemical description of the oxidation of two thiol groups to disulfide.)

S-Thiolation, in general, inactivates enzymes. However, the inactivation of glycogen phosphorylase phosphatase prevents the inactivation by dephosphorylation of glycogen phosphorylase, and this increases glycogen breakdown to glucose. At the same time, S-thiolation inactivates glycogen synthase.

The effects of this mechanism on glycolysis are complex. Fructose 1,6-bisphosphatase is activated, thus increasing the switch from glycolysis to gluconeogenesis. Hexokinase is inhibited by GSSG, but glucose 6-phosphatase, which must be activated to permit glucose to escape from cells, is activated by GSSG. All these reactions would seem to increase glycolysis. However, pyruvate kinase, the enzyme involved in the last step in glycolysis, is inactivated to some extent by GSSG, principally by raising its K_m. There are conflicting reports on the effect of S-thiolation on other enzymes, so it is not clear whether or not S-thiolation is the mechanism underlying the "Pasteur effect." The activity of a number of other enzymes is increased by S-thiolation.

Glucose 6-phosphate, the controlling enzyme of the hexose monophosphate shunt, behaves like glucose 6-phosphatase in being activated by S-thiolation. Thus both these enzymes may have disulfide in their active centers or a disulfide bridge is required to maintain the correct conformation. Both of course react with the same substrate. Thus the effect of raised oxygen tensions is to activate the shunt.

An enzyme important in cholesterol synthesis, 3-hydroxy-3-methylglutaryl-CoA reductase, is inhibited by GSSG. It is too early to say if this might lead to a new rational treatment for hypercholesterolemia. Melatonin is synthesized at night, and the two enzymes serotonin-N-acetyltransferase and acetyl-CoA hydrolase involved in its synthesis are inactivated by GSSG, so it is tempting to suggest nocturnal histotoxic hypoxia may be the cause of the diurnal variation (Stupfel et al., 1974). A number of other enzymes such as adenylate cyclase and leucocyte collagenase are inhibited or activated, respectively, by GSSG.

The role of the GSH/GSSG ratio *in vivo* in metabolic regulation is difficult to quantify, since there will be competition between enzymes and on account of intracellular heterogeneity of oxygen distribution, there will be a corresponding heterogeneity of this ratio. Overall cell ratios are higher (300:1) in a hypoxic tissue such as liver compared with muscle (100:1).

Intracellular Oxygen Heterogeneity

In 1980 Benson et al. studied the heterogenous distribution of oxygen within a single mouse liver cell by observing the oxygen quenching of the fluorescence of pyrenebutyric acid (PBA). A liver cell stained with PBA was allowed to flatten actively on a cover slip until its diameter rose to 60 μm, giving it a thickness of about 1 μm. It was examined under an oil immersion 54x objective in a video microscope. The cell was illuminated with UV light 345 nm and the fluorescence was observed at 420 nm. The cell was exposed to various values of P_{O_2} and the image digitized at 1 μm^2 pixels.

The relationship between P_{O_2} and oxygen tension is given by the Stern Volmer relationship,

$$\frac{F_o}{F_x} = 1 + \alpha Kq \ P_{O_2}$$

where

F_o is the unquenched fluorescence

F_x is the fluorescence at $P_{O_2} = x$

α is the Bunsen solubility coefficient

Kq is the quenching constant.

At each pixel the value of αKq was calculated by linear regression. The distribution of αKq values is given in the figure. They range from essentially zero to a few pixels, not shown, with a value five times that in water. Since the quenching efficiency is about 100%, the high values can only be due to a high value of α. α in lipid is about five times that in water, so the pixels with a high αKq may contain a high level of lipid. When we doubled the lipid content of liver cells by feeding carbon tetrachloride, the number of pixels with a high value doubled. In addition, adipocytes treated the same way as the liver cells showed a majority of the pixels with very high values. There seemed to be an obvious explanation for the low or zero values—a reduction of Kq to zero by steric hindrance. Vaughn and Weber (1970) showed that the binding of PBA to bovine serum albumin reduces Kq to very nearly zero. Since mouse liver cells contain significant amounts of albumin, this protein might be responsible for low Kq values if it behaved the same way as BSA. This hypothesis was tested by measuring the Kq of various cell organelles and cytosol. The Kq of the

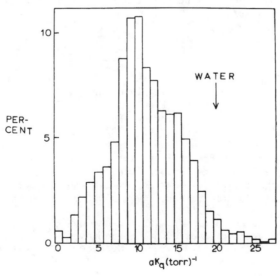

Histogram of αKq of 1 μm^2 pixels in a flat liver cell.

organelles was less than in water (Longmuir and Knopp, 1976) but no lower than 45%. It seemed more likely that albumin would be found in cytosol, and a systematic study of cytosol binding to PBA and its effect on Kq was undertaken (Weinbrecht, 1984). Cytosol stained with PBA was prepared by a number of methods. Cells were allowed to take up PBA before the extraction of cytosol by freezing, thawing, and centrifugation; PBA was added directly to cytosol and indirectly by dialysis. In each case the mole ratio PBA: albumin was kept below two to ensure essentially all the PBA was bound. The value of Kq in each preparation was almost identical, 30% that in water. The possibility of a low value of α was excluded by direct measurement (Knopp et al., 1983). The only other explanation for the failure of oxygen to quench fluorescence in some pixels is the exclusion of oxygen.

Although our measurements were made on non-respiring, starved cells—that is to say, cells which in bulk showed less than 1% of the respiratory activity of unstarved cells—it is quite possible that a few cells were actively respiring. The process of "flattening" on a cover slip is an active one prevented by respiratory inhibitors. In our preparation we measured the first cell to stop spreading, and may therefore have selected one of the few respiring cells capable of producing local anoxia.

Much of the volume of liver cells is free of respiratory enzymes, and it was assumed oxygen in this domain was unnecessary and irrelevant. However, the local oxygen tension determines the local GSH/GSSG ratio which, in turn, determines the activity of a number of enzymes. Here is a new non-respiratory regulatory role for oxygen.

SUMMARY

Many intracellular enzymes are activated or inactivated by S-thiolation. The extent of this depends on the local oxygen tension. Thus oxygen should not be considered as an enzyme poison for certain enzymes but as a regulatory of metabolic activity.

ACKNOWLEDGEMENT

I wish to thank Dr. H. Robert Horton for many helpful discussions on S-thiolation of enzymes.

REFERENCES

Benson, D. M., Knopp, J. A., and Longmuir, I. S., 1980, Intracellular oxygen measurements of mouse liver cells using quantitative fluorescence video microscopy, Biochim. Biophys. Acta, 591:187.

Chance, B., 1955, Ex cathedra statement, Faraday Society Discussion, Oxford.

Chance, B., Nishiki, K., and Oshino, N., 1976, Glutathione release, an indicator of hyperoxic stress, in: "Oxygen and Physiological Function," F. F. Jöbsis, ed., Professional Information Library, Dallas.

Hartley, B. S., 1970, Private communication.

Knopp, J. A., Longmuir, I. S., Pittman, J. L., and Knoeber, L., 1983, Microheterogeneity of fluorescence quenching constants in liver cells, Federation Proc., 42:2169.

Longmuir, I. S., and Knopp, J. A., 1972, A new method of measuring intracellular oxygen by fluorescence quenching, Federation Proc., 31:365.

Longmuir, I. S., and Knopp, J. A., 1976, Measurement of tissue oxygen with a fluorescent probe, J. Appl. Physiol., 41:598.

Steele, R. E., 1960, The polarographic measurement of oxygen in biological systems, J. Polarograph. Soc. IV:2.

Stupfel, M., Moutet, J. P., and Magnier, M., 1974, Periodes d'eclairement et de mortalite de la souris SPF male et femelle en hypoxia aigue, J. Physiol. (Paris), 69:209A.

Vaughn, W. M., and Weber, G., 1970, Oxygen quenching of pyrenebutyric acid fluorescence in water, Biochemistry, 9:464.

Weinbrecht, P., 1984, Interactions of pyrenebutyric acid with cellular components, M.S. thesis, North Carolina State University, Raleigh.

Ziegler, D. M., 1985, Role of reversible oxidation-reduction of enzyme thiols-disulfides in metabolic regulation, Ann. Rev. Biochem., 54: 305.

METHODS
AND
INSTRUMENTATION

BLOOD GAS ANALYSIS USING FLUORESCENCE AND ABSORPTION INDICATORS IN OPTICAL

SENSORS (OPTODES) WITH INTEGRATED EXCITATION AND FLUORESCENCE DETECTION ON

SEMICONDUCTOR BASIS

N. Opitz and D.W. Lübbers

Max-Planck-Institut für Systemphysiologie
Rheinlanddamm 201, 4600 Dortmund, FRG

In order to develop "Intelligent Fluorescence Optical Sensors"
(INFOS) for better stability and an improved microprocessor compatibility
we have combined the optical sensors, the so-called optodes (1, 2) with
integrated excitation source and fluorescence detection device on semiconductor basis. These INFOS can be applied for the analysis of various
substances such as oxygen, carbon dioxide, and hydrogen ion activities
(3, 4).

For the excitation of the fluorescent indicator two different
luminescent sources are used: light emitting diodes (LED's) and electro-
luminescent foils (ELF's). Both excitation sources have the advantage of
emitting "cold light", which brings about an improved thermostability of
the sensors. Moreover, LED's as light sources possess a high degree of
long-term stability and ELF's are useful for the irradiation of large
sensor areas resulting in increased signal-to-noise ratios (SNR's).
Fluorescence detection has been performed in both cases with photodiodes,
which, in combination with the ELF's, were arranged as diode arrays.

Fig. 1 schematically depicts an arrangement consisting of an optode
device connected to a blue light emitting diode (type LD 76, Knitter,
max. emission at 490 nm, half bandwidth $\pm$ 35 nm) and two photodiodes for
light detection (type BPX 91 B, Siemens, blue enhanced sensitivity).
Owing to the compact arrangement of this integrated sensor unit SNR is
improved due to the relatively broad acceptance angle of the fluorescence
emission by the detecting diodes as well as broad band excitation and
fluorescence detection. The fluorescence transillumination mode can
always be advantageously applied for analysis of gas mixtures and low
absorbing fluids, provided that fluorescence scattering by cells or
particles of the medium to be measured are negligible (5). With such a
device SNR's of up to about 5.000 enable the application of a number of
fluorescence indicators with a low oxygen sensitivity, but which, contrary
to indicators with a higher O_2 sensitivity, can be excited within the
visible spectral range so that simple and reliable excitation sources
like LED's and ELF's can be applied. Moreover, it seems noteworthy that a
low oxygen sensitivity of an indicator is often combined with an increased
long-term stability, because of an enhanced photostability of the fluorescence molecules (6). As an example for a low sensitive O_2 indicator we
used a benzopyrane derivative of BASF (Thermoplast-F-Rot 344, λ_{em}=570 nm).

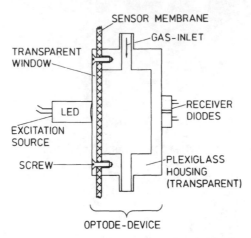

Fig. 1: Schematical representation of an optode device connected to a blue light emitting diode as excitation source and two photodiodes for fluorescence detection.

The absorption of this indicator overlaps well with the emission of blue and green LED's (450-530 nm). Fig. 2 shows an original recording of a calibration trace of this indicator using the sensor unit depicted in Fig. 1 during stepwise changes of the oxygen content within the measuring chamber of the optode. Due to the low oxygen sensitivity of the indicator corresponding to an overall quenching constant of $K=5,35 \times 10^{-5}$ $Torr^{-1}$, the calibration curve can be linearized (Fig. 3a) by plotting the relative fluorescence intensities, S, versus pO_2 : $S (pO_2) = S_O \cdot (1-K \cdot pO_2)$, with S_O = fluorescence intensity at zero pO_2 and K = overall quenching constant, although the fundamental process of fluorescence quenching by oxyen follows Stern-Volmer's hyperbolic equation (7). Therefore, in

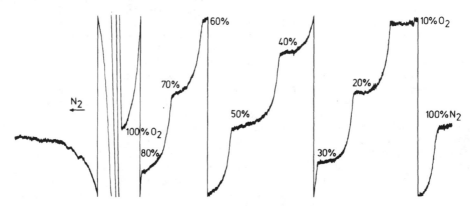

Fig. 2: Calibration trace of the O_2 indicator Thermoplast-F-Rot 344 using the sensor unit depicted in Fig. 1 (time axis from right to left).

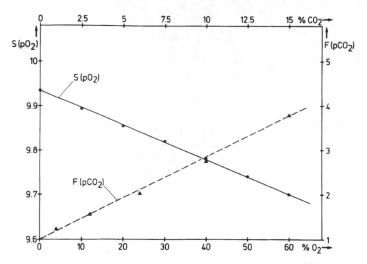

Fig. 3a: Linear calibration curve of the integrated fluorescence optical
O_2 sensor according to S $(pO_2) = S_o \cdot (1-K \cdot pO_2)$ using
the indicator Thermoplast-F-Rot 344 with a small overall quenching constant of $K=5,35x10^{-5}$ $Torr^{-1}$.

3b: Linearized calibration curve of the CO_2 sensor depicted in
Fig. 4 using phenol red dissolved in 10^{-2} M $NaHCO_3$ and
emulsified into a silicone rubber membrane.

contrast to O_2 indicators with a higher O_2 sensitivity such as pyrenebutyric acid (8), the precision of oxygen measurement within the complete
measuring range (0 - 100 % O_2) is independent of the actual pO_2 and
amounts in this case to approximately $\pm$ 2 Torr. Finally, as an advantageous consequence of the linear sensor characteristic (9), the kinetics of
back-and-forth reactions are symmetrical. Thus, response times for both
directions are identical, e.g. about 20 s till 90 % of final value with
measurements in gas mixtures. The complete measuring assembly showed an
overall drift of only 0.06 % per hour. The SNR amounted in this case to
approx. 5.000.

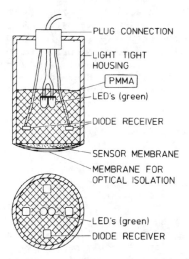

Fig. 4: Excitation and detection integrated optical sensor
embedded into polymethylmethacrylate (PMMA).

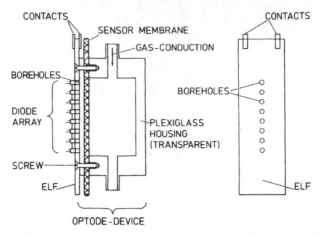

Fig. 5: Schematic representation of an optode with integrated electro-
luminescent foil (ELF) as excitation source and a photodiode
array as fluorescence detector.

For an improved optical coupling between sensor layers and miniaturized potometers LED's and photodiode detectors can be embedded into
different transparent polymers, i.e. polymethylmethacrylate (PMMA),
epoxy- or acrylic glass. Fig. 4 shows such an integrated sensor, where two
LED's (green) and four photodiodes have been enclosed into PMMA. Facing
the photodetectors and LED's, which in this case are both positioned at
the same side, the sensor membrane is coupled onto the PMMA surface via
an additional, optically isolating membrane, which prevents optical
interferences. The membrane system may be mechanically stabilized by a
built-in metallic network. This integrated optical sensor can be additionally equipped with electronical elements for impedance conversion and
signal amplification as well as sensor-specific signal correction, e.g.
for correction of temperature dependence. Fig. 3b represents a linearized
CO_2 calibration curve of the sensor (Fig. 4) performed with the pH-
sensitive absorption indicator phenol red, dissolved in 10^{-2} M $NaHCO_3$,
and emulsified into a silicone rubber membrane ($\sim 70 \mu m$) in order to
suppress the direct pH sensitivity, since silicone rubber membranes are
impermeable to ions. According to the Henderson-Hasselbalch equation the
steepness of the linearized calibration curve is mainly determined by the
bicarbonate concentration of the indicator buffer solution and the pK
value of the pH-sensitive weak electrolyte (5). Owing to a SNR of about
1.400 the precision of measurements within a pCO_2 range of 0-80 Torr is
better than $\pm$ 1 Torr.

If it is possible to use sensors with larger areas of sensor membranes, the ELF's are advantageously applied in connection with several
photodiode detectors, since large membrane areas result in increased
SNR's (up to about 2.000). Fig. 5 shows an arrangement where a blue
emitting ELF (Data Modul) has been prepared with eight concentric bore-
holes (diameter $\sim$ 1.2 mm) to obtain suitable windows for the radiation
receivers. A photodiode array, directly coupled onto the ELF, fits with
its eight photosensitive receiver layers into the window of the foil,
which is closely connected to a fluorescence optical sensor chamber. The
ELF (max. emission 460 nm) is operated at 400 Hz and 115 V with the aid

of an inverter (Data Modul). With such a compact device similar calibration curves as depicted in Fig. 3 have been recorded by means of the above mentioned indicator layers.

REFERENCES

1. D. W. Lübbers and N. Opitz, Optical fluorescence sensors for continuous measurement of chemical concentrations in biological systems, Sensors and Actuators 4:641-654 (1983).
2. N. Opitz and D. W. Lübbers, Electrochromic dyes, enzyme reactions and antigen-antibody interactions in fluorescence optic sensor (optode) technology, TALANTA (in press, 1987).
3. J. L. Gehrich, D. W. Lübbers, N. Opitz, D. R. Hansmann, W. W. Miller, J. K. Tusa and M. Yafuso, Optical fluorescence and its application to an intravascular gas monitoring system. IEEE Trans. Biomed. Eng. 33:117-231 (1986).
4. D.W. Lübbers, J. Gehrich and N. Opitz, Fiber optics coupled fluorescence sensors for continuous monitoring of blood gases in the extracorporeal circuit, in: "Life Support Systems, Extracorporeal Gas Exchange, Design and Techniques," A. Lautier, J. P. Gille, eds., Saunders, London (1986), pp. 94-108.
5. N. Opitz and D. W. Lübbers, Compact CO_2 gas analyzer with favourable singnal-to-noise ratio and resolution using special fluorescence sensors (optodes) illuminated by blue LED's, in: "Oxygen Transport to Tissue VI," Adv. Exp. Med. & Biol., Vol. 180, D. Bruley, H. I. Bicher, D. Reneau, eds., Plenum Press, New York-London (1984), pp. 757-762.
6. A. Rademacher, S. Märkle and H. Langhals, Lösliche Perylen-Fluoreszenzfarbstoffe mit hoher Photostabilität, Chem. Ber. 115:2927-2934 (1982).
7. O. Stern and M. Volmer, Über die Abklingzeit der Fluoreszenz, Z. Phys. 20:183-188 (1919).
8. N. Opitz and D.W. Lübbers, Increased resolution power in Po_2 analysis at lower Po_2 levels via sensitivity enhanced optical Po_2 sensors (Po_2 optodes) using fluorescence dyes, in: "Oxygen Transport to Tissue VI," Adv. Exp. Med. & Biol., Vol. 180, D. Bruley, H. I. Bicher, D. Reneau, eds., Plenum Press, New York-London, (1984) pp. 261-267.
9. N. Opitz, Untersuchungen zur Kinetik und Einstellzeit fluoreszenzoptischer Sensoren (Optoden) für die Blutgasanalyse, Biomed. Tech. 30:96-97 (1985).

METHODS OF QUANTITATING CEREBRAL NEAR INFRARED

SPECTROSCOPY DATA

M. Cope, D.T. Delpy, E.O.R. Reynolds*, S. Wray+, J. Wyatt*,
and P. van der Zee

Department of Medical Physics and Bio-Engineering
*Department of Paediatrics, +Department of Physiology
University College London, Shropshire House
11-20 Capper Street, London WC1E 6JA

INTRODUCTION

Non invasive infrared spectroscopy is a well established
technique for monitoring changes in the oxygenation status of tissues
(1). The technique has in particular been successfully employed to
monitor changes in cerebral blood and tissue oxygenation by observing the
absorption of haemoglobin and cytochrome aa3 respectively. Because of
the highly light scattering nature of the tissues studied, it has
normally not been possible to quantitate the observed changes.

We have been applying near infrared spectroscopy (nirs)
clinically to study the oxygenation of the brain of newborn infants in
the neonatal intensive care unit at UCH (2). These studies have been
performed in the transillumination mode with purpose built equipment (3).
By operating in transillumination mode we are able to define a minimum
path length for the light traversing the tissues. This permits upper
limits to be placed on any calculated values of nir derived data. We
propose that in certain circumstances, absolute quantitation of nir data
is possible. This is achieved by correlating measured variations in the
nir absorption with data from other physiological sensors during a small
known disturbance from a previous baseline.

EXPERIMENTAL BACKGROUND

The instrument used to monitor NIR absorption changes in the
neonate employs four semiconductor laser diodes as light sources
(wavelengths 775, 813, 847 and 904 nm). Light from the diodes is guided
to the head by a fibre optic bundle, and attached to a site equidistant
from the anterior fontanelle and the external auditory meatus. Light
emerging at the other side of the head is collected in a similar optode,
and lead to a photomultiplier tube detector. Changes in light absorption
at three of the wavelengths are converted to equivalent changes in
oxygenated haemoglobin (HbO_2), deoxygenated haemoglobin (HbR) and
oxygenated cytochrome aa3 (cytaa3). Total Haemoglobin (HbT) is obtained

from the sum of HbO$_2$ and HbR. This calculation is performed by a linear summation of absorption changes (in optical densities) multiplied by the following factors at each wavelength:

Multiplying Factors

	775 nm	847 nm	904 nm
HbO$_2$	-927	-618	+1831
HbR	+1433	-1744	+709
cytaa3	-84	+1339	-1107

The resulting concentration changes are expressed in µmol/l multiplied by optical path length in centimetres. The multiplying factors for haemoglobin were derived from measured absorption spectra for haemoglobin solutions. Those for cytochrome aa3 were obtained from in vivo spectra of (oxygenated-reduced) cytochrome aa3 (4). The assumption made in these calculations is that changes in concentration of the chromophores in the tissues produce a logarithmic change in transmitted light intensity. This relationship applies in clear (ie non scattering) absorbing media, but experimental data showing that it can be applied in the case of tissue is limited. Experimental studies of the attenuation of transmitted light in brain tissue (5,6) do however indicate a logarithmic fall in intensity with distance when measured through considerable thickness of tissue. A Monte Carlo simulation of light transport in brain tissue (7) also predicts that for transmitted light, a logarithmic relationship is obtained over a considerable range of absorption coefficient, and that this applies also for tissues having a large range of scattering coefficient. (It is interesting to note that the model predicts a non logarithmic response for the reflected light intensity over the equivalent range of absorption and scattering co-efficients. If these predictions are correct, then a simple linear addition cannot be employed when performing reflection mode nir spectroscopy).

To verify the predictions, a simple experiment has been performed using the nir instrument. The optodes were attached to opposite faces of a square transparent plastic container (5 cm side length), and the container filled with physiological saline. Aluminium Oxide particles (7 um diameter) were then added to produce a scattering medium. A magnetic stirrer was used to prevent particle settling. Sufficient scatterer was

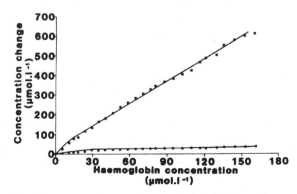

Fig. 1. Changes in calculated HbT and cytaa3 concentration in a scattering phantom as a function of added haemoglobin concentration.

added to produce an attenuation in transmitted intensity of approximately 5 OD (the attenuation of neonatal brain is approximately 1 OD/cm). Known aliquots of washed human red blood cells were then sequentially added to the container, and the resulting absorption changes recorded. These absorption variations were then converted to equivalent changes in total haemoglobin and cytochrome aa3 using the previously mentioned relationship. The results of this experiment are shown in Figure 1. It can be seen that the resulting relationship between HbT and true haematocrit, is in fact slightly non linear over this large range of haemoglobin concentration. This range is however greater than that observed in the human neonate. Over the physiological range of tissue haemoglobin concentration, the assumption of linear addition of absorbencies is justifiable within experimental error. The calculated cytochrome signal in this experiment should have remained constant, since there was no cytochrome present in the solution. It can be seen however that this signal did change slightly with haemoglobin concentration. The derived cytochrome signal was particularly sensitive to the initial rise in haemoglobin concentration, and virtually unaffected by later increases. This probably reflects the initial change in "effective" optical path length with wavelength, as the added absorber selectively attenuates the more highly scattered components of transmitted light.

This effect is again predicted by the Monte Carlo simulation (7) where maximum deviation from linearity is observed in circumstances of high scattering and low absorption. In practice, brain tissue always contains some absorber (haemoglobin or cytochrome), and therefore this initial part of the experimental curve does not represent a practical physiological situation. It is therefore justifiable to say that over the normal range of physiological variation, changes in blood volume will not affect the derived cytochrome signal. By repeating this experiment using a scattering medium which more closely mimics brain tissue, we hope to quantitate the overall non linearity of this response, and incorporate these results in the algorithm.

QUANTITATION OF CEREBRAL MIXED VENOUS SATURATION

Changes in HbO_2, HbR and HbT can be observed by tilting of an infant so that the head is raised or lowered with respect to the heart. Such a manoeuvre alters by a small amount, the hydrostatic pressure in both the cerebral arterial and venous compartments. Since the pressure change is small in comparison with mean arterial pressure, and the compliance of the blood vessels in the arterial system is low, it may reasonably be assumed that the blood volume change has occurred in the cerebral venous compartment. If this is so, then the mixed cerebral venous saturation (SvO_2) can be derived from the formula:

$$\% \ SvO_2 = \frac{\Delta HbO_2}{\Delta(HbO_2 + HbR)} \times 100$$

This calculation requires no assumption to be made regarding the mean optical path length through the tissues. The calculation is only valid if the cerebral blood flow and oxygen consumption remain constant during the manoeuvre. This is likely to be so if $PaCO_2$ and SaO_2 are stable during the tilting procedure. These conditions can be met by monitoring $PaCO_2$ with a transcutaneous electrode and SaO_2 with a pulse oximeter.

In practice, we have found the small tilt involved in the procedure (approximately 10°) does not disturb the infant, who will often sleep throughout the test. Figure 2 is a recording made during a head tilt on a full term infant with cystic encephalomalacia. In this infant

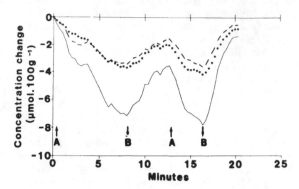

Fig. 2. Changes in HbO_2, HbR and HbT during tilting of
an infant 10° head up at A, back to level at B
and repeated. The infant was 9 weeks old
having been born at term and had developed
encephalomalacia following severe birth
asphyxia.

the mean saturation was 66%. In a normal preterm infant, of 28 weeks
gestation, studied at five weeks of age, the average SvO_2 was 38%.

QUANTITATION OF CEREBRAL BLOOD FLOW CHANGES

If arterial oxygen saturation is maintained at approximately
100%, then changes in HbR can be related to changes in cerebral blood
flow. This assumes that cerebral oxygen consumption remains constant,
which is likely to be true in normal preterm infants. In one infant
studied inthis way, cerebral blood flow was seen to increase linearly in
response to changes in $PaCO_2$. If SaO_2 equals 100%, and SvO_2 is assumed to
be 50%, then assuming arterial and venous compartments to be of equal
volume, HbR constitutes 25% of HbT. Under these circumstances, the blood
flow increases by approximately 9% per kPa. Figure 3 illustrates such a
change in blood flow.

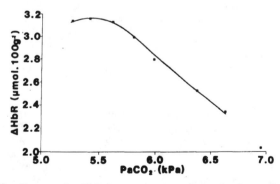

Fig. 3. Change in HbR in response to variation of $PaCO_2$
in a 2 day old infant born at 36 weeks of
gestation with congenital myopathy. The SaO_2
was in the range 98-99% during the manoeuvre.

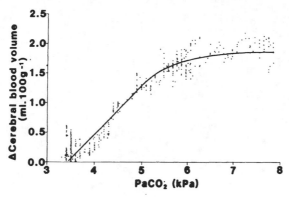

Fig. 4. Changes in cerebral blood volume with
increasing $PaCO_2$ in a 3 day old infant showing
a linear increase up to 5.5 kPa but no increase
thereafter. He had been born at 39 weeks of
gestation with Listeria septicaemia but had no
evidence of cerebral abnormality.

QUANTITATION OF CEREBRAL BLOOD VOLUME

The sum of the HbO_2 and HbR signals will provide information on
the instantaneous changes in cerebral blood volume. The cerebral blood
volume changes significantly in normal infants in response to changes in
$PaCO_2$. However in several infants with cerebral oedema, we have observed
a very limited response to changes in $PaCO_2$. Figure 4 shows the changes
in blood volume in an infant with Listeria Septicaemia, but no evidence
of cerebral abnormality.

It is possible to quantitate cerebral volume in some
circumstances by using additional information available on arterial
saturation. If the cerebral blood flow, blood volume and oxygen
consumption remain constant, then if SaO_2 is varied slightly, the same
change in SaO_2 will occur in all blood compartments in the brain. A
linear relationship will then be obtained between (HbO_2-HbR) and
percentage SaO_2 (Figure 5). The cerebral blood volume (CBV) can be
calculated from the slope of this relationship using the formula:

$$CBV \frac{\Delta(HbO_2 - HbR)}{\Delta \% SaO_2} \times 50$$

Applying this relationship to clinical data on neonates results
in volume estimates in the range of 3.0 to 17.0 ml/100g. (Brain tissue
specific gravity is assumed to be 1.05 g/ml).

DISCUSSION

The algorithm used to convert nir absorption changes to
concentration changes of HbO_2, HbR, etc, produces results expressed in
terms of umol/l tissue multiplied by optical path length. When applying
the quantitation techniques mentioned in this paper to clinical data, it
is necessary to choose a value for this path length (the exception to
this is the quantitation of cerebral mixed venous saturation, where no
path length estimate is required). By operating in transmission mode, we

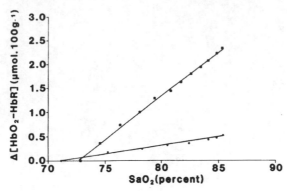

Fig. 5. Relationship between SaO_2 and [HbO_2 - HbR] at
two different values of $PaCO_2$ ■ = 7.8 kPa,
● = 3.5 kPa. Same infant as Figure 4.

are able to place a lower limit on the optical path, (ie the optode
spacing), and in this paper, that lower limit has been used. There is
however some measured data on which to base an estimate of the upper
limit for the path length, although as previously mentioned, it could
also be calculated from the Monte Carlo model (7). We can estimate an
upper limit by comparing the results obtained above for cerebral blood
volume with normal data for human volunteers measured by alternative
techniques. Although no data on baseline cerebral blood volume in human
neonates has been published, values in adult volunteers obtained by
various techniques have been within the range of 3 to 6 ml/100g (8,9).
The values derived from nir data on normal neonates, assuming a path
length equal to the optode spacing are in the range of 3.0-17.0 ml/100g.
This implies an average optical path of 1.5 to 5 times the optode
spacing, with a value of approximately 2.0 being most likely. If this is
true, then one could apply this path length to the quantitation of data
on cytochrome aa3.

CONCLUSION

Quantitation of near infrared spectroscopy data is normally not
possible due to uncertainties over the optical path length through the
tissues. We have shown that in some circumstances, quantitation is
possible by comparing nir signal changes with those obtained from
separate independent monitors of physiological variables. By comparing
the results obtained in this way on normal patients with the expected
normal values, it is possible to estimate the average optical path length
through brain tissue.

ACKNOWLEDGMENTS

This work was supported by grants from the Medical Research
Council, Science and Engineering Research Council, the Wolfson Foundation
and Hamamatsu Photonics KK.

REFERENCES

1. F.F. Jobsis, Noninvasive, infrared monitoring of cerebral and myocardial oxygen deficiency and circulatory parameters, Science, 198:1264-1267 (1977).
2. J.S. Wyatt, M. Cope, D.T. Delpy, S. Wray, E.O.R. Reynolds, Quantitation of cerebral oxygenation and haemodynamics in sick newborn infants by near infra red spectrophotometry, Lancet, 8515:1063-1066 (1986).
3. M. Cope, D.T. Delpy, A system for long term cerebral blood and tissue oxygenation measurement on newborn infants by near infra red transillumination, in: "Optical monitoring of metabolism in vivo", F.F. Jobsis, ed., Plenum Press (in press).
4. S. Wray, M. Cope, D.T. Delpy, J.S. Wyatt, E.O.R. Reynolds, Characterisation of the near infrared absorption spectra of cytochrome aa3 and haemoglobin for the noninvasive monitoring of cerebral oxygenation, Biochem. Biophys. Acta (submitted).
5. L.O. Svaasand, R. Ellingsen, Optical properties of human brain, Photochem. & Photobiol., 38:3:293-299 (1983).
6. L.O. Svaasand, R. Ellingsen, Optical penetration in human intracranial tumours, Photochem. & Photobiol., 41:1:73-76 (1985).
7. P. van der Zee, D.T. Delpy, Simulation of the point spread function for light in tissue by a Monte Carlo method, in: "Oxygen transport to tissue", vol. IX, I.A. Silver, ed., Plenum Press, New York (1987).
8. G. Ladurner, E. Zilkha, L.D. Iliff, G.H. Du Boulay, J. Marshall, Measurement of regional cerebral blood volume by computerised axial tomography, J. Neurol., Neurosurg. & Psychiatry, 39:152-158 (1976).
9. F. Sakai, K. Nakazawa, Y. Tazaki, K. Ishii, H. Hiro, H. Igarishi, T. Kanda, Regional cerebral blood volume and haematocrit measured in normal human volunteers by single photon emission computed tomography, J. Cereb. Blood Flow & Met., 5:207-213 (1985).

COMPUTED POINT SPREAD FUNCTIONS FOR LIGHT IN TISSUE

USING A MEASURED VOLUME SCATTERING FUNCTION

P. van der Zee, and D. T. Delpy

Department of Medical Physics and Bioengineering
University College London, Shropshire House
11-20 Capper Street, London WC1E 6JA

INTRODUCTION

Optical techniques are increasingly being used in the field of medicine in areas as diverse as surgery (for cutting and coagulation), cancer treatment (through photoradiation therapy) and blood flow monitoring (by laser doppler measurements). At University College, we are using the technique of near infrared spectroscopy (nirs) to monitor changes in cerebral blood and tissue oxygenation in newborn infants (1), and are investigating methods of optical imaging across the head (2). In all these applications, a detailed knowledge of light transport in tissue is required. For spectroscopy studies, in order to quantitate data, one needs to know the effective photon pathlength through the tissue, and a knowledge of the path also allows one to calculate the volume of tissue from which results are being obtained. In the case of imaging through tissue, data is required on the point spread function (PSF) for the tissue, both for the prediction of the image quality that could be obtained using various imaging schemes, and for use in image enhancement and reconstruction computations.

In order to provide the above data, we have developed a Monte Carlo program to model light transport in brain tissue. This model requires as its input, data on the scattering and absorption properties of brain tissue. Some data on absorption properties have been published in the literature (3,4), but few results are available on its scattering properties. Previously therefore we have used a calculated volume scattering function (VSF) derived using Mie scattering theory and applied to a range of cell sizes (5). We have now measured the VSF for adult rat brain, and incorporated this into the model. The model has been further improved by the inclusion of specular reflection and refraction at the tissue boundaries, and by the addition of the reflected PSF to the calculated parameters. The model has been used to simulate the PSF for light transmission through, and reflection from, a homogeneous slice of brain tissue. A general formula has also been derived to describe the transmitted PSF as a function of tissue scattering and absorption coefficients and thickness.

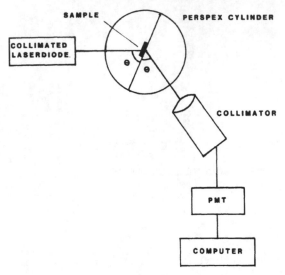

Figure 1.   Schematic diagram of system used to measure the volume scattering function (VSF).

MEASUREMENT OF BRAIN TISSUE VOLUME SCATTERING FUNCTION

The VSF describes the scattered light intensity as a function of scattering angle <u>for a single scattering event</u>. In order to measure this function accurately one must therefore ensure that multiple scattering does not occur within the tissue sample. This requires the use of very thin tissue samples, a consequence of which is extremely low light levels in the scattered beam. The system used to measure the VSF is shown in Figure 1. This consists of a goniometer with an angular resolution of approximately one minute of arc. The sample is held in a 65 mm diameter

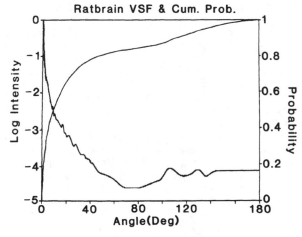

Figure 2.   Averaged VSF and cumulative probability for adult rat brain tissue at 783 mm.

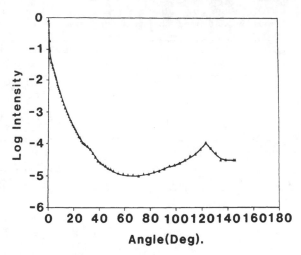

Figure 3. Averaged VSF for human red blood cells at
783nm. The sharp peak at 120° is an
instrumental artifact.

clear acrylic cylinder, which is manufactured from two halves, which are
separated by a 12.5 mm thick spacer. The spacer has a 10 mm hole in the
middle to contain the brain slice. The light source used was a 783nm
collimated laser diode, which produced a beam of 1mm diameter. A
photomultiplier was used to detect the light, its output being fed to a
computerised data collection system. With no sample in the cylinder, the
half width of the system response was 0.5°. Scattering measurements were
made on nine samples of brain tissue taken from freshly killed adult
wistar rats. Thin samples were taken from various areas of the brain.
and the results then combined to produce an average VSF for brain tissue.
This result is shown in Figure 2. Data for scattering angles greater
than 150° were not obtainable with this system, so the value at 150° has
been used. As expected, the VSF is strongly forward peaked. Figure 2
also shows the cumulative probability for the scattering angle (θ). This
is defined as:

$$P(\theta) = \frac{{}_0\int^{\theta} I(\theta) \sin\theta \, d\theta}{{}_0\int^{\pi} I(\theta) \sin\theta \, d\theta}$$

The cumulative probability is required by the Monte Carlo model to
generate the scattering angle at each interaction (6).

The system illustrated in Figure 1 has also been used to measure the
VSF for human blood. To do this, a spacer of 0.5 mm thickness was used,
and diluted blood with a haematocrit of 0.8%, was circulated through the
resulting sample cell using a peristaltic pump. The results of these
measurements can be seen in Figure 3.

THE MODEL

Some details of the Monte Carlo model have been discussed
previously (5). In the simulation, the path of individual photons is

Table 1. The range of absorption and scattering coefficient used in the Monte Carlo model.

Absorption Coefficient (mm^{-1})	Scattering Coefficient (mm^{-1})
1.0	4
0.4	2
0.167	1
0.105	0.5
0.077	0.333
0.0606	0.25
0.05	0.20
0.02	0.167

followed as they enter a homogeneous slice of tissue, perpendicular to the surface. Scattering is assumed to occur at discrete centres, and at each interaction, a new scattering length and scattering angle are determined by random numbers and the cumulative probabilities for scattering length and scattering angle. (The latter determined from the measured VSF for rat brain). Absorption is assumed to take place uniformly along the photon path. The photons are followed until they exit from the front or rear surface where specular reflection and refraction are taken into account. Data is stored on the photons exit coordinates, angle and total pathlength. The resulting data can be analysed and displayed as a function of any of the stored parameters.

RESULTS

Simulations have been performed for a homogeneous tissue slab having a thickness of 10 mm, and a refractive index of 1.4. A wide range of tissue absorption (mua) and scattering (mus) coefficients (Table 1) have

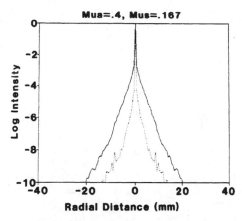

Figure 4. Transmitted and reflected PSF for tissue of relatively high absorption and low scattering coefficient. Intensities have been normalised and are plotted on a logarithmic scale.

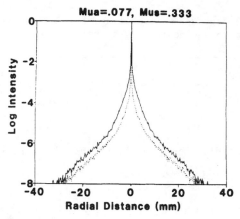

Figure 5. Transmitted and reflected PSF for tissue of medium absorption and scattering coefficient. Intensities have been normalised and are plotted on a logarithmic scale.

been modelled, encompassing the extremes of values quoted in the literature (3,4). For each combination of scattering and absorption coefficients, the model was run for a total of 100,000 photons. The data obtained has been analysed to give the transmitted and reflected PSF, and the total transmission and reflection. Figures 4, 5 and 6 show the reflected and transmitted PSF for different combinations of scattering and absorption coefficients. The transmitted PSF has been fitted to a

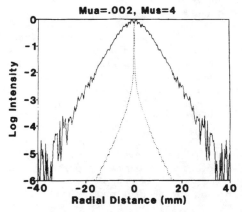

Figure 6. Transmitted and reflected PSF for tissue of relatively low absorption and high scattering coefficient. Intensities have been normalised and are plotted on a logarithmic scale.

Table 2.   General equation describing the PSF in terms of tissue
           absorption and scattering coefficient and thickness.

---

1) $I(r) = P1 \exp(-r^2 * P2)$ Gaussian term.

     $+ P3 \exp(-\sqrt{1 + (r/d)^2} * P4)/(1 + (r/d)^2)$ Diffusion term.

     $+ P5 \exp(-r * P6)$ Exponential Term.

     $+ K0$ Unscattered component.

   where r = radial distance from the centre, d = tissue thickness.

2) $K0 = \exp(-(mua + mus) * d)$

3) $P1 = 10^{-(6.7\ mua + 1.3\ mus - 1.7)}$

4) $P2 = 8.4\ mua + 2.3$

5) $P3 = 10^{((0.165 - 0.217\ mus)((1/mua) - 23) - 0.6)}$

6) $P4 = 1.12 + 0.72\ mus + 10\ mua$

7) $P5 = 10^{(-0.71\ mus + (0.068 * 1/mua) - 0.34)}$

8) $P6 = 0.48\ mua - 0.1\ mus + 0.81$

---

generalised equation (Table 2) using a non-linear least squares curve
fitting routine (7).   The parameters P1 to P6 in the equation were
subsequently fitted to functions of the absorption and scattering
coefficients and tissue thickness.   These equations are also given in
Table 2.

DISCUSSION

      The equations in Table 2 are the results of empirical fits to the
data generated by the Monte Carlo model.   They apply to a tissue
thickness of 10mm, and to the range of absorption and scattering
coefficients given in Table 1.   They can however be applied to tissue of
different thicknesses by suitable scaling of mua and mus.   Although the
equation is empirical, the choice of terms in the function was made on
the basis of a simplified analysis of the light transport problem.   The
equation therefore consists essentially of four terms, the general
characteristics of which are discussed below.

a)   The Gaussian term. This is thought to result from light which has
     undergone only a few small angle scatterings events (8).   Its
     amplitude (P1) decreases with increasing scattering and absorption
     coefficient.   The decrease with scattering is probably due to the
     increasing chance of multiple and large angle scattering.   The width
     of the gaussian term (P2) depends only upon the absorption
     coefficient in the range covered in this simulation.

b)   The Diffusion term. This applies to the high scattering region of
     light transport (9).   The behaviour of the amplitude of this term
     (P3) is more complex, but broadly speaking, it increases with
     increased scattering.   The width of the diffusion term (P4) is
     largely dependent upon the absorption coefficient.

c)   The Exponential term. The physical basis for the inclusion of this
     term is not fully understood, but it is required to fully describe
     the PSF over the whole range of absorption and scattering
     coefficients.   This term probably tries to fill the gap between the

small angle scattering and the diffusion regimes. Its parameters (P5 and P6) vary only slowly with absorption and scattering coefficients.

d) **The Unscattered component**. These are the photons that travel through the tissues without undergoing any scattering event.

CONCLUSION

We have simulated the transport of light through brain tissue using scattering data derived experimentally. From these simulations we have been able to derive a generalised formula describing the shape of the transmitted point spread function. Using this formula, it is possible to calculate the PSF for brain tissue of differing thicknesses, absorption and scattering coefficients without having to perform a computationally extended Monte Carlo calculation.

ACKNOWLEDGMENTS

This work was supported by grants from the Science and Engineering Research Council and Hamamatsu Photonics KK.

REFERENCES

1. J.S. Wyatt, M. Cope, D.T. Delpy, S. Wray, E.O.R. Reynolds, Quantification of cerebral oxygenation and haemodynamics in sick newborn infants by near infrared spectrophotometry, Lancet. 8515: 1063-1066 (1986).
2. S.R. Arridge, M. Cope, P. van der Zee, P.J Hillson, D.T. Delpy, Visualisation of the oxygenation state of the brain and muscle in newborn infants by near infrared transillumination, in: "Information Processing in Medical Imaging", S. L. Bacharach, ed., Martinus Nijhoff Ltd (1986).
3. L.O. Svaasand, D.R. Doiron, A.E. Profio, Light distribution in tissue during photoradiation therapy. Univ. South Calif. Inst. for Physics & Imaging Science, Report MISG 900-02 (1981).
4. L.O. Svaasand, R. Ellingsen, Optical properties of human brain, Photochem & Photobiol., 38:3:293-299 (1983).
5. P. van der Zee, D.T. Delpy, Simulation of the point spread function for light in tissue by a Monte Carlo method, in: "Oxygen Transport to Tissue", Vol IX, I. A. Silver, ed., Plenum Press (in press) (1987).
6. Tabulated data on the VSF and cumulative probability are not included in this paper but are available upon request from the authors.
7. Gaushouse, Non Linear Squares fit (Dec. 1965) Univ. of Wisconsin Computing Centre.
8. A.M. Whitman, M.J. Beran, Beam spread of laser light propagating in a random medium, J. Opt. Soc. Am., 60:12:1595-1602 (1970).
9. C.C. Johnson, Optical diffusion in blood, IEEE Trans. BME, 17:2: 129-133 (1970).

TEMPERATURE DEPENDENCE OF ENZYME OPTODES

AS EXEMPLIFIED BY THE GLUCOSE OPTODE

K.-P. Völkl*, N. Opitz, and D.W. Lübbers

*Physiologisches Institut der Universität Münster
Robert-Koch-Straße 28, D-4400 Münster, FRG
Max-Planck-Institut für Systemphysiologie
Rheinlanddamm 201, D-4600 Dortmund 1, FRG

Enzyme optodes, which were described as continuously measuring devices in 1980 (1), have been mainly used to measure glucose and lactate concentrations (2). The enzyme optode works according to the same principle as the enzyme electrode (3), but uses fluorescent indicators as sensors.

It is known that fluorescence, diffusion, and also enzymatic reactions are temperature dependent. Since in physiological experiments temperature may change, the temperature dependence of the enzyme optode was investigated. The experiments show that the temperature dependence of the optode influences range and accuracy of measurements.

METHODS

The glucose optode was prepared as described earlier in detail (1,2). Glucose oxidase E.C. 1.1.3.4 and catalase E.C. 1.11.1.6 were obtained by Boehringer (Mannheim, FRG), the fluorescence indicator of oxygen, pyrene butyric acid, by Eastman (Kodak, USA).

A gas mixing pump for N_2-O_2 mixtures (Wösthoff, Bochum, FRG) enabled the adjustment of N_2-O_2 ratios in steps of 1%. A cooling and heating installation permitted temperature changes between 280 and 330 K.

In order to avoid disturbances by reflected scattered light, the optode was mounted into an Aminco-Bowman fluorescence photometer (American Instruments Company) so that the angle between the surface of the optode and the exciting light was smaller than 0.78 rad. Fluorescence was excited at $\lambda = 342$ nm and the relative intensity of the emitted light was recorded at $\lambda = 395$ nm.

RESULTS

The quenching of oxygen is described by the Stern-Volmer equation (4).

$$I_O/I = 1 + k \cdot \alpha \cdot Po_2 \qquad (1)$$

I_O = fluorescence intensity without quencher, I = fluorescence intensity with quencher (oxygen), α = solubility coefficient for oxygen, k = overall quenching constant, Po_2 = oxygen pressure.

First, the temperature dependence of the indicator layer was investigated without glucose in the absence of oxygen during equilibration with nitrogen. The results of these experiments are presented in Fig. 1. The fluorescence intensity decreased with increasing temperature, T. Plotting the logarithms of the fluorescence intensity against 1/T (T in degree Kelvin) one obtained a straight line, which was described by the equation $\log I_O = 1113/T - 2.0$ with a correlation coefficient of r = 0.996.

In the presence of oxygen the relative fluorescence intensity behaved similarly. The relative fluorescence intensity during equilibration with oxygen was measured for each temperature, taking the fluorescence intensity during equilibration with nitrogen as the 100% value. Fig. 2 shows that the results can be also presented as a straight line: $\log I = 900.8/T - 1.54$ with a correlation coefficient of 0.985.

The ratio of the two investigated fluorescence intensities, I_O, and, I, decreased linearly with the increasing temperature (Fig. 3): $I_O/I = - 0.0122 \cdot T + 5.49$ (or rounded off $- 0.012 \cdot T + 5.4$)with the correlation coefficient of 0.99. An increase of the temperature from 283 to 335 decreases I_O/I by a factor of about 0.7.

Then, the temperature dependence of the glucose optode perfused with glucose-containing solutions was investigated in the temperature range between 290 and 330 K (17 - 57°C). The fluorescence signal of the glucose optode can be described by the following equation (2):

$$I_O/I = 1 + k \cdot \alpha \cdot Po_2 - k'[glucose] \qquad (2)$$

At different temperatures, the optode was first equilibrated with nitrogen and this intensity was used as the 100% value. Hereafter, the

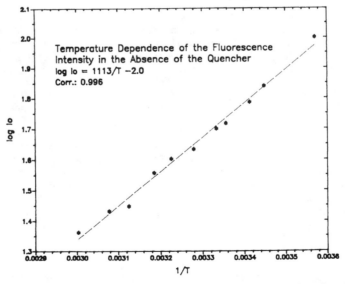

Fig. 1: Temperature dependence of the fluorescence intensity in absence of oxygen. Logarithms of the fluorescence intensity are plotted against 1/T. Corr = correlation coefficient, T = temperature in Kelvin. (n = 11).

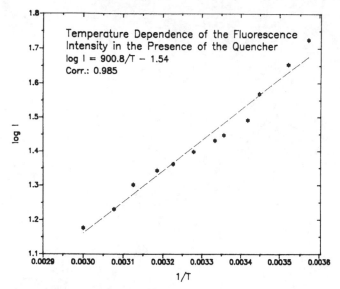

Fig. 2: Temperature dependence of the fluorescence intensity in the presence of the quencher oxygen. Logarithms of the fluorescence intensity are plotted against 1/T (for explanations see Fig. 1, n = 12).

optode was equilibrated with a constant oxygen partial pressure and the fluorescence intensity was measured at different glucose concentrations. According to eq. (2), drawing I_0/I vs glucose concentration yielded straight lines: the calibration curves of the glucose optode. By increasing temperature, the slope of the calibration curve increases. Temperature influenced diffusion processes as well as enzyme activity.

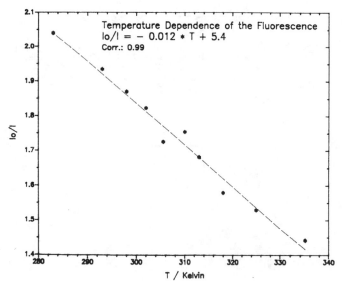

Fig. 3.. Ratio I_0/I vs temperature. The ratio I_0/I decreases linearly with the increasing temperature.

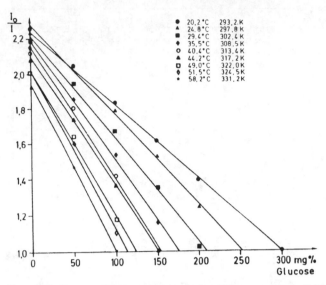

Fig. 4: Temperature dependence of the calibration curves of the glucose
optode (temperature in degrees Celsius and Kelvin).

The change of the slope with temperature could be described by a
straight line S = - 0.3 T + 80.3 (or rounded off 81) with a correlation
coefficient of 0.993 (Fig. 5). Between 293 and 331° K the slope increased
by a factor of about 2.5 and the measuring range decreased from 0 - 300 mg
% glucose at 293° K to 0 - 100 mg % glucose at 331° K.

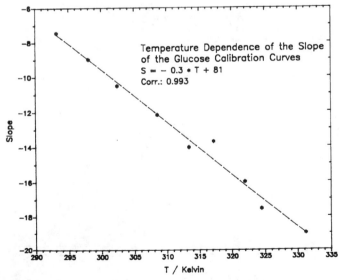

Fig. 5: Temperature dependence of the slope of the glucose calibration
curves (Corr.: Correlation coefficient).

DISCUSSION

Concerning the temperature dependence of fluorescence one has to consider both the influence upon excitation as well as on emission. Excitation is brought about by the absorption of radiation, which, according to the Lambert-Beer's law, depends on the indicator concentration, the thickness of the indicator layer, and the extinction coefficient of the indicator species. Temperature-dependent concentration changes caused by a variation of the layer thickness may be negligible, due to the limited temperature range investigated. In addition, temperature influences upon the extinction coefficient may be neglected as well, according to former investigations performed by Opitz et al. (5). Thus, the temperature course of fluorescence within the range of about 7 - 60° C (280 - 330° K) can be mainly restricted to the temperature dependence of the emission of radiation, i.e. the temperature dependence of the fluorescence quantum efficiency or, equivalently, the temperature dependence of the fluorescence lifetime, τ, whose reciprocal value is proportional to the sum of the transition probabilities of radiation (n_f) and radiationless desactivation (n_w):

$$\frac{1}{\tau} \sim n_f + n_w \qquad (3)$$

Since the transition probability for radiationless desactivation (n_w) can be caused by internal molecular conversion (n_i) and by fluorescence quenching ($n_k \cdot [O_2]$), where $[O_2]$ is the actual quencher concentration of oxygen, one can write:

$$\frac{1}{\tau} \sim n_f + n_i + n_k \cdot [O_2] \qquad (4)$$

with $[O_2] = \alpha \cdot Po_2$ follows:

$$\frac{1}{\tau} \sim n_f + n_i + n_k \cdot \alpha \cdot Po_2 \qquad (5)$$

In the absence of quencher molecules eq. (5) is reduced to

$$\frac{1}{\tau_o} \sim n_f + n_i \qquad (6)$$

For the ratio of the relative fluorescence intensity in the presence and absence of oxygen (I and I_o) follows:

$$\frac{I_o}{I} = \frac{\tau_o}{\tau} = \frac{n_f + n_i + n_k \cdot \alpha \cdot Po_2}{n_f + n_i} \qquad (7)$$

For $n_k/(n_f + n_i) = k$ one obtains eq. (1).

Since τ_o and, therefore, I_o decrease with increasing temperature, one can conclude from eq. (7) that the sum of n_f and n_i must increase. This increase is mainly due to an increase of desactivation processes by internal conversion.

The same may hold for the temperature dependence of the fluorescence intensity, I, since the temperature courses of I and I_o are quite similar, whereas the experimentally observed decrease of the quotient (I_o/I) may be mainly due to a decrease of the product ($n_k \cdot \alpha \cdot Po_2$). This result can be explained by the dominating temperature behaviour of the solubility coefficient of oxygen, α, since it can be assumed that n_k will increase with temperature.

Considering the above discussed temperature effects of the indicator layer in connection with the glucose optode the calibration curves of the glucose sensor should be parallel and reach smaller values for the axial intercept with increasing temperature. Fig. 5 shows that the slope of the calibration curves is also temperature dependent, which could be mainly explained by an increase of the velocities of the substrate diffusion as well as of the enzyme reaction. Thus, the decrease of I with increasing temperature - due to an increase of n_i - is surpassed by the increase of the velocities of the substrate diffusion and of the enzyme reaction within the temperature range investigated. Therefore, an increase of the temperature results for the glucose optode in a reduced range of measurement and, simultaneously, in a higher resolution of the measuring values.

REFERENCES

1. K.-P. Völkl, U. Grossmann, N. Opitz and D. W. Lübbers, The use of O_2-optode for measuring substances as glucose by using oxidative enzymes for biological applications, in: "Oxygen Transport to Tissue," A. G. B. Kovach, E. Dora, M. Kessler, and I. A. Silver, eds., Plenum Press, New York-London (1980), pp. 99-100.
2. K.-P. Völkl, Die Enzymoptode - eine neue Methode zur kontinuierlichen Messung von Substraten in biologischen Flüssigkeiten. Inauguraldissertation, Bochum (1981)
3. S. J. Updike and G. P. Hicks, The enzyme electrode, Nature 214:986 (1967).
4. O. Stern and M. Volmer, Über die Abklingungszeit der Fluoreszenz, Physik. Z. 20:183 (1919.
5. N. Opitz, H.-J. Graf and D. W. Lübbers, Oxygen sensor for the temperature range of 300 to 500 degrees Kelvin based on fluorescence quenching of indicator-treated silicone rubber membranes. Proc. 2nd Int. Meeting on Chemical Sensors, Bordeaux 1986, pp. 657-660.

A SIMPLE FIBRE OPTIC APPARATUS FOR MEASUREMENT OF THE OXY-HAEMOGLOBIN

BINDING ISOTHERM

M. McCabe, R. Hamilton[*], and D. Maguire[+]

Department of Chemistry and Biochemistry, James Cook University
of North Queensland, Townsville Q4811, Australia
[*]Department of Respiratory Medicine, Charing Cross Hospital
Medical School, London
[+]School of Sciences, Griffith University, Nathan Q4111
Australia

INTRODUCTION

We hesitate to report yet another variant of the automatic (polarographic) technique for the determination of the oxy-haemoglobin dissociation curve first described by Longmuir and his colleagues (Franco et al., 1962). Yet it remains true that this method has yet to find much favour since it requires either a considerable preparative effort in the need for a suitable fragmented sarcosomal fraction (Colman & Longmuir, 1963; McCabe, 1973), or (in the variant proposed by Imai et al., 1970) it requires a substantial and dedicating modification of a spectrophotometer (the cutting out of the bottom of the cell housing compartment).

We describe here a variant of the Imai method, namely the simultaneous measurement of spectra and pO_2 during a slow oxygenation, without the need to mutilate a spectrophotometer.

MATERIALS AND METHODS

A standard oxygen electrode system from Rank Bros., Bottisham, Cambs, UK was used as the reaction vessel. It was modified by the addition of two bundles of optic fibre arranged as shown in Figure 1. The fibres were plastic, 0.4 mm diam. Twelve of them were inserted into the respirometer cell by drilling a series of holes as shown in the figure and threading the optic fibre through both the walls of the water jacket and the reaction cell. The fibres were then fixed in place using an epoxy resin (which crept some way along the shank of the fibre). Finally the fibres were cut flush against the inner wall of the reaction vessel. In this way each fibre was optically aligned against another across the respirometer chamber. Ingoing and outgoing fibres were collected into two bundles which could be inserted into the cell housing of the spectrophotometer, either through a convenient gas port, or simply by replacing the lid with a black cloth. The terminus of each bundle was arranged into the light path of the spectrophotometer in the way shown in Figure 2. The steel plate provided a support and acted as a mask, preventing any stray light from "short circuiting" the optic fibres. During use the respirometer cell was also covered with a black cloth to minimise stray (background) light.

205

APPLICATION TO THE STUDY OF THE HAEMOGLOBIN–OXYGEN BINDING ISOTHERM

Haemoglobin preparation

Human blood (0.1 ml) obtained by finger prick was washed into saline (0.85%) and washed five times centrifuging at 1500 rpm for 5 min between each wash. Haemolysis was achieved by addition of 1.5 vol of distilled water and 0.5 vol toluene, followed by vigorous agitation. The resulting system was centrifuged at 1500 rpm for 10 min and the toluene layer which contained the cell debris was removed with a Pasteur pipette. The resultant haemoglobin solution was diluted to a 0.1% solution in 0.1 M sodium phosphate buffer at an appropriate pH.

Haemoglobin concentration was measured spectrally using the data of Benesch et al., (1973) as modified by Van Assendelft et al., (1975). The selection of a suitable wavelength for determination of the fractional saturation is dependent on the haemoglobin concentration. It was found that for haemoglobin concentrations of 0.1% a suitable wavelength was 560 nm. At this point the oxyhaemoglobin spectrum displays a distinct minimum while the deoxy spectrum has a large maximum.

Procedure

A continuous stream of high purity nitrogen gas was blown gently over the surface of the solution which was simultaneously stirred. Haemoglobin was judged deoxygenated when the recorded pO_2 remained a constant minimal value and the spectrum showed a single peak with no inflection at 560 nm.

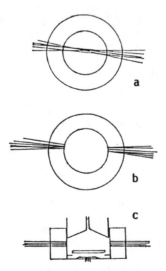

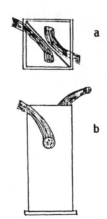

Fig. 1. (a) Plan of respirometer with optic fibres inserted through the chamber of the respirometer and the water jacket.
(b) As above after optic fibres cut from the chamber of the respirometer.
(c) Side view of optic fibres inserted into respirometer.

Fig. 2. (a and b) plan and end elavation respectively, showing the mounting of optic fibre bundles into the light path of the spectrophotometer. The terminus of the fibre bundles can be located into the light path by adjusting a steel wire stirrup (not shown) which is suspended from the angled steel plate. The plate stands on a 1 cm^2 base plate.

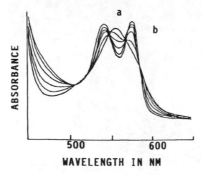

Fig. 3. Spectra of human haemoglobin at various stages
of reoxygenation; (a) prior to entry of any
oxygen, and (b) at the completion of oxygenation.

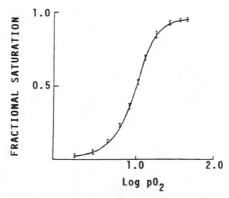

Fig. 4. A comparison of the complete equilibrium curve
determined by the present method and showing the
variation in six separate determinations (I)
superimposed on results obtained by Imai et al
(1970)

When complete deoxygenation was achieved (approx. 40 min), the N_2 was
turned off and air was allowed to diffuse back into the apparatus causing a
slow reoxygenation of the haemoglobin. Output from the spectrophotometer was
fed onto one arm of an X-Y recorder, while the polarographic signal due to
oxygen in solution was fed onto the other arm. In this way the recorder automatically generated the oxygen binding isotherm of the haemoglobin sample.
As an added check on events, the whole visible spectrum of the haemoglobin
sample was periodically recorded during the reoxygenation phase (see Figure 3).

Table 1. Comparison of P_{50}, n, and the Bohr effect ($\emptyset$) produced by the present method with accepted values reported in the literature

pH	log P_{50} This work	log P_{50} Literature	n This work	n Literature	$\emptyset$ This work	$\emptyset$ Literature
7.8	0.48	0.50[a] 0.48[b]	2.84	2.77[a] 3.11[b]	-0.74	-0.64[a] -0.71[b]
7.4	0.78	0.79[a] 0.74[b] 0.75[c]	2.90	2.88[a] 3.17[b]	-0.49	-0.51[a] -0.51[b]
7.4	0.97	0.98[a] 0.95[b] 0.96[c]	3.12	2.92[a] 2.98[b]	-0.29	-0.36[b] -0.31[b]
6.5	1.11	1.15[a] 1.09[b] 1.09[c]	2.96	3.00[a] 2.95[b]	+0.01	+0.15[b]
6.0	1.11	1.01[b]	3.19	3.17[b]		

(a) Tuchinda et al., (1975)
(b) Imai, 1968
(c) Imai et al., (1970)

RESULTS

The complete curve determined by the present method is compared with that obtained (under otherwise identical conditions) by the technique of Imai et al., (1970). The results are shown in Figure 4.

pH variation was determined over the range of pH from 6.0 to 7.8. Table 1 shows the results of the measured values of P_{50} and n (a measure of subunit interaction, see Hill, 1910). The table also shows the effects of temperature variation, the average temperature coefficients ($\Delta \log P_{50}/\Delta T$) are calculated for the change from 20° to 25°C and from 25° to 30°C respectively. The measured coefficient lies within the range of 0.018 to 0.029 quoted by Benesch et al.,(1969).

The heat of combination, Q, of the reaction was calculated from the equation of Roughton (1964), and the results are also shown in the table. Again it can be seen that the measured values are in close agreement with the values reported by Roughton (1964).

DISCUSSION

Before a new technique can be applied to investigational research it must be proven to produce results consistent with values obtained using currently accepted techniques. It is apparent from the literature that there are no universally accepted and standardised conditions for the measurement of the O_2-Hb binding isotherm. However, the technique of Imai et al., (1970) seems both elegant and well validated. For this reason the agreement of the present technique with that of Imai is reassuring.

The apparatus described here can be readily and cheaply assembled from standard commercially available (and inexpensive) equipment. Furthermore it has the advantage that the spectrophotometer selected for use need not be altered in any way.

REFERENCES

Benesch, R.E., Benesch, R., and Yu, C.I., 1969, Oxygenation of hemoglobin in the presence of 2:3 DPG; Effect of temperature, pH, ionic strength and hemoglobin concentration. Biochemisry, 8:2567-2581.

Benesch, R.E., Benesch, R., and Yung, S. 1973, Equations for the spectrophotometric analysis of hemoglobin mixtures, Analyt. Biochem, 55: 245-248.

Colman, C.H. and Longmuir, I.S., 1963, Automatic Registration of oxyhaemoglobin dissociation curves, J. Appl. Physiol., 18:420-423.

Franco, C.H., Longmuir, I.S., and McCabe, M., 1962, New methods for the continuous registration of oxygaemoglobin dissociation curves, J. Physiol. (London), 161, p.54.

Hill, A.V., 1910, J. Physiol., (London), 40 IV.

Imai, K., 1968, Oxygen equilibrium characteristics of abnormal hemoglobin Hiroshima, Arch. Biochem. Biophys., 127:543-547.

Imai, K., Morimoto, H., Kotani, M., Watari, H., Hirata, W., and Kuroda, M., 1970, Improved method for automatic measurement of the oxygen equilibrium curve of hemoglobin, Biochim. Biophys. Acta., 200:189-196.

McCabe,M., 1973, A micromethod for the study of the oxygen-haemoglobin equilibrium, Trans. Biochem. Soc., 1:882-884.

Roughton, F.J.W., 1964, Transport of oxygen and carbon dioxide, in: "Handbook of Physiology", Section 3, Vol. Amer. Physiol. Soc., Washington, D.C.

Tuchinda, S., Nagai, K., and Lehmann, H., 1975, Oxygen dissociation curve of haemoglobin, Portland, FEBS Letters, 49:390-391.

ABSORBANCE PROFILE OF RED BLOOD CELL SUSPENSION IN VITRO AND IN SITU

T. Tamura*, O. Hazeki, M. Takada* and M. Tamura

*Research and Developement, Shimadzu Corporation
Kyoto, Japan
Biophysics Div., Research Institute of Applied Electricity
Hokkaido Univ. Sapporo, Japan

INTRODUCTION

In order to achieve non-invasive monitoring of the oxygenation state of hemoglobin, there are many problems that must be solved. (e.g. selection of appropriate measuring wavelengths, establishment of the calibration curve etc.)

One of the major difficulties is that we need to use many rats to investigate each of these problems. To overcome this, we tried working with red blood cell suspensions in a cuvette instead of rats, and have obtained satisfactory results.

We measured the absorbance changes of hemoglobin in relation to changes in hematocrit values in vitro and in situ. To eliminate possible interference from cyt aa$_3$, two wavelengths, 750nm and 780nm, were used. At these two wavelengths there was a linear relationship between the absorbance change and the hematocrit value. The absorbance ratios between these wavelengths ($\Delta O.D._{750}/\Delta O.D._{780}$) in vitro showed a value very close to that in situ.

METHOD

We measured absorbance changes in red blood cell suspensions with changes in hematocrit values in vitro. For this we used an optical cuvette with a 1 cm light path, illuminated with diffused light from a fiber-optic light guide. The transmitted light was fed via a second light guide to a photodetector (Fig. 1).

Our in vivo model was the perfused rat head which was transilluminated by a light guide as shown in Figure 1. The efferent light was fed to a photodetector system via another light guide. The head was purfused with red cell suspensions of different hematocrit.

By applying the optical coefficients determined from the in vitro experiments, we were able to measure the amount of oxy-hemoglobin and deoxy hemoglobin in rat head in vivo during changes in the O$_2$ concentration of the inspired gas.

SYSTEM

The block diagram of our system is shown in Fig. 1.

(1) Optical System

    (1) Light Source: 250W halogen lamp
    (2) Wavelength: 750nm, 780nm (Filter)
    (3) Illuminating light guide: bundle 10mm dia.
    (4) Detecting light guide: bundle 10mm dia.
    (5) Detector: photomultiplier (R712)

(2) Preparation of Rat (measurement in situ)

    (1) Male Wistar rats (200 ~ 240g) anesthetized with urethane
        (0.8g/kg i.p.)
    (2) The Trachea was cannulated to control the conditions of the
        inspired gas.
    (3) The Carotid artery was cannulated to permit infusion of the red
        blood cell suspension into the rat head.
    (4) The Vena cava was also cannulated in order to collect infused
        blood samples.

(3) Red Blood Cell Suspension (measurement in vitro)

    We used blood from pigs.
        oxy-hemoglobin samples : Red cells were washed 3 times in
                                saline bubbled with 100% O_2: O_2 satu-
                                ration was over 98%
        deoxy-hemoglobin samples: Red cells were washed in saline con-
                                  taining a small amount of $Na_2S_2O_4$ and
                                  bubbled with N_2: O_2 saturation was
                                  below 2%

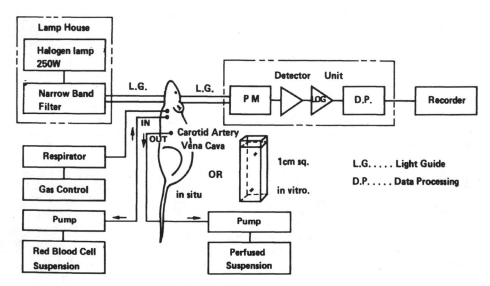

Fig. 1 Block Diagram of our system

RESULTS AND DISCUSSION

(1) Dilution of Blood in Rat Head under Anoxic Conditions

After ventilating the rat with N_2, we cut the jugular vein and infused an 0.1ml/bolus of oxygen-free saline into the carotid artery. Then the absorbance changes at 750nm and 780nm were measured across the head. Their mutual ratio ($\Delta O.D._{750}/\Delta O.D._{780}$) was 1.25 (Fig. 2).

(2) Absorbance Changes of Red Blood Cell Suspension in vitro

We measured the absorbance changes of the completely deoxygenated red blood cell suspension in vitro against changes in hematocrit values. We found a linear relationship between the absorbance change and hematocrit value in the region of Hct 3 ∿ 15% (Fig. 3-(A)). According to Twersky's theory, the change of O.D. (Optical Density) in the range of Hct 3 ∿ 15% is mainly affected by absorption by red blood cells and not by light scattering. There was also a linear relationship between $\Delta O.D._{750}$ and $\Delta O.D._{780}$, and the absorbance ratio ($\Delta O.D._{750}/\Delta O.D._{780}$) was 1.22. (Fig. 3-(B))

Under anoxic conditions the absorbance ratios between these two wavelengths in situ are very close to those in vitro.

We performed the same experiment under the aerobic conditions in vitro and also found a linear relationship between O.D. and Hct in the same range of Hct as under the anoxic conditions. (Fig. 3-(A)). The absorbance ratio ($\Delta O.D._{750}/\Delta O.D._{780}$) was 0.93. (Fig. 3-(B))

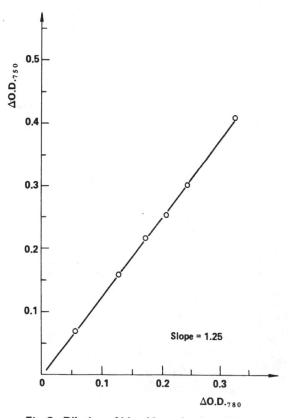

Slope = 1.25

Fig. 2   Dilution of blood in rat head under anoxic conditions

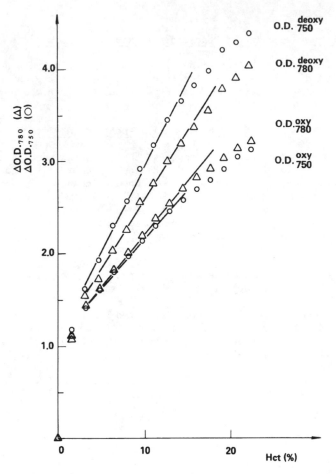

Fig. 3-(A) Correlation between Hct and O.D. changes of oxygenated and deoxygenated Red blood cell suspension in vitro.

(3) Basic Equations

From a linear relationship between the hematocrit and the O.D. change,

$$\Delta O.D._{750} = k_1 \Delta[HbO_2] + k_1' \Delta[Hb] \qquad (1)$$
$$\Delta O.D._{780} = k_2 \Delta[HbO_2] + k_2' \Delta[Hb] \qquad (2)$$

where k_1, k_2, k_1' and k_2' are absorption coefficients at 750nm and 780nm and $\Delta[HbO_2]$ and $\Delta[Hb]$ indicate the change of the amount of oxy-Hb and deoxy-Hb.

From equations (1), (2)

$$\Delta[HbO_2] = K_1 \{\Delta O.D._{750} - (k_1'/k_2') \ \Delta O.D._{780}\} \qquad (3)$$
$$\Delta[Hb] = K_2 \{\Delta O.D._{750} - (k_1/k_2) \ \Delta O.D._{780}\} \qquad (4)$$
$$\Delta([HbO_2] + [Hb]) = K_3 \{\Delta O.D._{750} - (k_1'-k_1)/(k_2'-k_2) \ \Delta O.D._{780}\} \qquad (5)$$

where $K_1 = k_2'/(k_1 k_2' - k_1' k_2)$, $K_2 = -k_2/(k_1 k_2' - k_1' k_2)$
$K_3 = (k_2' - k_2)/(k_1 k_2' - k_1' k_2)$

To obtain the coefficient $(k_1' - k_1)/(k_2' - k_2)$, we measured the difference of absorbance of deoxy-hemoglobin and oxy-hemoglobin with changes of hematocrit in vitro. At the given wavelengths, the slope was 2.16. (Fig. 4)

214

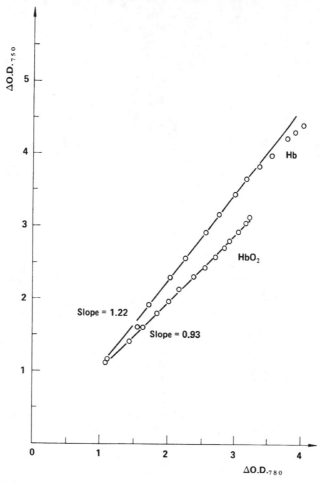

Fig. 3-(B) Linear relationship between O.D. changes 750nm and 780nm in vitro

From the result of the in vitro measurements given above we can find the relative values of the optical coefficients. (Table 1) So, equations (3) ∿ (5) become as follows:

$$\Delta[HbO_2] = -\Delta O.D._{750} + 1.22\Delta O.D._{780} \qquad (3)'$$
$$\Delta[Hb] = 0.76\Delta O.D._{750} - 0.71\Delta O.D._{780} \qquad (4)'$$
$$\Delta([HbO_2] + [Hb]) = -0.24\Delta O.D._{750} + 0.51\Delta O.D._{780} \qquad (5)'$$

Table 1. Relative values of the optical coefficient

	oxy	deoxy
750mm	0.71	1.22
780mm	0.76	1

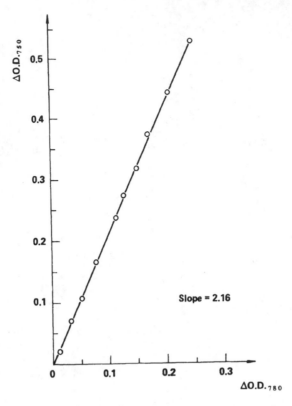

Fig. 4   Linear relationship between the O.D. changes of oxy-minus
deoxy Hb at 750nm and 780nm in vitro

(4)   Application to in vivo Measurements

To check the coefficients and equations derived from in vitro measurements, we measured the absorbance changes and calculated the parameters,
$\Delta[HbO_2]$ and $\Delta[Hb]$, with changes in O_2 concentration of the inspired gas.
These parameters showed very consistent changes as shown in Fig. 5.   For
example, oxy-hemoglobin decreased and deoxy-hemoglobin increased in line
with the decrease of O_2 concentration.   Finally, when we cut the jugular
vein of the rat to reduce the circulating blood volume under anoxic
conditions in which O_2 saturation of the blood was considered to be zero,
the parameter $\Delta[Hb]$ showed considerable decrease but the parameter $\Delta[HbO_2]$
remained constant.

The results of these experiments confirm the coefficients and
equations and all our assumptions.

Thus it is practicable to use a suspension of red blood cells in vitro
instead of living tissue, to make quantitative measurements of the oxygenation state of hemoglobin in situ.

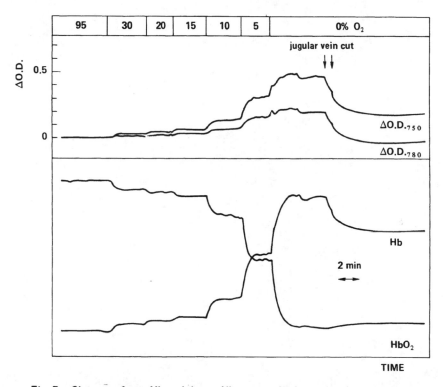

Fig. 5 Changes of oxy-Hb and deoxy-Hb content in the rat head
according with the aerobic to anoxic changes of the ventilate conditions

ESTIMATION OF THE TRANSFER COEFFICIENTS OF OXYGEN AND CARBON MONOXIDE IN THE BOUNDARY OF HUMAN AND CHICKEN RED BLOOD CELLS BY A MICROPHOTOMETRIC METHOD

I. Shibuya, K. Niizeki and T. Kagawa

Department of Physiology, Yamagata University School of Medicine, 990-23 Yamagata, Japan

A microphotometric method to measure the reaction rates of O_2 and CO was developed by Mochizuki et al. (1973) for the purpose of measurements in a single red blood cell (RBC). To date, the rates of oxygenation, deoxygenation, Bohr-on and Bohr-off shift, and replacement reaction of O_2 by CO have been studied by this method (Mochizuki et al., 1973; Tazawa et al., 1974; Tazawa and Mochizuki, 1976; Tazawa et al., 1976). However, since the reaction rates were much influenced by the thickness of the liquid layer remaining around the RBC, the definite reaction processes of O_2 and CO with RBCs have not yet been established. In the present study, to clarify the kinetics of the diffusion and reaction processes of O_2 and CO in the RBC and the RBC boundary, we measured reaction rates of O_2 and CO in human and chicken RBCs and computed numerical solutions of partial differential equations of diffusion of O_2 and CO. Then, by comparing the two results, we estimated the diffusion rate factor across the boundary layer including the RBC membrane, i.e., the transfer coefficients of O_2 and CO (ηo_2 and ηco).

Furthermore, the pulmonary diffusing capacity for CO (DLco) has been widely used in clinical evaluation of the degree of pulmonary diffusion impairment, because of the difficulty in estimating the pulmonary diffusing capacity for O_2 (DLo$_2$). To our knowledge, however, there are only a few reports on the quantitative comparison of the kinetics of the diffusion and reaction of O_2 and CO. Savoy et al. (1980) measured DLo$_2$ and DLco simultaneously, and found that DLco was greater than DLo$_2$. However, DL is influenced not only by the diffusion but also by the reaction with hemoglobin. In this study, we made sequential measurements of ηo_2 and ηco in both isolated RBCs and RBCs in the capillary. Then, we compared o_2 and co in both experimental conditions to clarify whether or not the diffusion rate across the boundary including the RBC membrane differs between O_2 and CO.

MATERIALS AND METHODS

1. Microphotometric apparatus

The microphotometric apparatus developed by Mochizuki et al. (1973) was partially modified and used in the present study. Figure 1 shows a schematic illustration of the device. Light emitted from the tungsten lamp (C) was condensed through lenses (G) and the light, with a wave length shorter than 470 nm, was reflected by a dichroic mirror (H). To avoid photodissociation of carboxy hemoglobin (COHb), the light projected onto the RBCs in the small reaction cuvette (K) had a wave length of between 400 to 435 nm, this was achieved by using a short pass filter (I) and narrowing the beam to a spot of about 7-10 µm with an iris diaphragm (J). The transmitted light from the RBC was magnified by an objective (L) and divided into two by a half mirror (N). The beams were introduced into photomultipliers (P) through two interference filters (402 and 416.5 nm). The change in O_2 and CO saturation of hemoglobin was measured by multiplying the difference of transmission at 402 and 416.5 nm with the I-V converter (Q) and differntial amplifier (R).

In the experiments on the CO reaction, a rotating sector (F) with a duty ratio of 2/15 was used.

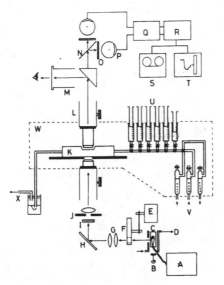

Fig. 1　Schematic illustration of microphotometric apparatus.
A:D.C. power supply, B:ventilator, C:tungsten light source, D:water perfusion, E:motor, F:sector, G:lenses, H:dichroic mirror, I:filter (400-435nm), J:iris diaphragm, K:reaction cuvette, L:objective (x40), M: ocular (x10), N:half mirror, O:filters (402 and 416.5nm), P:photomultipliers, Q:I-V converter, R:differential amplifier, S:data recorder, T:X-T recorder, U:syringes for gas injection, V:humidifier, W:water bath, X:gas outlet.

2. Preparation of blood samples

Experiments were made on both human RBC suspensions and chicken RBC in the chorioallantoic capillaries. Human blood was freshly drawn from healthy male subjects aged from 22 to 49 years and suspended in 26 mM-bicarbonate solution. About 10-20 µl of the blood were dropped onto the observation window of the air-tight reaction cuvette and used in the experiments. To avoid movement and desiccation of the blood sample, nylon mesh of about 1 mm lattice and wet paper were put in the reaction cuvette.

Chorioallantoic capillaries were prepared according to Tazawa and Ono (1974) and Tazawa et al. (1974). In brief, the chorioallantoic membrane was removed from a chicken embryo of 16 days old with the attached inner membrane and stuck to the observation window. Then, exess blood was soaked up with blotting paper and the inner membrane, only, carefully removed.

3. Gas mixtures

In the experiments on O_2, gas mixtures containing 0, 1.48, 3.12, 5.73, 14.27, or 94.4% O_2 and 5.6% CO_2 (N_2 balance) were sequentially injected into the reaction cuvette with glass syringes (U) after being humidified with distilled water at 37 °C (V).

The experiments on CO followed the O_2 experiments. First, a gas mixture containing ca. 0.4% CO, 14.3% O_2 and 5.6% CO_2 was introduced into the reaction cuvette to observe the reaction rates of CO with RBCs. Then, a gas mixture containing 10% CO, 14.3% O_2 and 5.6% CO_2, was followed by 100% CO gas to obtain maximal saturation of COHb.

4. Determination of dissociation curves for O_2 and CO

To compute the numerical solutions of differential equations for the diffusion and reaction of O_2 and CO with RBCs, the equations for the dissociation curves for O_2 and CO must be obtained. From the measured saturation of O_2 (So_2) and CO (Sco), dissociation curves for O_2 and CO were determined in both human and chicken RBCs.

The dissociation curve for O_2 is given (Hill, 1910) by:

$$So_2 = k \cdot Po_2^n / (1 + k \cdot Po_2^n). \tag{1}$$

Thus, from the linear relationship between Po_2 and $So_2/(1-So_2)$, k and n values were determined by the method of least squares.

If the amount of deoxyhemoglobin can be disregarded, the following equation (Douglas et al., 1912) satisfies:

$$\frac{[COHb]}{[O_2Hb]} = \frac{Pco}{Po_2} \cdot M, \tag{2}$$

where M is the partition coefficient. In addition, $[O_2Hb]$ is given by:

$$[O_2Hb] = [HB]_{total} - [COHb]. \tag{3}$$

Eliminating $[O_2HB]$ from Eqs. (2) and (3), Sco is given by:

$$Sco = \frac{M \cdot Pco}{M \cdot Pco + Po_2}. \tag{4}$$

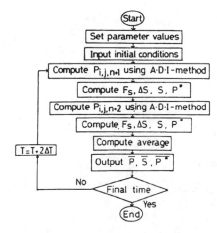

Fig. 2 A flow chart for solving the differential equation of diffusion for the O_2 and CO reaction in a disk model of the RBC.

Table 1. Parameters used in the computation of the reaction rates.

Parameters	Human RBC	Chicken RBC
Radius of RBC (c)	3.5×10^{-4} cm	5.0×10^{-4} cm
Thickness of RBC (d)	1.6×10^{-4} cm	2.0×10^{-4} cm
O_2 capacity of RBC (N)	0.435 ml/ml,RBC	0.335 ml/ml,RBC
k in O_2 dissociation curve	3.92×10^{-4}	1.43×10^{-4}
n " " " "	2.44	2.53
M in CO dissociation curve	200	
Diffusion coefficient (D) of O_2	0.46×10^{-5} cm^2/sec	
Solubility (α) of O_2	0.31×10^{-4} ml/ml,RBC·Torr	
Diffusion coefficient of CO	0.49×10^{-5} cm^2/sec	
Solubility of CO	0.23×10^{-4} ml/ml,RBC·Torr	

We measured Sco at several Pco values and a constant Po_2 of 100 Torr and determined the value of M.

4. Computation of the reaction rates of O_2 and CO with the RBC

The rates of diffusion of O_2 and CO in human and chicken RBCs were computed according to Kagawa and Mochizuki (1982) in a disk-shaped model. Figure 2 shows the outline of the computation and Table 1 summarizes the parameter values used. The values for dissociation curves for O_2 and CO were estimated in this study. The other parameter values for human and chicken RBCs were taken from Kagawa and Mochizuki (1982) and Romanoff (1960), respectively.

According to Kagawa and Mochizuki (1982), the partial differential equation for diffusion and reaction of O_2 and CO in RBCs is given by:

$$\alpha \frac{\partial P}{\partial t} = \alpha D \left(\frac{\partial^2 P}{\partial r^2} + \frac{1}{r} \cdot \frac{\partial P}{\partial r} + \frac{\partial^2 P}{\partial z^2} \right) - Fs\,(P - P^*), \qquad (5)$$

where r and z are distances in the radial and perpendicular direction, respectively, and P and P* are partial pressure and back pressure of O_2 or CO, respectively. Fs is the velocity factor for the combination of O_2 or CO with hemoglobin and is given by the following equations
For Fs of O_2:

$$Fs = 2.09\,(1 - So_2)^{2.02}, \qquad (6)$$

and for CO:

$$Fs = \frac{5}{Po_2 - 5}\,(1 - Sco). \qquad (7)$$

Initial conditions are as folows:

$$\left(\frac{\partial P}{\partial r} \right)_{r=0} = \left(\frac{\partial P}{\partial z} \right)_{z=0} = 0. \qquad (8)$$

Boundary conditions are as follows:

$$\alpha D \left(\frac{\partial P}{\partial r} \right)_{r=c} = \eta\,(Pr{=}c - Pg), \qquad (9)$$

and

$$\alpha D\left(\frac{\partial P}{\partial z}\right)_{z=\pm d} = \eta(Pz=\pm d - Pg), \qquad (10)$$

where η is the diffusion rate across the boudary layer including the RBC membrane, i.e. the transfer coefficient, and Pg is the partial pressure of O_2 and CO in the gas phase.

Changes in O_2 or CO saturation, ΔS, during time increment (Δt) are calculated as:

$$\Delta S = Fs\ (P - P^*)\cdot\Delta t/N. \qquad (11)$$

Then, saturation at the next time increment is:

$$S = S + \Delta S \qquad (12)$$

From the S value, back pressures of O_2 and CO are calculated as follows:
For O_2:

$$P^* = \left(\frac{S}{k(1 - S)}\right)^{1/n}, \qquad (13)$$

and for CO:

$$P^* = \frac{Po_2\cdot S}{M(1 - S)}. \qquad (14)$$

With Δt of 0.1 msec, Po_2 and Pco were computed from Eq. (5) by using the alternating-direction implicit (A.D.I) method developed by Douglas and Rachford (1956) to obtain spatial averages of P, P*, and S ($\bar{P}$, $\bar{P^*}$ and $\bar{S}$) up to 15 sec.

The above computation was performed using FORTRAN programs with a VAX-11/750 (Digital Equipment).

RESULTS

1. O_2 dissociation curve of the human RBC

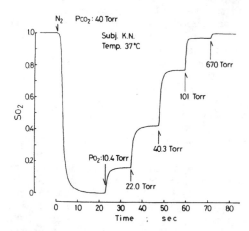
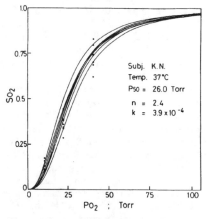

Fig. 3   An example of O_2 reaction with human RBC. Pco_2 was kept at 40 Torr.

Fig. 4 O_2 dissociation curve determined from 8 data on the human RBC.

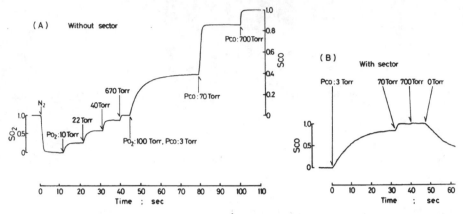

Fig. 5 (A) An example of O_2 and CO reactions measured in the human RBC without the rotating sector. (B) An example of CO replacement reaction measured with the rotating sector.

Figure 3 shows a typical example of the reaction of O_2 with human RBCs. From the level of maximal saturation, the RBC was completely deoxygenated by injection into the reaction cuvette of the gas mixture containing 5.6% CO_2 in N_2. Then, So_2 sequentially increased with increasing Po_2. In this example, half time of deoxygenation and oxygenation from 22 to 40 Torr Po_2 were 3.0 and 0.8 sec, respectively. From the data, the equation for the O_2 dissociation curve was obtained. Figure 4 shows the O_2 dissociation curve determined in RBC of subject K.N. at 37 °C. Some heterogeneity in the dissociation curve was observed in each measurement. On the average, P_{50}, n, and k values were estimated to be 26.3 Torr, 2.44, and 3.92×10^{-4}, respectively. We used these values in computation of numerical solutions of differential equations for the diffusion of O_2.

2. CO dissociation curve of the human RBC

Following the measurement of oxygenation, the replacement reaction of O_2 by CO was observed. Figure 5(A) shows an example of the replacement reaction, in which the rotating sector was not used. The first half shows the measurement of oxygenation, and the latter half, the replacement reaction by CO. After the maximal oxygenation level was reached, the reaction rate of CO was observed by injecting a gas mixture of 3 Torr Pco and 100 Torr Po_2. Equilibrium was reached by about 30 sec, and the half time of the replacement reaction was 3.5 sec. After equilibration, a gas mixture of 70 Torr Pco and 100 Torr Po_2 was introduced, followed by pure CO gas. The Sco at 3 and 70 Torr Pco was 0.40 and 0.86, respectively, suggesting that photodissociation of COHb occurred. Then, after the replacement of CO gas by CO-free O_2 gas, the replacement reaction of O_2 by CO was observed again by using the rotating sector. Figure 5(B) shows the result of the replacement reaction with the sector. The rate of the replacement reaction was almost similar to that obtained without the sector, however, the Sco at 3 and 70 Torr Pco was increased to 0.85 and 0.98 by use of the sector. Figure 6 shows the CO dissociation curves calculated with changing M values from 1 to 200 and measured data obtained in four conditions. The open and closed triangles indicate the data measured by without dichroic mirror, filter and sector and those with only the dichroic mirror, respectively. The estimated M value was 2 to 5, in contrast, the M value has been reported to be about 200 (Sendroy et al, 1929; Allen and Root, 1957). Therefore, in these condition, the affinity of CO to hemoglobin was reduced to 1/50 to 1/100 by photodissociation. When the short pass filter was used, as

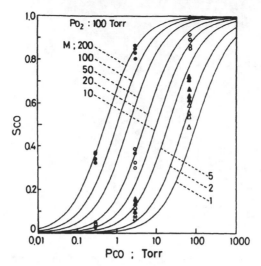

Fig. 6 CO dissociation curves calculated with changing M values from 1 to 200, and the relationship between Sco and Pco measured in four conditions. Open triangles: without dichroic mirror, filter and sector; closed triangles: with dichroic mirror; open circles: with dichroic mirror and short pass filter; closed circles: with dichroic mirror, short pass filter, and rotating sector.

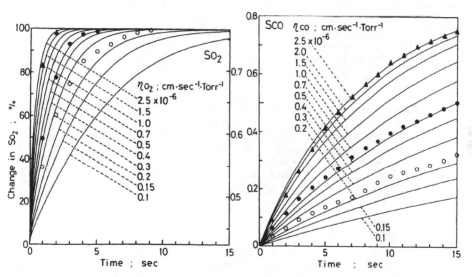

Fig. 7 Numerical solutions and typical data for O_2 diffusion in the human RBC.

Fig. 8 Numerical solutions and typical data for CO diffusion in the human RBC.

indicated by the open circles in Fig. 6, the M value was increased to 10 to 20. Furthermore, when both the short pass filter and the rotating sector were used, the M value was about 200, suggesting photodissociation of COHb was almost completely excluded. Therefore, in this study, we measured the CO reaction with the short pass filter and rotating sector.

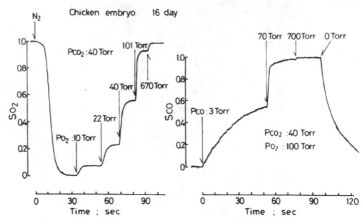

Fig. 9 Typical examples of the O_2 and CO reactions measured
sequentially in the chorioallantoic capillary. Left panel:
oxygenation reaction; right panel: replacement reaction by CO.

3. Numerical solutions and determination of ηo_2 and ηco
        Figures 7 and 8 show the numerical solutions for O_2 and CO
diffusion in the disk model RBC, which were computed with changing η from
0.1 to 2.5 x10^{-6} cm·sec^{-1}·Torr^{-1}. For O_2 diffusion, initial and final
Po_2 were 22 and 40 Torr, respectively, and Pco_2 was kept constant at 40
Torr. For CO diffusion, initial and final Pco were 0 and 3 Torr,
respectively. The Po_2 and M values were 100 Torr and 200, respectively.
The closed triangles and open and closed circles show three typical sets
of data. The same symbol indicates sequential measurements in the same
RBC. The profiles of the data all agreed well with the numerical
solutions. The estimated ηo_2 values were 2.0, 0.5, and 0.27, and the ηcc
values 2.5, 0.5, and 0.24 x10^{-6} cm·sec^{-1}·Torr^{-1}, showing good agreement.

4. O_2 and CO reaction in RBCs in the chorioallantoic capillary
        Figure 9 shows typical examples of the O_2 and CO reactions measured
sequentially in the chorioallantoic capillary from a 16 day old chicken
embryo. The rates of oxygenation and replacement reaction by CO were
slower than those obtained in the human RBC. The O_2 affinity of
hemoglobin was smaller than that of human hemoglobin. The mean P_{50} was
33.8 Torr from 11 measurements. This value coincides well with that
reported by Tazawa et al (1976). In the CO replacement reaction,
photodissociation of COHb was excluded by using the rotating sector.

5. Numerical solutions and η values in the chorioallantoic capillary
        Figures 10 and 11 show the numerical solutions for O_2 and CO
diffusion in the chicken RBC, which were computed with changes in the η
value from 0.05 to 2.5x10^{-6} cm·sec^{-1}·Torr^{-1}. Initial and final Po_2 and
Pco values were the same as those in the computations in the human RBC.
The closed circles and bars are mean and SD of the 11 measurements, which
agreed well with the numerical solutions. The estimated ηo_2 and ηco both
ranged from 0.1 to 0.4 x10^{-6} cm·sec^{-1}·Torr^{-1} and averaged 0.17x10^{-6}
cm·sec^{-1}·Torr^{-1}, showing good agreement.

DISCUSSION

        In the present study, we measured sequentially the reaction rates
of O_2 and CO in both the human RBC and the chicken RBC in the
chorioallantoic capillary, with a microphotometric device. In addition,
we calculated the numerical solutions for the partial differential

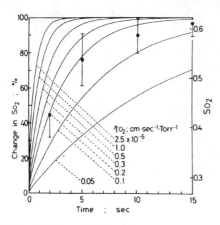 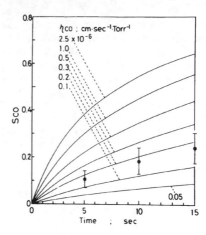

Fig. 10 Numerical solutions and mean ± SD of measured data for O_2 diffusion in the chorioallantoic capillary (n=11).

Fig. 11 Numerical solutions and mean ± SD of measured data for CO diffusion in the chorioallantoic capillary (n=11).

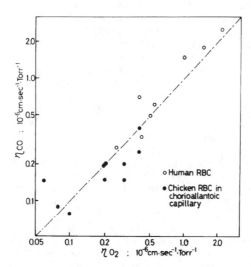

Fig. 12 Relationship between η_{O_2} and η_{CO} in the experiments on human and chicken RBCs. The open and closed circles are the data for the human RBC and the chicken RBC in the chorioallantoic capillary, respectively.

equations of O_2 and CO diffusion. From the measured and calculated reaction rates, we estimated the diffusion rates across the boundary layer, including the RBC and capillary membranes for O_2 and CO, i.e., η_{O_2} and η_{CO}. The relationship between η_{O_2} and η_{CO} estimated in the human and chicken RBC are shown in Fig. 12. The two η values for both types of RBC showed the same linear relation, suggesting the resistance of the RBC and capillary membrane is comparable for O_2 and CO. If the diffusion across the boundary layer of the RBC satisfies Graham's law, the diffusion rate would be proportional to the solubility and inversely proportional to the square root of the molecular weight. From this viewpoint, Meyer et al. (1981) stated that DL_{O_2} would be 1.23 times DL_{CO}. However, our present results indicate that Graham's law may not be satisfied as far as the boundary layer of the RBC concerned.

Savoy et al. (1980) and Meyer et al.(1981) measured DLo_2 and $DLco$ and compared the two DL values. However, $DLco$ is influenced not only by diffusion but also by reaction with hemoglobin. Consequently, $DLco$ changes widely with changes in Po_2 (Uchida et al., 1986). On the other hand, the ηco value is independent of Po_2. It depends only on the resistance of the RBC and capillary membrane. Therefore, in comparing the diffusion processes for O_2 and CO in the RBC and capillary membrane, comparison between η values will be advantageous.

SUMMARY

The reaction rates of O_2 and CO with the human and chicken red blood cell (RBC) were measured by using a microphotometric apparatus. In the experiments on the human RBC, a small amount of RBCs were put in an air-tight reaction cuvette. Gas mixtures containing various concentrations of O_2 and CO were sequentially injected into the cuvette and the change in O_2 and CO saturation of hemoglobin was measured from the change in transmission of the RBCs at 402 and 416.5 nm. The reaction rate of CO with RBCs was significantly influenced by photodissociation of carboxyhemoglobin (COHb). To eliminate this, a short-pass filter (400 to 435 nm) and a sector (100 Hz) were used. By comparing the measured reaction rates of O_2 and CO with the theoretical rates obtained from the numerical solutions of the partial differential equations of the diffusions of O_2 and CO, the transfer coefficients of O_2 and CO (ηo_2 and ηco) in the RBC boudary, including the RBC membrane and water layer around the RBC, were estimated. Both the values showed good agreement, ranging from 0.3 to 2.5 x 10^{-6} cm·sec^{-1}·$Torr^{-1}$. Furthermore, the chorioallantoic capillary of chicken embryo was used for the measurements of the reaction rates of O_2 and CO with RBC through the capillary membrane. The reaction rates of O_2 and CO in the chorioallantoic capillary were slower than those obtained in the human RBC. By comparing the measured reaction rates and the numerical solutions, the ηo_2 and ηco in the boundary, including the capillary membrane, plasma, and RBC membrane, were estimated. These two values ranged from 0.1 to 0.4x10^{-6} cm·sec^{-1}·$Torr^{-1}$ and showed good agreement. These results suggest that the diffusion rates for O_2 and CO across the capillary and RBC membrane are similar.

We aknowledge Prof. Dr. Masaji Mochizuki for his helpful advice on the measurements and theoretical computations, and Prof. Dr. Hiroshi Tazawa for his technical advice on the experiments on the chorioallantoic capillary. We are grateful to Prof. Dr. Katsuhiko Doi for his helpful cooperation in this study. This study was supported by Grant-in-Aids for Scientific Research from the Ministry of Education, Science and Culture of Japan (No.61770063 and 62770062).

REFERENCES

Allen, T. A. and Root, W. S. (1957) Partition of carbon monoxide and oxygen between air and whole blood of rats, dogs and men as affected by plasma pH. J. Appl. Physiol. 10:186–190.
Douglas, C. G., Haldane, J. S., and Haldane, J. B. S. (1912) The laws of combination of hemoglobin with CO and O_2. J. Physiol. 44: 275–304.
Douglas, J., Jr. and Rachford, H. H., Jr. (1956) On the numerical solution of the heat conduction problems in two and three space variables. Trans. Am. Math. Soc. 82: 421–439.

Hill, A. V. (1910)   The possible effects of the aggregation of the molecules of hemoglobin on its dissociation curve.   J. Physiol. 40: 4P.

Kagawa, T. and Mochizuki, M.   (1982)   Numerical solution of partial differential equation describing oxygenation rate of the red blood cell.   Jpn. J. Physiol. 32: 197–218.

Meyer, M., Scheid, P., Riepl, G., Wagner, H.-J., and Piiper, J.   (1981) Pulmonary diffusing capacities for O_2 and CO measured by a rebreathing technique.   J. Appl. Physiol. 51: 1643–1650.

Mochizuki, M., Tazawa, T., and Ono, T.   (1973)   Microphotometry for determining the reaction rate of O_2 and CO with red blood cells in the chorioallantoic capillary.   In: Oxygen Transport To Tissue. ed. by Bruley, D.F. and Bicher, H.I.   Plenum Publ. New York, pp. 997–1006.

Romanoff, A. L. (1960)   The avian embryo.   Structural and functional development.   Macmillan Company, New York.

Savoy, J., Michoud, M.-C., Robert, M., Geiser, J., Haab, P., and Piiper, J.   Comparison of steady state pulmonary diffusing capacity estimates for O_2 and CO in dogs.   Respir. Physiol. 42: 43–59.

Sendroy, J., Jr., Liu, S. H., and Van Slyke, D. D.   (1929)   The gasometric estimation of the relative affinity constant for carbon monoxide and oxygen in whole blood at 38 C.   Am. J. Physiol. 90: 511

Tazawa, H. and Mochizuki, M.   (1976)   Rates of oxygenation and Bohr shift of capillary blood in chick embryos.   Nature, 261: 509–511.

Tazawa, H. and Ono, T.   (1974)   Microscopic observation of the chorioallantoic capillary bed of chicken embryos.   Respir. Physiol. 20: 81–89.

Tazawa, H., Ono, T., and Mochizuki, M.   (1974)   Reaction velocity of carbon monoxide with blood cells in the chorioallantoic vascular plexus of chicken embryos.   Respir. Physiol. 20: 161–170.

Tazawa, H., Ono, T., and Mochizuki, M.   (1976)   Oxygen dissociation curve for chorioallantoic capillary blood of chicken embryo.   J. Appl. Physiol. 40: 393–398.

Uchida, K., Shibuya, I., and Mochizuki, M.   (1986)   Simultaneous measurements of cardiac output and pulmonary diffusing capacity for CO by a rebreathing method.   Jpn. J. Physiol. 36: 657–670.

COLOR ANALYSIS METHOD FOR STUDYING OXYGEN TRANSPORT IN HEMOGLOBIN

SOLUTIONS USING AN IMAGE-INPUT AND -PROCESSING SYSTEM

Masafumi Hashimoto, Ryuji Hata, Itiro Tyuma,
Akio Isomoto*, and Mitsuro Uozumi**

Department of Physico-chemical Physiology, Osaka University
Medical School, Osaka 530, *Department of General Education
Kinki University, Higashiosaka 577, and **Osaka Prefectural
Institute of Public Health, Osaka 537, Japan

INTRODUCTION

Since Scholander (1960) and Wittenberg (1959) found that the diffusion of oxygen in aqueous solution is facilitated in the presence of hemoglobin, its dynamic behavior has been investigated extensively by many
researchers (Hemmingsen and Scholander, 1960; Hemmingsen, 1962; Wittenberg,
1966; Kutchai and Staub, 1969; Spaan, Kreuzer, and Wely, 1980). The phenomena were observed not only in the steady state of oxygen transport but
also in the transient state to equilibrium. Scholander and Wittenberg
wetted a Millipore membrane with hemoglobin solution and estimated the membrane permeability to oxygen using a gas analysis technique. On the other
hand, Spaan et al. measured the oxygenation rate of hemoglobin by spectrophotometry when oxygen was flushed onto the surface of the thin-layered
hemoglobin solution. In both cases, the enhanced transport of oxygen by
hemoglobin was observed.

Although Scholander and Hemmingsen (1960) and Enns (1964) proposed
a "bucket-brigade" model, i.e., the enhanced transport occurs because oxygen exchange takes place when hemoglobin molecules collide, the following
model is now accepted. Oxygen and oxyhemoglobin diffuse independently at
rates of transport which are in proportion to their concentration gradients, with their individual diffusion coefficients; thus the total oxygen
transport will be the sum of diffusion of oxyhemoglobin and of unbound
oxygen. Based on this model, partial differential equations can be obtained with respect to time and position, in which the diffusion phenomena,
dissociation and association of both molecules are expressed. To analyze
the experimental data, the differential equations were integrated analyt-
ically (Snell, 1965; Wyman, 1966; Kutchai, Jacquez, and Mather, 1970;
Kreuzer and Hoofd, 1970; Kreuzer and Hoofd, 1972) and numerically by a
computer (Moll, 1968; Kreuzer and Hoofd, 1976), and the solutions obtained
under some specified boundary conditions and parameter values, agreed with
the results of Scholander and Wittenberg. In most cases, however, the
authors made no attempt to clarify the course of the gradients of oxygen
and oxyhemoglobin or only assumed the existence of linear profiles. Moreover, the experimental data were confined to the gross behavior of the
transport for a restricted range of the reaction space, and did not provide information about the concentration profiles in the space. Thus,

the validity of the equations was not directly verified. This prompted
us to develop a method for observing oxygen transport in a position-sensitive way and for obtaining the concentration profiles. We have now succeeded in establishing the reliable method using an image-input and -proc-
essing system including a microscope.

Our system is composed of a 3-tube video camera and a microcomputer-
aided color image analyzer, which resolves and quantizes an image into 512
× 480 pixels and the color of each pixel into several stages of brightness
of red (R), green (G), and blue (B) components. Since the R, G, and B
brightness values indicate the intensities of light coming into the camera
in the corresponding wave bands, we were able to introduce a spectrophotometric principle from the Beer-Lambert law and then estimate the oxygen
saturation of hemoglobin at each pixel from the values of R, G, and B
brightness based on analogy with the spectrophotometric method (Experiment
I). We measured the oxygen transport in hemoglobin solutions in glass
capillary tubes using this method (Experiment II). The results indicated
the usefulness of the method and provided information about the nature of
facilitated diffusion, particularly the course of the concentration gradients of oxyhemoglobin.

MATERIALS AND METHODS

Hemoglobin Sample

Hemoglobin was obtained from human erythrocytes using Drabkin's method
(1946) and was dissolved in 0.1 M (pH 7.0) potassium phosphate buffer. The
oxygen saturations were changed using a tonometric procedure and the oxygen saturation and hemoglobin concentration were determined with a spectrophotometer (Hitachi 320L). All experiments were carried out at room temperature (20°C).

Experimental System

Experiment I. Hemoglobin solutions at different concentrations and
oxygen saturations were photographed with a 3-tube video camera (Ikegami
Tsushinki Co., Tokyo, Model ITC-730). The solution was placed in a stand-

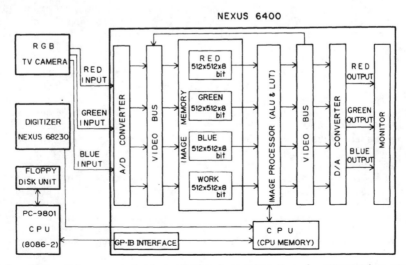

Fig. 1. Schematic presentation of the image-input and -processing system.

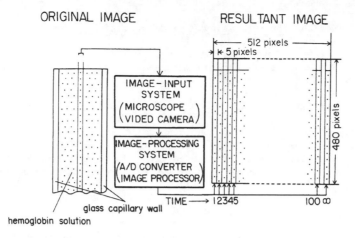

ORIGINAL IMAGE RESULTANT IMAGE

512 pixels
5 pixels

IMAGE-INPUT
SYSTEM
(MICROSCOPE
VIDED CAMERA)

IMAGE-PROCESSING
SYSTEM
(A/D CONVERTER
IMAGE PROCESSOR)

480 pixels

TIME ⟶ 12345   100 ∞

glass capillary wall

hemoglobin solution

Fig. 2. Schematic presentation of the system and procedure of Experiment II.

ard optical cell with a light path length of 1 cm and set on the optical
axis of the video camera with a white screen background and oblique illu-
mination from a white light bulb (100V-60W). The image analyzer used was
a Nexus 6400 (Kashiwagi Res. Corp., Tokyo) which resolves and quantizes an
image obtained via the video camera into 512×480 pixels and the color of
each pixel into 256 ($=2^8$) stages of R, G, and B brightness, then stores the
digital image information of $512 \times 480 \times 3 \times 8$ bits in R-, G-, and B-frame
memories addressed for each pixel and, if necessary, in a floppy disk of
1 Mbyte. A personal computer (NEC, Tokyo, Model PC-9801-F2) was connected
to the Nexus 6400 through a GP-IB interface to allow programmed processing
commands to be input into the image analyzer, and a digitizer (Nexus 68230)
was used to indicate the position or range of hemoglobin color to be proc-
essed. When the sample objects were photographed, the gamma-correction
channel of the video camera was switched off, to have the input values of
R, G, and B brightness proportional to the corresponding light energies
coming into the video camera (Fig. 1).

Experiment II. A 3-tube video camera (Ikegami Tsushinki Co., Tokyo,
Model ITC-350M) and the Nexus 6400 system described in the previous section
were connected with a microscope (Olympus Inc., Tokyo, Model BH-2). A
deoxyhemoglobin solution prepared by flushing pure nitrogen gas (99.99%)
over a hemoglobin solution was put in a glass capillary tube with a 1 ml
syringe connected anaerobically to one end of the tube. Oxygen began to
diffuse into the hemoglobin solution from the other end of the tube which
was open to air. The experiment was started when a portion of the oxygenated hemoglobin was discarded by operating the syringe and a new surface
of the deoxyhemoglobin solution appeared at the open end of the tube. The
5-pixel width and 480-pixel length images along the center axis of the tube
were entered in the R-, G-, and B-frame memories at intervals of 2.9 seconds, thus arranging 100 images of 5-pixel width from left to right with
time, and the image at about 20 minutes after the beginning of the experiment at the right end. The resultant image was transformed using a built-
in function of the image processor as the R, G, and B values of any pixel
were replaced with the median of the neighboring 5×5 pixels, which allowed us to reduce noise in the image. Thus, using an algorithm for estimating the oxygen saturation of hemoglobin from the R, G, and B brightness
values, we were able to know the saturation at any desired time and position of the experimental process (Fig. 2).

Computer Algorithm for Estimating Oxygen Saturation of Hemoglobin

Since the R, G, and B brightness values indicate the intensities $[I(i), i = R,G,B]$ of the light transmitted through an object in the corresponding wave bands, we can employ the Beer-Lambert law:

$$- \log I(i)/I_0(i) = \varepsilon(i)D, \quad i = R,G,B, \qquad [1]$$

where $I_0(i)$ is the intensity of incident light; $\varepsilon(i)$ is the absorption coefficient; D is the density of the object defined by the concentration multiplied by the effective light path length. This gives the following expression for the optical densities of two different color components:

$$- \log I(i)/I(j) + \log I_0(i)/I_0(j) = D[\varepsilon(i) - \varepsilon(j)], \qquad [2]$$
$$i = R,G,B, \quad j = R,G,B, \quad i \neq j.$$

Since $\log I_0(i)/I_0(j)$ has a constant value determined by the incident light (equal to zero in the case of white incident light), this equation indicates that $\log I(i) - \log I(j)$ is linear to D regardless of the luminosity, and in the xy-coordinate space defined as $x = \log I(G) - \log I(B)$ and $y = \log I(R) - \log I(G)$, the hue of the object's color is represented by the direction from the point $[x = \log I_0(G) - \log I_0(B) = C_1$ and $y = \log I_0(R) - \log I_0(G) = C_2$, where C_1 and C_2 are constants] and the density by the distance from that point. In this space, the R, G, and B values of the color of hemoglobin solution at the oxygen saturation Y $(0 \leq Y \leq 1)$ are indicated by point S, the xy values of which can be expressed with parameters D and Y:

$$x = DY[\varepsilon_1(B) - \varepsilon_1(G)] + D(1 - Y)[\varepsilon_0(B) - \varepsilon_0(G)] - C_1 \qquad [3]$$

$$y = DY[\varepsilon_1(G) - \varepsilon_1(R)] + D(1 - Y)[\varepsilon_0(G) - \varepsilon_0(R)] - C_2 \qquad [4]$$

where ε_1 and ε_0 are the absorption coefficients of oxy- and deoxy-hemoglobin, respectively. When Y varies from 0 to 1 at fixed D, S moves from S_0 to S_1 along the line $S_0 S_1$ in proportion to the value of Y (S_1 and S_0 denote S for oxy- and deoxy-hemoglobin, respectively). Thus, we can readily obtain the Y values from the actual locations of the points S, S_1 and S_0 in the space (Hashimoto et al., 1987).

RESULTS

Experiment I

The values of $\log I(R) - \log I(G)$ at various oxyhemoglobin concentrations (represented by ΔA_{588nm}) are shown in Fig. 3. The plots of the three experiments under slightly different lighting conditions show straight lines for hemoglobin concentrations below 0.83×10^{-4} M (ΔA_{588nm} = 0.80), having similar slopes but different ordinal intercepts. Assuming that the effective light path length does not change during a series of experiments, the results indicate that $\log I(R) - \log I(G)$ is linearly related to D with a slope independent of the luminosity. The same relationships but different slopes were found for $\log I(G) - \log I(B)$ versus oxyhemoglobin concentration, $\log I(R) - \log I(G)$ versus deoxyhemoglobin concentration, and $\log I(G) - \log I(B)$ versus deoxyhemoglobin concentration.

When the oxygen saturation was changed at a fixed hemoglobin concentration, the representative points in the xy-coordinate space also moved

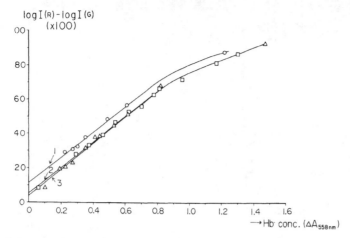

Fig. 3. Relationship between log I(R) - log I(G) and oxyhemoglobin concentration. The three experimental series under different lighting
conditions are shown.

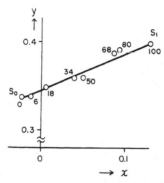

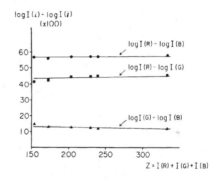

Fig. 4. Distribution of points
corresponding to various
oxygen saturations in the
xy-coordinate space. The
numbers beside the symbol
o indicate the percentage
fractional saturation.

Fig. 5. Experimental relationship
between log I(i) - log I(j)
and z value. The luminosity was changed by controlling the distance between
the sample and the light
bulb.

on a straight line, from point S_0 to S_1 where deoxy- and oxy-hemoglobin
appeared in the space, respectively, in proportion to their values of oxygen saturation (Fig. 4).

Figure 5 shows the experimental relationship between log I(i) - log
I(j) (i = R,G,B, j = R,G,B, i ≠ j) and z [= I(R) + I(G) + I(B)], when the
luminosity was changed. As the z value represents luminosity, the results
indicate that the values of log I(i) - log I(j) are independent of the
luminosity.

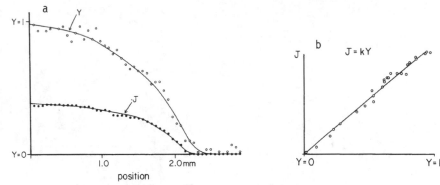

Fig. 6. One of the results obtained from the R, G, and B values of brightness of the 5-pixel width images. (a) Profiles of oxygen saturation (Y) and flux of oxygen (J). (b) Relationship between J and
Y at each position in (a). The J values are shown by an arbitrary unit.

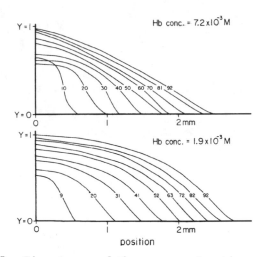

Fig. 7. Time course of the oxygen saturation profile.

Experiment II

Figure 6a shows the oxygen saturation profile at a specific time in
our experiment (open circles). The data points for oxy- and deoxy-hemoglobin were obtained from the R, G, and B brightness values of the 5-pixel
width images at the right end and the left end, respectively. Each point
indicates the oxygen saturation obtained from the R, G, and B brightness
values of the corresponding graphic element; the solid curve was drawn by
eye to be at the center of the plots. The hemoglobin at the surface exposed to air must have been fully saturated with oxygen, but it could not
be observed because of shading by the tube edge.

Figure 7 shows the time course of the oxygen saturation profile. The
profile at any time appeared to have a common shape with a plateau and a

cliff, so that the fairly clear-cut diffusion front could be readily detected by its color. It moved as fast in the center of the capillary tube as it did near the wall.

We can obtain the oxyhemoglobin concentration profile by multiplying the values of oxygen saturation by the hemoglobin concentration. By numerically integrating the oxyhemoglobin concentration curve from an arbitrarily indicated position to the diffusion front in the diffusing direction, we can obtain the sum of the oxygen flux up to the time at that position, the differential of which with respect to time (2.9-second interval) gives the flux of oxygen at that time and position. Figure 6a shows the flux against their position (closed circles), and Fig. 6b shows the relationship between flux and oxygen saturation at each position. This relationship leads to the empirical formula $J = kY$, where J is the flux, Y is the oxygen saturation of hemoglobin, and k is a constant.

DISCUSSION

The present study showed that the image-input and -processing system is very useful and versatile for quantitative analysis of hemoglobin color to measure oxygen saturation. The method was developed based on the spectrophotometric principle of the Beer-Lambert law. The linear relationship between hemoglobin concentration and the difference between the logarithms of the brightness values of the two color components confirmed its applicability, enabling us to estimate the oxygen saturation of hemoglobin by knowing the R, G, and B brightness values for oxy- and deoxy-hemoglobin. Luminosity (the intensity of incident light) was able to be ignored because the difference of the logarithms of the brightness values of the two color components is independent of luminosity as ascertained by our theory and experiments. This allowed us to analyze the hemoglobin color even in microscopic images which were not uniformly lighted. Our 3-tube video camera and image analyzer can be considered as a three-wavelength spectrophotometer sensitive to the two-dimensional spatial distribution of an object's color. Thus, the method could be used to study the oxygen transport in a hemoglobin solution.

As shown in the experimental results mentioned above, we are now able to describe the diffusion behavior of oxygen quantitatively; the flux of oxygen at any point of the diffusion path can be estimated by measuring the distribution of oxygen saturation of hemoglobin giving the empirical formula $J = kY$. The distribution of oxygen saturation does not show a linear gradient of unbound oxygen since the plateau level does not always correspond with full saturation. Further, the distribution does not reflect the sigmoidal oxygen equilibrium curve of hemoglobin. Therefore, Fick's diffusion model cannot explain our experimental results. The relation, $J = kY$, suggests an unidirectional transport, such as convection or electrophoresis. These are unlikely, however, in the absence of a disorder of the diffusion front and an electric field. Another unidirectional transport mechanism was proposed by Enns (1964), who assumed that oxygen is transported similarly to heat conduction. Heat conduction itself like diffusion, is not however, an unidirectional phenomenon in a sense that it occurs in proportion to temperature gradient. Questions remain about the phenomena, and to elucidate these problems an investigation including the reaction of oxygen and hemoglobin is in progress.

SUMMARY

A method for quantitative analysis of hemoglobin color to estimate the oxygen saturation was developed. The method uses an image-input and

-processing system composed of a 3-tube video camera and a digital image analyzer. Using the system connected to a microscope, facilitated diffusion of oxygen in hemoglobin solutions was observed and analyzed in a position-sensitive way. The results confirmed its applicability to this study and gave information about the diffusion mechanism expressed by the empirical formula $J = kY$, where J is the flux of oxygen, Y is the oxygen saturation of hemoglobin, and k is a constant.

ACKNOWLEDGMENT

The authors thank Dr. Y. Ueda of Kansai College of Acupuncture Medicine for a gift of the hemoglobin sample.

REFERENCES

Drabkin, D. L., 1946, The crystallographic and optical properties of the hemoglobin of man in comparison with those of other species, J. Biol. Chem., 164:703-727.

Enns, T., 1964, Molecular collision-exchange transport of oxygen by hemoglobin, Proc. Nat. Acad. Sci., 51:247-252.

Hashimoto, M., Hata, R., Isomoto, A., Tyuma, I., and Fukuda, M., 1987, Color analysis method for estimating the oxygen saturation of hemoglobin using an image-input and processing system, Anal. Biochem., 162:178-184.

Hemmingsen, E., and Scholander, P. F., 1960, Specific transport of oxygen through hemoglobin solutions, Science, 132:1379-1381.

Hemmingsen, E., 1962, Accelerated exchange of oxygen-18 through a membrane containing oxygen-saturated hemoglobin, Science, 135:733-734.

Kreuzer, F., and Hoofd, L., 1970, Facilitated diffusion of oxygen in the presence of hemoglobin, Respir. Physiol., 8:280-302.

Kreuzer, F., and Hoofd, L., 1972, Factors influencing facilitated diffusion of oxygen in the presence of hemoglobin and myoglobin, Respir. Physiol., 15:104-124.

Kreuzer, F., and Hoofd, L., 1976, Facilitated diffusion of CO and oxygen in the presence of hemoglobin or myoglobin, Adv. Exp. Med. Biol., 75:207-215.

Kutchai, H., and Staub, N., 1969, Steady-state hemoglobin-facilitated O_2 transport in human erythrocytes, J. Gen. Physiol., 53:576-588.

Kutchai, H., Jacquez, J., and Mather, F., 1970, Nonequilibrium facilitated oxygen transport in hemoglobin solution, Biophys. J., 10:38-54.

Moll, W., 1968/1969, The influence of hemoglobin diffusion on oxygen uptake and release by red cells, Respir. Physiol., 6:1-15.

Scholander, P. F., 1960, Oxygen transport through hemoglobin solutions, Science, 131:585-590.

Snell, F. M., 1965, Facilitated transport of oxygen through solutions of hemoglobin, J. Theoret. Biol., 8:469-479.

Spaan, J. A. E., Kreuzer, F., and van Wely, F. K., 1980, Diffusion coefficients of oxygen and hemoglobin as obtained simultaneously from photometric determination of the oxygenation of layers of hemoglobin solutions, Pflügers Arch., 384:241-251.

Wittenberg, J. B., 1959, Oxygen transport —— a new function proposed for myoglobin, Biol. Bull., 117:402-403.

Wittenberg, J. B., 1966, The molecular mechanism of hemoglobin-facilitated oxygen diffusion, J. Biol. Chem., 241:104-114.

Wyman, J., 1966, Facilitated diffusion and the possible role of myoglobin as a transport mechanism, J. Biol. Chem., 241:115-121.

CLINICAL EVALUATION OF CONTINUOUS VENOUS OXYGEN

SATURATION MONITORING DURING ANESTHESIA

Sho Yokota, Masako Mizushima, Osamu Kemmotsu, Shigeo Kaseno, Tomomasa Kimura, Yukiko Goda, and Kazuo Sasaki

Department of Anesthesiology, Hokkaido University School of Medicine, Sapporo 060, Japan

INTRODUCTION

A new pulmonary artery flow-directed balloon catheter combined a fiberoptic reflectometry system (Oximetrix, Opticath) provides a continuous digital display of mixed venous oxygen saturation ($S\bar{v}O_2$) together with the usual intermittent blood sampling, and pressure and flow measurement capabilities of the thermodilution pulmonary catheter. Some reports[1-3] have described a good correlation between on-line $S\bar{v}O_2$ and hemodynamic changes, although one concluded that there was no significant correlation between $S\bar{v}O_2$ and cardiac index[4]. This study therefore was designed to evaluate this fiberoptic reflectometry system in patients during anesthesia by comparing in vivo $S\bar{v}O_2$ measurements with those obtained in vitro $S\bar{v}O_2$ using a Radiometer ABL-300. We hoped to be able to correlate on-line $S\bar{v}O_2$ measurements with intermittently measured and derived oxygen transport variables.

METHODS

Fourteen adult patients were studied, who were scheduled for elective major abdominal surgery and had indications for pulmonary artery catheter insertion. An informed consent was obtained from each patient at the time of the pre-anesthesia visit. Anesthesia was induced by intravenous administration of thiamylal and the trachea was intubated with the aid of succinylcholine. Anesthesia was maintained with enflurane/nitrous oxide in oxygen and respiration was adjusted to keep normocapnia. A fiberoptic oxymetric catheter (Opticath) was inserted through the right internal jugular vein and correct positioning of the catheter in the pulmonary artery was confirmed by continuous monitoring of the pressure waves. The right radial artery was cannulated for both arterial pressure monitoring and blood gas analysis. Continuous monitoring of arterial, central venous and pulmonary arterial pressures was made together with measurement of end-tidal CO_2, $S\bar{v}O_2$, arterial oxygen saturation (SaO_2) by pulse-oxymetry, ECG and cardiac output (CO), during the whole procedure. Serial measurements of hemodynamic variables, together with arterial and venous blood gas analyses were made following induction of anesthesia, 5 minutes before and after incision of the skin and anytime during the procedure when on-line SvO_2 values increased or decreased by 5% or more. The following parameters of oxygen transport were caluculated by derivation from arterial and venous contents (CaO_2 and $C\bar{v}O_2$) measured with a Radiometer ABL-300:

Oxygen delivery $(\dot{D}O_2)$ = CO x CaO_2, Oxygen consumption $(\dot{V}O_2)$ = CO x $(CaO_2 - C\bar{v}O_2)$, Oxygen extraction ratio (O_2ER) = $(CaO_2 - C\bar{v}O_2)/CaO_2$.

RESULTS

Regression analysis of the 94 paired in vivo and in vitro SvO_2 measurements showed a good correlation (r = 0.885, p < 0.001) (Fig. 1). There was a significant positive correlation between increases or decreases in on-line $S\bar{v}O$ values of more than 5% and corresponding changes in cardiac index (r = 0.849, p < 0.001) (Fig. 2). There were 21 instances during study when $S\bar{v}O_2$ values changed by 5% or more, decreasing 8 times and increasing 13 times. There was a significant negative correlation between on-line $S\bar{v}O_2$ values and O_2ER (r = -0.733, p < 0.001) (Fig. 3). However, regression analysis of on-line $S\bar{v}O_2$ values and $\dot{D}O_2$ or $\dot{V}O_2$ showed poor correlations.

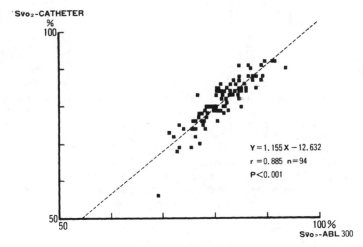

Fig. 1. $S\bar{v}O_2$ as calculated from ABL-300 plotted against value obtained from oxymetry system.

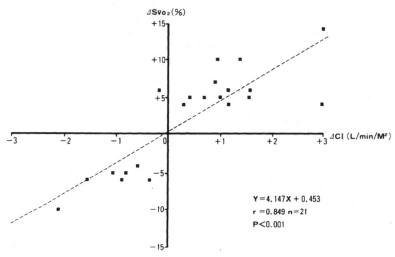

Fig. 2. Correlation of on-line $S\bar{v}O_2$ changes with corresponding changes in cardiac index (CI).

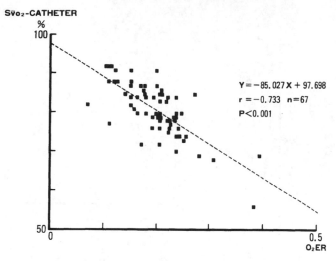

Fig. 3. $S\bar{v}O_2$ vs Oxygen extraction ratio (O_2ER)

DISCUSSION

These data indicate that during anesthesia this fiberoptic reflectometry system provides accurate, on-line and in vivo $S\bar{v}O_2$ values in the
range of 60% or more. These data are accordance with those of previous reports[1,4]. Our data also show that changes of $S\bar{v}O_2$ values closely reflect
changes of cardiac index as reported by Waller, et al[1]. Sudden decreases
of on-line $S\bar{v}O_2$ can warn of deleterious changes in hemodynamic variables
and/or oxygen transport variables. Fig. 4 demonstrates a sharp decrease
of $S\bar{v}O_2$ caused by shivering during recovery from anesthesia, which indicates
VO_2 was greatly increased. $S\bar{v}O_2$ was restored to normal levels after intravenous administration of diazepam and oxygen by face mask.

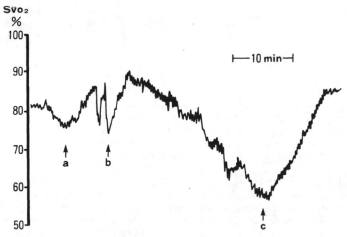

Fig. 4. A sudden drop of $S\bar{v}O_2$ seen by postoperative
shivering in a 55 yrs-old female.
(a: suction, b: extubation, b-c: shivering)

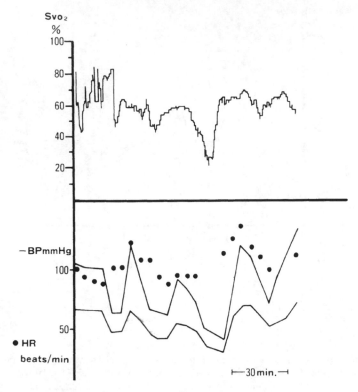

Fig. 5. Changes of on-line $S\bar{v}O_2$ values during massive
bleeding in liver resection of a 73 yrs-old male.

Fig. 5 shows a simultaneous tracing of $S\bar{v}O_2$, arterial blood pressure, and heart rate. Extreme changes of $S\bar{v}O_2$ and blood pressure can be observed by massive bleeding and blood transfusion. These results demonstrate that this system permits us to identify sudden hemodynamic changes as soon as they occur in a patient, by providing continuous monitoring of $S\bar{v}O_2$. It also shows that the system enhances the efficacy of patient management by permitting immediate assessment of response to therapeutic interventions[2].

According to Fick's principle, $S\bar{v}O_2 = SaO_2 - \dot{V}O_2/CO$ when hemoglobin values are relatively constant. In clinical anesthesia such as ours, oxygenation of arterial blood is almost complete with an SaO_2 of nearly 100% while $\dot{V}O_2$ is relatively stable during the anesthetized state. Therefore, it is reasonable to conclude that $S\bar{v}O_2$ reflects cardiac output. However, this is not always true as seen in fig. 4, in cases where $\dot{V}O_2$ is increased by shivering or excitement during recovery from anesthesia. In our study, we showed that $S\bar{v}O_2$ correlates well with O_2ER.

O_2ER can be simplified to $1 - S\bar{v}O_2/SaO_2$ by removing cardiac output and this equation can be expressed as $1 - S\bar{v}O_2$ since arterial saturation is usually close to 100%. $S\bar{v}O_2$ can be expected to correlate well with O_2ER. Furthermore, our data indicate that continuous $S\bar{v}O_2$ monitoring can provide online information not only about hemodynamic state but also on oxygen transport.

In conclusion, continuous monitoring of $S\bar{v}O_2$ during anesthesia is clinically valuable, since 1) on-line values for $S\bar{v}O_2$ were closely related to those obtained in vitro from a Radiometer ABL-300, 2) there was a significant correlation between increases or decreases by 5% or more in $S\bar{v}O_2$ values and corresponding changes in cardiac index, 3) there was a significant correlation between $S\bar{v}O_2$ and percentage oxygen extraction from blood, and 4) continuous monitoring of $S\bar{v}O_2$ can provide an indication of deterioration of cardiopulmonary function.

SUMMARY

Clinical evaluation of continuous $S\bar{v}O_2$ monitoring during anesthesia was made in 14 surgical patients utilizing a fiberoptic reflectometry system combined with a pulmonary artery flow-directed balloon catheter. On-line values for $S\bar{v}O_2$ by the system were closely related to those obtained in vitro from a Radiometer ABL-300. There was a good correlation between changes of in vivo $S\bar{v}O_2$ and corresponding changes in cardiac index. We also observed that there was a significant correlation between $S\bar{v}O_2$ and oxygen extraction ratio. Our data indicate that continuous $S\bar{v}O_2$ monitoring during anesthesia can provide on-line information not only about hemodynamic state but also on oxygen transport.

REFERENCES

1. J. L. Waller, J. A. Kaplan, D. I. Bauman, et al: Clinical evaluation of a new fiberoptic catheter oximeter during cardiac surgery. Anesth Analg 61:676 (1982).
2. H. Birman, A. Hag, E. Hew, et al: Continuous monitoring of mixed venous oxygen saturation in hemodynamically unstable patients. Chest 86:753 (1984).
3. J. M. Gore and K. Sloan: Use of continuous monitoring of mixed venous saturation in the coronary care unit. Chest 86:757 (1984).
4. P. L. Baele, J. C. McMichan, H. M. Marsh, et al: Continuous monitoring of mixed venous oxygen saturation in critically ill patients. Anesth Analg 61:513 (1982).

CONTINUOUS MEASUREMENT OF OXYGEN DELIVERY AND OXYGEN CONSUMPTION IN AWAKE LIVE ANIMALS

Setsuo Takatani*, Hiroyuki Noda*, Hiromasa Kohno**,
Hisateru Takano*, and Tetsuzo Akutsu*

*National Cardiovascular Center, Research Institute
 Suita, Osaka, Japan
*Terumo Inc., Research and Development, Fuji, Shizuoka
 Japan

INTRODUCTION

The spectrophotometric reflection method can provide accurate and rapid measurement of hemoglobin oxygen saturation (SO_2) by utilizing optical absorption differences of oxy- and deoxy-hemoglobin (Figure 1, Van Assendelft, 1970) and scattering characteristics at the red cell-plasma interface. Based on the reflection technique, various oximeters have been developed including the fiber-optic catheter (Cole et al., 1972;Baele et al., 1982), catheter-tip oximeters (Yee et al., 1977; Schmitt et al., 1986), and finger tip pulse oximeters (Mendelson et al., 1983). Though these systems can provide continuous and accurate measurement of hemoglobin oxygen saturation in various clinical settings at the bed-side, there is not yet a clinically reliable in vivo oximeter that can provide simultaneous measurements of both hemoglobin content ([Hb]) and oxygen saturation. Thus, when measurement of cardiac output or oxygen consumption is required in the clinical setting, the blood is analyzed for its hemoglobin content and then the oxygen content derived. Also, the measurement of SO_2 is affected by the variations in the hematocrit which exist from person to person and which occur, for example, during cardio-pulmonary bypass, due to hemodilution. Calibration is thus necessary as the hematocrit shift occurs.

In this research, which involves continuous measurement of oxygen delivery and oxygen consumption in awake live animals, an optical sensor that can provide both hemoglobin content and oxygen saturation was designed and developed. Following the in vitro characterization, the sensors were incorporated into artificial hearts to measure continuously both arterial and mixed venous SO_2 and [Hb] and to derive oxygen consumption in awake live animals. This paper reports the design, in vitro testing and initial application of the sensors for continuous monitoring of oxygen delivery and consumption in the awake live animals.

THEORY OF MEASUREMENT

SO_2 Measurement
Based on the conventional dual wavelength reflection method, the ratio of the reflectances at the near-infrared (800nm) and red (665nm) wavelengths is used in the linear regression equation to derive SO_2 as follows;

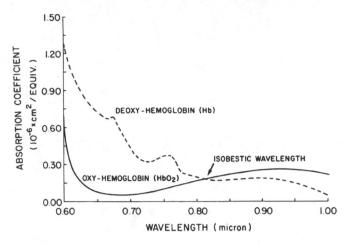

Figure 1. Absorption coefficients of oxy- and deoxy-hemoglobin in the red and near-infrared regions.

$$SO_2 = A + B \times (R800/R665) \tag{1}$$

where A and B are the constants that depend on physiological characteristics of the blood and on sensor characteristics, and R800 and R665 are reflectances at the 800 and 665nm wavelengths. Referring to Figure 1, 665nm is where the absorption difference between the oxy- and deoxy-hemoglobin is the largest and 800nm is the reference wavelength at which oxygenation does not affect optical absorption and is called the isobestic wavelength. This empirical relation was derived by Polanyi and Hehir(1962) in application to the fiber-optic catheter oximeter. However, due to the non-linear effect of scattering occurring at the interface of red blood cell and plasma, changes in the hematocrit shift the relation between the ratio and SO_2, thus yielding a family of curves. Also, the nonlinearity increases when the SO_2 level decreases below 60%, mainly due to increased absorption at the red wavelength. Thus, in order to increase the accuracy and to compensate for the variation in hematocrit, a higher order approximation and estimation of hematocrit is necessary to adjust the constants A and B in equation (1). Though a dedicated micro-computer may be used to follow hematocrit variation, such method may be tedious. In this research, based on a theoretical study using the three dimensional photon diffusion theory, an empirical relation has been proposed to compensate for the hematocrit effect and nonlinear effect by adding a compensation constant C in the denominator of the ratio term i.e. R800/(R665+C) (Takatani et al., 1987). Figure 2 shows the theoretical result based on the photon diffusion theory demonstrating that the addition of the compensation term in the denominator will eliminate the hematocrit-dependent shift of the curves as well as the non-linear effect at the lower SO_2 values. The constant term C added in the denominator acts to augment the reflectance changes at the red wavelength. Since the optical absorption at the red wavelength, particularly at the lower SO_2 levels, is considerably higher than that at 800nm, when the hematocrit decreases, its effect on the net reflectance at the red wavelength is much higher, thus shifting the curves. The value of C depends on the sensor geometry and the optimum value is close to the reflectance value of the red wavelength at low SO_2

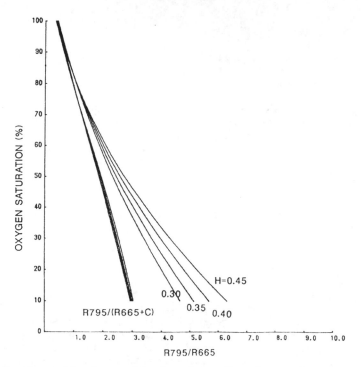

Figure 2. Theoretical results based on the photon diffusion theory demonstrating that the addition of constant C in the denominator eliminates the hematocrit-dependent shift and linearizes the SO_2 vs. ratio curves.

levels. Thus, its effect can be neglected when the SO_2 is high, but as the SO_2 gets smaller its effect becomes large enough to shift the slope of the curve to that given for high SO_2 values, thus linearizing the curve. Thus, by addition of the constant term C, although overall sensitivity may be lost, the scattering variation due to alteration in the hematocrit and the nonlinear effect at the lower SO_2 values can be minimized. Also, the continuous measurement of the hematocrit and adjustment of the constants A and B in equation (1) are not required.

Hemoglobin Content Measurement

The optical absorption at the isosbestic wavelength of the oxy- and deoxy-hemoglobin does not vary with the changes in the SO_2 (Figure 1), but the net reflectance varies when the concentration of the scattering particles changes. Hence, the reflectance at this wavelength can be utilized to obtain the concentration of scattering particles. The result based on the photon diffusion theory indicates that the reflectance at the 800nm isosbestic wavelength is a non-linear function of the hematocrit or hemoglobin content (Takatani et al., 1987; Takatani et al., 1980; Takatani et al., 1976). Using the second or third order equation, the hemoglobin content can be estimated as follows;

$$[Hb] = C_1 + C_2 \times R800 + C_3 \times R800^2 \tag{2}$$

where C_1, C_2, and C_3 are the constants that depend on the transducer and physiological characteristics of blood.

Oxygen Delivery and Oxygen Consumption Measurement

When the SO_2 and [Hb] are known, the oxygen content (OC) of whole blood can be derived as.

$$OC = 1.35 \times [Hb] \times SO_2 \qquad (3)$$

where 1.35 is the maximum oxygen binding capacity of 1gm hemoglobin and OC is expressed as the volume % or cc of oxygen/100ml of blood. Since at the pressure of room air the oxygen dissolved in the plasma is low in comparison to that chemically combined with hemoglobin, it was neglected in the analysis. Then, the oxygen delivered (OD) to the peripheral tissues and oxygen consumption ($\dot{V}O_2$) of the whole body can be derived when the cardiac output (CO) is known, as follows:

$$OD = OC_a \times CO \qquad (4)$$

$$\dot{V}O_2 = (OC_a - OC_v) \times CO \qquad (5)$$

where OC_a and OC_v are the oxygen content of arterial and venous blood, respectively.

DESCRIPTION OF THE OPTICAL SYSTEM

Optical Sensor

Figure 3 shows the block diagram of the optical sensor that was designed and fabricated (Takatani et al., 1987). The top of the TO-5 can was cut open and a specially designed substrate with light emitting diode (LED) and photodiode chips were inserted and wire connections were made between the chips and the external pins. The wavelengths of the LEDs used were 665 and 795nm and an SGD-040B photodiode (EG&G,USA) was used as a photodetector. The second wavelength of 795nm is very close to the isosbestic wavelength of 800nm and at this wavelength reflectance variation due to SO_2 is negligible. An isolator was placed between the LED and photodiode chips to prevent direct coupling effect. After the wiring was completed, the chip surface was covered with clear epoxy resin (Hysol), followed by polishing of its surface to minimize surface scattering. The separation between the LED and photodiode was about 3 mm, a distance based on photon diffusion analysis to ensure that the detected

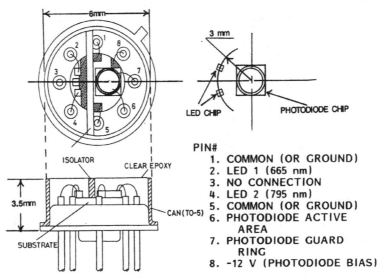

PIN#
1. COMMON (OR GROUND)
2. LED 1 (665 nm)
3. NO CONNECTION
4. LED 2 (795 nm)
5. COMMON (OR GROUND)
6. PHOTODIODE ACTIVE AREA
7. PHOTODIODE GUARD RING
8. -12 V (PHOTODIODE BIAS)

Figure 3. Schematic drawing of a hybrid type optical sensor showing the layout of LED and photodiode chips.

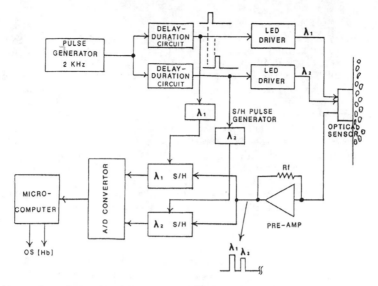

Figure 4. Schematic drawing of the reflection oximeter system.

light is free of surface scattering and also to ensure that the detector captures the diffused light coming from the deeper layer of the medium.

Instrumentation System

Figure 4 shows the instrumentation system. The 2 KHz pulse from the internal pulse generator is fed into the delay-duration circuit where two LED-driver pulses of 20 micro second duration separated by 20 micro second are generated. The LEDs are pulsed in order to obtain higher intensity output and also to eliminate DC drift problems. The reflected light from the whole blood is first pre-amplified, followed with gain stage amplification. Then the signal enters the sample/hold circuits where the reflectance at each wavelength is synchronously sampled and held. The sampled DC signal is then digitized for computer analysis. Prior to computation of the SO_2 and [Hb], signal averaging is done to increase signal-to-noise ratio; sampling rate and number of averages can be entered from the key-board at the time of experiments. In general, a sampling interval of 25msec is used and 20 data points are averaged to yield one data point for every 0.5 sec.

EXPERIMENTAL METHOD

In Vitro Evaluation

Prior to in vivo experiments, the sensor was evaluated in vitro using goat, bovine and human whole blood. The red blood cell size ranged from 30, 50 to 87 cubic micron in goat, bovine and human, respectively (Didisheim, 1985). Since the scattering is a function of wavelength-to-cell size, these three species presented a different scattering medium to test the hypothesis of SO_2 and [Hb] measurement. Fresh blood of each species was circulated through the optical sensor using the heart lung machine where blood PO_2 was varied by changing the mixing ratio of the N_2 and O_2 gases ventillated through the hollow fiber oxygenator. The blood pH was buffered at around 7.4 by balancing the gases with 5% CO_2. The hematocrit was varied by diluting the blood with physiological saline. At various values of SO_2, blood samples were obtained and their SO_2 and [Hb] were analyzed using a Hemoximeter OSM2.

In Vivo Evaluation

Initially, the sensor was evaluated in acute experiments using adult goats weighing around 30-40Kg. A pair of pneumatic artificial hearts (Takatani et al., 1984) were utilized to bypass the natural heart. The purpose of bypassing the natural heart by artifcial hearts is to separate the heart and peripheral organs, to control the pump output precisely at desirable levels, and to obtain independent responses of the peripheral circulatory system at various levels of oxygen delivered. The optical sensor was fixed in a tube so as to fit in the port of the artificial heart. Figure 5 shows the experimental set-up. Under gas anesthesia, the bypass circuit was prepared; the right bypass pump drained the blood from the right atrium and returned it to the pulmonary artery, while the left pump circulated blood from the left atrium to the aorta. The natural heart was fibrillated by electrical shock, allowing the artificial hearts to control the entire circulation. Thus, the optical sensor of the right pump measured the mixed venous SO_2 and [Hb], while that of the left determined arterial SO_2 and [Hb]. In the experiments, respiratory condition , pump flow and hematocrit were intentionally altered to evaluate the responses of both sensors in terms of [Hb] and SO_2.

After acute evaluation was completed, long term experiments were carried out to continuously monitor oxygen delivery and consumption in awake live animals whose circulation was entirely supported by a pair of pneumatic artificial hearts. Experimental procedure was the same as in the acute case. However, after completion of the bypass circuit, the chest was closed and the animal was returned to the cage and allowed to recover from the anesthesia. Postoperatively, hemodynamic variables including left and right atrial pressures, and pulmonary and systemic arterial pressures were continuously monitored using fluid filled catheters connected to the P-50 pressure transducers. The electromagnetic flow probes were mounted on the outflow graft of each pump to provide left and right pump flows. The optical sensor positioned in the outflow port of each pump provided arterial and mixed venous [Hb] and SO_2, from which data, in combination with hemodynamic data, oxygen delivery and consumption of the animal was continuously computed on-line using a microprocessor.

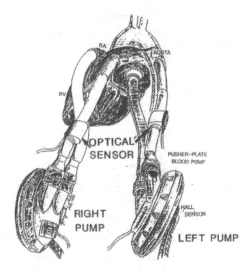

Figure 5. Bi-ventricular bypass arrangement showing how two artificial hearts are connected and where the optical sensors are mounted.

In Vitro Study

In the three species, the modification of the ratio term by addition of a constant minimized hematocrit-dependent shift of curves and also linearized the curve (Takatani et al., 1987). The optimum value of the constant C was dependent on the red blood cell size and increased from 0.02 to 0.09 in goat to human. Over the hematocrit range of 25-45%, the standard errors improved from about 8% to 2%. As for hemoglobin content measurement, reflectance at 795nm revealed a nonlinear relationship and the second order approximation yielded standard errors of about 0.3-0.4 gm% in the three species. The correlation coefficient between the oxygen content derived by the sensor and that from the Hemoximeter was 0.9971 (Y=1.008 X - 0.1147).

In Vivo Study

Figure 6 shows the results obtained from the acute in vivo experiments when arterial SO_2 and [Hb] were varied to evaluate the responses of the arterial and venous sensors. The standard deviations

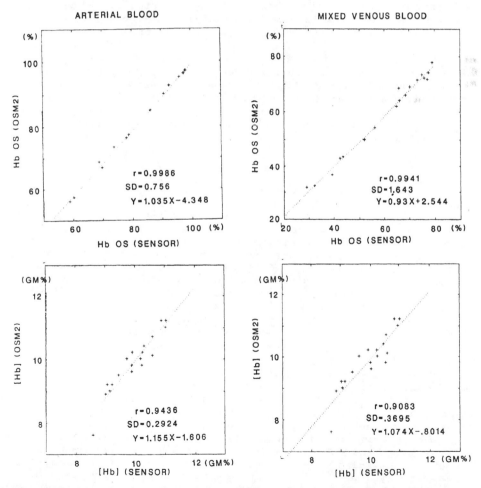

Figure 6. Correlation plots between the measurements obtained by the prototype optical sensors vs. the Hemoximeter OSM2 measurements in the in vivo experiments. Arterial SO_2 (top left), arterial [Hb] (bottom left), mixed venous SO_2 (top right), and mixed venous [Hb] (bottom right).

of errors in estimation of SO_2 were 0.75% (arterial) and 1.6% (venous), and of [Hb] was 0.3 gm% for both sensors.

To date, long term experiments have been performed in a total of three animals with an experimental duration of 3, 6 and 40 days. The 3-day animal died due to bleeding from the site of anastomosis inside the chest. In the 6-day animal, the outflow connector of the left pump was accidentally disconnected and the animal died of bleeding the night of 6th postoperative day. The last animal lived for a duration of 40 days; various hemodynamic studies including exercise and drug loading tests were carried out under controlled pump output (these studies will be reported later). On the 36th day, however, the left pump inflow valve (Bjork Shiley disk valve) failed and the animal was sacrificed on the 40th day. The mixed venous SO_2 and oxygen consumption data are presented here.

Figure 7 shows the changes in the arterial and mixed venous SO_2, [Hb], pump flows, and oxygen consumption of the 6-day animal on the 0th, 2nd and 6th postoperative days (POD). On the 0th POD, mixed venous SO_2 varied with slight movement of the animal; with the animal lying it was around 50-60% and it declined momentarily to about 10-15% during changes of posture, and returned to about 30-40% during standing. The oxygen consumption ranged around 5-6cc/min/Kg with the animal lying, momentarily increasing to about 15-20 cc/min/Kg and returning to 8-10 cc/min/Kg during standing. On the 2nd POD as the animal recovered from anesthesia and surgery, she became a little calmer and the mixed venous SO_2 became more stable. On the 6th POD, mixed venous SO_2 became very stable, and oxygen consumption during lying decreased to about 4-5 cc/min/Kg, transiently increasing to about 8-10 cc/min/Kg during changes of posture from lying to standing and returning to about 7-8 cc/min/Kg during standing. One interesting finding was that [Hb] increased by about 5% with the decrease in the mixed venous SO_2.

DISCUSSION

This paper gives a description of an optical sensor for continuous measurement of both SO_2 and [Hb] and its initial application to continuous monitoring of oxygen delivery and consumption in the awake live animals. In the measurement of SO_2, the addition of a constant in the denominator of the ratio term minimized the effect of hematocrit variation on the SO_2 value and improved the linearity, although overall sensitivity was somewhat reduced. This approach was verified in the blood of three different species where red blood cell size varied from 35, 50 to 87 cubic micron. Secondly, the hemoglobin content was derived from the reflectance at the isosbestic wavelength of 795nm. The second order approximation was shown to yield a fairly accurate prediction of the hemoglobin content of blood in the three different species. Thus, the computation of the oxygen content of whole blood became possible without requiring blood samples.

In the in vivo evaluation of the sensor, the sensor was mounted in a tube and inserted in series with the artificial hearts. this allowed continuous monitoring of both SO_2 and [Hb] of arterial and mixed venous blood and the derivation of oxygen consumption as a function of various physiological parameters. Thus, continuous monitoring of oxygen consumption became possible for the first time in awake live animals.

The mixed venous SO_2 varies in response to changes in 1) respiratory function, 2) cardiac function, 3) hemoglobin content, and 4) metabolic status or oxygen consumption. As demonstrated in this study, the mixed venous SO_2 varied, sensitively reflecting the status of the animal. It changed quite rapidly in response to movement of the animal. In particular, its variation was somewhat erratic directly after the surgery. This may be caused by the oxygen debt developed during the operation due to surgical stress and anesthetic effect. As shown in

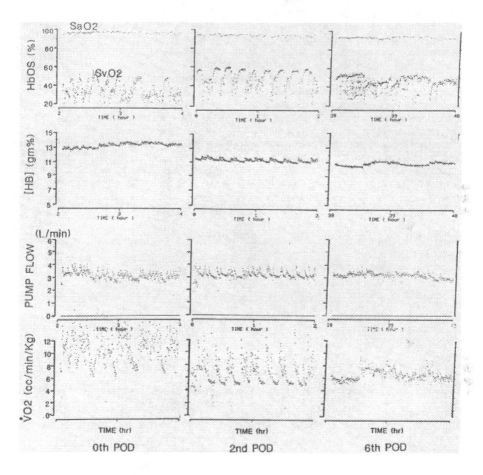

Figure 7. Arterial and mixed venous SO_2 (top), arterial and mixed venous [Hb] (second from the top), pump flow (third from the top), and oxygen consumption (bottom) of an awake live animal, supported by the artificial hearts, on the 0th, 2nd, and 6th postoperative day. Each trace is over a 2 hour duration.

Figure 7, the mixed venous SO$_2$ instantaneously dropped to a lower level when the animal tried to change its posture from standing or lying to other postures, followed by recovery to the intermediate level. This may be caused by the activity in the muscles during a change of posture; this would require more oxygen and would widen the A-V oxygen difference since pump flow would not increase to meet the increased oxygen demand. Another hypothesis could be that the blood pooled in the venous bed during lying could be released during movement and instantaneously lower the mixed venous saturation. Concerning the venous saturation of the skeletal muscle, Kramer et. al. (1939) electrically stimulated the muscle and measured the venous saturation. The venous saturation declined to 10-15% for each stimulation due to high energy utilization. Thus, the delayed action seen in this experiment may be explained from the instantaneous use of high oxygen in the skeletal muscle of the legs which lowered the mixed venous saturation. Figure 8 summarizes the changes in the mixed venous saturation and oxygen consumption during changes of the posture. There was approximately a minute delay before the stable mixed venous saturation was achieved after movement of the animals. This occurred during both standing and lying. The main cause may be due to increased oxygen use in the active muscles of the legs.

With regard to the [Hb], the decrease in the mixed venous saturation was associated with a slight increase in [Hb] level. This finding was confirmed through analysis of blood samples by the Hemoximeter OSM2. During movement, the blood pooled in the venous bed may flow or decreased venous saturation may trigger release of the blood from the spleen or the liver. This finding should be confirmed in animals with normal hearts.

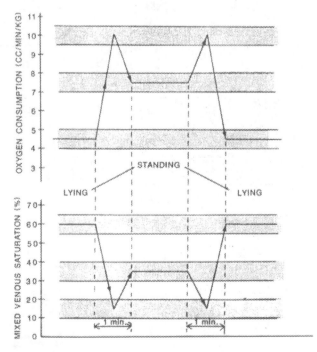

Figure 8. Changes in the mixed venous SO$_2$ (bottom) and oxygen consumpiton (top) of an awake live animal during changes in posture from lying to standing and from standing to lying. During changes of posture, it took about 1 minute before the mixed venous SO$_2$ reach a steady state level. Usually, the mixed venous SO$_2$ instantaneously decreased to lower levels due to increased oxygen use in the leg muscles. The decrease was somewhat lower during standing than lying.

The increased level of oxygen consumption directly after the surgery was correlated with the increased level of blood lactate. Following surgery the blood lactate level was high at around 30-40 mg% and returned to the pre-operative level of about 3-4 mg% after 6-7 hours. This finding is common in most of the animals that undergo heart surgery and may be specifically related to the oxygen debt developed during the operation, and thus to repay this debt more oxygen is required and this widens the A-V oxygen difference immediately after the surgery.

Further study is currently underway using the same animal model to evaluate mechanisms of circulatory regulation based on the concept of oxygen delivery to and oxygen consumption by, the peripheral tissues.

SUMMARY

For continuous measurement of both SO_2 and [Hb] of whole blood, a hybrid type optical sensor was developed, tested in vitro and applied to the continuous measurement of arterial and mixed venous SO_2 and [Hb] of awake live animals whose circulation was entirely supported by a pair of artificial hearts. The purpose of using artificial hearts was to control the cardiac function precisely and to obtain independent responses of the peripheral circulatory system. Both sensors in the arterial and venous circuits functioned satisfactorily for a duration of 40 days; no blood clots were observed around the sensor and the mixed venous SO_2 varied sensitively in response to changes in the respiratory status, pump output, [Hb], and oxygen consumption, thus showing that it could be used as an indicator to evaluate the adequacy of oxygen delivery to peripheral tissues.

ACKNOWLEDEGEMENT

This research was partially supported by the grants in aid from the Ministry of Education under #60480324(Principal Investigator: Setsuo Takatani), from the Ministry of Health and Welfare, and from Terumo Inc, Shizuoka, Japan.

REFERENCES

Baele, P.L., McMihan, J.C., March, H.M., et al., 1982, Continuous monitoring of mixed venous oxygen saturation in critically-ill patients, Anesth. Analog., 6:513.

Cole, J.S., Martin, W.E., Cheung, P.W., and Johnson, C.C., 1972, Clinical studies with a solid state fiberoptic oximeter, Amer. J. Card., 29:383-388.

Didisheim, P., 1985, Comparative hematology in the human, calf, sheep and goat, ASAIO J., 8(3):348.

Kramer, K., Obal, F., and Quensel, W., 1939, Unterschungen uber den Muskelstoffwechsel des Warmerbluters. III. Mitteilung. Die Saurerstoffaufnahme des Muskels wahrend rhythmischer Tatigkeit. Pflugers Arch ges Physiol, 241:717.

Mendelson, Y., Cheung, P.W., Neuman, M.R., Fleming, D.G. and Cahn, S.D., 1983, Spectrophotometric investigation of pulsatile blood flow for transcutaneous reflectance oximetry, Oxygen Transport to Tissue, 4:93-102.

Polanyi, M.L. and Hehir, R.M., 1962, In vivo oximeter with fast dynamic response, Rev. Sci. Instru., 33(10):1050-1054.

Schmitt, J.M., Meindl, J.D. and Mihm, R.G., An integrated circuit-based optical sensor for in vivo measurement of blood oxygenation, IEEE Trans. Biomed. Engr., BME-33(2):98-107.

Takatani, S., Cheung, P.W., and Ko, W.H., 1976, Estimation of hemoglobin concentration of whole blood using infrared reflectance, Proceedings of the 30th Annual Conference on the Engineering in Medicine and Biology, 19:171.

Takatani, S., Cheung, P.W., and Ernst, E.A., 1980, Noninvasive tissue reflectance oximeter: An instrument for measurement of tissue hemoglobin oxygen saturation in vivo, Annals of Biomed. Engr., 8:1-15.

Takatani, S, Tanaka, T., Takano, H., Nakatani, T., Taenaka, Y., Umezu, M., Matuda, T., Iwata, H., Noda, H., Nakamura, T., Seki, J., Hayashi, K., and Akutsu, T., 1984, Development of a high performance implantable total artificial heart system, Life Support Systems, 2(suppl 1):249.

Takatani, S., Kohno, H., Noda, H., Takano, H., and Akutsu, T., 1987, A miniature hybrid reflection type optical sensor for measurement of hemoglobin content and oxygen saturation of whole blood, IEEE Trans Biomed Engr, (Submitted for publication)

Van Assendelft, P.W., 1970, Spectrophotometry of Hemoglobin Derivatives, Assen, The Netherlands, Royal Vangorcum Ltd.

Yee, S., Schibli, E., and Krishnan, V., 1977, A proposed miniature red/infrared oximeter suitable for mounting on catheter tip, IEEE Trans. Biomed. Engr., BME-24(2):195-197.

ESTIMATION OF THE OXYGEN TRANSPORT TO TISSUES FROM AN IN VITRO MEASUREMENT

OF WHOLE BLOOD PASSAGE TIME AND VENOUS BLOOD P_{O_2} IN DIABETICS

Y. Kikuchi, T. Koyama*, N. Ohshima, and K. Oda**

Institute of Basic Medical Sciences, University of Tsukuba
Ibaraki 305, *Research Institute of Applied Electricity
Hokkaido University, Sapporo 060, **Sapporo City Hospital

INTRODUCTION

The flow properties of blood have recently attracted much attention among the many factors involved in oxygen transport to tissues. For example, altered blood flow properties, especially an impaired deformability of red cells, have been reported with the suggestion of their possible relevance to hypoxic disturbances in tissues in several diseases including diabetes mellitus (Schmid-Schonbein and Volger, 1976; McMillan, 1976; McMillan et al., 1978) and occlusive arterial diseases (Ehrly, 1976; Reid et al., 1976). Furthermore, rheological therapies, such as hemodilution and drugs which may ameliorate impaired red cell deformability, have been extensively tried in the latter case to improve peripheral circulation and hence oxygen supply to tissues (Ehrly, 1976; Marcel, 1979). So far, however, no quantitative studies have been made to assess the complications for oxygen transport to tissues in vivo of these modifications of blood flow properties.

In the present study an attempt was made, with particular reference to diabetics, to establish on a quantitative basis whether or not such impairments in blood flow properties do significantly impede the oxygen delivery to tissues; in other words, whether these may account for tissue hypoxia. For this purpose, the filterability of diabetic whole blood samples was reexamined by means of a modified Nuclepore filtration test in which incorporated considerable improvements in reliability and quantitativity (Kikuchi et al., 1983). Differences in blood passage time through the filter reflect alterations of blood flow properties particularly pertinent to the passage of blood through capillary vessels and hence are directly related to possible changes in capillary transit time in vivo. Oxygen diffusion from capillary blood to tissues was calculated using the Krogh (1919) tissue cylinder model; oxygen partial pressures in venous blood (PvO_2) derived from such calculations were compared with actual PvO_2 values. Alterations in blood flow properties and their possible consequences for oxygen transport to tissues were thus assessed in individual patients.

The variability of tissue cylinder dimensions in diabetics is discussed and on the basis of the present measurements and calculations it is suggested that increases in tissue cylinder radius, i.e. decreases in capillary density, exist in a certain proportion of the patients.

MATERIALS AND METHODS

Subjects.   Studies were made on eight in-patients with diabetes mellitus (mean age 49 years, range 24-70; mean duration of diabetes 11 years, range 1-24; mean fasting blood glucose 127 mg/dl, range 80-191; mean HbA_1 9.5 %, range 6.7-12.4) and 8 healthy volunteers with no family history of diabetes (mean age 41 years, range 31-54; matching of age, sex, and smoking habits with the patients was not possible).

Nuclepore filtration test.   A 9.5 ml venous blood sample, withdrawn into 0.5 ml heparin solution (1,000 I.U./ml), was taken by venepuncture from an antecubital vein from each of the subjects, who rested for 10 minutes prior to blood sampling.   Syringes containing blood samples were sealed with rubber stoppers and gently rotated in a water bath thermostatted at 37° C.   A sample volume of 0.6 ml was used for each determination of the Nuclepore filtration rate.   Blood was transferred into a 0.5 ml syringe connected to the inlet of the filter holder and caused to flow through the 5 µm Nuclepore filter by application of 10 cmH_2O negative pressure to the outlet; the time at which the sample meniscus passed each 0.05 ml graduation mark on the syringe was recorded from 0.5 to 0 ml with an electronic timer.

The filters were repeatedly flushed with saline before every measurement to remove air bubbles from the pores; this process is essential for making reproducible measurements.   A fresh filter was used for every blood sample, the pore density of each being calibrated before use by determination of the saline passage time.   The filtration apparatus used in the present study has been described in detail by Kikuchi et al. (1983).

Hematocrit (microhematocrit, 11,000 G, 5 min, in duplicate) and PO_2 (Radiometer BGA-3) were determined using a small portion of each whole blood sample.   All the measurements were carried out within 30 minutes after blood sampling.

Calculation of PO_2 by the tissue cylinder model.   The oxygen diffusion from capillary blood to surrounding tissues was caculated using the Krogh tissue cylinder model.   The equations and parameters such as dimensions of the tissue cylinder and oxygen consumption rate of the tissue used in the present calculation are given in the Appendix.

RESULTS

Blood passage times and hematocrits obtained for the diabetics and the healthy subjects are shown in Figure 1.   The patients with complications and those receiving insulin therapy are denoted by different symbols.   The figures on the data points show PO_2 values of the blood samples, i.e. PvO_2 values of the subjects.   In contrast to the healthy cases, the diabetics showed a wide scatter in the values for blood passage time, hematocrit and also PvO_2.

A nomogram between PvO_2, capillary transit time, and hematocrit obtained from the present calculation of the oxygen diffusion in the tissue cylinder is shown in Figure 2.   The oxygen consumption rate of the tissue was taken to be an average of that for skeletal muscle (at rest) and for skin (values were obtained from Finch and Lenfant, 1972; Ganong, 1983), since the blood samples were taken from the antecubital vein.   The solid curves give a nomogram for cases in which the oxygen affinity of hemoglobin is normal (PO_2 at which hemoglobin is half-saturated with oxygen, P_{50}, is assumed to be 27 mmHg).   The dashed curves show cases in which the oxygen affinity of hemoglobin is increased to give a P_{50} value of 21 mmHg (Ditzel, 1976).   The hemoglobin-oxygen dissociation curves corresponding to these P_{50} values are shown in the Appendix.

In healthy subjects with PvO_2 and hematocrit values of 44.6 ± 1.9 mmHg (mean ± S.D.) and 41.6 ± 2.6 %, respectively, the nomogram gives a mean capillary transit time of 3.1 ± 0.3 sec, a value close to reported values for skeletal muscles (Honig et al., 1977) and also for retina (Kohner, 1976).   Similarly, capillary transit times of 4.8 ± 1.7 sec are obtained in the diabetics from their PvO_2 and hematocrit values (35.1 ± 8.0 mmHg and 39.1 ± 4.8 %, respectively).   The latter estimates, however, become 3.0 ± 1.2 sec if P_{50} is assumed to be 21 mmHg.   On the other hand, values of 11.0 ± 0.7 sec

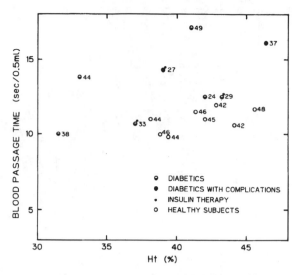

Figure 1.   Blood passage time and hematocrit values obtained for the healthy subjects and diabetics.   The diabetic patients with complications, or on insulin therapy are denoted by different symbols.   The numbers on the data points give the PO_2 values of the blood samples, i.e. PvO_2 values of the subjects.

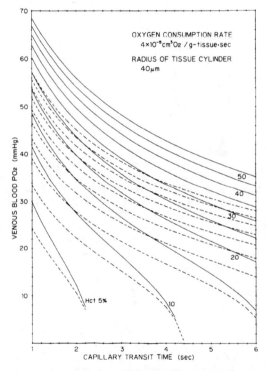

Figure 2.   A nomogram which gives PvO_2 as a function of capillary transit time and hematocrit.   The solid curves are for blood in which the oxygen affinity of hemoglobin is normal (P_{50} = 27 mmHg) and the dashed curves for blood with an increased oxygen affinity (P_{50} = 21 mmHg).   The curves are given for every 5 % change in hematocrit value.

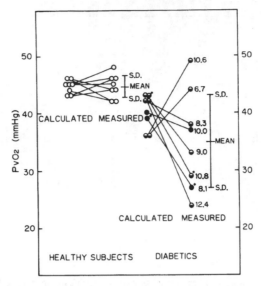

Figure 3. Comparisons between calculated and measured PvO_2's in the healthy subjects and diabetics. The figures given for the diabetic cases show HbA_1 contents.

and 13.4 $\pm$ 2.3 sec were obtained for the in vitro blood passage time in those healthy and diabetic subjects, respectively (Figure 1).

It appears of interest to see what values are expected for PvO_2 if the capillary transit time is assumed to be closely related to the observed in vitro blood passage time. To this end, PvO_2 values were obtained from the nomogram by assuming, as a first step, a linear relation between the capillary transit time and the blood passage time. Figure 3 shows the measured PvO_2 values and the values thus estimated (denoted as calculated PvO_2 values). The linear relation assumed was such that the means of the calculated and the measured PvO_2 values coincided with each other in the healthy subjects. Thus, the high degree of consistency shown for these subjects merely reflects the small scatter in their blood passage time and PvO_2 values. However, the diabetics exhibit considerable discrepancies when the calculations are made using the same model parameters and the same relationship between the two passage times as were used for the healthy subjects. Although such discrepancies in the diabetics might be a result of over-simplification, it still appears that the calculations and comparisons of PvO_2 values within the present framework could give some insight into the microcirculatory derangements in the patients; this is discussed later.

DISCUSSION

Several studies have reported an impaired red cell deformability and increased intercellular interaction, i.e. an increased tendency for red cells to aggregate, in diabetes mellitus (Schmid-Schonbein and Volger, 1976; McMillan, 1976; McMillan et al., 1978). The observed increases in blood passage time in the diabetics (Figure 1) could be due to these probable changes in the flow properties of their red cells. However, some researchers have denied an impairment in red cell deformability and proposed that the impaired filterability of whole blood samples of diabetics should be attributed to the white cells (Stuart et al., 1983). This problem has been assessed using red cell-plasma suspensions poor in white cells; the results still suggested that anomalies exist in the red cell component and these might be associated with stronger interactions between plasma components and the red cell surfaces, associated with increased glycosylation (to be published elsewhere).

Taking advantage of the analogy between the in vitro pore passage and in vivo capillary passage of whole blood, the present study attempted to analyze the oxygen supply to tissues taking into account the observed whole blood passage times and hematocrits. However, as already noted, calculations using the parameters determined for the healthy subjects failed to give consistency between the calculated and measured values of PvO_2 in the diabetics (Figure 3). Nevertheless, from such discrepancies it appears possible to make the following speculation as to the microcirculation in the patients.

Physiological adjustments can be expected to act to maintain the capillary circulation despite detrimental changes such as the present impairment of blood flow properties. Hence a normal PvO_2 is still possible even if reduced red cell deformability is observed. This might be the case in the two patients who showed normal PvO_2's in spite of their reduced red cell deformability. However, the regulatory mechanisms might themselves be impaired by diabetic derangements including autonomic neuropathy. In this case the capillary circulation will become more dependent on physical factors, and this might account for the findings in the two patients who showed PvO_2's slightly lower than the calculated values. In the other four cases, however, a much longer capillary transit time than expected from its in vitro index must be assumed in order to explain their extremely low PvO_2 values.

Although a very slow microcirculation might occur in diabetics, it appears more reasonable to assume the following two possibilities in the last four patients; an increased oxygen affinity of hemoglobin (Ditzel, 1976) or an increase in the dimensions of the tissue cylinder. The former possibility, however, appears less probable, since diabetic whole blood has been shown to have a normal oxygen affinity because of a compensatory increase in 2,3-DPG (Samaja et al., 1982). In addition, there seemed to be no correlation between PvO_2 and HbA_1 in these patients (Figure 3). On the other hand, the latter possibility is suggested by the high incidence of capillary degeneration in diabetics. It is obvious that the larger the tissue volume to which a capillary supplies oxygen, the larger is the arterio-venous difference in PO_2. Capillary degeneration, i.e. a decrease in capillary density, is expressed as an increase in tissue cylinder radius in the present model. The tissue cylinder radius necessary to give consistency between measured and calculated PvO_2's in the present model can be determined. Values of the tissue cylinder radius thus obtained are shown in Figure 4. The four cases of the patients showed radii considerably larger than the normal average value in agreement with the above discussion. Thus, it seems likely that such an estimation of the tissue cylinder radius could be valuable for detecting microcirculatory derangements in diabetics.

In the present study the venous blood oxygen tension was used as an index of the oxygen transport to tissues. It must be noted here that a PvO_2 lower than normal does not necessarily mean a reduction in the amount of oxygen supplied to tissues. It can be easily shown that a fall in venous PO_2 results when the same amount of oxygen is removed from the capillary blood despite a reduced inflow of oxygen. It also holds in the case of an increased oxygen affinity of hemoglobin that the same amount of oxygen is delivered to tissues; but a decrease of the same amount in the oxygen content of the blood results in a larger decrease in PO_2 at values higher that P_{50}. A lower PvO_2 does, however, indicate a situation in which tissue hypoxia or anoxia occur more readily should the capillary circulation stagnate occasion-

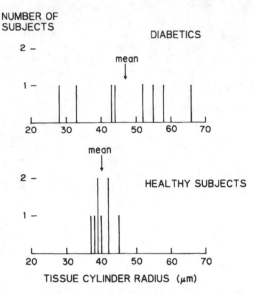

Figure 4. Values of the tissue cylinder radius estimated from the measured PvO_2's for the healthy subjects and diabetics.

ally. In this respect, it may be further suggested that impaired red cell deformability can have greater effects than an increased oxygen affinity of hemoglobin, because the former may reduce the oxygen inflow, i.e., the amount of oxygen available while the latter does not affect the oxygen content itself in blood.

A similar comment should be made concerning hemodilution. Capillary transit time will probably decrease with decreasing hematocrit as is indicated by the almost linear relationships between the blood passage time and hematocrit (Kikuchi et al., 1983). No alteration in the oxygen transport to tissues is produced by hemodilution so long as increases in blood flow velocity cancel out decreases in hematocrit. However, although a reduced hematocrit may give the same PvO_2, it does not mean the same capaity to supply oxygen to tissues. As is clear from the nomogram given in Figure 2, the lower the hematocrit, the more rapid the decreases in tissue PO_2 on the venous side during stasis. As Figure 5 shows, the emergence of the lethal corner is especially rapid in tissues with high oxygen consumption rates such as renal, retinal and nervous tissues. In view of this result, it appears that the optimal hematocrit is not that with the highest ability to transport oxygen to tissues (this factor would be independent of hematocrit as discussed above) but that with the capacity to maintain the oxygen supply to tissues despite fluctuations in the capillary circulation which may often occur.

ACKNOWLEDGMENT

The present study was supported by a Grant-in-Aid for Scientific Research (No. 62570379) from the Ministry of Education, Science and Culture of Japan.

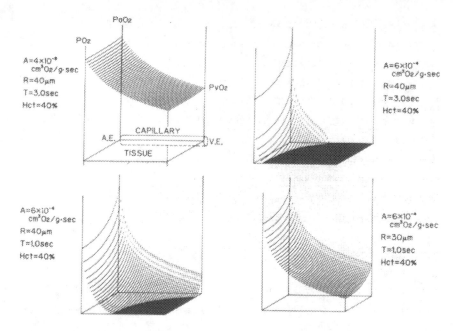

Figure 5. Examples of PO_2 profiles in the tissue cylinder calculated for given values of the capillary transit time, oxygen consumption rate, and tissue cylinder radius. A.E. = arterial end. V.E. = venous end. Lethal corner is shaded black.

APPENDIX

Equations and boundary conditions that were used in the present calculation are as follows (Bruley and Knisely, 1970; Leonard and Jorgensen, 1974). Capillary equation

$$\left[1+\frac{NKnP^{n-1}}{C_1(1+KP^n)^2}\right]\frac{\partial P}{\partial t}=D_1\left[\frac{\partial^2 P}{\partial r^2}+\frac{1}{r}\frac{\partial P}{\partial r}\right]+D_1\frac{\partial^2 P}{\partial x^2}-V\left[1+\frac{NKnP^{n-1}}{C_1(1+KP^n)^2}\right]\frac{\partial P}{\partial x} \qquad (1)$$

tissue equation

$$\frac{\partial P}{\partial t}=D_2\left[\frac{\partial^2 P}{\partial r^2}+\frac{1}{r}\frac{\partial P}{\partial r}\right]+D_2\frac{\partial^2 P}{\partial x^2}-\frac{A}{C_2} \qquad (2)$$

boundary conditions

$r = r_1 : D_1 c_1 (\partial P/\partial r)|_{blood} = D_2 c_2 (\partial P/\partial r)|_{tissue}$, $P|_{blood} = P|_{tissue}$;

$r = r_2 : \partial P/\partial r = 0,$

where

A: oxygen consumption rate of the tissue, 4×10^{-5} cm^3 O_2/cm^3 tissue.sec

for the forearm (muscles and skins) and 7×10^{-5} cm^3 O_2/cm^3 tissue.sec

for the whole body (Ganong, 1983; Finch and Lenfant, 1972);

c_1: oxygen solubility coefficient in blood, 3.4×10^{-5} cm^3 O_2/cm^3 blood.mmHg;

c_2: oxygen solubility coefficient in tissue, 2.9×10^{-5} cm^3 O_2/cm^3 tissue.mmHg;

D_1: oxygen diffusion constant in blood, 1.1×10^{-5} cm^2/sec;

D_2: oxygen diffusion constant in tissue, 1.7×10^{-5} cm^2/sec;

K: constant characterizing the hemoglobin dissociation curve, 0.0002;

n: constant characterizing the hemoglobin dissociation curve, 2.6 for
   P_{50} = 27 mmHg and 2.8 for P_{50} = 21 mmHg;

N: oxygen capacity of blood at saturation, 0.2 cm^3 O_2/cm^3 blood at Hct
   45 % (hemoglobin 15 g/dl).

P: oxygen partial pressure (mmHg);

r: radial axis of coordinates;

r_1: radius of the capillary, 2.5×10^{-4} cm;

r_2: radius of the tissue cylinder, 40×10^{-4} cm (Honig et al., 1971);

t: time (sec);

V: blood flow velocity in the capillary (cm/sec);

x: longitudinal axis of coordinates (cm).

Figure 6 shows hemoglobin dissociation curves given by $KP^n/(1 + KP^n)$ for the values of K and n given above. The length of the tissue cylinder was assumed to be 0.1 cm (Honig et al., 1977) and hence a blood transit time of 1 sec corresponds to a blood flow velocity of 0.1 cm/sec. The oxygen tension of arterial blood, PaO_2, was assumed to be 95 mmHg in the present calculation.

In capillary-tissue systems in vivo the longitudinal variations in P are much smaller than the variations in P along the radial direction r; the diffusion of oxygen longitudinally can be neglected in comparison with the radial diffusion.

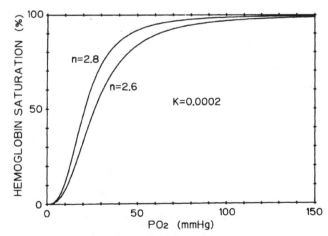

Figure 6. Hemoglobin-oxygen dissociation curves given by the Hill equation $KP^n/(1 + KP^n)$. The two curves correspond a P_{50} of 27 mmHg and a P_{50} of 21 mmHg.

At steady states, i.e. $\partial P/\partial t = 0$, the equation (2) is readily integrated.

$$P = P\bigg|_{r_1} - \frac{Ar_1^2}{4D_2C_2}\left[1 - \frac{r^2}{r_1^2}\right]\bigg|_{} - \frac{Ar_2^2}{2D_2C_2}\ln\frac{r}{r_1} \qquad (3)$$

Equation (1) can be numerically integrated by taking finite differences in x and r. The radial changes in P in the capillary blood are also very small and can be neglected. Then equation (1) can be rewritten as follows with the use of the boundary conditions and equation (3).

$$V\left[1 + \frac{NKnP^{n-1}}{C_1(1+KP^n)^2}\right]\frac{\partial P}{\partial x} = \bigg|\frac{2}{r_1}\frac{D_2C_2}{C_1}\frac{\partial P}{\partial r}\bigg|_{r_1\text{tissue}} = \frac{A}{C_1}\left[1 - \frac{r_2^2}{r_1^2}\right] \quad (4)$$

The numerical integration of this equation can be easily carried out.

Examples of calculated PO_2 profiles in the tissue cylinder are shown in Figure 5. It is shown that although the capillary blood flow corresponding to a capillary transit time of 3.0 sec is sufficient to supply oxygen to resting muscle or skin, it is too slow to oxygenate, for example, brain tissues (Ganong, 1983; Finch and Lenfant, 1972). Even if the capillary transit time is reduced to 1.0 sec, these tissues, especially on the venous side are still insufficiently supplied with oxygen so long as the tissue cylinder radius is 40 μm.

Another nomogram which would be useful for a whole body estimation (the oxygen consumption rate of the tissue is taken to be that of the whole body) is given in Figure 7.

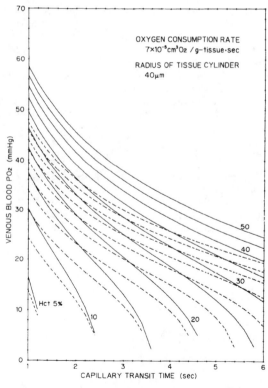

Figure 7. A nomogram in the same form as in Figure 3 but for the tissue oxygen consumption rate of the whole body.

REFERENCES

Bruley, D.F., and Knisely, M.H., 1970, Hybridsimulation-oxygen transport
    in the microcirculation, Chem. Eng. Progr. Symp. Ser., 66:22-32.
Ditzel, J., 1976, Oxygen transport impairment in diabetes, Diabetes,
    25:832-838.
Ehrly, A.M., 1976, Improvement of the flow properties of blood: a new
    therapeutical approach in occlusive arterial disease, Angiology,
    27:188-196.
Finch, C.A., and Lenfant, C., 1972, Oxygen transport in man, N. Engl. J.
    Med., 286:407-415.
Ganong, W.F., 1983, Review of Medical Physiology, Los Altos, Lange Medical
    Publications, 488.
Honig, C.R., Frierson, J.L., and Nelson, C.N., 1971, O_2 transport and VO_2
    in resting muscle: significance for tissue-capillary exchange, Am. J.
    Physiol., 220:357-363.
Honig, C.R., Feldstein, M.L., and Frierson, J.L., 1977, Capillary lengths,
    anastomoses, and estimated capillary transit times in skeletal muscle,
    Am. J. Physiol., 233:H122-H129.
Kikuchi, Y., Arai, T., and Koyama, T., 1983, Improved filtration method
    for red cell deformability measurement, Med. Biol. Eng. Comput.,
    21:270-276.
Kohner, E.M., 1976, The problems of retinal blood flow in diabetes,
    Diabetes, 25:839-844.
Krogh, A., 1919, The number and distribution of capillaries in muscles
    with calculations of the oxygen pressure head necessary for supplying
    the tissue, J. Physiol., 52:409-415.
Leonard, E.F., and Jorgensen, S.B., 1974, The analysis of convection and
    diffusion in capillary beds, Annual Rev. Biophys. Bioeng., 3:293-339.
Marcel, G.A., 1979, Red cell deformability: physiological, clinical and
    pharmacological aspects, J. Med., 10:409-416.
McMillan, D.E., 1976, Plasma protein changes, blood viscosity, and diabetic
    microangiopathy, Diabetes, 25:858-864.
McMillan, D.E., Utterback, N.G., and La Puma, J., 1978, Reduced erythrocyte
    deformability in diabetes, Diabetes, 27:895-901.
Reid, H.L., Dormandy, J.A., Barnes, A.J., Lock, P.J., and Dormandy, T.L.,
    1976, Impaired red cell deformability in peripheral vascular disease,
    Lancet, 1:666-668.
Samaja, M., Melotti, D., Carenini, A., and Pozza, G., 1982, Glycosylated
    haemoglobins and the oxygen affinity of whole blood, Diabetologia,
    23:399-402.
Schmid-Schonbein, H., and Volger, E., 1976, Red-cell aggregation and
    red-cell deformability in diabetes, Diabetes, 25:897-902.
Stuart, J., Kenny, M.W., Aukland, A., George, A.J., Neumann, V., Shapiro,
    L.M., and Cove, D.H., 1983, Filtration of washed erythrocytes in
    atherosclerosis and diabetes mellitus, Clin. Hemorheol., 3:23-30.

OXYGEN TENSION MEASUREMENTS USING AN OXYGEN POLAROGRAPHIC ELECTRODE SEALED IN

AN IMPLANTABLE SILASTIC TONOMETER: A NEW TECHNIQUE

J. Rabkin[*], R. Alena[**], J. Morse[**],
W.H. Goodson, III[*], and T.K. Hunt[*]

[*]Department of Surgery, University of California, San Francisco, and [**]Biogenesis, Inc., San Francisco, California

Reliable measurements of tissue oxygen tension have proven useful in the clinical management of patients (1). A monitor of the consuming organ directly measures the adequacy of perfusion for meeting the needs of the tissue. However, the current, "open-end" tonometer, consisting of an implanted Silastic (silicone) tube through which electrodes are inserted intermittently, is sufficiently cumbersome and time-consuming to preclude its routine clinical use in patient management (2). Furthermore, repeated, separate measurements are required to follow trends. We have designed and tested a new oxygen probe incorporating a three-electrode polarographic system sealed within a Silastic tube which eliminates the difficulties encountered with the open tonometer system.

MATERIALS AND METHODS

The probe consists of an electrode subassembly fitted into a tonometer (fig. 1). The electrode subassembly has a platinum cathode, a platinum anode and a reference electrode sealed in a 22-ga Luerhub intravenous cannula (Abbott Hospitals, Inc., North Chicago, Il.). The tonometer subassembly consists of a 5-cm length of medical grade Silastic tubing (Dow-Corning Corp., Midland, Mich.)(outer diameter 1.3 mm, inner, 0.8) sealed with Silastic cement and filled with the electrolyte solution containing a stabilizing additive. The electrodes are each connected to a short cable attached to the Luerhub. The assembly is then sealed and sterilized. When ready for use the cable is connected to a small, dedicated picoammeter which supplies the polarizing potential and monitors the probe output current, displayed directly as millimeters of mercury of oxygen tension (Biogenesis, Inc., San Francisco, Calif.)(fig. 2). The electrode is calibrated to 150 mm Hg at the tissue temperature in a waterbath bubbled with air to ensure saturation. It can also be calibrated at 37° C and adjustment made for the actual tissue temperature. The electrode is inserted into the tissue using a 14-ga breakaway needle (Luther Medical Products, Costa Mesa, Calif.) or peelaway catheter assembly (Cook Critical Care, Bloomington, Ind.) and is secured in place with a sterile transparent plastic dressing.

Performance Characteristics

The output of the electrode is approximately 10 namps. The 95-percent response to a step change in the partial pressure of oxygen is approximately

267

Rev 20 Probe Dimensions

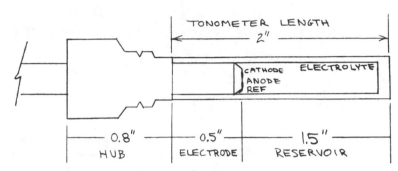

TONOMETER LENGTH
2"

CATHODE
ANODE
REF
ELECTROLYTE

0.8"
HUB

0.5"
ELECTRODE

1.5"
RESERVOIR

Fig. 1

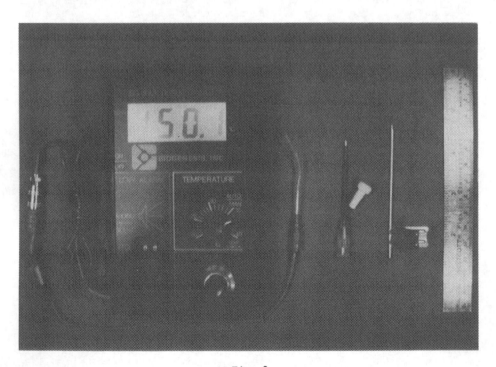

Fig. 2

three minutes at 37° C (fig. 3). Temperature variation on the output of the electrode over the physiologic range is approximately three percent/°C. After a 30-60 minute warmup period the electrode is generally stable within ± 10 percent under conditions of use for up to 48 hours.

The new device has been tested in vivo in both small and large animals in parallel with our standard open tonometer system (fig. 4). The monitoring devices were placed in the subcutaneous tissue of the abdominal wall which contains a vascular bed physiologically responsive to changes in vascular volume. The electrode systems tracked one another with a maximal offset of approximately 15 percent. The initial response time to step changes in inspired oxygen tension and to ten-percent blood volume manipulation was less than four minutes and often within one minute.

DISCUSSION

Many indirect parameters are used to monitor the hemodynamics of postoperative patients. Peripheral tissue perfusion is the ideal measurement.

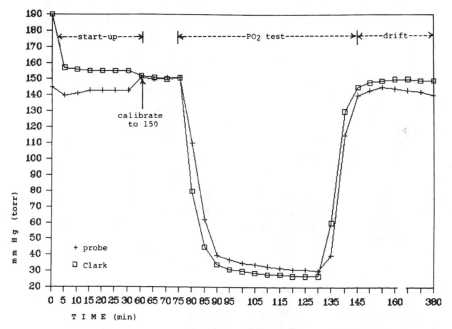

Fig. 3 In-vitro test compares a Clark-type oxygen sensor (Yellow Springs Instruments, Yellow Springs, Ohio) and the three-electrode probe. Both stabilized quickly and showed low drift. The monitors were set to 150 mm Hg at one hour. The calibration chamber was bubbled with nitrogen to bring the pO2 from 150 to 30 torr in about 15 minutes. The Clark sensor followed the change immediately; the three-electrode probe lagged behind by about two minutes. They tracked the pO2 levels within three percent. After the pO2 test the probes were allowed to run for two hours at room air and 37° C to test the drift rate.

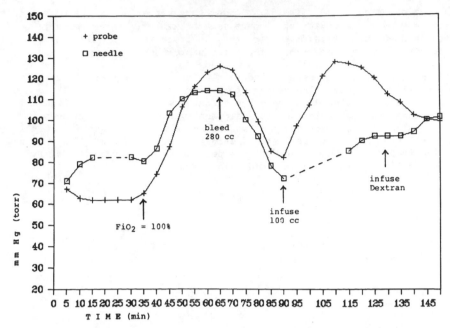

Fig. 4 This in-vivo test in a sheep compares the original needle electrode method with the new three-electrode probe. Both came to a baseline rapidly and showed a response to inspired oxygen. After bleeding induced a volume deficit, the PO2 values dropped. Reinfusion of the shed blood followed by Dextran resulted in an increase of the PO2 values. The graph of the needle electrode shows two portions where the output was unreliable due to loss of signal. Flushing fresh electrolyte through the Silastic tube restored operation. The three-electrode probe exhibited no loss of signal during this experiment.

Tissue oxygen tension reflects tissue perfusion; it can be used to calculate perfusion grossly and, more accurately, to identify changes (3). Tissue oxygen tension necessarily varies between tissues. During times of stress normal protective mechanisms sacrifice perfusion to some tissue beds in order to sustain others.

Perfusion to subcutaneous tissue is sacrificed early during circulatory stress (4). It is, therefore, a sensitive indicator during situations of hemodynamic compromise. Furthermore, it has been shown, in hypovolemic dogs, that the fall in subcutaneous oxygen tension parallels the rapid fall in perfusion to the gut, as well as to the skin and subcutaneum, as indicated by radiolabeled microsphere dilution techniques (unpublished data). These characteristics, along with the accessibility of the tissue, make monitoring of subcutaneous tissue oxygen tension acceptable as an indicator of circulatory homeostasis, as well as a specific indicator of subcutaneous perfusion.

Clinical monitoring of tissue oxygen tension has been a difficult task, however. Many techniques have been described, but all suffer distinct disadvantages when used in a clinical setting. Of the three to have gained ac-

270

ceptance recently, conjunctival oximetry, transcutaneous oximetry and pulse oximetry, only the first adequately reflects the unaltered phsyiology of the underlying tissue.

Transcutaneous oximetry utilizes a heated electrode, to increase local perfusion by dilating the underlying capillary bed and to facilitate transcutaneous migration of gases by changing the skin lipid structure, both of which destroy normal homeostasis. In the specific situation of monitoring downstream from a vascular obstruction, however, transcutaneous oximetry has demonstrated its value as a predictor of healing (5,6). Pulse oximetry, although an excellent indicator of oxyhemoglobin saturation, does not monitor tissue gas levels. Conjunctival oximetry does measure underlying tissue gas tensions (7). Unfortunately, its placement within the orbit, with the associated discomfort and theoretical risk to the eye, makes it unacceptable in many cases to the patient and clinician alike. Furthermore, it can only monitor perfusion centrally in a tissue bed likely to be spared the decrease in perfusion accompanying mild hypovolemia. A modification of the conjunctival electrode has allowed for intraoperative intraabdominal tissue gas monitoring (8). This may prove useful in the acute operative setting, but has little value otherwise in the clinical management of patients.

Subcutaneous polarography has been found to be a useful monitor of tissue oxygen tension (1), which has been shown to fall during smoking (unpublished data) and hemodialysis (9). Jonsson, et al., have found a direct relation between tissue oxygen tension and collagen accumulation (unpublished data). An ongoing prospective clinical trial has identified an increased incidence of wound infections in patients with low subcutaneous tissue oxygen tension (unpublished data).

Unlike microelectrodes, which, when inserted directly into the tissue, can be "poisoned" by tissue proteins affecting their calibration, the oxygen-permeable Silastic tonometer protects the electrodes, bathing them in fluid whose gas tension equilibrates with the surrounding tissue, providing a mean integrated value. Although invasive, the method minimally alters the natural physiology, limiting the impact of monitoring on the observed parameter. The inert materials are nonreactive and no heat is applied. The tonometer has been well tolerated by patients in several hundred trials, with minimal discomfort and no episodes of clinical infection or mechanical breakdown.

The current system requires intermittent insertion of the polarographic needle electrodes and Ag/AgCl reference electrode into the subcutaneously implanted Silastic tube. The tube is filled with isotonic saline to complete the circuit. A polarographic potential of 0.7 V DC is applied and the current measured. The current is a linear function of the pO_2 at the polarographic electrode. This method provides accurate readings, but the difficulty of application and calibration limits its utility in a clinical setting. Although simple in theory, the method is laborious for both patient and clinician. The equipment must be monitored while in use, during which time the patient must be immobile. Even without any accompanying discomfort some subjects are unable to cooperate, particularly if they are disoriented. This makes the open system useless in some extreme circumstances because of repeated mechanical problems preventing a stable baseline from being reached.

The goal of the clinical probe design effort was to develop an electrochemical system that did not suffer from the drift problems of the Ag/AgCl reference electrode and to perfect a low-cost manufacturing process for production. The original method uses a two-electrode system. The new clinical probe has the three-electrode system used by electrochemists for analysis of various concentrations of materials in solution. It is more precise and more flexible.

271

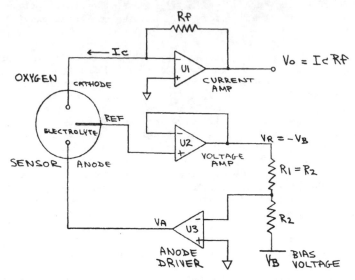

Fig. 5 Circuit diagram, three-electrode system

The circuit diagram of the three-electrode system shows the third electrode, which acts as an anode (fig. 5). In the two-electrode setup, in contrast, the reference electrode functions as the anode. Separating the reference and anode electrodes is preferable for two reasons. First, because the reference electrode functions only as a voltage sensor it draws very little current and so its potential is not a function of the current through the sensor. Consequently, because the current requirements of the reference electrode are sustantially reduced, a smaller electrode or a different type of reference electrode can be used.

In the three-electrode probe the cathode is held at ground potential and the current measured by the current amp. The reference electrode maintains a definite potential between it and the electrolyte solution. This potential is monitored via the voltage amplifier. The polarographic bias potential VB is applied to one input of a voltage-summing anode-driving amplifier. The other input is connected to the output of the reference voltage amplifier (VR). The closed loop feedback configuration of the circuit, in conjunction with the three-electrode cell, requires that the anode assume a potential that results in VB equaling VR. That is, the polarographic bias potential VB appears between the cathode and reference electrode, but the anode passes the current rather than the reference electrode. These advantages of the three-electrode configuration together with a suitable reference electrode have resulted in a miniature oxygen probe with the superior drift and performance characteristics described above.

An additional requirement for the new probe was that it be easy to use and resistant to mechanical difficulties. With our new sealed electrode mechanical effects on the monitor are minimal. Patients' activities will not be impaired by the monitoring. Similarly, the electrode may be allowed to stabilize unsupervised and consulted intermittently by existing staff to identify trends. This simplification will allow the device to be cost-effective by reducing the manpower required to operate it.

SUMMARY

Our new method is comparable to the standard, "open-end" tonometer system and incorporates all its abilities to monitor oxygen tension and perfusion. The new device has the advantage of being completely sealed, providing a true, continuous direct measurement over several days, with freedom of movement for the patient without operator attention and with improved stability of the electrode, which minimizes the drift artifacts previously encountered. These improvements will facilitate routine clinical monitoring of tissue oxygen tension.

REFERENCES

1. N. Chang, W.H. Goodson, III, F. Gottrup, et al., Direct measurement of wound and tissue oxygen tension in postoperative patients, Ann. Surg. 197:470 (1983).

2. F. Gottrup, R. Firmin, N. Chang, et al., Continuous direct tissue oxygen measurement by a new method using an implantable Silastic tonometer and oxygen polarography. Am. J. Surg. 146:399 (1983).

3. F. Gottrup, R. Firmin, J. Rabkin, et al., Directly measured tissue oxygen tension and arterial oxygen tension assess tissue perfusion, Crit. Care Med. (in press, 1987).

4. T.K. Hunt, B.H. Zederfeldt, T.K. Goldstick, et al., Tissue oxygen tensions during controlled hemorrhage, Surg. Forum 18:3 (1967).

5. T.R.S. Harward, J. Volny, F. Golbranson, et al., Oxygen inhalation induced transcutaneous pO2 changes as a predictor of amputation level, J. Vasc. Surg. 2:220 (1985).

6. P.T. McCollum, V.A. Spence and W.F. Walker, Oxygen inhalation induced changes in the skin as measured by transcutaneous oxymetry, Br. J. Surg. 73:882 (1986).

7. W.C. Shoemaker, S. Fink, C.W. Ray, et al., Effect of hemorrhagic shock on conjunctival and transcutaneous oxygen tensions in relation to hemodynamic and oxygen transport changes, Crit. Care Med. 12:949 (1984).

8. H.B. Kram and W.C. Shoemaker, Method for intraoperative assessment of organ perfusion and viability using a miniature oxygen sensor. Am. J. Surg. 148:404 (1984).

9. J. A. Jensen, W.H. Goodson, III, R. Omachi, et al., Subcutaneous tissue oxygen tension falls during hemodialysis, Surgery 101:146 (1987).

TISSUE OXYGEN UPTAKE FROM THE ATMOSPHERE BY A NEW, NONINVASIVE

POLAROGRAPHIC TECHNIQUE WITH APPLICATION TO CORNEAL METABOLISM

B. W. Brandell[1], T. K. Goldstick[2,1], T. A. Deutsch[3], and J. T. Ernest[4]

[1]Dept. of Biomedical Engineering, Northwestern University
Evanston, Illinois, USA
[2]Dept. of Chemical Engineering, Northwestern University
Evanston, Illinois, USA
[3]Dept. of Ophthalmology, Rush-Presbyterian-St. Luke's
Medical Center, Chicago, Illinois, USA
[4]Dept. of Ophthalmology, University of Chicago, Chicago
Illinois, USA

INTRODUCTION

Corneal O_2 uptake is important because most of the cornea's metabolic requirements are derived from atmospheric O_2. With the widespread use of contact lenses, research on the effect of this O_2 transport mechanism on corneal metabolism and physiology has taken on new importance (Hill and Fatt, 1964; Klyce and Beuerman, 1985; Holden, et al., 1985b). There also is evidence that corneal metabolism is altered after surgery (Chaston and Fatt, 1982; Kwok, 1985) and with diabetes mellitus (Graham, et al., 1981).

The first attempt at quantifying corneal O_2 flux noninvasively involved mounting a stirred, saline-filled chamber of known volume on a scleral contact lens (Hill and Fatt, 1963). The chamber PO_2 was monitored by a Clark O_2 electrode. When mounted on the eye, the scleral contact lens provided a tight seal and exposed the fluid of the chamber directly to the cornea. The rate of change of PO_2 in the chamber was directly proportional to the flux through the anterior surface of the cornea. Using this method, Hill and Fatt reported an oxygen uptake rate of 4.8 $\mu l/cm^2 \cdot hr$ for the human cornea _in vivo_. While the method provided an absolute indication of O_2 uptake, the procedure involved keeping the sensor on the eye for about 20 minutes, resulting in discomfort to the subject. For this reason, the technique was abandoned in favor of a faster, less traumatic method.

That new procedure involved placing the plastic membrane of an unmodified, standard, blood-gas Clark O_2 electrode in direct contact with the cornea of the eye (Hill and Fatt, 1964). The electrode was pressed against the cornea of the eye for just a few seconds, greatly reducing the discomfort of the experimental subjects. Because the 90% response time of the electrode typically used in these experiments was itself on the order of 10 seconds and because the corneal O_2 uptake disturbed the diffusion field in the membrane, the electrode output could not be correlated with the PO_2 on the membrane surface. Because of this problem, most results reported using this technique characterize corneal O_2 uptake by the time it takes the electrode current to fall from a value corresponding to 140 torr to a value corresponding to 40 torr (Benjamin and Hill, 1986). The technique only indicates changes in O_2 uptake by the same eye and cannot be used to compare uptake rates between eyes. Although the shortcomings of this method have been acknowledged by the original researchers (Jauregui and Fatt, 1971; Fatt, 1978), some work has been reported which ignores these faults and assumes that the electrode current maintains its relationship to PO_2 during the measurement, and that this PO_2 reflects the O_2 content of the electrode membrane (Jauregui and Fatt, 1972; Holden, et al., 1985a). By estimating the solubility and thickness of the electrode membrane, Holden, et al. claimed to have measured an absolute value of corneal O_2 flux of 3.31 ($\mu l/cm^2 \cdot hr$) on intact corneas of anaesthetized New Zealand white rabbits.

The present sensor, Figures 1a and 1b, combines the benefits of the two techniques described above. It has the ability to determine an absolute value of corneal O_2 flux, as with the original sensor, but with nearly the speed of measurement obtained using direct application of a Clark electrode to the cornea. The saline-filled chamber used in the original experiments (Hill and Fatt, 1963) is replaced by a thin membrane material of high O_2 solubility and diffusivity, approximating an O_2 reservoir. This reservoir is attached directly to the electrode membrane, eliminating the scleral contact lens. The reservoir size is such that, when the sensor is placed on the cornea, the rate of decrease of PO_2 is slow enough so that the electrode correctly gives the PO_2 in the reservoir, but fast enough so that the determination of corneal O_2 flux

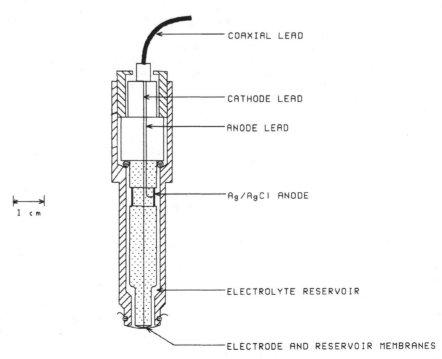

Figure 1a.   Schematic of the corneal O_2 uptake sensor. A magnified sketch of the cathode tip region of the sensor is shown in Figure 1b.

requires only a few minutes on the cornea. This is only possible when the diffusion coefficient of O_2 in the reservoir is high enough so that the O_2 is always well-mixed (i.e. the PO_2 profile in the reservoir is essentially flat). This is the key to our new design.

METHODS

The sensor consists of a reservoir mounted on top of the 12 μm polypropylene membrane of a standard blood-gas electrode (model E5047, Radiometer, Copenhagen). The reservoir is made of either porous polytetraflouroethylene (GORE-TEX, Gore and Associates, Bethesda, Maryland) or silicone rubber (Silastic, Dow Corning, Midland, Michigan). The high porosity (80%-90%) and open pore structure allow the diffusion properties of GORE-TEX to be approximated by those of air. Silicone rubber also behaves as a reservoir because of its high O_2 solubility and diffusivity. Analytical and computer models were used to demonstrate that the O_2 in GORE-TEX and silicone rubber reservoirs remained well-mixed during conditions approximating those during application to the cornea. Both materials used were in a thickness range of 60-200 μm. They provided an excellent barrier to aqueous penetration and had good compressive strength. While several mechanical and chemical methods of attaching the reservoir to the electrode membrane were tested, best results were obtained when the natural adhesive force between the materials was used. When pressed against the cornea, the deformation of the corneal surface provided a seal around the periphery of the sensor and prevented any leakage of ambient O_2 into the reservoir. Since the E5047 measures PO_2 just outside its membrane, it continuously monitors the level of O_2 in the reservoir. The sensor was calibrated by exposing it to air and then to pure nitrogen. Because the electrode membrane was exposed to the same environment (i.e. the reservoir) during both calibration and application to the cornea, no calibration error was introduced during the experiment.

A "phantom eye" was constructed to test different reservoir materials and to calibrate the sensor before any animal experiments. The phantom eye consisted of an O_2 impermeable tube, 5 cm in diameter, with one end sealed, and the other covered by a polypropylene "phantom cornea",

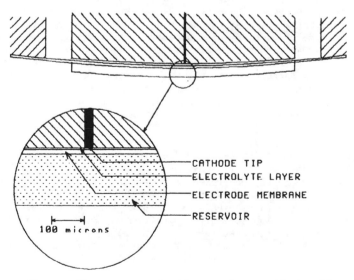

Figure 1b.   Close up of the sensor cathode tip region.

12 μm thick. The interior of the phantom eye was flushed with calibration gases and thus maintained at a known PO_2. The transmissibility of the phantom cornea, $(Dk/L)_{pc}$, was measured independently. Transmissibility is defined here as the O_2 diffusion coefficient (D)-solubility (k) product divided by the membrane thickness (L). With a PO_2 difference of 150 torr between the atmosphere and the inside of the phantom eye, a known flux of O_2 of 2.25 $\mu l/cm^2 \cdot hr$ was established across the phantom cornea, approximating that expected to be found during in vivo rabbit experiments.

The phantom eye was used to experimentally verify the suitability of the materials as reservoirs. The sensor, with the material in question mounted as a reservoir, was applied to the phantom eye. For the phantom cornea, the time constant was on the order of one second. Thus, when the sensor was placed against the phantom cornea and the reservoir material was such that the PO_2 remained well-mixed in the reservoir and decreased very slowly, the PO_2 in the phantom cornea could always be approximated by a series of steady state profiles (straight lines whose slope approached zero with time). In other words, the PO_2 profile in the phantom cornea was pseudo steady state. A theoretical analysis of such a phantom cornea/reservoir system showed that the PO_2 in the reservoir was a monoexponentially decaying function of time with a time constant much larger than that of the phantom cornea. This type of response was demonstrated experimentally. Therefore the material indeed behaved as if it were a reservoir of O_2. Figure 2 shows the results of such an extended application fitted with a monoexponential. The dimensionless PO_2 is defined as:

$$(P(t)_{res} - P_o)/(P_1 - P_o),$$

where $P(t)_{res}$, or P, is the PO_2 in the reservoir as measured by the electrode, P_o is the PO_2 inside the phantom eye (0 torr in all experiments), and P_i is the PO_2 in the reservoir at the moment of first application to the phantom cornea (room PO_2 in all cases). Both GORE-TEX and silicone rubber exhibit the necessary behavior to be considered reservoir materials. Although any GORE-TEX in the thickness range 60-200 μm would probably have worked, the best results were obtained with a pore size of 0.2 μm, a porosity of 80%, and a thickness of 75 μm. For the silicone rubber reservoir, the best results were obtained with a thickness of 125 μm. Figure 2 shows a reasonable agreement with the simple, pseudo steady state model. Much of the discrepancy which occurred at long times (low reservoir PO_2) was undoubtedly caused by leakage into the reservoir caused by radial diffusion from the sensor periphery and electrode drift over two hours. Nonetheless in the usual three minute application, as will be discussed below, agreement was excellent between the data and a monoexponential (see Figure 3). On the phantom cornea, the sensor recorded a time constant with both reservoir materials of approximately 20 minutes.

Experiments were conducted to determine the corneal O_2 flux of seven rabbits. The animals were killed by lethal intravenous injection and the right eye was immediately proptosed. The anterior chamber was cannulated and perfused with balanced salt solution (BSS) at a flow rate of 10 $\mu l/min$, a PO_2 of 55 torr, a PCO_2 of 40 torr, and approximately normal intraocular pressure. The corneal temperature was allowed to fall to room temperature. The anterior surface was kept moist with BSS during the entire experiment. The sensor, with reservoir membrane intact, was calibrated using room air and pure nitrogen to determine the relationship between electrode current and reservoir PO_2.

The sensor was then applied for three minutes to the phantom cornea resulting in a sensor output response as shown in Figure 3. Any application of the sensor initially introduced largely unpredictable transients in electrode current due to temperature imbalances and changes in electrode geometry induced by hydrostatic pressure effects. For this reason, the first minute of data after application was never used. The final two minutes of sensor current data were fitted either to a straight line (GORE-TEX) or a monoexponential (silicone rubber) to give an extrapolated value of the initial rate of change of PO_2 in the reservoir. From the known phantom cornea flux and calculated initial PO_2 time derivative, the proportionality constant between the rate of change of PO_2 in the reservoir and the O_2 flux from the reservoir was determined.

After removing the sensor from the phantom eye it was allowed to equilibrate in air until its current output stabilized at its room air value. Then it was applied to the rabbit cornea for three minutes, resulting in a sensor output response as shown in Figure 4. Again, the first minute of data was not used. A fit yielded the initial rate of change of the reservoir PO_2. The corresponding O_2 flux into the cornea from the reservoir was calculated using the proportionality constant determined from the phantom eye experiment. The rabbit eye was allowed several minutes to recover between experiments. An average of six experiments per rabbit were conducted.

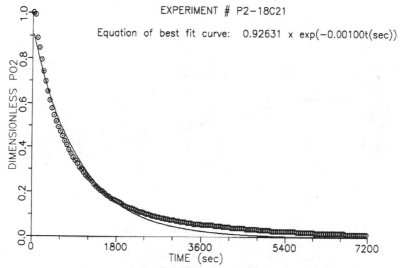

Figure 2. Long-term application of a sensor, fitted with a 0.2 μm pore size GORE-TEX reservoir, to the phantom cornea. The graph shows that GORE-TEX behaves as a well-stirred reservoir when applied to the phantom eye. The points are experimental and the curve is the best fitting monoexponential found by nonlinear regression using the previous calibration nitrogen current as P_o, and fitting for $P(t)_{res}$ at t=0 (i.e. P_i).

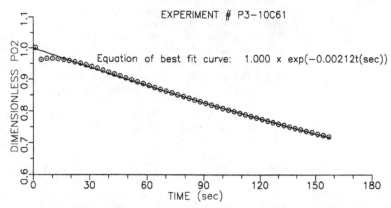

Figure 3. Three minute application of a sensor, fitted with a silicone rubber reservoir, to the phantom eye. The points are experimental and the curve is the best fitting monoexponential found by nonlinear regression excluding the first 60 sec of data. The nitrogen calibration current was used for P_o, and P_i was found by fitting for $P(t)_{res}$ at t=0.

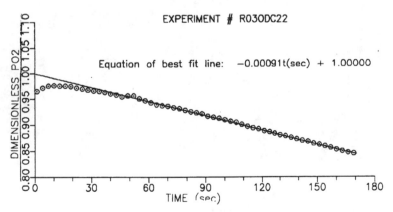

Figure 4. Three minute application of a sensor, fitted with a 0.2 μm pore size GORE-TEX reservoir, to the cornea of the proptosed right eye of a previously killed rabbit. The corneal temperature was approximately $20^{\circ}C$. The points are experimental and the line is the regression line excluding the first 60 sec of data. The nitrogen calibration current was used for P_o, and P_i was found by fitting for $P(t)_{res}$ at t=0.

RESULTS

Initially it was necessary to prove that the PO_2 profile in the reservoir was in fact essentially flat (i.e., the O_2 in the reservoir was well-mixed by diffusion alone). The only possible way to do this was with a computer model of the cornea/sensor system. However, since the flux from the reservoir at the corneal interface could be assumed constant at a known value and the flux at the electrode interface could be assumed to be zero, a simple unsteady state mathematical model (Carlslaw and Jaeger, 1959) represented the O_2 diffusion process in the reservoir quite well. The PO_2 distribution in a silicone rubber reservoir as a function of time after application to the cornea is shown in Figure 5. The profiles with GORE-TEX would have been even flatter. For the reservoir materials used in this study then, the PO_2 profiles were always flat to less than 0.5%. This confirms that the rate of decrease was slow enough so that the reservoir could always be considered well-mixed.

The results of typical experiments involving application of the sensor to the phantom cornea and to the rabbit cornea are shown in Figures 3 and 4, respectively. A monoexponential fit to the data of Figure 3 was used to determine the initial time rate of change of $P(t)_{res}$. The

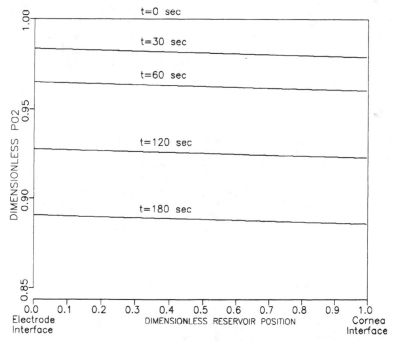

Figure 5. Analytical, unsteady state solution approximating the PO_2 distribution in a silicone rubber reservoir as a function of time after application to a cornea with a constant flux of 2 $\mu l/cm^2 \cdot hr$. The PO_2 profiles indicate that O_2 remains "well-stirred" by diffusion only.

relationship between the initial rate of change and the initial flux out of the reservoir is given by:

$$(Vk/A)_{res} \cdot [dP/dt]_{t=0} = -(Dk/L)_{pc} \cdot (P_i - P_o) \qquad (1)$$

where V is the volume and A is the area of the reservoir, and $(Dk/L)_{pc}$ is the phantom cornea transmissibility, which was measured independently. The only unknown in equation (1), $(Vk/A)_{res}$, can then be calculated. The value of $(Vk/A)_{res}$ gave us the necessary information to measure the rabbit corneal flux.

The rabbit data in Figure 4 were fitted to a straight line for simplicity. The application period in this case was much shorter than the time constant of the exponential decay. The calculated values of initial change in reservoir PO_2, $[dP/dt]_{t=0}$, and the value for $(Vk/A)_{res}$ previously determined using equation (1) were used to determine the flux, qO_2, into the cornea of the rabbit by:

$$qO_2 = (Vk/A)_{res} \cdot [dP/dt]_{t=0} \qquad (2)$$

Values for qO_2 are given in Table I. There was no difference in the results obtained with GORE-TEX and silicone rubber reservoirs on the same rabbit.

A theoretical determination of corneal O_2 flux can be made based on published values of thickness, Dk, and QO_2 (the O_2 consumption rate per unit volume of tissue, i.e., ml O_2 (STP)/ml·hr) of the three layers of the cornea, and the PO_2 of aqueous humor posterior to the cornea (Freeman and Fatt, 1972; Freeman, 1972). The values of Dk and QO_2 must be adjusted to compensate for the temperature difference between our experiments ($20^{\circ}C$) and that of Freeman ($37^{\circ}C$). Adjusted to $20^{\circ}C$, we calculated a value of qO_2 of 2.3 $\mu l/cm^2 \cdot hr$. This agrees reasonably well with the average in Table I, 1.7 $\mu l/cm^2 \cdot hr$.

DISCUSSION

The results above indicate that this method shows promise as a technique for determining corneal O_2 flux noninvasively. The method also may provide a way to determine the consumption rate of the various layers of the cornea in vivo. The sensor allows simultaneous measurement of both corneal O_2 flux and PO_2 at the anterior surface. With this "extra" boundary information, it may be possible to determine the forcing function (i. e. consumption) through inverse mathematics. The authors are currently exploring methods of solution to this inverse problem and any necessary modification to the device which may be required to make it possible.

The clinical application of this measurement technique would require several technical advances over the current device. Temperature control would have to be included to assure that the sensor is maintained at the same temperature as the human cornea in vivo. An accurate prior knowledge of $(Vk/A)_{res}$ would be desirable without the need for experimental determination on a phantom eye. The sensor should also be constructed so that minimal mechanical deformation occurs upon application to the cornea. All of the improvements above will also result in a shorter residence time on the cornea and less discomfort to the patient. After these improvements have been made, it is intended to use this device in a study of corneal O_2 uptake of diabetic patients.

Table I. Experimental Values of Corneal Oxygen Flux for Seven Rabbits.

Rabbit	Flux ($\mu l/cm^2 \cdot hr$)	Standard Deviation ($\mu l/cm^2 \cdot hr$)
R02OD	1.75	0.10
R03OD	1.88	0.35
R04OD	0.84	0.73
R05OD	2.56	0.27
R06OD	1.40	0.56
R07OD	1.37	0.35
R08OD	1.06	0.14
AVERAGE	1.67	

SUMMARY

A standard Clark electrode has been modified to continuously monitor the PO_2 in a thin, disk shaped reservoir membrane mounted on the electrode membrane surface. In vitro tests were conducted to determine the proportionality constant between the rate of change of reservoir PO_2 and the flux of O_2 out of the reservoir. The device was then used to determine the corneal O_2 uptake on proptosed eyes of previously sacrificed rabbits. Our average measured uptake at $20^\circ C$, 1.7 $\mu l/cm^2 \cdot hr$, agrees with the value 2.3 $\mu l/cm^2 \cdot hr$ calculated from a diffusion analysis of the cornea utilizing literature values of the parameters at $37^\circ C$ for the three component layers of the cornea when they are adjusted for temperature to $20^\circ C$.

ACKNOWLEDGEMENTS

This research was supported by U. S. National Eye Institute (NIH) grants EY-06697 to Greenmark, Incorporated, Niles, Illinois, USA, EY-07041, and the Louise C. Norton Trust. We are grateful for the advice and assistance provided by Dr. Bernard Marker, Hang Duk Roh and Gwynne E. Rowley.

REFERENCES

Benjamin, W. J., and Hill, R. M., 1986, Human corneal oxygen demand: the closed-eye interval, Graefe's Arch Clin Exp Ophthalmol, 224:291.

Carslaw, H. S., and Jaeger, J. C., 1959, "Conduction of Heat in Solids", 2nd ed., p. 112, Oxford University Press, Oxford.

Chaston, J., and Fatt, I., 1982, Corneal oxygen uptake under a soft contact lens in phakic and aphakic eyes, Invest Ophthalmol Vis Sci, 23:234.

Fatt, I., 1978, "Physiology of the Eye, An Introduction to the Vegetative Functions", Butterworths, Boston.

Freeman, R. D., 1972, Oxygen consumption by the component layers of the cornea, J Physiol, 225:15.

Freeman, R. D., and Fatt, I., 1972, Oxygen permeability of the component layers of the cornea, Biophys J, 12:237.

Graham, C. R., Jr., Richards, R. D., and Varma, S. D., Oxygen consumption by normal and diabetic rat and human corneas, Ophthalmic Res, 13:65.

Hill, R. M., and Fatt, I., 1963a, Oxygen uptake from a reservoir of limited volume by the human cornea in vivo, Science, 142:1295.

Hill, R. M., and Fatt, I., 1963b, Oxygen depletion of a limited reservoir by the human conjunctiva, Nature, 200:1011.

Hill, R. M., and Fatt, I., 1964, Oxygen deprivation of the cornea by contact lenses and lid closure, Am J Optom Arch Am Acad Optom, 41:678.

Holden, B. A., Sulonen, J., Vannas, A., Sweeney, D. F., and Efron, N., 1985a, Direct in vivo measurement of corneal epithelial metabolic activity using a polarographic oxygen sensor, Ophthalmic Res, 17:168.

Holden, B. A., Sweeney, D. F., Vannas, A., Nilsson, K.T., and Efron, N., 1985b, Effect of long-term extended contact lens wear on the human cornea, Invest Ophthalmol Vis Sci, 26:1489.

Jauregui, M. J., and Fatt, I., 1971, Estimation of oxygen tension under a contact lens, Am J Optom and Arch Am Acad Optom, 48:210

Jauregui, M. J., and Fatt, I., 1972, Estimation of the in vivo oxygen consumption rate of the human corneal epithelium, Am J Optom Arch Am Acad Optom, 49:507.

Klyce, S. D., and Beuerman, R. W., 1985, The effects of contact lenses on the normal physiology and anatomy of the cornea: symposium summary, Eye Res, 4:719.

Kwok, S. L., 1985, Effect of epithelial cell injury on anterior corneal oxygen flux, Am J Optom Physiol Opt, 62:642.

OXYGEN REACTION VESSELS

N. Okutani*, B. Hagihara*, M. McCabe**, W. Ohtani***,
N. Negayama***, M. Nishioka*, T. Nakai* and T. Morita***

* The Hagihara Medical Technology Research Institute
  72 Shimoyamaguchi, Yamaguchicyo, Nishinomiya, Hyogo
** James Cook University, Townsville, Q. 4811. Australia
*** Otsuka Electronics Co.,Ltd., 26-3 3Chome Shodai
  Tajika, Hirakata, Osaka, Japan

INTRODUCTION

Two types of oxygen electrodes have been used in biological
research, i.e., a "separated system" in which the cathode and the anode
are placed in different parts of the subject to be measured and a
"combined oxygen electrode system" (Clark electrode) in which the
cathode and the anode are put together in an electrolyte layer behind a
thin hydrophobic oxygen permeable membrane[1].

The "separated" oxygen electrode was first applied by Chance and
Williams[2], to the measurement of oxygen consumption by mitochondrial
suspensions. Much information on mitochondrial respiration was obtained
through the simultaneous use of this method and spectrophotometry.
However, their method allowed the introduction of atmospheric oxygen to
the reaction mixture. In order to prevent oxygen introduction,
Longmuir[3] and Hagihara[4] introduced different types of rotating
electrodes.

The Clark type electrode has various advantages such as complete
protection of the cathode by a hydrophobic membrane to provide high
stability. Therefore, various attempts were made to apply this
electrode to oxygen reaction vessels.

Imai et al[5] incorporated this electrode into a spectrophotometer
cuvette. Hamilton et al[6] described electrode systems with a stirring
device.

Recent improvements in oxygen electrode technology have resulted in
the manufacture of transcutaneous blood oxygen electrodes [7,8,9]. All
such electrodes are characterised by a small flat design(7-10mm thick;
16-20mm in diameter), which makes them suitable as the base plate for
oxygen reaction vessels. Additionally these transcutaneous oxygen
electrodes contain a temperature control system which can be used to
maintain the temperature of the reaction medium. These characteristics
enable construction of reaction vessels with volumes (1-2ml) far less
than that of conventional reaction vessels (volume=5-10ml) with rod

shaped oxygen elctrodes attached to the side. In further attempt to make
the vessel smaller, we incorporated the temperature control system in the
metal sleeve covering the cell instead of in the electrode part itself.

Using such a construction, the final volume of the vessel was
reduced to 0.1-0.5ml. In this paper, we mainly describe the construction and a representative application of this newly designed vessel.

APPARATUS

The construction of three representative types of oxygen reaction
vessels are depicted in Figs. 1 and 2. Fig. 1a shows the construction
of the most widely used conventional reaction vessel. A rod-shaped
Clark electrode is attached on the side wall of the glass reaction cell.
The cell is immersed in a constant temperature bath which may or may not
be connected to a circulating water bath. This design limits the ability
to attach spectrophotometric apparatus and other installments for
simultaneous measurements. As with all other oxygen reaction vessels,
the top inlet of the cell must be made very narrow in order to prevent
the introduction of atmospheric oxygen into the reaction medium. But
unless the side walls become narrow very gradually, there is a tendency
for air bubbles to become trapped on the side walls, especially near the
electrode-cell wall boundary where air bubbles adhere easily. In order
to set the oxygen electrode adequately in the cell wall and to prevent
entrapment of air bubbles, the size of the cell must be made considerably
large(usually, the cell must have a diameter more than 2.5 times the
outer diameter of the electrode). Thus there is a lower limit on the
vessel volume which depends on the electrode size. In the case of a 10mm
dia. electrode, the vessel volume cannot be made much smaller than 10ml.

With the design depicted in Fig. 1b, which uses a flat transcutaneous electrode as the cell base, the cell can be made into a
gradually narrowing cylinder. Due to this construction, the diameter of
the cell can be made much smaller than the outer diameter of the
electrode. Thus, even in order to prevent entrapment of air bubbles, the
cell height need not be very high. This makes it possible to greatly
reduce the size of the reaction vessel. Considering such factors as the
heat maintenance of the reaction medium and smooth rotation of the
magnetic stirring bar, the optimal volume of the cell is about 1ml, or
about 1/10th of the conventional reaction vessel. Since the electrode -
cell wall boundary is placed on the bottom of the vessel, removal of air
bubbles at the boundary is accomplished simply by stirring the medium
before an experiment. Thus, even though the volume of this cell is much
smaller than that of the conventional cell, accurate measurements can be
made with high reproducibility.

Fig. 2 shows further improved versions of the oxygen reaction
vessels. The flat oxygen electrode used as the cell base here was
designed specifically for the use in reaction vessels and does not have
the temperature control system which is characteristic of transcutaneous
oxygen electrodes. This enables increased ease of construction as well
as improvement of electrode characteristics. The temperature control
system is incorporated in the metal sleeve surrounding the glass cell
wall, providing more uniform and stable temperatures of the reaction
medium and enabling smaller inner volumes of the reaction vessel compared
to the design shown in Fig. 1b.

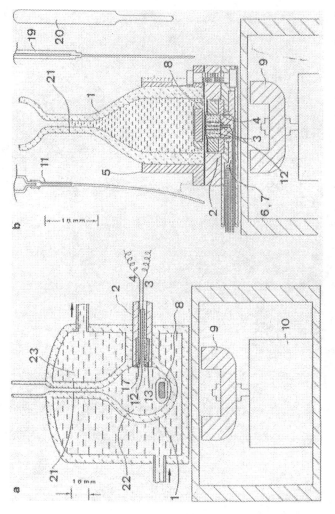

Fig. 1 Construction of the conventional oxygen reaction vessel (a) and the vessel using a trans-cutaneous PO₂ electrode (b) (refer to Fig. 2 for the description of the vessel componets).

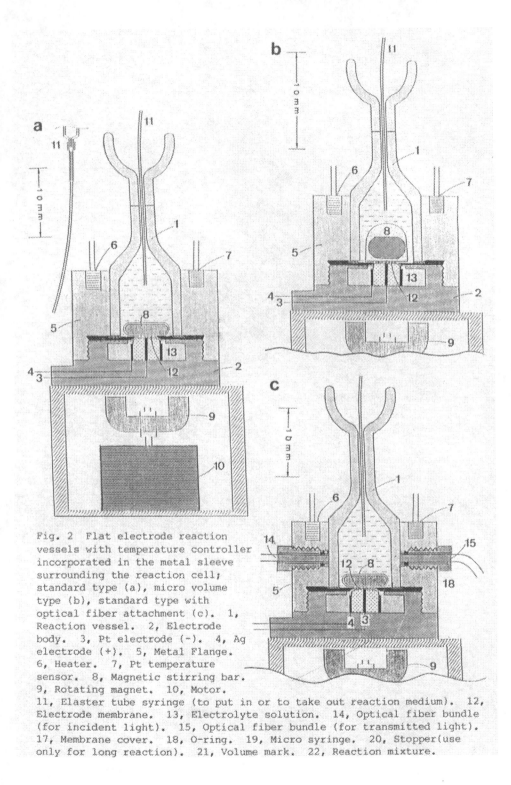

Fig. 2 Flat electrode reaction
vessels with temperature controller
incorporated in the metal sleeve
surrounding the reaction cell;
standard type (a), micro volume
type (b), standard type with
optical fiber attachment (c). 1,
Reaction vessel. 2, Electrode
body. 3, Pt electrode (-). 4, Ag
electrode (+). 5, Metal Flange.
6, Heater. 7, Pt temperature
sensor. 8, Magnetic stirring bar.
9, Rotating magnet. 10, Motor.
11, Elaster tube syringe (to put in or to take out reaction medium). 12,
Electrode membrane. 13, Electrolyte solution. 14, Optical fiber bundle
(for incident light). 15, Optical fiber bundle (for transmitted light).
17, Membrane cover. 18, O-ring. 19, Micro syringe. 20, Stopper(use
only for long reaction). 21, Volume mark. 22, Reaction mixture.

Fig. 2a shows the standard type of this improved reaction vessel with an inner volume of 0.5ml. The type shown in Fig. 2b enables the reduction of the inner volume to 0.1ml with the use of an enlarged magnetic stirring bar. Fig. 2c shows a standard improved vessel with two optical fiber attachments for delivering incident light and receiving transmitted light respectively for simultaneous optical measurements with a spectrophotometer.

APPLICATION: MEASUREMENT OF EQUILIBRATION OF HAEMOGLOBIN WITH OXYGEN

The oxygen concentration of Haemoglobin solutions (42uM Haemoglobin in 0.2M pH7.0 phosphate buffer; temp.=20°C) in the reaction cell was changed by sending gas mixtures of differing ratios of O_2 to N_2 to the small gas phase at the top of the solution.

Spectra were measured instantaneously by a photodiode array spectrophotometer (MCPD-110, Otsuka Electronics Co.,Ltd.) using a halogen lamp as the light source.

The variation of the oxygen concentration as measured and recorded using the oxygen electrode at the bottom of the reaction vessel is shown in Fig. 3a. Before beginning the oxygen trace, the reaction medium was equilibrated with air. Pure N_2 was then sent to the solution until the partial pressure of O_2 in the solution decreased to 0mmHg. 4% O_2, 8% O_2, and 21% O_2(air) were then sent successively to the vessel. After the O_2 concentration approached a constant value with air, a trace amount of sodium dithionite was added in order to confirm the 0 oxygen level. During this process, spectra were measured at the times indicated by lines B, C, D, E, F and G (Fig. 3b).

Summary

Transcutaneous oxygen electrodes were modified to be more suitable as a component of oxygen reaction vessels. The temperature control system was removed from the transcutaneous electrode to decrease the thickness and improve the stability. The temperature control system was incorporated in the metal sleeve surrounding the glass reaction vessel to shorten the distance between the magnetic stirrer and stirring bar, enabling smooth stirring with a short magnetic bar. With these modifications, we have succeeded in reducing the vessel volume to about 0.5ml, or two to four times smaller than reaction vessels incorporating unmodified transcutaneous electrodes (vessel volume=1-2ml) and about twnty times smaller than reaction vessels using rod-shaped Clark electrodes (vessel volume about 10ml).

In another vessels modified as above, two optical guides were connected to the metal sleeve for irradiating the solution and receiving transmitted light simultaneously to enable simultaneous measurements of oxygen concentration absoption spectra. The relationship between oxygen concentration and absorption spectra of Hb is described as an application of this vessel.

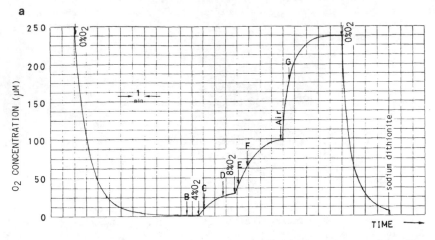

Fig. 3a Plot of oxygen concentration variation as measured by the oxygen reaction vessel shown in Fig. 2c.

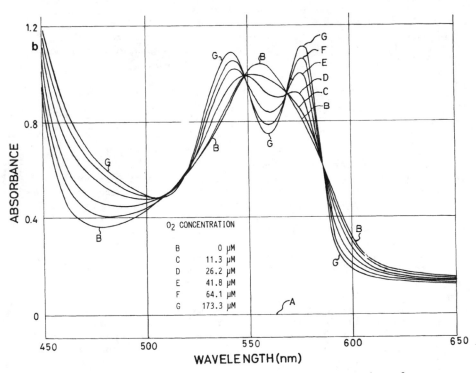

Fig. 3b Spectra of haemoglobin at various concentration of oxygen indicated by lines B to G in Fig. 3a.

REFERENCES

1. L.C. Clark, Jr., Monitor and control of blood and tissue oxygen tensions, Trans. Amer. Soc. Art. Int. Org. 2:41 (1956).
2. B. Chance and G.R. Williams, Respiratory enzymes in oxidative phosphorylation, J. Biol. Chem. 217:383 (1955).
3. I.S. Longmuir, Respiration rate of bacteria as function of oxygen concentration, Biochem. J. 57:81 (1954).
4. B. Hagihara, Techniques for the application of polarography to mitochondrial respiration, Biochim. Biophys. Acta 46:134 (1961).
5. K. Imai, H. Morimoto, M. Kotani, H. Watari, W. Hirata and M. Kuroda, An improved method for automatic measurement of the oxygen equilibrium curve of hemoglobin, Biochim. Biophys. Acta 200:189 (1970).
6. R. Hamilton, D. Maguire and M. McCabe, A versatile micro stirrer and oxygen electrode system for a spectrophotometer cuvette, Anal. Biochem. 93:386 (1979)
7. A. Huch, R. Huch, B. Arner and G. Rooth, Continuous transcutaneous oxygen tension measured with a heated electrode, Scand. J. Clin. Lab. Invest. 31:269 (1973).
8. P. Eberhard, K. Hammacher and W. Mindt, Methode zur Kutanen Messung des Sauerstoff partialdrickes, Biomed. Tech. 18:216 (1973).
9. B. Hagihara, T. Fukai, Y. Hachino, K. Nakayama, F. Ishibashi, A. Ohta, S. Ohminato, H. Takemura, T. Hasegawa, K. Kurachi, Y. Okada, T. Sugimoto, K. Nomura and K. Yoshida, A new tc PO_2 electrode and its application to adults, in "Birth Defects 15", (1979).

HEMOGLOBIN OXYGEN
AFFINITY

COMPARISON BETWEEN P_{50} MEASURED BY AUTOMATIC ANALYZER ON MICROSAMPLES IN

BUFFER SOLUTION AND BY TONOMETRIC METHOD ON WHOLE BLOOD

Makoto Koizumi

Pulmonary Division
Aiiku Hospital
Sapporo, Japan

INTRODUCTION

To evaluate the position of the oxyhemoglobin dissociation curve(ODC) in oxygen transport is important for patients with hypoxemia. That is, moving ODC rightside means advantage to more oxygen supply to the body in certain circumstances. One of the major factors to shift ODC is 2,3-diphosphoglycerate(2,3-DPG), but because responses to hypoxia are different among individuals, Pa_{CO_2} and blood pH are involved and affect 2,3-DPG synthesis.

There are several methods to determine P_{50}, which is oxygen pressure at the oxygen saturation of 50 percent. One is using a dual wavelength spectrophotometric method on microsamples in buffer solution, and another is tonometric method on whole blood. The purpose of this study is to compare P_{50} measured by automatic analyzer with P_{50} by tonometric method. Especially the question is directed to CO_2 gas. Is it able to evaluate P_{50} of patients with hypercapnia and compensated pH by automatic analyzer?

MATERIAL

Twenty patients with hypoxemia in stable stage were chosen. Fourteen patients were chronic obstructive pulmonary disease (COPD). The others were pulmonary fibrosis, pulmonary infarction, bronchiolitis, tuberous sclerosis, Kartagener's syndrome, mucoviscidosis and unknown anemia. Arterial blood was drawn from brachial artery. Blood gases showed; pH 7.389 ± 0.043, Pa_{CO_2} 50.0 ± 10.7 Torr, Pa_{CO_2} 67.9 ± 14.7 Torr, Hb 14.0 ± 2.3 g/dl (Mean $\pm$ S.D.)

METHOD

Automatic analyzer (Imai type apparatus) utilizes a dual wavelength. $40 \mu l$ of whole blood was suspended into 5 ml of buffer solution1 Po_2 was monitored by Po_2 electrode made by Industry Laboratory. pH of solution was continuously monitored and adjusted 7.40. Effect of CO_2 was checked at first and was found that there was a little increase or the same of P_{50} after correcting pH altered by mixing CO_2 into solution with NaOH as P_{50} measured by without CO_2 mixing. Then measuring P_{50} was done without CO_2.

Tonometric method used three cuvettes. Each cuvette contained 1 ml

of whole blood and was shaked in the water bath. Each blood was mixed
with gases (O_2; 3%, 3.5%, 4%: CO_2; 5.6%) for 20 minutes and measured
saturation of Hb with CO-oximeter (IL 282). Each blood gas was also
analyzed by IL 813 and corrected P_{O_2} by using equation ($\Delta P_{O_2} = 0.0013 \times$ B.E.
-0.5ΔpH). P_{50} was calculated by Hill's plot at pH 7.40 and P_{CO_2} 40 Torr.

RESULTS

1) There is tendency but not significant relationship between P_{50} measured by automatic analyzer and tonometric method (Figure 1).

2) There is significant relationship between P_{50} and 2,3-DPG except for
patients with hypercapnia and compensated pH imbalance (Figure 2).

3) In patients with hypercapnia and compensated pH imbalance, P_{50} measured by tonometric method showed higher than P_{50} measured by automatic
analyzer. P_{50} measured by automatic analyzer showd in patients with
hypercapnia and compensated pH than in control group (Figure 3).

4) P_{50} measured by tonometric method showed higher in patients with
hypercapnia and compensated pH than in control group. But 2,3-DPG level
showed lower in patients with hypercapnia and compensated pH than in
control group. (Table 1).

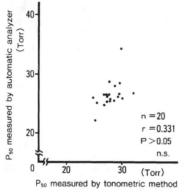

Fig. 1 Relationship between P_{50} measured by tonometric method and
P_{50} measured by automatic
analyzer.

Fig. 2 Relationship between 2,3-DPG
and P_{50} (In a group with hypoxemia except for patients with
compensated pH in hypercapnia).

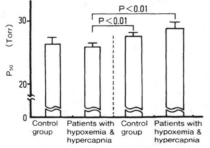

Fig. 3 Comparison between two P_{50}
measurements in control group
and patients group.

Table 1 Patients group shows higher
P_{50} but lower 2,3-DPG than
normal control.

	normal subjects	patients with hypercapnia and compensated pH	P
n	12	6	
Pao_2 (Torr)	90.2±8.9	58.0±6.0	$P < 0.001$
$Paco_2$ (Torr)	40.9±1.8	58.0±9.8	$P < 0.001$
pH	7.380±0.016	7.366±0.018	n.s.
2,3-DPG (μmol/gHb)	14.1±1.5	11.3±1.9	$P < 0.01$
P_{50} (Torr)	27.6±0.6	28.8±0.8	$P < 0.001$

(mean ± 1 S.D.)

DISCUSSION

P_{50} measured by automatic analyzer was affected by solution. Buffer solution was adjusted pH 7.40 and there is possibility to change intra erythrocytic environment by buffer system. Therefore in this system P_{50} might be representative of 2,3-DPG level, so in figure 3 P_{50} showed lower.

P_{50} measured by tonometric method in patients with hypercapnia and compensated pH showed higher even though 2,3-DPG level remained low. This mechanism of increase in P_{50} is thought to be an effect of increased bicarbonate and pH related factors to compensate for the pH imbalance.

Therefore it requires careful consideration to evaluate P_{50} in patients with hypercapnia with compensated pH in hypoxemia using automatic analyzer.

INFLUENCES OF CARBON MONOXIDE ON THE BINDING OF OXYGEN, CARBON DIOXIDE, PROTON AND 2,3-DIPHOSPHOGLYCERATE TO HUMAN HEMOGLOBIN

K. Yamaguchi, M. Mori, A. Kawai and T. Yokoyama

Department of Medicine, School of Medicine, Keio University
Tokyo 160, Japan

INTRODUCTION

Pioneering work on the effect of CO on O_2 equilibrium with hemoglobin (Hb) was performed by Douglas and Haldane (1912). Haldane (1912) also presented a theoretical procedure for the calculation of the position of the O_2 dissociation curve at a given amount of carboxyhemoglobin (HbCO). Although their studies were followed later by many investigators (Roughton and Darling, 1944; Okada et al., 1976; Yamaguchi et al., 1981), the analyses were mainly focussed on the competitive replacement reaction between O_2 and CO through Hb. On the other hand, several authors (Hlastala et al., 1976; Okada et al., 1976; Zwart et al., 1984) aimed to assess the effect of CO on allosteric interactions induced by ligands including hydrogen ion, i.e. the fixed acid Bohr effect as well as the molecular CO_2, i.e. the carbamino effect. Unfortunately however, their results were not in full agreement especially as regards the saturation dependence of the fixed acid Bohr factor. Furthermore, little work has been done to estimate a quantitative influence of CO on the effect provided by 2,3-diphosphoglycerate (DPG) within the red blood cells.

Recently, we developed a spectrophotometric method for accurate determination of three Hb derivatives including reduced Hb, HbO_2 and HbCO based on their isosbestic points. Using the new technique, the effects of CO on allosteric interactions mediated through the hydration of CO_2, i.e. CO_2 Bohr effect and intracellular DPG were systematically investigated.

MATERIALS AND METHODS

Fresh blood was repeatedly drawn from two of the authors into heparinized syringes, immediately placed on ice for storage and used on the same day. The concentration of Hb in the blood sample was measured by the cyanomethemoglobin method (Drabkin and Austin, 1935) and that of methemoglobin with the method described by Evelyn and Malloy (1938). Effective Hb concentration was calculated from the difference between the concentration of total Hb and that of methemoglobin. Hematocrit was determined with a microcentrifuge. DPG content was examined by an enzymatic analysis (no. 665, Sigma chemical, St. Louis, MO).

Alteration of HbCO and DPG levels

When no alterations in the concentration of HbCO and of DPG were required, blood was kept on ice until use.

To vary the HbCO concentration in the blood sample, a part of it was completely carboxygenated by means of equilibration with a gas mixture containing 50% CO and 6% CO_2 in N_2 at 37°C for 30 min. Thereafter, it was anaerobically mixed with a deoxygenated blood equilibrated with a gas mixture consisting of 6% CO_2 in N_2 so that the saturation of Hb with CO, SCO in the specimen was adjusted to either 0, 10, 15, 20, 40 or 50%. In this case, the DPG concentration in the blood remains at a near normal level because of the short equilibration time.

To decrease the DPG content in red blood cells, the blood was incubated at 37°C for 6 hrs. The sample thus prepared was divided into two parts, the one being deoxygenated and the other carboxygenated following the method as described above. These were mixed at varied ratios, resulting in a DPG content at half the normal level and in various SCO values.

Determination of O_2 dissociation curves

The blood specimens prepared as above were diluted at 1:100 into isotonic buffer solution having the same electrolyte composition as human serum, and were equilibrated at 37°C for 40 min in a Farhi tonometer with gas mixtures containing varied O_2 concentrations from 1 to 9%. Thereby, the sum of SO_2 and SCO, defined as S_T, ranged from 20 to 90% depending on the O_2 fraction in the gas mixture and the coexisting HbCO content. In the following, the symbols SO_2 and SCO are used, respectively, for the concentration of HbO_2 and HbCO relative to the effective Hb concentration. To alter the pH of the suspension, CO_2 concentration in the equilibrating gas was varied from 4 to 9%, resulting in pH values of 7.2, 7.4 and 7.6.

The equilibrated suspension was anaerobically introduced into a small glass chamber (18 μl in volume), the temperature of which was maintained at 37°C with a circulating water bath system. The SO_2 and SCO values in the specimen were determined by means of a dual-wavelength spectrophotometric

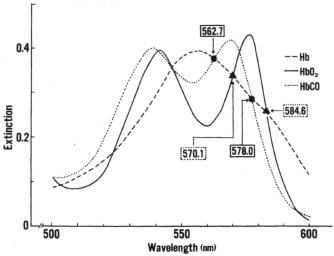

Fig. 1.   Spectral properties of reduced Hb, HbO_2 and HbCO.
●: isosbestic points between reduced Hb and HbCO.
▲: isosbestic points for reduced Hb and HbO_2.

300

method (US-501, Unisoku Scientific Instruments, Osaka). The extinction differences at 562.7 and 578.0 nm, being the isosbestic points between reduced Hb and HbCO, were observed as a measure of SO_2, while those at 570.1 and 584.6 nm, similarly the isosbestic points of reduced Hb and HbO_2 were monitored for determining SCO (Fig. 1). Repeated measurements showed little change in SCO of diluted red cell suspensions when an O_2 concentration of less than 9% was used as the equilibrating gas. The extinction difference at 100% SO_2 was obtained using the red cell suspension equilibrated with 50% O_2 and 6% CO_2 in N_2, whereas that of 100% SCO was measured from the suspension tonometered with 50% CO and 6% CO_2 in N_2. The extinction difference for the saturation of Hb without O_2 and CO, i.e. S_T equal to zero, was determined by adding 10 mmol/l of sodium dithionite to the red cell suspension.

PO_2, PCO_2 and pH of the suspension were observed, after each run, with electrodes (IL 13, Instrumentation Laboratories, Lexington, MA) thermostatted at 37°C. O_2 and CO_2 electrodes were calibrated by using isotonic buffer solution equilibrated with known fractions of O_2 and CO_2 in an IL 237 tonometer (Instrumentation Laboratories, Lexington, MA).

One O_2 dissociation curve was generally constructed from six different points of SO_2, and a total of 45 dissociation curves were studied at varied SCO, pH and DPG concentrations. Experimental results were fitted by applying the Hill equation (Hill, 1910), the slope of which, the n value, was calculated by least-squares regression. Based on the simulated curves, the CO_2 Bohr factor, B, at given SO_2, SCO and DPG concentration was estimated. Similarly, the effect of DPG on O_2 dissociation curves in the presence of CO, DPGF, was appraised at fixed SO_2, SCO and pH. These two coefficients are expressed as:

$$B = (\Delta log PO_2 / \Delta pH)_{SO_2, SCO, DPG} \tag{1}$$

$$DPGF = (\Delta log PO_2 / \Delta Z(DPG))_{SO_2, SCO, pH} \tag{2}$$

where Z(DPG) is a molar ratio of DPG relative to effective Hb.

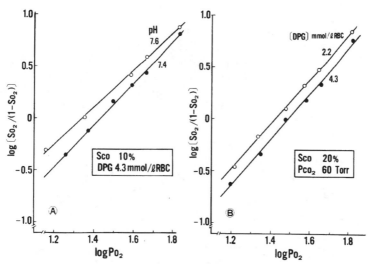

Fig. 2. Hill plots of experimental results.
   A. O_2 dissociation curves in the presence of CO at different pH.
   B. O_2 dissociation curves in the presence of CO at different DPG concentrations.

RESULTS

Total Hb and methemoglobin concentrations were averaged at, respectively, 20.6 mmol/(1RBC) and 0.2 mmol/(1RBC), resulting in an effective Hb concentration of 20.4 mmol/(1RBC). Mean hematocrit was 0.47, while DPG concentration in fresh blood was 4.3 mmol/(1RBC) and a corresponding value of Z(DPG)
was found to be 0.21. The blood samples incubated for 6 hrs showed an averaged DPG concentration of 2.2 mmol/(1RBC) and Z(DPG) of 0.11, both being
about half the normal values.

As shown in Fig. 2, the experimental results were well fitted by the
Hill equation, indicating that n value for O_2 dissociation curve decreased
with increasing concentration of coexisting HbCO.

CO_2 Bohr effect

The CO_2 Bohr factor, B was plotted as a function of S_T (Fig. 3). The
Bohr effect was consistently larger at a low S_T than a high S_T. When SCO was
more than 40%, the effect at a given S_T was augmented as increasing SCO. On
the other hand, no significant difference in effect was demonstrable for SCO
less than 20%, in which the representative CO_2 Bohr factor, defined at 50%
S_T, was −0.48.

The influence of intracellular concentration of DPG on the CO_2 Bohr
effect in the presence of CO is depicted in Fig. 4, showing that the effect
was more prominent at a low DPG concentration. For a SCO below 20%, the relative change of the CO_2 Bohr effect in the presence of CO due to diminishing
DPG concentration appeared to be approximately the same as in the absence of
CO.

Effect of DPG

The quantitative effect of DPG on O_2 affinity in the presence of CO was
estimated on the basis of the coefficient, DPGF defined in eq. 2. In contrast

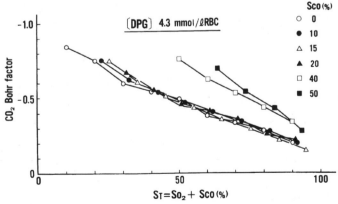

Fig. 3.   Effect of CO on the CO_2 Bohr factor at normal DPG
content.

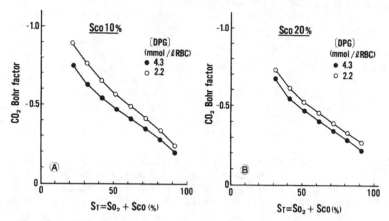

Fig. 4. Effect of DPG ON THE CO_2 Bohr factor at varied Sco.

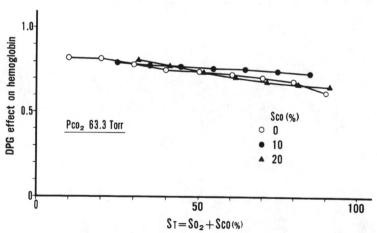

Fig. 5. Influence of CO on allosteric effect induced by DPG.

to the CO_2 Bohr factor, DPGF was tended to be fairly constant for S_T ranging from 20 to 90% (Fig. 5). The influence of CO on DPGF was not evident when the coexisting SCO was less than 20%.

When the pH in red cell suspension was changed from 7.2 to 7.6 by altering PCO_2 but keeping the base-excess at a constant level, little dependence of DPGF on the saturation was demonstrable (Fig. 6). Furthermore, the coefficient at pH 7.6 seemed to be nearly identical to that at pH 7.2. Thus, the averaged value for DPGF under a constant base-excess was estimated at 0.85, fairly independent of degree of saturation.

DISCUSSION

The accuracy of the O_2 dissociation curve of diluted red cell suspensions was evaluated by comparing the measured SO_2 values in the absence of CO (at pH 7.4 and 37°C) with those obtained from the standard curve studied with whole blood (Severinghaus, 1979). The relative difference in SO_2 does not exceed 5%.

Reliable determination of O_2 dissociation curves in the presence of CO is only possible if no alteration in HbCO content occurs during the measurements. The binding of CO to Hb was shown to be strong enough to avoid a significant change in SCO during the equilibration of the suspension by means of a gas mixture containing O_2 less than 9% (see methods). The finding is qualitatively consistent with those reported by Hlastala et al. (1976) and Zwart et al. (1984).

The conventional spectrophotometric technique involves the difficulty of distinguishing HbO_2 from HbCO. In the present study, however, this limitation was wholly overcome by using isosbestic points observed for reduced Hb and HbCO as a measure of SO_2 and those for reduced Hb and HbO_2 as a measure of SCO, respectively (Fig. 1). The present method may also eliminate the artifacts produced by light scattering in the fluid containing red cell particles (cf. Yamaguchi et al., 1985) and thus allows us to estimate SO_2 and SCO values accurately in a diluted red cell suspension.

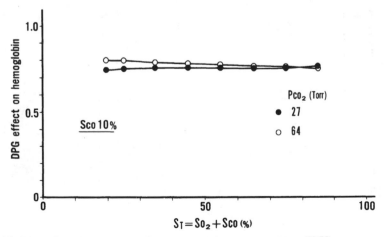

Fig. 6.   Influence of CO_2 on allosteric effect caused by DPG at fixed Sco and base-excess.

Influence of CO on the CO_2 Bohr effect

The CO_2 Bohr factor, B at 50% saturation was -0.48 for a normal DPG level, falling within the range of -0.48 to -0.54 determined by other authors with various techniques (cf. Hlastala and Woodson, 1983). The saturation dependency of the CO_2 Bohr effect is in good agreement with the recent work of Hlastala and Woodson (1983). Additionally, the influence of intracellular DPG concentration on the Bohr effect also seems to be consistent with the findings reported by the above authors.

In the case of the coexistent HbCO content being high (i.e. $SCO \geq 40\%$), the value of B at a given S_T became more negative as SCO increased, whereas it showed no significant difference at low HbCO levels (i.e. $SCO \leq 20\%$; Fig. 3). This may imply that at a high SCO there is appreciable difference, but at a low SCO virtually no difference, between the heme-heme interaction induced by O_2 and by CO. This finding is qualitatively, but not quantitatively, compatible with the analyses performed by Zwart et al. (1984) who showed that at SCO less than 50% the heme-heme interaction for CO is practically identical to that for O_2. Based on the present experimental results, it is probably safer to conclude that the allosteric interaction caused by O_2 is difficult to differentiate from that caused by CO for a SCO below 20%.

The CO_2 Bohr effect is the sum of allosteric interactions induced both by the carbamino reaction and by the change of proton activity surrounding Hb molecules, mediated through the hydration of CO_2. In contrast, the fixed acid Bohr effect is due to the change in proton activity alone. According to the work done by Hlastala et al. (1976), the fixed acid Bohr effect in the presence of CO at SCO below 25%, representing a maximum at intermediate saturation but falling off at both high and low saturations, is not significantly different from that in the absence of CO. Joining their and our experimental results, the carbamino effect induced by molecular CO_2 should be manifested primarily at low saturations and may not be affected by the presence of CO in SCO range from 0 to 20%.

To summarize briefly, CO combines with Hb practically in the same manner as O_2 at least for SCO values less than 20%. In this case, an identical coefficient can be used to describe the CO_2 Bohr effect on HbO_2 and HbCO. To express dependence of the allosteric effect provided by CO_2 hydration on the saturation and intracellular DPG concentration, experimental data obtained at SCO below 20% were fitted by polynomial regression, giving the empirical equation as follows:

$$B = -0.77 - 1.28 \cdot Z + 1.86 \cdot Z^2 + (0.36 + 1.17 \cdot Z + 11.50 \cdot Z^2) \cdot S_T$$
$$+(0.05 + 2.15 \cdot Z - 20.19 \cdot Z^2) \cdot S_T^2 \tag{3}$$

where B is the CO_2 Bohr factor related to both O_2 and CO, while Z indicates Z(DPG), i.e. the molar ratio of DPG to effective Hb. The equation is generally valid in circumstances where S_T ranges from 20 to 90% and the portion occupied by SCO is less than 20%.

Influence of CO on allosteric effect of DPG

As clearly shown in Fig. 5, less dependency of the DPG effect on HbCO concentration was demonstrated for a SCO ranging from 0 to 20%, again supporting the simple idea that the heme-heme interaction for O_2 was virtually not different from that for CO at a low SCO level. Thus, an identical coefficient may be applied to estimate the DPG effect on HbO_2 and on HbCO for SCO up to 20%.

The saturation dependency of the allosteric effect provided by DPG was not demonstrable for a range of S_T from 20 to 90% (Figs. 5 and 6). Although a clear explanation of the results can not be given, our findings appear to be consistent with those reported by Arturson et al. (1974) who have examined the effect of DPG on O_2 affinity in the absence of CO at varied PCO_2 and pH, and have shown the trend qualitatively similar to ours.

The allosteric effect of DPG was little influenced when the pH surrounding Hb molecules was changed by altering PCO_2 (Fig. 6). This is presumably
explicable from the reverse effects imposed by molecular CO_2 and proton activity on the binding of DPG to Hb. A low PCO_2 enhances the DPG binding to
Hb due to less competition at the N-terminal valine of both β-chains, thus
augmenting the effect of DPG on the affinity of heme ligands. On the other
hand, low PCO_2 increases the pH within the red blood cell, resulting in the
reduction of the DPG effect since the binding constant between Hb and DPG
may be lowered with increasing pH (Garby and Verdier, 1971).

Physiological significance

The experimental data obtained in the present study provide more accurate factors for calculating the effects of various allosteric ligands on
O_2 and CO dissociation curves in the blood. In pulmonary and systemic capillaries, SO_2 as well as the pH along them vary successively with the gas exchange of O_2 and CO_2. When a small amount of CO is added to an inspired gas,
interactions among allosteric ligands become more complicated. In such circumstances, factors describing the CO_2 Bohr and DPG effects observed under
conditions of O_2 and CO coexistence are indispensable for estimating the
displacements of PO_2 and PCO in the blood due to successive gas exchange in
the capillaries. The relationship is expressed by:

$$(\Delta \log PO_2)_{SO_2, SCO} = (\Delta \log PCO)_{SO_2, SCO} = B \cdot \Delta pH + DPGF \cdot \Delta Z(DPG) \qquad (4)$$

where B is given in eq. 3 and DPGF equals 0.85. The equation is generally
validated when HbCO concentration is relatively low, i.e. SCO below 20%.

SUMMARY

In an attempt to estimate the influences of CO on the CO_2 Bohr effect
and the 2,3-diphosphoglycerate (DPG) effect linked to the reversible binding
of O_2 to the hemoglobin molecule (Hb), O_2 dissociation curves of human blood
in the presence of CO were investigated at 37°C over a DPG concentration
ranging from 2.2 to 4.3 mmol/(1RBC) and a pH range of 7.2 to 7.6. The sample
with a low DPG concentration was made by incubating whole blood for 6 hrs,
whereas the saturation of Hb with CO, SCO in the sample was adjusted by anaerobically mixing completely carboxygenated blood with that free of O_2 and
CO so as to give the final SCO at either 0, 10, 15, 20, 40 or 50%. The blood
samples thus prepared were diluted at 1:100 in isotonic buffer solution and
were equilibrated with gas mixtures containing O_2 ranging from 1 to 9% and
CO_2 from 4 to 9%. The SO_2 and SCO values of diluted red cell suspensions were
examined by means of a dual-wavelength spectrophotometric method based on the
isosbestic points for reduced Hb, HbO_2 and HbCO. The extinction difference
at 562.7 and 578.0 nm was monitored as a measure of SO_2, while that at 570.1
and 584.6 nm was recorded for determining SCO.

Fitting the experimental results by the Hill equation, the coefficients
describing allosteric interaction induced by CO_2 hydration and by DPG in the
presence of both O_2 and CO were calculated for the total saturation, S_T defined as the sum of SO_2 and SCO, ranging from 20 to 90%. The CO_2 Bohr effect
was consistently larger at a lower S_T. At SCO above 40%, the effect at a
given S_T was augmented with increasing SCO, but this was not demonstrated
for the SCO below 20%. The CO_2 Bohr effect was also dependent on the intracellular concentration of DPG, the effect being more prominent at a low DPG
concentration.

The allosteric effect of DPG on O_2 affinity was practically independent
of S_T between 20 and 90%, and was little influenced by CO so long as the coexisting SCO was lower than 20%. In circumstances where base-excess was kept
constant, the effect of DPG was likely to be insensitive to the change of pH
mediated by altering PCO_2.

306

The present analyses may indicate that the manner of CO combination with Hb differs appreciably from that of O_2 for a high SCO level, but is virtually the same for a low SCO.

REFERENCES

Arturson, G., Garby, L., Robert, M., and Zaar, B., 1974, The oxygen dissociation curve of normal human blood with special reference to the influence of physiological effector ligands, J. Clin. Lab. Invest., 34:9.

Douglas, C. B., Haldane, J. S., and Haldane, J. B. S. , 1912, The laws of combination of hemoglobin with carbon monoxide and oxygen, J. Physiol. Lond., 44:275.

Drabkin, D. L., and Austin, J. H., 1935, Spectrophotometric studies. II. Preparations from washed blood cells, nitric oxide hemoglobin and sulf-hemoglobin, J. Biol. Chem., 254:1178.

Evelyn, K. A., and Malloy, H. T., 1938, Microdetermination of oxyhemoglobin, methemoglobin, and sulfhemoglobin in a single sample of blood, J. Biol. Chem., 126:655.

Garby, L., and de Verdier, C-H., 1971, Affinity of human hemoglobin A to 2,3-diphosphoglycerate: Effect of hemoglobin concentration and of pH, Scand. J. Clin. Lab. Invest., 27:345.

Haldane, J. B. S., 1912, The dissociation of oxyhemoglobin in human blood during partial CO-poisoning, J. Physiol. Lond., 45:XXII.

Hill, A. V., 1910, The possible effects of aggregation of the molecules of haemoglobin on its dissociation curve, J. Physiol. Lond., 40:4P.

Hlastala, M. P., Mckenna, H. P., Franada, R. L., and Detter, J. C., 1976, Influence of carbon monoxide on hemoglobin-oxygen binding, J. Appl. Physiol., 41:893.

Hlastala, M. P., and Woodson, R. D., 1983, Bohr effect data for blood gas calculations, J. Appl. Physiol.: Respirat. Environ. Exercise Physiol., 55:1002.

Okada, Y., Tyuma, I., Ueda, Y., and Sugimoto, T., 1976, Effect of carbon monoxide on equilibrium between oxygen and hemoglobin, Am. J. Physiol., 230:471.

Roughton, F. J. W., and Darling, R. C., 1944, The effect of carbon monoxide on the oxyhemoglobin dissociation curve, Am. J. Physiol., 141:17.

Severinghaus, J. W., 1979, Simple accurate equations for human blood O_2 dissociation computations, J. Appl. Physiol.: Respirat. Environ. Exercise Physiol., 46:599.

Yamaguchi, K., Kawashiro, T., Yokoyama, T., 1981, Mathematical representation of the CO-Hb dissociation curves of whole blood under various O_2 tensions, Prog. Resp. Res., 16:158.

Yamaguchi, K., Nguyen-Phu, D., Scheid, P., and Piiper, J., 1985, Kinetics of O_2 uptake and release by human red blood cells studied by a stopped-flow technique, J. Appl. Physiol.: Respirat. Environ. Exercise Physiol., 58:1215.

Zwart, A., Kwant, G., Oeseburg, B., and Zijlstra, W. G., 1984, Human whole-blood oxygen affinity: Effect of carbon monoxide, J. Appl. Physiol.: Respirat. Environ. Exercise Physiol., 57:14.

EVIDENCE THAT CHANGES IN BLOOD OXYGEN AFFINITY MODULATE OXYGEN DELIVERY:

IMPLICATIONS FOR CONTROL OF TISSUE PO_2 GRADIENTS

Robert D. Woodson

Department of Medicine
The University of Wisconsin-Madison
Madison, WI 53705 USA

INTRODUCTION

Oxygen flux to respiring tissue depends upon the oxygen diffusion gradient between blood in the microvascular network and sites of oxygen utilization. Current knowledge of how this gradient is regulated is quite incomplete. The magnitude of the gradient is determined by a number of morphological and functional parameters, one of which is blood oxygen affinity. Experiments examining control of oxygen delivery have traditionally made use of induced changes in parameters determining systemic oxygen transport (F_IO_2, [Hb], and/or flow), with secondary effects on the pattern of PO_2 values along the capillary. Changes in blood oxygen affinity, by contrast, allow one to observe effects of a change in PO_2 downstream in the capillary while PO_2 upstream and oxygen transport (CaO_2 x flow) are not primarily affected.

MODEL

Fig. 1 shows a model of PO_2 in tissue as a function of O_2 delivery along the capillary. For purposes of simplicity, the model assumes the traditional geometry where a single vessel supplies a cylinder of tissue having homogeneous oxygen consumption by radial diffusion without interactions from adjacent capillaries. The curves show expected tissue PO_2 values at that perpendicular distance from the microvessel where PO_2 is 10 Torr lower than in the capillary and when in vivo blood P_{50} is 36 Torr ("control") and 20 Torr ("low P_{50}"), respectively. The figure shows that tissue PO_2 falls to zero at this distance when ~94% of O_2 is released (control) and ~83% (low P_{50}). Fig. 2 shows model predictions at that distance where tissue PO_2 is 20 Torr lower than blood PO_2. In this case, tissue PO_2 falls to zero when about 80% and 40%, respectively, of oxygen has been delivered.

RESULTS AND DISCUSSION

If these predicted differences in tissue PO_2 are of real importance in terms of providing a sufficient pressure head for oxygen diffusion, one would expect some effect when P_{50} is acutely changed. Figs. 3 and 4 show data taken from the literature for the brain and heart, respectively, when the organ is freely perfused. Each point is the mean of a separate study. In the case of both organs, a rise in P_{50} evokes a decrease in flow and a

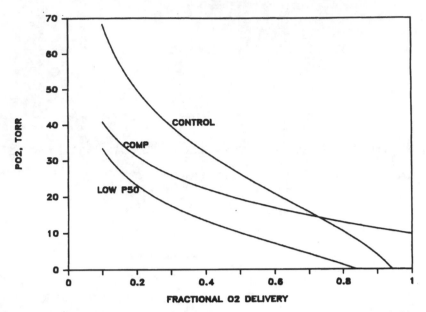

Fig 1.　　Tissue PO_2 as a function of theoretical fractional O_2 delivery when the　capillary-tissue PO_2 gradient is 10 Torr.

fall in P_{50} produces an increase in flow. This relationship between P_{50} and cerebral or coronary flow is apparently an intrinsic property of these organs in that it is observed with intact and with isolated organs.

The predictive impact of this flow response to a P_{50} reduction is shown in Figs. 1 and 2. The curve labelled "comp" assumes a flow increment of 100% in combination with the 16 Torr reduction in P_{50}. In the first portion of the capillary, tissue PO_2 is lower with the flow increment than under control conditions but by definition more than sufficient. More importantly, PO_2 is well maintained at the end of the capillary and is actually somewhat higher than in the control condition.

One implication of these flow responses to P_{50} changes is that the associated shift in tissue PO_2 profiles could be obligatory for maintenance of oxygen flux. Thus, if the flow increase in response to a P_{50} decrease is obligatory ·for sufficient O_2 diffusion, prevention of the flow response should impair organ function and oxygen consumption. Alternatively, if the response to a decrease in P_{50} were acting to furnish the heart or brain with an O_2 pressure head well in excess of that minimally needed for diffusion, blocking the increase should have no effect on organ function.

One study has looked at this possibility in brain. When flow to brain was set at the normal level for this species and held constant, a drop in P_{50} caused a decrease in VO_2 and abnormal function (Woodson et al., 1982). According to this study it would appear that tissue PO_2 profiles in brain are regulated rather tightly, as a drop in the capillary-to-tissue pressure head of 9 Torr decreased oxygen consumption by 17% and a drop of 12–13 Torr decreased oxygen consumption by about 25%.

A similar circumstance may occur in the heart. Available studies have largely been carried out at subphysiological values for coronary

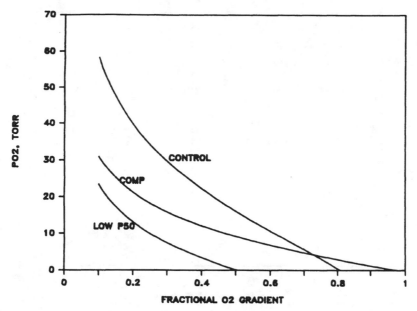

Fig 2.   Tissue PO_2 as a function of theoretical fractional O_2 delivery
         when the capillary-tissue PO_2 gradient is 20 Torr.

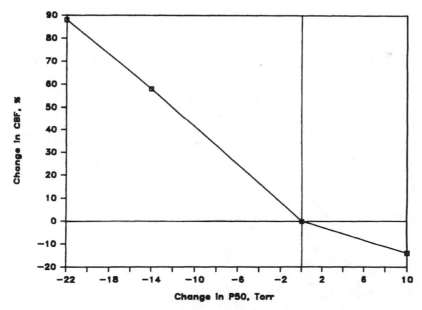

Fig 3.   Effect of a P_{50} change on cerebral blood flow (CBF). Each point
         is the mean from a study in the literature (Woodson and Auerbach,
         1982; Wade et al., 1980; Koehler et al., 1983).

311

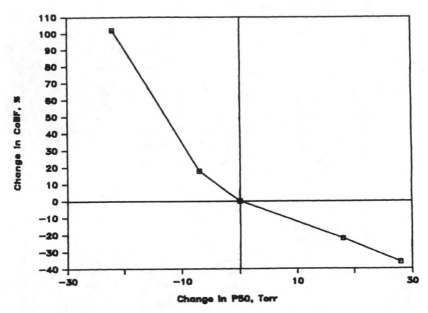

Fig 4.   Effect of a P_{50} change on coronary blood flow (CoBF). Each
            point is the mean from a study in the literature (Woodson and
            Auerbach, 1982; Stucker et al., 1985) or from a paper presented
            at this symposium.

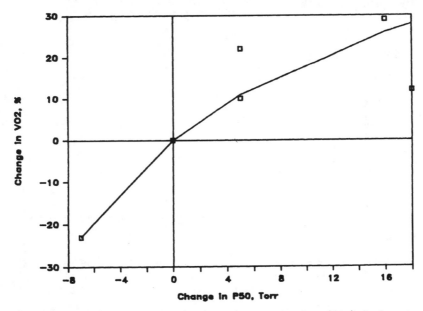

Fig 5.   Effect of a P_{50} change on oxygen consumption (VO_2) in hearts with
            constant coronary flow. Each point is from a study in the liter-
            ature (Gross, Warltier, and Hardman, 1977; Valeri et al., 1980;
            Apstein et al., 1985) or from a paper presented at this symposium.

flow. Fig. 5 shows means of changes in P_{50} with resultant effects on VO_2 in several literature studies. These data demonstrate that a decrease in P_{50} is reflected by a change in VO_2, implying that tissue PO_2 gradients, at least under the conditions of study, are in a range such that the gradient is a limiting factor.

CONCLUSION

A simple model examining possible effects of changes in blood oxygen affinity on tissue PO_2 profiles suggests that a decrease in P_{50} might require compensation to sustain oxygen consumption. Data from the literature suggest that this is the case for heart and brain, where an increase or decrease in P_{50} produces a reciprocal change in blood flow during free perfusion. The model indicates that this change in perfusion significantly sustains downstream PO_2 when P_{50} is reduced. When this compensation is blocked, changes in VO_2 or function may result.

REFERENCES

Apstein, C. S., Dennis, R. C., Briggs, L., Vogel, W. M., Frazer, J., and Valeri, C. R., 1985, Effect of erythrocyte storage and oxyhemoglobin affinity changes on cardiac function, Am. J. Physiol., 248 (Heart Circ. Physiol., 17):H508.

Gross, G. J., Warltier, D. C., and Hardman, H. F., 1977, Effect of ortho-iodo sodium benzoate on hemoglobin-oxygen affinity in normal and ischemic myocardium, J. Pharmacol. Exp. Ther., 203:72.

Koehler, R. C., Traystman, R. J., Rosenberg, A. A., Hudak, M. L., and Jones, Jr., M. D., 1983, Role of O_2-hemoglobin affinity on cerebro-vascular response to carbon monoxide hypoxia. Am. J. Physiol., 245 (Heart Circ. Physiol., 14):H1019.

Stucker, O., Vicaut, E., Villereal, M.-C., Ropars, C., Teisseire, B. P., and Duvelleroy, M. A., 1985, Coronary response to large decreases of hemoglobin-O_2 affinity in isolated rat heart, Am. J. Physiol., 249 (Heart Circ. Physiol., 18):H1224.

Valeri, C. R., Yarnoz, M., Vecchione, J. J., Dennis, R. C., Anastasi, J., Valeri, D. A., Pivacek, L. E., Hechtman, H. B., Emerson, C. P., and Berger, R. L., 1980, Ann. Thorac. Surg., 30:527.

Wade, J. P. H., Du Boulay, G. H., Marshall, J., Pearson, T. C., Ross-Russell, R. W., Shirley, J. A., Symon, L., Wetherley-Mein, G., and Zilkha, E., 1980, Cerebral blood flow, haematocrit and viscosity in subjects with a high oxygen affinity haemoglobin variant, Acta Neurol. Scandinav., 61:210.

Woodson, R. D., and Auerbach, S., 1982, Effect of increased oxygen affinity and anemia on cardiac output and its distribution, J. Appl. Physiol.: Respirat. Environ. Exercise Physiol., 53:1299.

Woodson, R. D., Fitzpatrick, Jr., J. H., Costello, D. J., and Gilboe, D. D., 1982, Increased blood oxygen affinity decreases canine brain oxygen consumption, J. Lab. Clin. Med., 100:411.

EFFECTS OF HIGH O_2 AFFINITY OF BLOOD ON OXYGEN CONSUMPTION ($\dot{V}O_2$) OF DOG

GRACILIS MUSCLE AT VARYING O_2 DELIVERY

H. Kohzuki, Y. Enoki, Y. Ohga, S. Shimizu, and S. Sakata

Second Department of Physiology
Nara Medical University
Kashihara, Nara 634, Japan

INTRODUCTION

Theoretical cosiderations indicate that high O_2-affinity blood, i.e. a left-shift of oxygen dissociation curve (ODC), is disadvatageous for oxygen transport to tissue in normoxia (Willford et al., 1982). This is supported by the fact that the hemoglobinopathies with high O_2 affinity induce erythrocytosis. It has been suggested that this erythropoietic change is one of the compensatory responses in living system that cope with tissue hypoxia. The term "tissue hypoxia" means that oxygen content or oxygen partial pressure (PO_2) limit aerobic metabolism of tissue, or decrease in oxygen consumption (Honig, 1977). The question arises here whether oxygen consumption ($\dot{V}O_2$) of peripheral tissue decreases or not, when the compensatory mechanisms, such as the erythrocytosis or increase in regional blood flow, are not induced.

To clarify this point, we investigate the relationship between a decrease in P_{50} (PO_2 at half oxygenation of blood) and $\dot{V}O_2$ of skeletal muscle. The dog resting gracilis muscle was perfused alternately with normal and high O_2 affinity blood at constant arterial O_2 content (CaO_2) and at various blood flows, i.e. various levels of O_2 delivery (CaO_2 x blood flow). As previously reported for rat skeletal muscle (Kolář and Janský, 1984) the resting $\dot{V}O_2$ was delivery-independent above a critical O_2 delivery rate of 0.40 ml/min·100g, but was delivery-dependent below this critical level.

MATERIALS AND METHODS

The preparation of the gracilis muscle. Mongrel dogs of either sex weighing 3.2-6.4kg (n=3), were anesthetized with pentobarbital sodium (30 mg/kg body weight, i.v.). The animals were ventilated with room air through a cuffed endotracheal tube by means of a respirator (IGARASHI, Model B-2) at a tidal volume of 15ml/kg and a frequency of 20/min. The right gracilis muscle was surgically isolated from the surrounding tissues and all vessels from the femoral artery and vein supplying the gracilis were preserved but the other vessels were ligated and cut. A glass T cannula (catheter volume: 0.54 or 0.71ml) for the venous line was inserted into the femoral vein proximal to the origin of the main gracilis vein. An arterial glass T cannula (catheter diameter: 2.5-3.0mm) was placed in

315

the femoral artery distal to the origin of the main gracilis artery. A thermostat-controlled IR lamp kept the surface of the muscle at 37°C. Prior to peristaltic pump-perfusion, the isolated gracilis muscle was flushed with Ringer-Krebs-Bülbring solution containing 4% BSA (Sigma). Blood reservoirs for high and normal O_2-affinity blood were maintained at 4°C so as to prevent excess lactate production. The temperature of inflowing blood was kept at 37°C by a dual-tube heat exchanger (volume: 3.5ml). The steady state perfusion pressure when the flow rate was artificially changed, was measured by a mercury manometer. Blood flows were determined by weighing the venous effluents collected for 5min in a tared vessel.

The blood for perfusion. Mongrel dogs of either sex, 6.2-8.6kg body weight (n=3), were anesthetized with pentobarbital sodium (30mg/kg body weight, i.v.). The animals were artificially ventilated as stated above. After cannulation of the right carotid artery and heparinization (5,000 IU), exsanguination was started. The heparinized blood (10,000 IU) was centrifuged for 5min at 3,000rpm. The packed red cells were separated from plasma and the "buffy coat" was removed. The plasma was stored at 4°C and before use was centrifuged for 5min at 3,500rpm to remove debris. The packed red cells were stored at 4°C for 24h with 30mM sodium cyanate in buffered saline (pH 7.4). The carbamylated blood was washed twice with Ringer-Krebs-Bülbring solution equilibrated with $95\%O_2$-$5\%CO_2$ gas mixture (pH 7.4) and then washed with the same solution containing 4% BSA. Finally, the washed red cells were resuspended in the original plasma to obtain a 45% hematocrit. The buffer solution used in the present experiment was passed through a filter of 0.45μm pore size (Millipore, HA). The base excess of the blood was adjusted to zero with 7% bicarbonate. The blood-gas equilibration was performed with a bubble oxygenator (JUNKEN) for 12-15 min at 37°C. The blood gas values of the equilibrated blood in reservoirs at 4°C did not change during the course of experiment (3-4h).

The measurements of the blood gases and oxygen consumption. The gracilis muscles were perfused alternately with the blood of lower P_{50} (n= 3) and with the blood of normal O_2-affinity. In one experiment, the perfusions were made at 9 different flow rates. Each type of perfusion was continued for 20 to 30min. SO_2 was measured with OSM-2 (Radiometer), and pH, PCO_2 and PO_2 with a blood gas analyzer (Radiometer, BMS-Mk2). VO_2 was calculated according to the Fick principle and the O_2 delivery was presented as the product of the arterial oxygen content (CaO_2) and the blood flow. In the calculation for the blood oxygen content from Hb concentration, Hüfner's factor, theoretically being 1.39, was corrected for the presence of such inactive hemoglobins as HbCO and Hb^+. Only the Hb^+ content was actually measured by a sensitive difference spectrum method (Enoki et al., unpublished), since the HbCO content was fairly constant (0.3%) throughout the animals (Shimizu et al., 1986). The means of the methemoglobin concentrations in control and low P_{50} blood which were prepared by refrigerated incubation with cyanate for 24h, were 0.90±1.07 and 1.66±1.65% (n=3), respectively. This difference was not statistically significant. Hematocrit (Ht) was measured with a microhematocrit centrifuge (KUBOTA, KH-120) and Hb concentration by the cyanmet-Hb method. The mean corpuscular hemoglobin concentration (MCHC) was calculated from Hb concentration and Ht. The oxygen dissociation curve (ODC) and P_{50} were determined by the micro-tonometric and single point method (Kohzuki et al., 1983). At the end of each experiment, both the perfused muscle and the corresponding muscle from the contralateral leg were excised and weighed. The perfused muscles weighed 20±6% more than the contralateral non-perfused muscles (means±SD; 12.7±7.6g). The contralateral muscle weight was used to calculate the blood flow, O_2 delivery and muscle VO_2 on 100g basis. All experimental data were presented as means±SD. For the statistical comparison, the unpaired t-test was used.

Table 1. Hematological and blood gas data on high O_2-affinity blood and normal blood for perfusion in reservoir

		Normal affinity blood	High affinity blood
n		3	3
Ht	(%)	43.5±2.2	44.6±1.3
Hb	(g/dl)	16.1±0.5	16.4±0.3
MCHC	(g/dl)	37.0±2.7	36.7±1.3
P_{50}	(Torr)	37.1±3.0	19.0±1.4*
SO_2	(%)	95.8±1.4	97.4±1.5
pH		7.43±0.03	7.43±0.05
PCO_2	(Torr)	33.4±1.5	32.5±2.0
PO_2	(Torr)	104.4±5.0	107.2±3.3
CaO_2	(g/dl)	21.56±0.76	22.11±0.56

Values are means±SD. High O_2-affinity blood was prepared by refrigerated storage at 4°C with cyanate. Ht, hematocrit; Hb, hemoglobin; MCHC, mean corpuscular hemoglobin concentration; P_{50}, PO_2 at half oxygenation of blood; SO_2, O_2 saturation; PO_2 and PCO_2, partial pressure O_2 and CO_2; CaO_2, arterial O_2 content. *; Difference between normal blood and high affinity blood is statistically significant (P=0.05).

RESULTS

    Table 1 presents the data on hematological and blood gas parameters for the two types of blood used for the perfusions. These values agree with each other, except the mean P_{50} being 19.0 Torr for high O_2-affinity blood and 37.1 Torr for normal affinity blood. Figure 1 shows a representative pattern of the pressure-flow relationship in the gracilis muscle when perfused with the blood of normal and high O_2-affinity, together with the O_2 delivery to $\dot{V}O_2$ relationship. We can see an autoregulation, albeit weak, of the blood flow. The difference in oxygen affinity per se did not seem to induce any apparent change in this pressure-flow relationship.

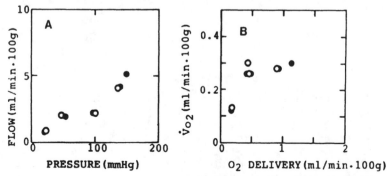

Fig. 1 A and B. Typical pattern of the relationships between $\dot{V}O_2$ and O_2 delivery (B), and between perfusion pressure and blood flow (A) in the perfused dog gracilis muscle. O_2 delivery is the product of blood flow and arterial O_2 content. Results for the perfusion with normal O_2-affinity blood (○) and with high O_2-affinity blood (●).

Resting $\dot{V}O_2$ was almost constant above a critical O_2 delivery level but decreased below it. For example, blood flows at the perfusion pressures 49 and 55 mmHg, respectively, for normal and high O_2-affinity blood perfusion were 1.87 and 1.99 ml/min·100g, respectively (Fig. 1 A). Blood flows at pressures of 137 and 140 mmHg for the normal and high affinity perfusions were 4.08 and 4.14 ml/min·100g, respectively (Fig. 1 A).

Figure 2 shows the relationship between the $\dot{V}O_2$ normalized to the maximum $\dot{V}O_2$ in each perfusion and O_2 delivery in three perfusions. $\dot{V}O_2$ remains nearly constant and delivery-independent over an O_2 delivery range from 0.40 to 2.00 ml/min·100g, while it turned out to be less and delivery-dependent below a critical O_2 delivery of 0.40 ml/min·100g. The following three points were concluded: 1) The mean maximal $\dot{V}O_2$ above the critical delivery level for the high O_2-affinity blood was 0.25±0.03 ml/min·100g and did not differ statistically from that for control perfusion, 0.26±0.03 ml/min·100g(Table 2). 2) Critical O_2 delivery was the same, 0.40 ml/min·100g, for both kinds of blood. 3) Consequently, the maximal O_2 extraction ratio (0.63) by skeletal muscle, represented by a slop of the $\dot{V}O_2$ vs O_2 delivery in the delivery-dependent portion, was the same for both kinds of blood. Figure 3 shows oxygen partial pressure of the venous effluent (PVO_2) as a function of the O_2 delivery during perfusion with normal and lower P_{50} blood. PVO_2 for the higher O_2-affinity blood was significantly lower than that for the normal affinity blood over the whole O_2 delivery range. The "critical PVO_2" which was defined as PVO_2 at the critical O_2 delivery

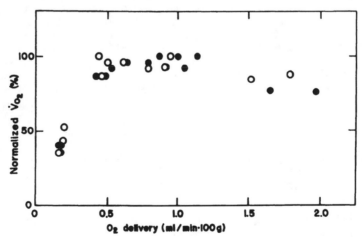

Fig. 2. Normalized $\dot{V}O_2$ as related to O_2 delivery in the resting dog gracilis muscles perfused alternately with the normal ($\bigcirc$; P_{50}= 37.1±3.0 Torr) and high O_2-affinity dog blood prepared by refrigerated storage with cyanate ($\bullet$; P_{50}=19.0±1.5 Torr). $\dot{V}O_2$ changes from O_2 delivery-dependent to -independent at the critical O_2 delivery level, 0.40 ml/min·100g muscle. Plateau $\dot{V}O_2$, critical O_2 delivery and maximal O_2 extraction ratio in the perfusion with the high O_2-affinity blood were not different from those with normal blood.

Table 2. Influence of high O_2-affinity blood on oxygen consumption of resting dog gracilis muscle

O_2 delivery(ml/min·100g)	$\dot{V}O_2$ (ml/min·100g)	
	Normal affinity blood	High affinity blood
0.00 – 0.40	0.12±0.02 (3)	0.10±0.02 (3)
0.40 – 0.90	0.26±0.03 (5)	0.25±0.01 (6)
0.90 – 1.80	0.25±0.03 (4)	0.25±0.04 (5)
0.40 – 1.80	0.26±0.03 (9)	0.25±0.03 (11)

Values are means±SD. O_2 delivery indicates a product of blood flow and arterial O_2 content. Numbers in parentheses indicate number of observations collected at each O_2 delivery range for each perfused gracilis.

was 33 Torr for the control perfusion and 20 Torr for the high affinity perfusion. Thus the "critical PVO_2" value was evidently influenced by the change in O_2 affinity of circulating blood.

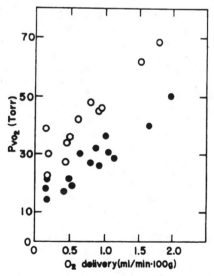

Fig. 3. PO_2 value of venous effluent (PVO_2) in relation to the O_2 delivery in the perfusions as presented in Figure 2. O, normal blood; ●, high affinity blood.

DISCUSSION

Criticisms of the present experimental procedures. Our isolated
muscle preparation ensured a stable pressure-flow and $\dot{V}O_2$-O_2 delivery
relationship for 3 to 4h. An improvement in this series of perfusions was
to preserve the perfusion blood in the reservoir at 4°C, which made it
possible to maintain fairly constant arterial gas and lactate values during
the perfusion period. One more important point was to correlate the $\dot{V}O_2$ to
the O_2 delivery. As experimentally demonstrated by Kolář and Janský (1984),
O_2 delivery is dependent on a number of variables such as blood flow, PaO_2,
SaO_2, and Ht and is the most decisive factor for $\dot{V}O_2$. The maximal and
constant $\dot{V}O_2$ values in the present results compared well with the previously
published values in the autoperfused gracilis muscle (Durán and Renkin,
1974).

Influence of elevated O_2 affinity of blood on the $\dot{V}O_2$-O_2 delivery
relationship in resting skeletal muscle. Increasing the O_2 affinity of
blood (P_{50}=19.0 Torr) did not affect aerobic metabolism even if blood flow
or O_2 delivery was reduced to quite a low level near the critical O_2
delivery. Furthermore, the high affinity blood did not induce a change in
pressure-flow relationship of the muscle when compared with normal blood.
Critical PVO_2 has often been considered the threshold at which $\dot{V}O_2$ begins
to decrease due to tissue hypoxia or anoxia (Lübbers, 1980). If this is
the case, the decrease in critical PVO_2 observed in the present results may
indicate the involvement of such compensatory mechanisms as capillary
recruitment in the high O_2 affinity perfusion.

Comparison with previous results by others. Theoretical cosiderations
predict that in normoxia, high affinity blood is disadvantageous for oxygen
transport because of reduction in PVO_2 leading to tissue hypoxia (Willford
et al., 1982). However, to date there has been no clear-cut experimental
proof for this notion. Yhap et al. (1975) and, Harken and Woods (1976)
reported the depression of $\dot{V}O_2$ in canine hindlimb perfused with stored canine
blood as compared with that perfused with fresh blood, although they did
not present their blood gas values. Ross and Hlastala (1981) found no decrease in $\dot{V}O_2$ of canine gracilis muscle perfused with low P_{50} blood at low
and high flow rates, but there was a reduction after perfusion with stored
blood. They interpreted the fall in $\dot{V}O_2$ as due to some factor other than
O_2-affinity. Our present results support their opinion. Recently, Van der
Plas et al. (1986) disproved the disadvantageous effect of low P_{50} hemoglobin on the function and $\dot{V}O_2$ of rat liver. Woodson et al. (1982) reported
results from dog brain perfused with moderately anemic dog blood of average
Ht and CaO_2 of 27% and 11.4 ml/dl, respectively. The blood for perfusion
was prepared by incubation with cyanate at 37°C for 9h. In view of our
recent results (Shimizu et al., 1986) it is highly probable that a considerable fraction of the hemoglobin was converted to inactive methemoglobin
in this experiment. Apstein et al. (1985) also found a reduction in $\dot{V}O_2$
of isolated rabbit heart perfused with DPG-depleted human RBC (stored
blood for 22-29 days) at Ht 30% and CaO_2 13.5 ml/dl. In this case, the
perfusion blood was also rather anemic and the O_2 extraction ratio of 24-
31% was lower than that of in vivo heart muscle. Possibly the measured
$\dot{V}O_2$ would originally be at a lower level than normal.

However it might be argued that the reported reductions of $\dot{V}O_2$ could
be caused by factor(s) other than the high O_2 affinity of the blood.

If high affinity blood does not affect the steady state $\dot{V}O_2$, PVO_2 or
$P\bar{V}O_2$ will have to decrease upon elevation of the O_2 affinity, which, in
turn, will provoke such physiological regulatory reactions as increased
erthropoietin production, myogenic change of organ blood flow, and
respiratory modulation through the respiratory chemoreceptors. Actually,

Lechermann and Jelkmann (1985) found a higher level of plasma erythropoietin in normoxic rats exchange-transfused with low P_{50} blood. Woodson and Auerbach (1982) also observed an increased blood flow to heart and brain of rats transfused with low P_{50} blood and suggested that it occurred via a vascular myogenic regulation mechanism to maintain normal tissue or capillary PO_2. Our results indicate that blood flow was not altered by differences in the O_2 affinity of the blood. It is quite possible that PO_2 sensing and regulation of the blood flow in brain and heart are different from that in skeletal muscle. Birchard and Tenney (1986) reported that the hypoxic ventilatory response was not influenced by high O_2 affinity blood and the results were consistent with the hypothesis of PaO_2 sensing by a peripheral chemoreceptor.

SUMMARY

To clarify the influence of high O_2-affinity blood on oxygen consumption by peripheral tissue, we perfused the dog gracilis muscle alternately with normal and low P_{50} blood at constant CaO_2 and varying blood flows. $\dot{V}O_2$ was plotted against O_2 delivery which was a product of CaO_2 and blood flow. $\dot{V}O_2$ of the resting dog gracilis muscle was O_2 delivery-dependent below a critical O_2 delivery of 0.40 ml/min·100g, but became delivery-independent above this level. This relationship was essentially the same as that observed in rat skeletal muscle (Kolář and Janský, 1984), and was independent of the O_2 affinity of blood, both qualitatively and quantitatively, although the PVO_2 was significantly lower in the perfusions with the high affinity blood than in those with normal blood. The average PVO_2 value by which the delivery-dependent and -independent $\dot{V}O_2$ regions were separated decreased from 33 Torr in the normal perfusions to 20 Torr in the high O_2 affinity perfusions. In conclusion, the elevation of blood O_2 affinity does not induce any reduction in $\dot{V}O_2$, and thus hypoxia, in resting skeletal muscle under normoxic conditions when O_2 delivery is not below the critical level.

REFERENCES

Apstein, C.S., Dennis, R.C., Briggs, L., Vogel, W.M., Frazer, J., and Valeri, C.R., 1985, Effect of erythrocyte storage and oxyhemoglobin affinity changes on cardiac function, Am. J. Physiol., 248: H508.
Birchard, G.F., and Tenney, S.M., 1986, The hypoxic ventilatory response of rats with increased blood oxygen affinity, Respir. Physiol., 66: 225.
Durán, W.N., and Renkin, E.M., 1974, Oxygen consumption and blood flow in resting mammalian skeletal muscle, Am. J. Physiol., 226: 173.
Harken, A.H., and Woods, M., 1976, The influence of oxyhemoglobin affinity on tissue oxygen consumption, Ann. Surg., 183: 130.
Honig, C.R., 1977, Hypoxia in skeletal muscle at rest and during the transition to steady work, Microvasc. Res., 13: 377.
Kohzuki, H., Enoki, Y., Sakata, S., and Tomita, S., 1983, A simple microtonometric method for whole blood oxygen dissociation curve and a critical evaluation of the "single point" procedure for blood P_{50}, Jpn. J. Physiol., 33: 987.
Kolář, F., and Janský, L., 1984, Oxygen cosumption in rat skeletal muscle at various rates of oxygen delivery, Experientia, 40: 353.
Lechermann, B., and Jelkmann, W., 1985, Erythropoietin production in normoxic and hypoxic rats with increased blood O_2 affinity, Respir. Physiol., 60: 1.
Lübbers, D.W., 1980, Tissue oxygen supply and critical oxygen pressure, in : "Oxygen transport to tissue", Kovách, A.G.B., Dóra, E., Kessler, M.,

and Silver, I.A., ed., Akadémiai Kiadó, Budapest.

Ross, B.K., and Hlastala, M.P., 1981, Increased hemoglobin-oxygen affinity does not decrease skeletal msucle oxygen consumption, J. Appl. Physiol., 51: 864.

Shimizu, S., Enoki, Y., Kohzuki, H., Ohga, Y., and Sakata, S., 1986, Determination of Hüfner's factor and inactive hemoglobin in human, canine, and murine blood, Jpn. J. Physiol., 36: 1047.

Van der Plas J., De vries-van Rossen, A., Bleeker, W.K., and Bakker, J.C., 1986, Effect of coupling of 2-nor-2-formylpyridoxal 5´-phosphate to stroma-free hemoglobin on oxygen affinity and tissue oxygenation. Studies in the isolated perfused rat liver under conditions of normoxia and stagnant hypoxia, J. Lab. Clin. Med., 108: 253.

Willford, D.C., Hill, E.P., and Moores, W.Y., 1982, Theoretical analysis of optimal P_{50}, J. Appl. Physiol., 52: 1043.

Woodson, R.D., and Auerbach, S., 1982, Effect of increased oxygen affinity and anemia on cardiac output and its distribution, J. Appl. Physiol., 53: 1299.

Woodson, R.D., Fitzpatrick, Jr J.H., Costello, D.J., and Gilboe, D.D., 1982, Increased blood oxygen affinity decreases canine brain oxygen consumption, J. Lab. Clin. Med., 100: 411.

Yhap, E.O., Wright, C.B., Popovic, N.A., and Alix, E.C., 1975, Decreased oxygen uptake with stored blood in the isolated hindlimb, J. Appl. Physiol., 38: 882.

EFFECTS OF DIFFERENCES OF OXYGEN AFFINITY ON CIRCULATORY RESPONSE TO HYPOXIA

Arisu Kamada, Akihiko Suzuki, Yasushi Akiyama, Shuichi Inaba, Kumi Dosaka, Fujiya Kishi, and Yoshikazu Kawakami

First Department of Medicine, Hokkaido University School of Medicine, Sapporo, Japan

INTRODUCTION

It is well known that tissue hypoxia is an important prognostic predictor in patients with respiratory failure especially in acute exacerbation. It can occur for a number of reasons such as hypoxemia, peripheral circulatory disturbance, malfunction of the hemoglobin, and disturbance of oxygen supply to the tissue. Compensatory responses to acute hypoxia occur in various organs and the magnitude of such response is also an important factor in prognosis. In the circulatory system, cardiac blood flow and output together with heart rate and systemic blood pressure, all increase although the magnitude of these responses may be affected by age, the nature of the disease, severity, and so on. Whether differences in oxygen supply to the tissue under stable conditions affect the magnitude of response is as yet unknown.

P_{50}, (oxygen tension at which the hemoglobin is 50% saturated with oxygen) is recognized as an expression of the affinity of the hemoglobin for oxygen, as well as one of the factors affecting oxygen supply to the tissue. In this study we tried to investigate whether differences of P_{50} affect the circulatory response to acute hypoxia as a model of acute exacerbation.

SUBJECTS AND METHODS

Studies were made on eighteen male patients with chronic obstructive pulmonary disease (COPD). They were diagnosed as COPD on the basis of their history, physical examinations, chest roentgenogram, pulmonary function studies and so on. They were studied in a supine position. Right cardiac catheterization was performed via the cubital vein. Cardiac output was measured by the thermodilution method and arterial blood samples were obtained from the brachial artery. Minute ventilation was analyzed breath by breath with a hot wire flowmeter. End-tidal PO_2, and PCO_2, were measured continuously by a mass spectrometer. S_aO_2 was monitored by a finger tip oximeter. P_aO_2 and P_aCO_2 were varied independently by changing the inspiratory gas composition. (Kawakami et al. 1981) Blood gases were analyzed with an IL-1303 analyzer (Instrumentation Laboratory, USA). Initially, control values for blood gases and hemodynamic parameters were established with the patients breathing room air.

323

P_{50} was calculated from control mixed venous blood by method of Aberman (1975). Subjects were divided into two groups according to low (< 26.6 torr) or high (> 26.6 torr) P_{50}. Isocapnic hypoxia was then induced progressively (final P_aO_2 = 45 torr) and maintained for 10 minutes. Blood gases and hemodynamic parameters during hypoxia were measured at that time.

Results before and after hypoxia were compared using Student's paired t test. Results between groups were compared using Student's unpaired t test, and a p value < 0.05 was considered to be statistically significant. Data are presented as mean ± SEM.

RESULTS

Mean values of hemodynamic measurements and blood gases in each group are shown in Table 1. The low P_{50} group includes eleven subjects with a mean P_{50} of 25.8 ± 0.2 torr. The high P_{50} group includes seven subjects with a mean P_{50} of 27.5 ± 0.2 torr. No differences could be detected in respect of age, height, and weight.

Comparison between groups

Comparing control values (while breathing room air) P_{50} and S_aO_2 differed between the two groups, but there were no difference in respect of pH, P_aCO_2, P_aO_2 or any hemodynamic parameters. Likewise a comparison of values after hypoxia, showed no differences either. (Table 1.)

Effects of hypoxia

In the low P_{50} group C_aO_2, $C_{\bar{v}}O_2$, COD (coefficient of oxygen delivery which is calculated as $C_aO_2/C_aO_2-C_{\bar{v}}O_2$) decreased (18.9 ± 0.9 to 16.3 ± 0.7, 14.1 ± 0.7 to 12.0 ± 0.6 and 4.0 ± 0.2 to 3.6 ± 0.1 respectively), but $C_aO_2-C_{\bar{v}}O_2$ did not change. Arterial pH and P_aCO_2 remained unchanged while P_aO_2, S_aO_2, $P_{\bar{v}}O_2$, and $S_{\bar{v}}O_2$ decreased. (Table 1., Fig. 1)

Pulmonary vascular resistence (PVR) and mean pulmonary arterial pressure (mPA) increased after hypoxia, (210.2 ± 33.1 to 265.1 ± 34.2 and 16.5 ± 1.2 to 22.7 ± 1.5, respectively) (Table 1.) Heart rate increased (73.6 ± 2.8 to 81.6 ± 2.9) but stroke volume (61.5 ± 5.5 to 62.3 ± 4.2) and cardiac output (4.4 ± 0.3 to 5.1 ± 0.3) did not change (Table 1., Fig. 2.).

In the high P_{50} group C_aO_2, $C_{\bar{v}}O_2$, COD also decreased (18.0 ± 0.7 to 15.3 ± 0.6, 13.5 ± 0.7 to 11.1 ± 0.6 and 4.0 ± 0.2 to 3.7 ± 0.2 respectively), and $C_aO_2-C_{\bar{v}}O_2$ did not change. Arterial pH and P_aCO_2 remained constant, while P_aO_2, S_aO_2, $P_{\bar{v}}O_2$, and $S_{\bar{v}}O_2$ decreased. (Table 1., Fig. 3)

PVR and mean PA pressure also increased after hypoxia, (238.6 ± 28.0 to 280.7 ± 21.8 and 18.7 ± 1.1 to 24.1 ± 1.7, respectively) (Table 1.).

Cardiac output (4.6 ± 0.3 to 5.4 ± 0.3) and heart rate (65.7 ± 2.3 to 78.1 ± 2.9) increased, whereas cardiac output did not change in the low P_{50} group. Stroke volume (71.1 ± 5.0 to 69.0 ± 3.2) did not change (Table 1., Fig. 4.).

Table 1. Hemodynamic measurements and blood gases in each group

		LOW P_{50}		HIGH P_{50}	
		ROOM AIR	HYPOXIA	ROOM AIR	HYPOXIA
n		11		7	
P_{50}	torr	25.8±0.2	22.7±1.5 §	27.5±0.2 *	24.1±1.7 §
mPA	torr	16.5±1.2	265.1±34.2 §	18.7±1.1	280.7±21.8 §§
PVR	dynes·sec·cm^{-5}	210.2±33.1	265.1±34.2 §	238.6±28.0	280.7±21.8 §§
CO	L/min	4.4±0.3	5.1±0.3	4.6±0.3	5.4±0.3 §§
HR	beats/min	73.6±2.8	81.6±2.9 §	65.7±2.3	78.1±2.9 §§
SV	ml/beat	61.5±5.5	62.3±4.2	71.1±5.0	69.0±3.2
CaO_2	vol%	18.9±0.9	16.3±0.7 §	18.0±0.7	15.3±0.6 §
$C\bar{v}O_2$	vol%	14.1±0.7	12.0±0.6 §	13.5±0.7	11.1±0.6 §
COD		4.0±0.2	3.6±0.1 §	4.0±0.2	3.7±0.2 §
pH		7.40±0.00	7.42±0.00 §	7.39±0.00	7.43±0.01 §
$PaCO_2$	torr	40.8±1.2	38.4±0.8 §	44.2±1.8	40.6±0.8 §
PaO_2	torr	77.5±3.6	45.9±1.9 §	71.0±3.2	44.4±2.1 §
SaO_2	%	95.4±0.5	82.8±1.9 §	93.2±0.8 *	79.3±1.9 §

All values are mean ± SEM
§ : different from ROOM AIR value, $p < 0.05$
* : different from LOW P_{50} value, $P < 0.05$

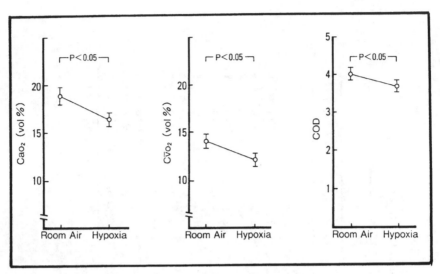

Fig. 1. Effects of hypoxia on oxygen transport in the low P_{50} group.

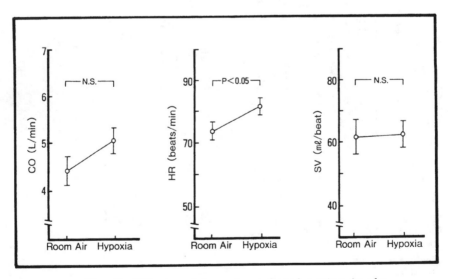

Fig. 2. Effects of hypoxia on hemodynamics in the low P_{50} group.

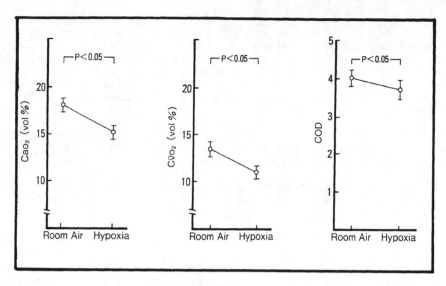

Fig. 3. Effects of hypoxia on oxygen transport in the
high P_{50} group.

DISCUSSION

Chronic respiratory failure is caused by various diseases including
COPD, interstitial pneumonia, pulmonary thromboembolism and so on. When

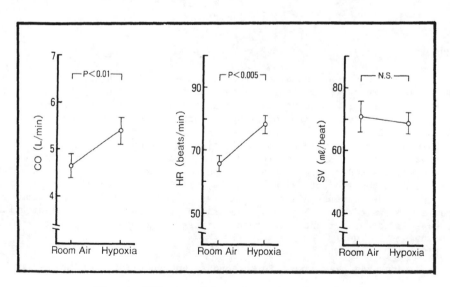

Fig. 4. Effects of hypoxia on hemodynamics in
the high P_{50} group.

subjects in such circumstances fall into acute exacerbation, hypoxemia occurs immediately. As a response to hypoxemia, ventilation increases in the respiratory system, while in the circulatory system, heart rate, cardiac output, and systemic blood pressure increase via peripheral chemoreceptor reflex, and blood flow redistribtion occurs. Compensation for hypoxemia is made by integration of these responses, but the magnitude is varied among individuals.

This study was made to determine whether differences of P_{50}, i.e., differences in oxygen supply to the tissue, during stable conditions affect the circulatory response to acute hypoxia.

There were no differences between pH, P_aCO_2, and P_aO_2, but 2,3-DPG, ATP and other parameters which might affect P_{50} were not measured. So the reason for differences in P_{50} observed in this study is unclear.

In healthy subjects, exercise or hypoxic load increases cardiac output mainly due to increases in heart rate, but stroke volume also increases. However in COPD patients, exercise or hypoxic load increases cardiac output only while heart rate and stroke volume do not change. (Miyamoto, 1984) These results in the high P_{50} group are consistent with this, but those in the low P_{50} group are not. This may be due to differences in the heart-rate response to hypoxia. Subjects whose heart-rate response to hypoxia is diminished have lowered oxygen transport. (Miyamoto, 1984)

In this study parameters of oxygen transport such as C_aO_2, C_vO_2, COD, and $C_aO_2-C_vO_2$ showed no difference. The reason may be due to small difference of P_{50} but P_{50} itself has a narrow range in vivo and it would be difficult to compare low and high P_{50} groups over a wide range.

CONCLUSIONS

1) Circulatory response to hypoxia is well preserved in the high P_{50} group but deteriorated in the low P_{50} group.
2) This suggests that patients with low P_{50} cannot compensate adequately to hypoxia during acute exacerbation.

SUMMARY

Studies were made on eighteen male patients with chronic obstructive pulmonary disease (COPD) to investigate whether differences of P_{50} affect the circulatory response to acute hypoxia as a model of acute exacerbation. Subjects were divided into two groups according to low (< 26.6 torr) or high (> 26.6 torr) P_{50}. Isocapnic hypoxia was induced progressively (final P_aO_2 = 45 torr) and maintained for 10 minutes. Blood gases and hemodynamic parameters were measured before and after hypoxia. Results before and after hypoxia and results between groups were compared. The low P_{50} group includes eleven subjects with a mean P_{50} of 25.8 ± 0.2 torr. The high P_{50} group includes seven subjects with a mean P_{50} of 27.5 ± 0.2 torr. Comparison between groups showed no significant differences. In the high P_{50} group cardiac output and heart rate increased, whereas cardiac output did not change in the low P_{50} group.

We conclude that circulatory response to hypoxia is well preserved in the high P_{50} group but deteriorated in the low P_{50} group, and this suggests that patients with low P_{50} cannot compensate adequately to hypoxia during acute exacerbation.

ACKNOWLEDGMENT

This work was supported in part by the Ministry of Health and Welfare Grants in Aid for Special Disease Survey and Research Project in "Respiratory Failure".

REFERENCES

Aberman, A., Cavanilles, J.M., Weil, M.H., and Shubin, H., 1975, Blood P_{50} calculated from a single measurement of pH, PO_2, and SO_2, J. Appl. Physiol., 38:171.

Kawakami, Y., Asanuma, Y., Yoshikawa, T., and Murao, M., 1981, A control system for arterial blood gases, J. Appl. Physiol., 50:1362.

Miyamoto, K., 1984, Pathophysiological studies on hypoxic circulatory responses in health and chronic obstructive pulmonary disease, J. Jpn. Soc. Intern. Med., 70:1461.

INCREASED VENTILATORY RESPONSE TO ACUTE HYPOXIA WITH HIGH Hb-O_2 AFFINITY INDUCED BY Na-CYANATE TREATMENT IN THE RAT

Yasuichiro Fukuda, Toshio Kobayashi[1], Hiroshi Kimura[2] and Ryoko Maruyama

Department of Physiology II, Internal Medicine[1] and Chest Medicine[2], School of Medicine, Chiba University
1-8-1 Inohana, 280 Chiba, Japan

INTRODUCTION

The characteristic affinity of hemoglobin (Hb) for O_2 has an important functional significance in determining the amount of O_2 delivered to tissues in systemic hypoxic conditions. Animal species with a high Hb-O_2 affinity have been shown to be more resistant to hypoxic hypoxia compared with those species which have a low O_2 affinity (Dawson and Evans, 1966; Eaton et al.,' 1974; Hall, 1966; Turek et al., 1973). Furthermore the influence of altered Hb-O_2 affinity on the tolerance to acute hypoxia has been evaluated by artificial modification of Hb-O_2 affinity in experimental animals (Eaton et al., 1974; Teisseire et al., 1979). Eaton et al. (1974) demonstrated that increased rather than decreased Hb-O_2 affinity permitted survival at greatly reduced environmental pressure (hypobaric hypoxia). In their study Hb-O_2 affinity was increased (P_{50} was decreased) by carbamylation of Hb with sodium cyanate (NaOCN). The nature of this protective action of high Hb-O_2 affinity, however, has not been well clarified. Acute systemic hypoxic hypoxia produces respiratory stimulation which is followed by respiratory depression leading eventually to death. The former is initiated by stimulation of peripheral arterial chemoreceptors, and the latter results from a direct depressant action of brain tissue hypoxia on the respiratory neural mechanism in the lower brain stem (Cherniack et al., 1970/1971; Cherniack et al., 1977; Koepchen et al., 1977; Lahiri, 1976; Lahiri and Gelfand, 1981; Morrill et al., 1975). In the present study, we have sought to define the mechanism of the protective action of high Hb-O_2 affinity produced by NaOCN treatment, especially the protection of respiratory regulating mechanism during acute hypoxia in the halothane-anesthetized spontaneously breathing rat.

MATERIAL AND METHODS

Male rats of the Wistar strain were obtained at a body weight of about 340-360g. In the NaOCN group, the animal was treated subcutaneously with 50mg/kg NaOCN (0.1ml/100g body weight) daily for a period of about 2 weeks. Control animals received 0.9% NaCl solution (0.1ml/100g body weight) for the same period. The animal was anesthetized with halothane (induction 2.5%, during operation 1.0-1.2%, maintenance during experiment 0.8%) and was placed on a heating pad in the supine position. The rectal temperature was kept constant at 37°C throughout the experiment. The trachea was cannulated with a polyethylene tubing, and which was connected to a respiratory gas circuit (Fig. 1). The circuit consisted of three flowmeters with needle valves for mixing N_2, O_2 and CO_2 gases at a desired ratio, a reservoir bottle (2 liters) and a halothane vaporizer (Dräger, Vapor, Germany). The base-line total gas flow through the circuit was kept constant at 600ml/min. The respiratory flow was measured by a pneumotachograph and the breath-by-breath tidal volume, inspiratory and expiratory times (Ti, Te) were obtained electronically from the pneumotachograph output (Fig. 1). End-tidal P_{O_2} and P_{CO_2} (P_{ETO_2}, P_{ETCO_2}) were monitored by an expiratory gas analyzer (San-Ei Instrument, 1H21, Tokyo) the expiratory gas being sampled through a fine stainless steel tube placed in the tracheal cannula (Fukuda et al., 1982). A polyethylene catheter for blood pressure measurement was inserted into the femoral artery. In some experiments, Pa_{O_2} was directly measured by a Clark type O_2 electrode placed in the femoral artery together with simultaneous measurement of arterial Hb-O_2 saturation (Sa_{O_2}) by an oximeter (Minolta, SM-32, Tokyo) placed on the tail artery.

Ventilatory parameters, Pa_{O_2}, P_{ETO_2} and P_{ETCO_2} in normoxic conditions

Ventilatory parameters such as respiratory flow, tidal volume, Ti, Te, respiratory frequency, minute ventilation (ATPS), Pa_{O_2}, P_{ETO_2} and P_{ETCO_2} were measured during the breathing of a gas mixture containing 20-21% O_2 in N_2.

Progressive hypoxia test with intact carotid sinus nerve

The reservoir bottle was first filled with a gas mixture with an O_2 concentration of more than 30% (inspiratory P_{O_2} above 200mmHg). The O_2 in the bottle was then gradually replaced with N_2, decreasing inspiratory P_{O_2} almost linearly from the hyperoxic level (above 200mmHg) down to about 30-50 mmHg within 10 min. When hypoxic stimulation caused hyperventilation, CO_2 gas was injected into the inspiratory line in order to maintain P_{ETCO_2} at a constant level as measured in the hyperoxic condition (isocapnia). In some experiments, a Hb-O_2 dissociation curve was obtained in in vivo condition.

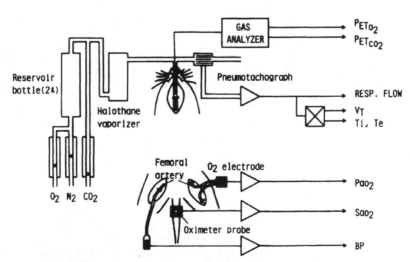

Fig. 1. Experimental set-up. A small resistance head for the pneumotachograph was placed in the respiratory gas circuit. A Clark type O_2 electrode was placed between the proximal and distal cut ends of the femoral artery through which heparinized blood flowed. V_T, tidal volume; Ti, Te, inspiratory and expiratory times; Resp. flow, respiratory flow.

Table 1   Ventilatory parameters during the breathing of normoxic gas

	Body weight (g)	Minute ventilation (ml*/100g/min)	Tidal volume (ml*/100g)	Respiratory frequency (breaths/min)	$P_{ET_{CO_2}}$ (mmHg)
Control (n=9)	394 ± 35	30.8 ± 3.5	0.280 ± 0.043	113 ± 8.1	43.1 ± 2.1
NaOCN (n=9)	372 ± 43	28.4 ± 2.1	0.337 ± 0.032	85 ± 8.7	42.0 ± 2.7
Difference	NS	NS	p<0.01	p<0.001	NS

Values are mean ± SD. *, ATPS; p, statistically significant
difference between control and NaOCN groups (t-test); NS,
statistically not significant.

Progressive hypoxia test after bilateral carotid sinus nerve section

The above procedure was repeated after bilateral section of the carotid
sinus nerve. In the rat, there are no functional aortic bodies and carotid
sinus nerve section entirely eliminates the peripheral chemoreceptor
afferent activity in the anesthetized condition. The ventilatory responses
to the progressive hypoxia test were compared before and after carotid sinus
nerve section with or without administration of NaOCN.

RESULTS

1. Ventilatory parameters in normoxic conditions

NaOCN administration for 2 weeks resulted in a reduced gain of body
weight although the difference between control and NaOCN groups was not
statistically significant (Table 1). Table 1 shows ventilatory parameters
and $P_{ET_{CO_2}}$ values during the breathing of normoxic gas. An interesting
observation was that NaOCN administration produced significant changes in
breathing pattern, i.e., a low breathing frequency with a large tidal volume
but without large changes in minute ventilation and $P_{ET_{CO_2}}$. Our previous
study showed that the $P_{ET_{CO_2}}$ value measured by the present technique accorded
well with Pa_{CO_2} (Fukuda et al., 1982). In contrast, there was a large
arterial to end-tidal (mean alveolar) P_{O_2} difference in the anesthetized rat
(Fig. 2). The average (± SD) Pa_{O_2} and $P_{ET_{O_2}}$ in control and NaOCN rats were
73.6 ± 8.6 and 117 ± 6.1 mmHg (control, n=36) and 75.2 ± 8.8 and 118 ± 6.5
mmHg (NaOCN, n=30) respectively. The difference between control and NaOCN
groups was not significant. The large arterial to end-tidal P_{O_2} difference
is probably due to relatively large physiologic 'shunts' or venous admixture
in the lung. The difference became smaller in hypoxia (Fig. 2). The relationship between Pa_{O_2} and $P_{ET_{O_2}}$ was, however, nearly equivalent in both
groups of rats indicating that NaOCN administration did not initiate large
changes in O_2 diffusion or ventilation-perfusion ratio in the lung.

2. Arterial Hb-O_2 saturation, Pa_{O_2} and ventilatory response during progressive hypoxia test

An in vivo Hb-O_2 dissociation curve obtained during a progressive hypoxia test at a maintained $P_{ET_{CO_2}}$ (43-45 mmHg) is presented in Fig. 3. Since

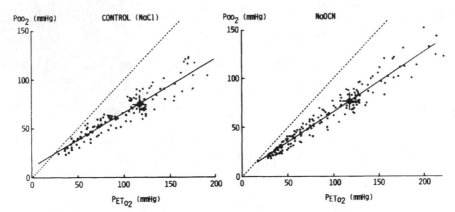

Fig. 2. Relationship between $P_{ET_{O_2}}$ and Pa_{O_2}. Large filled circle with cross indicates mean $\pm$ SD value of $P_{ET_{O_2}}$ and Pa_{O_2} during the breathing of normoxic gas. Values were not significantly different between control and NaOCN groups. Dotted line is the line of identity.

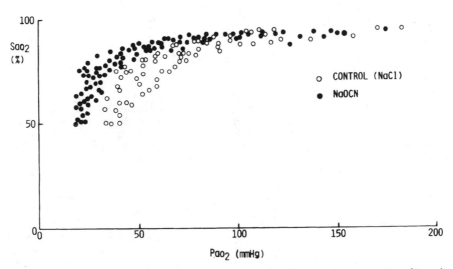

Fig. 3. Relationship between Pa_{O_2} and arterial Hb-O_2 saturation (Sa_{O_2}) measured simultaneously during progressive hypoxia test at a maintained $P_{ET_{CO_2}}$.

the progressive hypoxia test was terminated when the ventilation reached the maximum level and then became depressed, Sa_{O_2} could not be reduced to less than 50%. The estimated P_{50} value (P_{O_2} at 50% saturation) was about 34mmHg in the control rats. In the NaOCN group the Hb-O_2 dissociation curve was displaced to the left and the P_{50} value was about 20mmHg. The results indicated an increased Hb-O_2 affinity after NaOCN administration. Figures 4 and 5 illustrate typical examples of changes in ventilatory parameters during a progressive hypoxia test. The P_{ETo_2} was used conventionally as an index of the level of hypoxia. A decrease in P_{ETo_2} from hyperoxic to normoxic levels induced a small increase in ventilation, and at P_{ETo_2} levels below normoxia there was a sharp increase in respiratory frequency with increased tidal volume. A further decrease in P_{ETo_2} (below 50mmHg in control rat), however, depressed ventilation with an accompanying reduction in respiratory frequency mainly due to prolongation of expiratory time with a maintained large tidal volume. The extent of maximum increase in minute ventilation, tidal volume and respiratory frequency are shown in Fig. 6. The threshold P_{ETo_2} for ventilatory depression defined as the P_{ETo_2} at which the maximum increase in ventilation was attained was 51.7 $\pm$ 5.4 mmHg (mean $\pm$ SD, n=8) in the control rats (Fig. 6). Increases in minute ventilation, tidal volume and frequency were all significantly larger in the NaOCN group (Figs. 4, 5, 6), the threshold P_{ETo_2} for ventilatory depression was decreased to 28.4 $\pm$ 5.0 mmHg (mean $\pm$ SD, n=8) (Fig. 6). Thus the ventilatory response to acute hypoxia was clearly augmented and ventilatory depression during severe hypoxia was inhibited after administration of NaOCN.

3. Ventilatory responses to progressive hypoxia after bilateral section of the carotid sinus nerve

Ventilatory parameters during the breathing of hyperoxic gas after bilateral section of the carotid sinus nerve (CSN) are listed in Table 2. The differences between control and NaOCN groups were similar to those in CSN-intact animals in normoxic condition (compare Table 1 and 2), i.e., slow respiratory frequency with a relatively large tidal volume in the NaOCN group though the difference in tidal volume was not statistically significant. After bilateral section of the CSN, ventilation was gradually decreased even at the normoxic range during the progressive hypoxia test

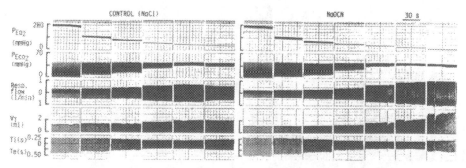

Fig. 4. Typical examples of changes in ventilatory parameters during progressive hypoxia test. V_T, tidal volume; Ti and Te, inspiratory and expiratory times. Respiratory frequency was decreased in severe hypoxia.

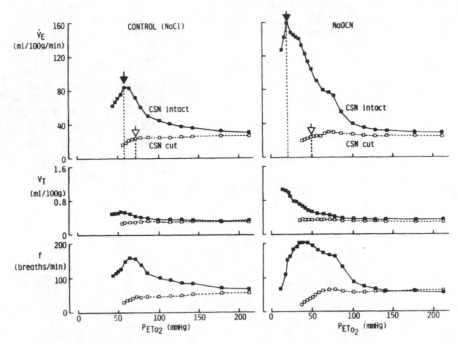

Fig. 5.   Typical examples of ventilatory response to hypoxia.
CSN, carotid sinus nerve; filled arrow with dotted line,
threshold P_{ETO_2} for ventilatory depression in CSN intact
rat; open arrow with dotted line, threshold P_{ETO_2} for 10%
reduction in ventilation after CSN section; $\dot{V}_E$, minute
ventilation; V_T, tidal volume; f, respiratory frequency.

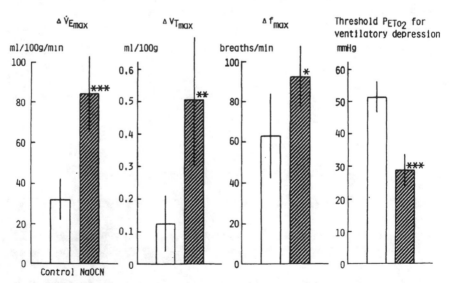

Fig. 6.   The extent of maximum increase in ventilation during hypoxia
test and threshold P_{ETO_2} for ventilatory depression. The height
of each column represents mean ($\pm$SD) value obtained from 8 rats.
Statistically significant difference between control and NaOCN
groups, * p<0.02, ** p<0.01, *** p<0.001 (t-test).

Table 2   Ventilatory parameters after carotid sinus nerve section

| | During the breathing of hyperoxic gas | | | | Threshold PET_{O_2} for 10% reduction in ventilation during hypoxia |
| | Minute ventilation | Tidal volume | Respiratory frequency | PET_{CO_2} | |
	$(ml^*/100g/min)$	$(ml^*/100g)$	(breaths/min)	(mmHg)	(mmHg)
Control (n=8)	26.1 ± 2.4	0.291 ± 0.034	90 ± 12.4	44.5 ± 1.7	80.2 ± 16.8
NaOCN (n=8)	25.7 ± 3.4	0.340 ± 0.088	73 ± 11.6	42.3 ± 3.7	53.0 ± 8.2
Difference	NS	NS	$p < 0.05$	NS	$p < 0.01$

Values are mean $\pm$ SD. *, ATPS; p, statistically significant difference between control and NaOCN groups (t-test); NS, statistically not significant.

(Fig. 5). The threshold PET_{O_2} for 10% reduction in ventilation during progressive hypoxia was, however, significantly lower in the NaOCN than in the control group (Fig. 5, Table 2). The protective effect of NaOCN treatment, therefore, appeared even in the absence of carotid chemoreceptor afferent activity.

DISCUSSION

The present experiment demonstrated that an increase in Hb-O_2 affinity following administration of NaOCN augments the ventilatory response to acute systemic hypoxia caused by inhalation of low Po_2 gas, and inhibits the ventilatory depression due to severe hypoxia. The smaller gain of body weight during NaOCN treatment may result from toxic side effects (Teisseire et al., 1979). The mechanism of changes in breathing pattern during the breathing of normoxic gas is obscure but it may be due to some alteration in the neural regulatory mechanism of the respiratory pattern. Clearly the ventilatory effect of hypoxia is the sum of the chemoreceptor-mediated excitatory effect and a direct depressant effect on the respiratory neurons (Lahiri, 1976; Lahiri and Gelfand, 1981; Morrill et al., 1975). The increased ventilatory response to hypoxia in the NaOCN-treated rat may be due to an inhibition of the depressant effect of hypoxia. Since ventilatory depression was caused mainly by reduction in respiratory frequency, due to prolongation of expiratory time, the process of expiratory-inspiratory phase switching (inspiratory onset) seems to be susceptible to hypoxia. Koepchen et al. (1977) reported that inspiratory neurons were more susceptible to hypoxia than were expiratory neurons. Our unpublished observations showed that the carotid chemoreceptor afferent activity is increased even in severe hypoxia which depresses ventilation. In the NaOCN group, the reduction in respiratory frequency during severe hypoxia was inhibited and the increase in tidal volume was greatly augmented. All these protective actions of NaOCN treatment seems to be related to a high Hb-O_2 affinity. The amount of O_2 that would be released from Hb in severe hypoxic condition, may be sufficient to prevent the strong reduction in brain tissue Po_2 and maintain the respiratory phase switching mechanism. The fact that the rise in blood lactate concentration during acute hypoxia is small in the NaOCN-treated rat (Teisseire et al., 1979) supports the idea that aerobic metabolism can be better maintained in animals with low rather than high P_{50} values in severe hypoxic condition. Hemoglobin concentration and O_2 content of the blood are also increased in the NaOCN-treated rat (Teisseire et al., 1979). Recently Birchard and Tenney (1986) described that the ventilatory responses to hypoxia expressed as a function of Sa_{O_2} were significantly larger in the NaOCN-treated rat. The mechanism by which NaOCN administration induces a beneficial effect, however, may not result from high Hb-O_2 affinity alone. In the NaOCN-treated animal the tissue or cells may be exposed to low Po_2 (tissue hypoxia) even during the breathing of normoxic or hyperoxic gas because of high Hb-O_2 affinity. Thus the animals might have become acclimatized to such conditions during treatment with NaOCN, with some accompanying functional or morphological changes in the respiratory regulating mechanism, as in animals acclimatized to high altitude. There is also a possibility that increased afferent stimulatory activities from the peripheral arterial chemoreceptors overcome the direct depressant action of hypoxia on the central nervous structures in the NaOCN-treated rat. Evidence that the protective action still occurred after bilateral section of the CSN indicates that the respiratory neural structures in the lower brain stem become less susceptible, or become tolerant, to hypoxia in the NaOCN-treated animals.

SUMMARY

The effects on the ventilatory response to acute hypoxia of increasing the Hb-O_2 affinity by NaOCN administration were studied in the halothane anesthetized spontaneously breathing rat. Increases in ventilation during the progressive hypoxia test were significantly augmented, and ventilatory depression occurring in severe hypoxia was clearly inhibited in the NaOCN-treated rat. Beneficial effects of NaOCN treatment probably result from the protection of respiratory regulating mechanism from functional deterioration in severe hypoxia.

ACKNOWLEDGEMENT

We are indebted to Ms. M. Iino for preparing photographs and Dr. Y. Enoki for helpful discussion.

REFERENCES

Birchard, G.F., and Tenney, S.M., 1986, The hypoxic ventilatory response of rats with increased blood oxygen affinity, Respir. Physiol., 66: 225.

Cherniack, N.S., Edelman, N.H., and Lahiri, S., 1970/1971, Hypoxia and hypercapnia as respiratory stimulants and depressants, Respir. Physiol., 11: 113.

Cherniack, N.S., Kelson, S.G., and Lahiri, S., 1977, The effects of hypoxia and hypercapnia on central nerveous system output, in: "Respiratory Adaptation, Capillary Exchange and Reflex Mechanisms," A.S. Paintal and P. Gill-Kumar, eds., Vallabhbhai Patel Chest Inst., Delhi.

Dawson, T., and Evans, J.V., 1966, Effects of hypoxia on oxygen transport in sheep with different hemoglobin types, Am. J. Physiol., 210: 1021.

Eaton, J.W., Skelton, T.D., and Berger, E., 1974, Survival at extreme altitude: protective effect of increased hemoglobin-oxygen affinity, Science, 183: 743.

Fukuda, Y., See, W.R., and Honda, Y., 1982, Effect of halothane anesthesia on end-tidal Pco_2 and pattern of respiration in the rat, Pflügers Arch., 392: 244.

Hall, F.G., 1966, Minimal utilizable oxygen and the oxygen dissociation curve of blood of rodents, J. Appl. Physiol., 21: 375.

Koepchen, H.P., Klussendorf, D., Barchert, J., Lessman, D.W., Dinter, A., Frank, C.H., and Sommer, D., 1977, Type of respiratory neuronal activity pattern in reflex control of ventilation, in: "Respiratory Adaptations, Capillary Exchange and Reflex Mechanisms," A.S. Paintal P. Gill-Kumar, eds., Vallabhbhai Patel Chest Inst., Delhi.

Lahiri, S., 1976, Depressant effect of acute and chronic hypoxia on ventilation, in: "Morphology and Mechanisms of Chemoreceptors," A.S. Paintal, ed., Vallabhbhai Patel Inst., Delhi.

Lahiri, S., and Gelfand, R., 1981, Mechanisms of acute ventilatory responses, in: "Regulation of Breathing, Part II," F. Hornbein, ed., Marcel Dekker, New York - Basel.

Morrill, C.G., Mayer, J.R., and Weil, J.V., 1975, Hypoxic ventilatory depression in dogs, J. Appl. Physiol., 38: 143.

Teisseire, B.P., Soulard, C.D., Heigault, R.A., Leclerc, L.F., and Laver, M.B., 1979, Effects of chronic changes in hemoglobin-O_2 affinity in rats, J. Appl. Physiol., 46: 816.

Turek, Z., Kreuzer, F., and Hoofd, L.J.C., 1973, Advantage or disadvantage of a decrease of blood oxygen affinity for tissue comparing man and rat, Pflügers Arch., 342: 185.

PHARMACOLOGICAL EFFECTS ON HEMOGLOBIN-OXYGEN AFFINITY IN VITRO AND IN PATIENTS WITH CHRONIC OBSTRUCTIVE PULMONARY DISEASE

Fujiya Kishi, Yoichi Nishiura, Arisu Kamada and
Yoshikazu Kawakami

First Department of Medicine, Hokkaido University
School of Medicine, Sapporo, Japan

INTRODUCTION

The presence of chronic hypoxia in peripheral tissue in patients with chronic obstructive pulmonary disease (COPD) is one of the major factors leading to a poor prognosis (Kawakami et al., 1983). If hemoglobin-oxygen affinity can be reduced by pharmacological agents, oxygen delivery to the tissues should be improved in mild hypoxemia. We examined the effects of three agents on hemoglobin-oxygen affinity in vivo and in vitro. The first is pentoxifylline which is a xanthine derivative and has the effect of improving capillary circulation and increasing intracellular ATP (Ohshima and Sato, 1981; Stefanovich, 1975). The second was coenzyme Q_{10} (CoQ_{10}) which improves oxygen utilization in ischemic heart muscles (Mochizuki et al., 1980). The third was calcium hopantenate which increases glucose utilization in brain tissues (Tamura et al., 1986).

METHODS

In vitro study

Freshly drawn heparinized venous blood obtained from healthy volunteers was used for the in vitro experiment. Immediately after collection, 1mM glucose was added to the blood and the sample tube was shaken at 72 rpm on a reciprocating shaker in a water maintained at 37°C. A control sample was taken from the preincubated blood to measure the hemoglobin-oxygen dissociation curve, erythrocyte 2,3-diphosphoglycerate(2,3-DPG), hemoglobin concentration (Hb), ATP and glucose. This preincubated blood was divided into four tubes. To one was added saline solution as a placebo and to the others, three concentrations of each agent. The samples were then incubated for 4 hours.

Pentoxifylline was studied at 3, 15 and 75μg/ml, CoQ_{10} at 1, 20 and 100μg/ml, and a calcium hopantenate was 0.01, 0.1 and 1mg/ml. Hemoglobin-oxygen affinity was measured with a Hemox-Analyzer(TCS Co. USA) and expressed as P_{50} (the partial pressure of oxygen at which the hemoglobin is 50% saturated).

In vivo study

Seven patients(five males and two females)with clinically stable

COPD were given pentoxifylline 300mg per day orally for 4 weeks. Their age was 62 ± 10yrs(mean$\pm$SD), %VC (percent of predicted vital capacity) 74 ± 4%, FEV_1% (a ratio of forced expiratory volume in one second to forced vital capacity)54 ± 12%, pH 7.37 ± 0.01, $Paco_2$ 44 ± 2 torr, Pao_2 73 ± 4 torr.

Statistical evaluation was by unpaired t-test for the in vitro study and paired t-test for the in vivo study.

RESULTS

1. In vitro study

1) Pentoxifylline at 3 and 15ug/ml significantly increased P_{50} from the placebo(27.0 ± 0.3 Torr, mean$\pm$SE) to 28.2 ± 1.7 and 28.1 ± 1.3Torr respectively($p < 0.05$) after 4 hour incubation (Fig.1). With 75µg/ml pentoxifylline, P_{50} increased slightly to 27.8 ± 1.4 Torr ($0.05 < p < 0.1$). 2,3-DPG was slightly higher($0.05 < p < 0.1$) with pentoxifylline of 3 (11.15 ± 4.12 µmol/gHb) and 75µg/ml (10.62 ± 2.32 µmol/gHb) than placebo (8.91 ± 0.78 µmol/gHb) (Fig.2). Hb and ATP were not changed.

2) In CoQ_{10}, P_{50}, 2,3-DPG and Hb were unchanged compared to the placebo after four hours incubation (Table 1).

3) Four hours incubation with calcium hopantenate caused no change in P_{50}, 2,3-DPG, Hb,ATP or glucose when compared to the placebo (Table 2).

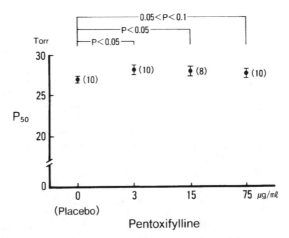

Fig.1. Effects of pentoxifylline on P_{50} in vitro. Mean$\pm$SEM. Parentheses show numbers of experiment.

2. In vivo study

After administration of pentoxifylline , P_{50} increased significantly from 29.0±0.6 to 30.4±0.6 Torr (p<0.05) and 2,3-DPG increased slightly from 11.67±0.56 to 14.33±1.12μmol/gHb(p<0.1) (Fig. 3). Arterial blood gases showed no change between control and after administration of pentoxifylline in patients with COPD (Fig. 4).

DISCUSSION

Hemoglobin-oxygen affinity has an important role in the process of tissue oxygenation. Pendelton et al.(1972) and Oski et al.(1979) observed that when propranolol, a blocking agent for the beta-adrenergic receptor, was added to a suspension of intact human erythrocytes, there was a rightward shift of the hemoglobin-oxygen dissociation curve. Schrumpf et al.(1977) found a decrease of hemoglobin-oxygen affinity during propranolol administration to patients with coronary artery disease . Since propranolol may cause bronchospasm, one must be careful in its administration to patients with COPD. In chronic respiratory failure with COPD, the greater the hypoxemia , the more 2,3-DPG increases . This rise of 2,3-DPG increases the P_{50} and leads to a compensatory improvement oxygen delivery to the peripheral tissues . However, in mild hypoxemia there is no evidence of increased 2,3-DPG or P_{50}. It is therefore useful to reduce hemoglobin-oxygen affinity by pharmacological agents in mild hypoxemic patients for whom oxygen therapy is not available .

Pentoxifylline is one of the xanthine derivatives known to improve capillary circulation by increasing the ability of erythrocyte deformability, besides which has the effect of raising intracellular ATP(Ohshima and Sato, 1981; Stefanovich,1975). Soliman et al.(1987) reported that pentoxifylline improved tissue oxygenation after anesthesia and during an operation in rats.

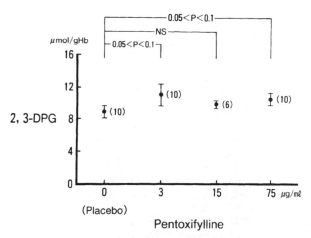

Fig.2. Effects of pentoxifylline on 2,3-DPG in vitro. Mean±SEM. Parentheses show numbers of experiment.

Table 1. Effect of coenzyme Q_{10} in vitro (n=7). Values show mean±SEM.

	Pre-incubation	After 4 hours incubation			
		Placebo	1µg/ml	20µg/ml	100µg/ml
P_{50} (Torr)	28.5±0.9	29.3±1.2	29.5±1.3	29.6±1.4	30.1±1.3
2,3-DPG (µmol/gHb)	9.23±1.06	9.78±1.53	10.09±1.26	10.61±1.67	9.83±2.02
Hb (g/dl)	14.1±1.0	14.0±1.0	13.9±1.0	13.9±0.9	14.0±1.0

Table 2. Effect of calcium hopantenate in vitro(n=7). Values show mean± SEM.

	Pre-incubation	After 4 hours incubation			
		Placebo	0.01mg/dl	0.1mg/dl	1mg/dl
P_{50} (Torr)	27.5±1.0	28.0±1.1	28.4±1.0	28.6±1.0	28.8±1.1
2,3-DPG (µmol/gHb)	11.95±0.42	11.40±1.17	12.20±0.80	12.36±1.04	11.77±1.49
Hb (g/dl)	13.5±1.2	13.5±1.3	13.5±1.2	13.5±1.2	13.4±1.2
ATP(mg/dl)	450±23	459±16	451±32	433±30	445±41
Glucose (mg/dl)	108±12	61±16	58±16	63±20	62±17

In this experiment pentoxifylline has also been shown to reduce hemoglobin-oxygen affinity . It is suggested because of the slight increase in 2,3-DPG that the mechanism for this is anaerobic glycolysis in erythrocytes. Other mechanisms that may be responsible for the increase of P_{50} could be related to changes in Hb, ATP and pH, but Hb and ATP were not significantly changed. Although pH decreased from 7.38 in the preincubation state to 7.30 after 4 hours incubation, there were no differences between placebo and pentoxifylline treated incubates in vitro studies. The in vivo study in patients with COPD has also disclosed a decrease of hemoglobin-oxygen affinity after 4 weeks administration of pentoxifylline. The amount of pentoxifylline, 300mg per day oral administration, is assumed to give about 3µg/ml blood level, a concentration equal to that in vitro. 2,3-DPG has been shown to increase slightly in vivo. As arterial blood gases did not change from pretreatment to after 4 weeks of treatment, the shift of hemoglobin-oxygen affinity may be explained by changes in anaerobic erythrocytic glycolysis.

Neither CoQ_{10} nor calcium hopantenate produced any changes of

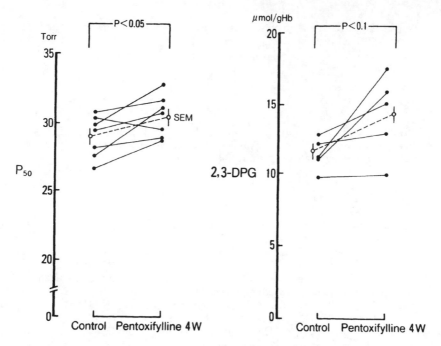

Fig.3. Effects of pentoxifylline on P_{50} (left panel) and 2,3-DPG (right panel) in patients with chronic obstructive pulmonary disease. Open circles show mean values.

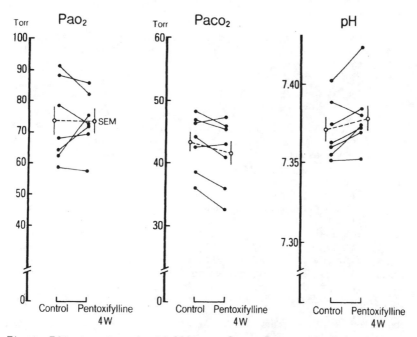

Fig.4. Effects of pentoxifylline on Pao_2, $Paco_2$ and pH in patients with chronic obstructive pulmonary disease. Open circles show mean values.

hemoglobin-oxygen affinity or 2,3-DPG, although it has been suggested that they affect glucose metabolism in the ischemic state. This indicates that 2,3-DPG is an important factor in changing hemoglobin-oxygen affinity.

CONCLUSION

These results suggest that pentoxifylline reduces hemoglobin-oxygen affinity by increasing 2,3-DPG and improves oxygen delivery to the tissues in vitro and patients with COPD.

SUMMARY

We examined the effects of pentoxifylline, coenzyme Q_{10} (CoQ_{10}) and calcium hopantenate on hemoglobin-oxygen affinity in vivo and in vitro. In vitro study, pentoxifylline at 3 and 15µg/ml significantly increased P_{50} from the placebo(27.0 ± 0.3 Torr, mean$\pm$SE) to 28.2 ± 1.7 and 28.1 ± 1.3Torr respectively(p< 0.05) after 4 hours incubation. With 75µg/ml pentoxifylline, P_{50} increased slightly to 27.8 ± 1.4 Torr ($0.05 < p < 0.1$). 2,3-DPG was slightly higher ($0.05 < p < 0.1$) with pentoxifylline of 3 (11.15 ± 4.12 µmol/gHb) and 75ug/ml (10.62 ± 2.32 µmol/gHb) than placebo. Other agents, 1-100µg/ml of CoQ_{10} and 0.01-1mg/ml of calcium hopantenate, neither P_{50} nor 2,3-DPG, Hb,ATP and glucose were changed compared to placebo. In vivo study, after administration of pentoxifylline 300mg per day orally for 4 weeks to seven patients with COPD, P_{50} increased significantly from 29.0 ± 0.6 to 30.4 ± 0.6 Torr (p<0.05) and 2,3-DPG increased slightly from 11.67 ± 0.56 to 14.33 ± 1.12µmol/gHb(p<0.1). These results suggest that pentoxifylline reduces hemoglobin-oxygen affinity by increasing 2,3-DPG and improves oxygen delivery to the tissues in patients with COPD.

This work was supported in part by the Ministry of Health and Welfare Grant-in-Aid for Special Disease Survey and Research Project in Respiratory Failure.

REFERENCES

Kawakami, Y., Kishi, F., Yamamoto, H., Miyamoto, K., 1983, Relation of oxygen delivery, mixed venous oxygenation, and pulmonary hemodynamics to prognosis in chronic obstructive pulmonary disease. N Engl J Med., 308:1045.

Mochizuki, S., Feuvray, D., Ishikawa, S., Saso, F., Yoshiwara, T., Ozasa, H., Shimada, T., Saito, N., Abe, M and Neeley, J.R., 1980,Energy metabolism and funktion during ischemia and with reperfusion in rat hearts: Effects of substrates, hormone and coenzyme Q_{10}, in : "Biomedical and Clinical Aspects of Coenzyme Q, Vol 2, " p.377, Yamamura,Y., Folkers, K and Ito, Y.,ed., Elsevier/North-Holland Biomedical Press.

Oski, F.A., Millar, L.D., Delivoria-Papadopoulous, M., Manchester, J.H. and Schelburne, J.C., 1972, Oxygen affinity in red cells, Changes induced in vivo by propranolol, Science, 175:1372.

Pendelton, P.G., Newman, D.J., Sherman, S.S., Bran,E.G. and Maya, W.E., 1972, Effect of propranolol upon the hemoglobin-oxygen dissociation curve, J Pharmacol Exp Ther., 180:647.

Schrumpf, J.D., Sheps, D.S., Wolfson, S., Aronson, A.L., Cohen, L.S., 1977, Altered hemoglobin-oxygen affinity with long-

term proranolol therapy in patients with coronary artery
disease,<u>Am J Cardiol.</u>, 40:76.

Soliman, M.H., O'Neal, K. and Waxman,K., 1987, Pentoxifylline
improves tissue oxygenation following anesthesia and
operation, <u>Crit Care Med.</u>, 15:93.

Stefanovich, V., 1975, Beeinflussung des ATP-Gehaltes der
Erythrozyten durch Pentoxifyllin, <u>Med Welt.</u>, 26:1882.

Tamura,J., Hashimoto, K., and Kuriyama, K., 1986. Effect of calcium
hopantenate on energy metabolism in the experimental
ischemic brain (in Japanese). <u>Folia pharmacol japon.</u>,
88:26p.

CENTRAL NERVOUS
SYSTEM

MEASUREMENT OF RAT BRAIN SPECTRA

B. Hagihara[*], N. Okutani[*], M. Nishioka[*], N. Negayama[**]
W. Ohtani[**], S. Takamura[**], and K. Oka[**]

[*]The Hagihara Medical Technology Reserch Institute
72 Shimoyamaguchi, Yamaguchi-cho, Nishinomiya, Hyogo, Japan
[**]Otsuka Electronics Co., Ltd.
26-3 3chome Shoudaitajika, Hirakata, Osaka, Japan

INTRODUCTION

The proper functioning of the human brain is strongly dependent
on a sufficient supply of blood oxygen. Thus the monitoring of cerebral
blood oxygen levls is of clinical importance in a wide variety of
medical fields such as anaesthesiology, brain surgery, pediatrics and
haemotology. But the direct monitoring of cerebral blood oxygen levels
has been difficult to achieve using conventional means. For example,
although PO_2 analysis of blood sampled from the carotid artery is
often practiced, the values obtained do not necessarily represent the
PO_2 of the brain itself.

The application of absorption and reflection spectrophotometry
to living tissues and organisms have resulted in oximeters and other
apparatus for continuous, non-invasive monitoring of blood oxygen
levels(Hamaguri, 1987; Emson et al, 1962). But most oximeters are
applicable only to parts of the body where short light paths are
available, such as the fingertip and the ear lobe(Polanyi and Ostrowski,
1971), and are difficult to apply to the brain where direct optical
monitoring of blood oxygen is hampered by the skull and surrounding
tissues.

Jobsis(Jobsis, 1977) used the relatively good transmission of
light in the near infrared range(700-1300nm) through deep tissues to
develop methods for probing inner organs such as the brain. This wavelength range is suited for monitoring blood oxygen levels since it
includes the absorption peaks for deoxyhaemoglobin(768nm) and oxidized
cytochrome aa_c(840nm) as well as the isobestic point for deoxy- and
oxy-haemoglobin(805nm).

In many of the previous works using near infrared spectrophotometry to probe the brain, the optical paths were taken between opposite
locations on the surface of the brain. Since these paths include the
surface tissue, the transmitted or reflected light contain information
not only of the brain itself but also of the surrounding tissue as well.
In addition, the skull scatters as extremely large amount of light,
lowering the S/N ratios for optical measurements. Thus much of the
previous work was limited to two- or three-wavelengths methods which do

not give accurate information when there are changes in the blood flow
and movement of probe attachments.

A method of monitoring blood oxygen levels of the brain optically
by attaching optical probes to the orifices of both ears and using an
"ear-to-ear" path to measure time courses at two wavelengths and whole
absorption spectra from 600 to 1100nm of the brain stem is described
here. By using this optical path, not only is the distance between the
light probes shortened, but it is also possible to avoid thick sections
of the skull and overlying tissues. Since the brain stem is essential
to the function of the rest of the brain, it is an appropriate "target"
for monitoring of oxygen saturation levels. Furthermore, since the
light probes are inserted in the ear orifices, greater stability of
measurement is attained without the use of firm pressure as is required
when the probes are attached to the surface of the head. These
advantages lead to a considerable improvement in the S/N ratio and
blood oxygen levels of the brain, unperturbed by pressure applied by the
probes, may be determined from spectra.

EXPERIMENTAL

A diagram of the system used in making optical measurements of the
rat brain in-vivo is shown in Fig. 1. The rat, anaesthesized with
Nembutal(50mg per kg body weight), was kept in a small closed box in
which the internal gas content was adjusted by sending N_2 and O_2
(both containing 5% CO_2) in controlled ratios. The PO_2 of the
inspired gas was monitored by means of a temperature controlled oxygen
electrode connected to a recorder. The rat head was firmly held at
three points, namely, the orifices of both ears and the incisors.

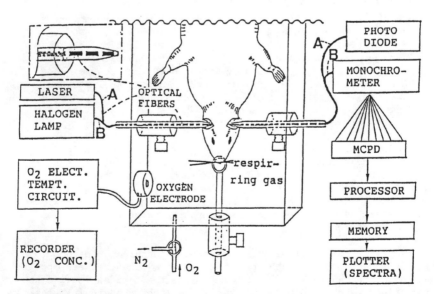

Fig. 1. System used for making optical measurements of the rat brain.
A. Arrangement for measurements of time courses.(laser as light source)
B. Arrangement for measurements of absorption spectra.(white light)

Incident light from a light source was irradiated by means of an optical fiber bundle attached to the middle ear(ie. inside the ear orifice) on one side of the head(see enlarged figure at the upper left corner). The light mainly passes through the pons, and partially through the mesencephalon, the diencephalon, the cerebellum and the medulla oblongata. The transmitted light was received and sent to a detection system by means of another optical fiber bundle attached to the middle ear on the opposite side of the head.

For monitoring time courses, alternative pulses from two semiconductor lasers of different wavelengths(805 and 870nm) were irradiated on the rat. A photodiode was used as the detector. The resulting signals were separated and then sent to a chart recorder. In the case of spectral measurements, light from a white-light source (halogen lamp) was used and a photodiode array spectrophotometer(MCPD-110, Otsuka Electronics Co., Ltd.) was used as the detection system. Scanning of wavelengths is eliminated by the combination of grating and a photodiode array in this spectrophotometer and spectra with wide wavelength ranges are instantaneously recorded in the memory.

Plasma clearance experiments using an indicator dye, indocyanine green(ICG), were performed to consider the stability of measurements. ICG was dissolved in distilled water and injected via the vein(tail vein in the case of the rat) in 50mg per kg body weight doses. In a second set of experiments, the PO_2 inside the box was varied and oxygen saturation levels of the brain stem were determined from the in-vivo spectra.

RESULTS

Stability of Measurements. The time couse of transmitted light intensity of the rat brain as measured at the ear-to-ear location during an ICG experiment is shown in Fig. 2. At both wavelengths, the intensities drop sharply as ICG is introduced and gradually return to the initial levels as the ICG is removed by the liver. Although the main figure traces seem to be broadened by noise, magnifications of a portion of the course prior to ICG injection show that this is due to the heart pulse(fast cycle; 4.8Hz) and the respiration rate(slow cycle; 0.6Hz).

Difference spectra measured at various times after ICG injection for three different cases(human fingertip, rabbit ear stem, rat ear-to-ear) are compared in Fig. 3. The reference spectrum in each is the absorption spectrum measured prior to ICG injection. The variation of the absorption values at 880nm normalized by the highest absorbance value for each case is 26% for the rabbit ear stem, 11% for the human fingertip and less than 5% for the rat ear-to-ear measurements. Since there is practically no absorption by ICG at 880nm, variations of absorbance values at this wavelength are due to the movement of the probes and changes in the blood flow due to changes in the pressure applied by the probes. These and the above results show that high stability is provided by the ear-to-ear method for optical measurements.

Determination of oxygen saturation levels. Fig. 4. shows spectra of rabbit erythrocyte in an oxygen reaction cell(Okutani et al,this volume) measured at various oxygen concentrations. The oxygen concentrations shown were monitored by the electrode incorporated in the reaction cell. Since the peak at 768nm is due to deoxy-haemoglobin only, the peak height at this wavelength is proportional to the ratio of

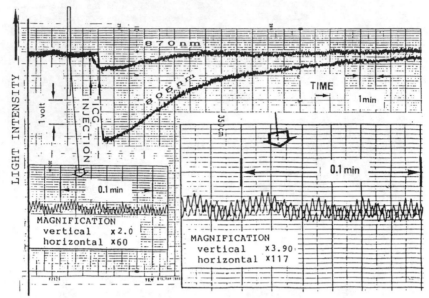

Fig. 2. Time courses of transmitted light intensity from the rat brain during an ICG experiment. ICG(1mg per kg body wheight) was injected from the tail vein at the time indicated by the two arrows (shift in position is due to the offset of the pen recorder).

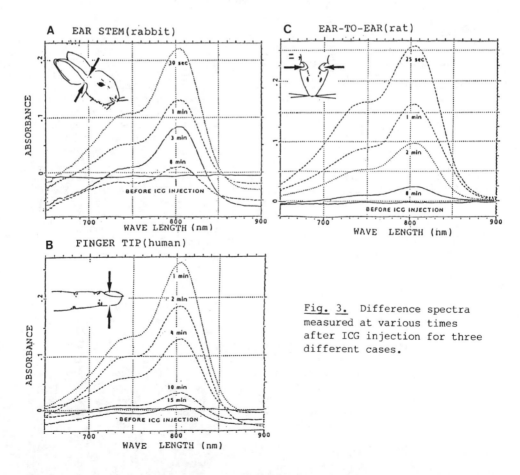

Fig. 3. Difference spectra measured at various times after ICG injection for three different cases.

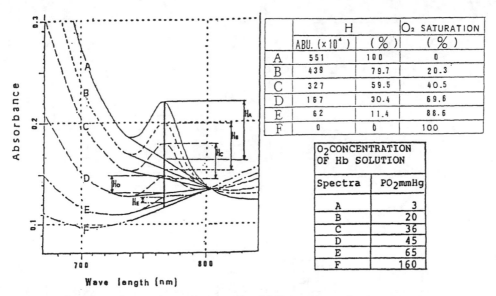

	H		O_2 SATURATION
	ABU. ($\times 10^4$)	(%)	(%)
A	551	100	0
B	439	79.7	20.3
C	327	59.5	40.5
D	167	30.4	69.6
E	62	11.4	88.6
F	0	0	100

O_2CONCENTRATION OF Hb SOLUTION	
Spectra	PO$_2$mmHg
A	3
B	20
C	36
D	45
E	65
F	160

Fig. 4. Absorption spectra of rabbit erythrocyte in an oxygen reaction vessle at various oxygen concentrations and table of oxygen saturations calculated from the spectra.

the amount of deoxy-haemoglobin to the total amount of haemoglobin present in the sample:

peak height(768nm) α deoxy Hb/(oxy Hb + deoxy Hb)

At 0% oxygen saturatin, the amount of deoxy-haemoglobin is equal to the total amount of haemoglobin. Thus if the peak height at 768nm at 0% oxygen saturation is H_A, then the oxygen saturation when the peak height is H(provided that there are no changes in the blood flow), is given by:

O_2 saturation(%) = oxy Hb/(oxy Hb + deoxy Hb) = $100 - H/H_A \times 100$

A conventional method of determining the peak height is to substract the absorbance value at the isobestic point(805nm) from the value at 768nm. But since the absorption spectra of blood in living tissues contain contributions from light scattering and absorption by tissue pigments, peak heights and thus oxygen saturations determined as above will be different for subjects of different skin coloration and optical path lengths even if the actual oxygen saturations are the same. In order to determine peak heights less dependent on these effects, a line tangent to the trough of the blood spectra near 744nm from the isobestic point was determined and the distance from this line to the point in the spectra at 768nm was taken as the peak height.

A suitable reference may be chosen in order to further lessen the effects of the extraneous contributions on the determination of the peak height. Fig. 5. shows spectra of the rat brain taken using the reflection spectrum of a brown ceramic plate waith a violet-tint as the reference. The spectrum used for determing 0% oxygen saturation was measured when the rat was asphyxiated by sending pure N_2 in the closed box and waiting for 10 seconds after the PO$_2$ in the box reached OmmHg. With this reference, the in-vivo spectra have nearly the same shape as the in-vitro spectra, showing that the effects of the extraneous

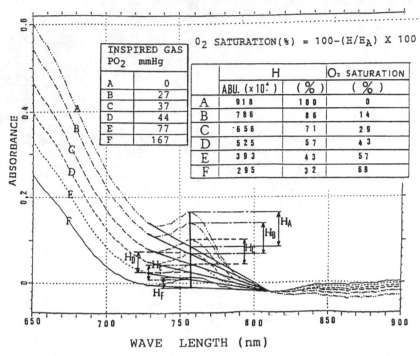

Fig. 5. Absorption spectra of rat brain in-vivo taken when gases of various oxygen concentrations were inspired.

contributions have been lessened and that thus the oxygen saturation may be determined more accurately.

DISCUSSION

The above results indicate that the method of using the "ear-to-ear" path is quite effective for optical measurements as it provides high stability and S/N ratios. The use of a photodiode array spectro-photometer makes possible the measurment of spectra in short time intervals and detailed information on changes in the brain in-vivo may be gained. Unlike two- or three-wavelengths methods, measurement of whole spectra will clearly reveal changes in blood flow, probe movements and differences in absorbance values due to light scat-tering, absorption by tissue pigments and use of different references, thus making it possible to reduce the effects these may have on the determination of oxygen saturation.

The method of graphycally determining peak heights and using suitable references are examples of possible means for determining oxygen saturations more accurately from whole spectra. The selection of a suitable reference could be facilitated with more information on the scattering and absorption properties of tissues. Although changes in the blood flow were not considered here, corrections for this effect should be possible with the knowledge of relative extinction coeffecients of oxy-haemoglobin and deoxy-haemoglobin at given wave-lengths and oxygen saturation. These topics and applications to humans and other subjects are presently being investgated.

SUMMARY

A method of optically monitoring oxygen saturation of blood
haemoglobin in the rat brain stem was developed. Optical fiber bundles
were attached to the orifices of both ears of the rat to irradiate
incident light from one ear and receive transmitted light from the
other ear. Absorption spectra were measured using a white-light source
and a photodiode array spectrophotometer. Stable measurements of
optical time cources and absorption spectra were made using this "ear-
to-ear" path. Oxygen saturation levels were calculated from spectra by
using a suitable reference and an improved method of determining
spectral peak heights.

REFERENCES

Emson, Y., Briscoe, W. A., Polanyi, M. L., and Cournand, A., 1962, In
    vivo studies with an intravascular and intracardiac reflection,
    J. appl. Physiol., 17:552.
Hamaguri, K., 1987, Oximeter, Japan Patent 62-16646 and Non invasive
    oximeter, Japan Patent 62-16647.
Jobsis, F. F., 1977, Non invasive infrared monitoring of cerebral and
    myocardial oxygen sufficiency and circulatory parameters, Science,
    198:1264.
Okutani, N., Hagihara, B., McCabe, M., Ohtani, W., Negayama, N.,
    Nishioka, M., Nakai, T., and Morita, T., Oxygen reaction vessels,
    in this volume.
Polanyi, M. L., and Ostrowski, D. S., 1971, Blood oxygenation and pulse
    rate monitoring appratus, US Patent 3,628,525.

THE SIMULTANEOUS MEASUREMENTS OF TISSUE OXYGEN CONCENTRATION AND ENERGY

STATE BY NEAR-INFRARED AND NUCLEAR MAGNETIC RESONANCE SPECTROSCOPY

M. Tamura*, O. Hazeki*, S. Nioka, B. Chance, and D.S. Smith**

Department of Biochemistry and Biophysics and ** Department
of Anesthesia, School of Medicine, Univ. of Pensylvania
Philadelphia, and * Biophysics Division, Research Institute
of Applied Electricity, Hokkaido Univ., Sapporo

Introduction

There is no doubt that tissue oxygen concentration is the critical
determinant in the energy metabolism of living tissues. The concept of
critical oxygen concentration, which was introduced to emphasize this,
has been defined as the oxygen concentration at which the respiration
rate of the tissue starts to obey first order kinetics rather than
zero-order with respect to oxygen. The critical oxygen concentration
values reported for isolated mitochondria and tissue show
discrepancies. The tissue oxygen gradient can explain the variation
between the two.

During this decade, tissue energy state has been monitored noninvasively by phosphorous nuclear magnetic resonance (P-NMR)
spectroscopy, where the ratio of phosphocreatine to creatine has been
successfully used as the energy indicator, as well as the ATP:ADP.
Tissue oxygen concentration has been also monitored by several optical
method, where hemoglobin, myoglobin and cytochromes have been used as
the optically active oxygen indicator. Recent advances in near-
infrared photometry have extended these techniques to normal
circulatory conditions in situ.

This paper describes the first attempts to apply those two non-
invasive techniques in situ and to correlate the brain energy state and
its oxygen concentration under various conditions, to yield an energy-
oxygen diagram.

Materials and Methods

Isoflurane anesthetized dogs and cats were fully instrumented with
respiratory and electrophysiological non-interfering transducers.
Changes in the tissue oxygen concentration were achieved by changing
the oxygen concentration in the inspired gas (FiO_2).

^{31}P-NMR spectra were measured by a 12 inch bore 2T Oxford magnet
and a Phospho-energetics 280. The degree of hemoglobin oxygenation in
the brain was continuously monitored by near-infrared photometry

together with the redox change of cytochrome aa₃ copper. A multi-channel time-sharing spectrophotometer, fitted with the 4 m of flexible light guide, was used for the measurement. Simultaneous measurements by NMR and transmission optics were done by setting the end of the light guide on the surface of the exposed skull bone very close to the surface coil of NMR.

Results

Fig. 1. summarizes the changes in the ratio of phosphocreatine to inorganic phosphate, PCr/Pi, and the absorbance change at 780-830 nm in relation to the degree of hemoglobin oxygenation, 700-730 nm, in the dog brain. The PCr/Pi (3,1) started to fall at 90% deoxygenation of hemoglobin, achived by respiration with 10% oxygen in the inspired gas. PCr/Pi remained constant above 10% hemoglobin oxygenation. The absorbance change at 780-830 nm gave two phases. The straight line portion (above 10% oxygenation of hemoglobin) was the absorbance change due to hemoglobin oxygenation. The marked deviation at the ordinate (below 10% oxygenation of hemoglobin) was due to the reduction of the copper in cytochrome aa₃. The dashed line in Fig. 1 (45° straight line) shows the absorbance at 780-830nm, due to the hemoglobin oxygenation only.

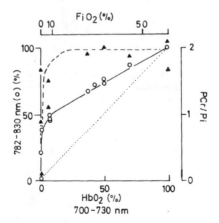

Fig. 1. The relationship between hemoglobin oxygenation (700-730 nm) PCr/Pi, and absorbance change at 780-830 nm in the blood circulated dog brain. (▲) PCr/Pi. (o). Absorbance at 780-830 nm.

Fig. 2. shows the relationship between PCr/Pi and hemoglobin oxygenation of cat brains with normal blood-circulation and with fluorocarbon substitution. In Fig. 2-A (blood circulated) PCr/Pi remained constant at above 8% oxygenation of hemoglobin, and then started to fall sharply at lower oxygenation. The absorbance at 780-830 nm deviated markedly from the straight line below 8% hemoglobin oxygenation. The fluorocarbon substituted cat (Fig. 2-B) gave a similar relationship between PCr/Pi and hemoglobin oxygenation. The deviation in the 780-830 nm absorbance at the ordinate was much more pronounced in the fluorocarbon substituted cat than in the normal one.

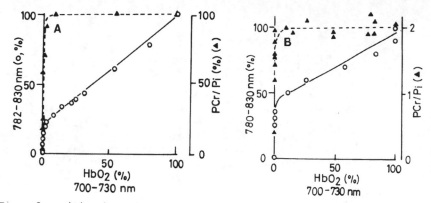

Fig. 2 . (A). The relationship between hemoglobin oxygenation, PCr/Pi and absorbance change at 780-830 nm in the blood circulation, at brain. (B). Fluorocarbon-substituted cat brain.

Fig. 3. shows the profile of hemoglobin oxygenation and tissue energy state, PCr/Pi, in the normal dog brain, where oxygenation level of hemoglobin was taken as 100% with respiration of pure oxygen. PCr/Pi started to fall at 10% Fi O_2. About 90% of hemoglobin was deoxygenated under the condition. Our near-infrared measurement picks up mainly the oxygenation state of venous blood, as shown in Fig. 4. The superior sagittal sinus was cannulated and pO_2 of the venous blood was measured by the gas analyzer. The oxygenation curve observed in the dog brain in situ is very close to that of blood.

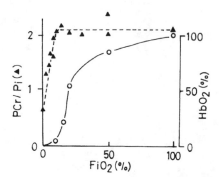

Fig. 3. The relationship between FiO_2, hemoglobin oxygenation and PCr/Pi of blood circulated dog brain.

Discussion

The critical hemoglobin oxygenation was 8 and 10% for cat and dog brains at approximately 10% FiO_2, where PCr/Pi started to fall. Critical FiO_2, however, was higher in fluorocarbon substituted cats than in controls. Our fluorocarbon substituted cats had a hematocrit of less than 5%, but the relationship between PCr/Pi and hemoglobin oxygenation was unchanged. This suggests that the oxygen gradient between the blood vessels and the interior of the cell seems independent of blood volume. Since our optical measurement picks up mainly the venous oxygen concentration (Fig. 4), $PvO_2 \sim 10$mmHg was critical for the brain.

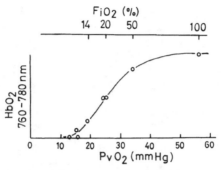

Fig. 4. The relationship between hemoglobin oxygenation and PvO_2 at sagittal sinus. Blood circulated dog brain.

The deviation in the absorbance change at 780-830 nm was due to the overlap of the absorbance change of copper in cytochrome aa_3 with that of hemoglobin. This was verified by the results of the fluorocarbon substitution (Fig. 2-B). Thus, the fall of PCr/Pi paralleled the reduction of cytochrome aa_3. The energy-oxygen diagram presented here can be applied more widely to various tissues and to various conditions such as the impairment of the circulatory system.

Summary

The oxygenation and energy states of brain tissues were measured simultaneously by near-infrared photometry and nuclear magnetic resonance spectroscopy in situ. In both cat and dog, the critical hemoglobin oxygenation was 10%, below which the ratio of phosphocreatine (PCr) to inorganic phosphate (Pi) started to fall. The fall of PCr/Pi paralleled the reduction of copper in cytochrome aa_3. The separation of the cytochrome aa_3 signal from that of hemoglobin by our optical method was confirmed by the substitution of blood by fluorocarbon solution. The energy-oxygen diagram (PCr/Pi against hemoglobin oxygenation, HbO_2) was the same in normal- and fluorocarbon substituted cats, but energy curve shifted to the right in the latter when PCr/Pi plotted against the inspired oxygen, FiO_2.

FIBER OPTIC SURFACE FLUOROMETRY/REFLECTOMETRY AND 31-P-NMR

FOR MONITORING THE INTRACELLULAR ENERGY STATE IN VIVO

A. Mayevsky* S. Nioka and B. Chance

*Life Sciences Dept., Bar-Ilan University, Ramat-Gan 52100
Israel and Biochemistry/Biophysics Dept.
University of Pennsylvania, Phila., PA 19104

INTRODUCTION

In vivo monitoring of intracellular events was and is a great challenge to many investigators. One of the main approaches used for many years in various animal organs is the optical monitoring of the redox state of members of the mitochondrial respiratory chain in vivo. In the past few years, phosphorus-31 nuclear magnetic resonance (^{31}P-NMR) spectroscopy has been developed as a practical tool for monitoring organ energy metabolism in vivo including brain, heart or muscle.

The purpose of the present article is to describe in detail the basic principles of the fiber optic surface fluorometry technique and to describe the combination between this optical technique and the ^{31}P-NMR spectroscopy approach.

Various members of the respiratory chain exhibit different optical properties in the reduced and oxidized forms, thus enabling the non-invasive monitoring of various organs in vitro as well as in vivo. Chance and Williams (1955) defined various metabolic stages of the mitochondria in vitro which depend upon the availability of oxygen, substrate and phosphate acceptor (ADP).

Using the autofluorescence of the mitochondrial NADH, the excitation and emission spectra of NADH in a suspension of rat liver mitochondria were recorded. The intensity of the emission spectrum depends upon the metabolic state of the mitochondria. The maximum intensity recorded in the resting state, state 4, decreased during the transition to the active state, state 3, in which the electron transport chain was the limiting factor. No significant increase in NADH was recorded by exposing the state 4 mitochondria to anoxic conditions (state 5). Other members of the respiratory chain, such as flavoproteins, cytochromes b and c, also provide good signals for state 4-3-4 transition.

The first observation showing a correlation between NADH fluorescence and physiological activity in vivo was described by Chance and Jöbsis 959). They observed the changes in fluorescence in frog sartorius muscle following a twitch. The decrease in fluorescence was interpreted as a state 4 - state 3 transition.

Chance and co-workers (1962) described for the first time the measurement of NADH in an anesthetized exposed rat brain cortex, using surface fluorometry. They studied the effect of anoxia on the oxidation-reduction state of NADH and correlated it to the electrical activity of the cortex. The principle of the method is that the exposed brain is illuminated by 366 nm to excite the NADH, and the emitted fluorescence light is measured in the range of 430-480 nm. The same approach was later applied to the in vitro system as well as to the brain (Jobsis, 1972).

The measurement of NADH fluorescence is affected by different factors such as absorption, reflectance and scattering. These factors can affect the excited as well as the emitted light from the NADH. Oxydeoxy hemoglobin transition has a very small effect on the NADH-emitted light (Chance et al., 1973). However, by using a blood-perfused tissue such as the brain, hemodynamic alternation or swelling of the tissue may occur under different conditions, such as anoxia. In order to compensate and correct the NADH fluorescence measurement, Aubert, Chance and keynes (1964) recorded a reflectance signal together with the fluorescence emission excited at the same wavelength from slices of the electric organ of the electric fish. Later on, Jöbsis et al. (1971) used the 366 nm wavelength to measure reflectance as well as fluorescence from the surface of exposed cat brain. Kobayashi et al. (1971) used 620 nm reflectance measurement in order to correct for NADH fluorescence signal. Harbig and Reivich (1976) employed combined reflectance at 366 nm and 450 nm fluorescence, using the Ultrapak apparatus. The ratio of correction used by Jöbsis (1971) and Harbig (1976) was in the range of 0.7-1.5:1.

In 1972, we introduced optical fibers into the surface fluorometry/reflectometry replacing the old "rigid" optical system used until then and, as a result, measurement of NADH simultaneously with reflectance signal at 366 nm from the brain of an <u>awake</u> animal became possible (Chance et al., 1973; 1974; Mayevsky and Chance, 1973).

During the last decade, this technique has been developed, improved and applied to many experimental setups in brain research, as well as in other organs in vivo (Kedem et al., 1981; Mayevsky and Chance, 1982).

During the past few years, phosphorus-31 nuclear magnetic resonance (^{31}NMR) spectroscopy has been developed to a practical tool for in vivo noninvasively monitoring of brain energy metabolism (Chance et al., 1978; Gordian and Radda, 1981). Very recently, the dog brain was used as a model for studying the ^{31}P-NMR responses to and recovery from hypoxia (Hilberman et al., 1984). In order to interpret the results obtained in vivo by the ^{31}P-NMR technique, it is desirable to compare it to another in vivo monitoring system since results obtained by standard biochemical assay techniques are not directly comparable. The development and usage of the fiber optic surface fluorometry/reflectometry for monitoring brain NADH redox state have been reviewed (Mayevsky, 1984). This approach opened up the possibility to monitor brain NADH while the animal was located inside the NMR magnet.

METHODS

Fluorometer/Reflectometer

The basic principles of the fluorometers/reflectometers developed have been described previously (Chance et al., 1975; Mayevsky, 1978) and very recently we described the new multichannel model (Mayevsky and Chance, 1982).

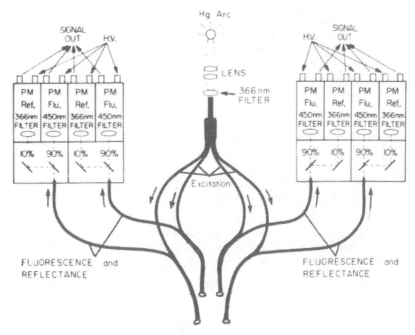

Fig. 1. Four-channel DC fluorometer/reflectometer for studying the NADH redox state simultaneously from 4 different organs or 4 separate spots in the same organ, such as the brain.

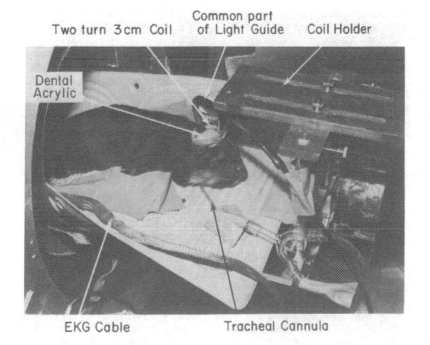

Fig. 2. Combination between the light guide and the radiofrequency (RF) coil connected to the puppy brain in the ^{31}P-NMR magnet.

The light source in all fluorometers used was a 100W water or air-cooled mercury arc having a strong 366 nm filter (Corning 5860, 7-37 plus 9782, 4-96). The 4-channel fluorometer is shown schematically in Fig. 1. Using the experimental set-up described previously (Mayevsky, 1978) we obtained scans of the emission spectra from rat brains using fiber optic attachment to the fluorometer, as previously described by Chance et al. (1962). Furthermore, we compared two sources of excitation light, namely, the usual mercury arc and the laser (Linonix Model 400 PS) which excites at 324 nm. The results indicate that the location of the peak of the emission spectrum is identical in normoxia and anoxia and is close when the laser is used. We also used the laser as a source in a few in vivo studies and found that the intensity of the light did not harm the brain during several hours of measurements. The intensity of energy on the tissue after passing through the filters and the light guide was 0.03 mW. For excitation of NADH, a 366 nm light is used and therefore only UV transmitting fibers are usable to build the light guides. Fibers from Schott (Mainz, W. Germany) and from the American Optical Corporation (Mass., USA) were used. Very recently, new UV fibers have appeared (Math Assoc. Inc.) having a core diameter of 200-300u and preliminary experiments show qualitatively similar results.

The usual arrangement of the excitation and emission bundle of fibers are a Y-shaped light guide used in the time sharing, as well as most DC fluorometer/reflectometers.

In the Y-shaped light guide where the light is split at the entrance to the photomultipliers measuring the fluorescence and reflectance. We have arbitrarily chosen a 90:10 mirror providing adequate fluorescence and reflectance signals, respectively.

^{31}P-NMR spectroscopy

In the present study, we correlated the ^{31}P-NMR spectra to the NADH redox state monitored from the brain of puppies located inside the magnet and exposed to hypoxic conditions. The studies were done using a phosphoenergetics 260-80 NMR Spectrometer (Phosphoenergetics, Inc., Phila, PA, USA) with a 31-cm bore, horizontal Oxford superconducting magnet (Oxford Instruments, Oxford, UK). Details regarding the spectrometer, animal holder and probe design were very similar to those published recently (Hilberman, 1984). Intramitochondrial NADH redox state was measured by a DC surface fluorometer/reflectometer (Mayevsky and Chance, 1982) connected to the brain by a 3-meter flexible UV light guide. The puppy was anesthetized and a 9-mm hole was drilled in the parietal bone. The dura mater was removed gently and a plexiglass holder with the light guide was implanted above the brain and fastened to the skull with dental acrylic cement. The puppy was ventilated by a Harvard rodent respirator via a tracheal cannula. The coil (2-turn 3-cm diameter) was placed on the skull around the cemented light guide and fixed in place by the coil holder. Fig. 2 shows the technical details of the puppy connected to the monitoring system.

Animal Operation

Rats and gerbils were anesthetized by IP injection (0.3ml/100 gr) of Equithesin [each ml contains: pentobarbital (9.72 mg); chloral hydrate (42.51 mg); magnesium sulfate (21.25 mg); propylene glycol (44.34% w/v); alcohol (11.5%) and water]. The animal was placed in a head holder a midline incision of the skin was made. An appropriate hole (2 mm in diameter) was drilled in the parietal bone and the cannula (light guide holder) was placed epidurally. Four stainless steel screws were used as

ECoG electrodes or for holding the cannula to the skull by dental acrylic cement.

In most of our studies, the dura mater remained intact and the optical signals measured showed the same responses are compared with brains in which the dura was removed. In rats, gerbils and guinea pigs, the dura mater remained intact, but in chickens, rabbits or puppies, the dura is so thick that measurement of NADH was possible only after its removal. The 4-channel fluorometer was used in the rat and gerbil model. In the rat, we monitored NADH in brain, liver, kidney and testis simultaneously. In the gerbil, we monitored four points on the same brain hemisphere.

Calibration procedure

In order to decrease the variation between animals, a standard procedure for signal calibration was applied. The reflectance and fluorescence signals obtained from the photomultipliers (RCA 931B) were calibrated to a standard signal (0.5V), as recently described in detail (Mayevsky and Chance, 1982), by variation of photomultiplier dynode voltage obtained from the high voltage power supply. The standard signal (0.5V), used to calibrate the recorder, was set to give a half-scale deflection on the recorder (2.5 cm) with the pen resting at midscale. The gain was increased as required by a factor of 2 or 4 to give 50% or 25% of the full scale correspondingly. The changes in the fluorescence and reflectance signals were calculated relatively to the calibrated signals under normoxic conditions. This type of calibration is not absolute, but provides reliable and reproducible results from various animals and also between various laboratories using this approach.

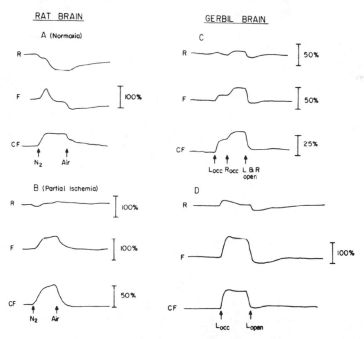

Fig. 3. Brain metabolic and hemodynamic responses to hypoxia in the rat (A,B,) and to ischemia in the gerbil brain (C,D). R - reflectance; F - fluorescence; CF - corrected fluorescence; L_{occ} - left carotid artery occlusion; R_{occ} - right carotid artery occlusion.

RESULTS AND DISCUSSION

Figure 3 shows typical responses of the rat brain to anoxia (A,B) and a gerbil brain to ischemia (C,D). Under anoxia (A), oxygen availability decreased and NADH was increased (CF). The reflectance trace shows the typical autoregulation response to the lack of oxygen, namely, an increase in blood volume led to a decrease in the amount of the reflected light (R). The correction ratio used for the fluorescence signal (1:1) led to a clear anoxic response in the CF trace. Under partial ischemia induced by ligation of the two common carotid arteries in the rat (B), the reflectance trace shows a minimal change under the anoxic episode, and the uncorrected fluorescence trace (F) was identical to the corrected one (CF). It seems that the brain was unable to increase its blood flow due to carotid ligation.

Under partial or complete ischemia induced in the gerbil brain by unilateral or bilateral carotid artery occlusion, the reflectance trace shows a very small change and therefore the artifacts introduced into the fluorescence measurement are minimal (C,D).

Figs. 4 and 5 show the response of four various body organs to hypoxia and anoxia of an anesthetized rat ventilated with a respirator. Two signals from each organ are presented, namely, reflectance (R) and corrected fluorescence (CF) of the brain (B), liver (L), kidney (K) and the testis (T).

By comparing the response of the four organs to lower pO_2, a few significant points are to be stressed. First, the magnitude of the decrease in the R signal is not affected so much by the level of pO_2, namely, that the autoregulation mechanisms taken place, in order to increase blood volume to the organ under hypoxia, is activated even by 10% O_2 breathing.

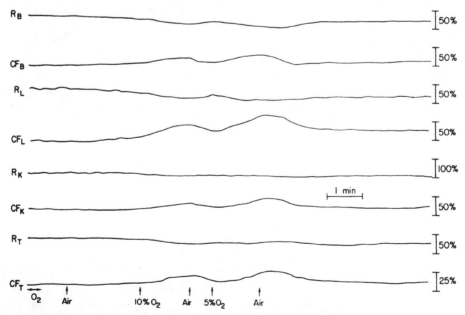

Fig. 4. Reflectance (R) and corrected fluorescence (CF) responses to hypoxia measured from the brain (R_B, CF_B), liver (R_L, CF_L), kidney (R_K, CF_K) and testis (R_T, CF_T). of an anesthetized rat.

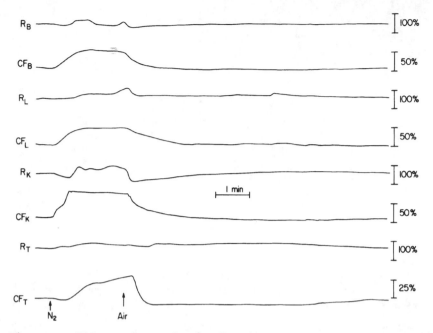

Fig. 5. Effect of anoxia (100% N_2) on the reflectance (R) and corrected fluorescence (CF) measured from the brain, liver, kidney and testis.

Due to the increase in blood volume in the measuring site, the blood will absorb more light and the reflectance signal will decrease. In the testis, the R response was minimal and is probably due to the small vascularity in the measurement site. The liver and kidney also showed a uniform response of the R under the three conditions.

The correlation between the ^{31}P-NMR spectroscopy and NADH redox state measured from the puppy brain exposed to hypoxia (5% O_2) is shown in Fig. 6. The puppy was located inside the magnet, as shown in Fig. 2, and respirated with 100% O_2. Lowering the FiO_2 to 5% led to increased NADH levels (CF) and as the mitochondrial activity was partially inhibited, the PCr/Pi ratio decreased during the hypoxic episode.

Since the correction factor is the main problem in monitoring NADH in vivo, a more detailed discussion on this question was done regarding the fiber optic approach (Mayevsky, 1984). Four main factors may affect the measurement of NADH fluorescence: 1) movement artifacts; 2) changes in the oxygenation level of the blood; 3) changes in the absorption properties of the tissue monitored; and 4) blood volume changes in the tissue under observation.

To summarize the NADH fluorescence correction problem we can say that the main factors affecting the NADH measurements under most conditions are the blood volume changes, and to a lesser extent, the change in the absorption properties of the tissue. Movement artifacts and oxy-deoxy transition of Hb effects can probably be neglected.

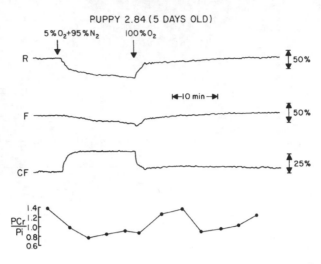

Fig. 6. Correlation between the hemodynamic, metabolic brain activities and the phosphocreatine to inorganic phosphorus ratio in a 5-day-old puppy.

SUMMARY

Various members of the respiratory chain exhibit different optical properties in the reduced and oxidized forms, thus enabling the non-invasive monitoring of various organs in vitro as well as in vivo. Since the pioneering work of Chance, Cohen, Jobsis and Schoener in 1962, many groups of investigators adopted their approach in monitoring NADH oxidation reduction states in vivo for the brain as well as for other body organs.

In 1972, we introduced flexible, optical fibers into the surface fluorometry replacing the usual "rigid" optical system used by other groups. During the last decade, this technique has been developed, improved and applied to many experimental setups in brain research and very recently was combined with ^{31}P NMR spectroscopy for the puppy and the adult dog brain in vivo.

In our system, the effects of movement artifacts and changes in blood oxygenation are negligible while the effects of tissue absorption or blood volume changes are considerable and could be minimized by subtraction of the reflectance signal from that of the fluorescence (1:1 ratio) providing the corrected fluorescence signal.

Acknowledgement. This work was supported by the NIH Grant NS 22881 and Ben Franklin Partnership's Advanced Technical Center of SE Penna, and the Dr. Jaime Lusinchi Research Center in Applied Life Sciences at Bar-Ilan University, Israel.

REFERENCES

Aubert, X., Chance, B., and Keynes, R.D., 1964, Optical Studies of biochemical events in the electric organ of Electrophorous, Proc. Roy. Soc. B., 160:211-245.

Chance, B., and Williams, G.R., 1955, Respiratory enzymes in oxidative phosphorylation, I. Kinetics of oxygen utilization, J. Biol. Chem., 217:383-393.

Chance, B., and Jobsis, F.F., 1959, Changes in fluorescence in a frog sartorius muscle following a twitch, Nature, 184:195-196.

Chance, B., Cohen, P., Jobsis, F., and Schoener, B., 1959, Intracellular oxidation reduction states in vivo, Science. 137:499-508.

Chance, B., Legallis, V., and Schoener, B., 1962, Metabolically linked changes in fluorescence emission spectra of cortex of rat brain, kidney, and adrenal gland, Nature, 195;1073-1075.

Chance, B., and Schoener, B., 1962, Correlation of oxidation-reduction changes of intracellular reduced pyridine nucleotide and changes in elkectroencephalogram of the rat in anoxia, Nature, 195:956-958.

Chance, B., Oshino, N., Sugano, T., and Mayevsky, A., 1973, Basic principles of tissue oxygen determination from mitochondrial signals, in: Internat. Symp. Oxygen Transport to Tissue, Adv. Exp. Med. Biol, Vol.37A, Plenum Publ., NY, pp. 277-292.

Chance, B., Mayevsky, A., Goodwin, C., and Mela, L., 1974, Factors in oxygen delivery to tissue, Microvasc. Res., 8:276-282.

Chance, B., Legallais, V., Sorge, J., and Graham, N., 1975, A versatile time sharing multichannel spectrophotometer, reflectometer and fluorometer, Anal Biochem., 66498-514.

Chance, B., Nakase, Y., Bond, M., Leigh, J.S., Jr. and MacDonald, G., 1978, Detection of 31_p nuclear magnetic resonance signals in brain by in vivo and freeze-trapped assays, Proc. Natl. Acad. Sci. USA., 75:4925-4929.

Eke, A., Hutiray, Gy and Kovach, A.G.B., 1979, Induced hemodilution detected by reflectometry for measuring microregional blood flow and blood volume in cat brain cortex, Am. J. Physiol., 236:759-768.

Gadian, D.G. and Radda, G.K., 1981, NMR studies of tissue metabolism, Ann. Rev. Biochem., 50:69-83.

Harbig, K., Chance, B., Kovach, A.G.B., and Reivich, M., 1976, In vivo measurement of pyridine nucleotide fluorescence from cat brain cortex, J. Appl. Physiol., 41:480-488.

Hilberman, M., Subramanian, V.H. Haselgrove, J., Cone, J.B., Egan, J.W., Gyulai, L., and Chance, B., 1984, In vivo time-resolved brain phosphorous nuclear magnetic resonance, J. Cereb. Blood Flow Metabol., 4:334-342.

Jobsis, F.F., O'Connor, M., Vitale, A. and Verman, H., 1971, Intracellular redox changes in functioning cerebral cortex; I. Metabolic effects of epileptiform activity, J. Neurophysiol., 34:735-749.

Jobsis, F.F., 1972, Oxidative metabolism at low PO2, Fed. Proc., 31:1404-1413.

Kedem, J., Mayevsky, A., Sonn, J. and Acad, B., 1981, An experimental approach for evaluation of the O_2 balance in local myocardial regions in vivo, Quart. J. Exp. Physiol., 66:501-514.

Kobayeshi, S., Kaede, K., Nishiki, K. and Ogata, E., 1971, Microfluorometry of oxidation-reduction state of the rat kidney in situ, J. Appl. Physiol., 31:693-696.

Kobayeshi, S., Nishiki, K., Kaede, K. and Ogata, E., 1971, Optical consequences of blood substitution on tissue oxidation-reduction state microfluorometry, J. Appl. Physiol., 31:93-96.

Mayevsky, A., Chance, B., 1973, A new long-term method for the measurement of NADH fluorescence in intact rat brain with implanted cannula, in: Internat. Symp. on Oxygen Transport to Tissue, Adv. Exp. Med. Biol., Vol. 37A, Plenum Publ., New York, pp.239-244.

Mayevsky, A., 1978, Shedding light on the awake brain, in: Frontiers in Bioenergetics: From Electrons to Tissues, Vol. II. P.L. Dutton, J. Leigh and A. Scarpa, eds., Academic Press, New York, pp.1467-1476.

Mayevsky, A. and Chance, B., 1982 Intracellular oxidation reduction state measured in situ by a multichannel fiber-optic-surface fluorometer, Science, 217:537-540.

Mayevsky, A., 1984, Brain NADH redox state monitored in vivo by fiber optic surface fluorometry, Brain Res. Rev., 7:49-68.

Rosenthal, M. and Jobsis, F.F., 1971 Intracellular redox change in functioning cerebral cortex, II. Effects of direct cortical stimulation, J. Neurophysiol., 34:750-762.

Sundt, T.M. and Anderson, R.E., 1975 "Reduced nicotinamide adenine dinucleotide fluorescence and cortical blood flow in ischemic and nonischemic squirrel monkey cortex, 1., Adnimal preparation, instrumentation, and validity of model, Stroke, 6:270-278.

Sundt, T.M. Jr., Anderson, R.E. and Sharbrough, F.W., 1976, Effect of hypocapnia, hypercapnia and blood pressure on NADH fluorescence, electrical activity and blood flow in normal and partially ischemic monkey cortex, J. Neurochem., 27:1125-1133.

THE EFFECT OF HYPERBARIC OXYGEN ON CEREBRAL HEMOGLOBIN OXYGENATION AND DISSOCIATION RATE OF CARBOXYHEMOGLOBIN IN ANESTHETIZED RATS: SPECTROSCOPIC APPROACH

Ryuichiro Araki*, Ichiro Nashimoto* and Takehito Takano**

*Department of Hygiene, Saitama Medical School, Saitama 350 -04, Japan. **Department of Public Health and Environmental Science, Faculty of Medicine, Tokyo Medical and Dental University, Tokyo 113, Japan

INTRODUCTION

Hyperbaric oxygen (HBO) has been regarded as an effective treatment for various ischemic diseases because of its ability to provide a large quantity of oxygen to the peripheral tissues (Boerema et al., 1960). However, there is little direct evidence to demonstrate that HBO actually increases oxygen tension in ischemic tissue *in vivo* in spite of its remarkable efficacy in the clinical situation (Kawamura et al., 1978).

Near-infrared (NIR) spectroscopy has been proposed for continuous and non-invasive monitoring of hemoglobin (Hb) oxygenation and redox state of cytochrome aa_3 in the brain (Jöbsis, 1977, 1979; Kariman and Burkhart, 1985; Ferrari et al., 1986a, 1986b). By using this technique, we examined the effect of HBO on Hb oxygenation and redox state of cyt. aa_3 in the brain of anesthetized rats.

Treatment for carbon monoxide (CO) poisoning is another target for clinical application of HBO. The blood level of COHb is considered to be a reliable indicator of the patient's condition although it has been a controversial issue whether formation of the CO complex of myoglobin and/or cyt. aa_3 is involved in CO toxicity (Goldbaum et al., 1975; Piantadosi et al., 1985, 1987). Because of the isolation of the patient inside a pressure chamber, hyperbaric oxygen therapy places restrictions on medical procedures such as blood sampling. In an attempt to overcome this problem we have used visible light reflectance spectroscopy on anesthetized rats to determine whether this method would be practicable for non-invasive, continuous monitoring of COHb during hyperbaric oxygen therapy.

MATERIALS AND METHODS

Animals. Male albino rats of the Wistar strain weighing 230–280 g were used in this study. The animals were anesthetized with sodium pentobarbital (initial 30 mg/kg i.p., 25 mg/kg/hr), and then immobilized with pancronium bromide (0.5 mg/kg/hr). The trachea was cannulated and

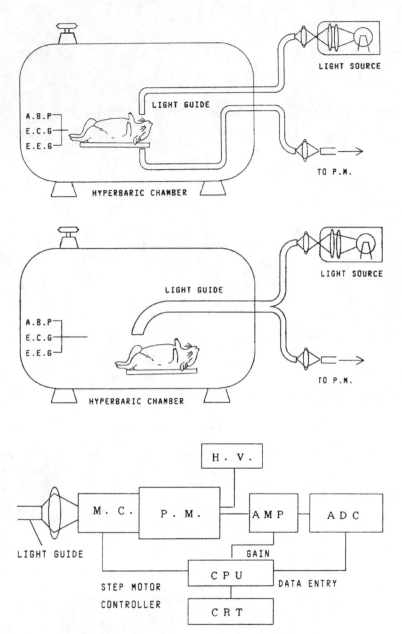

Figure 1. Experimental setup used in this study. *Top*; setup for transmittance measurement of the rat brain. *Middle*; Setup for surface reflectance measurement of the rat groin. *Bottom*; block diagram of the optical system. M.C.; monochromator (Model H-20IR, Jobin-Yvon). P.M.; photomultiplier (Model R-712, Hamamatsu Photonics Ltd.). H.V.; high voltage supplier. AMP; voltage-controlled amplifier. ADC; A/D converter (Model ANALOG-PRO I, CANOPUS Electronics Ltd.). CPU; 8086-based computer (Model PC-9801M2, NEC).

artificial respiration at a rate of 480 ml/kg/min was instituted with a constant volume respirator (Harvard). Catheters were placed in the femoral artery and vein for measurement of arterial blood pressure and infusion of substances. ECG, EEG and heart rate were monitored continuously.

NIR and visible spectrophotometry. Transmittance NIR spectra of the rat brain was measured with a computer-assisted scanning spectrophotometer. Light from a 12 V-100 W tungsten-iodine lamp was condensed as a light spot of 2 mm diameter and guided onto the top of the shaved head of the animal via a light guide through the wall of a hyperbaric chamber. Another light guide was inserted into the mouth of the animal at an angle of 45° to detect light transmitted through the brain. The transmitted light was then fed into a photomultiplier (Model R-712; Hamamatsu Photonics Ltd.) through a monochromator (Model H-20IR; Jobin-Yvon). The monochromator was equipped with a high-speed stepping motor. An 8086-based computer (Model PC-9801M2; NEC) including a stepping motor controller (Model AB98-06V; Adtec System Science) and an A/D conveter (Model ANALOG PRO I, CANOPUS Electronics Ltd.) was used for control of the optical system. Spectra in the range from 700 to 1000 nm were measured in 600 msec, and accumulation of 16 spectra were performed to obtain high signal-to-noise ratio. A reference spectrum was measured prior to the experiment, and difference spectra against the reference one were displayed on a CRT at 10 sec intervals. Real-time calculations of dual wavelength traces were also performed and results were displayed on the CRT.

Visible spectra of the surface reflectance from the rat groin was measured with the same apparatus used for the transmittance measurement. Light from the lamp was guided onto the shaved right groin of the animal via a Y-shaped light guide, and the reflected light was then fed back into the optical system. The spectra from 400-700 nm were measured. Fig. 1 summarizes the experimental setup and gives a block diagram of the optical system.

RESULTS AND DISCUSSION

Fig. 2 shows the effect of N_2 breathing on transmittance NIR spectra of the rat brain. When the inspired gas was switched from room air to nitrogen, a spectral change occurred within 20 sec (spectrum No. 62) and reached a maximum within 1 min. The spectral change obviously indicated deoxygenation of Hb. In addition, a delayed decrease in optical absorption around 830-840 nm was also observed (spectrum No. 64-66). While 40 sec of N_2 breathing induced maximum spectral change in the ranges from 700 to 800 nm and from 900 to 1000 nm (spectrum No. 64), the decrease in absorption around 830-840 nm was incomplete at this time (spectrum No. 64 -66). This delayed change might be attributable to reduction of cyt. aa_3.

Fig. 3 demonstrates HBO-induced hyperoxygenation of Hb in the rat brain. Pressure was not raised above 4 ATA in order to avoid the risk of acute oxygen poisoning. Since HBO elicits arterial vasoconstriction which causes decrease in blood flow (Kawamura et al., 1978), it has been controversial whether HBO actually increases tissue oxygen tension effectively. The present data, however, clearly demonstrated that HBO caused pressure-dependent Hb hyperoxygenation in the rat brain. On the other hand, absorption around 830-840 nm was affected only a little by HBO. By using reflectance spectrophotometry, Hempel et al. (1977) have observed HBO-induced oxidation of cyt. aa_3 in cerebral cortex of the anesthetized cat. This discrepancy might be due to the different responses of cerebral cortex and subcortex to HBO. In the ischemic brain which was obtained by ligation of the bilateral common carotid arteries, HBO improved tissue

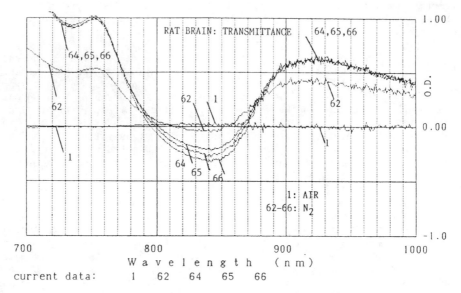

Figure 2. Effect of N_2 breathing on transmittance NIR spectra of the rat brain.

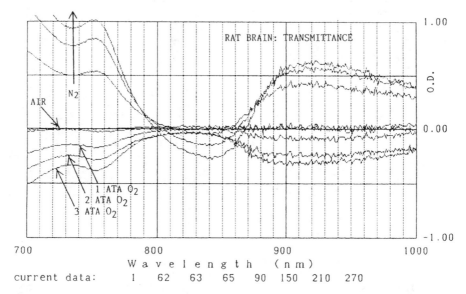

Figure 3. Effect of HBO on Hb oxygenation in the rat brain. The inspired gas was switched to O_2, and then animals were pressurized with the hyperbaric chamber.

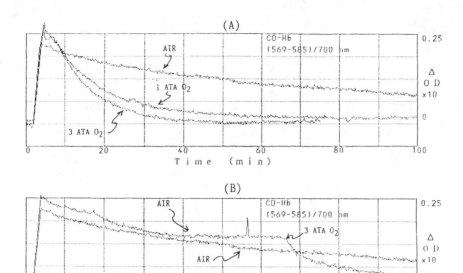

Figure 4. Effect of HBO on dissociation of COHb in the CO-poisoned rats.
COHb content was determined from visible reflectance spectra of the
surface of the groin. *A*; the animals were subjected to 1 ATA air, 1 ATA
oxygen, and 3 ATA oxygen immediately after inhalation of 1% CO for 2 min.
B; the CO-poisoned rat was subjected to 3 ATA oxygen after 1 hr of air
breathing at 1 ATA.

oxygenation in regard to both Hb oxygenation and the oxidation-reduction
state of cyt. aa_3 (data not shown).

Another application of optical measurement to the living body in a
hyperbaric environment is shown in Fig. 4. Surface reflectance spectrophotometry was used instead of transmittance spectrophotometry in order
to simplify the measurement system. It was confirmed that the wavelength
pair of 569-585 nm was not affected by oxygenation-deoxygenation of Hb.
Absorption at 700 nm was used to compensate partly for interference by
changes in blood flow (Araki et al., 1985). After 2 min of 1% CO
breathing, 40-50% of COHb was formed. The animals were then subjected to
1 ATA air, 1 ATA oxygen, and 3 ATA oxygen immediately after inhalation of
CO (Fig. 4A). The halftime for disappearance of COHb was 70 min with 1
ATA air, and shortened to 16 and 13 min with 1 ATA and 3 ATA oxygen,
respectively. The delayed application of HBO also accelerated the rate of
COHb dissociation (Fig. 4B). Tight correlation between the optical signal
and COHb content as determined from blood samples has been reported
previously (Araki et al., 1985). These results demonstrated the
usefulness of optical monitoring in a hyperbaric environment.

SUMMARY

By measuring near-infrared transmittance spectra, we examined the effect of HBO on cerebral Hb oxygenation in normal and ischemic brain in the anesthetized rat. The oxygenation state of Hb was around 80% in the rat brain under 1 ATA air breathing. HBO did not induce further cerebral Hb oxygenation above 2 ATA in control animals but improved tissue oxygenation in the ischemic brains. The oxidation-reduction state of cyt. aa_3 in the normal brain was not affected by HBO. In the ischemic brain, however, HBO prevented ischemia-induced reduction of cyt. aa_3. Non-invasive optical monitoring of COHb with visible reflectance spectrophotometry was also examined. HBO markedly accelerated dissociation of COHb. Tight correlation was found between the optical signal and COHb content determined from blood samples. These results demonstrated the usefulness of optical monitoring *in vivo* under hyperbaric conditions.

REFERENCES

Araki R., Gotoh Y., and Nashimoto I., 1985, Optical detection of various hemoglobin derivatives in circulating blood in the anaesthetized rat: Noninvasive, real-time observation by visible reflectance spectrophotometry (in Japanese), *Jap. J. Hyg.*, 40:227.

Boerema I., Meyne N. G., Brummelkamp W. K., Bouma S., Mensch M. H., Kamermans F., Hanf M. S., and Aalderen W. V., 1960, Life without blood (A study of the influence of high atmospheric pressure and hypothermia on dilution of the blood), *J. Cardiovasc. Surg.*, 1:133-146.

Ferrari M., Marchis C. D., Giannini I., Nicola A. D., Agostino R., Nodari S., and Bucci G., 1986a, Cerebral blood volume and hemoglobin oxygen saturation monitoring in neonatal brain by near IR spectroscopy, *Adv. Exp. Med. Biol.*, 200:203-211.

Ferrari M., Elietta, Z., Giannini I., Sideri G., Fieschi C., and Carpi A., 1986b, Effects of carotid artery compression test on regional cerebral blood volume, hemoglobin oxygen saturation and cytochrome-c-oxidase redox level in cerebrovascular patients, *Adv. Exp. Med. Biol.*, 200:213-221.

Goldbaum L. R., Ramirez R. G., and Absalon K. B., 1975, What is the mechanism of carbon monoxide toxicity?, *Aviat. Space. Environ. Med.*, 46:1289-1291.

Jöbsis F. F., 1977, Noninvasive, infrared monitoring of cerebral and myocardial oxygen sufficiency and circulatory parameters, *Science*, 198:1264-1267.

Jöbsis F. F., 1979, Oxidative metabolic effects of cerebral hypoxia, *Adv. Neurol.*, 26:299-318.

Hempel F. G., Jöbsis F. F., LaManna J. L., Rosenthal M. R., and Saltzman H. A., 1977, Oxidation of cerebral cytochrome aa_3 by oxygen plus carbon dioxide at hyperbaric pressures, *J. Appl. Physiol.*, 43:873-879.

Kariman K. and Burkhart D. S., 1985, Heme-copper relationship of cytochrome oxidase in rat brain *in situ*, *Biochem. Biophys. Res. Commun.*, 126:1022-1028.

Kawamura M., Sakakibara K., and Yusa T., 1978, Effect of increased oxygen on peripheral circulation in acute, temporary limb hypoxia, *J. Cardiovasc. Surg.*, 19:161-168.

Piantadosi C. A., Sylvia A. L., Saltzman H. A., and Jöbsis-Vandervliet F. F., 1985, Carbon monoxide-cytochrome interactions in the brain of the fluorocarbon-perfused rat, *J. Appl. Physiol.*, 58:665-672.

Piantadosi C. A., Sylvia A. L., and Jöbsis-Vandervliet F. F., 1987, Differences in brain cytochrome responses to carbon monoxide and cyanide *in vivo*, *J. Appl. Physiol.*, 62:1277-1284.

BRAIN TISSUE TEMPERATURE: ACTIVATION-INDUCED CHANGES

DETERMINED WITH A NEW MULTISENSOR PROBE

J.C. LaManna, K.A. McCracken, M. Patil, and O. Prohaska

Departments of Neurology, Physiology/Biophysics, and
Biomedical Engineering, Case Western Reserve University
Cleveland, Ohio 44106, U.S.A.

INTRODUCTION

Because of its high oxidative metabolic rate, the brain is a
significant heat producer. As with all mammalian tissue, the operating
temperature range of the brain must be kept at a nearly constant level
for normal function. Brain function is especially sensitive to
increases in temperature (Cabanac, 1986). The temperature of the brain
tissue is determined by the heat conduction properties of the tissue,
the heat produced by metabolism, and the heat and heat-transfer capacity
of the blood flowing through the tissue. It has always been thought
that one of the functions of the cerebral circulation was to act as a
cooler of the brain, and some mammalian species have a carotid rete
system to maximize this cooling effect (Baker, 1979; Cabanac, 1986).
While there is data on the larger animals such as dog, cat, and primate
which indicate that significant cooling of the brain during physical
excercise occurs (Baker, 1979; Hayward, 1966), there is only sparse
information concerning the relationships of metabolism and blood flow in
the CNS of smaller mammals such as rodents.

With the availability of new sensor types and arrangements through
methodological advances in thin film deposition techniques, it has
become possible to explore some of the relationships between brain
tissue temperature, and function-driven changes in metabolism and blood
flow in the rat.

METHODS

Experiments were performed on 250-350 g Wistar rats. Rats were
anesthetized with either chloral hydrate (400 mg/kg i.p., supplemented
i.v. as needed) or with brevital, for the placement of arterial and
venous catheters and tracheostomy prior to nitrous oxide/halothane (70%/
2%, balance O_2) anesthesia. Rats were paralyzed and placed on a
positive pressure ventilator. The arterial catheter was used for
continuous blood pressure monitoring and recording, as well as
intermittent sampling for blood gases and pH, and the respirator
adjusted to result in normal range blood gases. Body temperature was
monitored and kept near 37°C by a rectal probe and hot water heating
pad. Rats were placed in a head holder and immobilized. The skin
overlying the top of the skull was retracted and a burr hole about 5 mm

diameter was made over the parietal cortex, leaving the dura intact. A 5 mm length of large diameter (1 cm) polyethylene tubing was attached with acrylic adhesive to the bone surrounding the opening. Vacuum grease and bone wax were used to ensure a tight seal between the tubing and the bone. The tubing thus formed a reservoir over the dura which was then filled with mineral oil in order to minimize the effects of atmospheric exposure of the cerebral cortex.

Brain tissue temperature, oxygen tension, and electrical potential were measured at each of two depths with a thin-film multiple sensor probe (Prohaska et al, 1986; 1987; Prohaska, 1987). Figure 1 depicts the sensor arrangement on the needle-shaped glass substrate. The glass needle was inserted through the dura overlying the parietal cortex so that the individual sensors were positioned as shown. With this positioning, one temperature sensor was within the cerebral cortex (T_B) and one was in the striatum (T_A). The temperature sensors were thin-film, 0.25 micron Germanium thermistors with a sensitivity of 100mV/$^\circ$C, resolution of .001 $^\circ$C, and a response time of 5 msec. Both of the tissue temperature sensors and the deep rectal temperature probe were calibrated between 20-40 $^\circ$C against a mercury thermometer before each experiment.

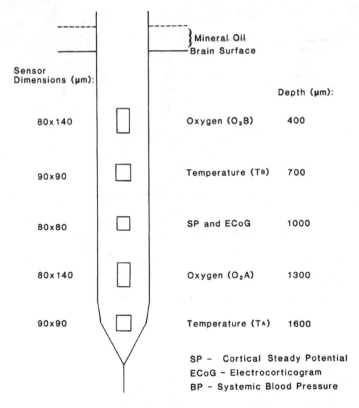

Figure 1: Schematic representation of the multisensor probe used in these studies. Sensor dimensions are shown, in microns, on the left and depth below the brain surface on the right. The glass substrate is itself 500 microns wide and 100 microns thick. Symbols in parentheses and at bottom right explain the labels that appear on the following figures.

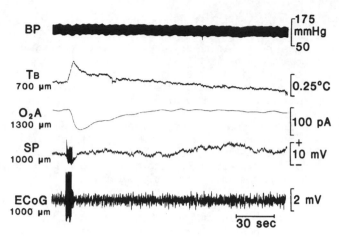

Figure 2: Transient increase in brain tissue temperature in response to direct cortical electrical stimulation (10 Vhom, 20 Hz, 0.5 msec, 4 sec). Symbols as given in Figure 1.

The Au-Ag/AgCl oxygen sensors and the Ag/AgCl potential sensors were of the chamber type. The Gold working electrode of the oxygen sensor was 20 microns in diameter. The chamber design means that the electrodes were not in direct contact with the tissue. Amplified signals from the probe were displayed on a chart recorder together with arterial blood pressure.

RESULTS AND DISCUSSION

Our goal in the studies reported was to test the responsiveness of the multisensor probe under the types of experimental conditions that have been used in the past to examine tissue metabolism and blood flow. The first experimental condition that we observed was the baseline tissue temperature soon after insertion of the probe. In each rat, regardless of anesthetic, tissue temperature at both sensor sites was cooler than core. The mean difference in 8 rats was 1.2 °C at the cortical site and 1.8 °C at the deeper site. These differences from core temperature were significant at the $p < 0.01$ level by the paired t-test.

We then examined the response of the tissue to direct cortical electrical stimulation in the chloral hydrate-anesthetized rats. One to 10 second trains of 0.5 msec duration pulses at 10-20 Hz were delivered through 0.1 mm diameter stainless steel surface electrodes positioned within a few mm of the probe. Typical negative steady potential shifts (SP) were observed through the potential sensor (see Figure 2). Increases in tissue temperature up to 0.5 °C always accompanied electrical stimulation that produced SP shifts. These temperature transients were present at both depths and were of about the same size although the response from the deeper site was usually slightly smaller and slower. As shown in the figure, tissue oxygen tension near the deeper site decreased with electrical stimulation, but the tissue oxygen tension near the cortical site usually increased (not shown). The tissue temperature responses were slower than the potential changes and faster than the tissue oxygen tension changes. The time course of the tissue temperature changes was similar to, and reminiscent, of the time course of the changes seen in blood volume and extracellular potassium ion concentration in response to the same type of stimulation (LaManna et al, 1987). When spreading depression, as a result of too vigorous electrical stimulation was produced, large and long-lasting increases in tissue temperature were also consistently observed.

The effect of pentylenetetrazol (PTZ)-induced seizure activity on deep brain temperature is shown in Figure 3 from a rat under nitrous oxide/halothane anesthesia. In this figure, a series of seizure bursts are shown accompanied by large and prolonged increases in temperature. Tissue temperature remained elevated even inter-ictally in this rat.

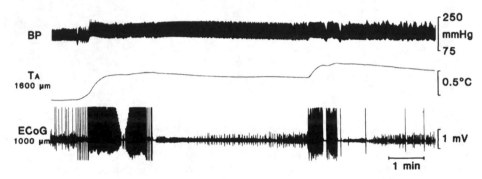

Figure 3: Prolonged increases in brain tissue temperature during PTZ-induced seizures. Symbols as in Figure 1.

All of the previous examples of tissue temperature responses occurred during experimental conditions which would be expected to increase both metabolic rate and blood flow. Thus, from the above results it was not possible to distinguish between two mechanisms to explain the observed temperature increases. Tissue temperature increases could be due to increased metabolic heat production, but since the absolute tissue temperature is less than core temperature, and since the rat has no carotid rete system (Baker, 1979), then the increases in tissue temperature may be indicating passive heating of the tissue by the increased flow into the tissue of slightly warmer blood. We performed two additional experiments to determine if either or both of these mechanisms contributed to the findings. The first experiment was designed to increase blood flow primarily, using a vasodilating stimulus that should have only minor acute effects on metabolism. This was easily done by ventilating chloral hydrate-anesthetized rats with an oxygen mixture containing 5% CO_2. The result is shown in Figure 4. There was an increase in tissue temperature at both sensor sites. This suggests that increasing blood flow warms the tissue. That this was not the only possible explanation for tissue temperature rise seen with increased neuronal activity was shown by the following, somewhat more complicated protocol.

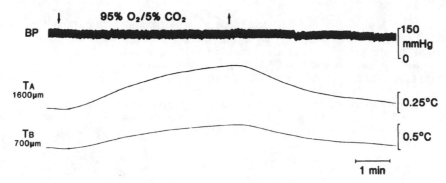

Figure 4: Increase in brain tissue temperature in response to hypercapnia (between arrows). Note that T_B was recorded at about half the sensitivity of T_A, so that the magnitude of the response was about the same at both sites. Symbols as in Figure 1.

When a rat is made acutely and severely hypoxic by ventilation for 1-2 minutes with 100% Nitrogen, tissue blood flow is greatly, if not maximally, increased and at the same time metabolic heat production from oxygen consumption is greatly diminished. If oxygen is readmitted to the respirator, then there will be an initial period of time where blood flow remains at maximum (i.e., unchanged) but oxidative metabolism increases from near zero to near maximum. The tissue temperature changes under these conditions are shown in Figure 5. With the onset of hypoxia from nitrogen breathing, blood pressure rises, brain tissue oxygen content falls to a very low plateau level, but tissue temperature rises drastically. Tissue temperature remains elevated even after systemic blood pressure begins to fall. When oxygen is readmitted to the respirator, tissue oxygen content rises to hyperoxic levels. The rise in tissue temperature is even larger under these conditions than during the hypoxic phase. We interpret the first increase in temperature as due to an increased tissue blood flow and the second as due to maximally stimulated oxidative metabolism.

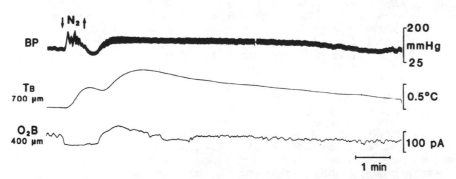

Figure 5: Changes in brain tissue temperature during severe acute hypoxia and reoxygenation induced by ventilation with 100% nitrogen (N_2) for the period of time indicated by the arrows on the blood pressure trace.

The last experimental condition we examined was comparing the rate of brain tissue cooling in rats under chloral hydrate and nitrous oxide/halothane anesthesia, after complete cessation of blood flow due to an intravenous KCl injection. Examples are shown in Figure 6. The rate of cooling in these brains should be dependent only on the ambient temperature differential. Comparable rates of cooling were observed under both anesthetics. The initial temperature drop in these rats occurred at a rate of about 0.5 °C/minute at an initial temperature difference of about 10 °C with ambient.

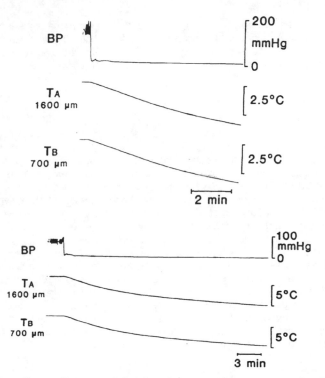

Figure 6: Rates of brain tissue cooling at two depths below the surface following cessation of blood flow and metabolism produced by cardiac arrest-induced total cerebral ischemia in a nitrous oxide/halothane anesthetized rat (top) and a chloral hydrate anesthetized rat (bottom).

In conclusion, it would seem that in the anesthetized rat brain, resting metabolic heat production is low and brain tissue temperature is below core and, in the absence of a carotid rete, therefore also below systemic arterial blood temperature. With increases in blood flow, the tissue is heated – tissue temperature approaches blood temperature. If this is indeed the situation in the anesthetized rat, it would imply that the warming effect of increased blood flow is through an increase in the volume of tissue that is actively blood perfused, i.e., through capillary recruitment. This phenomenon would also explain why thermal methods for blood flow determination are qualitative at best, and then only when differential pulse systems are used (Ferrari et al, 1982).

Metabolic stimulation increases tissue temperature as well. The temperature increases due to metabolism are in the same range as that produced by blood flow changes. The local change in tissue temperature with electrical activation due to increased blood flow and increased metabolism would affect energy of activation calculations of metabolic reactions and might explain the previously reported, somewhat higher than expected, Q_{10} values for electrically stimulated cerebral oxidative metabolism (LaManna et al, 1980).

SUMMARY

Local brain tissue oxygen tension, temperature, and electrical potential were continuously and simultaneously measured at each of two different depths in anesthetized, paralyzed rat brain. Brain tissue temperature increases up to 1°C were recorded in response to direct electrical stimulation, spreading depression, PTZ-induced seizures, hypercapnia, and hypoxia. An increase in brain tissue temperature was also recorded during reoxygenation after hypoxia. Thus, we have shown that, in this preparation, increases in either blood flow or oxidative metabolism lead to transient warming of the brain.

REFERENCES

Baker, M.A., 1979, A brain-cooling system in mammals, Sci. Am., 250:130-139.

Cabanac, M., 1986, Keeping a cool head, NIPS, 1:41-44.

Ferrari, M., Giannini, I., Carta, F., Argiolas, L., Carpi, A., 1982, Quantitative measurements of tissue blood flow by fast pulse heated thermistors, Physiol. Chem. Phys., 14:553-560.

Hayward, J.N., 1966, Cerebral cooling during increased cerebral blood flow in the monkey, Proc. Soc. Exp. Biol. Med., 124:555-557.

LaManna, J.C., Rosenthal, M., Novack, R., Moffett, D. F., Jobsis, F.F., 1980, Temperature coefficients for the oxidative metabolic responses to electrical stimulation in cerebral cortex, J. Neurochem., 34:203-209.

LaManna, J.C., Sick, T.J., Pikarsky, S.M., Rosenthal, M., 1987, Detection of an oxidizable fraction of cytochrome oxidase in intact rat brain, Am. J. Physiol., 253:C477-C483.

Prohaska, O.J., Olcaytug, F., Pfunder, P., Dragaun, H., 1986, Thin-film multiple electrode probes: Possibilities and limitations, IEEE Trans. B.M.E., 33:223-229.

Prohaska, O.J., Kohl, F., Goiser, P., Olcaytug, F., Urban, G., Jachimowicz, A., Pirker, K., Chu, W., Patil, M., LaManna, J.C., and Vollmer, R., 1987, Multiple chamber-type probe for biomedical application, Transducers 87, Proceedings of the 4th Conference on Solid-state Sensors and Actuators, Japan Institute of Electrical Engineers, Tokyo, pp. 812-815.

Prohaska, O.J., 1987, Thin-film micro-electrodes for in vivo electrochemical analysis, in: "Biosensors," A.P.F. Turner, I. Karube, and G.S. Wilson, eds., Oxford University Press, New York, pp. 377-389.

ACTIVE AND BASAL CEREBROMETABOLIC RATE FOR OXYGEN (CMRO2)

AFTER COMPLETE GLOBAL BRAIN ISCHEMIA IN RATS

Edwin M. Nemoto, John A. Melick, and Peter Winter

Department of Anesthesiology and Critical Care Medicine
University of Pittsburgh School of Medicine
Pittsburgh, Pennsylvania 15261

INTRODUCTION

The ultimate goal of this study is to devise a method for prognostic assessment of the viability of the brain in prolonged coma or in the vegetative state. The method we are attempting to develop is based upon the following hypotheses and observations: (1) The capacity of the brain to support spontaneous electrical activity (i.e., synaptic transmission) is the essence of viability; (2) severe neurologic dysfunction occurs not only as a result of irreversible neuronal necrosis but also, a potentially reversible failure of synaptic transmission in viable neurons (i.e., quiescent neurons); (3) active CMRO2 (ACMRO2) is O_2 consumed to generate energy to support spontaneous brain electrical activity (i.e., restoration of neuronal ionic gradients after depolariztion) while basal CMRO2 (BCMRO2) provides the energy needed to maintain ionic gradients and energy-requiring anabolic reactions to maintain the viability of the neuron in the resting state;[1,2] and (4) barbiturates, namely, thiopental, blocks spontaneous brain electrical activity and therefore, ACMRO2, while BCMRO2 is unaffected.[3] Thus, the compartmentation of total CMRO2 into ACMRO2 and BCMRO2 after cerebral insults will allow estimates on the viability of the brain (ACMRO2) and neuronal necrosis (BCMRO2).

METHODS AND MATERIALS

Male, Wistar albino rats weighing 300 to 350 g had free access to food and water up to the time of the studies. Anesthesia was induced with 5% halothane in 70% N_2O/30% O_2 and maintained via face mask on 2% halothane/ 70% N_2O/30% O_2. Femoral artery and vein catheters (PE-50) were inserted and the trachea cannulated with PE-240 catheters. The rats were paralyzed with 0.2 mg Pavulon, IV every hour and mechanically ventilated (Harvard rodent ventilator) on 0.5% halothane/70% N_2O/30% O_2.

The skin overlying the calvarium was reflected laterally via a midline incision and the temporal muscle reflected ventrolaterally after blunt dissection from the skull. The rats' heads were then fixed in a stereotoxic device (David Kopf, Inc.) with the sharpened ear bars fixed on the temporal bone instead of the ear canal to minimize pain. Lidocaine jelly (5%) was applied to all exposed tissue and bone surrounding the calvarium.

Via craniectomies, a short-bevel 30 ga needle was inserted into the torcula for sampling of cerebral venous blood and a 10 μm (tip diam) platinum microelectrode was inserted into the superior saggital sinus 0.5 cm posterior to the frontal suture. The microelectrode was polarized at +250 mV via a Transidyne chemical microsensor and the output recorded on a Linseis recorder.

Arterial blood pressure and rectal temperature were both continuously recorded. Arterial pressure was monitored via a Statham strain gauge connected to a Honeywell preamplifier and recorded on the Linseis recorder. Rectal temperature was monitored via a Yellow Spring telethermometer and maintained at 38.0 ± 0.5 $^{\circ}$C by an automatic temperature controller and heated water blanket. Arterial blood gas and pH measurements were made with a Model 165 Corning trielectrode blood gas machine. Arterial and cerebral venous O_2 content was measured with an IL model 282 Co-oximeter.

Following all surgical preparations, halothane was discontinued and the rats were ventilated on 70% N_2O/30% O_2 for 30 min to allow stabilization of physiological variables: MAP, 100 to 125 torr; PaO_2, > 100 torr; $PaCO_2$, 35 to 40 torr; pHa, 7.35-7.40; and BE, $\pm$ mEq/l.

At the end of the stabilization period, 1% H_2 gas was added to the inspired gas mixture for a 10 min equilibration or until a steady level of H_2 was recorded in the superior saggital sinus. H_2 was abruptly discontinued and the H_2 clearance monitored until baseline was attained. After the first 5 min of the clearance, arterial and cerebral venous blood samples were obtained for arterial blood gas analyses and O_2 content measurements. All blood samples withdrawn were replaced with an equal volume of blood from a donor rat.

Following the measurement of CBF and CMRO2 on 70% N_2O/30% O_2 the rats were equilibrated on 1% H_2. A bolus of sodium thiopental (Pentothal, Abbott Laboratories, Inc.) 90 mg/kg was infused in doses of 45 mg/kg x 2 spaced 1 min apart while titrated IV infusion of norepinephrine (Levophed, Breon, Inc.) was used to maintain MAP at 100 torr. Continuously recorded EEG over the frontoparietal cortex verified an isolelectric EEG after infusion. Within 20 secs after the administration of the Pentothal, H_2 was discontinued to begin H_2 clearance measurement of CBF with arterial and cerebral venous samples obtained as previously described. In this manner, total CMRO2 and BCMRO2 values are obtained. ACMRO2 is then the difference, total CMRO2 minus BCMRO2. These procedures were repeated in different rats to obtain total CMRO2, ACMRO2 and BCMRO2 measurements two times preischemia with the measurements spaced 30 min apart.

Prior to fixation of the rat in the stereotoxic device, a miniature tourniquet designed for the rat was loosely wrapped around the neck. To induce complete global brain ischemia, Arfonad in 2 mg doses was injected IV (without exceeding a total of 6 mg) to decrease MAP to 50 torr with PEEP at 15 cm H_2O. When a stable MAP of 50 torr was attained, the tourniquet was inflated to 1500 torr and MAP controlled by manipulation of PEEP. Toward the end of the ischemic episode, MAP was restored by decreasing PEEP 2 to 5 cm H_2O and titrated infusion IV, of norepinephrine to MAP of 100 torr. At the precise end of ischemia, the tourniquet was deflated.

Postischemia, arterial blood gases and pH and MAP were tightly controlled. Base deficits were corrected with IV $NaHCO_3$. CBF and CMRO2 compartments were measured at 1, 3 and 6 h postischemia by the methods previously described. It is important to note that after the measurement of BCMRO2 following thiopental infusion, the study in that particular rat was terminated. Thus, measurements of ACMRO2 and BCMRO2 at the various times pre- and postischemia required that the rats be used only up to the point of

BCMRO2 measurement. Thus, different rats were used for CMRO2 compartment measurements made at the different time points of C1, C2, 1, 3 and 6 h.

CBF was calculated by the $T\frac{1}{2}$ method using a λ value of 1.0 for H_2. CMRO2 was calculated as the product CBF x A-VO2. The data were analyzed by ANOVA for unpaired data and tested for significance with the Newmann-Keuls test. A P-value of 0.05 or less was considered statistically significant.

RESULTS

Preischemic (C1 and C2) total CMRO2 was about 6 ml/100 g/min with active and basal CMRO2 values ranging between 2 and 4 mls/100 g/min in a nearly 50:50 distribution between the two compartments (Table 1).

Total CMRO2 in the sham ischemic group increased (P < 0.05) by about 30% to 8 ml/100 g/min at 1 and 3 h postischemia. Thirty min of ischemia suppressed this rise in CMRO2 resulting in a lower (P < 0.05) value at 1 h post-insult compared to the corresponding value in the sham group. After 15 min of ischemia, total CMRO2 was generally less depressed compared to 30 min ischemia with a marked increase in total CMRO2 at 6 h.

The postinsult depression in total CMRO2 in the 30 min ischemia compared to the sham group or the activation of CMRO2 in the sham group was primarily due to differences in active CMRO2 at 1 and 3 h (Table 1). However, a difference in basal CMRO2 was also observed at 1 h between the sham and 30 min ischemia groups. In the 15 min ischemia groups, the marked increase in total CMRO2 after 6 h was due to an increase in basal rather than active CMRO2.

In the rat brain under 70% N_2O/30% O_2 anesthesia in an unstressed state, active and basal CMRO2 each represents about 50% of total CMRO2 (Table 2). In the sham ischemia group, it remains essentially the same for up to 6 h. After 30 min of ischemia, however, ACMRO2 was depressed (P < 0.05) at 3 h postinsult and BCMRO2 increased (P < 0.05) compared to the preischemic values and the corresponding values in the sham group. As previously noted, the large increase in total CMRO2 noted at 6 h after 15 min of ischemia was attributable to a 50% increase in BCMRO2 and a 50% decrease in ACMRO2.

DISCUSSION

Our findings of an approximately 50% distribution of the total O_2 consumed by the brain between the active and basal components corroborate the findings of earlier reports.[1,2] As discussed extensively by Astrup,[1,2] the active CMRO2 component is primarily associated with the energy requiring processes associated with spontaneous brain electrical activity whereas basal CMRO2 is probably related to the energy necessary for the maintenance of the cellular metabolic apparatus and maintaining transmembrane ionic gradients. It has also been clearly established that the barbiturate anesthetic thiopental, will only abolish ACMRO2 without influencing BCMRO2 even at multiples of the anesthetic dose.[3] This property is unique to the barbiturates and perhaps other anesthetics, but not to halothane which will decrease BCMRO2 at higher anesthetic doses.[4]

Our findings on the changes in ACMRO2 and BCMRO2 indicate that compartmentation of CMRO2 may be an effective method for prognosis in prolonged coma and the vegetative state. However, more work is required to clearly establish guidelines and baselines of values associated with the various states of coma or brain dysfunction. Recently, Sari et al[5] published a report on the use of thiamylal and its effect on cerebral A-VO2 difference as a measure of the responsivity or viability of the brain of patients suffering various cerebral insults and reported that the magnitude of the

Table 1. Cerebrometabolic Rate for O_2 (CMRO2) Compartments after Global Ischemia in Rats

CMRO2 Compartment		C1	C2	CMRO2 (ml/100 g/min)	1H	3H	6H
TOTAL	X	6.3	5.4	SHAM	8.1*	8.6*	5.0
	SD	1.6	3.2		3.6	4.4	1.2
	n	7	4		9	7	5
	X			30' ISCH	5.5+	5.9	5.5
	SD				1.7	2.2	1.2
	n				5	7	6
	X			15' ISCH	7.4	6.1	10.5*+$
	SD				1.9	1.3	3.4
	n				3	3	5
ACTIVE	X	2.6	3.1	SHAM	3.7	4.7	2.8
	SD	1.3	2.2		3.2	3.7	0.4
	n	7	4		9	7	3
	X			30' ISCH	2.5	1.9+	2.4
	SD				2.4	2.1	1.5
	n				5	7	6
	X			15' ISCH	3.0	2.7	3.0
	SD				1.6	0.5	2.9
	n				3	3	5
BASAL	X	3.7	2.3	SHAM	4.4*	3.9	2.7
	SD	1.9	1.2		1.8	2.7	1.2
	n	7	4		9	7	3
	X			30' ISCH	3.0+	4.0	3.1
	SD				1.6	2.1	0.9
	n				5	7	6
	X			15' ISCH	4.4	3.4	7.5+$
	SD				0.7	0.9	2.0
	n				3	3	5

*P < 0.05 compared to preischemic value.

+P < 0.05 compared to corresponding value in sham group.

$P < 0.05 compared to corresponding value in the other ischemic group.

Table 2. Active CMRO2 (ACMRO2) and Basal CMRO2 (BCMRO2) as Percent of Total CMRO2 after Complete Global Ischemia in Rats.

CMRO2 COMPARTMENT		C1	C2	CMRO2 (% of Total)	1H	3H	6H
ACMRO2	$\overline{X}$	42.8	55.0	SHAM	40.0	52.6	48.6
	SD	22.1	18.3		25.6	20.0	6.3
	n	7	4		9	7	3
	$\overline{X}$			30' ISCH	40.4	30.3[**]	40.5
	SD				29.2	30.1	21.0
	n				5	7	6
	$\overline{X}$			15' ISCH	38.9	44.2	25.6[*+]
	SD				12.1	4.2	18.1
	n				3	3	3
BCMRO2	$\overline{X}$	57.2	44.9	SHAM	60.0	47.4	51.4
	SD	22.1	18.3		25.6	19.9	6.3
	n	7	4		9	7	3
	$\overline{X}$			30' ISCH	59.6	69.7[*+]	59.5
	SD				29.2	30.1	21.0
	n				5	7	6
	$\overline{X}$			15' ISCH	61.0	55.8	74.4[*+]
	SD				12.1	4.2	18.1
	n				3	3	5

[*] $P < 0.05$ compared to preischemic value.

[+] $P < 0.05$ compared to corresponding value in sham group.

response in the cerebral A-VO2 was directly related to recovery. This abbreviated version of our method lends further support to our findings and proposal for the assessment of brain viability.

An equally exciting aspect of our findings relates to the mechanisms of postinsult hyper- or hypometabolism. We have always assumed that post-insult hypermetabolism was attributable to excessive neuronal activation resulting from excessive neurotransmitter release, accumulation or hyper-sensitivity of receptors. However, our results show an increase in BCMRO2 after ischemia which would suggest that postinsult hypermetabolism could represent energy consumption used to restore ion gradients and re-pair damaged cellular metabolic components. The distinction is important because the inhibition of hypermetabolism due to excessive neurotrans-mitter activation of neurons may be beneficial, but not if the hyper-metabolism is due to cellular repair.

REFERENCES

1. J. Astrup, L. Symon, N. M. Branston, and N. A. Lassen, Cortical evoked potential and extracellualr K^+ and H^+ at critical levels of brain is-chemia, Stroke 8:51 (1977).

2. J. Astrup, Energy-requiring cell functions in the ischemic brain. Their critical supply and possible inhibition in protective therapy, J. Neurosurg. 56:482 (1982).

3. J. D. Michenfelder, The interdependency of cerebral functional and metabolic effects following massive doses of thiopental in the dog, Anesthesiology 41:231 (1974).

4. J. D. Michenfelder, and R. A. Theye, In vivo toxic effects of halothane on canine cerebral metabolic pathways, Am. J. Physiol. 229:1050 (1975).

5. A. Sari, Y Matayoshi, A. Yonei, H. Ogasahara, T. Nonoue, K. Yokota and S. Yamashita, Cerebral arteriovenous oxygen content difference during barbiturate therapy in patients with acute brain damage, Anesth. Analg. 65:1196 (1986).

ACKNOWLEDGEMENT

Supported in part by the American Heart Association, Grant No. 84-1138.

AUTOREGULATION OF INTRACELLULAR PO$_2$ IN APLYSIA GIANT NEURONS AND
RESPONSE OF NEURONAL ACTIVITY FOLLOWING CHANGES IN EXTRACELLULAR PO$_2$

W. Erdmann[*], C.F. Chen, S. Armbruster[*] and
B. Lachmann[*]
[*] Department of Anaesthesiology, Erasmus University
Rotterdam, The Netherlands
Department of Physiology, International University
Miami, U.S.A.

INTRODUCTION

The abdominal ganglion of the marine gastropode, aplysia californensis, is a valuable preparation for cellular neuron studies. The size
of some cells range between 500 μ and 1 mm, and thus allow intracellular
measurements of various parameters with microelectrodes, the majority of
the cells are located near the surface of the ganglion and can readily
be visualized under the dissecting microscope. The majority of the cells
are defined according to their functional behaviour divided into groups
named with letters (e.g. R, L), the cells in the groups are numbered (e.g.
R3, R4, R5, R6).
Some of the neurons have single cell pacemaker activity, which is interesting because of its function in temporal organization and information
processing in the nervous system. These neurons (e.g. R3-R6) spontaneously
discharge spikes at a regular rate under constant environment, others
discharge in bursts (e.g. L4).
After penetration of an electrode into the cell, spontaneous activity
ceases and returns with a recovery delay of 1 HR, and more (e.g. R6,
firing rate 0,5 Hz to 1.6 Hz) at room temperature (20°C).
The effects of oxygen on the excitability of neurons have been studied in
various groups of animals including hypoxic and hyperoxic tolerance
(Chalozinitis and Sugaya, 1958; Eccles et al. 1966; Chalozinits, 1968;
Kerkut and York, 1969; Steefin 1975). However, the difficulty in recording intracellular PO$_2$ (IPO$_2$) as opposed to extracellular PO$_2$ (EPO$_2$) has
prevented gathering of essential information concerning the effects of
IPo$_2$ on the bioelectric phenomena.
The objective of the research projected for this model applying a gold
microelectrode system for simultaneous PO$_2$ and action potential measurement (Kunke et al., 1972; Erdmann et al., 1973; Erdmann and Krell, 1976)
was:

1. To determine the intracellular PO$_2$ in relation to extracellular PO$_2$
   to get some insight into the properties of the cell membrane as diffusion barrier.
2. To determine hypoxic and hyperoxic tolerance of the bioelectric phenomena.

METHODS

The spontaneously active pacemaker neuron R6 of the abdominal ganglion of 12 gastropodes (aplysia californensis) were dissected and pinned to a flat plate. The plate was brought into a flat chamber (Fig. 1) with a saline reservoir and a constant and changeable gas flow across the surface of the saline.

Both EPo_2 and IPo_2 were recorded polarographically by employing gold microelectrodes electrochemically etched to tip diameters of 1 μ covered with a collodium membrane and resistances of 100 MΩ to 1000 MΩ. As reference electrodes served Ag-Ag Cl electrodes inserted into KCl-filled glass microelectrodes. Measurement was performed at a polarisation voltage of - 700mV with both recording and reference electrodes as seperate amplification circuits as well intra and extracellularly to avoid interference of the applied voltage with neuronal activity. KCl-filled glass microelectrodes were used to record the bioelectric function (Chen, Erdmann and Halsey 1978).

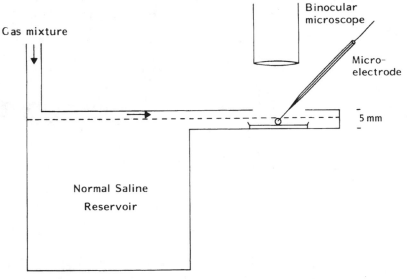

Fig.1: Measurement chamber with saline reservoir.
The oxygen concentration of the constant flow gas mixture (Oxygen-Nitrogen) can be feed back controlled via extracellular PO_2 measurement with a second electrode controlling the oxygen flow regulating electronic valve. Extracellular PO_2-electrode and both reference electrodes, extra - as well as seperate intracellular, are not indicated in this schematic drawing of the experimental set up.

RESULTS

At extracellular PO$_2$ values of $\pm$ 20 mmHg intracellular PO$_2$ showed stable values of between 4.5 and 8 mmHg PO$_2$ as differences among individual cells. Electrical spontaneous cellular activity was recorded in form of intermittant spike intervals (ISI) ranging from 1 to 3 sec ISI. Acute changes to hypoxic values of extracellular PO$_2$ were followed by an immediate decrease of intracellular PO$_2$ which recovered to the original value in 1 to 2 minutes as long as extracellular PO$_2$ was not decreased to below $\pm$ 8 in some and $\pm$ 10 mmHg PO$_2$ in other cells. Further decrease resulted in a permanent decrease of intracellular PO$_2$ and increase of ISI combined with irregularity after an up to 1 minute lasting transient decrease of ISI (transient increase of spike frequency). Acute changes to hyperoxic values of extracellular PO$_2$ were followed by an increase of intracellular PO$_2$ which decreased to the original value slowly with a time delay of up to 10 minutes as long as extracellular PO$_2$ was not increased to values above 42 mmHg in all cells. This readjustment capability of the cells to hyperoxic conditions showed a large variability from cell to cell with some cells even being capable to compensate for extracellular PO$_2$ values of above 60 mmHg PO$_2$. Intermittent spike interval was increased during hyperoxic intracellular PO$_2$ conditions and restored concomittant with the PO$_2$ returning to the original (Fig. 2 and 3).

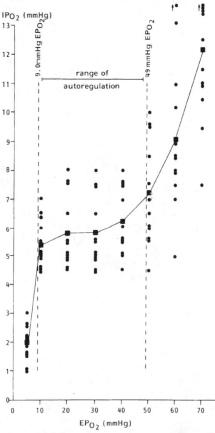

Fig.2: Intracellular PO$_2$ (IPO$_2$) in R6 of the abdominal ganglion of aplysia californensis plotted against extracellular PO$_2$ (EPO$_2$). Between 10 mmHg and 50 mmHg EPO$_2$ the IPO$_2$ is kept fairly constant (autoregulative changes of the diffusion resitance of the cell membrane?).

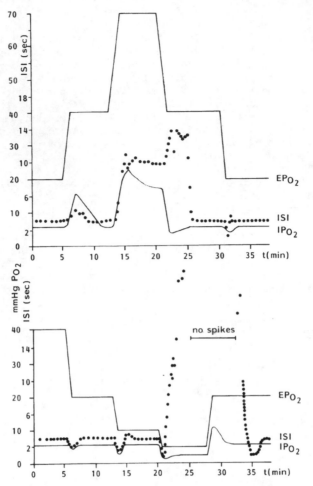

Fig.3.: Responses of IPO$_2$ to changes of EPO$_2$ and effects on bioelec-
tronic function of aplysia giant neurons (R6, spontaneously
discharging spikes). The intermittant spike interval (ISI) is
registered. ISI changes are closely related and very sensitive
to IPO$_2$ changes, but not to changes of EPO$_2$. Regarding only
the intracellular PO$_2$ there is no hypoxic-hyperoxic tolerance
of cellular neurogenic function to be seen.

DISCUSSION

   The results of the described experiments as a follow up investigation of findings described by Chen, Erdmann and Halsey (1978) show clearly
that bioelectric responses to changes in PO$_2$ correspond closely to the
IPO$_2$ where the oxygen is actually utilized but not to the EPO$_2$ from
where the oxygen has to diffuse into the cell for oxydative metabolism.
It is possible that mitochondrial activity is higher near the very inner
neuromembrane where a steep oxygen decrease occurs over a distance
of just a few microns after the measuring tip has penetrated through the
membrane than in the rest of the neuron where a nearly homogenous oxygen
partial pressure is seen (Fig.4). However, this mitochondrial, probably
highly metabolising, layer is so thin that the electrode after bulking

of the membrane before acute penetration immediately passes through into the deeper parts of the cell with relatively homogenous PO_2 values (= IPO_2).

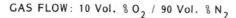

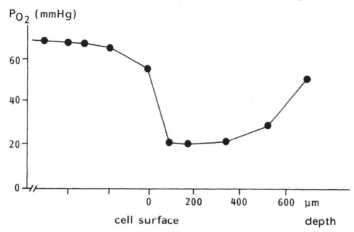

Fig.4: Oxygen partial pressure profile of a pacemaker neuron of aplysia californensis. After a steep decrease of the oxygen partial pressure during penetration through the cell membrane and the immediate inner layer of the neurogenic membrane intracellular PO_2 (IPO_2) shows no further significant changes inside the cell.

Another important question arising when discussing PO_2 dependent neurogenic activity is the one of PO_2 related changes of oxygen consumption. This topic could so far not be studied because of methodological difficulties with the current experimental set up. Previous studies employing metabolic blockers, DNP and ouabain (Chen, Erdmann and Halsey, 1978) with uncoupling of oxidative phosphorylation (DNP) or blocking ATP hydrolysis (ouabain), and thus depletion of cellular ATP did not bring further insight into this aspect.

Under consideration that this study still leaves many question open following statements can so far be concluded: it is very questionable if there exists any hypoxic - hyperoxic tolerance of biolectric phenomena of neurons to changes in intracellular PO_2. However, extracellular changes of PO_2 do not reflect on intracellular PO_2 in relatively broad range of PO_2 changes, with variability among different cells. The so far described tolerance to changing oxygen supply conditions of the cells seems to be based on the capacity of the cell membrane to autoregulate intracellular PO_2 on a steady state value (autoregulative changes of the diffusion characteristics of the cell membrane for oxygen?).

REFERENCES

Chalazonitis N, Sugaya E. (1958). Effets anoxiques sur l'autoactivite electrique des neurons geants d'Aplysia. C.R. Acad. Sci. 247: 1495-1502.

Chalazonitis N. (1968). Intracellular pO2 control on excitability and synaptic activiability in aplysia and helix identifiable giant neurons. Ann. N.Y. Acad. Sci. 147: 421-459.

Chen CF, Erdmann W, Halsey JH. (1978). The sensitivity of aplysia giant neurons to changes in extracellular and intracellular pO2. Adv. Exp. Med. Biol. 94: 691-696.

Eccles RM, Loyning Y, Oshima T. (1966). Effects of hypnoxia on the monosynaptic reflex pathway in the cat spinal cord. J. Neurophysiol. 29: 315-332.

Erdmann W, Kunke S, Krell W. (1973). Tissue-pO2 and cell function. An experimental study with multimicroelectrodes in the rat brain. In: Oxygen Supply. Kessler, Bruley, Clark, Lübbers, Silver, Strauss (eds.). Urban & Schwarzenberg, München-Berlin-Wien.

Erdmann W, Krell W. (1976). Measurement of diffusion parameters with noble metal electrodes. Adv. Exp. Med. Biol. 75: 225-228.

Kerkut GA, York B. (1969). The oxygen sensivity of the electrogenic sodium pump in snail neurons. Comp. Biochem. Physiol. 28: 1125-1134.

Kunke S, Erdmann W, Metzger H. (1972). A new method for simultaneous pO2 and action potential measurement in microareas of tissue. J. Appl. Physiol. 32: 436-438.

Steffin H. (1975). An oxygen dependent current in aplysia neurons: an approach to the question of cellular hypoxic adaptation. Comp. Biochem. Physio. 52A: 691-699.

BRAIN PROTECTION AGAINST OXYGEN DEFICIENCY BY NIZOFENONE

Hiroshi Yasuda, Noriyoshi Izumi, Masato Nakanishi
and Yutaka Maruyama

Research Laboratories, Yoshitomi Pharmaceutical
Ind. Ltd., Chikujo-gun, Fukuoka-ken 871, Japan

INTRODUCTION

Oxygen deprivation is one of the most damaging conditions affecting the brain. When the oxygen supply to the brain is deficient, oxidative energy metabolism ceases and all related metabolic activities are drastically changed, resulting in disorder of cerebral function and tissue deterioration. Much effort has been devoted to finding suitable agents which would enable the brain to withstand oxygen deficiency (Hossmann, 1982).

Barbiturates have been reported to be effective in protecting the brain from hypoxia or ischemia in various animal experiments (Michenfelder et al., 1976; Smith, 1977), but their clinical effectiveness is still controversial and their undesirable side-effects on cardiorespiratory functions limit their use (Abramson et al., 1986).

We have developed a basic experimental screening system for evaluating cerebral protective agents (Nakanishi et al, 1973; Yasuda et al., 1978; Izumi and Yasuda, 1983), and have found a potent, novel cerebroprotective compound, nizofenone fumarate (nizofenone, Fig. 1), with little significant side-effects (Yasuda et al., 1978, 1979, 1981; Tamura et al., 1979; Ochiai et al., 1982; Izumi and Yasuda, 1983).

This paper describes the amelioration by nizofenone of cerebral metabolic disorder following tissue oxygen deprivation and discusses its action mechanism.

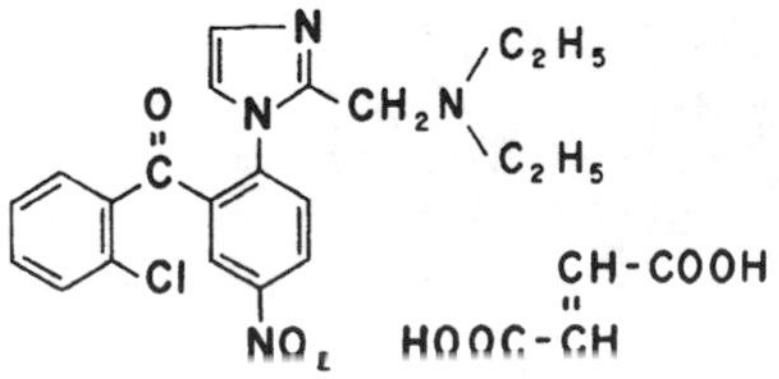

Fig. 1. Chemical structure of nizofenone.

MATERIALS AND METHODS

Animals

Male ddY mice (20-25 g) and infant ICR mice (4-6 g, 9 days old) were used. Animals were housed at 24 ± 1°C.

Histotoxic anoxia induced by KCN injection

Male adult mice (10 per group) were given nizofenone intra-peritoneally, followed 30 min later by an intravenous injection of a lethal dose of KCN (2.5 mg/kg) dissolved in physiological saline solution. The mortality rate of each group was compared with that of the control group.

Cerebral energy metabolism during histotoxic anoxia

The animals were rapidly frozen in liquid nitrogen at measured time intervals (0.5, 1, 2, and 4 min) after the KCN injection. The frozen brains were dissected, weighed and crushed to powder in liquid nitrogen. The powder was mixed with 0.3 ml methanol containing 0.1 M HCl, and 1 ml 3 M $HClO_4$ at -10°C. Three ml 3 mM EDTA was added, and the entire contents homogenized and centrifuged at 13,000 g for 10 min at 0°C. The supernatant was adjusted with 2 M $KHCO_3$ to pH 6.5-6.8, and the precipitated $KClO_4$ was removed by centrifugation. The supernatants were stored at -70°C for analysis.

Cerebral energy metabolism during complete ischemia

Infant ICR mice were given nizofenone intraperitoneally (1 mg/kg) and decapitated 30 min later. The severed heads were rapidly frozen in liquid nitrogen at 0.5, 1 and 2 min following decapitation, and the brains treated for analysis as described above.

Analysis

Brain high energy phosphate compounds, glucose and lactate were determined by the enzymic fluorometric method of Lowry and Passonneau (1972) as previously reported (Yasuda et al., 1979, 1983).

RESULTS

Brain protection against histotoxic anoxia caused by KCN

The protective effect of nizofenone against KCN-induced histotoxic cerebral anoxia was examined in mice (Table 1). Mice injected with KCN (2.5 mg/kg i.v.) fell into a coma with con-vulsions and stopped breathing about 1 min later. The mortality rate of the untreated control animals was 80 % or more. Pre-treatment with nizofenone resulted in a dose-dependent decrease in the mortality rate, with significant protection observed at as low a dose as 0.3 mg/kg (i.p.). At 10 mg/kg, all of the animals used revived from the coma after several min and no apparent behavioral disorders were observed the next day. The dose of nizofenone for reducing the mortality rate to one-half was estimated to be approximately 0.2 mg/kg.

Table 1. Effect of Nizofenone on Mortality Induced by KCN in Mice

	Dose mg/kg i.p.	No. of mice survived/used	Mortality rate (%)
Control	–	2/10	80
Nizofenone	0.1	5/10	50
	0.3	9/10**	10
	1	9/10**	10
	10	10/10**	0

**: p<0.01
Mice were intraperitoneally treated with nizofenone 30 min prior to the injection of KCN (2.5 mg/kg i.v.).

Amelioration of cerebral metabolic disorder induced by KCN

Fig. 2 shows the changes in cerebral high energy phosphate stores, glucose and lactate concentrations in mice during histotoxic cerebral anoxia induced by KCN. In the control animals, the cerebral phosphocreatine (P-creatine) stores rapidly decreased to 29 % of the normal level of 3.08 ± 0.09 µmol/g 1 min after the KCN injection, and to 2 % 1 min later, and thereafter remained at this low level. In the nizofenone-treated animals, the cerebral P-creatine concentration decreased to a minimum of 58 % of the normal level 2 min after the KCN injection, and recovered to 87 % 2 min later.

The cerebral ATP concentration in the control animals was also completely depleted 4 min after the KCN injection, while the ATP concentration in the nizofenone-treated animals stayed at 88 % or more of the normal level of 2.83 ± 0.03 µmol/g.

The cerebral glucose concentration in the control animals rapidly decreased to about one-fifth of the normal level 1 min after the KCN injection and slowly decreased to near zero within the next 3 min. In the nizofenone-treated animals, the depletion rate was significantly lower, and the glucose concentration recovered to almost the normal level 4 min after the KCN injection.

The cerebral lactate concentration increased to about 5 times the normal value of 1.78 ± 0.22 µmol/g 1 min after the KCN injection in the untreated control group. The rate of accumulation of cerebral lactate during anoxia was significantly lower in the nizofenone-treated group. The lactate level had returned almost to normal 30 min after the injection (data not shown).

The cerebral energy charge potential (ECP), which is an index of the energy state of the tissues, was calculated in order to evaluate any imbalance between the rate of production of ATP and the rate of utilization of ATP in the tissues. In the control animals, the cerebral ECP decreased to 22 % of the normal value of 0.899 ± 0.005 2 min after the KCN injection. The ECP in the nizofenone-treated animals remained at near normal levels.

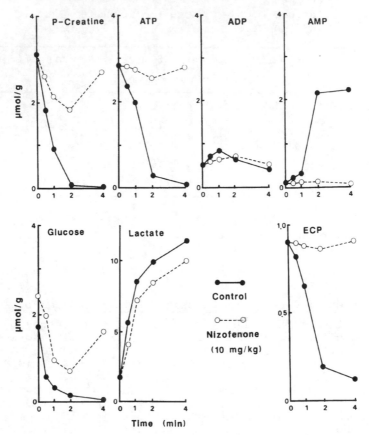

Fig. 2. Effect of nizofenone on cerebral metabolic disorder
induced by KCN injection. Each circle represents
the mean value for 6 animals.

Attenuation of cerebral metabolic disorder during ischemia

Fig. 3 shows the changes in cerebral high energy stores, and
in glucose and lactate concentrations in infant mice during complete global ischemia following decapitation. For the nizofenone-
treated group, the rate of depletion of cerebral P-creatine, ATP
and glucose was significantly lower than for the untreated control group. Nizofenone treatment also reduced significantly the
rate of accumulation of cerebral lactate.

DISCUSSION

Nizofenone dose-dependently decreased the mortality rate of
mice injected with KCN and provided significant protection at
0.3 mg/kg. This result coincides with the results of the basic
screening tests (Table 2), indicating that nizofenone has a
potent and consistent cerebral protective action. Its excellent
cerebroprotective effect has been ascertained by Tamura et al.
(1979), who demonstrated that nizofenone reduces the size of
cerebral infarctions following chronic middle cerebral artery
occlusion in cats.

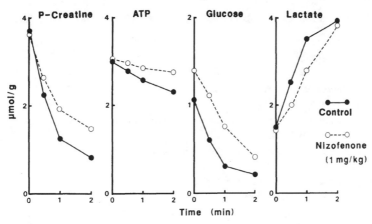

Fig. 3. Effect of nizofenone on cerebral metabolic changes
during complete ischemia following decapitation.
Each circle represents the mean value for 6 animals.

Nizofenone ameliorated KCN-induced cerebral metabolic disorder (Fig. 2). It lowered the rate of depletion of cerebral P-
creatine and glucose concentrations and also lowered the rate of
accumulation of lactate, while keeping cerebral ATP and ECP at
nearly their normal levels. In addition, in the nizofenone group
these metabolic changes returned to normal, but in the control
group they were irreversible.

Additional biochemical evidence that nizofenone has the
ability to decrease cerebral energy consumption and ameliorate
cerebral metabolic disorder was obtained from another experiment
in which cerebral oxygen deficiency was induced by complete ischemia following decapitation (Fig. 3).

The present results coincide with the previous findings
that nizofenone reduces the cerebral energy demand or $CMRO_2$
(Yasuda et al., 1979; Ochiai et al., 1981). This action could
contribute to its cerebral protective effect: with reduced cerebral energy demand, less ATP is required so that the cerebral
ATP concentration can be maintained at higher levels during oxygen deficiency. Indeed, in the nizofenone-treated group higher,
nearly normal, levels of ATP and ECP were maintained throughout
the KCN-induced cerebral anoxia (Fig. 2), whereas the levels in
the control group were irreversibly depleted.

The lower cerebral lactate accumulation in the nizofenone-
treated group indicates that there is less lactic acidosis to
cause tissue damage.

Free fatty acids (FFA) in the brain are normally present in
small amounts but accumulate during oxygen deficiency (Bazan,
1970). This FFA accumulation is accompanied by a decrease in
some membrane phospholipids and is considered to be one of the
injurious reactions to brain damage (Yoshida et al., 1980;
Siesjö, 1981). In addition, cerebral reoxygenation after ischemia
propagates peroxidative reactions with free or esterified polyunsaturated fatty acids in membranes, which enhance the tissue
damage (Yoshida et al., 1980; Chan and Fishman, 1980; Siesjö,
1981). Nizofenone has been reported to attenuate the cerebral FFA
accumulation (Yoshida et al., 1980; Nemoto et al., 1982; Yasuda

Table 2. Effective Dose of Nizofenone in Various Models

Model	Route		Parameter	Effective dose
Complete ischemia (decapitation)	Mouse	i.p.	Gasping time	0.1 mg/kg
Global ischemia (c.a. ligation)	Mouse	i.p.	Survival time	0.3
Hypobaric hypoxia (low pressure)	Mouse	i.p.	Survival time	0.2
Normobaric hypoxia (4 % oxygen)	Mouse	i.p.	Survival time	0.3
Asphyxic hypoxia (asphyxia)	Rat	i.p.	EEG	0.2
Histotoxic hypoxia (KCN)	Rat	p.o.	Mortality	0.3

et al., 1985) and prevent peroxidative membrane disintegration (Yoshida et al., 1980; Yasuda et al., 1981). These properties of nizofenone could also contribute to its cerebral protective effect.

Prostacyclin is a potent vasodilator and inhibitor of platelet aggregation, while thromboxane A_2 possesses vasocon-stricting and platelet aggregating properties. This prosta-cyclin-thromboxane A_2 balance may be perturbed in certain patho-logical states including cerebral vasospasm, transient ischemic attacks and the 'no-reflow phenomenon' observed after ischemia. Prostacyclin has recently been reported to have some therapeutic benefit in the treatment of ischemic stroke (Gryglewski et al., 1983). Nizofenone has been demonstrated to stimulate prostacyclin synthesis while inhibiting thromboxane A_2 formation (Yasuda et al., 1984). This action of nizofenone may also contribute to its cerebral protective effect.

In the present study, we were able to demonstrate that nizofenone decreases cerebral energy consumption and so increases the resistance to anoxia or ischemia. This finding suggests that this drug may be effective in attenuating human brain damage following oxygen deficiency, which occurs with strokes, cardiac arrest, or during thoracic, cardiac or neurosurgery. In fact, a clinical study by Saito et al. (1983) has shown that nizofenone improves the functional outcome following subarachnoid hemorrhage which often leads to vasospasm-induced late-ischemic attacks. Furthermore, a recent large scale double-blind, controlled trial by Ohta et al. indicated that this drug is of value in protecting the brain from ischemic damage (Ohta et al., 1986).

We conclude that nizofenone can provide cerebral protection against oxygen deficiency mainly by reducing cerebral energy con-sumption. Another possible mechanism of action is its protection of membranes from peroxidative degradation caused by ischemia and/or post-ischemia.

SUMMARY

The cerebral protective effect of nizofenone against tissue oxygen deficiency was investigated in mice. Treating mice with nizofenone resulted in a dose-dependent decrease in the rate of KCN-induced mortality and significant protection was observed at a dose as low as 0.3 mg/kg (i.p.). The cerebroprotective action of nizofenone was also demonstrated biochemically: Nizofenone (10 mg/kg i.p.) ameliorated KCN-induced anoxic disorder of cerebral energy metabolism, characterized by irreversible depletions of cerebral high energy phosphate stores and glucose concentrations and a marked accumulation of lactate, while keeping the cerebral energy charge potential (ECP) close to its normal value. Nizofenone showed similar effects in another experiment in which cerebral metabolic disorder was induced by complete ischemia following decapitation.
These findings suggest that nizofenone has a considerable cerebroprotective action against oxygen deficiency.

REFERENCES

Abramson N.S. and Brain Resuscitation Clinical Trial I Study Group, 1986, Randomized clinical study of thiopental loading in comatose survivors of cardiac arrest, New Eng. J. Med., 314: 397.
Bazan N.G., 1970, Effects of ischemia and electroconvulsive shock on free fatty acid pool in the brain, Biochim. Biophys. Acta, 218: 1.
Chan P.H. and Fishman R.A., 1980, Transient formation of superoxide radicals in polyunsaturated fatty acid-induced brain swelling, J. Neurochem., 35: 1004.
Gryglewski R.J., Nowak S., Kostka-Trabka E., Kusmiderski J., Dembinska-Kiec A., Bieron K., Basista M. and Blaszczyk B., 1983, Treatment of ischemic stroke with prosta-cyclin, Stroke, 14: 197.
Hossmann K.A., 1982, Treatment of experimental cerebral ischemia, J. Cereb. Blood Flow Metabol., 3: 275.
Izumi N. and Yasuda H., 1983, A convenient cerebral ischemia model using mice, J. Cereb. Blood Flow Metabol., 3 (Suppl. 1): 391.
Lowry O.H. and Passonneau J.V., 1972, "A flexible system of enzymatic analysis", Academic Press, New York.
Michenfelder J.D., Milde J.H. and Sundt T.M.Jr., 1976, Cere-bral protection by barbiturate anesthesia, Arch. Neurol., 33: 345.
Nakanishi M., Yasuda H. and Tsumagari T., 1973, Protective effect of anti-anxiety drugs against hypoxia, Life Sci., 13: 467.
Nemoto E.M., Shiu G.K., Nemmer J. and Bleyaert A.L., 1982, Attenuation of brain free fatty acid liberation during global ischemia: a model for screening potential thera-pies for efficacy?, J. Cereb. Blood Flow Metab., 2: 475.
Ochiai C., Asano T., Tamura A., Sano K., Fukuda T. and Nakamura T., 1981, An experimental study on the mecha-nism of the protective action of pentobarbital and Y-9179 against cerebral ischemia, Neurol. Med. Chir. (Tokyo), 21: 303.

Ochiai C., Asano T., Takakura K., Fukuda T., Horizoe H. and
Morimoto Y., 1982, Mechanisms of cerebral protection by
pentobarbital and nizofenone correlated with the course
of local cerebral blood flow changes, Stroke, 13: 788.
Ohta T., Kikuchi H., Hashi K. and Kudo Y., 1986, Nizofenone
administration in the acute stage following subarachnoid hemorrhage. Results of a multi-center controlled
double-blind clinical study, J. Neurosurg., 64: 420.
Saito I., Asano T., Ochiai C., Takakura K., Tamura A. and
Sano K., 1983, A double-blind clinical evaluation of
the effect of nizofenone (Y-9179) on delayed ischemic
neurological deficits following aneurysmal rupture,
Neurol. Res., 5: 29.
Siesjö B.K., 1981, Cell damage in the brain: a speculative
synthesis, J. Cereb. Blood Flow Metab., 1: 155.
Smith A.L., 1977, Barbiturate protection in cerebral hypoxia, Anesthesiology, 47: 285.
Tamura A., Asano T., Sano K., Tsumagari T. and Nakajima A.,
1979, Protection from cerebral ischemia by a new
imidazole derivative (Y-9179) and pentobarbital: A
comparative study in chronic middle cerebral artery
occlusion in cats, Stroke, 10: 126.
Yasuda H., Shuto S., Tsumagari T. and Nakajima A., 1978,
Protective effect of a novel imidazole derivative
against cerebral anoxia, Arch. Int. Pharmacodyn. Ther.,
233: 136.
Yasuda H., Nakanishi M., Tsumagari T., Nakajima A. and
Nakanishi M., 1979, The mechanism of action of a novel
cerebral protective drug against anoxia. I. The effect
on cerebral energy demand, Arch. Int. Pharmacodyn.
Ther., 242: 77.
Yasuda H., Shimada O., Nakajima A. and Asano T., 1981, Cere-
bral protective effect and radical scavenging action,
J. Neurochem., 37: 934.
Yasuda H., Nakanishi M. and Izumi N., 1983, Cerebral pro-
tection by pentetrazol against hypoxia, in: "Current
Problems in Epilepsy", Baldy-Moulinier M., Ingvar D.H.
and Meldrum B.S. ed., John Libbey, London, pp. 292-296.
Yasuda H., Ochi H. and Tsumagari T., 1984, Stimulation of
prostacyclin synthesis by nizofenone, Biochem.
Pharmacol., 33: 2707.
Yasuda H., Kishiro K., Izumi N. and Nakanishi M., 1985,
Biphasic liberation of arachidonic and stearic acids
during cerebral ischemia, J. Neurochem., 45: 168.
Yoshida S., Inoh S., Asano T., Sano K., Kubota M., Shimazaki
H. and Ueta N., 1980, Effect of transient ischemia on
free fatty acids and phospholipids in the gerbil brain:
lipid peroxidation as a possible cause of postischemic
injury, J. Neurosurg., 53: 323.
Yoshida S., Inoh S., Asano T., Sano K., Kubota M., Shimazaki
H., Ueta N., Yasuda H. and Shimada O., 1980, Brain free
fatty acids in transient cerebral ischemia and the
effect of metabolic depressants, Brain Nerve, 32: 931.

THE INFLUENCE OF THE CALCIUM ANTAGONISTS FLUNARIZINE AND VERAPAMIL ON CEREBRAL BLOOD FLOW AND OXYGEN TENSION OF ANESTHETIZED WFS-RATS

Hermann P. Metzger   and   Y. Savas

Dept. of Physiology, Medizinische Hochschule Hannover
P.O. Box 61o 18o, 3ooo Hannover, Fed. Rep. of Germany

INTRODUCTION

The beneficial influence of calcium antagonists in the treatment of coronary heart disease, hypertension, etc., has been examined from various points of view in both, chronic and acute animal experiments (Henry, 1980; Holmes et al., 1984). However, there is little information concerning the effects of calcium antagonists on cerebrocortical surface PO_2 (sPO_2) and cerebral blood flow (CBF). In contrast to the generally accepted animal models simulating ischemia, hypertension, arteriosclerosis, etc., experiments have been performed near the metabolic equilibrium of the normal rat. This knowledge might be helpful in finding an adequate therapy for the ischemic patient and in learning more about the unexpected side-effects of the calcium antagonists.

The influence of two structurally completely different molecules, verapamil and flunarizine, on sPO_2 and CBF has been studied. In addition, the time-responses of CBF and sPO_2 to the infusion of calcium antagonists have been analysed in order to find specific reactions. The results of the analysis were then compared with the action of three different pharmacological molecules known to influence the systemic blood pressure (MAP): sodium-nitroprusside for lowering of MAP, norepinephrine for MAP elevation, and dihydroergotamine for slightly increasing and stabilising MAP.

MATERIAL AND METHODS

Animals and anesthesia: Wistar-Frömter rats (N=56, 180-250 g b.w.) were anesthetized with Ketavet (30 mg/100 g b.w.; ketamine hydrochloride, Parke-Davis, München, FRG) and Rompun (1.2 mg/100 g b.w.; xylazine hydrochloride, Bayer Leverkusen, FRG). Following intubation and relaxation the animals were artificially ventilated by using positive inspiratory pressure with a respiratory rate of 70/min and a tidal volume of 1.5 - 2.2 ml depending upon the body weight of the animals. Hyperventilation ($PaCO_2$ = 28 mm Hg) was induced in some animals in order to enhance the vasodilatory effect of the calcium antagonists. During cooling with ringer-lactate (37°C) a small window measuring 5x5 mm^2 was drilled into the skull cap between sagittal, frontal, and lambdoidea sutures; the bones were carefully removed using fine forceps.

PO_2 and H_2 surface electrode construction: Miniaturized polarographic electrodes used to measure surface PO_2 (sPO_2) and H_2 clearance were especially designed to fit the dimensions of the cranial window of the rat. Six wires made of gold (15 um diameter) for sPO_2 measurements and four wires made of platinum-iridium (100 um diameter) for H_2 clearance measurements were soldered onto small printed plates, connected with fine teflon-insulated wires and inserted into a specially produced flexible cable sheath. The electrode arrangement was cast into a plexi-glass mould with artificial resin (Hysol, Dexter Corp., New York). A silicone rubber tube was used to protect the cable against moisture. Miniaturized pico-amperemeter amplifiers were fixed in position directly adjacent to the rat's head. The signals obtained from the latter were amplified and plotted on a six-channel potentiometric recorder (Linseis Corp., Selb, FRG). A fine thermistor was inserted into the body of the electrode to continually monitor the temperatur of the animal during experiments on the exposed brain surface as well as for calibration purposes in ringer-lactate.

The CBF data were calculated as mean values from 5 different hydrogen washout curves for each animal. The curves were plotted on semilogarithmic paper; the slope of the straight line constructed through the interval of the first two minutes was used for CBF calculation. A factor of 0.69 was taken as the H2-distribution coefficient between blood and tissue.

RESULTS

In order to increase the vasodilatory effect of the calcium antagonists verapamil and flunarizine on the cerebral vascular smooth muscle, six animals were hyperventilated inducing hypocapnic vasoconstriction. After 10 minutes of artificial ventilation $PaCO_2$ decreased from 38.5 to 26.7 mm Hg while pHa remained almost constant at 7.38. PaO_2 increased from 91.7 to 102.3 mm Hg; the base deficit lay between 6 - 7.7 mmol/l. Nevertheless, mean arterial pressure (MAP) remained almost constant during the course of hyperventilation (MAP$\pm$SD = 85$\pm$ 5 mm Hg). Surface PO_2 remained constant during the course of hyperventilation. Under these conditions, the vasodilatory effect was not enhanced with either a bolus application or with an infusion of calcium antagonists. Therefore, during the course of the other experiments (N=50) the animals were adjusted according to a normal acid-base status.

Flunarizine application: A bolus injection of flunarizine (N=8) caused a considerable drop of MAP from 85 to 69 mm Hg while surface PO_2 showed a slight decrease from 22.5 to 19 mm Hg. In a second group of experiments (N=5) the MAP also showed an obvious fall from 85 to 56 mm Hg, whereas the sPO_2 remained constant. The heart rate diminished from 271$\pm$9 to 260$\pm$11/min (N=13) after flunarizine administration.

Infusion of flunarizine (0.31 mg/kg·min) induced a slow MAP decrease corresponding to the effective plasma concentration; in two cases there was a decrease as far as 50 mm Hg. The mean sPO_2 curves calculated from the individual time-courses of sPO_2 showed a slight and insignificant increase during the 16-minute period of infusion (Fig. 1).

In contrast to sPO_2, the time-course of cerebral blood flow (CBF) in response to flunarizine infusion was characterized by a considerable increase in all cases examined (N=7), (Fig. 2). Extremely high CBF values, in the 3 - 3.5 ml/g·min range, were observed in some of the experiments (N=3) whereas in others (N=4) CBF amounted to 1.2 - 1.5 ml/g·min.

Verapamil infusion: In response to a two-minute application of the drug, the "specific pharmacological effect" of verapamil can clearly be seen (Fig. 3). Following the beginning of verapamil infusion a rise in sPO_2 occured corresponding to the time-dependent increase of the verapamil plasma concentration. When the infusion was discontinued, sPO_2 and MAP immediately approximated their initial levels with the same kinetics as were observed prior to infusion. If the verapamil infusion was extended over a 30-minute period, the mean sPO_2 calculated from 5 individual electrode-readings showed a clear sPO_2 elevation in three cases and a slight sPO_2 increase in one. A maximum sPO_2 rise of 50% was obtained from the experimental data of all animals from an initial sPO_2 in the range 20 - 35 mm Hg (Fig. 4). The mean arterial blood pressure showed a decrease from 90 to 51 mm Hg during the course of infusion. In one example, sPO_2 increased slightly while MAP remained almost constant. In this case the infusion probably was not completely effective. Cerebral blood flow showed a maximum increase of 2oo-3oo% of the initial values. However, the absolute values were somewhat smaller in comparison with experiments on the flunarizine group. Nevertheless, almost the same effects have been observed in relation to the relative CBF and sPO_2 increments. In some of the experiments CBF showed a pronounced increase while in some of the others it was elevated only slightly (Fig. 5).

Control experiments: In order to study the influence of the volume effects on CBF and sPO_2 values, equivalent amounts of ringer-lactate were infused during a 60-minute interval (N=9 animals). Initial CBF values were in the range of 0.6 - 1.5 ml/g·min and did not change during the whole course of the experiments (Fig. 6).

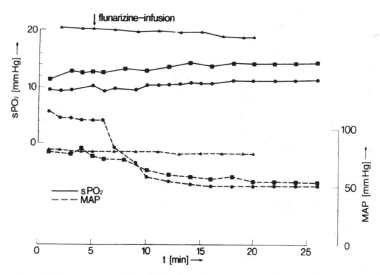

Fig. 1. Cerebral surface PO_2 (sPO_2) and mean arterial pressure (MAP) in response to flunarizine infusion (0.31 ml/kg·min). Each curve summarizes the data from 5 different electrode registrations per animal.

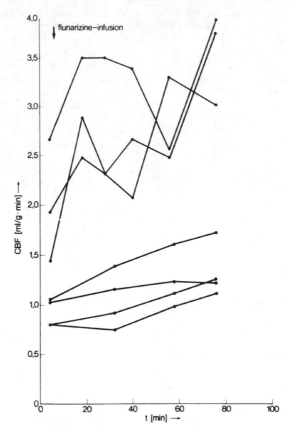

Fig. 2. Effects of flunarizine infusion on CBF measured by H_2 clearance. Each curve was calculated as mean value from 5 different electrode-registrations per animal.

Experiments with substances affecting blood pressure (non-calcium antagonists): A bolus application of the beta- and alpha-sympathicomimetic n-ethyl-norepinephrine caused a pronounced MAP and parallel sPO_2 increase which disappeared completely within a few minutes. Sodium nitroprusside, on the other hand, induced a fall of systemic pressure but kept sPO_2 constant. A similiar MAP and sPO_2 time-course was observed in response to injection of the alkaloid dihydroergotamine which is known to prevent blood volume storage within the venous system. The CBF and sPO_2 responses to different drugs were transient and initial values were restored within 10-20 minutes after application in all cases.

DISCUSSION

The helpful effects of the calcium antagonists in bringing the disturbed microcirculation towards its normal level have already been examined for a variety of conditions by preparing single blood vessels (Triggle, 1982), perfusing the isolated brain (Krieglstein and Weber, 1986) and studying whole animals by determining organ-blood supply and cardiac

414

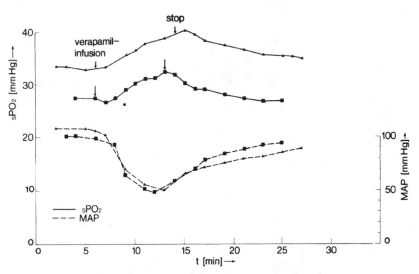

Fig. 3. "Specific pharmacological effect" of verapamil infusion (7.5 ug/kg·min) on sPO$_2$ and MAP. Note the symmetric time course between the beginning and end of infusion.

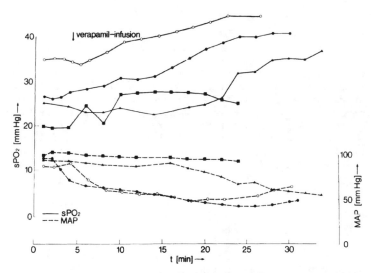

Fig. 4. sPO$_2$ of cerebral cortex and MAP in response to verapamil infusion (7.5 ug/kg·min). Each curve represents the mean of 5 different electrode-registrations per animal.

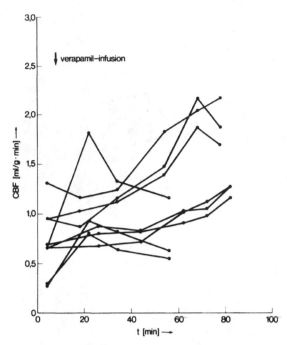

Fig. 5. Effects of verapamil infusion on CBF measured by means of H_2 clearance. Each curve was calculated as mean value from five different electrode-registrations per animal.

functions (Flaim and Kanda, 1982). Calcium antagonists affect potential-dependent and/or receptor specific channels of the plasmalemma reducing the entrance of calcium ions into the cell and preventing intracellular calcium overload. As a consequence, myocytes and vascular smooth muscle cells are protected against energy consuming contractions. Moreover, the active calcium transport into the mitochondria and the sarcoplasmic reti-culum has to be taken into consideration, if the cytosolic calcium con-centration is excessively disturbed such as in ischemia. Recently, a subgroup of calcium antagonist molecules have been identified as having the capability of selectively blocking the intracellular receptor molecule calmodulin which is known to be located within the vascular smooth muscle (Hidaka et al., 1980). Calmodulin activates light chain-kinase by forming the activated calcium-calmodulin complex which is necessary for the phosphorylation of light chain myosin. Some evidence exists that flunarizine characterized to be highly lipophilic belongs to the group of calmodulin inhibitors. On the other hand, verapamil with less lipophilic and more hydrophilic properties (also found in nifedipine and diltiazem) serves as a slow channel blocker causing pronounced smooth muscle relaxa-tion but with little calmodulin inhibition.

Flunarizine, a calmodulin inhibitor and verapamil, a calcium channel blocker have been compared. A noticeable difference concerning the kinetic effects of the two molecules has not yet been evaluated from the

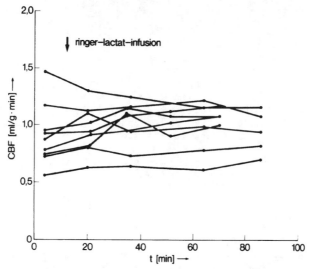

Fig. 6. Control experiments: Infusion of equal amounts of ringer-lactate. Each curve represents the mean of 5 different electrode-registrations from the same animal.

experiments. The reactions are unspecific and, unfortunately, do not allow any further speculation into the molecular mechanisms concerning CBF and sPO_2 on the superficial grey matter of the rat.

It is interesting to note that both molecules, flunarizine and verapamil cause a pronounced CBF increase despite a considerable fall in systemic pressure. A remarkable flow redistribution exists even within the normal animal due to cerebral vasodilatation. A slight improvement of cerebral sPO_2 has been observed for verapamil but not for flunarizine. The difference between the molecules with respect to local oxygen tension is not clear. A fall in systemic arterial pressure has been observed for both calcium antagonists – flunarizine and verapamil. The MAP decrease might be caused by a lowering of heart minute volume and by a reduced peripheral resistance.

In summarizing the results, it is concluded that the improvement of the cerebrocortical oxygen supply by vasodilatation of vascular smooth muscle is compensated at least in part by the marked drop in systemic pressure. However, flunarizine and verapamil cannot provide improvement in the therapy for the ischemic patient. The necessity for further development of brain specific calmodulin inhibitors and calcium channel blockers is derived from the data. This generation of molecules may have the quality of minor efficiency within other microcirculatory systems of the organism.

SUMMARY

The beneficial influence of calcium antagonists in restoring the disturbed circulation and metabolism towards its normal level has been described for a variety of circulatory disorders (arteriosclerosis, ischemia, hypertension). To date, however, little information exists concerning the physiological effects of the blocking molecules on cerebral blood flow (CBF) and oxygen tension at the organ surface (sPO$_2$). White rats (Wistar-Frömter strain, N=56, 180-250 g bw) were anesthetized with ketamine-xylazin, and artificially ventilated until normal acid-base status was achieved. Polarographic multi-wire O$_2$ and H$_2$ electrodes (15 um diameter, 1 g weight) were balanced on the open brain through a cranial window drilled into the skull-cap in order to perform surface PO$_2$ and hydrogen clearance measurements.

Two different calcium antagonists, verapamil and flunarizine, were tested in order to verify structural differences of the molecules. Infusion of verapamil (7.5 ug/kg·min) and flunarizine (0.3 mg/kg·min) induced an increase in CBF by 55%/h (verap.) and by 62%/h (flun.) respectively while the controls (infusion of equal amounts of ringer-lactate) remained constant, ranging between 0.5 and 2.7 ml/g·min (n=25 measurements). Surface PO$_2$ improved distinctly in response to verapamil (39% increase) but was uneffected by flunarizine (less than 1%). Both drugs, however, lowered MAP by 18% (verap.) and by 33% (flun.) respectively probably due to peripheral vasodilatation and to the lowering of heart minute volume. In comparison with flunarizine the small MAP change after verapamil has resulted in the rise in CBF and sPO$_2$. The considerable MAP drop in response to flunarizine, on the other hand, compensated the improved CBF increase in part and led to an almost constant sPO$_2$ level.

REFERENCES

Flaim, S. F., and Kanda, K., 1982, Comperative pharmacology of calcium
    blockers based on studies of cardiac output distribution, in:
    "Calcium Blockers", F. Flaim and R. Zelis, eds., Urban and
    Schwarzenberg, Baltimore-München.
Henry, P. D., 1980, Comperative pharmacology of calcium antagonists
    nifedipine, verapamil and diltiazem, Am. J. Cardiol., 46:1047.
Hidaka, H., Yamaki, T., Naka, M., Tnaka, T., Hayashi, H., and
    Kobay-ashi, R., 1980, Calcium regulated modulator protein
    interacting agents inhibit stimulated protein kinase and ATPase,
    Mol. Pharmacol., 17:66.
Holmes, B., Broyden, R. N., Heed, R. C., Speight, T. M., and Avery, G. S.,
    1984, Flunarizine: a review of its pharmacodynamic and pharmaco-
    kinetic properties and therapeutic use, Drug, 27:6.
Krieglstein, J., and Weber, J., 1986, Calcium entry blockers protect
    brain energy metabolism against ischemic damage, Adv. Exp. Med.
    Biol., 200:243
Triggle, D. J., 1982, Biochemical pharmacology of calcium blockers, in:
    "Calcium Blockers", F. Flaim and R. Zelis, eds., Urban and
    Schwarzenberg, Baltimore-München.

CARDIOVASCULAR
SYSTEM

VENTRICULAR SYSTOLIC PRESSURE-VOLUME AREA (PVA) AND

CONTRACTILE STATE (Emax) DETERMINE MYOCARDIAL OXYGEN DEMAND

Hiroyuki Suga, Yoshio Yasumura, Takashi Nozawa, Shiho Futaki, Nobuaki Tanaka, and Masaaki Uenishi

Dept. of Cardiovascular Dynamics, National Cardiovascular Center Research Institute, Suita, Osaka, 565 Japan

INTRODUCTION

Cardiac contraction requires a continuous oxygen supply via the coronary circulation. The oxygen demand of the heart per unit mass under physiological conditions is high, around 8 ml/min/100g, comparable to that of active skeletal muscle. Cardiac oxygen demand varies widely depending on cardiohemodynamic, neural (sympathetic and vagal), and humoral (endocrinological and pharmacological) conditions (Gibbs, 1978). When, as in normal, coronary oxygen supply (= coronary flow x arterial oxygen content x maximum oxygen extraction fraction) exceeds cardiac oxygen demand, cardiac contractions can continue without being restricted by the supply. However, when the coronary oxygen supply is limited abnormally, the heart cannot maintain normal contractions and the cardiac contractile state is depressed (Weber et al., 1980). In fact, an acute decrease in coronary perfusion decreases cardiac contraction immediately. For a better understanding of the balance between cardiac oxygen supply and demand, factors determining the cardiac oxygen demand must be fully elucidated. The cardiac oxygen demand of ongoing contractions can be quantified by cardiac oxygen consumption (Vo_2) under aerobic conditions, i.e., non-ischemic (Weber et al., 1980). In this sense, we can use cardiac oxygen demand and Vo_2 interchangeably under conditions of sufficient coronary oxygen supply. However, the value for cardiac oxygen demand expressed as Vo_2 of the ongoing contraction under decreased coronary perfusion conditions may be smaller than that obtained from the heart with a sufficient coronary oxygen supply (Weber et al., 1980).

Early in this century, Evans and Matsuoka (1915) investigated how cardiac pressure and volume loads affected cardiac Vo_2. Their results showed that for a given external work load the heart consumes more oxygen for a higher pressure load and less oxygen for a greater volume or flow load. This characteristic of cardiac energetics has been firmly established (Sarnoff et al. 1958; Suga et al., 1982). This background has generated the contemporary concept that systolic pressure is one of the most important determinants of cardiac Vo_2 (Gibbs, 1978). In contrast, external work per se has long been considered as a secondary determinant of cardiac oxygen demand and Vo_2.

Sonnenblick et al. (1965) proposed the maximum speed (Vmax) of myocardial shortening as an index of contractile state. Vmax increases with positive inotropic agents such as catecholamines and Ca^{2+}. Positive inotropic agents not only enhance myocardial contraction but also augment cardiac Vo_2 (Graham et al., 1968): for a given systolic pressure (or force), Vo_2 is increased by positive inotropic agents. This is called the oxygen wasting effect. This background has yielded the current view that cardiac contractility is another important determinant of Vo_2 (Gibbs, 1978). Besides these two determinants (pressure or force, and contractility), Vo_2 per min increases in proportion to heart rate (Boerth et al., 1969). Basal metabolism requires a stable Vo_2 per min although it increases slightly with positive inotropic agents (Klocke et al., 1965). Generation of the membrane action potential requires a minute fraction of Vo_2 (Klocke et al., 1966).

Despite the elucidation of these determinants of cardiac Vo_2 (Gibbs, 1978), one cannot yet reliably predict cardiac Vo_2, under a variety of cardiac conditions, from cardiohemodynamic parameters including cardiac output, arterial pressure, heart rate, and contractility (Rooke and Feigl, 1982).

With this background in mind, we have been searching for a new way by which we can quantitatively relate left ventricular Vo_2 with cardiohemodynamic variables and ventricular contractility (Suga et al., 1981a, 1981b, 1982, 1983a, 1983b, 1984a, 1984b, 1984c, 1985, 1986, 1987). We consider that our efforts so far have been successful to this end. Therefore, on this occasion, we have summarized the advances, made over the last 10 years, in our research to facilitate the understanding of determinants of cardiac oxygen demand.

METHODS

Excised, Cross-Circulated Heart Preparation

We used the left ventricle of the excised dog heart metabolically supported by cross circulation with a support dog (Suga et al., 1981a, 1986). It was independent of neural control. The left ventricle was fitted with a thin balloon which was connected to a custom-made volume servo pump (Suga et al., 1981a, 1983a). With this pump, we could control precisely and measure accurately the left ventricular volume. Left ventricular pressure was measured with a miniature pressure gauge inside the balloon. The temperature of the heart was kept at $35-37^{0}C$. Blood gas and pH were corrected to normal. Heart rate was fixed by electric pacing. Mean coronary perfusion pressure was usually 65-100 mmHg, which was equal to the systemic arterial pressure of the support dog. Coronary arterial oxygen content was 12-17 vol%. The heart maintained normal contractions over 4-6 hours. For details, see the References.

Cardiac Oxygen Consumption

Coronary flow was measured with an electromagnetic flowmeter on the coronary venous draining tube from the right heart. Coronary arteriovenous oxygen content difference (AVOD) was continuously measured with an oximeter (Waters Instrument 600A, or Erma PWA-200) or an AVOX oxygen content difference meter (Shepherd and Burger, 1977), calibrated against a Lex-O_2-Con (Lexington Instrument) oxygen content analyzer. The product of AVOD and coronary flow yielded cardiac Vo_2 per min. This was divided by heart rate to determine cardiac Vo_2 per beat in the steady state. Since the right ventricle was kept collapsed and mechanically unloaded, we assumed that the right ventricular Vo_2 was constant regardless of left ventricular Vo_2 which varied with left ventricular loading conditions.

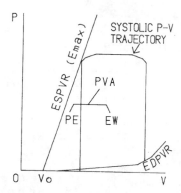

Fig. 1   Pressure-volume area (PVA) in the pressure-volume (P-V) diagram. PVA is the area circumscribed by the end-systolic P-V relation (ESPVR) line, end-diastolic P-V relation (EDPVR) curve, and the systolic P-V trajectory. PVA consists of external work (EW) and potential energy (PE). Emax is the slope of ESPVR.

Pressure-Volume Area (PVA)

As a new approach to the problem, we utilized a new measure of the total mechanical energy (TME) generated by left ventricular contraction (Suga, 1979; Suga et al., 1984b). The measure is designated as left ventricular systolic pressure-volume (P-V) area (PVA) (Suga, 1979), as shown in Fig. 1. PVA is the area in the P-V diagram circumscribed by the end-systolic (ESPVR) and end-diastolic P-V relation (EDPVR) curves and the systolic P-V trajectory. PVA consists of external mechanical work (EW) and mechanical potential energy (PE). EW is the area under the systolic P-V trajectory. PE is the area under the ESPVR which can be assumed to be practically linear in the dog left ventricle (Suga and Sagawa, 1974; Suga, 1979). The concept of PVA is based on a time-varying elastance model of the left ventricle (Suga and Sagawa, 1974; Suga, 1979). PVA represents the mechanical energy required for the time-varying elastance to increase from an end-diastolic low level to an end-systolic high level (Suga, 1979). The dimensions of PVA are mmHg ml, convertible to J where 1 mmHg ml = 1.33 x 10^{-4} J. For more details of the Methods, refer to the References.

Experimental Protocols and Statistics

See the References.

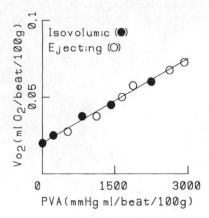

Fig. 2   Regression line of Vo$_2$ on PVA of isovolumic (●) and ejecting (O) contractions in a dog left ventricle in a stable contractile state.

RESULTS

Loading Conditions (Preload, Afterload, Mode of Contraction)

Fig. 2 shows an example of the regression of Vo$_2$ on PVA in a given contractile state in a dog left ventricle. Open circles represent the data of ejecting contractions and closed circles those of isovolumic contractions at different end-diastolic volumes. PVA of an ejecting contraction consists of external work (EW) and potential energy (PE). PVA of an isovolumic contraction consists only of PE. Despite this difference, Vo$_2$s of both ejecting and isovolumic contractions correlated linearly and closely with their PVAs in the same way. Similar results were always obtained in many other dog left ventricles (Suga et al., 1981a, 1981b, 1982, 1983a, 1983b, 1984a, 1984b, 1984c, 1986).

We confirmed the load-independence of the Vo$_2$-PVA relationship under a variety of contractions where the P-V trajectories deviated widely from a normal rectangular loop during both systolic and diastolic phases. As long as the magnitude of PVA was maintained, alterations in its shape, which changed the relative magnitudes of EW and PE, did not significantly affect Vo$_2$ (Suga et al., 1981b, 1984a, 1984b, 1984c).

From these results, we obtained an empirical equation of Vo$_2$ = A x PVA + D, where A and D are constants. D is the Vo$_2$-axis intercept and equal to Vo$_2$ of unloaded contraction. When Vo$_2$ and PVA are expressed respectively in ml O$_2$/beat/100g left ventricle and mmHg ml/beat/100g, A = 1.8 x 10^{-5} with a coefficient of variation (CV) of about 0.25 and D = 0.017 with a CV of about 0.20 on the average in more than 50 hearts (Suga et al., 1981a, 1981b,

424

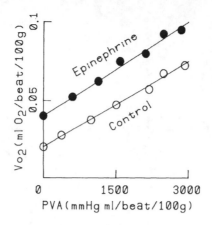

Fig. 3   Regression lines of Vo_2 on PVA of contractions in the control contractile state and in an enhanced contractile state produced by epinephrine, in a dog left ventricle.

1983a, 1983b, 1984c). A was reproducible in a given heart but varied between different hearts.

When PVAs were equal for ejecting and isovolumic contractions, Vo_2s were also equal, but peak systolic pressure, peak systolic wall force, and their time integrals were variably smaller in ejecting contractions than in isovolumic contractions (Suga et al., 1981b, 1987). This result shows that PVA correlates better with Vo_2 than do any of these conventional indices of Vo_2.

Heart Rate

Vo_2 per min increases with heart rate. However, the effect of heart rate on Vo_2 per beat is controversial (Boerth et al., 1969). We varied heart rate between 80 and 220 beats/min by atrial or ventricular electric pacing. At a constant heart rate, the Vo_2-PVA relation was linear regardless of loading conditions. The Vo_2-PVA relation was shifted little by heart rate changes (Suga et al., 1983a). We have not determined whether the Vo_2-PVA relation would shift with heart rate changes below 80 beats/min because of experimental difficulties.

Contractility

Fig. 3 shows a representative example of the effect of enhanced

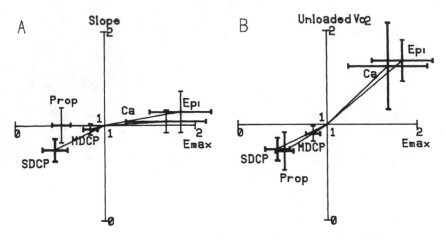

Fig. 4   Relative changes in the slope of the Vo_2-PVA relation (A) and in the unloaded Vo_2 (or Vo_2-axis intercept) (B) against relative changes in the ventricular contractility index, Emax, when the ventricular contractile state was enhanced by epinephrine (Epi) and Ca^{2+} and depressed by propranolol (Prop) or by mildly and severely decreased coronary perfusion pressure (MDCP and SDCP). Means $\pm$ SD from many dog left ventricles.

contractile state on the Vo_2-PVA relation. We enhanced ventricular contractility by epinephrine (Suga et al., 1983b), norepinephrine (unpublished), isoproterenol (unpublished), dobutamine (Nozawa et al., 1987), Ca^{2+} (Suga et al., 1983b), acetylstrophanthidin (unpublished), and paired pulse stimulation (unpublished). These positive inotropic interventions elevated the Vo_2-PVA relation without significantly changing its slope (Suga et al., 1983b, 1984c, 1986). Conversely, negative inotropic agents such as propranolol, verapamil, and pentobarbital lowered the Vo_2-PVA relation without changing its slope (unpublished).

Fig. 4 shows relative changes in the slope of the Vo_2-PVA relation (Panel A) and unloaded Vo_2, or the Vo_2-axis intercept of the Vo_2-PVA relation, (Panel B) against relative changes in Emax produced by epinephrine, Ca^{2+}, and propranolol. Emax is an index of left ventricular contractile state, defined as the slope of ESPVR as shown in Fig. 1. The sensitivity of the slope to changes in Emax was small. The sensitivity of the unloaded Vo_2 to Emax changes was high. The unloaded Vo_2 seems to change largely linearly with Emax in our study as well as in that of Burkhoff et al. (1985).

In accordance with these results, we modified the empirical equation to Vo_2 = A x PVA + B x Emax + C. Fig. 5 shows this relation as a plane surface in a three-dimensional diagram with Vo_2, PVA, and Emax coordinates. The slope of the plane surface along the PVA coordinate at a fixed Emax represents coefficient A, and the slope of the plane along the Emax coordinate at a fixed PVA represents coefficient B. Constant C is the height of the plane surface at zero PVA and Emax, corresponding to basal metabolism per beat (BM). B x Emax + C represents the height of the plane as a function of Emax at zero PVA and is equal to D in the first equation. When the dimensions of Emax were mmHg/(ml/100 g), B had a mean value of 0.0024, tending to be higher for catecholamines than Ca^{2+} though not statistically significant (Suga et al., 1985). The mean value of C was 0.014 ml O_2/beat/100g, comparable to the directly measured basal metabolic Vo_2 converted to a per-beat value (Suga et al., 1983b). The surface was drawn

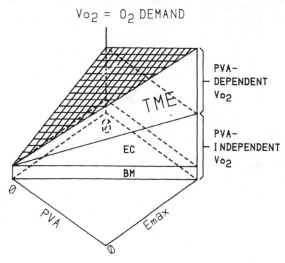

Fig. 5   Three-dimensional surface of the Vo_2-PVA-Emax
relation: Vo_2 or oxygen demand = A x PVA + B x Emax
+ C, where PVA = systolic pressure-volume area
equivalent to total mechanical energy (TME), Emax =
end-systolic ventricular elastance as an index of
contractility. A x PVA is oxygen consumption or demand
for mechanical energy. B x Emax is oxygen consumption
or demand for excitation-contraction coupling (EC). C
is oxygen consumption or demand for basal metabolism
per beat (BM).

triangular because maximum PVA at a maximum end-diastolic volume decreases
with decreases in Emax and PVA is zero when Emax is zero.

Low Coronary Perfusion

To decrease coronary oxygen supply, we decreased coronary perfusion
pressure from the control level (mean 80 mmHg) to mildly, and severely,
decreased levels (50 and 30 mmHg). At each coronary perfusion pressure
level, we produced ejecting contractions at different end-diastolic volumes
and obtained the Vo_2-PVA relation (unpublished). Heart rate was fixed by
pacing.

The mildly decreased coronary perfusion pressure decreased Emax by 17%
and lowered the Vo_2-PVA relation by 10% in the Vo_2-axis intercept without
changing the slope. However, the severely decreased coronary perfusion
pressure decreased Emax by 56% from control and lowered the Vo_2-PVA relation
by 26% from control in the Vo_2-axis intercept, with a significant, 24%,
decrease in the slope (unpublished). Fig. 4 shows relative changes in the
slope of the Vo_2-PVA relation and the unloaded Vo_2 against relative changes
in Emax produced by mildly, and severely, decreased coronary perfusion

pressure. The decreases in unloaded Vo_2 with decreased Emax under decreased coronary perfusion pressure were in virtually the same proportion as the increases produced by epinephrine, Ca^{2+}, and propranolol described above, and shown in Fig. 4B. However, the relative decrease in the slope with decrease in Emax under severely decreased coronary perfusion pressure was disproportionately large as shown in Fig. 4A. These changes were reproducible when coronary perfusion pressure was restored to normal.

With decreased coronary perfusion pressure, coronary flow also decreased and AVOD increased. These changes accompanied decreases in Emax. These responses are opposite to those accompanying propranolol treatment where coronary perfusion pressure and flow remained unchanged and AVOD decreased. Although we measured coronary arterial and venous lactate concentrations, the former was always higher than the latter even at the severely decreased coronary perfusion pressure. This result indicates that the heart as a whole did not become ischemic despite a possibility of regional ischemia. This situation suggests a delicate balance or slight imbalance between coronary oxygen supply and cardiac oxygen demand by the depressed contractions of this heart.

We also decreased and increased coronary perfusion pressure while producing ejecting contractions at the same end-diastolic volume. Emax and Vo_2 changed in proportion to coronary perfusion pressure and flow. This result indicates the possibility that decreased coronary oxygen supply decreases cardiac oxygen demand by depressing contractility.

KCl Arrest

The excised, cross-circulated heart was arrested by continuously infusing KCl into the coronary circulation and Vo_2 was measured. The average value of Vo_2 in the control was 1.5 ml O_2/min and it was not significantly increased by epinephrine and Ca^{2+} (Suga et al., 1983b). This indicates the independence of the constant C, i.e., Vo_2 of basal metabolism, from Emax. However, decreasing the coronary perfusion pressure from 80 mmHg to 30 mmHg under KCl arrest decreased Vo_2 from 1.5 ml O_2/min to 1.0 ml O_2/min (unpublished). These results indicate that, whereas Vo_2 for basal metabolism is independent of Emax when coronary oxygen supply is normal, it decreases when coronary oxygen supply is limited. Therefore, Fig. 5 must be slightly modified in the low Emax range to incorporate this result to describe cardiac oxygen demand under a limited coronary oxygen supply.

DISCUSSION

Our experimental results reviewed here show that left ventricular oxygen consumption (Vo_2) and oxygen demand vary as an explicit function of the total mechanical energy (TME) measured by PVA and ventricular contractility measured by Emax. Fig. 5 represents this relationship clearly. The sum of the Vo_2 components for basal metabolism (BM) and excitation-contraction coupling (EC) can be assumed to be practically independent of PVA and hence is called PVA-independent Vo_2. The excess Vo_2 above the PVA-independent Vo_2 is a linear function of PVA or TME and hence is called PVA-dependent Vo_2. The Vo_2-PVA-Emax surface is a plane so long as coefficient A in the empirical equation remains unchanged despite changes in Emax as with epinephrine, Ca^{2+}, propranolol, etc. We consider that the Vo_2-PVA-Emax surface as shown in Fig. 5 can be taken to represent cardiac oxygen demand for an ongoing steady-state contraction as an explicit function of PVA and Emax. This contention is tenable when the coronary oxygen supply is adequate.

When values are obtained under a limited coronary oxygen supply, the

Vo_2-PVA-Emax plane as shown in Fig. 5 cannot always be considered to represent cardiac oxygen demand, as a function of PVA and Emax, under a sufficient coronary oxygen supply. The reason is that the observed PVA and Emax are related to weakened contractions in a contractile state depressed as a consequence of insufficient cardiac oxygen supply. The true oxygen demand capability of this heart may be obtained when coronary oxygen supply is restored to a sufficient level.

Although we give mean values for A, B, and C in the Results, they varied widely among different dog hearts (Suga et al., 1981a, 1983a, 1983b, 1984c, 1985; Burkhoff et al., 1985). The coefficients of variation of A, B, and C were 0.2-0.3 (Suga et al., 1985). Hence, cardiac oxygen demand varies among different hearts even when their PVA and Emax are given. We have not yet succeeded in identifying the factors which produce these interindividual variations of A, B, and C in dog left ventricles.

SUMMARY

We have briefly reviewed the recent advances in the understanding of primary determinants of cardiac oxygen consumption (Vo_2). We focused on our own experimental findings that Vo_2 is a function of left ventricular systolic pressure-volume area (PVA), as a measure of total mechanical energy generated by contraction, and the left ventricular contractility index (Emax). We conclude that the cardiac oxygen demand of the ongoing contractions is equivalent to Vo_2 and is a precise function of PVA and Emax whether coronary oxygen supply is normal or mildly limited in the excised, cross-circulated dog left ventricle. However, when the coronary oxygen supply is severely limited, Vo_2 may not always be equivalent to the oxygen demand capability shown by the heart when its coronary oxygen supply is restored.

ACKNOWLEDGMENT

Partly supported by Grants-in-Aid for Scientific Research (C59570047, B61480102) from the Ministry of Education, Science and Culture, and Research Grants for Cardiovascular Diseases (60A-1 and 60C-3) from the Ministry of Health and Welfare of Japan

REFERENCES

Boerth, R. C., Covell, J. W., Pool, P. E., and Ross, J., 1969, Increased myocardial oxygen consumption and contractile state associated with increased heart rate in dogs, Circ. Res., 24:725.

Burkhoff, D., Yue, D. T., Franz, M., Oikawa, R., Schaefer, J., and Sagawa, K., 1985, Influence of contractile state on myocardial oxygen consumption (abstract), Circulation, 72 (Supp III):III-298.

Evans, C.L., and Matsuoka, Y., 1915, The effect of various mechanical conditions on the gaseous metabolism, and efficiency of the mammalian heart, J. Physiol., 49:378.

Gibbs, C. L., 1978, Cardiac energetics, Physiol. Rev., 58:174.

Graham, T. P., Covell, J. W., Sonnenblick, E. H., Ross, J., and Braunwald, E., 1968, Control of myocardial oxygen consumption: Relative influence of contractile state and tension development, J. Clin. Invest., 47:375.

Klocke, F. J., Kaiser, G. A., Ross, J., and Braunwald, E., 1965, Mechanism of increase of myocardial oxygen uptake produced by catecholamines, Am. J. Physiol., 209:913.

Klocke, F. J., Braunwald, E., and Ross, J., 1966, Oxygen cost of electrical activation of the heart, Circ. Res., 18:357.

Nozawa, T., Yasumura, Y., Futaki, S., Tanaka, N., Igarashi, Y., Goto, Y., and Suga, H., 1987, Relation between oxygen consumption and pressure-volume area of in situ dog heart, Am. J. Physiol., in press.

Rooke, G. A., and Feigl, E. O., 1982, Work as a correlate of canine left ventricular oxygen consumption, and the problem of catecholamine oxygen wasting, Circ. Res., 50: 273, 1982

Sarnoff, S. J., Braunwald, E., Welch, G. H., Case, R. B., Stainsby, W. N., and Macruz, R., 1985, Hemodynamic determinants of oxygen consumption of the heart with special reference to the tension-time index, Am. J. Physiol., 192:148.

Shepherd, A. P., and Burger, C. G., 1977, A solid-state arteriovenous oxygen difference analyzer for flowing whole blood, Am. J. Physiol., 232:H437

Sonnenblick, E. H., Ross, J., Covell, J. W., Kaiser, G. A., and Braunwald, E., 1965, Velocity of contraction as a determinant of myocardial oxygen consumption, Am. J. Physiol., 202:931.

Suga, H., 1979, Total mechanical energy of a ventricle model and cardiac oxygen consumption, Am. J. Physiol., 236:H498.

Suga, H., and Sagawa, K., 1974, Instantaneous pressure-volume relationships and their ratio in the excised, supported canine left ventricle, Circ. Res., 35:117.

Suga, H., Hayashi, T., and Shirahata, M., 1981a, Ventricular systolic pressure volume area as predictor of cardiac oxygen consumption, Am. J. Physiol., 240:H39.

Suga, H., Hayashi, T., Suehiro, S, Hisano, R., Shirahata, M., and Ninomiya, I., 1981b, Equal oxygen consumption rates of isovolumic and ejecting contractions with equal systolic pressure volume areas in canine left ventricle, Circ. Res., 49:1082.

Suga, H., Hisano, R., Hirata, S., Hayashi, T., and Ninomiya, I., 1982, Mechanism of higher oxygen consumption rate: pressure-loaded vs. volume-loaded heart, Am. J. Physiol., 242:H942.

Suga, H., Hisano, R., Hirata, S., Hayashi, T., Yamada, O., and Ninomiya, I., 1983a, Heart rate independent energetics and systolic pressure-volume area in dog heart, Am. J. Physiol., 244:H206.

Suga, H., Hisano, R., Goto, Y., Yamada, O., and Igarashi, Y., 1983b, Effect of positive inotropic agents on the relation between oxygen consumption and systolic pressure volume area in canine left ventricle, Circ. Res., 53:306.

Suga, H., Goto, Y., Yamada, O., and Igarashi, Y., 1984a, Independence of myocardial oxygen consumption from pressure-volume trajectory during diastole in canine left ventricle. Circ. Res., 55:734.

Suga, H., Yamada, O., Goto, Y., 1984b, Energetics of ventricular contraction as traced in the pressure-volume diagram, Fed. Proc., 43:2411.

Suga, H., Yamada, O., Goto, Y., and Igarashi, Y., 1984c, Oxygen consumption and pressure volume area of abnormal contractions in canine heart, Am. J. Physiol., 246:H154.

Suga, H., Igarashi, Y., Yamada, O., and Goto, Y., 1985, Mechanical efficiency of the left ventricle as a function of preload, afterload, and contractility, Heart Vessels, 1:3.

Suga, H., Igarashi, Y., Yamada, O., and Goto, Y., 1986, Cardiac oxygen consumption and systolic pressure volume area, Basic Res. Cardiol., (Supp 1):39.

Suga, H., Goto, Y., Nozawa, T., Yasumura, Y., Futaki, S., and Tanaka, N., 1987, Force-time integral decreases with ejection despite constant oxygen consumption and pressure-volume area in dog left ventricle, Circ. Res., 60: in press.

Weber, K. T., Janicki, J. S., and Fishman, A. P., 1980, Aerobic limit of the heart perfused at constant pressure, Am. J. Physiol., 238:H118.

CARDIAC METABOLISM AS AN INDICATOR OF

OXYGEN SUPPLY/DEMAND RATIO

Kazuo Ichihara and Yasushi Abiko

Department of Pharmacology
Asahikawa Medical College
Asahikawa, Japan

INTRODUCTION

As long as one is alive, the heart must continuously beat to pump
blood into the organs and tissues in the whole body. As a result, a large
amount of energy is required to maintain cardiac contraction. Under
aerobic conditions, the energy is produced by oxidation of substrates, such
as glucose and fatty acids. Oxygen delivered by coronary arterial blood
is consumed by the heart at the rate of more than 40 liters a day. When
coronary stenosis occurs, the myocardium becomes deficient in oxygen and,
hence, in energy for its contraction. Thus, it is important to know an
oxygen supply/demand ratio in the ischemic myocardium, in order to study
the pathophysiology of ischemia of the heart. We (Abiko et al., 1979)
have tried to evaluate an antianginal or anti-ischemic effect of a drug by
using a change in cardiac metabolism as an indicator of the oxygen
supply/demand ratio in the ischemic myocardium. In the ischemic
myocardium, where myocardial metabolism shifts from aerobic to anaerobic in
type, drugs can prevent ischemic injury by switching the metabolism back
from anaerobic to aerobic. It has been known that oxygen deficiency
accelerates the rate of glycolysis through activation of
phosphofructokinase (PFK) to produce adenosine triphosphate (ATP)
anaerobically. Our data, however, indicate that ischemia inhibits the
glycolytic pathway at the level of PFK (Ichihara and Abiko, 1982).
Because there is no coronary flow in the ischemic myocardium, glycolysis
may be inhibited to prevent accumulation of metabolic end-products such as
lactate. Weishaar et al. (1979) have suggested that one can estimate the
severity of myocardial injury or the energy state of the ischemic
myocardium by studying glycolytic flux through the PFK reaction; as the
deleterious effect of ischemia becomes worse, the inhibition of the
glycolytic flux at the PFK stage becomes more severe. They (Weishaar et
al., 1979) measured the myocardial levels of glucose-6-phosphate (G6P),
fructose-6-phosphate (F6P) and fructose-1,6-diphosphate (FDP), and
calculated the ratio of ([G6P]+[F6P])/[FDP] as an index of glycolytic flux
at the PFK stage. The present study, therefore, was undertaken to examine
the effect of several kinds of anti-ischemic drugs on changes in the
([G6P]+[F6P])/[FDP] ratio during ischemia.

METHODS

Mongrel dogs of either sex weighing 8 to 15 kg were anesthetized with sodium pentobarbital (30 mg/kg, iv.), and endotracheally intubated and ventilated with a Harvard respirator. A left thoracotomy was performed between the 4th and 5th ribs to expose the left ventricular wall. After the heart was suspended in a pericardial cradle, a main trunk of the left anterior descending coronary artery was dissected free from the adjacent tissues at a point distal to the first diagonal branch, and was loosely encircled with a silk thread ligature. Ischemia was initiated by ligating the coronary artery. An ischemic region of the myocardium was assessed by visible cyanosis and the elevation of the ST segment of ECG, which was recorded by a wire electrode attached on the surface of the left ventricular wall.

After control observations had been completed, the dogs were divided into 8 approximately equal groups and received either saline or a drug at the dose level shown in Table 1. All the drugs were injected intravenously. Five min after drug injection, the ligature placed around the coronary artery was tied in half of the animals receiving drugs and also in half of animals receiving saline. After 3 min of coronary ligation, a full thickness sample of the myocardium was taken from the center of the ischemic area (ischemia). An equivalent sample was taken from the control animals, in which the coronary artery was not tied (nonischemia). The samples were immediately compressed and frozen with clamps previously chilled in liquid nitrogen in such a way that the endocardial and epicardial portions of the myocardium could be taken separately. The endocardial frozen tissue sample was used to determine the levels of G6P, F6P and FDP in neutralized perchloric acid extract according to the standard enzymatic procedures (Bergmeyer, 1974). The ratio of ([G6P]+[F6P])/[FDP] was calculated from the concentration of hexose phosphates in order to assess the rate of glycolytic flux through PFK reaction (Weishaar et al., 1979).

Table 1. Changes in the Myocardial Levels of
G6P, F6P and FDP during Ischemia

	G6P	F6P	FDP
Saline-treated			
Nonischemia	212± 37[a]	38± 9	113±28
Ischemia	1488±301	353±74	78±16
Propranolol-treated (1 mg/kg)			
Nonischemia	35± 6	6± 2	32± 6
Ischemia	171± 50	36± 9	32± 6
Carteolol-treated (100 µg/kg)			
Nonischemia	89± 35	20± 7	43± 9
Ischemia	350±107	77±27	20± 3
Nadolol-treated (1 mg/kg)			
Nonischemia	131± 21	37± 4	83±10
Ischemia	1098±212	273±42	111±22
Nifedipine-treated (10 µg/kg)			
Nonischemia	163± 22	29± 3	63± 7
Ischemia	799±208	142±45	30± 5
Diltiazem-treated (100 µg/kg)			
Nonischemia	170± 49	60±14	78±23
Ischemia	795±118	178±33	98±17
Verapamil-treated (100 µg/kg)			
Nonischemia	150± 19	24± 3	79± 9
Ischemia	487±101	101±29	41±17
Flunarizine-treated (1 mg/kg)			
Nonischemia	113± 29	27± 7	55±20
Ischemia	620±112	148±30	87±17

[a]Values are means±SE (µmol/g wet tissue)

RESULTS

    Changes in the levels of G6P, F6P and FDP induced by ischemia, and
effects of anti-ischemic drugs on the change, are shown in Table 1.   In
the saline-treated heart, ischemia increased the levels of G6P and F6P,
whereas it decreased the level of FDP.   In the drug-treated heart,
ischemia increased the levels of G6P and F6P, the increased levels of these
hexose monophosphates being smaller than those in the saline-treated heart.
The effect of ischemia on the level of FDP, however, showed a wide
variation in the drug-treated heart.   In the carteolol-, nifedipine- or
verapamil-treated heart, the FDP level decreased during ischemia, while it
increased in the nadolol-, diltiazem-, or flunarizine-treated heart.   The
level of FDP did not change during ischemia in the propranolol-treated
heart.   The ratio of ([G6P]+[F6P])/[FDP] calculated from the values of
hexose phosphate concentrations is illustrated in Fig. 1.

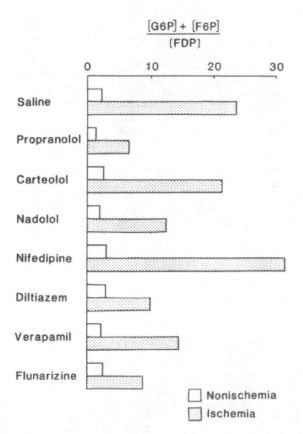

Fig. 1 Effects of anti-ischemic drugs on changes in the ratio of
        ([G6P]+[F6P])/[FDP] caused by ischemia.   Ischemia was induced by
        ligating the left anterior descending coronary artery for 3 min.

This ratio was increased by ischemia in the saline-treated heart from 2.2 to 23.6.   This suggests that the glycolytic flux at the level of PFK was inhibited by ischemia.   Propranolol, nadolol, diltiazem, verapamil, and flunarizine attenuated the increase in the ratio of ([G6P]+[F6P])/[FDP] caused by ischemia.   Nifedipine and carteolol, however, did not prevent the ischemia-induced increase in the ratio.

DISCUSSION

The crossover plot analysis of glycolytic intermediates in the ischemic myocardium has revealed that there is a negative crossover point between F6P and FDP (Opie, 1976; Weishaar et al., 1979; Ichihara and Abiko, 1982).   Static measurements of the substrates and products of a reaction do not quantify the rate of glycolytic flux unless a direct measurement of glycolytic flux is completed (Newsholme and Crabtree, 1979).   In fact, the levels of G6P and F6P can be increased with unchanged flux rates, increased flux rates, or decreased flux rates.   Nevertheless, the negative crossover point obtained in the ischemic myocardium suggests an inhibition of glycolytic flux, because hydrogen ions which inhibit the PFK reaction (Ui, 1966) accumulate during ischemia (Ichihara et al., 1979), and because in vitro studies using the isolated perfused working rat heart show the inhibition of glycolytic flux during ischemia (Rovetto et al., 1975).   Whether rates of glycolytic flux are inhibited or not, the ratio of ([G6P]+[F6P])/[FDP] increased more than 10-fold during ischemia (Fig. 1).   This increase in ratio does not directly mean a decrease in the rate of glycolytic flux, but it may express the degree of change in hexose phosphate levels during ischemia.   Pretreatment with either β-adrenergic blocking agents (propranolol and nadolol), or calcium entry blocking agents (diltiazem, verapamil and flunarizine) prevented the increase in this ratio in ischemia.   This result indicates that these drugs reduce the effect of ischemia on the myocardium.   Carteolol and nifedipine, however, did not prevent the increase in the ratio of ([G6P]+[F6P])/[FDP], indicating that changes in the levels of hexose phosphates still occurred even in the presence of these drugs in the ischemic myocardium.   Although carteolol is a β-adrenergic agent, it has a sympathomimetic action (Yabuuchi and Kinoshita, 1974).   A potent coronary dilatory effect of nifedipine may cause a coronary steal phenomenon (Shulz, 1985).   These facts may be responsible for the results observed with carteolol and nifedipine in terms of ischemic myocardial metabolism.
In conclusion, some anti-ischemic drugs may improve the oxygen supply/demand ratio in the ischemic myocardium, since these drugs attenuate the increase in the ratio of ([G6P]+[F6P])/[FDP] due to ischemia.

SUMMARY

We evaluated the anti-ischemic effect of drugs by using the inhibition of glycolytic flux at the level of the phosphofructokinase (PFK) reaction, caused by ischemia, as an indicator of the oxygen supply/demand ratio in the ischemic myocardium.   Ischemia was induced by ligating the left anterior descending coronary artery in the open-chest dog.   After 3 min of coronary ligation, the ischemic myocardium was removed.   The endocardial portion of the myocardial sample was used to determine the levels of glucose-6-phosphate (G6P), fructose-6-phosphate (F6P) and fructose-1,6-diphosphate (FDP), and the ratio of ([G6P]+[F6P])/[FDP] was calculated in order to assess the rate of glycolytic flux at the PFK stage.   Either saline or drug (propranolol, 1 mg/kg; carteorol, 100 μg/kg; nadolol, 1 mg/kg; nifedipine, 10 μg/kg; diltiazem, 100 μg/kg; verapamil, 100 μg/kg; and flunarizine, 1 mg/kg) was injected intravenously 5 min before coronary ligation.   In the saline-treated heart, ischemia increased the levels of G6P and F6P, whereas it decreased the level of FDP. The ratio of ([G6P]+[F6P])/[FDP] was increased by ischemia from 2.2 to 23.6, suggesting the inhibition of glycolytic flux at the level of the PFK reaction. In the drug-treated heart, ischemia increased the levels of G6P and F6P, but the increases were smaller than those in the saline-treated heart. Pretreatment with propranolol, nadolol, diltiazem, verapamil, flunarizine attenuated the increase in the ratio of ([G6P]+[F6P])/[FDP] caused by ischemia. Carteolol and nifedipine, however, did not modify the increase in the ratio of ([G6P]+[F6P])/[FDP], indicating that changes in the levels of hexose phosphates still occur even in the presence of these drugs in the ischemic myocardium. In conclusion, some anti-ischemic drugs may improve oxygen supply/demand ratio in the ischemic myocardium, since they attenuate the increase in the ratio of ([G6P]+[F6P])/[FDP] due to ischemia.

REFERENCES

Abiko, Y., Ichihara, K., and Izumi, T., 1979, Effects of antianginal drugs on ischemic myocardial metabolism, in: "Ischemic Myocardium and Antianginal Drugs," M.M. Winbury and Y. Abiko, ed., Raven Press, New York.

Bergmeyer, H.U., 1974, "Methods in Enzymatic Analysis," Academic Press, New York.

Ichihara, K., and Abiko, Y., 1982, Crossover plot study of glycolytic intermediates in the ischemic canine heart, Japan. Heart J., 23: 817.

Ichihara, K., Ichihara, M., and Abiko, Y., 1979, Involvement of beta adrenergic receptors in decrease of myocardial pH during ischemia, J. Pharmacol. Exp. Ther., 209: 275.

Newsholm, E.A., and Crabtree, B., 1979, Theoritical principles in the approaches to control of metabolic pathways and their application to glycolysis in muscle, J. Mol. Cell. Cardiol., 11: 839.

Opie, L.H., 1976, Effects of regional ischemia on metabolism of glucose and fatty acids. Relative rates of aerobic and anaerobic energy production during myocardial infarction and comparison with effects of anoxia, Circ. Res., 38 (Suppl.): I 52.

Rovetto, M.J., Lamberton, W.F.N., and Neely, J.R., 1975, Mechanisms of glycolytic inhibition in ischemic rat hearts, Circ. Res., 37: 742.

Schulz, W., Jost, S., Kober, G., and Kaltenbach, M., 1985, Relation of antianginal efficacy of nifedipine to degree of coronary arterial narrowing and to presence of coronary collateral vessels, Am. J. Cardiol., 55: 26.

Ui, M, 1966, A role of phosphofructokinase in pH-dependent regulation of glycolysis, Biochim. Biophys. Acta, 124, 310.

Weishaar, R., Ashikawa, K., and Bing, R.J., 1979, Effect of diltiazem, a calcium antagonist, on myocardial ischemia, Am. J. Cardiol., 43: 1137.

Yabuuchi, Y., and Kinoshita, D., 1974, Cardiovascular studies of 5-(3-tert-butylamino-2-hydroxy) propoxy-3,4-dihydrocarbostyril hydrochloride (OPC-1085), a new potent β-adrenergic blocking agent, Japan. J. Pharmacol., 24: 853.

THE DIFFERENT CONTRIBUTIONS OF CORONARY BLOOD FLOW TO CHANGES IN MYOCARDIAL OXYGEN CONSUMPTION BETWEEN EXCISED AND IN SITU CANINE HEARTS

Shiho Futaki, Takashi Nozawa, Yoshio Yasumura,
Nobuaki Tanaka, Masaaki Uenishi, and Hiroyuki Suga

Department of Cardiovascular Dynamics, National Cardiovascular Center Research Institute, Suita, Osaka, 565 Japan

INTRODUCTION

In recent years, our research has been directed to the relation between cardiac dynamics and energetics. We have utilized myocardial oxygen consumption (MVO_2) as the energy input to the heart. Since MVO_2 is obtained as the product of coronary blood flow (CBF) and arteriovenous oxygen content difference (AVD), both the behavior of coronary vascular tone and the oxygen extraction capability of the myocardium may affect MVO_2 and consequently cardiac performance. In general, myocardial oxygen extraction rate is about 75% (Goodale and Hackel, 1953), and the heart is working nearly at its maximum capability of oxygen extraction even in the resting state. Several reports showed relative constancy of coronary venous oxygen content and AVD independent of changes in MVO_2 (Gerola, et al., 1959; Katz and Feinberg, 1958; Feigl, 1983). Therefore, the coronary arterial reserve to increase blood supply is especially important when oxygen demand increases by physical exercise, etc. It was also well documented that CBF closely correlates with MVO_2 with typical correlation coefficients from 0.8 to 0.9 (Belloni, 1979).

From another point of view, MVO_2 is the most important determining factor of CBF (Braunwald, et al., 1958). Some metabolic factors, rather than mechanical factors per se, induced by changes in mechanical loading conditions to the heart are assumed to control the regulation of vasomotor tone of the coronary vessels so as to match the oxygen supply to the demand (Belloni, 1979; Berne, 1963; Berne, 1980). However, this metabolic coronary vasoregulaion remains to be clarified in terms of the underlying mechanism and the mediating substances.

Moreover, it is questionable whether this situation is applicable to the excised heart. Suehiro et al. (1982) reported the relation between MVO_2 and CBF in excised hearts, and demonstrated the relative importance of AVD as a contributor to MVO_2. In the present study, we looked into our previous data obtained in three different series of experiments to see if there was any difference in the contribution of CBF to changes in MVO_2 between excised and in situ canine hearts. The purpose of this study is to elucidate a specific pattern of oxygen uptake in the excised heart. Thus, this is a retrospective study which is a by-product of our previous studies (Nozawa, et al., 1987a, 1987b; Yasumura, et al., 1987) concerning cardiac dynamics and energetics.

METHODS

Surgical Preparations

In three different series of experiments, one with excised cross-circulated hearts (EC group), one with right heart bypass preparations (RB group) and another with in situ heart preparations (IS group), adult mongrel dogs were anesthetized with ketamine hydrochloride (5 mg/kg, im) followed by sodium pentobarbital (25 mg/kg, iv), and heparinized (750 u/kg, iv).

EC group. A total of 13 pairs of smaller and larger dogs were used. Briefly, arterial and venous cross circulation tubes were inserted into the left subclavian artery and the right ventricle in the smaller dog (heart donor dog). The other ends were inserted into the common carotid arteries and the right jugular vein in the larger dog (support dog). After the heart lung section was isolated from the systemic circulation in the heart donor dog, bilateral pulmonary hili were ligated and cross circulation was started. The heart from the donor dog was excised from the chest, the left atrium opened and all the chordae tendinae cut. A thin latex balloon with an unstretched volume of 60 ml was introduced into the left ventricle through the mitral annulus, where the balloon was secured and connected to the volume servo pump. The balloon and the water housing of the pump were primed with water. This system enabled us to control and measure left ventricular volume accurately. Heart rate was fixed in each experiment by atrial pacing. A miniature pressure gauge (Konigsberg, P-7) was placed inside the apical end of the balloon to measure left ventricular pressure. Mean arterial pressure of the support dog which served as the coronary perfusion pressure of the excised heart was maintained above 60 mmHg by infusing 10% Dextran-40 solution, transfusing blood collected from the heart donor dog, or giving phenylephrine (5-10 mg, im) as needed. To prevent allergic reactions, diphenhydramine hydrochloride (30-60 mg, im) was administered to the support dog prior to cross circulation.

RB group. In another group of 7 dogs, right heart bypass preparation was made. Briefly, the chest was opened by median sternotomy under artificial ventilation. All the systemic venous return, drained via cannulae in the superior and inferior venae cavae, was pumped into the main pulmonary artery with a roller pump. The left ventricular stroke volume was changed over a broad range by controlling the speed of this pump. Heart rate was not fixed. Another cannula was placed into the right ventricle to drain the coronary venous return without contamination of systemic venous blood. A catheter-tip micromanometer (Millar, PC-470) was inserted into the left ventricle through the apex to measure left ventricular pressure. Aortic pressure was measured with a Statham pressure transducer through a water-filled catheter inserted into the aortic arch from the femoral artery. If systemic hypotension occurred, 10% Dextran-40 solution or blood from the other dog was infused or Angiotensin II was administered by intravenous drip as needed. In the last 4 cases (exp.4 - exp.7) of this series, hexamethonium bromide (30mg/kg) and atropine sulfate (0.5mg/kg) were infused intravenously for pharmacological denervation of the heart.

IS group. In another group of 8 dogs, the chest was opened at the right fifth intercostal space under artificial ventilation. Briefly, the pericardium was incised and the heart was suspended in the pericardial cradle. Complete AV block was produced by cauterization of the bundle of His from the right atrial appendage. The coronary sinus was cannulated from the right atrial appendage to drain the coronary venous return. The tube was fitted tightly at the atrial-sinus junction so as to prevent contamination of blood in the right atrium. The coronary venous blood collected into the reservoir was then returned to the right external jugular vein with a roller pump. Left ventricular pressure was measured with a catheter-tip micromanometer (Millar, PC-470) inserted via the right common carotid artery. Heart rate was changed by right ventricular pacing in incremental steps of 20/min from 80/min to 160/min.

Oxygen Consumption

CBF was measured in each experiment with an electromagnetic flowmeter (MFV-2100, Nihon Kohden) in the middle of the coronary venous drainage tube. Although a part of the coronary venous return into the ventricle(s) (Thebesian flow) was not included in the flow measurement, it is known to be only a few percent of the total coronary flow (Moir, et al., 1963). AVD was measured continuously by a spectrophotometric method with an AVOX system (Shepherd and Burgar, 1977), calibrated against a Lex-O_2 Con oxygen content analyzer. The product of CBF (ml/min) and AVD (ml O_2/100ml) divided by 100 gives MVO_2 (ml O_2/min). CBF and MVO_2 were normalized with respect to the left ventricular wet weight to give (ml O_2/min)/100gLV. In EC and RB groups, this represented the left ventricular MVO_2 almost exclusively, because the right ventricle was collapsed by continuous drainage of the coronary venous return. But in the IS group, this contains a considerable amount of the right ventricular MVO_2, since the right ventricle was not unloaded. To alter myocardial oxygen demand, the following interventions were added to the heart; in the EC group, left ventricular volume and the contraction mode were changed with the servo pump; in the RB group, left ventricular preload and consequently stroke volume were changed with the roller pump; in the IS group, heart rate was changed by ventricular pacing. After these interventions, it usually took a few minutes to attain a new cardiac steady state. Thus, we measured data only in the steady state after waiting for a few minutes.

Coronary Vascular Resistance

We assumed that coronary arterial pressure was equal to the arterial pressure measured in the middle of the cross circulation tube in the EC group, and to the aortic pressure measured in the aortic arch in the RB group. Coronary vascular resistance (CVR) (mmHg/[(ml/min)/100gLV]) was then calculated by mean arterial or aortic pressure divided by CBF.

In the IS group, however, arterial pressure was not measured. We used left ventricular peak pressure as a substitute for the arterial pressure to calculate the tentative coronary vascular resistance, which was not pertinent to the comparison with the data of the other groups.

Data Analysis

Our statistical analyses depended on Snedecor and Cochran (1967). The linear regression analysis was applied to the correlation of CBF with MVO_2 and that of AVD with MVO_2 in each experiment. The correlation of CVR with MVO_2 was also analyzed in each experiment. Mean value for the correlation coefficient (r) was estimated by z-transformation. Mean values for the slope of CBF-MVO_2 regression line among the three groups were compared by one-way analysis of variance, and the least significant difference (LSD) method was applied when F-test was significant. Mean values for the slope of CVR-MVO_2 regression line between the EC and RB groups were compared by Student t-test. P values smaller than 0.05 were considered statistically significant. Data are presented as mean $\pm$ SD.

RESULTS

The variations and percent changes of MVO_2, CBF, AVD, and CVR in each experimental group are shown in Table 1. In the EC and RB groups where the left ventricular mechanical load was changed, MVO_2 varied over comparable

Table 1. Variations and Percent Changes in MVO_2, CBF, AVD and CVR in three Experimental Groups

		EC group	RB group	IS group
MVO_2	low	6.77 $\pm$ 1.57	6.05 $\pm$ 1.91	7.11 $\pm$ 1.47
	high	12.13 $\pm$ 2.24	10.81 $\pm$ 3.61	9.56 $\pm$ 2.12
	%change	84 $\pm$ 39	79 $\pm$ 25	34 $\pm$ 6
CBF	low	120 $\pm$ 40	117 $\pm$ 48	83 $\pm$ 16
	high	161 $\pm$ 44	225 $\pm$ 112	105 $\pm$ 18
	%change	37 $\pm$ 14	104 $\pm$ 47	27 $\pm$ 9
AVD	low	5.7 $\pm$ 1.6	5.3 $\pm$ 1.9	8.7 $\pm$ 1.8
	high	8.6 $\pm$ 1.9	7.1 $\pm$ 2.3	9.4 $\pm$ 2.1
	%change	55 $\pm$ 20	38 $\pm$ 37	7 $\pm$ 6
CVR	low	0.498 $\pm$ 0.181	0.439 $\pm$ 0.234	(0.791 $\pm$ 0.159)
	high	0.661 $\pm$ 0.329	0.737 $\pm$ 0.376	(1.152 $\pm$ 0.226)
	%change	34 $\pm$ 19	73 $\pm$ 41	(46 $\pm$ 10)

Values are shown as mean $\pm$ SD. MVO_2; (ml O_2/min)/100gLV.
CBF; (ml/min)/100gLV. AVD; ml O_2/100ml. CVR; mmHg/[(ml/min)/
100gLV]. low; lowest value. high; highest value.
%change; (high − low) x 100 / low.

wide ranges. On the other hand, in the IS group where the heart rate was changed, MVO_2 varied within a narrow range that accompanied smaller changes in both CBF and AVD. The lower average value for AVD in the EC and RB groups compared with that in the IS group was probably an inevitable consequence of hemodilution in these more invasive preparations. In this regard, the IS group simulated more physiological in situ coronary and cardiac conditions than the RB group.

Table 2 shows the slopes and the correlation coefficients (r) of the CBF–MVO_2, AVD–MVO_2 and CVR–MVO_2 relations in each experiment. In the EC group, AVD correlated significantly with MVO_2 in all cases, but CBF did so only in 7 cases. In contrast, CBF correlated significantly with MVO_2 in all cases in the RB and IS groups, but AVD did so only in one of the RB group and in 4 of the IS group. (In 2 other cases in the RB and IS groups, AVD correlated with MVO_2 although inversely.) The mean r values of the CBF–MVO_2 and AVD–MVO_2 relations in each group are also shown in Table 2; 0.737 and 0.916 in the EC group, 0.922 and −0.309 in the RB group, and 0.991 and 0.866 in the IS group. However, the scatter of r of the CBF–MVO_2 relation was quite large in the EC group, from the lowest value of 0.008 in exp. 12 to the highest of 0.991 in exp. 9. This suggests a possible mixture of different subgroups in the EC group in regard to the sensitivity of the coronary vasculature to altered MVO_2. We therefore turned our attention to the CVR–MVO_2 relation. It was noted that, although CVR correlated inversely with MVO_2 in all cases of the RB and IS groups, CVR did not correlate with MVO_2 in 4 cases of the EC group.

The slope of the CBF–MVO_2 relation indicates the sensitivity of CBF to a change in MVO_2. As seen in Table 2 and Fig. 1, the mean value for this slope in the EC group was 4.12 $\pm$ 3.14 (ml/ml O_2). The mean values for this slope in the RB and IS groups were 20.45 $\pm$ 6.13 and 9.67 $\pm$ 4.17, and there

Table 2   Slopes and Correlation Coefficients (r)

exp.No.	n	CBF–MVO$_2$ slope	r	AVD–MVO$_2$ slope	r	CVR–MVO$_2$ slope	r
EC group							
1	9	5.45	0.987*	0.523	0.972*	−0.085	−0.987*
2	12	7.50	0.908*	0.456	0.917*	−0.034	−0.943*
3	8	3.19	0.505	0.557	0.855*	−0.031	−0.944*
4	8	8.04	0.971*	0.311	0.965*	−0.012	−0.974*
5	8	4.41	0.812*	0.356	0.968*	−0.007	−0.857*
6	12	5.86	0.959*	0.509	0.949*	−0.058	−0.986*
7	15	2.58	0.230	0.589	0.747*	−0.007	−0.350
8	14	−2.50	−0.203	0.643	0.847*	−0.001	−0.087
9	16	7.34	0.991*	0.306	0.979*	−0.043	−0.903*
10	17	4.86	0.608*	0.379	0.899*	−0.012	−0.690*
11	19	1.01	0.229	0.862	0.957*	−0.033	−0.881*
12	14	0.03	0.008	0.745	0.923*	−0.004	−0.449
13	12	5.74	0.311	0.640	0.631*	−0.046	−0.570
mean		4.12	0.737	0.529	0.916	−0.029	−0.820
SD		±3.14		±0.169		±0.025	
RB group							
1	14	26.77	0.936*	−0.238	−0.722*	−0.040	−0.895*
2	10	16.81	0.766*	0.166	0.663*	−0.018	−0.939*
3	6	28.42	0.924*	−0.196	−0.521	−0.066	−0.960*
4	5	23.16	0.997*	0.017	0.226	−0.094	−0.996*
5	5	13.99	0.966*	−0.285	−0.592	−0.124	−0.947*
6	5	21.26	0.968*	−0.065	−0.425	−0.024	−0.967*
7	5	12.77	0.989*	−0.096	−0.471	−0.125	−0.961*
mean		20.45	0.922	−0.100	−0.309	−0.070	−0.937
SD		±6.13		±0.157		±0.045	
IS group							
1	5	5.22	0.906*	0.486	0.955*	(−0.063	−0.950*)
2	4	9.65	0.997*	0.185	0.948	(−0.236	−0.992*)
3	5	6.30	0.958*	0.470	0.904*	(−0.202	−0.981*)
4	5	4.59	0.947*	0.547	0.935*	(−0.097	−0.992*)
5	5	13.94	0.999*	−0.146	−0.931*	(−0.219	−0.996*)
6	5	12.59	0.999*	−0.017	−0.303	(−0.148	−0.987*)
7	5	9.25	0.997*	−0.063	−0.687	(−0.144	−0.978*)
8	5	15.82	0.999*	0.125	0.958*	(−0.147	−0.998*)
mean		9.67	0.991	0.198	0.866	(−0.157	−0.986)
SD		±4.17		±0.272		(±0.060)	

n; number of data points.
slope; regression coefficient
r; correlation coefficient
mean r; estimated by z-transformation
*; P < 0.05

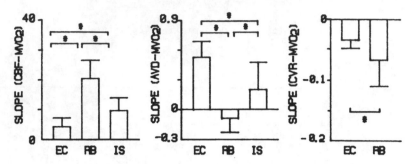

Fig. 1 Slopes of CBF-MVO$_2$, AVD-MVO$_2$, and CVR-MVO$_2$ relations. values are mean $\pm$ SD. *; P $<$ 0.05, significant difference by one-way analysis of variance and LSD method. See text for details.

were statistically significant differences among the three groups. Since CBF and AVD changed reciprocally, the mean values for the slope of the AVD-MVO$_2$ relation were also different among the three groups ;0.529 $\pm$ 0.619, -0.100 $\pm$ 0.157 and 0.198 $\pm$ 0.272 ((g/dl)/[ml O$_2$/100g LV]) in the EC, RB, and IS groups, respectively.

Similarly, the slope of the CVR-MVO$_2$ relation indicates the sensitivity of CVR to a change in MVO2. The mean value for this slope in the EC group was -0.029 $\pm$ 0.025 ([mmHg/(ml/100g LV)]/(ml O$_2$/100gLV)), which was significantly larger than that in the RB group (-0.070 $\pm$ 0.045).

DISCUSSION

Pressure-volume area (PVA), a specific area on the pressure-volume diagram, consists of external work and potential energy, and is considered as the total mechanical energy output of the left ventricle during one cardiac cycle (Suga, et al., 1981a, 1981b). We recently investigated the MVO$_2$-PVA relation in in situ hearts using right heart bypass preparations (Nozawa, et al., 1987). The MVO$_2$-PVA relation was closely fitted by a linear regression line with a similar slope value to that shown in excised cross-circulated hearts in the previous studies (Suga, et al., 1981a, 1981b). From the view point of intrinsic energy conversion efficiency, the efficiency from MVO$_2$ to PVA was not different from that in excised hearts. Nevertheless, the manner of oxygen delivery to the hearts was quite different as seen in the present study. When MVO$_2$ increased, AVD increased predominantly in excised hearts, whereas CBF did so in in situ hearts. We discuss this difference as follows.

The first consideration is that the extent to which the coronary circulation depends on left ventricular mechanics was obviously different in the EC group from that in the RB and IS groups. Since the coronary circulation of the EC group was supplied by the support dog, it was independent of left ventricular mechanics although elevated end-diastolic pressure or a shortened diastolic phase would diminish CBF. In contrast, the coronary circulation of the RB and IS groups perfused the myocardium which generated coronary perfusion pressure itself. Thus any interference with left ventricular mechanics to alter MVO$_2$ would simultaneously affect the coronary

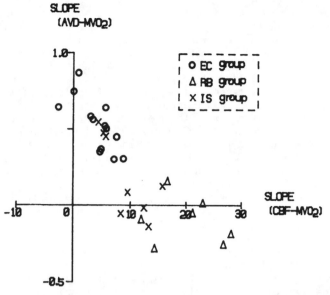

Fig. 2   Both slopes of CBF-MVO$_2$ and AVD-MVO$_2$ relations represent sensitivities or contributions of CBF and AVD to
changes in MVO$_2$. Each case scatters from left upward to right
downward in the order of EC, IS, and RB groups. See text for
details.

circulation, or coronary perfusion pressure. Altered coronary perfusion
pressure would necessarily affect CBF and probably re-affect MVO$_2$ by Gregg's
effect (Gregg, 1963). In fact, significant correlation was observed between
MVO$_2$ and mean aortic pressure in 4 cases of the RB group, but not in the EC
group. Therefore, it is fair to state that CBF and MVO$_2$ related more in-
timately to each other in the RB and IS groups than in the EC group through
the intermediary role of left ventricular mechanics which were also related
with both CBF and MVO$_2$.

The second consideration is that the different contribution of CBF to
MVO$_2$ can be accounted for in part by the different sensitivity of the coronary vasculature to changes in MVO$_2$. Inverse correlation of CVR to MVO$_2$ was
absent in 4 cases in the EC group. Furthermore, the slope of the CVR-MVO$_2$
relation in the EC group was significantly larger than that in the RB group.
This indicates impaired coronary vasodilatation along with the increased
MVO$_2$ in excised hearts. Thus some regulatory mechanism of coronary vascular
tone, and especially of metabolic vasoregulaion, must have been abolished by
excision.

Of course the effect of denervation cannot be excluded. Sarnoff et al.
(1958) noted that coronary vascular resistance frequently fell substantially, possibly due to denervation, between 30 minutes and 1 hour after
heart isolation was completed. But the absolute values for CVR in the EC
group were not necessarily lower compared with those in the RB group in the
present study. Moreover, even in pharmacologically denervated 4 cases in the
RB group, the tendency toward a greater contribution of CBF to MVO$_2$ was not
alleviated.

An alternative speculation is that the difference is due to the effect

443

of some substances such as adenosine, ATP, or tissue oxygen assumed to be the mediators of metabolic vasoregulation. However, we are on uncertain ground here since we have no evidence to support such an idea.

A summary diagram is presented (Fig. 2). The abscissa indicates the slope of the $CBF–MVO_2$ relation and the ordinate indicates that of the $AVD–MVO_2$ relation. The individual points represent each case of the EC, RB, and IS groups. As readily noticed, these points scatter diagonally from left upward to right downward in this diagram in the order of EC, IS, and RB groups. The more left and upward the point locates, the lesser is the contribution of CBF to MVO_2, and the greater is the contribution of AVD to MVO_2. Inversely, the more right and downward the point locates, the greater is the contribution of CBF to MVO_2. Interestingly, the IS group locates between the other two groups in this diagram in spite of the IS group being a more physiological preparation than the RB group. This is probably related to the intervention employed in the IS group to change MVO_2. The elevated heart rate will increase MVO_2, but will not so much increase CBF concomitantly because of the shortened diastolic phase during one cardiac cycle.

In summary, we observed the different contributions of CBF to changes in MVO_2 among three different preparations. CBF changed predominantly in in situ hearts so as to match with the oxygen demand, whereas AVD did so in excised hearts. Metabolic vasoregulation of the coronary artery was more or less impaired in excised hearts. The effect of denervation may be partially responsible for this disparity.

SUMMARY

The correlations of $CBF–MVO_2$ and $AVD–MVO_2$ were analyzed by linear regression analysis in three groups of canine heart preparations. CBF correlated well with MVO_2 in all cases of in situ heart preparations, and the sensitivity of CBF to changes in MVO_2 was high. In contrast, AVD, rather than CBF, correlated well with MVO_2 in all cases of excised hearts, and the sensitivity of CBF to changes in MVO_2 was low. The different manner of oxygen delivery in excised hearts from that in in situ hearts can be attributed to the characteristics of its coronary perfusion and the impaired mechanism of metabolic vasoregulation in the coronary vasculature.

REFERENCES

Belloni, F. L., 1979, The local control of coronary blood flow, Cardiovasc. Res., 13: 63.

Berne, R. M., 1963, Cardiac nucleotides in hypoxia: possible role in regulation of coronary flow, Am. J. Physiol., 204: 317.

Berne, R. M., 1980, The role of adenosine in the regulation of coronary blood flow, Circ. Res., 47: 807.

Braunwald, E., Sarnoff, S. J., Case, R. B., Stainsby, W. N., and Welch, G. H., 1958, Hemodynamic determinants of coronary flow: Effect of changes in aortic pressure and cardiac output on the relationship between myocardial oxygen consumption and coronary flow, Am, J.Physiol., 192(1): 157.

Feigl, E. O., 1983, Coronary physiology, Physiol. Rev., 63: 1.

Gerola, A., Feinberg, H., and Katz, L. N., 1959, Role of catecholamines on energetics of the heart and its blood supply, Am. J. Physiol., 196(2): 394.

Goodale, W. T., and Hackel, D. B., 1953, Myocardial carbohydrate metabolism in normal dog, with effects of hyperglycemia and starvation, Circ. Res., 1: 509.

Gregg, D. E., 1963, Effect of coronary perfusion pressure or coronary flow on oxygen usage of the myocardium, Circ. Res., 13: 497.

Katz, L. N., and Feinberg, H., 1958, The relation of cardiac effort to
    myocardial oxygen consumption and coronary flow, Circ. Res., 6: 656.
Moir, T. W., Eckstein, R. W., and Driscol, T. E., 1963, Thebesian drainage
    of the septal artery, Circ. Res., 12: 212
Nozawa, T., Yasumura, Y., Futaki, S., Tanaka, N., Igarashi, Y., Goto. Y.,
    and Suga, H., 1987a, Relation between oxygen consumption and pressure-
    volume area of in situ dog heart, Am. J. Physiol., 253: (in press).
Nozawa, T., Yasumura, Y., Futaki, S., Tanaka, N., Uenishi, M., and Suga, H.,
    1987b, Relation between oxygen consumption and pressure-volume area
    accounts for the Fenn effect in dog left ventricle, (in preparation).
Sarnoff, S. J., Case, R. B., Welch, G. H., Braunwald, E., and Steinby, W.
    N., 1958, Performance characteristics and oxygen debt in a nonfailing,
    metabolically supported, isolated heart preparation, Am. J. Physiol.,
    192(1): 141.
Shephard, A. P., and Burgar, C. G., 1977, A solid-state arteriovenous oxygen
    difference analyzer for flowing whole blood, Am. J. Physiol., 232:
    H437.
Snedecor, G. W., and Cochran, W. G., 1967, "Statistical Methods", Iowa State
    Univ. Press, Ames.
Suehiro, S., Suga, H., and Ninomiya, I., 1982, Interrelation between myocar-
    dial oxygen consumption, coronary blood flow and arteriovenous oxygen
    difference in the isolated cross-circulated canine heart, (in Japanese
    with English abstract), Resp. and Circ., 30: 1141.
Suga, H., Hayashi, T., and Shirahata, M., 1981a, Ventricular systolic
    pressure-volume area as predictor of cardiac oxygen consumption, Am.
    J. physiol., 240: H39.
Suga, H., Hayashi, T., Shirahata, M., Suehiro, S., and Hisano, R., 1981b,
    Regression of cardiac oxygen consumption on ventricular pressure-
    volume area in dog, Am. J. Physiol., 240: H320.
Yasumura, Y., Nozawa, T., Futaki, S., Tanaka, N., Goto, Y., and Suga, H.,
    1987, Dissociation of pressure-rate product from myocardial oxygen
    consumption in dog, Jpn. J. Physiol., (to be published).

MYOCARDIAL CAPILLARY FLOW PATTERN AS DETERMINED

BY THE METHOD OF COLOURED MICROSPHERES

W. John Reeves and Karel Rakusan

Department of Physiology, School of Medicine
Faculty of Health Sciences, University of Ottawa
Ottawa, Ontario, Canada

INTRODUCTION

In a paper presented to the ISOTT Meeting in 1973 D.F. Bruley stated
that "one of the greatest problems facing both the theoretician and the
experimentalist is the establishment of a meaningful geometry and flow
pattern" (Bruley, 1973). Surprisingly, in the 14 years since this
statement was made no technique has emerged which would allow for the
accurate analysis of capillary flow pattern. The determination of
capillary flow pattern, whether concurrent, countercurrent or any other
proposed geometry, has proven especially difficult for the heart with its
inherent contractile activity.

With no method available for the establishment of myocardial flow
pattern, the theoretician finds himself in the position of having to base
his modelling of oxygen transport to tissue on assumption. The incorporation of any assumption, whether anatomical or physiological, in the
modelling process must bring into question the relevance and accuracy of
any obtained results. This is due to the fact that any mathematical model,
no matter how sophisticated, stands or falls with the quality of its input
data (see reviews by Kreuzer, 1982; Fletcher, 1978; Middleman, 1972;
Leonard and Jorgensen, 1974).

The search for an accurate geometry and flow pattern has produced a
number of excellent research techniques. An incomplete listing of those
which are applicable to myocardium include: silicone elastomer perfusion
techniques (e.g. Bassingthwaighte et al., 1974), photomicrographs of
corrosion casts (e.g. Potter and Groom, 1983; Tomanek et al., 1982) and
intravital microscopic techniques (e.g. Bourdeau-Martini and Honig, 1973;
Steinhausen et al., 1978). The use of these techniques has provided the
data necessary to characterize the myocardial capillary network as exhibiting a parallel alignment corresponding to the orientation of muscle cells.
In addition, it appears that the capillary system exhibits no definable
beginning or end but is instead sequentially reinforced and drained, by
transverse arterioles and collecting venules, respectively.

The results of the in vitro techniques portray a geometry which
invites speculation as to the nature of myocardial flow pattern. This
tendency must be resisted due to the fact that changes in the contractile

state, driving pressure, vasomotion, etc. may result in temporal variability of flow pattern. In contrast, intravital microscopic techniques allow for the actual visualization of the myocardial microcirculation. With this technique Steinhausen et al. (1978) reported that on the surface of the rat myocardium, blood flow in 56% of capillaries was concurrent, with the remaining capillaries exhibiting a countercurrent flow pattern. Similar techniques applied to the sartorius muscle indicate that of 149 observations, 83% were concurrent and 17% countercurrent (Tyml et al., 1981). However, intravital microscopic techniques require manipulation of the tissue itself which may in fact alter the flow pattern of the region under study. In addition, these observations are limited to the most superficial layers of the epimyocardium, which is in no way representative of the rest of the organ, whose environment and thus flow is affected to a greater degree by the contractile activity of the heart.

Theoretical analysis of flow pattern through the application of appropriate geometric models has lead to conflicting results. Wieringa (1985) developed a stochastic three-dimensional network model which when applied to myocardium resulted in 57% of situations being concurrent and 43% of situations being countercurrent. Grunewald and Sowa (1977), utilizing the microcirculatory unit (MCU) model applied to skeletal muscle at rest, concluded that the concurrent MCU 1 and asymmetric MCU 16 models accounted for 61% and 39% of cases respectively, while the countercurrent MCU 19 Model was absent.

In an effort to resolve the inconsistencies of past results as well as to provide a method capable of transmural analysis of myocardial flow pattern we have developed a new technique based on the serial injection of non-radioactive microspheres. This technique, by providing a permanent record of flow activity during the period of microsphere injection allows for both qualitative and quantitative information to be obtained concerning myocardial capillary geometry and flow pattern.

METHODS

Sprague-Dawley rats of either sex (weighing 200-350 g) were placed in a flow through system and anesthetized using 2.5% halothane in oxygen, delivered at 500 ml/min. After surgical preparation a midline incision was made from cricoid cartilage to sternal notch. Blunt dissection and subsequent retraction of strap muscles exposed both carotid arteries and trachea. The left carotid artery was dissected free from the vagus nerve then catheterized using a bevelled PE-50 polyethylene catheter. This catheter was connected to a Statham pressure transducer P25Db which allowed for a continual written recording of arterial pressure using a 4 channel Grass 7D polygraph. A tracheostomy was performed, an intracath #16 was introduced and then connected to a Harvard rodent respirator delivering 2% halothane in oxygen at 500 ml/min. The right carotid was then dissected from the vagus nerve and a left ventricular PE-10 catheter, stabilized by a metal stylet, subsequently advanced into the aortic root. Gentle manipulation was used to position the catheter within the ventricular cavity without damaging the myocardium. The success of this proceedure was established through visualization of appropriate ventricular pressure recordings.

Serial injections of microsphere aliquots, each containing approximately 10 million, 12 μm, non-radioactive coloured microspheres (obtained from SRP, 435 N. Roxbury Drive, Beverly Hills, Ca. 90210, U.S.A.) were made into the left ventricle. Each 1 μm aliquot contained a different colour of microsphere suspended in a .01% Tween solution. This along with vortexing minimized any aggregation prior to injection. Each injection was carried out over a two minute period and was quickly followed by a 0.5 ml saline flush. A timed 10-minute interval elapsed before any subsequent injections were carried out.

After all injections were completed the heart was quickly removed, and after removal of apex and base quickly frozen in liquid N_2. The heart was allowed to reach an equilibrium temperature of -20°C in a cryostat (IEC Minotome Microtome-Cryostat) before the preparation of 70 μm thick longitudinal sections. These slices were then fixed (5% acetic acid, 95% absolute alcohol) for 1 minute and then stained with periodic acid-Schiff. This method results in the formation of three distinct microsphere aggregation patterns within the myocardial circulation;

(1) single sphere
(2) aggregates of one colour only
(3) aggregates containing more than one injected colour

Aggregates which contain more than one injected colour allow for the production of a "flow vector" directed from the position of a latter injected coloured microsphere to the position of a previously injected coloured microsphere (see Fig. 1). Analysis of "flow vectors" obtained from neighbouring capillaries provides the information necessary to determine the flow pattern existing between the capillaries during the period of microsphere injection. This method of analysis was expanded in our study to a range of five intercapillary distances from a reference capillary with its associated "flow vector".

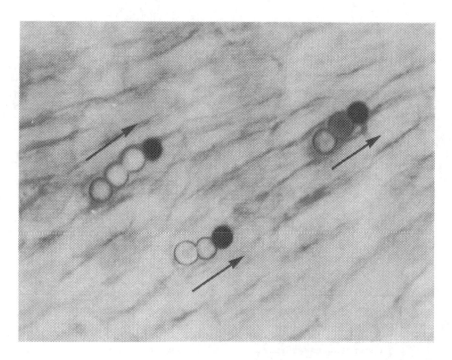

Figure 1. Sequential injection of black and blue (which appear clear) 12 μm non-radioactive microspheres allows for the production of capillary "flow vectors" within the myocardium.

RESULTS/DISCUSSION

In order to determine the nature of flow pattern existing within the myocardium, one requires the production of "flow vectors" in a suitable number of neighbouring capillaries. This requires the injection of more than 30 million microspheres, which although providing sufficient "flow vector" density, also presents a practical limitation on the method itself. This results from the fact that the aggregation of microspheres within a capillary may affect flow in the neighbouring vicinity and thus alter the observed flow pattern.

Aggregates lodged within the capillary network appear to be preferentially located in either the arterial ends of the capillaries or occasionally in the transverse arterioles. When lodged within the capillaries one can clearly visualize the parallel nature of the network as well as numerous examples of "Y" branching, "H" anastomosis and short capillary loops. In cases where large numbers of microspheres were aggregated within a transverse arteriole one could see that these vessels often cross the parallel capillary network on a diagonal, resulting in a staggered arrangement of arteriolar inflows to the capillary network. When numerous "flow vectors" can be visualized to have originated from a given transverse arterial, the arteriolar inflows to the capillaries often branch in opposite directions which supports the geometric concept of alternating transverse arterioles and collecting venules (Fig. 2).

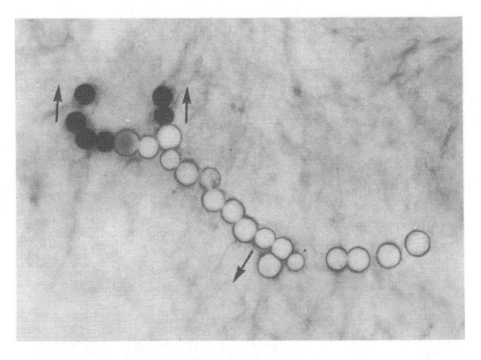

Figure 2. Aggregation of 12 μm microspheres within a transverse arteriole and its capillary branches provides qualitative information concerning the terminal vascular bed.

The apparent inability of microspheres to lodge on the venular side of the capillary network may hide the presence of long capillary loops which reverse their direction further along in the capillary network. This may present a second limitation to our technique in that these loops, if present, display a countercurrent flow pattern which escapes our analysis. This would then result in an overestimation in the percentage of concurrent flow by our technique.

Table 1. Results of "flow vector" analysis for rat myocardium.

Number of Intercapillary Spacings	1	2	3	4	5
Number of Flow Observations	343	424	394	308	322
Number of Concurrent Flow Patterns	334	392	366	272	276
Number of Countercurrent Flow Patterns	14	32	28	36	46
Percentage of Concurrent Flow Patterns	96	92	93	88	86

Quantitative results of vector flow pattern analyses over a range of five intercapillary distances are presented in Table 1. The number of observations for each intercapillary distance is in excess of 300 and provides the basis for the determination of capillary flow pattern in myocardium. These results are schematically presented in Fig. 3, which shows the percentage of concurrent blood flow as a function of inter-capillary distance. The obtained results were consistent for all animals studied, and in addition appeared to exhibit no transmural variation.

Our results (96% concurrent flow) appear to contradict previous experimental results obtained for rat myocardium by establishing this flow pattern to be essentially concurrent in nature. The small percentage of countercurrent observations appear to be the result of short capillary loops. The fact that this percentage of countercurrent flow patterns increases as one moves away from a reference capillary may be indicative of the presence of distinct capillary bundles, each exhibiting an internal concurrent flow pattern, which may be countercurrent to neighbouring capillary bundles (see Skalak and Schmid-Schonbein, 1986).

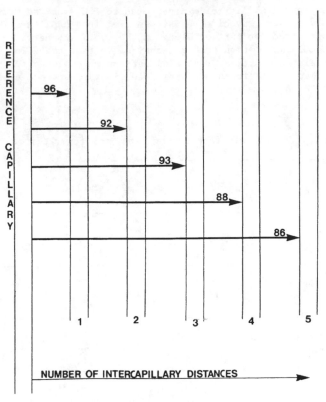

Figure 3. Percentage of concurrent flow patterns as a function of intercapillary distance.

SUMMARY

In summary, it can be said that in spite of possible limitations, this technique provides previously unattainable data on capillary flow pattern in the myocardium. Qualitative and quantitative results indicate that the following anatomical and physiological concepts should be incorporated in any further modelling of oxygen transport to myocardium:

(1) Concurrent flow pattern
(2) Presence of short capillary loops
(3) Presence of capillary bundles
(4) Staggered arrangements of arteriolar inflow to capillary network

ACKNOWLEDGEMENTS

Supported by the Medical Research Council of Canada.

The authors wish to thank Miza Bosc-Davie, Richard Seymour, Jimmy Gao, Ching-Ju Kuo for expert technical assistance, as well Anita Bouchard and Elizabeth Kucharczyk for typing and editing of the manuscript.

REFERENCES

Bassingthwaighte, J.B., Yipintsoi, T., Harvey, R.B., 1974, Microvasculature of the dog left ventricular myocardium, Microvasc. Res., 7:229.

Bourdeau-Martini, J., Honig, C.R., 1973, Control of intercapillary distance in rat heart; effect of arterial CO_2 and pH, Microvasc. Res., 6:286.

Bruley, D.F., 1973, Mathematical considerations for oxygen transport to tissue, in: "Oxygen Transport To Tissue", D.F. Bruley and H.I. Bicher eds., Plenum Press, New York and London.

Fletcher, J.E., 1978, Mathematical modeling of the microcirculation, Math. Biosci., 38:159.

Grunewald, W.A. Sowa, W., Capillary structure and O_2 supply to tissue, Rev. Physiol. Biochem. Pharmacol., 77: 149.

Kreuzer, F., 1982, Oxygen supply to tissues: The Krogh model and its assumptions, Experientia, 38:1415.

Leonard, E.F., Jorgensen, S.B., 1974, The analysis of convection and diffusion in capillary beds, A. Rev. Biophys. Bioengng., 3:293.

Middleman, S. 1972, "Transport phenomena in the cardiovascular system", Wiley-Interscience, New York/London/Sydney/Toronto.

Potter, R.F., Groom, A.C., 1983, Capillary diameter and geometry in cardiac and skeletal muscle studied by means of corrosion casts, Microvasc. Res., 25:68.

Skalak, T.C., Schmid-Schonbein, G.W., 1986, The microvasculature in skeletal muscle. IV. A model of the capillary network, Microvasc. Res., 32:333.

Steinhausen, M., Tillmanns, H., Thederan, H., 1978, Microcirculation of the epimyocardial layer of the heart, Pflugers Archiv, 378:9.

Tyml, K., Ellis, C.G., Safranyos, R.G., Fraser, S., Groom, A.C., 1981, Temporal and spatial distributions of red cell velocity in capillaries of resting skeletal muscle, including estimates of red cell transit times, Microvasc. Res., 22:14.

Wieringa, P.A., 1985, "The Influence of the Coronary Capillary Network on the Distribution and Control of Local Blood Flow," (Ph.D. Thesis) Technische Hogeschool, Delft, The Netherlands.

IMPROVING CARDIOCIRCULATORY PARAMETERS AND OXYGEN TRANSPORT CAPACITY BY SHIFTING THE pH TO THE ALKALINE SIDE

B. Lachmann, B. Jonson[x] and W. Erdmann

Department of Anesthesiology, Erasmus University, 3000 DR Rotterdam, Postbus 1738, The Netherlands, and Department[x] of Clinical Physiology, University of Lund, Sweden

INTRODUCTION

Because mammalian blood pH averages 7.40 at 37°C, most acid-base therapy is directed at maintaining pH at this value. Furthermore, it is well known that increased concentration of hydrogen ions has a marked negative ionotropic effect on the heart. Apart from the investigations of Rahn and Reeves (1) and Becker et al (2), concerning the influence of alkaline pH on cardiocirculatory parameters during hypothermia in vivo, studies concerning the influence of alkaline pH during normothermic conditions are not available. Therefore, the present study was carried out to determine the role of pH management on cardiac output (CO) and circulatory parameters.

METHODS

Using 9 beagle dogs (9.5-11 kg bw) under neurolept analgesia and muscular paralysis, studies were made on the effects of pH changes from 7.15-7.75 on CO, hemodynamics, arterial and venous blood gases, and transcutaneous carbon dioxide tension (TcPCO2). In order to minimize the effects of spontaneous or artificial ventilation on CO, all animals were ventilated by high frequency ventilation (12 Hz) using the HF-unit from the Servo-ventilator system, in connection with the Servo-ventilator 900 C (Siemens-Elema). CO was measured by thermodilution using the Edwards lung water computer 9310. TcPCO2 was monitored by a cutan-PCO2 monitor (Kontron) and PO2 was measured continuously using an intravascular PO2 monitor (Kontron).

pH changes, in steps from 0,1-0,2 in both directions (two or three times), were induced by infusions of 0,5-n HCL or 1-n NaHCO3. Body temperature and ventilator settings were kept constant during the protocol(fig. 1).

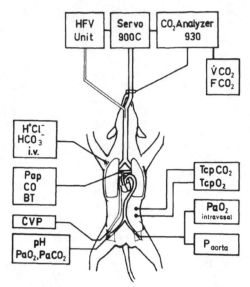

Fig.1: Experimental setup.
Pap - pulmonary artery pressure, CO-cardiac output, BT-body temperatur, CVP-central venous pressure, $TcpCO_2$-transcutaneous CO_2-tension, $TcpO_2$-transcutaneous O_2-tension, $\dot{V}CO_2$-CO_2-minute production, FCO_2-expiratory CO_2-fraction, HFV - high frequency ventilator, Servo 900 C - ventilator (Siemens-Elema, Stockholm, Sweden.

RESULTS

Figure 2 demonstrates the cardiac output from two dogs in which the pH changes were made in both directions. There is an almost linear correlation between pH and cardiac output. The left drawing of figure 2 demonstrates the results of one dog, the right of the second dog. In both animlas the cardiac output reached its maximum between a pH of 7.62 and 7.66 (indicated by arrow). The maximal cardiac output (fig. 3) and mean arterial blood pressure (fig. 4) in the different animals, was reached at a pH between 7.58 and 7.72 (mean 7.64).

The increase in cardiac output always led to an improvement in $P\bar{v}O2$ (fig.5) and to a decrease in $C(a-\bar{v})O2$ (fig.6). The difference between PaCO2 and tcPCO2 (fig.7) was always lower than 6 mmHg during pH increase, whereas H^+ infusion resulted in significantly decrease cardiac output, decreased $P\bar{v}O2$ and increased tcPCO2/PaCO2 differences.

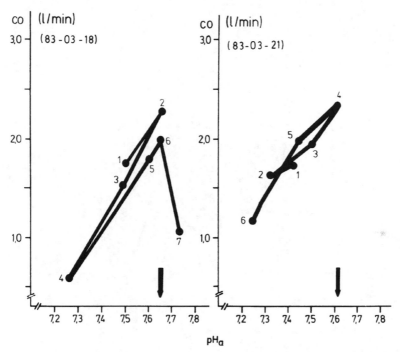

Fig.2: Two typical figures of cardiac output in relation to pH.
The numbers indicate the direction of pH changes either by
bicarbonate or by hydrochlorid acid.

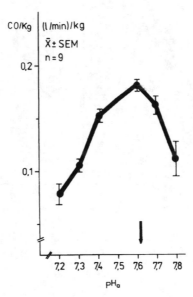

Fig.3: Cardiac output from 9 dogs (including 123 single measurements) in relation to pH. The highest cardiac output was found at pH of 7.62.

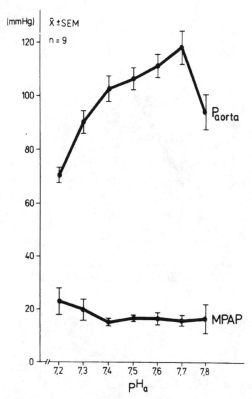

Fig.4: Mean arterial pressure and mean pulmonary artery pressure
in relation to pH. Same animals as in figure 3.

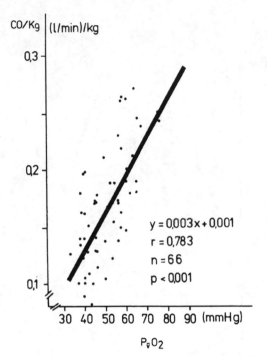

Fig.5: Regression line of mixed venous oxygen tension ($P\bar{v}O_2$) in relation to cardiac output.

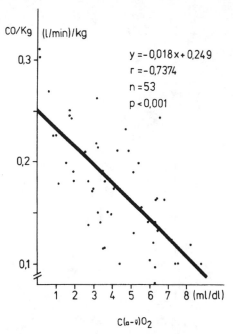

CO/Kg (l/min)/kg

$y = -0,018x + 0,249$
$r = -0,7374$
$n = 53$
$p < 0,001$

$C(a-\bar{v})O_2$

Fig.6: A regression line of arterial-venous oxygen content differences in relation to cardiac output.

CONCLUSIONS

It can be concluded that pH values of 7.3 often accepted during anaesthesia and intensive care in emergency situations, are not high enough for an optimal cardiac performance. To manage circulatory complications, optimal pH settings (7.50-7.65) may be of considerable value. When reaching such pH values by massive bicarbonate infusion, one has to prevent the resulting respiratory acidosis by optimal ventilator settings.

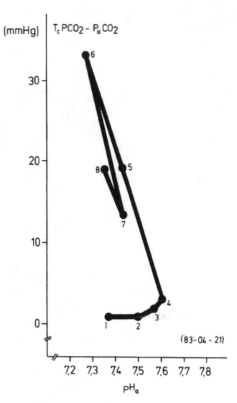

Fig.7: Differences between transcutaneous and arterial CO_2-tension.
The numbers indicate the direction of pH changes.
Note: pH-changes to the alkaline side have no effect on this
difference.

REFERENCES

1.Rahn, R; Reeves, RB. Hydrogen ion regulation: from the amazon to the
    operating room. 1982. In: Prakash O. (ed.). Applied Physiology in
    Clinical Respiratory Care. Martinus Nijhoff, The Hague/Boston/
    London. p.p. 1-15.
2.Becker, H, Vinten-Johansen, J, Buckberg GD. 1981. Myocardial damage caused
    by keeping pH 7.40 during systemic deep hypothermia. J. Thorac.
    Cardiovasc. Surg. 82: 810-820.

MYOCARDIAL OXYGEN TENSION DURING LEFT VENTRICULAR BYPASS

Yoshinori Mitamura, Takeo Matsumoto, and Tomohisa Mikami

Department of Medical Electronics, Research Institute of Applied Electricity, Hokkaido University, Sapporo 060, Japan

INTRODUCTION

In a small number of patients undergoing cardiac surgery, the heart is unable to supply sufficient cardiac output to maintain normal circulation at the end of the procedure. Many types of mechanical devices to assist circulation have been developed for these patients and some of them have been applied clinically. Circulatory assist devices have been shown to be effective in supporting the circulation in the presence of cardiogenic shock, in resting the failing heart and also in salvaging the ischemic heart. Evidence of their action includes reduction in myocardial oxygen consumption (Pennock et al., 1974), improvement of the ST level in the infarcted heart preparation (Pennock et al., 1976; Laks et al., 1977) and an increase in the diastolic pressure-time index/tension-time index (DPTI/TTI) ratio (Hughes et al., 1975). Although myocardial oxygen and carbon dioxide tensions (PmO_2 and $PmCO_2$) reflect the metabolic state of the myocardium more directly than the myocardial oxygen consumption, the ST level in the epicardial ECG or the DPTI/TTI ratio, the myocardial gas tension has not been used so far to evaluate the effect on the heart of left ventricular assist devices. It is the purpose of this study to investigate the effect of left ventricular bypass on myocardial oxygen tension.

MATERIALS AND METHODS

Left Ventricular Assist Device(LVAD)

A pneumatically driven diaphragm-type LVAD was developed as shown in Fig. 1. The LVAD consists of an inflow cannula, pump body, and outflow cannula. The pump cases are made of epoxy resin (Epicote 828, Hardner Epomate N-002). The blood-contacting surface of the pump housing is coated with segmented polyurethane (TM-3). The smooth blood-contacting diaphragm within the pump cases is also made of segmented polyurethane. Ceramic heart valves (#25) are employed (Mitamura et al., 1986). The maximum stroke volume of the pump is around 40 ml. The inflow and outflow cannulae are made of polyvinylchloride and are coated with segmented polyurethane. The LVAD is driven by the multi-mode artificial heart driving system (Takahashi et al., 1978) on counterpulsation using the ECG signal.

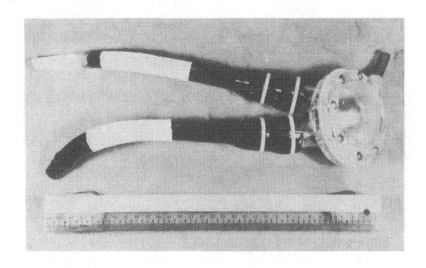

Fig. 1. Pneumatically driven diaphragm-type left ventricular
assist device.

Myocardial PO2 (PmO2) and PCO2 (PmCO2) Measurement

The experimental arrangement is shown in Fig. 2. Six mongrel dogs
weighing between 10.4 Kg and 19 Kg were used in this study. The dogs were
anesthetized with intravenous sodium pentobarbital. Respiration was
controlled by a mechanical respirator. Left thoracotomy was performed

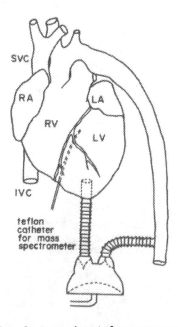

Fig. 2. Experimental arrangement
for myocardial PO2 and
PCO2 measurement.

through the 7th intercostal space. The outflow cannula was anastomosed to the descending aorta. A polyvinyl ring with an attached polyester sewing cuff was sutured to the apex of the left ventricle. A circular knife was then used to excise a section of the myocardium to allow insertion of the inflow cannula into the sinus of the left ventricle. After connection of the tubings to the pump was completed, and residual air in the pump removed, pumping was started. Two electrodes were sutured in place: one in the myocardium and the other in the intercostal muscle. PmO_2 and $PmCO_2$ were measured continuously by a mass spectrometer (Scientific Research Instruments Corp. MS-8) employing a flexible stainless steel sampling catheter 1 mm in diameter encased in a diffusible Teflon membrane. The sampling catheter was gently inserted into the left ventricular myocardium parallel to the left anterior descending coronary artery. A small nick was made in the epicardium and the Teflon diffusion probe was advanced into the subendocardial wall obliquely to the epicardial surface. A constant position was assured by securing the probe with a 3-O silk suture. Arterial and left ventricular catheterizations were performed through the femoral and carotid artery, respectively. Pressures were measured by strain gauge-type pressure transducers. The drive pressure of the LVAD was also measured with a strain gauge-type pressure transducer. All measurements were recorded on an 8-channel pen recorder.

The drive condition of the LVAD was switched back and forth between LVAD counterpulsation and pump-off (control) every 20 minutes.

Myocardial Oxygen Consumption (MOC) Measurement

Thirteen dogs (15-25 Kg) were used in this study. The dogs were classified into two groups: left ventricle-to-aorta bypass (LV-A; N=6) and left atrium-to-aorta bypass (LA-A; N=7). The experimental arrangement is shown in Fig. 3. Under intravenous anesthesia and controlled respiration, right thoracotomy was performed through the 4th intercostal space. An extracorporeal coronary blood flow loop was completed between the coronary sinus and the jugular vein using a flow-probe (6 mm) and T-shaped connector interposed in the silicone tube. After closure of the right

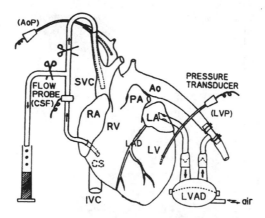

Fig. 3. Experimental arrangement for myocardial oxygen consumption measurement in the LA-A bypass preparation.

465

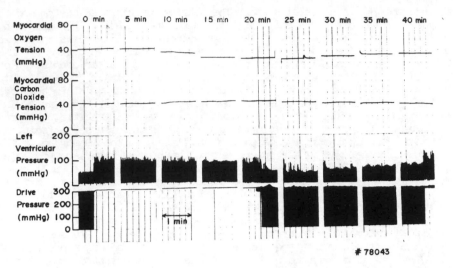

Fig. 4. Changes in PmO2 and PmCO2 during LVAD counterpulsation and in the control in the LV-A bypass preparation.

chest, left thoracotomy was performed and the LVAD implanted between the left ventricle and the aorta, or the left atrium and the aorta. Silicone catheters for mass spectrometry were inserted into the coronary sinus cannula and the brachial artery to measure PO2 and PCO2 continuously. During left ventricular bypass and in the control (pump-off), arterial and venous blood was sampled. The blood gases were analyzed by a blood gas analyzer and the hematocrit (Hct) was obtained by centrifugation. Oxygen saturation (SO2) was calculated from the Severinghaus nomogram. Coronary sinus blood flow (CSBF) was measured by collecting coronary sinus blood through the T-connector. MOC was calculated as 1.34 x Hct/3 X (SaO2 - SvO2) X CSBF.

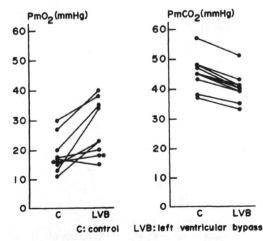

Fig. 5. PmO2 and PmCO2 during LVAD counterpulsation and the pump-off period.

RESULTS

PmO2 and PmCO2 Measurement

Changes in PmO2 and PmCO2 during LVAD counterpulsation following pump-off are exemplified in Fig. 4. With the pump-off, PmO2 decreased and PmCO2 increased gradually from their original level. After the start of pumping the left ventricular pressure was reduced significantly and the aortic pressure was increased. Maximum left ventricular pressure was lower than minimum aortic pressure. Total left ventricular bypass was obtained with pump-on in the LV-A preparations. PmO2 increased and PmCO2 decreased gradually toward the original level during LVAD counterpulsation. Ten myocardial gas tension measurements during LVAD counterpulsation (total left ventricular bypass) and the pump-off period are summarized in Fig. 5. PmO2 increased during left ventricular bypass except in one case. PmCO2 decreased during LVAD counterpulsation in all cases. PmO2 was 18.3 ± 1.9 mmHg (Mean $\pm$ S.E.; N=10) in the pump-off and increased to 26.4 ± 3.0 mmHg (N=10) during left ventricular bypass (p < 0.005, paired t-test). PmCO2 was 45.3 ± 1.8 mmHg (N=10) with pump-off and 40.2 ± 1.5 mmHg (N=10) with pump-on (p<0.005, paired t-test).

Myocardial Oxygen Consumption (MOC) Measurement

Continuous measurements disclosed an immediate change in CSBF at the time of pump-on and pump-off, while the level of blood gases remained almost constant (Fig. 6). In the LV-A bypass group, MOC was 1.99 ± 0.15 ml/min/100g (N=10) in the pump-off and decreased to 1.17 ± 0.05 (N=10) during left ventricular bypass(p<0.001, paired t-test). CSBF decreased from 24.1 ± 2.74 ml/min/100g (N=10) in the pump-off to 16.3 ± 1.69 (N=10) during left ventricular bypass (p<0.001, paired t-test). Arterial-venous oxygen content difference (A-VDo2) was 8.74 ± 0.63 vol% (N=10) in the

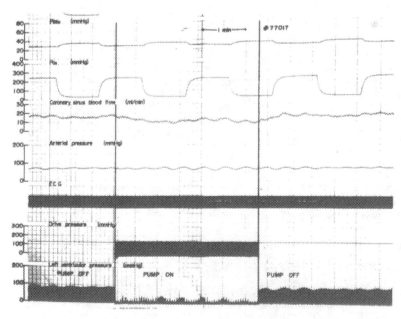

Fig. 6. Hemodynamics during LVAD counterpulsation. PO2 and PCO2 indicate alternately, values for arterial blood and for coronary sinus blood.

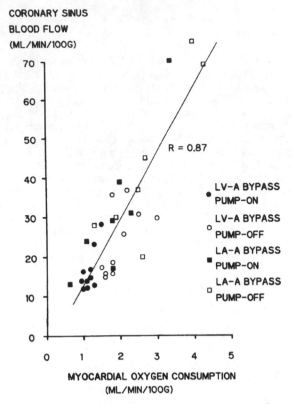

Fig. 7. Relationship between coronary sinus blood flow and myocardial oxygen consumption.

pump-off and 7.62 $\pm$ 0.56 (N=10) during left ventricular bypass (p<0.01, paired t-test). In the LA-A bypass group, CSBF was 43.4 $\pm$ 8.0 ml/min/100g in the pump-off and 31.9 $\pm$ 7.2 during LV bypass (N=7; p<0.05). MOC was reduced significantly from 2.74 $\pm$ 0.41 ml/min/100 g in the pump-off to 1.85 $\pm$ 0.34 during bypass (N; p<0.02). Arterial-venous oxygen content difference was 6.84 $\pm$ 1.04 vol% in the pump-off and 6.19 $\pm$ 0.79 during LV bypass (N=7).

The relationship between CSBF and MOC is shown in Fig. 7. All measurements with pump-on and pump-off in both LV-A bypass and LA-A bypass groups are plotted. CSBF increases with the increase of MOC. No significant correlation was observed between CSBF and mean arterial pressure (Fig. 8).

DISCUSSION

For temporary support of the failing post-cardiotomy heart, several types of mechanical circulatory assist devices have been applied clinically. Beneficial effects of left ventricular bypass on the normal heart have been demonstrated as evidenced by a reduction of myocardial oxygen consumption, improvement of the ST level in the epicardial ECG and an increase of DPTI/TTI ratio. Left ventricular bypass is effective in

468

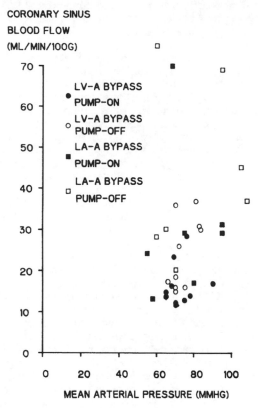

CORONARY SINUS
BLOOD FLOW
(ML/MIN/100G)

LV-A BYPASS
PUMP-ON

LV-A BYPASS
PUMP-OFF

LA-A BYPASS
PUMP-ON

LA-A BYPASS
PUMP-OFF

MEAN ARTERIAL PRESSURE (MMHG)

Fig. 8. Relationship between coronary sinus
blood flow and mean arterial pressure

assisting the heart by reducing myocardial metabolic requirements or
improving myocardial blood flow, and thereby improving the oxygen
supply/demand ratio in the myocardium. Reduced myocardial work implies
a decreased myocardial oxygen demand, however it does not guarantee an
improved oxygen supply to the myocardium. Reduction of myocardial oxygen
consumption itself does not always mean the improvement of myocardial
metabolism. Interpretation of the epicardial ST level measurement relies
on the evidence that ST segment elevation directly reflects the extent of
myocardial ischemia. However, an epicardial electrode at times fails to
detect the presence of underlying ischemia, as evidenced by significant
changes in local myocardial gas tensions and intramyocardial ST segment
voltage. This lack of sensitivity may reflect the inability of the
epicardial electrodes to sense ischemic changes in deeper myocardial
layers. Evaluation by the DPTI/TTI ratio relies on the evidence that DPTI
reflects myocardial oxygen supply and TTI oxygen demand. However,
applicability of the parameters DPTI and TTI is rather limited.
Intramyocardial gas tensions reflect the balance between oxygen supply and
consumption in local regions of myocardium more directly. Therefore
PmO_2 and $PmCO_2$ are more appropriate indices of the effects of left
ventricular bypass than myocardial oxygen consumption, epicardial ECG or
DPTI/TTI. Although various polarographic electrode methods have allowed
measurement of tissue PO_2, measurement of PCO_2 within tissues has been
less successful. In this study a mass spectrometric technique was used.

This technique has made it possible to continuously monitor the changes in both PmO2 and PmCO2. PmO2 increased and PmCO2 decreased significantly in normally perfused myocardium (Fig. 4 and 5) during LVAD counterpulsation. This is direct evidence that left ventricular bypass improves the myocardial oxygen supply/demand ratio.

One of the beneficial effects of counterpulsation is diastolic augmentation. During counterpulsation, blood is ejected into the artery from assist pump during diastole of the natural heart and maximum arterial pressure appears during diastole. This increased diastolic pressure, i. e. coronary perfusion pressure, may increase coronary circulation. However, as shown in this study CSBF decreased during counterpulsation, while coronary perfusion pressure was increased. The level of coronary blood flow is primarily set by the rate of myocardial oxygen consumption. Hence, as shown in Fig. 7, a linear relationship was observed between CSBF and MOC. Although CSBF decreased during counterpulsation, there are several pieces of evidence which account for the improvement of the myocardial oxygen supply/demand ratio during counterpulsation. First, the myocardial CSBF/MOC ratio was significantly ($p < 0.05$) higher during counterpulsation (13.7 ± 1.04; N=10) than in the control (12.1 ± 1.13; N=10) in the LV-A bypass group. Similar results were found in Pennock's experiments (Pennock et al., 1974). The CSBF/MOC ratio was 9.9 in the pump-off and 13.8 during complete left ventricular decompression. Secondly, a diffusion model for myocardial PO2 suggests that a rise in mean tissue PO2 will occur unless compensated for by a increase in intercapillary distance for proportional decrements in MOC and coronary blood flow. Consider the effect on myocardial PO2 if capillary density was constant and MOC halved. A corresponding halving of coronary blood flow would maintain coronary PVO2 constant. Since the diffusion of oxygen from each capillary would also be halved, the gradient for PO2 from the capillary to the tissue would be decreased, and, for the same capillary PO2, the mean tissue PO2 would increase. The transmural gradient of coronary blood flow is another factor. The distribution of coronary blood flow across the heart wall is influenced by counterpulsation. The flow in the endocardial half (ENDO) of the left ventricular wall is approximately equal to the flow in the epicardial half (EPI) when DPTI:TTI is greater than 0.8. When DPTI:TTI falls below 0.8, the ratio of ENDO:EPI flow falls below 1.0 indicating relative underperfusion of the subendocardium (Buckberg et al., 1972). During counterpulsation in the LV-A bypass group, the DPTI:TTI ratio is much higher than 1.0, because TTI is nearly zero as shown in Fig. 4. Therefore, a uniform distribution of coronary blood flow across the heart wall is expected.

The results of this study suggest that left ventricular bypass is effective in increasing the oxygen supply/demand ratio in the myocardium, while both MOC and CSBF decrease during left ventricular bypass.

SUMMARY

The effect of an assist pump on the heart was evaluated on the basis that myocardial gas tensions reflect the balance between oxygen supply and consumption in local regions of the myocardium.
A left ventricular assist device (LVAD) was implanted in dogs between the left ventricle (LV group), or the left atrium (LA group) and the aorta. The assist pump was driven in a counterpulsation mode. A silicone catheter for mass spectrometry was inserted into the myocardium to monitor continuously myocardial oxygen and carbon dioxide tensions (PmO2, PmCO2). Coronary sinus blood flow (CSBF) was measured by completing an extracorporeal flow loop between the coronary sinus and the jugular vein.
During LVAD counterpulsation, the LV pressure was reduced and diastolic

arterial pressure (i.e. coronary perfusion pressure) was increased. Total LV bypass was obtained in the LV group, and partial LV bypass in the LA group.

In the LV group, PmO2 was 18.3+1.9mmHg with pump-off and increased to 26.4+3.0 during LV bypass (N=10, p<0.005). PmCO2 was 45.3+1.8mmHg with pump-off and decreased to 40.2+1.5 during bypass (N=10; p<0.005).

CSBF significantly decreased during LV bypass from a control value of 24.1+2.74 ml/min/100g to 16.3+1.69 in LV group (N=10; p<0.001). It was 43.4+8.0 ml/min/100g in the pump-off and 31.9+7.2 during LV bypass in the LA group (N=7; p<0.05). Myocardial oxygen consumption (MOC) was significantly reduced during LV bypass (from 1.99+0.15ml/min/100g to 1.17+0.05 in the LV group (N=10; p<0.001) and from 2.74+0.41 to 1.85+0.34 in the LA group (N=7; p<0.02)). Arterial-venous oxygen content difference was 8.74+0.63 vol% in the pump-off and 7.62+0.56 during LV bypass in the LV group (N=10; p<0.001). In the LA group, the difference was 6.84+1.04 vol% in the pump-off and 6.19+0.79 during LV bypass (N=7).

The oxygen supply/demand ratio in local regions of myocardium was improved by LVAD, as evidenced by the increased PmO2. The coronary blood flow decreased during LV bypass, while coronary perfusion pressure was increased. CSBF was primarily set by MOC.

The results suggest that the LVAD counterpulsation is effective in improving the myocardial oxygen supply/demand ratio.

REFERENCES

Buckberg, G. D., Fixler, D. E., Archie, J. P., and Hoffman, J. I. E., 1972, Experimental subendocardial ischemia in dogs with normal coronary arteries, Cir. Res., 30:67.

Hughes, D. A., Igo, S. R., Daly, B. D. T., Migliore, J. J., and Norman, J. C., 1975, Effects of an abdominal left ventricular assist device on myocardial oxygen supply/demand ratios in normally perfused and ischemic bovine myocardium, Ann. Thorac. Surg., 19:301.

Laks, H., Ott, R. A., Standeven, J. W., Hahn, J. W., Blair, O. M., and Willman, V. L., 1977, The effect of left atrial-to-aortic assistance on infarct size, Circulation, 56:11.

Mitamura, Y., Mikami, T., Yuta, T., Matsumoto, T., Shimooka, T., Okamoto, E., Eizuka, N., and Yamaguchi, K., 1986, Development of a fine ceramic heart valve for use as a cardiac prosthesis, Trans. Am. Soc. Artif. Intern. Organs, 32:444.

Pennock, J. L., Pierce, W. S., Prophet, G. A., and Waldhausen, J. A., 1974, Myocardial oxygen utilization during left heart bypass, Arch. Surg., 109:635.

Pennock, J. L., Pierce, W. S., and Waldhausen, J. A., 1976, Quantitative evaluation of left ventricular bypass in reducing myocardial ischemia, Surgery, 79:523.

Takahashi, M., Mitamura, Y., Yamamoto, K., Mikami, T., Nakamura, T., Nishiura, K., and Sakakibara, K., 1978, Multi-mode artificial heart driving system, Jap. J. Artif. Organs, 7:141.

MYOCARDIAL OXYGEN TENSIONS DURING ISCHAEMIA IN FLUOROCARBON DILUTED PIGS

N.S. Faithfull*, M. Fennema and W. Erdmann

Departments of Anaesthesia, *University of Manchester, UK
and Erasmus University, Rotterdam, The Netherlands

INTRODUCTION

Fluorocarbons are inert compounds with a very high solubility for oxygen which over the last few years have been available in emulsified form in the oxygen transporting plasma substitute Fluosol-DA 20%. This preparation has undergone clinical trials in the USA and Japan in which it has been used to treat severe anaemia in Jehovah's witness patients who have refused blood transfusions on religious grounds (Tremper et al, 1982; Ohyanagi et al, 1984).

A valuable potential clinical application of fluorocarbon emulsions lies in their ability to support the ischaemically hypoxic microcirculation. Haemodilution with FDA can cause marked increases in myocardial oxen tension (PmO2) in the most ischaemic areas of an experimentally induced myocardial infarct (Faithfull et al, 1986) and further studies indicated that prior haemodilution with FDA could delay changes in PmO2 and suggested that this was due to radical redistribution of blood flow in the areas of myocardium bordering the infarct (Faithfull et al, 1986). This paper reports in detail the pattern of PmO2 changes occurring after induction of myocardial ischaemia following prior haemodilution with FDA.

METHODS

The model used in this study has been described more fully elsewhere (Faithfull et al, 1986a, Faithfull et al, 1986b). Briefly, two groups of five juvenile Yorkshire pigs (body weight 23.5-27 kg) were anaesthetized with an intraperitoneal induction dose of thiopentone 30 mg kg-1. The trachea was intubated and the lungs were ventilated with 66% nitrous oxide in oxygen and 1.0-1.5% halothane. Muscular relaxation was obtained using a continuous infusion (0.5 mg/kg/hr) of pancuronium bromide. Systemic arterial pressure was monitored via left femoral artery, and pulmonary artery pressure, pulmonary capillary wedge pressure and cardiac output were obtained using a 7-French gauge thermodilution Swan-Ganz catheter inserted via the left femoral vein. Central venous pressure was monitored via a catheter in the right femoral artery. All the above pressures were transduced and displayed on a Grass 7D polygraph and ink writing oscillograph.

Arterial and mixed venous blood-gas and acid-base state, haemoglobin concentration and oxyhaemoglobin saturation were regularly estimated using a Radiometer ABL1 acid/base laboratory and a Radiometer OSM1 cooximeter. In cases where FDA had been administered, readings from the cooximeter were inaccurate and the oxyhaemoglobin saturation was calculated from the computer subroutine described by Kelman (1966).

A thoracotomy was performed, the pericardium was opened and a ligature was loosely placed around the left anterior descending coronary artery (LAD) at the junction of its middle and lower thirds as it crossed the anterior surface of the left ventricle. To assess the effects of haemodilution on PmO2 in both the ischaemic and non-ischaemic areas, four steel-protected gold microelectrodes 200 microns in diameter were inserted approximately 3 mm into the myocardium. These were placed 1 cm apart and positioned so that two were in and two were outside the expected infarct area. A silver reference electrode was inserted into a muscle of the thoracic wall. Outputs from the polarised electrodes were amplified and recorded on a flat bed paperchart recorder. The electrodes were calibrated prior to the experiments and recalibrated at the termination of the studies. Any electrode showing a drift of more than 10 percent per hour was discarded. PmO2 values, corrected for temperature and drift, were calculated every minute for 15 minutes following LAD occlusion and every 5 minutes from 15 to 60 minutes.

One group of animals were bled 20 ml/kg body weight and the volume deficit was replaced by FDA. The other group served as controls and received no treatment. The LAD was occluded after the animals had been ventilated with 0.5% halothane in 100% oxygen for at least 30 minutes to ensure stabilisation of the preparations and allow maximal oxygen transport in the fluorocarbon group.

Cardiovascular and oxygenation variables were calculated using a Phillips P2000 microcomputer and statistical analysis was performed with an Apple IIe micro. Statistical significance was assessed using the paired Student's t tests and the nul hypothesis was rejected at a p value of less than 0.05.

RESULTS

Before LAD occlusion, cardiac output was significantly higher and systemic vascular resistance was significantly lower in the FDA diluted group. Otherwise there were no significant differences in systemic cardiovascular parameters. There were no significant differences in whole body oxygen consumption (VO2) or flux (QO2). Following LAD occlusion, no significant changes in cardiovascular or oxygenation parameters were seen in either group, even though an obvious macroscopically visible area of infarction was created.

The oxygen microelectrodes are referred to thus: that furthest into the ischaemic area is termed the deep ischaemic electrode; that next to it is the border ischaemic electrode; the one in the "nonischaemic" area closest to the infarct is the border non-ischaemic electrode and the electrode furthest from the infarct is the deep non-ischaemic electrode. In figures 1, 2, 3 and 4 graphical analysis of the results are presented from the deep ischaemic, border ischaemic, border non-ischaemic and deep non-ischaemic electrodes respectively. In the figures percentage changes in PmO2 from pre-occlusion values are presented for both control animals and for those prediluted with FDA. PmO2 values are presented every five minutes after LAD occlusion up to 60 minutes and statistical significant differences from the pre LAD occlusion values are indicated - statistical analysis were performed on the pre-transformed PmO2 data.

474

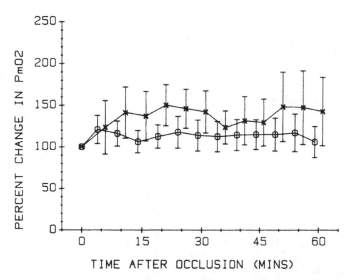

Fig.1 Percentage changes in myocardial oxygen tension (PmO2) in the deep ischaemic electrode following LAD occlusion in control animals (indicated by multiplication signs) and animals subjected to prior haemodilution with Fluosol-DA 20% (indicated by open circles).

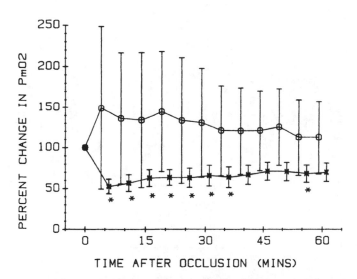

Fig.2 Percentage changes in myocardial oxygen tension (PmO2) in the border ischaemic electrode following LAD occlusion in control animals (indicated by multiplication signs) and animals subjected to prior haemodilution with Fluosol-DA 20% (indicated by open circles).

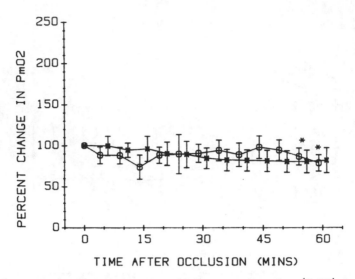

Fig.3 Percentage changes in myocardial oxygen tension (PmO2) in the border non-ischaemic electrode following LAD occlusion in control animals (indicated by multiplication signs) and animals subjected to prior haemodilution with Fluosol-DA 20% (indicated by open circles). Significant change from pre-occlusion values indicated by asterisks (* = p < 0.05)

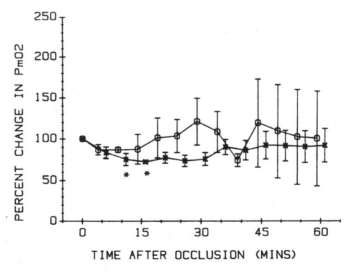

Fig.4 Percentage changes in myocardial oxygen tension (PmO2) in the deep non-ischaemic electrode following LAD occlusion in control animals (indicated by multiplication signs) and animals subjected to prior haemodilution with Fluosol-DA 20% (indicated by open circles). Significant change from pre-occlusion values indicated by asterisks (* = p < 0.05)

No significant changes were seen in PmO2 values at the deep ischaemic electrode (Fig. 1). PmO2s in the border ischaemic areas of the control animals were significantly decreased at all measurement points up to 35 mins and at 55 mins post occlusion; in the FDA pre-treated animals there were no significant changes (Fig. 2). In the FDA group one animal had extremely large increases in PmO2 (up to 500 percent). However even when these results were excluded there were still no significant decreases in PmO2 at the border ischaemic area in this group.

In the border non-ischaemic areas there were no changes in the control group though significant decreases were seen at 1, 55 and 60 minutes in the FDA group (Fig.3). Significant decreases in PmO2 were seen in control animals in the deep non-ischaemic electrode at 4, 6-12, 14 and 15 minutes though none were seen in the FDA pre-treated group (Fig 4).

DISCUSSION

Few studies have measured oxygen tension profiles in the myocardium and fewer still have looked at changes occurring in PmO2 during acute myocardial ischaemia. Those studies that have taken place have used polarographic needle electrodes (Moss 1968; van der Laarse 1978; Schuchhardt 1985; Faithfull et al 1986b); these are liable to artifacts caused by the movements of the heart and, due to the stirring effect, this results in increased current and exaggeration of tissue PO2. It has also been suggested that needle electrodes cause damage to the tissues and reduce blood circulation (Schuchhardt 1985), sometimes reducing oxygen tension due to vascular compression or displacement (Lund, 1985). Nevertheless, the inaccuracies introduced into polarographic needle electrode measurement would tend to remain fairly constant during the period of measurement and hence we are justified in comparing PmO2 changes in the two groups of animals in the present study.

Because halothane (and many other anaesthetic agents) can be detected by polarography this may have affected the results. However halothane induced errors occurring during polarographic measurement of oxygen tensions have been investigated by Severinghaus et al (1971) who concluded that, at the low concentrations used during anaesthesia, these were probably small in most instances. Because halothane concentration was kept constant for at least 1/2 hour before measurements were made, it is probable that any small error would be constant. Indeed, in the presence of a constant concentration of a second polarographically detectable substance in the tissues (halothane), the relative changes of the primarily detected substance (oxygen) may actually be underestimated (Faithfull et al, 1986a).

In previous experiments (Faithfull et al 1986b) we examined changes occurring in untreated pigs subjected to the present protocol and demonstrated very similar changes to those seen above. Only in the ischaemic border area did PmO2 decrease after LAD occlusion and it is therefore significant that PmO2 remained stable in this area in FDA treated animals. In these earlier studies we attributed ischaemic border area decreases in PmO2 to rapid shunting of blood into the centre of the infarct and the consequent occurrence of a myocardial steal. The present work suggests that, in the FDA group, blood was being effectively transferred from outside the ischaemic areas to provide tissue oxygenation via collateral channels. This mechanism was largely successful and it was only occasionally that decreases in PmO2 were observed in the border non-ischaemic area.

The effects of treatment with FDA could have been due to simple haemodilution as evidenced by cardiac output and decreases in systemic peripheral resistance. As far as we are aware no studies of PmO2 have been undertaken during simple haemodilution. However, Rude et al (1982) measured PmO2 in the dog using a mass spectrometer probe during administration of dextran to animals receiving 100 percent oxygen. Mean increses in PmO2 of 136 percent were seen suggesting that haemodilution with dextran may, by decreasing capillary transit time and improving flow through micro-collateral channels, sometimes be able to improve oxygen supply. Indeed, pre-ischaemic haemodilution with colloidal dextran has been shown by Cohn et al, (1975) to lessen subsequent ST elevation but the observations were not continued beyond 20 minutes of ischaemia. The significance of these findings are open to question as no details of cardiac work were reported and it is quite possible that oxygen supply /demand kinetics were dissimilar in the different groups. Our own work has shown that dextran is ineffective in improving PmO2 when haemodilution is performed after LAD occlusion.

PFC emulsions have a very low viscosity which, unlike that of blood, hardly alters as shear rate decreases in the microcirculation (Naito and Yokoyama, 1978). This implies a higher tissue blood flow for the same trans-capillary pressure and, under conditions of limiting oxygen flux, improved tissue oxygenation. Kloner and Golgar (1985) have maintained that the small PFC particles may bypass obstructions in the circulation caused by endothelial blebs and sludged red cells. Others (Faithfull et al, 1984; Smith et al, 1985) have maintained that the particles can penetrate deep into hypoxic tissue beds, not only bypassing sludged cells, but reoxygenating them and reversing the red cell membrane stiffening that occurs under conditions of hypoxia and acidosis (Schmid-Schonbein, Weiss and Ludwig, 1973). Hence the vicious circle of sludged standstill can be broken.

Whatever the mechanism of improvement of oxygenation of ischaemically hypoxic tissues, there is little doubt that PFC emulsions are effective in this respect and recent work (Faithfull and Cain, 1987 in press) has shown that haemodilution with FDA can substantially increase oxygen extraction and reduce critical oxygen delivery levels during progressive haemorrhagic shock even when breathing ambient air. It has been suggested that this is not only due to improved microcirculatory perfusion, but may also be due to improved diffusion of oxygen from the red cell into the tissues.

In conclusion we have demonstrated that haemodilution with FDA can radically alter the pattern of PmO2 changes occurring following subsequent LAD occlusion and the induction of a small myocardial infarction.

SUMMARY

Previous work by the authors has shown that, following ligation of the left anterior descending coronary artery (LAD), myocardial oxygen tension (PmO2) in expected areas of maximal ischaemia is maintained at the expense of ischaemic border zones of the infarct area. Post-ischaemic haemodilution with the fluorocarbon containing plasma substitute Fluosol-DA 20% (FDA) could significantly improve PmO2 and pre-ischaemic haemodilution can delay myocardial ischaemia. We now present an analysis of the pattern of PmO2 changes to be seen when myocardial ischaemia is induced following prior haemodilution with FDA.
Two groups of juvenile Yorkshire pigs were anaesthetised with intraperitoneal thiopentone, intubated and ventilated with halothane,

nitrous oxide and oxygen. After placement of cardiovascular monitoring
lines, a thoracotomy was performed. The pericardium was opened and 4
steel-protected gold microelectrodes were placed in the terminal supply
area of the LAD in such a way that 2 electrodes were in the area of
myocardial ischaema to be produced. One group of pigs were bled (20
ml/kg) and the loss was replaced with equal volumes of FDA. The animals
were ventilated with halothane and oxygen and the terminal LAD was
ligated. Electrode outputs were recorded on a flat bed recorder and
analysed.

LAD occlusion in the control animals resulted in similar changes in
PmO2 to those described above. In contrast, significant decreases in
PmO2 situated at the border of the ischaemic area were markedly delayed
in the FDA pretreatment group, suggesting definite differences in
microcirculatory blood flow following ischaemia in this group and
influence on the previously reported myocardial steal effects (1).

REFERENCES

Cohn L.H., Lamberti, J.L., Florian, A., Moses, R., Vandevanter, S, Kirk,
    E. and Collins, J.J., 1975. Effects of hemodilution on
        myocardial ischemia. J. Surg. Res., 18: 523.
Faithfull, N.S. and Cain S.M., 1987 in press, Critical Oxygen Delivery
        levels during shock following normoxic and hyperoxic
        haemodilution with fluorocarbons or dextran. Adv. Exp. Med. Biol.
Faithfull, N.S., Erdmann, W., and Fennema, M., 1986a, Effects of
        haemodilution with fluorocarbons or dextran on oxygen tensions in
        the acutely ischaemic myocardium. Brit. J. Anaesth., 58:1031.
Faithfull, N.S., Fennema, M., Erdmann, W., Dhasmana, M.K. and Eilers,
        G., 1986b, Effects of acute myocardial ischaemia on
        intramyocardial oxygen tensions, Adv. Exp. Med. Biol., 200: 49.
Faithfull, N.S., Fennema M, and Erdmann W., 1987, Prophylactic treatment
        of myocardial ischaemia by prior haemodilution with fluorocarbon
        emulsions. Brit. J. Anaesth., 59:124P.
Kelman, G.R., 1966, Digital computer subroutine for the conversion of
        oxygen tension into saturation, J. Appl. Physiol. 21:1375.
Kloner, R.A. and Golgar, D.M., 1985, Overview of the use of
        perfluorochemicals for myocardial ischemic rescue, Int. Anesth.
        Clin., 23, 115.
Laarse, A., van der, 1978, On multiple polarographic measurement of
        myocardial oxygen tension, Thesis, University of Amsterdam.
Lund, N., 1985, Skeletal and cardiac muscle oxygenation, Adv. Exp. Med.
        Biol., 191: 37.
Moss, A.J., 1968, Intramyocardial oxygen tension, Cardiovasc. Res., 3:
        14.
Naito, R. and Yokoyama, K., 1978, Perfluorochemical Blood Substitutes,
        Technical Information Series no 5, Green Cross Corporation,
        Osaka.
Ohyanagi, H., Nakaya, S., Okamura, S. and Saito, Y., 1984, Surgical use
        of Fluosol-DA in Jehovah's witness patients. Artif. Organs, 8:10.
Rude, R.E., Golgar, D., and Khuri, S.F., 1982, Effects of intravenous
        fluorocarbons during and without oxygen enhancement on acute
        myocardial ischemic injury assessed by measurement of
        intramyocardial gas tensions, Am. Heart J., 103:986.
Schmid-Schonbein, H., Weiss, J., and Ludwig, H., 1973, A simple method
        for measuring red cell deformability in models of the
        microcirculation. Blut, 16: 369.
Schuchhardt, S., 1985, Myocardial oxygen pressure: Mirror of oxygen
        supply, Adv. Exp. Med. Biol., 191: 21.
Severinghaus, J.W., Weiskopf, R.B., Nishimura, M. and Bradley, A.F.,
        1971, Oxygen electrode errors due to polarographic reduction of
        halothane. J. Appl.Physiol. 31: 640.

Smith, A.R., v Alphen, W., Faithfull, N.S., and Fennema, M., 1985. Limb
    preservation in replantation surgery. <u>J. Plast. Reconst. Surg.</u>,
    75: 227.
Tremper, K.K., Friedman, A.E., Levine, E.M., Lapin, R., and Camarillo,
    D, 1982, The preoperative treatment of severely anaemic patients
    with a perfluorochemical oxygen transport fluid, Fluosol-DA, <u>New</u>
    <u>Eng. J. Med.</u>, 307: 278.

THE INFLUENCE OF OXYGEN TENSION ON MEMBRANE POTENTIAL AND TONE OF CANINE CAROTID ARTERY SMOOTH MUSCLE

J. Grote[1], G. Siegel[2], K. Zimmer[1] and A.Adler[3]

[1] Institute of Physiology I, University of Bonn
Nussallee 11, D-5300 Bonn 1, Germany
[2] Institute of Physiology, The Free University of Berlin
Arnimallee 22, D-1000 Berlin 33, Germany
[3] Department of Internal Medicine I, University of Munchen
Klinikum Grosshadern, D-8000 Munchen 70, Germany

INTRODUCTION

During arterial hypoxia the brain blood flow rate increases when oxygen tensions below approximately 60 mmHg are attained (Betz, 1972; Grote and Schubert, 1982; Kuschinsky, and Wahl, 1978). Tissue PO_2 measurements as well as tissue metabolite assays performed under the same conditions indicate the existence of pronounced tissue hypoxia and the beginning of hypoxic changes in brain metabolism (Grote, Zimmer and Schubert, 1981). Possible mediators of hypoxia-induced cerebral vasodilatation are various metabolic factors, the concentrations of which increase under conditions of insufficient brain oxygen supply. Among these vasodilating factors, potassium and hydrogen ions as well as adenosine seem to play an important role (Betz, 1972; Kuschinsky and Wahl, 1978; Rubio et al., 1985). Different additional vasoactive substances released at low oxygen tensions by the tissue or by the vessel wall itself, especially by the endothelium, are under investigation (Busse et al., 1984; Coburn et al., 1979; Detar, 1980; Eckenfels and Vane, 1972; Fay, 1971; Furchgott and Zawadzki, 1980; Pitman and Duling, 1973; Rubanyi and Vanhoutte, 1985; Rubio et al., 1975; Sparks, Jr. 1980; Vanhoutte, 1976).

MATERIALS AND METHODS

To study the effect of varying ambient oxygen tensions on the electro-mechanical properties of arterial smooth muscle cells, combined measurements of membrane potential and vascular tone were performed using an isolated carotid artery preparation suspended in Krebs solution. The oxygen tension of the Krebs solution was varied between approximately 550 mmHg and 20 mmHg whereas the CO_2 tension was maintained constant at a normal level of about 35 to 40 mmHg. The oxygen tension was recorded continuously in the bath solution close to the vessel strip with multiwire platinum microelectrodes (Grote et al., 1981). In some experiments intermittent tissue PO_2 measurements were performed in the superficial layers of the carotid artery preparation using the same microelectrodes. The membrane potential of the smooth muscle cells was recorded intracellularly with glass microelectrodes filled with 3 M KCl (Siegel et al., 1978; Siegel et al., 1984). The resistances of the electrodes ranged from 50 to 150 M and the tip potentials were −5 to −80 mV. At each PO_2 level at least 10 impalements were performed and the measured membrane potential values averaged. For determining smooth muscle tone, the artery segments were attached to an isometric force transducer (Siegel et al., 1978).

RESULTS AND DISCUSSION

Under control conditions the Krebs solution was equilibrated with a gas mixture containing 95% oxygen and 5% carbon dioxide resulting in a mean oxygen tension of about 550 mmHg. The corresponding mean membrane potential, recorded intracellularly was −64.4 ± 0.8 mV. Since the initial values of mechanical tension differed slightly in the various preparations, normalization to a uniform level was performed, as described previously (Siegel, 1986).

Figure 1. shows the results of a typical experiment. The decrease of oxygen tension in the Krebs solution produced by lowering the oxygen content of the equilibration gas caused a hyperpolarization of the smooth muscle cell membrane and a fall in smooth muscle tone. The first reactions of both parameters could be detected 1 to 2 minutes after each stepwise change in PO_2 whi e the total effect was attained after approximately 10 min. The application time of a single PO_2 level ranged between 20 and 30 min.

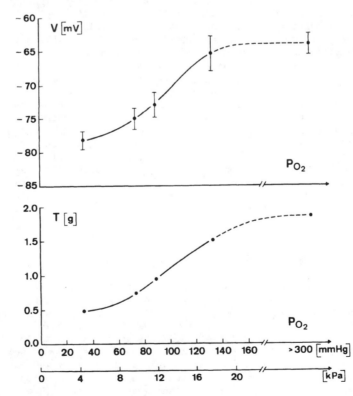

Figure 1. The effect of oxygen tension on membrane potential and smooth muscle tension determined in canine carotid artery segments.

The evaluation of membrane potential and mechanical tension recordings of six preparations resulted in the two mean dose-response curves given in Figure 2. When the oxygen tension was lowered, hyperpolarization and relaxation of the smooth muscle cells were observed as soon as PO_2 values below approximately 150 mmHg were reached. Further PO_2 decrease to values between 80 and 30 mmHg caused a strong hyperpolarization to a maximum of -74.8 ± 0.8 mV and a maximum fall in tension to 1 g at a mean PO_2 of 33 mmHg. Hyperpolarization of the membrane and reduction of the smooth muscle tone as a function of PO_2 ran parallel through all the experiments. When the oxygen tension was increased both cell reactions were reversed. During recent studies pronounced hypoxia with PO_2 values below approximately 20 mmHg produced a depolarization and subsequent contraction of vascular smooth muscle cells. Comparable results in tension measurements were found in canine coronary rteries by Rubanyi and Vanhoutte (1985).

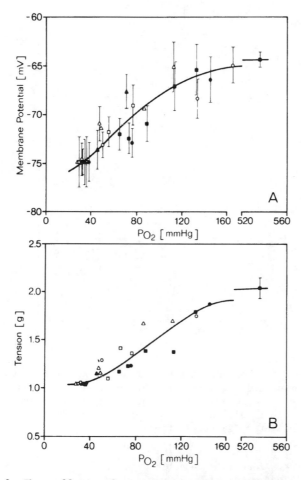

Figure 2. The effect of oxygen tension on membrane potential (A) and smooth muscle tension (B) determined in canine carotid artery segments. Mean values of six preparations $\pm$ SEM.

The present results do not give an explanation of the causal mechanisms inducing the changes in membrane potential and tone of the smooth musculature with varying PO_2. Changes in passive membrane properties such as an increase in K^+ permeability or a decrease of Na^+ permeability or active transport processes due to stimulation of the electrogenic outward Na^+ pump could be considered in relation to the hyperpolarization of the membrane (Detar, 1980; Sparks Jr., 1980). Since the hyperpolarization of smooth muscle cells induces a fall in intracellular Ca^{2+} activity (Siegel et al., 1978), a relaxation would be expected. Such changes in permeability or active transport could be a direct effect of hypoxia on the membrane or an indirect influence mediated by vasoactive substances like adenosine, eicosanoids and an endothelium-derived relaxing factor. When plotting smooth muscle tension against membrane potential recorded simultaneously, one obtains the hyperpolarized part of the activation curve (Fig.3). The identity of this activation curve with those found by Siegel and co-workers (Siegel et al., 1978; Siegel et al., 1984) by variation of the hydrogen or potassium ion concentration, suggests a change in the membrane permeability of the smooth muscle cells.

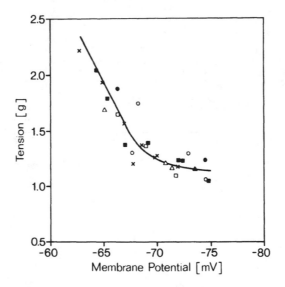

Figure 3. Dependency of smooth muscle tension on the membrane potential in isolated carotid artery segments (stationary activation curve). Changes of both parameters were induced by variations of PO_2 in untreated and indomethacin (10^{-5} M) preincubated (x) preparations.

Since the results resemble closely those of prostacyclin experiments (Siegel et al., 1986) the measurements were repeated following an application of indomethacin (10^{-5} M). Under control conditions the application of indomethacin resulted in a slight depolarization of the membrane from -65.0 ± 0.5 mV (n = 6) to -62.8 ± 0.6 mV (n = 6; P < 0.01) and a subsequent augmentation of the basal tension by 0.29 ± 0.08 g (Fig. 4). Similar changes in vascular smooth muscle tone caused by indomethacin have been reported by Rubanyi and Vanhoutte (1985). In the different preparations of the present experiments, indomethacin had no significant influence on the hypoxia-induced changes of membrane potential and mechanical force. Figure 4. summarizes the results whih clearly show an increasing hyper-polarization and relaxation with decreasing oxygen tension. Since it can be assumed that the indomethacin concentration applied leads to a complete blockade of prostacyclin synthesis (Rubanyi and Vanhoutte, 1985), the hyperpolarization of 9.1 mV and the subsequent fall in tension of 1.05 g can not be mediated by prostacyclin. These results confirm previous observations on the behaviour of smooth muscle tension of skeletal and cardiac muscle arteries during hypoxia (Broadley and Rothaul, 1986; Detar, 1980). The membrane potential and mechanical tension values determined simultaneously fit very well into the activation curve given in Figure 3, thus indicating that indomethacin does not affect the electro-mechanical properties of vascular smooth musculature during hypoxia. As found in experiments without blockade of prostacyclin synthesis, the hyperpolarization diminished and mechanical tension recovered at PO_2 values below approximately 15-20 mmHg.

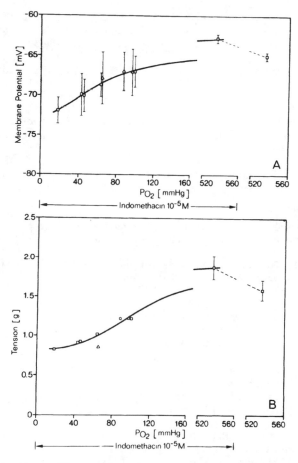

Figure 4. The effect of oxygen tension on membrane potential (A) and smooth muscle tension (B) determined in canine carotid artery segments preincubated with indomethacin (10^{-5} M. Mean values of three preparations.

SUMMARY

In vascular smooth muscle of canine carotid artery segments the decrease in oxygen tension from approximately 550 mmHg to approximately 30 mmHg leads to a dose-dependent hyperpolarization and relaxation of the vessel strip. At PO_2 values below approximately 15-20 mmHg depolarization and subsequent contraction were found. All changes of the parameters investigated were reversible. Indomethacin (10^{-5} M) had no significant effect on the hypoxia-dependent potential and tension changes. The hyperpolarized part of the activation curve was obtained by plotting tension versus membrane potential values .

REFERENCES

Betz, E., 1972, Cerebral blood flow: its measurement and regulation. Physiol. Rev., 52:595-530.

Bolton, T.B., 1979, Mechanisms of action of transmitters and other substances on smooth muscle, Physiol. Rev., 59:606-718

Broadley, K.J., and Rothaul, A.L., 1986, The coronary vasodilator mediator released by hypoxia and isoprenaline is not affected by cyclo-oxygenase inhibition. Prostaglandins, 31:295-306.

Busse, R., Forstermann, U., Matsud, H., and Pohl, U., 1984, The role of prostaglandins in the endothelium-mediated vasodilatory response to hypoxia. Pflugers Arch. Eur. J. Physiol., 401:77-83

Coburn, R.F., Grubb, S., and Aronson, R.D., 1979, Effect of cyanide on oxygen tension-dependent mechanical tension in rabbit aorta. Circ. Res., 44:368-378.

Detar, R., 1980, Mechanism of physiological hypoxia-induced depression of vascular smooth muscle contraction. Am. J. Physiol., 238:H761-H769.

Eckenfels, A., and Vane, J.R., 1972, Prostaglandins, oxygen tension and smooth muscle tone. Br. J. Pharmacol., 45:451-462.

Fay, F.S., 1971, Guinea pig ductus arteriosus. I. Cellular and metabolic basis for oxygen sensitivity. Am. J. Physiol., 221:470-479.

Furchgott, R.F., and Zawadzki, J.V., 1980, The obligatory role of endothelial cells in the relaxation of arterial smooth muscle by acetylcholine, Nature. 288:373-376.

Grote, J., Zimmer, K., and Schubert, R., 1981, Effects of severe arterial hypocapnia on regional blood flow regulation, tissue PO_2 and metabolism in the brain cortex of cats. Pflugers Arch. Eur. J. Physiol., 391:195-199.

Grote, J., and Schubert, R., 1982, Regulation of cerebral perfusion and PO_2 in normal and edematous brain tissue. In: "Oxygen Transport to Human Tissues", edited by J.A. Loeppky and M.L. Riedesel, pp. 169-178. Elsevier North Holland, Amsterdam, New York, Oxford.

Kuschinsky, W., and Wahl, M., 1978, Local chemical and neurogenic regulation of cerebral vascular resistance. Physiol. Rev. 22:656-689.

Pittman, R.N., and Duling, B.R., 1973, Oxygen sensitivity of vascular smooth muscle. I. In vitro studies. Microvasc. Res. 6:202-211.

Rubanyi, G.M., and Vanhoutte, P.M., 1985, Hypoxia releases a vaso-constrictor substance from the canine vascular endothelium. J. Physiol. (Lond.), 364:45-56.

Rubio, R., Berne, R.M., Bockman, E.L., and Curnish, R.R., 1975, Relationship between adenosine concentration and oxygen supply in rat brain. Am. J. Physiol., 228:1896-1902.

Siegel, G., Ehehalt, R., and Koepchen, H.P., 1978, Membrane potential and relaxation in vascular smooth muscle. in: "Mechanisms of Vaso-dilatation", edited by P.M. Vanhoutte, and I. Leusen, pp 56-72. S. Karger, Basel, Muchen, Paris, London, New York, Sydney.

Siegel, G., Grote, J., Zimmer, K., Adler, A., and Litza, N., 1986, Electro-physiological effects of hypoxia on vascular smooth muscle. in "Mechanisms of Vasodilatation", edited by P.M. Vanhoutte, Raven Press, New York, (in press).

Siegel, G., Stock, G., Schnalke, F., and Litza, B., 1986, The effect of iloprost on the electro-mechanical properties of vascular smooth muscle. Pflugers Arch. Eur. J. Physiol., 406:R41.

Siegel, G., Walter, A., Thiel, M., and Ebeling, B.J., 1984, Local regulation of blood floow. Adv. Exp. Med. Biol., 169:515-540.

Sparks, H.V., Jr., 1980, Effect of local metabolic factors on vascular smooth muscle. in: "Handbook of Physiology, Section 2: The Cardio-vascular System, Vol. II: Vascular Smooth Muscle", edited by D.F. Bohr, A.P. Somlyo, and H.V. Sparks, Jr., pp 475-513. American Physiological Society, Bethesda.

Vanhoutte, P.M., 1976, Effects of anoxia and glucose depletion on isolated veins of the dog. Am. J. Physiol., 230:1261-1268.

RESPIRATORY
SYSTEM

PULMONARY DIFFUSING CAPACITY FOR CARBON MONOXIDE AND NITRIC OXIDE

J. Piiper, K.-D. Schuster, M. Mohr, H. Schulz and M. Meyer

Abteilung Physiologie, Max-Planck-Institut für
experimentelle Medizin, D-3400 Göttingen, F.R.G.

INTRODUCTION

Carbon monoxide (CO) is widely used to determine the diffusive conductance or diffusing capacity (D) of the lungs. Its advantage over O_2 is the high affinity of hemoglobin for CO whereby P_{CO} in pulmonary capillary blood is low and the veno-arterial P_{CO} difference is minimal. Thus, the mean P_{CO} difference effective for alveolar-capillary CO transfer is close to alveolar Pco. But since the reaction of CO with hemoglobin is relatively slow, the diffusing capacity for CO includes a reaction component which cannot be easily assessed. The affinity of hemoglobin for nitric oxide (NO) is even higher (about 2000 times) than for CO and, more importantly, the association velocity constant of hemoglobin with NO is about 75 times higher than that with CO (1). Thus the diffusing capacity for NO ·may represent a closer estimate of the true pulmonary diffusing capacity, i.e. not complicated by reaction velocity. Moreover, a comparison with D_{CO} may provide information on the extent of reaction limitation in CO uptake.

For this purpose, simultaneous determinations of D_{NO} and D_{CO} were performed in anesthetized dogs using a rebreathing technique (2).

METHODS

The experiments were performed on 8 anesthetized supine dogs (mean body weight 20 kg), anesthetized with sodium pentobarbital, 20 mg/kg intravenously, paralyzed with alcuronium, 1 mg/h intravenously (initial dose 0.25 mg/kg), and artificially ventilated by a servo-controlled ventilator (3). Experiments were done in normoxia ($FIO_2 = 0.21$), in hypoxia ($FIO_2 = 0.11$) and in hyperoxia ($FIO_2 = 1.00$).

Rebreathing with a tidal volume of 0.5 l and frequency of 60/min was continued for 15 sec. The rebreathing mixture contained 0.06% NO, 0.06% CO (isotope $^{13}C^{18}O$) and 1% He in Argon. The partial pressures of He, CO_2, N_2, NO, $^{13}C^{18}O$ and $^{16}O^{18}O$ were continuously monitored by a respiratory mass spectrometer (4).

The stable isotope $^{13}C^{18}O$ (mass 31) was used in order to separate the CO signal from NO (mass 30) and N_2 (mass 28). N_2 was replaced by argon to prevent formation of NO from N_2 and O_2 at high temperature in the ion

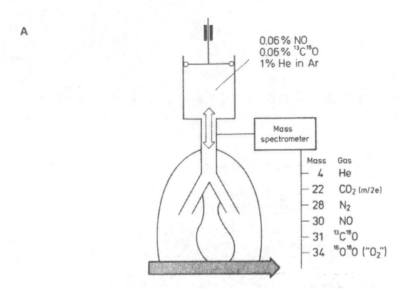

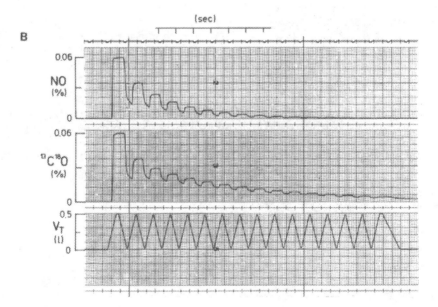

Fig. 1. Method for simultaneous estimation of D_{CO} and D_{NO}. A. Schema of set-up. B. Typical mass spectrometer record of NO, $^{13}C^{18}O$ and tidal volume (V_T) during rebreathing. For explanation, see Methods.

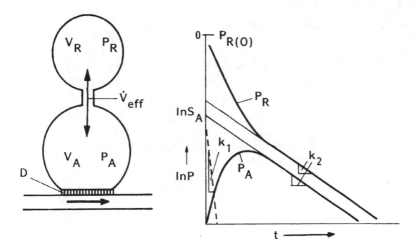

Fig. 2. Model for evaluation of partial pressure changes of NO and $^{13}C^{18}O$ in the lungs (P_A) and in the rebreathing bag (P_R). For explanations, see Methods.

source of the mass spectrometer. Corrections for the contribution by $^{16}O^{16}O$ to mass peaks 30 and 31 were achieved by measurement of the naturally ocurring, rare O_2 isotope $^{16}O^{18}O$ (mass 34). Helium was included to check the speed and completeness of intrapulmonary gas mixing.

A typical experimental record is shown in Fig. 1. The model and its application to determination of D is shown in Fig. 2. The time course of test gas partial pressures in the alveolar space (P_A) and in the rebreathing bag (P_R) followed a biexponential time course characterized by the rate constants, k_1 and k_2. The diffusing capacity was calculated as

$$ D = \frac{1}{R \cdot T} \cdot \frac{V_R}{S_A} \cdot \frac{k_2}{1 - k_2/k_1} \tag{1} $$

(R, gas constant; T, absolute temperature; V_R, rebreathing bag volume; S_A, intercept of the slow component of P_A). Also the alveolar volume (V_A) and the effective ventilation ($\dot{V}_{eff}$) could be determined from the test gas curves.

RESULTS

The following mean values (± SD) for D_{NO} and D_{CO}, in ml STPD/(min·Torr), were found:

	Hypoxia (11% O_2)	Normoxia	Hyperoxia (100% O_2)
D_{NO}	49.0 ± 11.0	48.1 ± 10.6	43.8 ± 9.2
D_{CO}	15.3 ± 3.4	13.9 ± 3.1	8.2 ± 2.4

These values reveal the following features:

(1) The D_{NO} values are higher than D_{CO} by a factor of 3.2 to 5.3.
(2) Both D_{NO} and D_{CO} decrease from hypoxia through normoxia to hyperoxia; but the relative changes are much larger for D_{CO} than for D_{NO}.

DISCUSSION

The much higher D_{NO} compared to D_{CO} is the prominent feature of the results. There are several possible explanations.

(1) The Krogh diffusion constant ratio NO/CO, K_{NO}/K_{CO}, is above unity due to higher solubility of NO in the air/blood barrier and in blood. But the experimental D_{NO}/D_{CO} ratio is much higher than the K_{NO}/K_{CO} ratio, 2.0, determined for water. Unfortunately, solubility of NO in tissue has not been determined.

(2) NO is a reactive gas. In presence of O_2 it is oxidized to NO_2. Loss of NO during rebreathing by chemical reactions (in gas phase and tissue) would accelerate its disappearance from gas phase and thus lead to an overestimation of D_{NO}. However, at least for gas phase, the reaction is much too slow to exert a significant effect.

(3) Model simulations of rebreathing have shown that in lungs with unequal distribution of D, $\dot{V}_A$ and $\dot{Q}$, the D_{NO}/D_{CO} ratio may be over-estimated or underestimated. However, high degrees of inhomogeneity unlikely present in normal lungs are needed for explanation of the observed D_{NO}/D_{CO} ratios.

(4) The D_{NO}/D_{CO} ratio is elevated due to slow reaction of CO with hemoglobin.

The following quantitative analysis is based on the factors (1) and (4), whereas the factors (2) and (3) are discarded.

In alveolar-capillary gas transfer of O_2, CO and NO, diffusion through various layers and chemical reaction with hemoglobin are involved. A highly simplifying approach is to consider three additive resistance terms in series:

$$R_{tot} = R_{diff(M)} + R_{diff(RBC)} + R_{reaction} \tag{2}$$

R_{tot} is the total resistance; $R_{diff(M)}$, resistance of the alveolar-capillary "membrane" and plasma; $R_{diff(RBC)}$, the resistance offered by the red blood cells; $R_{reaction}$, the resistance arising from non-instantaneous reaction with hemoglobin.

For application to the experimental data, the following assumptions are made. (1) Uptake of NO is not limited by reaction velocity ($R_{NO(reaction)} = 0$). (2) The $R_{diff(NO)}/R_{diff(CO)}$ ratio is equal to the Krogh diffusion constant ratio $K_{NO}/K_{CO} = 2.0$ (value determined for water).

This leads to the following interpretations.
(A) The (relatively small) observed changes of D_{NO} with the level of oxygenation,

$$\frac{D_{NO(hypoxia)}}{D_{NO(normoxia)}} = 1.11$$

$$\frac{D_{NO(hyperoxia)}}{D_{NO(normoxia)}} = 0.87$$

are attributed to true changes of diffusion conditions (due to changes in the number and/or surface area of perfused capillaries).

(B) The larger changes of D_{CO} with the oxygenation level are ascribed to both changes in diffusion conditions and in reaction velocity of CO with hemoglobin. The following values are obtained for the fractional resistance to CO uptake due to finite reaction rate Hb + CO:

in hypoxia, 0.38;

in normoxia, 0.41;

in hyperoxia, 0.64.

It is concluded that NO can be used for determination of pulmonary diffusive properties. In theory, it is better suited than CO. However, its chemical reactivity and toxicity set limits to its use, particularly in man.

REFERENCES

1. Gibson, Q.H., The kinetics of reactions between haemoglobin and gases, Progr. Biophys. 9: 1-53 (1959).
2. Hook, C. and M. Meyer, Pulmonary blood flow, diffusing capacity and tissue volume by rebreathing: Theory, Respir. Physiol. 48: 255-279 (1982).
3. Meyer, M. and H. Slama, A versatile hydraulically operated respiratory servo system for ventilation and lung function testing, J. Appl. Physiol. 55: 1023-1030 (1983).
4. Scheid, P., Respiratory mass spectrometry, in: "Measurement in Clinical Respiratory Physiology", G. Laszlo and M.F. Sudlow, eds., Academic Press, London, pp. 131-166 (1983).

RELATIONSHIP BETWEEN ALVEOLAR-ARTERIAL Po_2 AND Pco_2 DIFFERENCES AND THE CONTACT TIME IN THE LUNG CAPILLARY

Masaji Mochizuki and Tomoko Kagawa*

Geriatric Respiratory Research Center, Nishimaruyama Hospital, 064 Sapporo/Chuo-Ku, and *Department of Physiology, Yamagata University School of Medicine, 990-23 Yamagata, Japan

INTRODUCTION

As reported in previous papers (Kagawa and Mochizuki, 1987; Mochizuki et al., 1987), we have developed a method to evaluate the contact time (t_C) in the lung capillary, using the numerical solutions of simultaneous O_2 and CO_2 diffusions in the red blood cell (RBC) and the relation between the gas exchange ratio (R) and Pco_2 in rebreathing air. By comparing the t_C obtained with that determined from the D_LCO measurement, we could verify the validity of the contact time equation as well as that of the numerical solution (Mochizuki and Kagawa, 1986; Uchida et al., 1986; Shibuya et al., 1987). Our experimental evidence, cited above, suggested that the CO_2 equilibrium between alveolar air and capillary blood is not reached during the contact time. Thus, we attempted to calculate the alveolar-arterial Pco_2 difference ($P(A-a)CO_2$) by varying the t_C at various Po_2 and Pco_2 levels of venous blood.

From the results we could clarify the relationship between the $P(A-a)CO_2$ and t_C and its dependency on the venous Po_2 and Pco_2 ($P\bar{v}O_2$ and $P\bar{v}CO_2$). Furthermore, the influence of the t_C on the arterial-venous O_2 and CO_2 content differences ($C(a-v)O_2$ and $C(v-a)CO_2$) and on the R was clarified quantitatively.

COMPUTATION

The overall gas exchange process was computed by using a computer, VAX-11/750 (DEC) according to the program made in the previous study (Mochizuki, 1987). All the parameter values such as the size of RBC, the slope of the CO_2 dissociation curve, the buffer value, Donnan ratio and so on were exactly the same as those used by Mochizuki et al. (1987). The permeabilities of O_2 and CO_2 across the capillary wall were taken to be much greater than those across the RBC membrane, so that diffusivities across a diffusion barrier such as the alveolar membrane were entirely ignored. The time constant of the dehydration reaction of HCO_3^- in plasma due to carbonic anhydrase in the capillary wall was taken to be 0.1 sec. The alveolar Po_2 and Pco_2 (P_AO_2 and P_ACO_2) were taken to be 90 and 40

Torr, respectively, throughout the computation. The $P\bar{v}O_2$ was varied in the range of 27 to 39 Torr, and the $P\bar{v}CO_2$, in the range in which the blood gas exchange ratio, $C(v-a)CO_2$ to $C(a-v)O_2$, was between 0.7 and 1.0.

RESULTS

The changes in $C(a-v)O_2$ and $C(v-a)CO_2$ during the course of oxygenation with outward CO_2 diffusion are shown against the contact time in Fig. 1. The $P\bar{v}O_2$ and $P\bar{v}CO_2$ were taken to be 30 and 55 Torr,

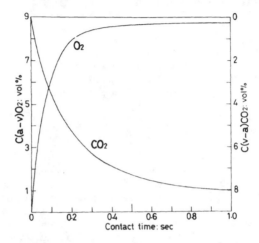

Fig. 1. Changes in arteriovenous O_2 and CO_2 content differences ($C(a-v)O_2$ and $C(v-a)CO_2$) against the contact time. $P\bar{v}O_2$ and $P\bar{v}CO_2$ were 30 and 55 Torr, and P_AO_2 and P_ACO_2 were 90 and 40 Torr, respectively.

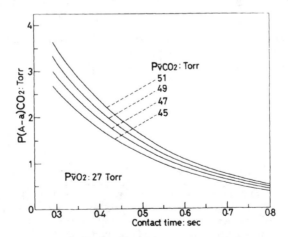

Fig. 2. Relationship between alveolar-arterial P_{CO_2} difference ($P(A-a)CO_2$) and the contact time. $P\bar{v}CO_2$ was varied from 45 to 51 Torr, while $P\bar{v}O_2$ was kept at 27 Torr.

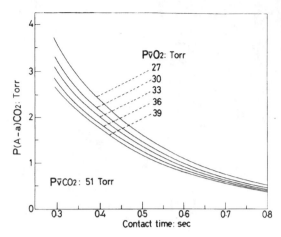

Fig. 3. Relationship between $P(A-a)CO_2$ and the contact
time under various values of $P\bar{v}O_2$ between 27 and
39 Torr. $P\bar{v}CO_2$ was constant at 51 Torr.

respectively. Because HCO_3^- ions enter the RBC following the CO_2 release,
and cause an increase in intracellular PCO_2, the change in intracellular
PCO_2 was significantly slower than the change in PO_2. The CO_2
equilibrium was not achieved even with the normal contact time, 0.7 sec
(Roughton, 1945; Uchida et al., 1986; Shibuya et al., 1987). Thus, the
intracellular PCO_2 in arterial blood was slightly higher than the P_ACO_2.
The half-time of $C(a-v)O_2$ was 50 msec, whereas that of $C(v-a)CO_2$ was 120
msec.

The $P(A-a)CO_2$ depended not only on the t_C, but also on the $P\bar{v}O_2$ and
$P\bar{v}CO_2$. Figure 2 shows the relationship between $P(A-a)CO_2$ and t_C at 4
different values of $P\bar{v}CO_2$ between 45 and 51 Torr, when the $P\bar{v}O_2$ was
maintained at 27 Torr. The $P(A-a)CO_2$ decreased exponentially as the
contact time was prolonged, and its magnitude increased linearly in
parallel with the $P\bar{v}CO_2$. The $P(A-a)CO_2$ also depended on the $P\bar{v}O_2$ through
the Haldane effect. Figure 3 shows the dependency of the $P(A-a)CO_2$ on
the $P\bar{v}O_2$, when the $P\bar{v}CO_2$ was kept at 51 Torr. The $P(A-a)CO_2$ increased as
the $P\bar{v}O_2$ decreased: the influence of the $P\bar{v}O_2$ was of a similar magnitude
to that of $P\bar{v}CO_2$. The $PaCO_2$ in Figs. 2 and 3 was described
quantitatively by the following equation:

$$PaCO_2 = P\bar{v}CO_2 \cdot (30/P\bar{v}O_2)^{0.33} \cdot \exp(-3.96 \cdot t_C - 0.712)$$
$$+ 40.4(t_C - 0.25)^{0.044}. \qquad (1)$$

As shown in Fig. 4, there was a linear relation between $C(v-a)CO_2$
and $P\bar{v}CO_2$, when the $P\bar{v}O_2$ was kept constant at 30 Torr. The $C(v-a)CO_2$ was
reduced as the t_C was shortened. Although the PO_2 dependency of the
$C(v-a)CO_2$ is not shown in Fig. 4, the $C(v-a)CO_2$ caused by the Haldane
effect was also reduced as the t_C shortened. The Haldane effect
coefficient, which was computed by dividing the difference between the
$C(v-a)CO_2$ at two $P\bar{v}O_2$ levels by the $C(a-v)O_2$ difference, was inversely
related to the t_C as given by the following equation in the $P\bar{v}O_2$ and
$P\bar{v}CO_2$ ranges of Figs. 3 and 4:

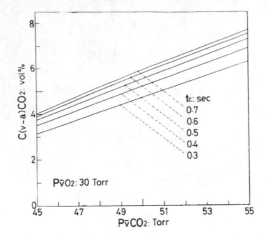

Fig. 4. Relationship between $C(v-a)CO_2$ and $P\bar{v}CO_2$, when the $P\bar{v}O_2$ was kept at 30 Torr. The parameter is the contact time.

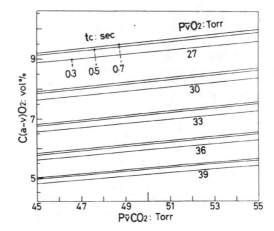

Fig. 5. Relationship between $C(a-v)O_2$ and $P\bar{v}CO_2$ at 5 different $P\bar{v}O_2$ values (27 - 39 Torr) and 3 different t_c (0.3, 0.5 and 0.7 sec).

$$HEC = 0.325 - 0.042/t_c. \tag{2}$$

The relation between $P\bar{v}CO_2$ and $C(v-a)CO_2$ was expressed by the following empirical equation:

$$P\bar{v}CO_2 = 2.5 \cdot C(v-a)CO_2 \cdot (P\bar{v}O_2/30)^{0.56} \cdot t_c^{-0.17}$$
$$+ 34 \cdot t_c^{-0.0278}. \tag{3}$$

As with $C(v-a)CO_2$, a linear relation was also observed between

$C(a-v)O_2$ and $P\bar{v}CO_2$. Figure 5 shows the $C(a-v)O_2$ plotted against the $P\bar{v}CO_2$ at 5 different $P\bar{v}O_2$ values between 27 and 39 Torr. The $C(a-v)O_2$ decreased as the t_C was shortened, but it increased in parallel with the $P\bar{v}CO_2$. The $C(a-v)O_2$ in Fig. 5 was tentatively described by the following equation:

$$C(a-v)O_2 = C(a-v)O_2^* \cdot \{1 - \exp(-10.8 \cdot t_C)\}, \qquad (4a)$$

where

$$C(a-v)O_2^* = 0.08 \cdot (P\bar{v}CO_2 - 50) + 27.15 - 0.854 \cdot P\bar{v}O_2$$
$$+ 0.755 \cdot 10^{-2} \cdot P\bar{v}O_2^{\,2}. \qquad (4b)$$

As a consequence of the interaction between O_2 uptake and CO_2 output by the RBC and its t_C-dependency, the R value also depends on the t_C. Figure 6 illustrates the R plotted against the $P\bar{v}CO_2$ at a constant $P\bar{v}O_2$ of 30 Torr, with different t_C values. The R significantly increased as the $P\bar{v}CO_2$ increased, but decreased with increases in the t_C. Thus, to keep the R constant at constant P_AO_2 and P_ACO_2 levels, the $P\bar{v}CO_2$ should change in accordance with the t_C value. Figure 7 shows the relation between $P\bar{v}O_2$ and $P\bar{v}CO_2$ which gives the constant R of 0.85 at 5 different t_C values of 0.3 to 0.7 sec. During exercise the t_C is reduced, in addition, the O_2 consumption rate increases together with the CO_2 output. In such a case, if the P_AO_2 and P_ACO_2 are maintained at 90 and 40 Torr, and the R is to remain constant at 0.85, the $P\bar{v}O_2$ and $P\bar{v}CO_2$ must change simultaneously within the region shown by the t_C lines. The relatonship shown in Fig. 7 was fitted by the following equation:

$$P\bar{v}CO_2 = 180 \cdot (t_C - 0.15)^{-0.296} \cdot P\bar{v}O_2^{\,x}, \qquad (5a)$$

where

$$x = -0.375 \cdot t_C^{-0.236}. \qquad (5b)$$

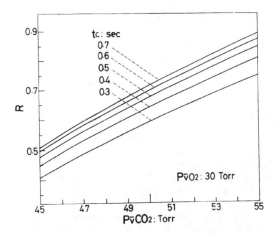

Fig. 6. Relationship between the gas exchange ratio (R) and $P\bar{v}CO_2$ illustrated for 5 different contact times, where the $P\bar{v}O_2$ was 30 Torr.

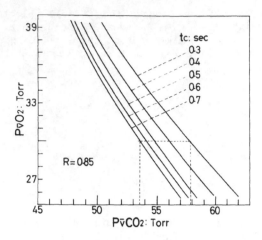

Fig. 7. Relationship between $P\bar{v}O_2$ and $P\bar{v}CO_2$ at 5 different contact times, where P_AO_2 and P_ACO_2 were 90 and 40 Torr, and the R was 0.85.

DISCUSSION

The Pco_2 in Figs. 1, 2 and 3 are, strictly speaking, those for intracellular Pco_2 in the capillary blood. When the exchange between the gas and blood phase ceases in the postcapillary, extracellular HCO_3^- enters the RBC and after the dehydration reaction, CO_2 molecules diffuse out of the RBC, enhancing the extracellular Pco_2. Therefore, the intracellular Pco_2 at the end capillary is somewhat different from the arterial Pco_2. Figure 8 shows the profiles of Pco_2, HCO_3^- content, Po_2 and O_2 saturation (So_2) and pH in intra- and extracellular fluid, when the gas exchange between blood and gas phase ceased at 0.3 sec. The $P\bar{v}O_2$ and $P\bar{v}CO_2$ were 27 and 61 Torr and P_AO_2 and P_ACO_2 were 90 and 40 Torr, respectively. The gas exchange ratio at the end capillary was 0.83, the value usually observed in the normal subject during exercise. In the postcapillary the Pco_2 is about 0.3 Torr higher than the intracellular Pco_2 at the end capillary. Therefore, the $PaCO_2$ at the shorter contact times of Figs. 1, 2 and 3 were slightly underestimated.

Up to date, it has been considered that hypercapnia is not caused by an impairment of diffusion, since CO_2 solubility in body fluid is about 20 times higher than that of O_2. However, the permeability of CO_2 across the RBC membrane is identical to that of O_2 (Niizeki et al., 1984), in addition, the CO_2 diffusion out of the RBC accompanies the inward HCO_3^- shift, which in turn increases intracellular Pco_2. Thus, the diffusion process of CO_2 becomes slower than the oxygenation rate, as seen in Fig. 1. For pulmonary diffusion impairment, two causes have generally been considered: one reason is the thickening of the diffusion layer around the RBC including the capillary wall, and the other is the shortening of the contact time. The capillary wall is permeable to CO_2 as reported by Schuster (1985), and hypercapnia may not occur even when there is a diffusion insufficiency for O_2 and CO. However, when the contact time is reduced, CO_2 equilibrium may not be achieved, and therefore, the $P(A-a)CO_2$ as well as the $P\bar{v}CO_2$ is inevitably increased. When neither the R nor the $P\bar{v}O_2$ is changed, the change in $P\bar{v}CO_2$ due to

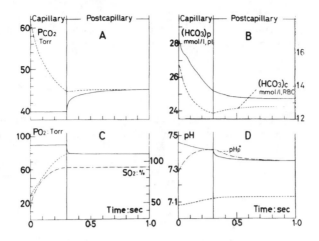

Fig. 8. Overall gas exchange profiles during oxygenation
accompanying outward CO_2 diffusion in normal
blood. The extracellular Po_2 was increased from
27 to 90 Torr, while the extracellular Pco_2 was
decreased from 61 to 40 Torr. The solid and
dashed lines show the profiles in plasma and RBC,
respectively. The interrupted line in graph C
illustrates the O_2 saturation (So_2). The
interrupted line in graph D represents the change
in pH in plasma corresponding to that of the
buffer-base of plasma protein (pH_p^*).

the shortening of the contact time can be evaluated from the relation of
$P\bar{v}CO_2$ to $P\bar{v}O_2$ illustrated in Fig. 7. For example, when the $P\bar{v}O_2$ is 30
Torr at R = 0.85 and t_c = 0.7 sec, the $P\bar{v}CO_2$ is expected to be 53.5 Torr,
as shown by the broken line. By reducing the contact time from 0.7 sec
to 0.3 sec, the $P\bar{v}CO_2$ will rise from 53.5 to 57.8 Torr. That is,
hypercapnia may occur without causing hypoxemia, because of the reduction
of the contact time.

Jones et al. (1967) reported that the $P(A-a)CO_2$ was about 2 Torr at
rest and it increased to 6 to 8 Torr during heavy exercise, suggesting
that the validity of our theoretical evidence. The above experimental
and theoretical values obviously suggest that the relatonship between
PaO_2, $PaCO_2$ and R, namely, the O_2/CO_2 diagram, is a function of the
contact time. When $P\bar{v}O_2$ = 39 Torr, $P\bar{v}CO_2$ = 48 Torr and t_c = 0.7 sec, the
$P(A-a)CO_2$ calculated from Eq. (1) becomes 0.35 Torr, which is acceptable
for drawing the O_2/CO_2 diagram. However, when the t_c is 0.3 sec or
thereabouts, some correction will be needed.

SUMMARY

From the numerical solutions of simultaneous O_2 and CO_2 diffusions
in the RBC, we calculated the $P(A-a)CO_2$ by varying the tc at various $P\bar{v}O_2$
and $P\bar{v}CO_2$ levels, whereas the P_AO_2 and P_ACO_2 were kept constant at 90 and
40 Torr, respectively. From the results we could clarify the
relationship between the $P(A-a)CO_2$ and the t_c and its dependency on $P\bar{v}O_2$

and $P\bar{v}CO_2$. Furthermore, the influence of the t_C on the $C(a-v)O_2$, $C(v-a)CO_2$ and R was clarified quantitatively.

REFERENCES

Jones, N. L., Campbell, E. J. M., McHardy, G. J. R., Higgs, B. E. and Clode, E. The estimation of carbon dioxide pressure of mixed venous blood during exercise. Clin. Sci. 32, 311–327, 1967.

Kagawa, T. and Mochizuki, M. Theoretical analyses for arterial-venous O_2 content difference and Haldane effect during rebreathing. Jpn. J. Physiol. 37, 267–282, 1987.

Mochizuki, M. and Kagawa, T. Numerical solution of partial differential equations describing the simultaneous O_2 and CO_2 diffusions in the red blood cell. Jpn. J. Physiol. 36, 43–63, 1986.

Mochizuki, M. Computer program for overall gas exchange in RBC, in "Overall Gas Exchange through Red Blood Cell" Ed. M. Mochizuki, pp. 535–574, Yamagata University School of Medicine. 1987.

Mochizuki, M., Shibuya, I., Uchida, K. and Kagawa, T. A method for estimating contact time of red blood cells through lung capillary from O_2 and CO_2 concentrations in rebreathing air in man. Jpn. J. Physiol. 37, 283–301, 1987.

Niizeki, K., Mochizuki, M. and Kagawa, T. Secondary CO_2 diffusion following HCO_3^- shift across the red blood cell membrane. Jpn. J. Physiol. 34, 1003–1013, 1984.

Roughton, F. J. W. The average time spent by the blood in the human lung capillary and its relation to the rates of CO uptake and elimination in man. Am. J. Physiol. 143, 621–633, 1945.

Schuster, K. D. Kinetics of pulmonary CO_2 transfer studied by using labeled carbon dioxide $C^{16}O^{18}O$. Respir. Physiol. 60, 21–37, 1985.

Shibuya, I., Uchida, K. and Mochizuki, M. Experimental analyses of pulmonary gas exchange in a standing position at rest and during treadmill exercise. Jpn. J. Physiol. 37, 303–320, 1987.

Uchida, K., Shibuya, I. and Mochizuki, M. Simultaneous measurement of cardiac output and pulmonary diffusing capacity for CO by a rebreathing method. Jpn. J. Physiol. 36, 657–670, 1986.

RELATIONSHIP BETWEEN CARDIORESPIRATORY DYNAMICS AND MAXIMAL AEROBIC CAPACITY IN EXERCISING MEN

Ryszard Grucza [*], Yoshimi Nakazono, and Yoshimi Miyamoto

Department of Information Engineering, Faculty of Engineering
Yamagata University, Yonezawa 992, Japan

[*] Department of Applied Physiology, Medical Research Centre
Polish Academy of Sciences, 00-730 Warsaw, Poland

INTRODUCTION

Some respiratory reactions show a relation to the physical fitness of
an individual indicating that respiratory mechanisms can play a role in
functional adaptation to exercise in man. Stegemann (1981) has reported
that trained and untrained persons have the same ventilation at rest but
that this is achieved with different ratios of frequency and tidal volume
when the respiratory center is stimulated at rest by CO_2. As was shown by
Yearg et al. (1985) the exercise equivalent for oxygen ($\dot{V}_E/\dot{V}_{O_2}$) decreases and maximal achievable ventilation ($\dot{V}_{Emax}$) increases in the course
of endurance training. It is also reported that at the onset of exercise
oxygen uptake and ventilation increase more rapidly toward the steady state
in trained than in untrained men (Hagberg et al., 1980). The latter finding has an important bearing on the estimation of the oxygen deficit
developing in man at the beginning of heavy exercise. It is not clear, however, whether the faster response of ventilation is a specific result of the
training applied, or whether it is a physiological reaction connected with
the level of maximal aerobic capacity ($\dot{V}_{O_2}max$) regardless of training. The
purpose of this study was, therefore, to test the dynamics of respiratory
response to exercise in relation to the individual level of $\dot{V}_{O_2}max$ in a
group of untrained men.

METHODS

Ten men in the range (±SD) of 22 ± 0.5 years in age, 61.4 ± 7.0 kg
in weight, and 170 ± 5cm in height participated in this study. The subjects
were university students. None of them had performed physical training during the 3 months preceding the experiment.
    The experimental protocol consisted of two physical exercises perfor-
med in the upright position on a bicycle ergometer (Lode, Groningen). The
subjects worked first at an intensity of 100 W for 10 min in order to estimate their maximum aerobic capacity ($\dot{V}_{O_2}max$) by an indirect method of
Åstrand-Ryhming (1954). The next day, the subjects exercised at an intensity of 50% $\dot{V}_{O_2}max$ for 10 min. The work started from nonpedaling rest at an
acoustic signal.
    The subjects breathed via a hot-wire type pneumotachograph (Minato,
type RF-2) with a small dead space of 10 ml. Respiratory frequency (f),

tidal volume (V_T), and minute ventilation ($\dot{V}_E$) were measured automatically, breath-by-breath, by a computer system (Miyamoto et al., 1981).
Stroke volume (SV) was continuously measured by an impedance cardiography
method (Nakazono and Miyamoto, 1985). For this purpose transthoracic
impedance was measured by an impedance plethysmograph (Nihon Kohden, type
RGA-5), and heart rate (HR) was monitored by an electrocardiogram (ECG).
Components of the transthoracic impedance (dz/dt and z_0) were sampled by
the computer system at the rate of 200 times·sec^{-1} when triggered by the R
wave of ECG. The sampling was repeated for 4 to 5 cardiac cycles in order to
obtain an average value. The averaging technique allowed the subjects to
breath spontaneously during the exercise. Cardiac output ($\dot{Q}$ = SV·HR) was
automatically calculated by the system. Digital data for f, V_T , $\dot{V}_E$, SV ,
HR , and $\dot{Q}$ were then converted to calibrated analog signals and continuously
recorded on an 8-channel chart recorder (San-Ei, Rectigraph 8k). Since
changes in the cardiac and respiratory responses exhibit an exponential
course from the beginning of an exercise a time constant was used to describe
the dynamics of these responses. The time constants (τ) were defined as the
time required to change from the control rest value to 63.2 % of the new
steady state level for each of the cardiorespiratory reactions. Steady state
levels were assumed to be established by the 7th min of exercise. During the
8th min the expired gas was collected into a Douglas bag and the oxygen
consumption of the subjects measured with an expired gas monitor (San-Ei,
type 1H21A). The data, presented as mean ± SD, were analysed by Student's
paired t-test and linear regression.

RESULTS

Maximum oxygen uptake ($\dot{V}_{O_2}$max) in the tested men was 2.41 ± 1.05
$l·min^{-1}$ (39.3 ± 6.2 $ml·kg^{-1}·min^{-1}$) . They performed exercise with a relative intensity of 50.6 ± 7.8 % $\dot{V}_{O_2}$max , which corresponded to an absolute
intensity of work of 90 ± 18 W.

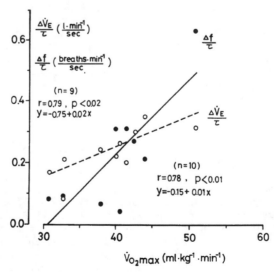

Fig. 1.   Dynamics of respiratory frequency ($\Delta f/\tau$) and minute
ventilation ($\Delta\dot{V}_E/\tau$) at the beginning of exercise in relation
to individual levels of maximum oxygen uptake ($\dot{V}_{O_2}$max).

The exercise caused an increase from 85 ± 10 to 132 ± 13 beats$\cdot$min^{-1} in HR, from 55 ± 11 to 73 ± 12 ml in SV, and from 4.5 ± 0.5 to 9.6 ± 1.6 l$\cdot$min^{-1} in $\dot{Q}$. Respiratory frequency (f) increased from 16 ± 4 to 20 ± 5 breaths$\cdot$min^{-1}, tidal volume (V_T) from 0.34 ± 0.11 to 1.53 ± 0.54 l, and ventilation ($\dot{V}_E$) from 5.1 ± 1.1 to 29.7 ± 6.1 l$\cdot$min^{-1} . In one subject no increase in respiratory frequency was observed.

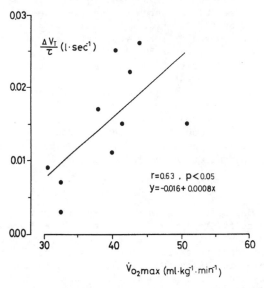

Fig. 2.   Dynamics of tidal volume ($\Delta V_T/\tau$) during the beginning of exercise in relation to individual levels of maximum oxygen uptake ($\dot{V}_{O_2}$max) .

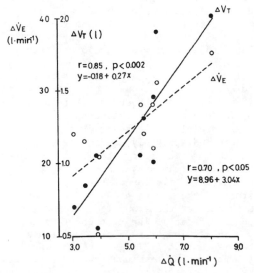

Fig. 3.   Relationship between increases in minute ventilation ($\Delta \dot{V}_E$) , tidal volume (ΔV_T) and cardiac output ($\Delta \dot{Q}$) during the steady state of exercise .

The time constants for the cardiac response to exercise were as follows: (HR) 45.8 ± 36.2 sec, (SV) 21.7 ± 16.3 sec, ($\dot{Q}$) 21.7 ± 10.1 sec. Time constants for the ventilatory response were : (f) 27.4 ± 28.7 sec, (V_T) 70.0 ± 23.2 sec, ($\dot{V}_E$) 69.5 ± 15.4 sec. The time course for ventilation was significantly longer than for heart rate ($p < 0.05$), stroke volume ($p < 0.001$) and cardiac output ($p < 0.001$). However, respiratory frequency increased relatively quickly with a time comparable to that for the increase in stroke volume and cardiac output.

Dynamics of the respiratory responses to exercise was significantly greater in subjects with higher aerobic capacity. Linear relationship between the dynamics of f , V_T, $\dot{V}_E$, and $\dot{V}_{O_2}$max are presented in Fig. 1 and Fig. 2. It can be estimated from the regression equation that subjects having a maximum oxygen uptake of 50 $ml \cdot kg^{-1} \cdot min^{-1}$ were able to increase respiratory frequency at a rate 0.43 $breaths \cdot min^{-1} \cdot sec^{-1}$, tidal volume at a rate 0.024 $l \cdot sec^{-1}$, and ventilation at a rate equal to 0.28 $l \cdot min^{-1} \cdot sec^{-1}$. The respective values for the subjects with lower $\dot{V}_{O_2}$max (40 $ml \cdot kg^{-1} \cdot min^{-1}$) were: 0.21 $breaths \cdot min^{-1} \cdot sec^{-1}$, 0.015 $l \cdot sec^{-1}$, and 0.22 $l \cdot min^{-1} \cdot sec^{-1}$.

Increases in tidal volume and minute ventilation were individually related to the increase in subjects' cardiac output during the steady state of exercise (Fig. 3). No relationship was found between respiratory frequency and cardiac output during this phase of exercise.

DISCUSSION

Ventilatory adaptation to submaximal exercise was faster in men with higher aerobic capacity. More rapid increase in $\dot{V}_E$ and $\dot{V}_{O_2}$ at the onset of constant work was previously reported as an effect of physical training on the respiratory response to exercise (Hickson et al.,1978; Hagberg et al., 1980). The training increases the level of oxidative enzymes in muscle cells, the density and number of their mitochondria, and the number and density of capillaries to the muscles (Grimby and Saltin, 1971; Yoshida et al., 1982). All these factors increase oxygen demand and utilization in the working organism. The higher oxygen demand can be realized only by improved cardiac performance and respiratory function.

Our results suggest that the dynamics of the cardiorespiratory response to exercise is strongly related to the actual level of $\dot{V}_{O_2}$max in subjects regardless of the training. This finding explain, to some extent, the individual differences in cardiovascular and respiratory function observed in exercising men. It seems that at least such factors as the level of $\dot{V}_{O_2}$max , body muscle mass, age, and previous physical activity of the subject should be taken into account when cardiorespiratory dynamics is analysed.

During transition from rest to exercise the response of the organism is probably structured to maintain homeostasis at the cellular level (Cooper et al., 1985). It can be assumed, therefore, that the increased supply of tissue O_2 is determined by the needs of the cells. The faster adjustment to the cardiorespiratory steady state during exercise can result in a smaller O_2 deficit and, probably, a smaller O_2 debt in a fitter man.

It was found in the present study that both respiratory frequency and tidal volume increased faster in subjects with higher $\dot{V}_{O_2}$max leading to a faster increase in minute ventilation at the onset of exercise. However, the mean time constant of the increase in the respiratory frequency (27.4 sec) was much shorter than that for the tidal volume (70.0 sec) and for ventilation (69.5 sec). The time constant for the respiratory frequency was only slightly longer than time constants for the stroke volume (21.7 sec) and cardiac output (21.7 sec). The fact that the time course for cardiac output preceded the time courses for respiration seems to support, indirectly, the concept of cardiodynamic hyperpnea postulated by Wasserman et al. (1974).

SUMMARY

Cardiorespiratory dynamics was tested in 10 men exercising with a relative intensity of 50% $\dot{V}_{O_2}$max for 10 min. Time constants for cardiac response (SV 21.7 sec, HR 45.8 sec, $\dot{Q}$ 21.7 sec) were shorter than those for the ventilatory response (f 27.4 sec, V_T 70 sec, $\dot{V}_E$ 69.5 sec). Respiratory dynamics was significantly related to the level of $\dot{V}_{O_2}$max exhibited by the subjects: (f) r = 0.79, p < 0.02 ; (V_T) r = 0.63, p < 0.05; ($\dot{V}_E$) r = 0.78, p < 0.01. It is concluded that in man the dynamics of the ventilatory response to exercise depend on the actual level of $\dot{V}_{O_2}$max in the individual.

This work was supported in part by the Foundation for Promoting Health Sciences (Kenko Kagaku Zaidan) grant No 61 101.

REFERENCES

Åstrand,P.O., and Ryhming, I., 1954, A nomogram for calculation of aerobic capacity (physical fitness) from pulse rate during submaximal work, J. Appl. Physiol., 7: 218 -221.

Cooper, Dan, M., Berry, C., Lamarra, N., and Wasserman, K., 1985, Kinetics of oxygen uptake and heart rate at onset of exercise in children, J. Appl. Physiol. 59: 211 - 217.

Grimby, G., and Saltin, B., 1971, Physiological effects of physical training, Scand. J. Rehab. Med. 3: 6 - 14 .

Hagberg, J. M., Hickson, R.C., Ehsani, A.A., and Holloszy, J.O., 1980, Faster adjustment to and recovery from submaximal exercise in the trained state, J. Appl. Physiol.: Respirat. Environ. Exercise Physiol. 48: 218 - 224.

Hickson, R.C., Bomze, H.A., and Holloszy, J.O., 1978, Faster adjustment of O_2 uptake to the energy requirement of exercise in the trained state, J. Appl. Physiol.: Respirat. Environ. Exercise Physiol. 44: 877 - 881.

Miyamoto, Y., Sakakibara, K., Takahashi, M., Tamura, T., Takahashi, T., Hiura, T., and Mikami, T., 1981, Online computer for assesing respiratory and metabolic function during exercise, Med. Biol. Eng. Comp. 19: 340 - 348.

Nakazono, Y., and Miyamoto, Y., 1985, Cardiorespiratory dynamics in men in response to passive work, Jpn. J. Physiol. 35: 33 - 43.

Stegemann, J., 1981, Effect of endurance training on autonomic function, in: " Exercise Physiology", J.S. Skinner, ed., Georg Thieme Verlag, Stuttgart - New York, pp. 293 - 294.

Wasserman, K., Whipp, B.J., and Castagna, J., 1974, Cardiodynamic hyperpnea: hyperpnea secondary to cardiac output increase, J. Appl. Physiol. 36: 457 - 464.

Yearg II, J.E., Seals, D.R., Hagberg, J.M., and Holloszy, J.O., 1985, Effect of endurance exercise training on ventilatory function in older individuals, J. Appl. Physiol. 58: 791 - 794.

Yoshida, T., Suda, Y., and Tekeuchi, N., 1982, Endurance training regimen based upon arterial blood lactate: effects on anaerobic threshold, Eur. J. Appl. Physiol. 49: 223 - 230.

OXYGEN UPTAKE AND SURFACTANT REPLACEMENT

B. Lachmann, S. Armbruster and W. Erdmann

Department of Anesthesiology, Erasmus University, 3000 DR
Rotterdam, Postbus 1738, The Netherlands

INTRODUCTION

Increased capillary permeability, associated with damage to the alveolar epithelium, can lead to derangement of pulmonary surfactant through
a variety of mechanisms. These include: washout of alveolar phospholipids
into the interstitial tissue and blood stream; inactivation by plasma components; surfactant depletion by foaming or ventilation with large tidal
volumes; disturbed synthesis, storage or release of surfactant secondary
to direct injury to type II cells by hypoxemia, acidosis and hypoperfusion
(for review see 1).

Besides these mechanisms, changes in alveolar configuration and size,
caused by an increase of fluid in the alveoli, is another important pathogenic factor. If fluid enters an alveolus, the radius of the alveolar
bubble becomes smaller and the negative pressure of the alveolar fluid
becomes even more negative. Therefore, once started in surfactant deficiency, alveolar flooding is a selfaccelerating process as the retractive
forse in the alveolar/air-liquid interface tends to increase with decreasing bubble radius.

Irrespective of the pathogenic mechanisms, a derangement of pulmonary
surfactant causes decreased lung distensibility, in addition to increasing the work of breathing and the oxygen consumption of respiratory
muscles. Other consequences are alveolar collapse (atelectasis) and
transudation of plasma into the alveoli with increased resistance to O_2
and CO_2 diffusion, and enhanced functional right to left shunt due to
perfusion of non-ventilated areas of the lungs. Finally, progressive hypoxemia and metabolic acidosis results secondary to increased production of
organic acids under anaerobic conditions (Fig.1).

The central role of surfactant deficiency and/or its inactivation by surfactant inhibitors in this context, is illustrated by recent studies in
animal ARDS models, e.g. surfactant deficiency by lung lavage,
capillary leakage syndrome after intravenous injection of an anti-lung
serum,which show that abnormalities in blood gases and lung mechanics
can be restored to normal by tracheal instillation of natural surfactant.

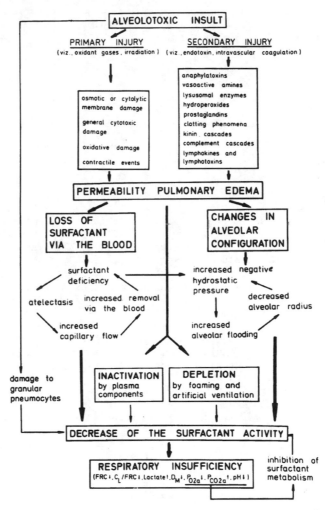

Fig. 1. Pathogenesis of adult respiratory distress syndrome with special reference to the surfactant system.

In experimental ARDS, induced by in vivo lung lavage, we were the
first to demonstrate that tracheal instillation of surfactant lipids results in striking improvement in gas exchange (2, 3) (Fig.2). Other
experiments (in animals receiving surfactant) have documented that the
improved blood gases are stable for at least five hours, whereas P_aO_2 in
control animals remains low, despite ventilation with PEEP and pure
oxygen (Fig.3).

Histologic lung sections from surfactant treated animals showed a uniform pattern of well-aerated alveoli, with only minimal intra-alveolar
oedema and hyaline membranes, whereas control animals ventilated with
the same respirator setting had extensive atelectasis and prominent
hyaline membranes.

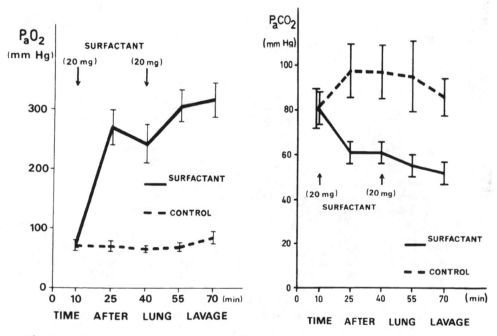

Fig.2: Effect of surfactant replacement on P_aO_2 and P_aCO_2 in guinea pigs
with severe ARDS induced by repeated lung lavage. The animals were
ventilated with pressure-controlled ventilation, 100% oxygen,
I/E ratio 1:1, frequency 20/min, insufflation pressure 28 cm H_2O
and PEEP 5 cm H_2O. Surfactant, made from bovine lungs, was administered twice via the tracheal cannula (arrows). Values are given as
mean $\pm$ SD. (From Lachmann et al. (2, 3) with permission).

These results indicate that the ventilator treatment per se is not harmful to the pulmonary parenchyma, provided that alveolar collapse is prevented by surfactant replacement and shear forces thereby avoided.

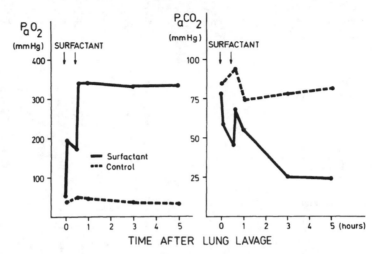

Fig.3.: Sequential recordings of P_aO_2 and P_aCO_2 in two adult guinea pigs subjected to repeated lung lavage, followed by ventilation for 5 h with pressure-controlled ventilation; 100% oxygen; I/E ratio, 1:2; frequency, 30/min; insufflation pressure, 28 cm H_2O; and PEEP, 6 cm H_2O. One animal received two tracheal instillations of natural surfactant (each dose 50 mg), 5 and 30 min after lavage (arrows). The improved blood gases, recorded after surfactant replacement, are stable throughout the period of observation.

One may argue that using the surfactant depleted lung model for surfactant replacement does not represent all changes typical for ARDS lungs. This is why we also used other animal models with severe respiratory failure and treated them with tracheal instillation of surfactant (bacterial pneumonia (4), respiratory failure due to free oxygen radicals (5), ARDS by intravenous injection of anti-lung serum = capillary leakage syndrome).

EXPERIMENTAL CAPILLARY LEAKAGE SYNDROME

Intravenous injection of anti-lung serum (ALS) in guinea pigs causes acute and fatal respiratory failure (6). Previous studies have shown marked decrease in thorax-lung compliance, increase in water content, decrease in phospholipid content and diminished surface activity in these lungs, as well as severe morphological damage, especially to the alveolar capillary membrane, of these lungs. Using this model - where the failure is clearly caused by damage to the capillary membranes, we could demonstrate that surfactant instillation also significantly improved blood gases and lung mechanics (7) in these animals (Fig.4).

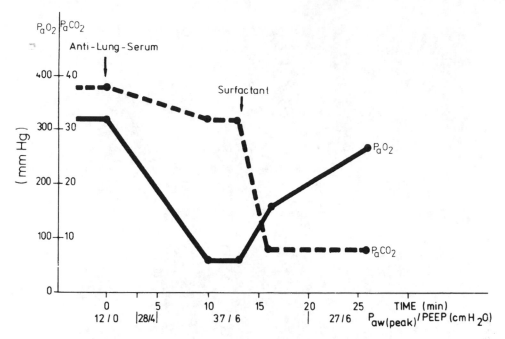

Fig.4: Course of P_aO_2 and P_aCO_2 in a guinea pig before and after anti-lung serum as well as after tracheal instillation of 3 ml surfactant (60 mg total phospholipids per ml). Paw - peak airway pressure, PEEP - positiv end - expiratory pressure.

SURFACTANT SUBSTITUTION IN CLINICAL ARDS

In a terminal patient with sepsis and severe ARDS (P_aO_2 of 19 mmHg, despite pressure controlled ventilation with an I:E ratio of 3:1; peak airway pressure of 48 cm H_2O; PEEP 12 cm H_2O; FiO_2 = 1), tracheal instillation of natural surfactant led, within a few hours, to a dramatic improvement in gas exchange (P_aO_2 from 19 mm Hg to 240 mm Hg; P_aCO_2 from 68 to 45 mmHg). These first clinical results already show that lungs from patients with severe RDS, superimposed with virus and bacterial pneumonia, can be re-aerated by tracheal instillation of exogenous surfactant (Fig. 5).

CONCLUSION

From these results we conclude that in the future surfactant replacement could be one of the most important therapeutic approaches for reducing the high mortality rate currently associated with ARDS due to a failure of oxygen uptake in the lung.

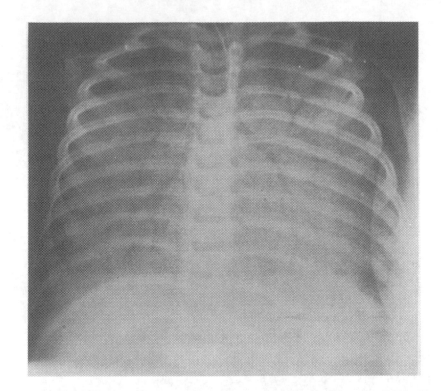

A

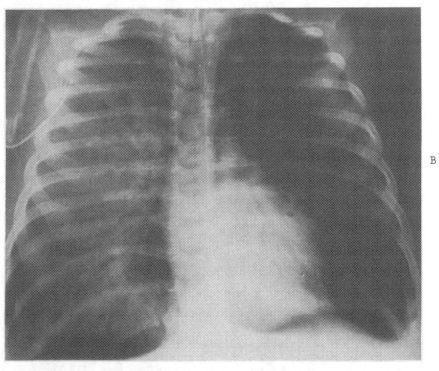

B

Fig.5: X-rays from a child with a severe ARDS immediatly before (A)
and 4 hours after surfactant replacement therapy (B). From
Lachmann (8) with permission.

REFERENCES

1. Lachmann B, Danzmann E. (1984) Acute respiratory distress syndrome.
   In: B. Robertson, LMG Golde, JJ Batenburg (eds). Pulmonary
   Surfactant. Elsevier, Amsterdam: 505-548.
2. Lachmann B, Fujiwara T, Chida S, Morita T, Konishi M, Nakamura K,
   Maeta H (1981). Improved gas exchange after tracheal instillation
   of surfactant in the experimental adult respiratory distress syndrome. Crit Care Med 9: 158.
3. Lachmann B, Fujiwara T, Chida S, Morita T, Konishi M, Nakamura K, Maeta
   H.(1983).Surfactant replacement therapy in the experimental adult
   respiratory distress syndrome (ARDS). In: Cosmi EV, Scarpelli EM
   (eds). Pulmonary Surfactant System. Elsevier, Amsterdam: 231-235.
4. Lachmann B, Bergmann K Ch. (1987). Surfactant replacement improves
   thorax-lung compliance and survival rate in mice with influenza
   infection. Am Rev Resp Dis 135: A6.
5. Lachmann B, Saugstad OD, Erdmann W. (1987). Effect of surfactant replacement on respiratory failure induced by free oxygen radicals. In:
   G. Schlag, H. Redl (eds) First Vienna Shock Forum. Alan Liss,
   New York: 305-313.
6. Lachmann B, Bergmann KC, Winsel K, Müller E, Petro W, Schäffer C, Vogel
   J (1975). Experimental respiratory distress syndrome after injection of anti-lung serum. III. Chronic experimental trial. Pädia-
   trie und Grentzgebiete 14: 211-233.
7. Lachmann B, Hallman M, Bergmann K. Ch. (1987). Respiratory failure following anti-lung serum: Study on mechanisms associated with surfactant system damage. Exp Lung Res 12: 163-180.
8. Lachmann B. (1987) The role of pulmonary surfactant in the pathogenesis
   and therapy of ARDS. In: J.L. Vincent (ed). Update in Intensive
   Care and Emergency Medicine. Springer-Verlag: 123-134.

NON - INVASIVE PULSE OXIMETRY DURING LUNG LAVAGE

S. Armbruster, B. Lachmann and W. Erdmann

Department of Anesthesiology, Erasmus University, 3000 DR
Rotterdam, Postbox 1738, the Netherlands

INTRODUCTION

Diagnostic lung lavages with or without bronchographic procedures
are investigational methods performed daily in pulmonary medicine. Less
commonly therapeuric lung lavages are performed as part of the treatment
of severe asthmatic status, complete mucus obstruction or in rare other
cases such as alveolar proteinosis. Therapeutic and diagnostic lung lavage often induces hypoxia. The present study was performed to investigate
whether on-line monitoring of oxygen saturation SpO2 could help to prevent
prolonged hypoxic episodes and thus render lung lavage more safe.

METHODS

Non-invasive SpO2 measurements were performed with a pulse oxymeter
(Biox 3700, Ohmeda, USA). On-line St-segment analysis (Mac PC, Marquette,
USA) was used as a parameter of life endangering oxygen supply deficiency.
21 Patients were studied during and after diagnostic lung lavage and bronchography. 5 Patients with alveolar proteinosis (3) respectively status
asthmaticus (2) were investigated during and after therapeutic lung lavage. All patients were anesthetized and paralyzed and ventilated with
pure oxygen.

RESULTS

DIAGNOSTIC LUNG LAVAGE AND BRONCHOGRAPHY

No patient had SpO2 decrease below 93% (fig. 1), nor did any one show
any hypoxic signs in the St-segment during the procedure. However, in the
recovery room where patients were returned to spontaneous room air respiration, all patients except one showed SpO2 values decreasing below 91%
with some values as low as 78% (fig. 2). In 6 of these hypoxic patients
ST-deprivation > 0,1 mV was seen at different SpO2 levels (fig. 3).

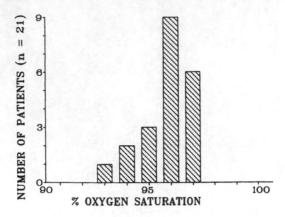

Fig. 1: SpO2 distribution during diagnostic lung lavage and bronchography.

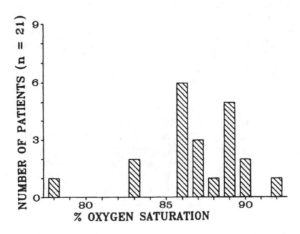

Fig. 2: SpO2 distribution in the recovery room following diagnostic lung lavage and bronchography in 21 patients breathing room air.

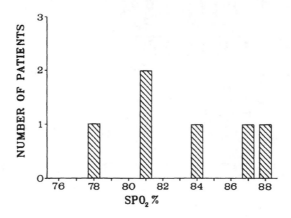

Fig. 3: ST-deprivation > 0,1mV in relation to SpO2 in 6 patients following diagnostic lung lavage and bronchography.

In two patients with coronary disease, severe ST-deprivation already occurred at 83%-88% (fig. 4) respectively at saturation. Where after,
with several minutes delay, the typical ST-elevation of a beginning infarction appeared, although saturation had been restored (fig. 5).

6 of the patients with SpO2 decrease in the recovery room received
anticoughing or sedatin drugs. In the other patients with SpO2 decrease
residual anesthesia must have been the cause.

SpO2 monitoring made it possible to immediately intervene and try to
reverse hypoxic periods and ST-segment alterations by appropriate O2-ap-
plication.

THERAPEUTIC LUNG LAVAGES

Oxygen saturation during aspiration of the lung lavage fluid decreased to values between 89% and 81%, depending on the anesthesia. However, this had no further side effects on the ST-segment; the periods of
suctioning were rather short (less than 1 minute). Nevertheless, SpO2
measurement in these cases is very important. Increased shunt was immediately detected in one patient, probably due to fluid transfer from the
treated lung to the untreated and ventilated lung. This relatively severe SpO2 decrease (to 77%) could be due to blockade of the small airways with fluid of high surface tension.

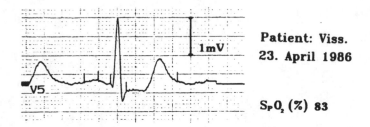

Patient: Viss.
23. April 1986

S_PO_2 (%) **83**

Fig. 4: ECG from a patient in the recovery room with an ST-deprivation
> 0,1 mV at SpO2 of 83%.

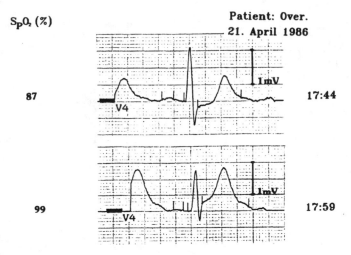

S_PO_2 (%)

Patient: Over.
21. April 1986

87   17:44

99   17:59

Fig. 5: ECG from a patient with known coronary disease in the recovery
room after bronchography. Note: ST-deprivation > 0,1 mV already
at SpO2 of 87% followed by an ST-elevation after 15 minutes de-
monstratin a beginning myocardial infraction.

RECOMMENDATIONS

To avoid the complications which may follow hypoxic episodes, especially after diagnostic and therapeutic lung lavage and bronchography,
pulse oximetry should be routinely applied. SpO2 values should be carefully controlled and kept above 90% in at least those patients with known
myocardial and coronary disease.

EFFECTS OF COMBINED RESISTIVE LOADING ON THE RESPIRATORY PATTERN AND

THE SUBJECT'S PERCEPTION

Zhang Li Fan, Wu Xing Yu, Jiang Shi Zhong, Zhang Rong
and Ma Mou Chao*

Department of Aerospace Physiology, The Fourth Military
Medical College, Xian, and *Institute of Psychology
Academia Sinica, Beijing, China

The respiratory regulatory system has a remarkable ability to adjust the ventilation to levels appropriate to changing chemical and/or mechanical loading conditions in order to preserve oxygen and carbon dioxide homeostasis. Compared with the large body of published work on the effects of chemical loading, there have been few quantitative studies on the effects and regulatory mechanisms of mechanical loading. The response to added mechanical loads is, however, of considerable practical importance.

External respiratory mechanical loads may be classified as elastic, flow resistive, pressure, threshold load etc. (Cherniack and Altose, 1981). In recent years, most studies on flow resistive load have focused on the effects of an added inspiratory resistance on respiratory pattern and ventilation, the mechanism of load compensation and the subject's perception of the added load to breathing. Great attention has been paid to research work on respiratory sensation with the aim of gaining further understanding of the neural and behavioral mechanisms involved in the conscious control of breathing (Altose and Cherniack, 1980). Recent work along these lines has demonstrated that respiratory sensation perceived with added inspiratory resistive loads follows Stevens' psychophysical power law and the perceived magnitude is directly related to the pressure generated by inspiratory muscles against resistive loads (Killian et al., 1982; Altose et al., 1982; Burdon et al., 1983). These results were obtained in research which was confined to inspiratory resistive loads. Reports concerning the effects of resistive loads applied either during expiration only (expiratory resistance, ER) or throughout the breathing cycle (combined resistance, CR) are scanty. Moreover, the researches on the combined resistance were mainly limited to one loading condition in which the added resistance applied to each phase was equal in physical intensity. In brief, the data on the effects of flow resistive load applied either to a single phase or throughout the breathing cycle are not systematic and are to some extent conflicting (Muza et al., 1984).

This paper describes the effects of combined flow resistive loads applied throughout the breathing cycle but with different inspiratory versus expiratory load ratios. Using a Fuzzy Set model for category judgement (Ma et al., 1986) and applying the concept of Just Noticeable Difference (JND) steps (Zhang et al., 1986) and equivalent sensation (Zhang et al., 1986), we succeeded in elucidating how the respiratory pattern and the subject's perception vary in accordance with both the total physical intensity of the resistance added and the ratio of inspiratory to expiratory load.

METHODS

Subjects. The experiments were performed on 53 healthy male subjects (staff personnel and college students) in a sitting position. All subjects gave normal spirometry values and none had a history of cardio-pulmonary disease.

Apparatus. Respiratory flow rate and oral mask cavity pressure were measured by a Fleisch-type pneumotachograph and a differential air pressure transducer (XY-1 model, Xian). The analog output was recorded by a data recorder (A-69, Magnescale Sony, Japan). A microcomputer (MDR Z-80, Beijing) was used for off-line computation and analysis of experimental data. The resistive tube of the pneumotachograph was inserted between the mouth piece, or the oro-nasal mask, and the two-way respiratory valve. Added resistance was provided by two resistance manifolds connected respectively to the inspiratory and the expiratory limb of the non-rebreathing valve. Each resistance manifold consisted of a series of resistors made of nylon net, arranged in a Lucite tube. Rubber stoppers were inserted in ports between the resistors to produce the desired load.

Psychophysical Scaling by Multistage Evaluation Scale (MES). A new kind of category scale, the Multistage Evaluation Scale (MES), was developed to estimate the respiratory sensation associated with added resistance to breathing. The MES answer sheet is shown in Table 1. The five graded categories of the respiratory resistive load sensation are as follows: (1) none (not noticeable); (2) light (noticeable, but no discomfort); (3) moderate (breathing with effort but tolerable); (4) heavy (difficulty in breathing, intolerable for any length of time); (5) severe (a sense of total suffocation). The first and second columns of Table 1 show, respectively, the subject's attitude (positive or negative) and the degree of certainty (or confidence). The subject, loaded with an inspiratory and/or an expiratory resistance, was first asked to select the answer which gave the closest description of his sensation from the resistive load. At the same time, he was asked to indicate the degree of certainty of the reaction by marking the appropriate box in that category. Next, he was required to indicate the degree of certainty in the two categories next to that which was the closest description of his sensation from the added load. Lastly, the attitude and certainty evaluation for the two remaining categories was completed by marking the appropriate boxes.

Table 1. MES answer sheet for respiratory resistive
load sensation ('✓' = subject's reaction)

Attitude	Certainty	None	Light	Moderate	Heavy	Severe
positive	very		✓			
positive	considerably					
positive	slightly	✓				
negative	slightly					
negative	considerably			✓		
negative	very				✓	✓

The following formula was used to quantify the sensory magnitude (F) elicited by a given added load with intensity u_i (i = 1, 2,..., n):

$$F(u_i) = \frac{1}{t}\left(\sum_{j=1}^{m} V_i y_{ij} w_i \Big/ \sum_{j=1}^{m} y_{ij} w_i\right), \quad i = 1, 2, \cdots, n$$

where t is a constant representing the intensity of the added resistance which may cause the most intolerable respiratory sensation. According to the result obtained by Pope et al. (1968) let t = 1100 $mmH_2O \cdot l^{-1} \cdot s$, so that the perceived magnitude is normalized. V_j is a term for the category on the sensory scale. y_{ij} is the weighting coefficient related to the degree of certainty. w_j is a set of moving weighting coefficients which reflect the relative importance of each category, r_j, (j = 1, 2, ..., m) on the scale. For further details see Ma et al. (1986). In addition, the Classical Category Scale (CCS) was used for comparison.

Determination of number of JND steps of respiratory sensation. The JND steps of respiratory sensation caused by either inspiratory or expiratory resistance in the range 10-500 $mmH_2O \cdot l^{-1} \cdot s$ were determined. The determination was repeated six to eight times on each subject for both inspiratory and expiratory load. Then both the individual and the group mean values of the threshold resistance corresponding to each step were obtained. The procedure for determining JND steps was as follows. The subject was seated upright in a dental chair and the apparatus and the experimenter were screened from his view by a curtain. Wearing a nose clip, he breathed quietly through the mouthpiece of the apparatus for several minutes at the beginning of each experiment. The desired resistance stimulus was presented by removing the rubber stopper from the appropriate port. The JND steps for the inspiratory and the expiratory phase were then determined by the differential threshold method (ascending series).

In order to estimate the perceived magnitude and record the respiratory flow rate and mouth pressure changes during different JND steps, each subject was asked to experience his individual mean threshold resistance for each JND step for 3 minutes. The physiological changes were recorded during the first two minutes and the respiratory sensation was estimated with the MES answer sheet in the third minute.

Construction of equivalent respiratory-resistive-load sensation contours. The equivalent respiratory-resistive-load sensation contours define the condition which could elicit a respiratory sensation with an intensity equivalent to that caused by the average inspiratory resistance of a given JND step. In this study, the group-averaged inspiratory resistances corresponding to the 2nd, 3rd and 4th JND steps above the basal load level respectively, were adopted as the reference resistance loads. Their resistive values were 37, 59 and 88 $mmH_2O\ 0.5 \cdot l^{-1} \cdot s$. When the relevant reference inspiratory load of a JND step was presented, the added resistance to the expiratory phase was kept at a basal level, $7mmH_2O\ 0.5 \cdot l^{-1} \cdot s$. The principle of the method employed to construct the equivalent sensation contours was that the subject, breathing from the circuit and experiencing various combined resisrive loads, was required to determine the intensity of four sets of combined resistances that could elicit a respiratory sensation equivalent to the reference inspiratory resistance of a certain JND step. The ratios of inspiratory versus expiratory resistance (IR:ER) of the four combined loads to be compared were about 3:1, 2:2, 1:3 and 0:4, respectively. At the end of each experiment, to confirm the detected equivalent relations and to record the respiratory flow rate and mouth pressure changes, the subject was asked, in the JND step sequence, to experience the reference resistance and all its tentative 'equivalent' combined resistances. During this period, any combined resistive load that was not in accord with the equivalent relationship could be readjusted and affirmed by the subject, using the same procedure. For further details see Zhang et al. (1986).

RESULTS

1. Comparison of the Physiological Effects of Added Resistive Loads Applied Separately to a Single Phase or Throughout the Breathing Cycle

Mask cavity pressure (P) and external respiratory work (W). Fig. 1, depicting the relationship between mask cavity pressure swing ($\tilde{P} = P_{Emax} + |P_{Imax}|$) and total resistance ($R_T$) added under three loading conditions, shows $\tilde{P}$ was a curvilinear function of R_T which exhibited a convexity, indicating that the P associated with the CR load was the largest among the three loading conditions.

The changes of other pressure or work rate indices against R_T, including the integral mean value of pressure $\bar{P}$, external respiratory workrate $\dot{W}$ and external respiratory work per litre of ventilation W/l, showed a trend similar to that of $\tilde{P}$. The above mentioned changes were confirmed by a 3 x 6 analysis of variance (ANOVA; P <0.01), which revealed that the main effects of factor A and B were highly significant (P <0.01). Moreover, there was a significant interaction between factor A and B (P <0.01) among the effects on indices $\tilde{P}$, $\bar{P}$, $\dot{W}$ and W/l.

Respiratory flow rate ($\dot{V}$). When the resistive load was applied separately to one respiratory phase, no matter which phase was impeded, $\dot{V}_{max}$ and $\tilde{V}$ of the impeded phase decreased as R_T increased; conversely those of the unimpeded phase increased. When the resistive load was applied throughout the breathing cycle (CR), neither flow rate nor $\dot{V}_{Imax}/\dot{V}_{Emax}$ ratio showed significant changes.

Respiratory timing. Under conditions of single phase resistive loading, the duration of the impeded phase lengthened as R_T increased.

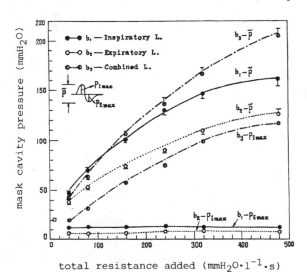

Fig. 1 The relationship between mask cavity pressure swing ($\tilde{P}$) and total resistance added (R_T) for different loading conditions (n = 37, the data for peak pressure, P_{max}, are also included. Note: Each point represents group mean, vertical bars represent SE.

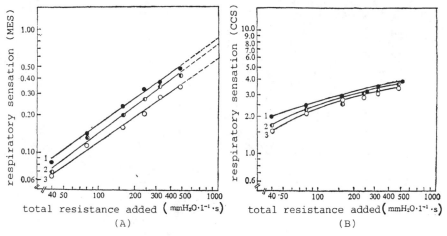

Fig. 2. Plots of grouped data of perceived magnitude of resistive loads
as a function of total added resistance (n = 37).
Note: (A) = Multistage Evaluation Scale; (B) = Classical Category
Scale; 1,inspiratory load; $Y = 0.0065 X^{0.7021}$; 2, combined load,
$Y = 0.0051 X^{0.7271}$; 3, expiratory load, $Y = 0.0057 X^{0.6645}$.

In contrast the duration of the unimpeded phase varied depending on which
phase was loaded. Expiratory time (T_E) was shortened with IR loads, but inspiratory time (T_I) was virtually unchanged with ER loads. Under the conditions mentioned the variation of the T_I/T_C ratio was due to the disproportionately small change of T_I compared with T_E. When CR loads were applied
both T_I and T_E lengthened proportionately, so that the T_I/T_C ratio remained
almost unchanged as R_T increased.

2. Comparison of the Subject's Perception to Added Resistance Applied either Separately to a Single Phase or Throughout the Breathing Cycle

Fig. 2 presents the relationship between R_T and the perceived magnitude estimated by either MES or CCS for the three loading conditions. From
Fig.2 it is apparent that the difference in the magnitude of the respiratory
sensation associated with both the intensity of the total resistance and the
phase of the breathing cycle that is loaded could be accurately assessed by
the MES (Fig. 2A). This was also confirmed by a 2-factor (6 x 3) analysis of
variance (ANOVA), which revealed that the main effects of both factors A
(R_T) and B (phase) were highly significant (P <0.01). In contrast to MES,
the perceived intensity obtained by use of CCS could reflect only the difference caused by factor A (P <0.01), not that caused by B (P >0.05, see
Fig. 2B). Furthermore, the psychophysical function obtained with MES could
be well expressed as a power function, $Y = aX^b$ (Fig. 2A), whereas the corresponding function with CCS was logarithmic (Fig. 2B).

Fig. 3, which gives a plot of the perceived magnitude estimated by MES
against the recorded swing in mask cavity pressure (P), shows that the
relationship between the physiological stimulus and the respiratory sensation assessed by MES exhibits a concavity, indicating more clearly the tendency of the respiratory sensation to intensify.

3. Just Noticeable Difference (JND) Steps of Respiratory Sensation

According to Fechner's law, the differential threshold, or Just
Noticeable Difference (JND) Step could be used as a standard unit in the

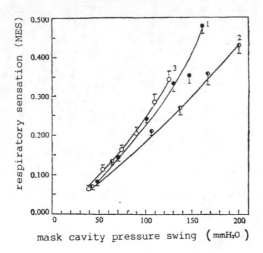

Fig. 3. Plots of grouped data of perceived magnitude (MES) of
resistive loads as a function of peak-to-peak mask cavity
pressure swing ($\widetilde{P}$), (n = 37). Note: see Fig. 2.

measurement of the subjective magnitude of a sense modality (Altose and
Cherniack, 1980). In Fig. 4 the perceived magnitude in JND units versus the
intensity of physical or physiological stimulus is plotted for inspiratory
and expiratory loads. It shows that the sensory continuum corresponding to
the range of added load (10-500 $mmH_2O \cdot 1^{-1} \cdot s$) consisted of 6 or 7 JND steps.
The data presented in Fig. 4 could be fitted to a logarithmic equation in
the form: $Y = b\ \ln(X+K)-a$, where Y is the estimated sensory intensity in JND
sensory units, X is the intensity of the physical or physiological stimulus,
and a, b, and K are parameters.

The JND step-perceived magnitude (MES) relationship, presumably
reflecting the discrimination-sensory magnitude relationship of respiratory
sensation is plotted and fitted by an exponential equation in the form:
$\hat{Y} = e^{a+bX}-K$, where $\hat{Y}$ is the estimated value of perceived magnitude, X is the
number of the JND step, and a, b, and K are parameters (see Fig. 5).

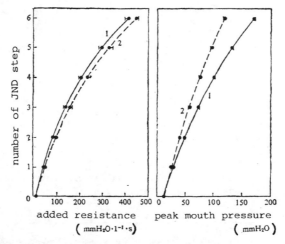

Fig. 4. The relationship between JND step and intensity of
physical (R_T) or physiological (P_{max}) stimulus.
Note: 1 = inspiratory load; 2 = expiratory load.

4. Equivalent Respiratory Resistive-Load-Sensation Contours

If the intensity of the physical (R) or physiological (P_{max}) stimulus from any combined resistance applied to the inspiratory and to the expiratory phase is plotted on, respectively, the horizontal and the vertical axis then the intensity of the combined resistance with a certain inspiratory to expiratory load ratio may be represented by a definite point in a plane.

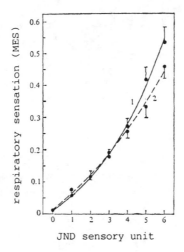

Fig. 5. The relationship between JND units and perceived magnitude (MES). Note: see Fig. 1. Fitted equation:
1. inspiratory phase: $\hat{Y} = e^{-1.8185 + 0.2429X} - 0.15$
2. expiratory phase: $\hat{Y} = e^{-1.0083 + 0.1295X} - 0.35$

The equivalent sensation contour can thus be drawn by a quadratic polynomial approximation to all such points representing combined loads which would elicit an equivalent resistive-load-sensation when applied. The equivalent respiratory resistive-load-sensation contours for the 2nd, 3rd and 4th JND step above the basal level, respectively, were drawn by defining P_{max} as the stimulus (see Fig. 6).

5. Effects on Respiratory Pattern of Combined Resistive Loads with Different IR/ER ratios

The relative changes of both inspiratory and inspiratory peak flow rates ($\dot{V}_{Imax}$ and $\dot{V}_{Emax}$) and the $\dot{V}_{Imax}/\dot{V}_{Emax}$ ratio during combined loadings with different IR/ER ratios but eliciting equivalent sensations at the 2nd, 3rd and 4th JND step above basal level, respectively, are presented in Fig. 7. Analysis of variance (3 x 5 ANOVA) confirmed that the relative changes of peak flow rates were related to both the number of the JND step (for $\dot{V}_{Imax}$, $P < 0.05$; for $\dot{V}_{Emax}$, $P < 0.01$) and the IR/ER ratio ($P < 0.01$); whereas the $\dot{V}_{Imax}/\dot{V}_{Emax}$ ratio was related to the IR/ER ratio ($P < 0.01$) only, and not to the JND step number ($P > 0.05$). The relationship between the $\dot{V}_{Imax}/\dot{V}_{Emax}$ ratio (Y) and the IR'/ER' ratio (Burdon et al., 1983) (X) was fitted by the linear regression equation, $\hat{Y} = 1.3046 - 0.4176 \ln X$ ($r = -0.9428$).

DISCUSSION

Two important aspects of this work should be emphasized. First, regarding the combined resistive loading as a general case, and the inspiratory loading as a special one, we studied systematically the effects of resistive load applied either to a single phase or throughout the breathing

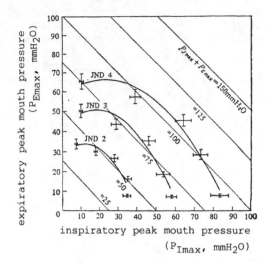

Fig. 6. Equivalent respiratory resistive-load-sensation contour
plotted by defining the peak mouth pressure (P_{max}) as
the stimulus. Note: n = 13; vertical bar represents SE.
JND 2: $Y = 28.4947 + 0.8985X - 0.0386X^2$
JND 3: $Y = 42.5551 + 0.9102X - 0.0255X^2$
JND 4: $Y = 58.6464 + 0.6782X - 0.0151X^2$

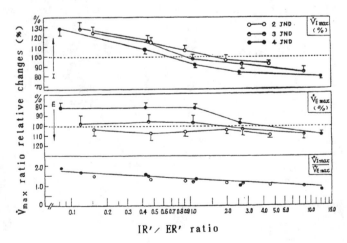

Fig. 7. Changes in respiratory flow rate in relation to the
number of the JND step and IR'/ER' ratio. (n = 13).
Note: see Fig. 1.

cycle. Second, adopting a new approach to studies on respiratory sensation, we were able to elucidate the physiological and psychological effects of combined resistive loads different in total intensity as well as in IR/ER ratio.

Recent studies suggest that the perceived magnitude of added inspiratory loads to breathing is directly related to factors associated with muscle force and indirectly to the added inspiratory loads. The inspiratory mouth peak pressure is regarded as an index of muscle force. Our results show that besides peak mouth pressure (mask cavity pressure), mask cavity pressure swing ($\overline{P}$), integrated mean of mask cavity pressure ($\overline{P}$), external respiratory work rate ($\dot{W}$) and external work per litre of ventilation (W/l) could also be treated as indices of respiratory muscle force. Our study was principally designed to compare a subject's perceptual performance during inspiratory, expiratory and combined loading. Our results show that when the intensity of total resistance (R_T) is considered, the perceived magnitude estimated by MES (S_{MES}) is arranged in the sequence IR>CR>ER (see Fig. 1A); for the perceptual performance plotted against the physiological stimulus, i.e., P or W against S_{MES}, the sequence is ER>IR>CR (see Fig. 3). Furthermore, applying the concept of JND steps and equivalent sensation, we have elucidated systematically the changes in respiratory flow rate and the subject's perception to combined resistive loads which differ not only in total resistance but also in IR/ER ratio (see Fig. 6 and 7).

We attribute our results to the methodology used in studying respiratory sensation. In the past two decades, studies of respiratory sensation have focused mainly on three aspects: 1. detection of the absolute and differential thresholds; 2. estimation of the perceived magnitude for suprathreshold loads by use of a category/ratio scale and elucidation of the functional relationship between sensory magnitude and the intensity of the physical or physiological stimulus; 3. investigation of the validity of Stevens' law in the study of respiratory sensation. In this work we have tried to estimate the perceived magnitude by the Multistage Evaluation Scale and elucidate the effects of combined resistance by determining the JND steps and constructing equivalent sensation contours. Let us discuss this in further detail.

Our results show that the perceived magnitude estimated by the use of the MES reflects accurately the difference in sensory magnitude caused by added loads with equal intensity but applied to a different respiratory phase. In addition, the psychophysical function obtained from the results of MES conforms to the Stevens' power law, whereas the Classical Category Scale lacks some of these attributes. We can now discuss the difference in methodological principle between the MES and the CCS. Firstly, the perception of added load seems to be a problem of possibility, rather than one of probability. It should be noted that only a weak connection exists between probability and possibility distribution. The MES based on multivariate logic can properly reflect the situation in which there are different degrees of reaction in more than one of the categories. The MES is suitable for the assessment of the fuzzy quality in psychology, like respiratory sensation and discomfort elicited by added resistive load. But the CCS is quite different. Here the category judgement is regarded as a random event and is based on 2-value logic. Secondly, using MES we can obtain a whole set of data at one go, including the subject's attitude and degree of certainty towards each sensory category. Using CCS, we obtain only a single numerical value for each try, the other data being neglected. Thirdly, with the MES the evaluation is conducted in two dimensions. The CCS, however, is a one-dimensional evaluation scale, which makes it relatively difficult to avoid subjective judgements.

There have been only a few studies concerning the applicability of the

Weber-Fechner's law to mechanical load detection. Investigations have shown that the detection of changes in the intensity of both resistive and elastic load is in accordance with Weber's law (Altose and Cherniack,1980). Our study was principally designed to discover whether we could determine the JND steps and apply Fechner's law to the study of respiratory sensation elicited by changes in intensity of the added resistive load. Our results show that the sensory continuum corresponding to either inspiratory or expiratory resistive loads, ranging from 10 to 500 $mmH_2O \cdot l \cdot s$ consisted of 6 or 7 JND steps. The results also show that the relationship between the number of the JND step $(\hat{Y})$ and either the physical or physiological stimulus (X) could be fitted by an equation, $\hat{Y} = b \ln(X+K)-a$ (a,b and K are parameters), indicating that the relationship is substantially in accordance with a logarithmic function as proposed by Fechner.

Adopting the concept of 'equivalent sensation' as used in the study of other sensory modalities, e.g., the sense of temperature, vibration, hearing etc., we obtained a set of average equivalent respiratory resistive-load-sensation contours. By means of these contours, the sensory magnitude could generally be predicted from both the intensity of the total resistive load added and the inspiratory versus the expiratory load ratio. The equal sensation contours presented in Fig. 6 show that in each defined equivalent sensation condition, the peak-to-peak mouth pressure swing $(\check{P})$, is higher during combined resistive loading than that during single-phase loading, indicating that the sensory threshold is higher during a combined load than a single-phase load. Fig. 6 also shows that physiological stimuli (mouth peak pressure) of equal amplitude elicit a stronger respiratory sensation when they occur in the expiratory phase than when they occur in the inspiratory phase. These results are consistent with the results shown in Fig. 3. Using a cross-modality matching method, Muza et al. (1984) have revealed a similar tendency as shown in their Fig. 3B.

SUMMARY

The physiological and psychophysical effects of combined flow resistive loads were studied systematically by using the Multistage Evaluation Scale (MES), determining the number of Just Noticeable Difference (JND) steps, and constructing equivalent resistive-load-sensation contours. During combined resistive loading, the relative changes of inspiratory and expiratory peak flow rate ($\dot{V}_{Imax}/\dot{V}_{Emax}$) were related to both the JND step number and the inspiratory versus expiratory resistance ratio (IR/ER). However, the ratio $\dot{V}_{Imax}/\dot{V}_{Emax}$ was related to the IR'/ER' ratio only. The subject's perception of combined loads differing both in intensity and IR/ER ratio could be well depicted by the equivalent resistive-load-sensation contours. Hence we could define a model for predicting the physiological and psychophysical effects of various resistive loading conditions.

REFERENCES

Altose, M.D. and Cherniack, N.S. (1980) Respiratory sensation and respiratory muscle activity, in: Advances in Physiological Sciences, Vol. 10, Respiration, eds I.Hutas and L.A. Debreczeni, pp. 111-119, Pergamon Press, Oxford.

Altose, M.D., DiMarco, A.F., Gottfried, S.B. and Strohl, K.P. (1982) The sensation of respiratory muscle force. Am. Rev. Resp. Dis. 126, 807.

Burdon, J.G.W., Killian, K.J, Stubbing, D.G. and Campbell, E.J.M. (1983) Effect of background loads on the perception of added loads to breathing. J. Appl. Physiol. 54, 1222.

Cherniack, N.S. and Altose, M.D. (1981) Respiratory responses to ventilatory loading, in: Regulation of breathing, Part I, ed. T.F. Hornbein, pp. 905-964, Marcel Dekker, New York.

Killian, K.J., Bucens, D.D. and Campbell, E.J.M. (1982) Effect of breathing patterns on the perceived magnitude of added loads to breathing. J. Appl. Physiol. 52, 578.

Ma, M.C., Zhang, L.F., Wu, X.Y. and Jiang, S.Z. (1986) A Fuzzy Set approach to the estimation of respiratory sensation. Acta Psychol. Sinica, 18, 8 (in Chinese).

Muza, S.R., McDonald, S. and Zechman, F.J.W. (1984) Comparison of subjects' perception of inspiratory and expiratory resistance. J. Appl. Physiol. 56, 211.

Pope, H., Holloway, R. and Campbell, E.J.M. (1968) The effects of elastic and resistive loading of inspiration on the breathing of conscious man. Respir. Physiol. 4, 363.

Zhang, L.F., Wu, X.Y., Wang, X.B., Tang, J.A. and Xiao, Z.Y. (1986) The difference threshold steps of respiratory resistive-load-sensation – a discriminatability scale. Chinese J. Appl. Physiol. 2, 197 (in Chinese).

Zhang, L.F., Wu, X.Y., Zhang, R. and Xiao, Z.Y. (1986) The equivalent respiratory resistive-load-sensation contours and their corresponding respiratory patterns in humans. Chinese J. Appl. Physiol. 2, 262 (in Chinese).

EFFECTS OF DOPAMINE AND DOBUTAMINE ON HEMODYNAMICS

AND OXYGEN TRANSPORT IN DOGS WITH PULMONARY EMBOLISM

Kimitaka Tajimi, Hiroyuki Tanaka,
Takeshi Kasai and Kunio Kobayashi

Trauma and Critical Care Center
Teikyo University School of Medicine
11-1, Kaga, 2-chome, Itabashi-ku, Tokyo 173   Japan

INTRODUCTION

Dopamine (DA) and dobutamine (DB) have been widely used as an inotropic agents in patients with circulatory and respiratory problems. In such patients, the most important therapeutic goal is to improve oxygen transport.

Although we have already reported the superiority of DB over DA on oxygenation in patients with acute respiratory failure, the choice of catecholamines to restore oxygen transport is still controversial.

The aim of this study is to compare the effects of DB with those of DA, on systemic and pulmonary circulation and on oxygen transport, in dogs with massive pulmonary embolism (PE).

MATERIALS AND METHODS

Twelve adult mongrel dogs, wighing 14.2±2.7 kg (mean±SD), were anesthetized with 25 mg/kg sodium pentobarbital and intubated with a 8.5 Fr endotracheal tube. A Swan-Ganz catheter was inserted into the pulmonary artery for measurements of mean pulmonary arterial (MPAP) and pulmonary capillary wedge (WP) pressures, and cardiac output (CO). The right femoral artery was cannulated with a polyethylene catheter for measurements of mean arterial pressure (MAP) and heart rate(HR), and for sampling of arterial blood. Right and left femoral veins were also cannulated with a polyethylene catheter for infusion of DA and DB, and Ringer-lactate solution.

CO was determined by the thermodilution method, using a cardiac output computer and taking the mean of three determinations obtained in rapid succession. Cardiac Index (CI), stroke volume index (SVI), systemic vascular resistance index (SVRI) and pulmonary vascular resistance index (PVRI) were calculated by the following formulae:

$$CI = CO/body \ surface \ area \ (l/min \cdot m^2)$$
$$SVI = CI \times 1000/HR \ (ml/beat \cdot m^2)$$
$$SVRI = (MAP - CVP) \times 79.9/CI \ (dyne \cdot sec/cm^5 \cdot m^2)$$
$$PVRI = (MPAP - WP) \times 79.9/CI \ (dyne \cdot sec/cm^5 \cdot m^2)$$

Arterial and mixed venous PO_2, SO_2, pH and Hgb were measured with a blood gas analyzer and co-oximeter.

PE was induced by injection of autologous muscle, which was removed from the femoral muscle and sliced into 2-mm cubes. Muscle cubes were slowly infused until there was a 100% increase in MPAP compared with the baseline value.

Throughout the experiment, animals breathed spontaneously.

After a stabilization period of more than 30 minutes, dogs were divided into two groups, a DB group and a DA group, and hemodynamic and blood gas baseline values obtained. Autologous muscle cubes were then infused to induce PE. The second determination of hemodynamic and blood gas parameters was made 30 minutes after termination of muscle cube infusion. Infusion of either DB or DA was begun immediately after the second measurement. The third, fourth and fifth determinations of hemodynamic and blood gas parameters were performed at 15, 30 and 60 minutes after the start of infusion of 10 μ/kg·min DB or DA infusion.

Data were analyzed by Student's t-test for paired and unpaired samples. Difference between means are considered statistically significant when P<0.05. The data are displayed as mean±SD.

RESULTS

PE significantly decreased SVI in both groups and there were no significant difference between the two groups. PE slightly increased HR and SVRI and significantly increased MPAP and PVRI in both groups, values for the two groups were not significantly different.

Infusion of either DA or DB significantly increased CI, SVI and HR, and significantly decreased SVRI, there were no significant difference between the two groups. Infusion of DB significantly decreased PVRI but DA had no significant effect. These hemodynamic changes are summarized in Table 1.

Changes if CaO_2, CvO_2, $PaCO_2$ and PvO_2 are summarized in Table 2. Infusion of DB significantly increased CaO_2 and CvO_2 from values at PE, DA however did not.

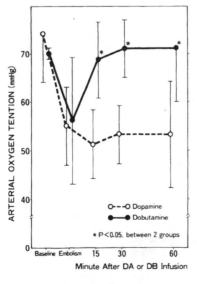

Fig. 1 Changes in PaO₂ by the infusion of dobutamine and dopamine

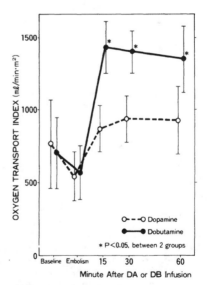

Fig. 2 Changes in oxygen transport by the infusion of dobutamine and dopamine

Table 1. Hemodynamic Responses with Dopamine or Dobutamine After Induction of Pulmonary Embolism

		Baseline	Embolism	Minutes After DA or DB Infusion		
				15	30	60
MAP (mmHg)	DA	140±17	139±18	148±30	150±31	149±20
	DB	137±25	138±22	144±26	149±34	144±34
CI (L/min·m²)	DA	5.50±1.99	4.69±1.65[a]	7.76±1.56[b]	7.86±1.36[b]	7.70±1.40[b]
	DB	5.12±1.56	4.41±1.08	8.36±1.16[b]	8.08±1.08[b]	7.81±0.93[b]
HR (beat/min)	DA	147±27	153±18	177±19	187±22	196±35
	DB	148±29	157±18	194±28[b]	198±17[b]	198±19[b]
SVI (mℓ/beat·m²)	DA	35.3±9.2	29.0±6.9[a]	45.2±10.2[b]	43.4±10.2[b]	40.7±10.4[b]
	DB	36.9±8.5	29.7±7.7[a]	43.3±11.2[b]	40.8±8.8[b]	39.8±5.9[b]
SVRI (dyne·sec/cm⁵·m²)	DA	2222±696	2555±679	1605±528[b]	1583±493[b]	1583±427[b]
	DB	2301±722	2662±945	1423±459[b]	1524±500[b]	1515±479[b]
MPAP (mmHg)	DA	17±6	29±10[a]	42±13	44±17	42±15
	DB	16±5	27±5[a]	30±7	28±7	28±7
PVRI (dyne·sec/cm⁵·m²)	DA	255±184	471±275[a]	465±220	474±231	477±188
	DB	307±151	566±167[a]	324±102[b]	323±100[b]	324±112[b]

a : P<0.05 between baseline and embolism
b : P<0.05 from embolism

Table 2. Changes in CaO_2, CvO_2, PvO_2 and $PaCO_2$ during DA and DB Infusion in Dogs with PE

		Baseline	Embolism	Minutes after DA or DB Infusion		
				15	30	60
CaO_2 (mℓ/dℓ)	DA	13.6±1.5	11.8±1.2	11.3±0.9	12.2±1.8	11.9±2.3
	DB	13.8±2.1	13.1±2.3	17.5±3.0[bc]	17.6±2.5[bc]	17.4±2.7[bc]
CvO_2 (mℓ/dℓ)	DA	10.9±1.3	8.4±1.2[a]	8.9±1.0	9.5±2.0	8.8±1.9
	DB	10.9±2.1	9.8±2.3	14.8±3.1[bc]	14.8±3.0[bc]	14.8±3.0[bc]
PvO_2 (mmHg)	DA	46±4	37±4[a]	41±4	42±4	40±5
	DB	43±3	39±7	50±5[bc]	51±5[bc]	50±4[bc]
$PaCO_2$ (mmHg)	DA	47±9	46±7	52±8	48±6	45±9
	DB	47±7	45±9	44±10	42±9	40±9

a : P<0.05 between baseline and embolism
b : P<0.05 from embolism
c : P<0.05 between the two groups

Figure 1 shows that PE decreased PaO_2, but it was restored to the baseline value by DB infusion. DA however did not improve PaO_2.

Changes in oxygen transport are shown in Figure 2. PE significantly decreased oxygen transport in both groups. Both DA and DB significantly increased oxygen transport compared with the values at PE. However, the increase in oxygen transport was greater with DB than DA and there were significant differences between two groups.

DISCUSSION

Several investigators have compared the hemodynamic effects of DA with those of DB. DA and DB have almost similar effects on systemic circulation in normal dogs.(Kosugi and Tajimi, 1985) In this study, CI and SVI were decreased by inducing PE with concomitant elevation in MPAP and PVRI. The ensuing acute pulmonary hypertension may cause hemodynamic deterioration by limiting right ventricular pump performance. This can be treated by increasing right ventricular contractility, enhancing right ventricular preload and reducing right ventricular afterload. In our study, DB significantly decreased PVRI, but DA did not affect it. Although the pulmonary circulatory reponses deffered between DA and DB, both restored the depressed CI and SVI. This indicates that increasing contractility is the main mechanism by which CI and SVI are improved in our PE model.

Our previous study indicates that pulmonary shunt during DB infusion is lower than that during DA infusion in patients with acute respiratory failure.(Tajimi et al., 1986) In this study, PaO_2 was decreased by induction of PE, which seems to be related to the changes in PVRI. DB improved PaO_2 to the baseline values with concomitant improvement in oxygen transport. On the other hand, DA did not improve PaO_2. Changes in PaO_2 after DA or DB infusion in both groups seem to be related to the changes in PVRI. DB infusion may be associated with a better ventilation-perfusion relationship than DA. Although both DA and DB increased CI and SVI to a similar level, DB improved oxygen transport more than DA.

We conclude that reduction in oxygen transport due to acute increase of PVRI with PE in dogs can be improved by infusion of DB and that DB may be the treatment of choice to restore oxygen transport in acute pulmonary hypertension.

REFERENCES

I. Kosugi and K. Tajimi: Effects of dopamine and dobutamine on hemodynamics and plasma catecholamine levels during severe lactic acid adidosis. Circ. Shock 17:95-102, 1985.

K. Tajimi, M. Watanabe, T. Kasai, K. Kobayashi: Effects of dopamine and dobutamine on pulmonary shunt. Abstracts of the 11th ANZICS at Tasmania, Australia, 1986.

HOME OXYGEN THERAPY

Niels Lund

Anesthesiology Department
University of Rochester
Rochester, New York

History records that oxygen was first used for treatment purposes in the United States on March 6, 1887, by Dr George E Holzapple. He used it for a teenager with pneumococcal pneumonia, generating oxygen from chlorate of potassium and black oxide of manganese, producing the life-giving gas in large test tubes heated by a lamp. Holzapple observed the resolution of cyanosis and the reduction in tachypnea in a matter of minutes.

The mid-60's brought us practical portable oxygen therapy for the first time. The role of oxygen in correcting cor pulmonale heart failure via a reduction in pulmonary hypertension along with the reversal of accompanying erythrocytosis was documented by Levine et al in 1967.

In the late 1980s, advances include the technique of transtracheal oxygen therapy, other semi-concealed oxygen delivery systems and conserving devices. Further studies have improved our knowledge of the type of patient most apt to benefit from home oxygen. Improvements in oxygen concentrator technology includes the possibility of a portable concentrator. Developments in home mechanical ventilation offer the opportunity of home care services for patients with both chronic lung diseases and neuromuscular disorders, who otherwise would either die or survive only in hospitals or special care facilities.

BASIS OF OXYGEN THERAPY

Hypoxemia

Those of us who are clinicians were taught in medical school that among COPD patients we would find "pink puffers" who were thought to suffer from emphysema, and "blue bloaters" who suffered from chronic bronchitis. The difference between these two types, we now know, is not necessarily that the blue bloater suffers a different disease, but that, in addition to being hypoxemic, he also suffers from significant pulmonary vascular resistance and pulmonary hypertension.

Arterial oxygen pressure (p_aO_2) is determined by the inspired oxygen pressure (p_iO_2), by the level of alveolar ventilation, and by the distribution of ventilation and perfusion in the lungs. Therefore, arterial hypoxemia occurs at high altitude because of the decreased p_iO_2, when hypoventilation increases the alveolar CO_2 pressure (p_ACO_2) [thereby decreasing the alveolar O_2 pressure (p_AO_2)], and when pulmonary or cardiac diseases alter ventilation/perfusion (V/Q) distribution.

The increase in p_aO_2, which results from therapeutically increasing p_AO_2, depends on the magnitude of the V/Q mismatch, and ranges from maximum improvement when there is no intrapulmonary shunting to no improvement when intrapulmonary shunting reaches 50% of cardiac output. The greatest benefit can be expected when there are few lung regions with very low V/Q ratios.

Oxygen Transport

Oxygen transport to the tissues involves multiple factors, which include the p_aO_2, hemoglobin concentration, oxyhemoglobin dissociation curve shape and position, cardiac output, and individual organ perfusion. When hemoglobin is deficient, it is usually taught that an increase in FiO_2 from 0.21 to 1.0 can increase dissolved oxygen to deliver approximately one-third of the resting tissue requirements. This is, in my opinion, doubtful (and will be touched upon later). I would prefer to increase circulating hemoglobin to at least 10g/100ml in order not to increase the FiO_2 to toxic levels.

The effect of a decrease in cardiac output on oxygen delivery can not be predicted by simple mathematic projection. Autoregulation of flow within and between organs provides partial compensation, although this may be impaired in disease (e.g. septicemia). In addition, a decrease in cardiac output usually causes a decrease in venous admixture, which offsets to a variable extent the adverse effect of a low mixed venous partial pressure of oxygen on this admixture.

Evidence of Hypoxia

Since clinically and routinely useful measurements of tissue oxygen pressures are not presently feasible, evidence of hypoxia must be based on assumptions derived from evaluations of oxygen delivery, of mixed venous pO_2, or of vital organ function (brain, heart, kidneys). We must not forget that hypoxemia may have a particularly deleterious effect on the respiratory muscles when the work of breathing is increased.

At p_aO_2 levels below 50 mm Hg, some degree of hypoxia can be assumed. At higher levels of p_aO_2, hypoxia due to defects in hemoglobin concentration or function or in perfusion may occur. The pO_2 of venous blood from an organ may be much higher than the pO_2 within many cells of that organ, which are distant from their arterioles and capillaries. Failure of autoregulation is another mechanism whereby venous pO_2 may not reflect tissue oxygen pressures. Mixed central venous pO_2, measured in the pulmonary artery, only grossly reflects total body tissue status and may be seriously misleading when interpreting what is happening in individual tissues and organs.

Effects of Hypoxia

A decreased oxygen supply impairs mitochondrial function and ultimately causes necrosis. Reversible anaerobic glycolysis occurs, and lactate/pyruvate ratio increases. Similar effects occur when oxygen utilization is impaired by, e.g., cyanide. Since oxygen diffuses through tissues by a pressure gradient mechanism, there is no easy way to define a specific tissue pO_2 at which cellular damage occurs.

In disease states, it is often difficult to separate the effects of hypoxia or hypercarbia on the whole patient from the symptoms or signs of the underlying disease. The effects of <u>acute</u> hypoxia have therefore been studied mostly in normal persons. In persons with acute hypoxemia, brain function is compromised in advance of malfunction of other vital organs. When the pO_2 approximates 55 mm Hg short-term memory is altered, and euphoria and impairment of judgment may occur. As hypoxemia worsens there is progressive loss of cognitive and motor functions, and loss of consciousness may occur at a p_aO_2 approximating 30 mm Hg. The heart initially responds to acute hypoxemia with tachycardia and increased stroke volume, both of which increase cardiac output and maintain oxygen delivery. As hypoxemia worsens, myocardial function begins to fail and disturbances of cardiac rhythm tend to occur. The lungs will respond to a decreased p_AO_2 with vasoconstriction and bronchoconstriction.

In <u>chronic</u> hypoxemic states, additional adaptive mechanisms, which tend to maintain oxygen transport, come into play. Increased ventilatory drive occurring at p_aO_2 levels below 55 mm Hg produces hypocapnia and, thus, increases alveolar and arterial pO_2; this change occurs extremely rapidly. An increase in erythrocyte 2,3-DPG causes a rightward shift of the oxyhemoglobin dissociation curve and increases oxygen availability to the tissues; this adaptation occurs over a period of hours to days. A slower adaptation to hypoxemia is the secretion of erythropoietin which stimulates the bone marrow and produces erythrocytemia.

CHRONICALLY HYPOXEMIC PULMONARY DISEASES

COPD may be considered the prototype of the chronic hypoxemic lung diseases, and most of the available data for efficacy of oxygen therapy come from studies on this patient group. Initial oxygen therapy studies in COPD showed that continuous supplemental oxygen for 4 to 8 weeks reduced elevated hematocrits, decreased pulmonary vascular pressures, and improved exercise tolerance.

Two hallmark studies were performed in the late 1970s: the British Medical Research Council Domiciliary study and the American Nocturnal Oxygen Therapy Trial (NOTT) study. The NOTT study compared nocturnal oxygen with continuous oxygen therapy, and the British study compared nocturnal oxygen with no oxygen therapy at all in COPD patients. The mortality rate was reduced in the nocturnal oxygen group compared with the no oxygen group in the British study, and was reduced by almost twofold in the continuous as compared with the nocturnal group in the NOTT study. Although the two groups were not exactly comparable, it appears that nocturnal oxygen is better than no oxygen and that continuous oxygen is better than nocturnal oxygen therapy. Both trials showed a reduction in hematocrit and reduced pulmonary vascular pressures; the latter changes were, however, not statistically significant. In addition, the NOTT study subgroups showing a high partial pressure of p_aCO_2, elevated hematocrit, elevated pulmonary artery pressure, or acidosis appeared to derive the most benefit from continuous as opposed to nocturnal oxygen.

Brain function is compromised in advance of malfunction of other organs. Neuropsychologic evaluation in the NOTT study showed improvement in most tests performed, and in the quality of life when all patients were considered, regardless of the number of hours of oxygen therapy.

INDICATIONS FOR OXYGEN THERAPY

While the scientific foundation for oxygen therapy, in carefully selected cases, is increasing, it is still incomplete.

Oxygen Therapy for Chronic Conditions

Adequate data for the efficacy of ambulatory oxygen therapy exist only for COPD. Before the initiation of long-term oxygen therapy, all four of the following conditions should be met:

1) An accurate, current diagnosis must have been established.

2) An optimal medical regimen prescribed by a physician knowledgeable in chest diseases must be in effect.

3) The patient should have recovered from any exarcebation, and should have been in a stable state for approx. 1 month. It is important to note that the need for stability before long-term oxygen therapy is begun does not exclude short-term (1 to 30 days) oxygen therapy, especially if the latter therapy allows the patient to be safely discharged from the hospital sooner.

4) Oxygen therapy has been shown to, or can reasonably be predicted to:

   a) improve the hypoxemia or evidence of tissue hypoxia, and

   b) provide overall clinical benefit.

Institution of Long-Term Oxygen Therapy

Long-term oxygen therapy should be considered only for those patients who have been on an optimal regimen for 30 or more days. When such patients have an arterial pO_2 of 55 mm Hg, or less, hypoxic organ dysfunction may be considered present, and long-term oxygen therapy may be prescribed.

Patients on optimal medical regimens with p_aO_2 values greater than 55 mm Hg, may have evidence of organ dysfunction such as secondary pulmonary hypertension, cor pulmonale, secondary erythrocytosis, and impaired Omentation or other central nervous system dysfunction. Such individuals should be considered for long-term oxygen therapy.

Smoking by patients receiving oxygen therapy has inherent safety risks, and some cases of oxygen therapy fires have been reported.

Dosage Guidelines

Oxygen should be given at a dosage to alleviate the hypoxemia (e.g. to increase the p_aO_2 to 60 to 80 mm Hg) and/or the deleterious effects of end-organ dysfunction. This goal can usually be accomplished by administering low-flow concentration of oxygen by, e.g., nasal cannulas at 4 l/min. The amount of oxygen in liters per minute should, of course, be adjusted to individual patient needs. Patients with continuous hypoxemia benefit maximally from continuous oxygen therapy 24 hrs/day, while patients with intermittent hypoxic states (e.g during exercise, sleep, or air travel) may benefit from intermittent therapy adjusted to relieve these hypoxemic periods.

HOME OXYGEN DELIVERY SYSTEMS

A wide variety of oxygen delivery systems are available for hospital and home use. With the event of DRGs in the USA, it is clear that more emphasis will be placed on home and outpatient care.

Oxygen for home use is relatively expensive when supplied in cylinders or liquid form. Oxygen concentrators, which may be less expensive, require periodic maintenance for maximum effectiveness and are not portable.

The devices that interface directly with the upper airway, such as catheters, cannulas, masks, are often uncomfortable when used for prolonged periods. They are also easily displaced or removed, and the FiO_2 delivered to the trachea is often unpredictable.

Up-to-date methods for delivering oxygen are:

1) Intermittent flow devices. which deliver oxygen only during inspiration triggered either by temperature or negative pressure.

2) Continuous flow reservoir cannulas which use a partial rebreathing of oxygen.

3) Transtracheal catheters which deliver oxygen continuously in the lower part of the trachea allowing the rebreathing of oxygen accumulated during exhalation in the upper part of airways. The intratracheal catheter is permanently implanted under local anesthesia. It is tunnelated subcutaneously and exits the skin below the breast. Thus, it is easily hidden by clothing, and forms a closed system.

Besides its aesthetic, a transtracheal catheter has other advantages:

1) It remains efficient whichever the patterns of breathing are (open mouth, speaking, exercise, sleep).

2) It may be coupled with an intermittent device allowing oxygen savings of up to 70%.

Blood gases with a transtracheal catheter are equivalent to those attained using at least twice as much oxygen delivered nasally. Infective exacerbations of disease become relatively infrequent as daily bronchial lavage improves pulmonary hygiene.

Home, Domiciliary and Portable Systems

There are three major sources of supplemental oxygen for home usage: oxygen cylinders, liquid oxygen and oxygen concentrators.

1) Oxygen Cylinders. The large cylinder most commonly used is the H cylinder with a capacity of 6,900 liters. At a flow rate of 2 l/min this will last for 56 hours. Smaller cylinders, e.g. size E (622 l), can be used for portability, and lasts 5.1 hours at 2 l/min flow. Light weight aluminum cylinders are available.

2) Liquid Oxygen. One cubic foot of liquid oxygen is the equivalent of 860 cubic feet (24,368 l) of gaseous oxygen. Reservoirs usually contain 40 to 90 pounds of oxygen and will last for 4 1/2 to 10 days with a continuous flow of 2 l/min. Liquid oxygen devices are provided with an ambulatory system. These 6 1/2 to 11 lb units are easily

filled from the larger reservoir and are carried like a shoulder bag. At flows of 1 to 3 l/min a patient may be mobile for 3 to 8 hours. Both the reservoirs and the portable units vent gaseous oxygen at a rate of at least 1 lb per day if the system is not used. Newer systems will not vent oxygen if a flow greater than 1 l/min is used continuously.

3) Oxygen Concentrators and Enrichers

a) Molecular sieve concentrators remove nitrogen and water from air. They force ambient air over a bed of inorganic sodium-aluminum silicate pellets which binds nitrogen, water vapor and trace gases such as carbon monoxide, carbon dioxide and all hydrocarbons. Two such beds are alternated; while one is used the other is back-flushed and revitalized. Newer models are capable of delivering 85% to 90% oxygen at flow rates of 1 to 4 l/min. This type of device requires supplemental humidification. Average weight 40 to 50 lb.

b) Membrane diffusion enrichers employ a semipermeable membrane that permits oxygen and water vapor to diffuse faster than other gases, primarily nitrogen. Membrane enrichers deliver 30% to 40% oxygen at flow rates of 1 to 10 l/min. Because of the relatively low oxygen concentration the gas flowrate must be doubled or tripled to equal oxygen delivery from a system that delivers 100% oxygen. An enricher system weighs approximately 110 lb. A new portable oxygen enricher was presented in 1987 by Akutsu and co-workers. They have developed a new polymer membrane characterized by an oxygen permeability approx. 20 times higher than that of poly (dimethylsiloxane), which had the highest permeability till now. Their enricher weighs 6.8 kg, so we now have a portable enricher.

Oxygen for air travel: commercial aircraft are pressurized to maintain atmospheres that are equivalent to 8,500 feet or less. All airlines require prior notice of air travel, and personal oxygen equipment is not allowed. Precise therapy usually cannot be guaranteed.

OXYGEN TOXICITY

The therapeutic use of oxygen entails the risk of adverse reactions, and its use should be based on an assessment of the potential toxicity versus therapeutic benefit.

Three categories of hazards have classically and clinically been associated with oxygen therapy:

1) Physical risks: fire hazard, tank explosions, trauma from catheters or masks, drying of mucous membranes

2) Functional effects: carbon dioxide retention, atelectasis

3) Cytotoxic manifestations, and,

4) Effects on microcirculatory distribution of blood flow.

Acute pulmonary syndromes associated with oxygen-induced toxicity include tracheobronchitis and ARDS, chronic syndromes are, e.g., bronchopulmonary dysplasia in the neonate and a similar but ill-defined syndrome in adults.

I do not intend to deal with functional effects in this presentation. Also, we all know the free radical theory of oxygen toxicity which states that an increased rate of generation of partially reduced oxygen products is responsible for the cytotoxicity of oxygen.

I would like to dwell for a short moment on the microcirculatory effects of increased oxygen levels. I am well aware of the fact that, in the COPD patient we are not usually dealing with increased arterial oxygen levels, and by increased I mean a p_aO_2 above, say, 135mm Hg. However, nobody knows whether lower than 135 mm Hg p_aO_2 levels in a COPD-patient adapted to very low everyday oxygen levels may be high for that specific patient, and that he may, therefore, experience oxygen toxicity, e.g., at a p_aO_2 of 100 mm Hg. Neither do we know whether the patient is overdosing his oxygen at home or not, thereby increasing his p_aO_2 to normally toxic levels of >130 mmHg.

Looking at the lungs first, it has been shown by several authors that increasing the FiO_2 to 1.0 will increase the intrapulmonary shunting, thus leading to a lower p_aO_2. Relatively recent data support the idea that an FiO_2 > 0.6 will increase intrapulmonary shunting. This is not to say that 100% oxygen should not be used acutely in emergency situations, but that it should never be used chronically.

We have data from humans (skeletal muscle and brain) that show abnormalities in tissue oxygenation when p_aO_2 is increased above 130 mm Hg. From animal studies we have data showing the same results of high arterial oxygen levels on liver, heart and other organs.

Apart from the effects of absolute arterial pO_2 levels, there is also a time factor involved, the extent of which we do not know yet. After acute exposure to high oxygen levels it takes a very long time for the tissue oxygen levels to come back to normal, actually we do not know how long!

In 1984 the American College of Chest Physicians published a report on oxygen therapy in which they state: (Oxygen) "Damage to organs other than the lung is not considered to be of major clinical importance in the adult". I must admit that I would also worry a little bit about my brain and heart!

Tolerance

Many factors appear to be involved in producing a tolerant state, including antioxidant defenses, age, nutrition, and hormonal influences. In animal studies increased tolerance to oxygen is strongly associated with elevated lung superoxide dismutase (SOD), catalase, glutathione peroxidase and other enzymes.

Deficiency of nutritional factors such as vitamins E and C, proteins (especially those rich in sulfur-containing amino acids), and some trace elements (e.g., selenium) decreases tolerance to oxygen in experimental studies.

Other factors known to have detrimental effects on tolerance in experimental oxygen toxicity include hyperthyroidism, elevated glucocorticoid levels and hypermetabolic states.

Oxygen also interacts with drugs and toxins to increase production of free oxygen radicals. Paraquat (the herbicide) is actively transported into lung cells and produces superoxide anions. Bleomycin, adriamycin, daunorubicin, and antibiotics that depend on quinoid groups for activity are all able to generate oxygen radicals.

The only effective treatment for oxygen toxicity is prevention reduce the FiO_2 to the lowest possible level for the shortest time necessary to achieve adequate tissue oxygenation.

REFERENCES

Anthonisen NR. Long-Term Oxygen Therapy. Ann Intern Med 1983;99:519-27.

Cooper CB, Waterhouse J, Howard P. Twelve year clinical study of patients with hypoxic cor pulmonale given long term domiciliary oxygen therapy. Thorax 1987;42:105-10.

Cry fire! Oxygen therapy. Respir Care 1976;21:1139-40.

Gjerde GE, Kraemer R. An oxygen therapy fire. Respir Care 1980;25:362-3.

Levine BE, Bigelow DB, Hamstra RD, et al. The role of long-term continuous oxygen administration in patients with chronic airway obstruction and hypoxemia. Ann Intern Med 1967;66:639-50.

Mathews PJ, Jr. Home Oxygen Delivery Systems. Curr Rev Respir Ther 1985;8:1-8.

McKeon JL, Saunders NA, Murree-Allen K. Domiciliary oxygen: rationalization of supply in the Hunter region from 1982-1986. Med J Austr 1987;146:73-7.

Medical Research Council Working Party. Long term domiciliary oxygen therapy in chronic hypoxic cor pulmonale complicating chronic bronchitis and emphysema. Lancet 1981;1:681-6.

Nocturnal Oxygen Therapy Trial Group. Continuous or nocturnal oxygen therapy in hypoxemic chronic obstructive lung disease: a clinical trial. Ann Intern Med 1980;93:391-8.

CAN BLOOD GAS ANALYSIS INDICATE WHEN MECHANICAL VENTILATION

SHOULD START IN PATIENTS WITH ACUTE MYOCARDIAL INFARCTION ?

Izumi Matsubara[*], Osamu Kemmotsu,[**] Ichiro Tedo[*],
Satoshi Gando[*], and Hirofumi Tsujinaga[*]

[*]Department of Critical Care Medicine, Sapporo
City Hospital, Sapporo, Japan
[**]Department of Anesthesiology, Hokkaido
University School of Medicine, Sapporo, Japan

INTRODUCTION

Pulmonary edema caused by heart failure is relieved by improvement of cardiac function and respiratory failure is improved by oxygen therapy in most cases. But in cases of severe hypoxemia due to decreased oxygen capacity in the lung, and when hypoxemia is not relieved by oxygen therapy alone, it is necessary to apply mechanical ventilation in order to improve alveolar ventilation and oxygenation in the lung. In patients with hypoxemia due to pulmonary edema caused by heart failure as a result of acute myocardial infarction, PaO_2, $PaCO_2$, and B.E. (base excess) were measured in each patient and the results used to help in deciding whether oxygen therapy with or without mechanical ventilation should be used.

METHODS

The subjects of this study were 69 patients with myocardial infarction who underwent blood gas analysis in the last three years. The patients were divided into two groups: group A received oxygen therapy only (n = 38) and Group B had oxygen with mechanical ventilation (n = 31). Group A included 33 males and 5 females with an average age 61.9 ± 11.3 years (mean ± SD). Group B included 26 males and 5 females with an average age 68.0 ± 11.4 years. Group B has been subdivided into surviving cases (n = 14) and fatal cases (n = 17).

The oxygen content of the breathing gas varied from patient to patient. The inhaled oxygen fraction, FiO_2, was 0.21 in room air while that given by venturi mask varied in different cases. With the nasal tube FiO_2 values were 0.28, 0.32, and 0.36 corresponding to 2.0, 3.0 and 4.0 ℓ/min. Mechanical ventilation was started in those patients (Group B) who failed to achieve an adequate PaO_2 on oxygen therapy alone.

Blood gas analysis was done 30 minutes after initiation of either oxygen alone or oxygen with mechanical ventilation. The t-test was applied to compare the two groups and when P values were less than 0.05 they were judged to be statistically significant.

RESULTS

1) Comparison between group A who received oxygen therapy only and group B who received both oxygen and mechanical ventilation

In patients with acute myocardial infarction, blood gas analysis was performed as they breathed room air, before either oxygen therapy or mechanical ventilation was started. The B.E. was -3.5 mEq/ℓ in group B, compared with -0.4 mEq/ℓ in group A (Figure 1). The $PaCO_2$ was equally low in both groups but PaO_2 was significantly lower in group B (62.6 mm Hg) than in group A (81.4 mm Hg) (Figure 1).

After oxygen therapy alone, patients in group A showed increases in PaO_2, but group B did not demonstrate any significant rise in PaO_2 (Figure 2). Figure 3 compares the ratio PaO_2/FiO_2 (P/F) values, which serve as an index of oxygenation capacity in the lung, in the two groups. The P/F value was still low even after oxygen therapy in both groups but was within the normal range in group A. On the other hand, the P/F value was extremely low (199.6) after oxygen therapy alone in group B.

In the cases undergoing evaluation of cardiac function by Swan-Ganz catheterization (26 cases in group A and 14 cases in group B), mean values were plotted according to the Forrester classification (Figure 4). Group A were in the H-I area and Group B were at the border between H-II and H-IV which indicates a decrease in cardiac function.

2) Comparison between the surviving and fatal cases in Group B

In group B (the 31 patients treated with oxygen and mechanical ventilation), PaO_2 was not improved by oxygen therapy alone and the P/F value fell. Of these cases 14 survived and 17 died. The blood gas analysis on admission of the fatal cases indicated significantly lower $PaCO_2$ due to hyperventilation and tended to show metabolic acidosis (Figure 5). The PaO_2 values in the two groups were not significantly different, being 64.5 mm Hg in the surviving cases and 62.1 mm Hg in the fatal cases (Figure 5). Figure 6 shows the value of blood gas analysis just before mechanical ventilation. There was no differences in PaO_2 between surviving and fatal cases but B.E. tended to be lower in the fatal cases ($p < 0.05$). The P/F values were low in both cases with no significant difference between them (Figure 6).

Just after mechanical ventilation, B.E. was still significantly lower in the fatal cases than in the surviving patients and the P/F values were improved in the surviving patients but not in the fatal ones (Figure 7). An increase in cardiac index and an improvement in the P/F value after mechanical ventilation were seen in the surviving patients but not in those who died (Figure 8).

DISCUSSION

Maneuvers to improve cardiac function are a first step in the treatment of pulmonary edema due to heart failure in acute myocardial infarction. However, it takes time to improve cardiac function in most cases, especially in the cases of acute myocardial infarction where damage to the heart muscle delays the restoration of cardiac contractility. In respiratory failure associated with pulmonary edema, PEEP (positive and expiratory pressure) is required with the aim of increasing the functional residual capacity and, in many cases, decreasing the pulmonary shunt. It is often reported that PEEP is effective in improving cardiac function as well as of oxygenation in the lung.

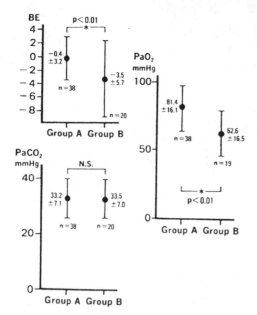

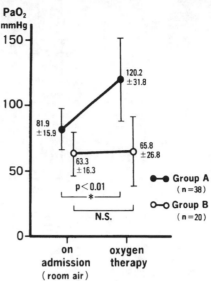

Fig. 1. Blood gas analysis on admission, (spontaneous respiration, room air).

Fig. 2. Changes in PaO_2 after oxygen therapy alone.

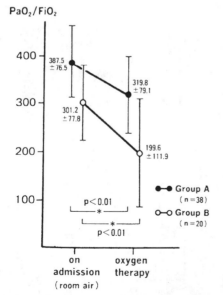

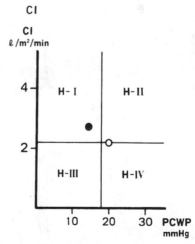

Fig. 3. Comparison of the P/F values, between oxygen therapy group and mechanical ventilation group.

Fig. 4. Comparison of cardiac index and PCWP between the oxygen therapy group and mechanical ventilation group.

● Group A   ○ Group B

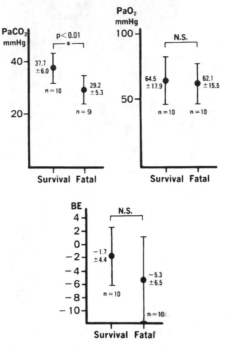

Fig. 5. In blood gas analysis on admission, surviving cases and fatal cases in mechanical ventilation.

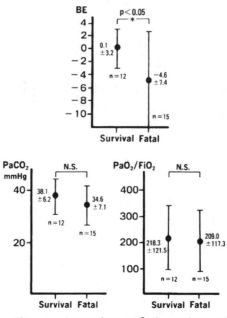

Fig. 6. Comparison of the values of blood gas analysis just before mechanical ventilation.

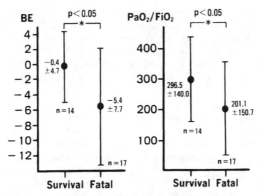

Fig. 7. Blood gas analysis just after
mechanical ventilation.

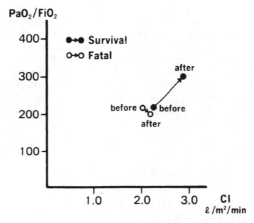

Fig. 8. Changes in the mean cardiac
index and P/F value before
and after mechanical ventilation.

In our study, the patients who were given mechanical ventilation were hypoxemic at the time of admission and the hypoxemia was not improved by oxygen therapy alone. It was concluded from the decrease in cardiac index that severe heart failure existed in these patients. Comparing the surviving and fatal cases, the fatal cases showed lower B.E. even after mechanical ventilation which suggests that cardiac function was not effectively improved. There was no difference in the P/F value between the surviving and non-surviving cases prior to mechanical ventilation but after mechanical ventilation, the P/F value was significantly increased in the survivors indicating an improvement in the oxygenation capacity in the lung. The cardiac index was improved in the surviving but not in the non-surviving patients. These results may indicate that cardio-pulmonary failure was not reversed, even by mechanical ventilation, in the fatal cases.

Therefore, patients with myocardial infarction who are severely hypoxemic on admission, and do not show improved oxygenation in the lung with oxygen therapy alone, should be given mechanical ventilation from an early stage of respiratory failure, to improve alveolar ventilation and oxygenation in the lung. When the P/F value, which indicates oxygenation capacity in the lung, is not increased by oxygen therapy alone, mechanical ventilation should be started. It would seem reasonable to assume that mechanical ventilation is indicated if the P/F value drops below 250.

Although there was no difference in the P/F value between the surviving and nonsurviving cases before mechanical ventilation, the P/F values rose to 296.5 in the survivors, which is significantly higher than the value of 201.1 in those patients who died. Therefore, improvement in the P/F value after mechanical ventilation seems to provide an index to predict the outcome of respiratory failure.

In acute myocardial infarction, the initial management is aimed at reducing cardiac load, by absolute bed rest and/or the administration of vasodilator drugs. While great efforts are made to improve cardiac function if cardiac failure develops, pulmonary care is often neglected. However, intensive respiratory management with mechanical ventilation often becomes a life-saving maneuver. In our experience, many cases showed no improvement of PaO_2, B.E. and P/F value with oxygen therapy alone and respiratory care with mechanical ventilation was necessary together with circulatory care.

CONCLUSION

When severe hypoxemia exists in myocardial infarction, the improvement of oxygenation capacity in the lung should be achieved by mechanical ventilation. Arterial blood gas analysis can indicate when the mechanical ventilation should be initiated. From our result, if the ratio of PaO_2/FiO_2 (P/F value) drops below 250, mechanical ventilation should be initiated.

SUMMARY

Arterial blood gas analysis of 69 patients with acute myocardial infarction were evaluated to provide a basis for respiratory care. Patients were divided into two groups: group A which received oxygen therapy only (n = 38) and group B which received oxygen therapy with mechanical ventilation (n = 31). The patients in group B were further divided into surviving cases (n = 14) and fatal cases (n = 17). On admission patients assigned to group B had lower PaO_2 values than those

placed in group A. In group B, there was no difference in the P/F value
before mechanical ventilation of the surviving and the fatal cases, but
the survivors demonstrated an improvement of the P/F value and an increase
in cardiac index after mechanical ventilation. It may be reasonable to
assume that a P/F value of less than 250 serves as an indicator for the
initiation of mechanical ventilation. An increase in the P/F value after
mechanical ventilation seems to be a valuable index to estimate prognosis
in respiratory failure.

REFERENCES

    Araki, Y., Hanamoto, S., Kho, M., 1986, Blood-Gas changes and
respiratory care in patients with acute myocardial infarction, Journal
of Intensive Care Medicine, 10 (No. 6):513.

    Räsänen, J., Nikki, P., and Heikkilä, J., 1984, Acute myocardial
infarction complicated by respiratory failure, Chest, 85:21.

    Räsänen, J., Väisänen, I.T., and Heikkilä, J., 1985, Acute
myocardial infarction complicated by left ventricular dysfunction and
respiratory failure. The effects of continuous positive airway
pressure, Chest, 87:158.

    Soma, K., and Kemmotsu, O., 1984, Management of pulmonary edema
based on differential diagnosis of cardiogenic and noncardiogenic
forms, Circulation Control, 5 (No. 2):167.

A STUDY OF OXYGEN THERAPY IN PATIENTS WITH CHRONIC RESPIRATORY FAILURE

T.Y. Ni, S.S. Hahn and S.P. Tsen

Research Section of Pulmonary Diseases
No. 2 Hospital, Harbin Medical University
Harbin, The People's Republic of China

INTRODUCTION

The purpose of oxygen therapy is to enhance the diffusing capacity of oxygen by raising the P_aO_2 so that S_aO_2, O_2 content and O_2 supply can be increased and hypoxemia relieved. But since oxygen therapy cannot replace ventilation how can hypercapnia be prevented? This paper reports studies on 245 cases of chronic respiratory failure due to chronic heart disease and elaborates the principle of oxygen therapy in such cases.

METHODS

1. The patients were grouped according to the severity of their hypoxemia with respiratory failure:
   (1) 1st group (64 cases): slight hypoxemia (P_aO_2: 51-60 mmHg). Oxygen was not administered.
   (2) 2nd group (80 cases): moderate hypoxemia (P_aO_2: 36-50 mmHg). These patients received controlled oxygen therapy by nasal tube for 2-4 weeks.
   (3) 3rd group (90 cases): severe hypoxemia (P_aO_2:<36 mmHg). These patients also received controlled oxygen therapy for 2-4 weeks, by nasal tube or Venturi mask.
   (4) 4th group (11 cases). These patients had moderate or severe hypoxemia but refused to accept oxygen therapy.

2. During and after oxygen administration, arterial blood gases were analysed with a BMS_3-MK_2 Radiometer. Measurements were made 5-6 times in the 1st and 4th groups during hospitalization, and 5-20 times in the 2nd and 3rd groups.

3. All 4 groups received almost the same general treatment including:
   (1) Intravenous infusion of antibiotics (Penicillin + Streptomycin, Erythromycin + Chloromycin, Leucomycin, Ampicillin, Amoxycillin or Cephatazidime etc).
   (2) Improvement of respiratory function by: removal of bronchial secretions using bronchodilators such as aminophylline + 5% glucose i.v., Salbutamol by mouth and aminophylline-co by mouth; moisturization of the respiratory tract with a nebulizer; mechanical ventilation if necessary.
   (3) Reduction of pulmonary hypertension by intravenous Regitine.

Table 1   Results of Oxygen Therapy with Fifteen Typical Patients

Case	Before O_2-therapy P_aO_2	Before O_2-therapy P_aCO_2	Methods of O_2 therapy	Comprehensive therapy	Resp. stimulants	During Hosp. P_aO_2	During Hosp. P_aCO_2	Before discharge P_aO_2	Before discharge P_aCO_2	Note
1	56.7	53.5	no	+	–			64.1	45	Slight HXA needs no O_2 therapy
2	58.5	56.8	no	+	Ventilator			69.9	47.4	Ventilator is beneficial to respiratory failure
3	49.4	57.1	24% O_2 (3 w.)	+	–			57.4	53.1	Moderate HXA needs O_2-therapy
4	40.1	74.2	24% O_2 (3 w.)	++	Ventilator			66.1	55	If P_aCO_2 is 70 mmHg, ventilator, powerful comprehensive therapy and resp. stimulants are indispensable
5	41.7	72.5	24% O_2	++	+	54.4	61.4	67.2	52.5	After O_2 administration for 3 weeks, P_aO_2↗, P_aCO_2↘
6	44.8	64.7	24-32% O_2 for 3 days	+	–	45.1	60.2			P_aO_2 goes up very slowly.
			42% O_2 for 3 days			47.5	57.1			A V–A shunt more than 30% is indicated
7	42.5	47.0	24% O_2 7 days	+	–	67.2	41.4			Very good results; suggests lung fibrosis
8	45.7	67.2	42% O_2 3 hrs	+	–	67.9	79.8			P_aCO_2 increases to 79.8, patient falls into CO_2-narcosis
9	39.4	65.4	24% O_2 3 hrs	+	–	48.4	69.1			P_aO_2 increases obviously, but P_aCO_2 goes up to 69.1
			28% O_2 4 hrs	+	R.S.	57.6	67.2			When powerful comprehensive therapy and resp. stimulants are given, P_aCO_2 drops.
10	35.7	74.1	24% O_2 4 hrs	++	–	47.2	73.7			When respiratory stimulants are given, P_aCO_2 drops.
			28% O_2 3 hrs	+++	+	57.6	67.2			
11	46.1	60.3	24% O_2 discontinuous	+	–	50.7	68.1			Result unsatisfactory
			24% O_2 continuous					58.4	51.7	When O_2 is given continuously, the result becomes satisfactory
12	32.9	60.7	24% O_2 10 days	+	–	43.4	65.2	36.4	64.0	Withdrawal of O_2 after brief O_2-inspiration, P_aO_2 drops
13	34.2	67.9	24% O_2 (2-3 w.)	+	–	52.4	57.7	67.8	51.4	Effect is better c̄ 3 weeks-O_2 therapy
14	32.1	65.7	24% O_2 (3-4 w.)	+	–	62.8	53.1	64.5	49.1	Effect is similar c̄ 4 weeks-O_2 therapy
15	37.4	64.5	24% O_2 (3 w.), then 12-15 hrs daily over 1 year	++	–			68.2	48.2	When respiratory failure is relieved, 12-15 hrs O_2-therapy is beneficial to patients to reduce pulm. hypertension & improve tolerance for exercise.

RESULTS

Typical results for 15 patients representing the 4 groups are summarized in Table 1.

1. The first group of patients, with slight hypoxemia, did not need any O_2-therapy. After effective general treatment, the alveolar ventilation improved and P_aO_2 went up (Case 1). Recovery could be hastened by the use of a ventilator (Case 2).

2. In patients with moderate to severe respiratory failure, O_2-therapy was essential (Cases 3 and 13).

3. When the concentration of the inspired O_2 was below 35%, this was referred to as 'controlled O_2-therapy' (Cases 3,7,13 and 14). If patients were given 32% O_2 for 3 days and the increase in P_aO_2 was less than 1/3 of the P_aO_2 before treatment, and the P_aCO_2 was below 65 mmHg, 38 or 40% O_2 was tried. If P_aO_2 still did not go up, this indicated that there might be a V-A shunt of more than 30% (Case 6).

4. If P_aCO_2 was over 60 mmHg before oxygen inhalation, the use of mechanical ventilation or respiratory stimulants was advisable and the bronchodilatory drugs had to be increased (Cases 9 and 10).

5. When P_aCO_2 was over 70 mmHg before oxygen-treatment, powerful general therapy, respiratory stimulants and mechanical ventilation were all indispensable (Cases 4 and 5).

6. Oxygen had to be given continuously day and night, otherwise the P_aO_2 fell when oxygen was withdrawn (Cases 11 and 12).

7. The course of O_2-therapy had to be continued for at least 3 weeks, otherwise the effect was unsatisfactory (Case 12).

8. After 3 weeks of continuous controlled O_2-therapy, O_2 was given for 12-15 hours daily in order to reduce pulmonary hypertension and improve tolerance to exercise (Case 15).

COMMENT

In the light of our experience we would make the following comments:

1. Indication index for O_2-therapy: Moderately hypoxemic patients should be given O_2-therapy since P_aO_2 values of 36-50 mmHg lie on the steep part of the oxyhemoglobin dissociation curve; this means that while P_aO_2 will rise in response to the administration of a low concentration of oxygen, it may fall rapidly if some crisis occurs. Thus moderate hypoxemia is an important indicator for O_2-therapy and we must not ignore it. For patients with severe hypoxemia, O_2 is indispensable, otherwise they will suffer irreparable harm.

2. Concentration of the inspired oxygen: the values for PiO_2 can be divided into 3 ranges: (1) low, 24-35%; (2) moderate, 36-60%; and (3) high, higher than 60%. Respiratory failure due to pulmonary heart disease is classified as ventilatory respiratory failure (Type II). In principle such patients should receive controlled O_2-therapy, i.e. a low level of O_2 is sufficient for ventilatory failure (Fig. 1). A low concentration of oxygen ensures that the P_aO_2 is sufficient to meet the physiological demands but does not undermine the chemical drive to the peripheral respiratory (sensors). If PiO_2 has been increased to 35% but P_aO_2 does not increase, or increases very little, there are only two possibilities: (1) a large V-A shunt exists in the lungs, or (2) the alveolar ventilation of the patient is under 1.5 l/min. Under these circumstances the most serious danger is the severity of the hypercapnia, rather than the hypoxemia. Owing to the different causes of hypoxemia in different patients, the effect of O_2-therapy varies. In our

research the best effect is seen in cases of hypoxemia due to lung fibrosis; next come cases of under ventilation and Va/Q mismatching; the worst cases are those with a V-A shunt.

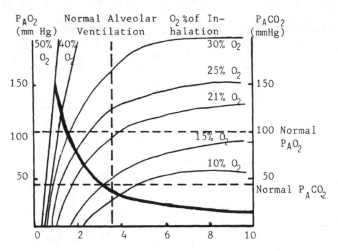

Fig.1. The influence of alveolar ventilation volume
changes on P_AO_2 and P_ACO_2.
—— Concentration of O_2 inhaled
— P_ACO_2 curve for different alveolar ventilation.

Why should we not use a high concentration of oxygen? The central chemical respiratory centre in patients with prolonged hypercapnia is narcotized (?depressed). Therefore, the respiration is being driven from the peripheral chemical sensors which depend on hypoxemic stimulation . Thus, at the start of the O_2 administration, the concentration of PiO_2 must not be high, otherwise the hypoxic stimulus will be eliminated and respiratory frequency will be reduced. Later, although the hypoxemic condition may have been eliminated, the patient will be under CO_2 narcosis. In our experience, if P_aCO_2 before O_2 inhalation is higher than 65 mmHg, PiO_2 must be below 35%; ideally 24% should be used at the beginning. So long as P_aO_2 is over

Table 2. P_AO_2 & P_AO_2-changes by Discontinuous Use of O_2

	Before O_2 inhalation	During O_2 therapy	Withdrawal of O_2
$P_A O_2$	70	110	50
$P_A CO2$	70	90	90
$P_A N2$	573	513	573
$P_A H2 O$	47	47	47
Total	760	760	760

55 mmHg the patient will not be in any danger but every effort should be made to decrease P_aCO_2 while P_aO_2 is rising.

3. Oxygen therapy must be given continuously day and night. From our study, so long as hypercapnia persists, O_2 must be administered continuously, otherwise the raised P_aO_2 will drop again (Case 11). Table 2 shows the blood-gas changes with the discontinuous use of O_2. If P_AO_2 is 70 mmHg before O_2 inhalation and P_ACO_2 is also 70 mmHg, then during the period of O_2 inspiration P_AO_2 will increase to 110 mmHg and P_ACO_2 will rise to 90 mmHg. After withdrawal of O_2, P_AO_2 will fall to 60 mmHg, while P_ACO_2 will remain at 90 mmHg. In principle, the lower the P_ACO_2 falls, the higher the P_AO_2 will be.

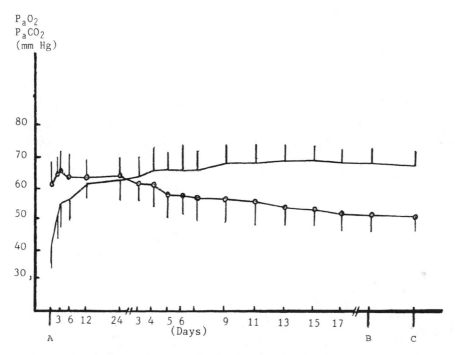

Fig.2. The dynamic changes of blood gases by O_2-therapy in 36 chronic respiratory failure patients.
•indicates P_aO_2; ⊘indicates P_aCO_2.
A, start of O_2-therapy; B, discontinuous O_2-therapy; C, three days after end of O_2-therapy.

4. The course of O_2 therapy: From Fig.2 we can see that during the initial period of O_2 inspiration, P_aO_2 increases rapidly owing to the difference between P_AO_2 and P_aO_2. After 3 days of O_2 administration, P_aO_2 does not rise as obviously as in the first 3 days. However, because of the effect of the general treatment, alveolar ventilation, alveolar exudate and interstitial edema are improved as are the ventilation and exchange function of the lungs and as a result P_aO_2 still rises but the rise is smaller than that during the initial period. During the next stage of O_2-therapy, owing to the effect of the general treatment, P_aO_2 still rises, slowly but definitely. After 2 weeks P_aO_2 rises very slowly. If O_2-therapy is stopped during this period the elevated P_aO_2 may drop. After 3 weeks P_aO_2 does not increase further. Therefore the course of O_2-therapy should not be shorter than 3 weeks.

5. Prolonged intermittent oxygen inhalation by convalescent patients has the advantage of decreasing pulmonary hypertension and it is beneficial in improving the tolerance to physical exercise. We have 3 such patients who inspired oxygen for 12-15 hours daily for more than a year. Their right sided heart failure has improved and they can look after themselves.

6. While those in the control group, who refused to accept oxygen-therapy, were given the same general treatment, P_aCO_2 dropped but P_aO_2 rose very, very slowly by less than 10 mmHg. This indicates that the use of oxygen therapy is essential for respiratory patients.

OTHER ORGANS
AND
TISSUES

ERYTHROCYTE AGGREGATION AS A DETERMINANT OF BLOOD FLOW:

EFFECT OF pH, TEMPERATURE AND OSMOTIC PRESSURE

Nobuji Maeda, Masahiko Seike, Kazunori Kon and Takeshi Shiga

Department of Physiology, School of Medicine, Ehime University

Shigenobu, Onsen-gun, Ehime, Japan 791-02

INTRODUCTION

Oxygen transport to tissues depends on (i) the oxygen transporting capacity of blood (mainly the hemoglobin content and the functional properties of hemoglobin regulated by various factors such as H^+, CO_2, 2,3-diphosphoglycerate and temperature) and (ii) the blood flow to tissue. Blood viscosity is one of the important factors affecting blood flow, and is influenced by erythrocyte deformation in the high shear regions and erythrocyte aggregation in the low shear regions (Chien, 1975).

Erythrocyte aggregation is induced in blood flowing at low shear rates, when macromolecules in plasma, such as fibrinogen and immuno-globulins, "interact" with the erythrocyte surface and then "bridge" between adjacent cells. The process is reversible, aggregations disintegrating at high shear rates. Since the erythrocyte aggregates increase the blood viscosity and thus decrease the blood flow, the inter-relation between the transit time of blood in the microcirculation and the ease of formation of aggregates is a rheologically important problem in oxygen transport. Thus, we measured the velocity of erythrocyte aggregation and examined the effect upon it of pH, temperature and osmotic pressure.

MATERIALS AND METHODS

(a) Erythrocytes and plasma: Fresh blood was obtained from healthy adult males (red cell type, O^+ or A^+) and heparinized. Blood was centri-fuged at 1200 x g for 5 min at 4°C. Plasma was collected and recentri-fuged at 15000 x g for 20 min at 4°C to remove platelets and any insoluble materials. Erythrocytes were washed twice with about 20 volumes of isotonic phosphate-buffered saline (42.6 mM Na_2HPO_4, 7.4 mM NaH_2PO_4, 90 mM NaCl, 5 mM KCl, 5.6 mM glucose, pH 7.4, 285 mOsm) or isotonic Hepes-buffered saline (50 mM Hepes, 120 mM NaCl, 5 mM KCl, 5.6 mM glucose, adjusted to pH 7.4 with NaOH, 285 mOsm) at 4°C.

(b) Measurement of velocity of erythrocyte aggregation: The rheoscopic apparatus (Schmid-Schönbein et al., 1969) (composed of a transparent cone-plate viscometer and an inverted microscope equipped with a temperature-controlled stage) combined with a video-camera (Sony, AVC 1150, Tokyo), an image analyzer (Luzex 450, Toyo Ink Co., Tokyo) and a computer (Hewlett Packard, HP-85, Palo Alto, CA) was used for the measurement of the velocity of erythrocyte aggregation (Shiga et al., 1983a; Maeda et al., 1984; Imaizumi et al., 1984; Maeda and Shiga, 1985; Maeda and Shiga, 1986; Maeda et al., 1986). A diagram of the apparatus is shown in Fig. 1.

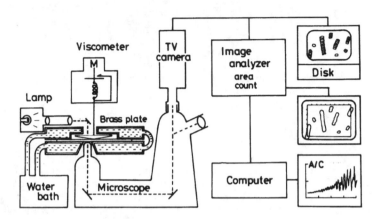

Fig. 1. Diagram of the apparatus for measuring the velocity of erythrocyte aggregation.

The washed erythrocytes were resuspended in isotonic phosphate-(or Hepes-) buffered saline containing 0.3-0.4 g/dl fibrinogen (from AB Kabi, Stockholm, Sweden; grade L, 90 % clottable) and 2-5 g/dl serum albumin (fatty acid free, fraction V from Miles Lab. Inc., Naperville, IL) or in the mixture of autologous plasma and isotonic phosphate-buffered saline (70:30, by volume). The final hematocrit was adjusted to 0.26 % to maximize the sensitivity and the reproducibility of the measurement. The suspension was immediately applied to the rheoscope, and erythrocyte aggregation observed at a constant shear rate (7.5 s^{-1}) and at a constant temperature (controlled by circulating water of a constant temperature through the microscope stage and the plate covering the cone, which were specially made of brass). The count of particles (i.e., single erythrocytes, one-dimensional aggregates (rouleaux) or three-dimensional aggregates) and the total area projected by particles in a frame of the video-image (200 x 150 μm^2) were consecutively encoded by the analyzer at an interval of approximately 1.3 s, and transferred to the computer. The velocity (v) of erythrocyte aggregation was expressed by the increase with the time of area/count (μm^2/min) (Shiga et al., 1983a; Shiga et al., 1983b; Imaizumi et al., 1984; Maeda et al., 1984; Maeda and Shiga, 1985; Maeda et al., 1986; Maeda and Shiga, 1986).

(c) Morphological observations: Erythrocytes suspended in various media were transferred to 15 vol of the same medium containing 1 % glutaraldehyde at constant temperature and further fixed with 1 % OsO$_4$. Their shapes were observed with a scanning electron microscope (Hitachi, S-500A, Japan) and then dimensions measured on photographs. The shapes of aggregates were also observed under the microscope and photographed.

RESULTS AND DISCUSSION

(I) Effect of pH on erythrocyte aggregation

Erythrocyte aggregation was observed in a medium containing 0.4 g/dl fibrinogen and 5 g/dl albumin or in 70 % autologous plasma at various pHs, as shown in Fig. 2. The velocity of erythrocyte aggregation increased linearly with increasing pH.

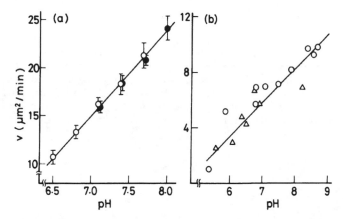

Fig. 2.　Effect of pH on the velocity of erythrocyte aggregation. Measured at 25 C in (a) 0.4 g/dl fibrinogen + 5 g/dl albumin in isotonic phosphate-(O) or Hepes-($\bullet$)buffered saline and (b) 70 % autologous plasma (different symbols from different experiments).

The aggregation of erythrocytes is affected not only by the properties of the bridging macromolecules, but also by the properties of the erythrocytes themselves (e.g., deformability, negative surface charge and shape). Thus, the shape of erythrocytes was examined at various pHs as shown in Table 1: with increasing pH of the medium, the diameter increased and the thickness and cell volume decreased. The cells at high pH were flattened. Since the binding of fibrinogen to the cell surface does not change at pH 6.5-8.0 (Rampling, 1984) and the surface negative charge of sialic acid (with pK=2.6) scarcely alters in the experimental range, the acceleration of erythrocyte aggregation is mainly due to the increased ratio of surface area to volume of cells at high pH.

(II) Effect of temperature on erythrocyte aggregation

The velocity of erythrocyte aggregation at various temperatures is shown in Fig. 3. In the medium containing 0.4 g/dl fibrinogen and 5 g/dl albumin, the velocity of erythrocyte aggregation increased with increasing temperature (Fig. 3a).

Table 1. Shape change of erythrocytes under various conditions

No.	pH	Temperature (°C)	Osmolarity (mOsm)	Diameter*	Thickness*	Cell volume**
1	7.4	25	285	100 + 5	100 + 8	100
2	6.1	25	285	81 + 5	137 + 14	120
3	6.7	25	285	96 + 6	120 + 8	111
4	7.1	25	285	98 + 6	107 + 11	102
5	7.8	25	285	104 + 5	88 + 9	93
6	7.4	2	285	91 + 5	120 + 11	N.D.
7	7.4	37	285	101 + 7	96 + 8	N.D.
8	7.4	25	189	84 + 7	151 + 29	119
9	7.4	25	221	82 + 6	124 + 16	115
10	7.4	25	338	101 + 6	99 + 11	94
11	7.4	25	430	107 + 7	84 + 7	85

Figures for each parameter represent percent + standard deviation, with the 100 percent value at pH 7.4, 25°C and 285 mOsm. *Diameter and maximum thickness were measured on photographs from scanning electron microscopy. **Cell volume was calculated from the cell count and the packed cell volume.

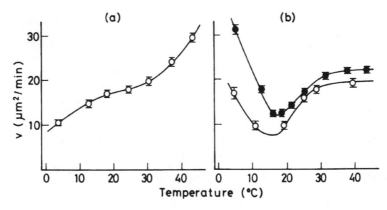

Fig. 3. Effect of temperature on erythrocyte aggregation. Measured at pH 7.4 ín (a) 0.4 g/dl fibrinogen + 5 g/dl albumin in isotonic phosphate-buffered saline and (b) 70 % autologous plasma (different symbols from different donors).

The temperature dependency of erythrocyte aggregation is interpreted as follows. (a) The change in medium viscosity: With decreasing temperature, the viscosity of medium increases, and the resultant increase of shear stress (under a constant shear rate) disintegrates the erythrocyte aggregates (Shiga et al., 1983a; Maeda and Shiga, 1985). (b) The change of interaction between erythrocytes and macromolecules: At low temperature, the flexibility of the erythrocyte membrane, which aids cell-to-cell adhesion, is impaired since the membrane rigidity increases (Shiga and Maeda, 1980). The change in the ratio of surface area to volume is also an important factor, since the cell diameter increases and the thickness decreases as the temperature rises (Table 1; see also Murphy, 1967). Furthermore the temperature-dependent alteration of molecular conformation of fibrinogen (Zyma et al., 1978) may affect the interaction between erythrocytes and fibrinogen.

In contrast, in 70 % autologous plasma, the velocity of erythrocyte aggregation was minimum at 15-18°C and increased at both higher and lower temperatures (above 30°C, the velocity saturated) (Fig. 3b). Thus the temperature dependency of erythrocyte aggregation in autologous plasma was quite different from that in the medium containing fibrinogen and albumin. The main reason for the difference must be the different composition of macromolecules (probably, in the globulin fraction) and other constituents. Furthermore, it should be noted that below 15-18°C the aggregates were mainly three-dimensional both in artificial medium containing fibrinogen and albumin and in 70 % autologous plasma, while above 15-18°C they were one-dimensional rouleaux in both media. Such a temperature-dependent shape change was not observed for immunoglobulin G-induced erythrocyte aggregation: the aggregates were three-dimensional in all temperature ranges (data not shown here). Shiga et al.(1983b) have observed that in the specific pathogen-free rats, the erythrocytes do not form three-dimensional aggregates in their own plasma, because of the low concentration of γ-globulin. Therefore, the difference in the shape of erythrocyte aggregates at low and high temperatures may result from the fibrinogen/γ-globulin ratio.

(III) Effect of osmotic pressure on erythrocyte aggregation

The effect of osmotic pressure on the fibrinogen-induced erythrocyte aggregation is shown in Fig. 4. With increasing osmotic pressure in the medium (adjusted with NaCl in the phosphate-buffered saline), the velocity of erythrocyte aggregation increased but was nearly constant in the range of 250-300 mOsm. Above 400 mOsm the velocity decreased. Morphologically, the erythrocyte aggregates were three-dimensional below 200 mOsm, while they were essentially one-dimensional rouleaux above 200 mOsm.

The phenomena are interpreted in terms of the change in various properties of erythrocytes (the shape change is shown in Table 1) and of the change in the ionic strength of medium, as follows. (a) In low osmolarity, the cells were nearly spherical (with increased thickness, decreased diameter and increased cell volume), thus the adhesion among such spherical cells by macromolecular bridging was random, and rouleaux were rarely formed. (b) As the osmolarity increased, the cells were flattened with an increase in diameter and a decrease in thickness, and the cell volume decreased (the diameter and the thickness were nearly constant around the isotonic condition). Thus, the increased ratio of surface area to volume accelerated the erythrocyte aggregation. (c) However, above 400 mOsm the membrane flexibility may be impaired owing to the increase of intracellular hemoglobin concentration. Echinocytosis, which inhibits

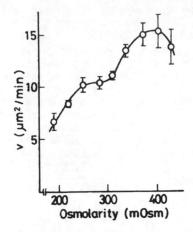

Fig. 4. Effect of osmotic pressure on
erythrocyte aggregation.
Measured in phosphate-buffered
saline containing 0.3 g/dl
fibrinogen + 5 g/dl albumin at
pH 7.4 and at 25°C.

erythrocyte aggregation, was not observed in this osmotic range. (d) The
increase in ionic strength in the medium decreases the electrostatic
repulsive force between negatively charged erythrocytes, and increases
erythrocyte aggregation (Chien, 1975; Jan and Chien, 1973). (e) The
binding of fibrinogen to erythrocytes increases at ionic strengths above
isotonicity (Rampling, 1984).

(IV) Physiological implications

The physiological consequence of erythrocyte aggregation is a
reduction of blood flow in the low shear region due to the increased blood
viscosity. As established in the present experiments, one of the
important factors leading to marked aggregation is the shape change of the
erythrocytes: the increased ratio of the surface area to the volume of the
cells accelerates erythrocyte aggregation, as observed at high pH and in
high osmolarity. However, the suppression of erythrocyte aggregation in
low pH may be related to the acceleration of oxygen transport by blood flow
to metabolically active tissues.

A change of temperature around 37°C did not affect the velocity of
erythrocyte aggregation in autologous plasma. However, in hypothermic
tissues exposed to a cold environment, the blood supply may be improved by
suppression of erythrocyte aggregation, in the temperature range of 15-
30°C. Since the viscosity of fluid increases at low temperature, the
suppressed erythrocyte aggregation is beneficial for the blood supply to
such hypothermic tissues. In contrast to this benefit, the accelerated
erythrocyte aggregation below 15°C may lead to pathological states, such as
chilblains. The change of erythrocyte aggregation under various osmotic

pressures may have serious implications in relation to the blood supply to tissues in the hydrated or dehydrated state.

Acknowledgement: The work was supported in part by a Grant-in-Aid for General Scientific Research of Japan and by a grant from the Ehime Health Foundation. The authors are indebted to Mr. D. Shimizu for the operating the scanning electron microscope and to Miss M. Sekiya for technical assistance.

SUMMARY
The effect of pH, temperature and osmotic pressure on velocity of erythrocyte aggregation was quantitatively examined with a rheoscope combined with a video-camera, an image analyzer and a computer, (a) in an artificial medium containing fibrinogen and albumin and (b) in diluted autologous plasma.
(1) With increasing pH of the medium, the velocity of erythrocyte aggregation increased.
(2) The velocity of erythrocyte aggregation in the artificial medium increased as the temperature rose. However, in 70 % autologous plasma the velocity was minimum at 15-18°C, increasing both above and below this temperature (above 30°C, the velocity saturated).
(3) The velocity of erythocyte aggregation decreased in hypotonic medium, while it increased in hypertonic medium (at osmotic pressures higher than 400 mOsm, the velocity decreased).
The mechanism of erythrocyte aggregation is discussed with special reference to the morphological changes produced by pH, temperature and osmotic pressure, and the implications of the phenomena for oxygen transport to tissues in (patho)physiological situations are considered.

REFERENCES

Chien, S., 1975, Biophysical behavior of red cells in suspensions, in: "The Red Blood Cell", D. M. Surgenor, ed., Vol. 2, 2nd edn., p. 1031-1133, Academic Press, New York.
Imaizumi, K., Imai, A., Maruyama, T., and Shiga, T., 1984, Inhibition of Ig G-, F(ab')$_2$- and myeloma protein-induced erythrocyte aggregation, by small Ig G-fragments, Clin. Hemorheol., 4: 431.
Jan, K-M., and Chien, S, 1973, Influence of the ionic composition of fluid medium on red cell aggregation, J. Gen. Physiol., 61: 655.
Maeda, N., Imaizumi, K., Sekiya, M., and Shiga, T., 1984, Rheological characteristics of desialylated erythrocytes in relation to fibrinogen-induced aggregation, Biochim. Biophys. Acta, 776: 151.
Maeda, N., and Shiga, T., 1985, Inhibition and acceleration of erythrocyte aggregation induced by small macromolecules, Biochim. Biophys. Acta, 843: 128.
Maeda, N., and Shiga, T., 1986, Opposite effect of albumin on the erythrocyte aggregation induced by immunoglobulin G and fibrinogen, Biochim. Biophys. Acta, 855: 127.

Maeda, N., Sekiya, M., Kameda, K., and Shiga, T., 1986, Effect of immunoglobulin preparations on the aggregation of human erythrocytes, Eur.
J. Clin. Invest., 16: 184.
Murphy, J. R., 1967, The influence of pH and temperature on some physical
properties of normal erythrocytes and erythrocytes from patients
with hereditary spherocytosis, J. Lab. Clin. Med., 69: 758.
Rampling, M. W., 1984, The binding of fibrinogen and fibrinogen degrada-
tion products to the erythrocyte membrane and its relationship to
haemorheology, Acta Biol. Med. Germ., 40: 373.
Schmid-Schönbein, H., Wells, R., Schildkraut, R., 1969, Microscopy and
viscometry of blood flowing under uniform shear rate (rheoscopy),
J. Appl. Physiol., 26: 674.
Shiga, T., Imaizumi, K,. Harada, N., and Sekiya, M., 1983a, Kinetics of
rouleaux formation using TV image analyzer. I. Human erythrocytes,
Am. J. Physiol., 245: H252.
Shiga, T., Imaizumi, K., Maeda, N., and Kon, K., 1983b, Kinetics of
rouleaux formation using TV image analyzer. I. Rat erythrocytes,
Am. J. Physiol., 245: H259.
Shiga, T., and Maeda, N., 1980, Influence of membrane fluidity on erythro-
cyte functions, Biorheology, 17: 485.
Zyma, V. L., Varets'ka, T. V., Svital's'ka, L. O., Demchenko, O. P., 1978,
Temperature effect on the conformational state of fibrinogen and its
derivatives, Ukr. Biokhim., 50: 459.

DEPENDENCE OF O_2 TRANSFER CONDUCTANCE OF RED BLOOD CELLS ON CELLULAR

DIMENSIONS

K. Yamaguchi, K.D. Jürgens, H. Bartels, P. Scheid and J. Piiper

Department of Medicine, School of Medicine, Keio University
Tokyo, Japan; Zentrum Physiologie, Medizinische Hochschule
Hannover, F.R.G.; Institut für Physiologie, Ruhr-Universität
Bochum, F.R.G.; Abteilung Physiologie, Max-Planck-Institut für
experimentelle Medizin, Göttingen, F.R.G.

INTRODUCTION

The experimental evidence for an advantageous effect of small red
blood cell (RBC) size on O_2 transfer kinetics was originally provided by
Holland and Forster (1966) who examined the initial O_2 uptake rate (k_c')
of RBC of varied size, using a stopped-flow technique, and concluded that
k_c' was inversely related to the RBC volume. Recently, however, several
authors (see Yamaguchi et al., 1985) have shown that O_2 uptake kinetics
measured in a rapid-mixing apparatus is significantly influenced by the
diffusion boundary layer around the cell which, it was predicted would
increase with increasing cell size. On the other hand, Coin and Olson
(1979), Vandegriff and Olson (1984) and our group (Yamaguchi et al., 1985)
have shown that the limitation imposed by the diffusion boundary layer
surrounding the RBC can be removed if the kinetics of O_2 release from RBC
are measured in a medium with a sufficiently high dithionite concentration.

The aim of this study was to investigate the O_2 release kinetics of
RBC of varied cell sizes in the presence of high dithionite concentrations. Based on the experimental results, the significance of dimensions
of RBC on O_2 transfer kinetics will be discussed.

MATERIALS AND METHODS

Fresh venous blood was drawn into heparinized syringes from nine
different species including man, llama (Lama lama), vicuña (Lama vicuna),
alpaca (Lama pacos), dromedary camel (Camelus dromedarius), domestic pygmy
goat (Capra hircus), muscovy duck (Cairina moschata), domestic hen (Gallus
domesticus) and turtle (Pseudemys scripta elegans). The blood samples were
immediately cooled to $4^{\circ}C$ for storage and used on the same day.

Various measurements on blood

The total hemoglobin (Hb) concentration of blood was determined by
the cyanomethemoglobin method (Drabkin and Austin, 1935). The Hb concentration in the RBC was estimated by dividing that of the blood by the

hematocrit. Since the avian and turtle RBC are nucleated, the cytoplasmic Hb concentration was calculated with a correction for the nucleus volumes, 19.6% in the duck (Gaehtgens et al., 1981), 21.6% in the hen (Abdalla et al., 1982) and 12% in the turtle (Saint Girons, 1970; Wintrobe, 1934). The size distribution and the numbers of RBC were measured with a Coulter counter and a multi-channel particle size analyzer. The RBC volume (mean corpuscular volume, MCV) was calculated from the hematocrit and the number of RBC. The O_2 dissociation curve including the CO_2 Bohr coefficient for the pH range between 7.1 and 7.7 was determined, at a given temperature, by equilibrating blood samples of each animal with a gas mixture containing appropriate fractions of O_2 and CO_2.

For the camelids, goat and man, the RBC dimensions including major and minor axes, plane surface area and circumference were measured using a computerized planimeter system. Approximating the RBC shape by a circular or oval disc, mean cell thickness and total surface area were also estimated. For the nucleated avian and turtle RBC, the values of mean cell thickness and total surface area were calculated using the measured MCV values and published data on major and minor axes and the ratio of total surface area to MCV (Gaehtgens et al., 1981; Sheeler, 1964; Wintrobe, 1934).

Stopped-flow measurements

The O_2 transfer kinetics were studied by spectrophotometric determination of the time course of hemoglobin O_2 saturation (So_2) after a rapid 1:1 mixing of RBC suspensions or Hb solutions with appropriate buffer solutions in a stopped-flow apparatus. A detailed description of the stopped-flow methodology has been presented elsewhere (Yamaguchi et al., 1985). The O_2 release kinetics with 40 mmol/l dithionite (after mixing) were usually investigated at 37°C and a fixed pH of 7.4 (RBC suspension) or 7.2 (Hb solution). Only for duck and hen RBC, were O_2 kinetics studied at 41°C and for turtle RBC, at 30°C.

Calculations

Reaction velocity constants. The apparent dissociation velocity was determined from So_2 and its rate of change, $d(So_2)/dt$, measured in Hb solution rapidly mixed with dithionite (40 mmol/l after mixing). The decrease of So_2 with time was nearly exponential and could thus be characterized by a So_2-independent dissociation velocity constant, k (1/sec):

$$k = \frac{-d(So_2)/dt}{So_2} \tag{1}$$

O_2 transfer conductance of RBC. The time course of So_2 of RBC was analyzed in terms of the specific O_2 transfer conductance, G (mmol/min/Torr/ml RBC) (Yamaguchi et al., 1985):

$$G = \dot{m}/(P_m - P_{eq}) \tag{2}$$

where $\dot{m}$ is the effective O_2 transfer rate per unit RBC volume. P_m and P_{eq} are the mean Po_2 in the surrounding medium and the Hb equilibrium Po_2 at a given mean So_2, respectively. In the case of the present study, P_m was zero because of the high concentration of dithionite present in the medium. P_{eq} was estimated from the O_2 dissociation curve which was determined as described above. Since the G values for O_2 release with 40 mmol/l dithionite changed little with time, and So_2 displayed a flat maximum, at least in the So_2 range of 0.3 to 0.7 (Nguyen-Phu et al., 1986; Yamaguchi et al., 1985, 1987 a, b), these G values were averaged and defined as the standard O_2 transfer conductance, G_{st}:

$$G_{st} = G \text{ at } 40 \text{ mmol/l dithionite } (0.3 > So_2 > 0.7) \qquad (3)$$

The specific O_2 transfer conductance of whole blood, θ_{st} (mmol/min/Torr/ml blood) was calculated from the G_{st} and the normal hematocrit, Hct, of each animal:

$$\theta_{st} = G_{st} \cdot Hct \qquad (4)$$

The standard O_2 transfer conductance of the single RBC, g_{st} (mmol/min/Torr) was obtained from the relationship:

$$g_{st} = G_{st} \cdot V \qquad (5)$$

where V indicates the average volume of single RBC, i.e. MCV.

Since O_2 transfer kinetics for the birds and the turtle were observed at temperatures other than 37°C, their values for k, G_{st}, θ_{st} and g_{st} were corrected, for comparison, to 37°C using the corresponding activation energies obtained for Hb solution and RBC of human blood (Yamaguchi et al., 1987b).

RESULTS

Hematological values and RBC dimensions (Table 1)

Cytoplasmatic Hb concentration was highest in camelid RBC. Correcting for the nucleus volume, the avian RBC showed a Hb concentration lower than that of camelids but significantly higher than that of human RBC, while turtle erythrocytes had nearly the same Hb concentration as human RBC. Erythrocyte counts increased with decreasing cell size.

Table 1. Hematological and morphological properties of blood

Species	Hct	Hb	RBC counts	(a) Equivalent diameter	Mean cell thickness	Surface area	MCV	S/V	(b) Sphericity index
	$(\ell \cdot \ell^{-1})$	$(mmol \cdot \ell RBC^{-1})$	$10^{10} \cdot (ml\ blood)^{-1}$	(μm)	(μm)	(μm^2)	(μm^3)	(μm^{-1})	
Llama	0.30	27.8	1.2	5.4	1.05	63	24	2.6	0.64
Vicuña	0.34	27.0	1.4	5.2	1.01	60	22	2.7	0.63
Alpaca	0.36	27.5	1.6	5.2	1.01	59	21	2.8	0.63
Camel	0.34	27.3	1.1	5.9	1.11	74	30	2.5	9.63
Goat	0.43	20.4	1.7	4.1	1.72	49	23	2.1	0.80
Human	0.47	19.8	0.5	8.0	1.75	143	87	1.6	0.67
Hen	0.33	23.7*	0.3	9.2	1.56	179	104	1.7	0.60
Duck	0.44	23.8*	0.3	10.6	1.75	236	155	1.5	0.59
Turtle	0.25	20.6*	0.08	13.9	2.15	398	327	1.2	0.58

* : corrected for nucleus volume. (a) : calculated from major and minor axes (not shown).
(b) : $4.84 \cdot (Volume)^{2/3}$/(surface area) given by Canham and Burton (1968). See text for details.

The RBC of the high-altitude camelids (llama, vicuña and alpaca) had the shape of ellipsoidal discs with a major axis of 7 μm, a minor axis of 4 μm, a thickness of 1 μm, a mean cellular volume of 22 μm³ and an approximated effective surface area of 61 μm². Camel RBC were similar in shape to those of their high-altitude relatives but of a larger size, 8 μm x 4 μm in the plane and 1.1 μm thick, with a mean volume of 30 μm³ and a surface area of 74 μm². Goat RBC were close to a circular disc in shape with a mean diameter of about 4 μm and a mean thickness of 1.7 μm. Their volume was nearly identical to that of high altitude camelids but their surface area, about 50 μm², was the smallest of all species investigated. For human RBC a thickness of 1.8 μm, a volume of 87 μm³ and an area of 143 μm² were obtained. The erythrocytes of the domestic hen and muscovy duck were ellipsoidal discs with mean thicknesses of, respectively, 1.6 and 1.8 μm, volumes of 104 and 155 μm³, and surface areas of 179 and 236 μm². Turtle RBC, also ellipsoidal in shape, were the largest cells investigated and had a mean thickness of 2.2 μm, a MCV of 327 μm³ and a surface area of 398 μm².

With the exception of goat cells, the ratio of surface area to volume, S/V, increased with decreasing cell size. Calculation of the sphericity index (Canham and Burton, 1968) revealed the goat RBC to be closest to a sphere with a value of 0.8. The ellipsoidal red cells have considerably lower sphericity indices ranging between 0.64 and 0.58. The lower values belong to the larger cells, indicating increasing flatness with increasing size.

O₂ transfer kinetics (Table 2, Figs. 1 and 2)

Oxyhemoglobin dissociation velocity constants, k, for the camelids and birds were found to range between 100 and 120 sec^{-1}, whereas those for man, goat and the turtle were about 200 sec^{-1}.

Table 2. O₂ kinetic data and estimated diffusion distance

Species	k	(a) G_{st}	(b) θ_{st}	(c) $g_{st} \cdot 10^{11}$	$Do_2 \cdot 10^6$	$\alpha_{O_2} \cdot 10^6$	(d) $Ko_2 \cdot 10^{11}$	Effective diffusion path
	(sec^{-1})				$(cm^2 \cdot sec^{-1})$	$(mM \cdot Torr^{-1})$		(μm)
Llama	119	0.55	0.17	1.4	6.3	1.6	1.00	0.27
Vicuña	124	0.60	0.20	1.4	6.7	1.6	1.06	0.27
Alpaca	110	0.58	0.21	1.3	6.4	1.6	1.02	0.28
Camel	106	0.41	0.14	1.3	6.5	1.6	1.03	0.36
Goat	191	0.42	0.18	1.1	10.2	1.5	1.48	0.40
Human	203	0.39	0.18	3.4	10.6	1.4	1.53	0.38
Hen	104	0.30	0.10	3.2	8.4	1.5	1.27	0.43
Duck	112	0.23	0.10	3.4	8.4	1.5	1.27	0.52
Turtle	218	0.20	0.05	6.4	10.1	1.5	1.47	0.55

(a) : $mmol \cdot min^{-1} \cdot Torr^{-1} \cdot ml\ RBC^{-1}$.   (b) : $mmol \cdot min^{-1} \cdot Torr^{-1} \cdot ml\ blood^{-1}$.   (c) : $mmol \cdot min^{-1} \cdot Torr^{-1}$
(d) : $mmol \cdot sec^{-1} \cdot cm^{-1} \cdot Torr^{-1}$.   See text for details.

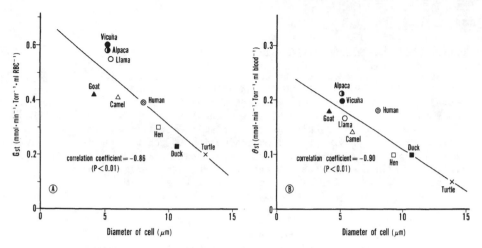

Fig. 1. Specific O$_2$ transfer conductance of packed RBC (A) and of whole blood (B) with varied cell size. Solid line: linear regression.

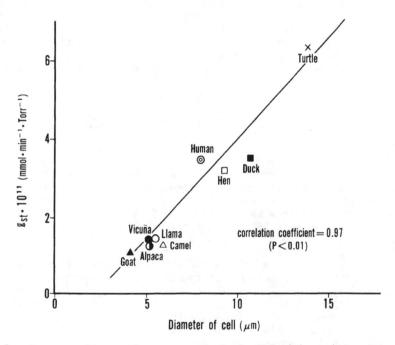

Fig. 2. O$_2$ transfer conductance of single RBC with varied cell size. Solid line: linear regression.

The standard O$_2$ transfer conductance of RBC, G$_{st}$, decreased as cell size increased. Specific O$_2$ transfer conductance of blood, θ_{st} (at the physiological Hct), showed qualitatively the same trend as G$_{st}$, while the standard O$_2$ transfer conductance of single RBC, g$_{st}$, tended to increase with increasing cell size.

DISCUSSION

Physiological factors determining O_2 transfer conductance

As extensively discussed by our group (Piiper et al., 1987; Yamaguchi et al., 1985; 1987 a, b), O_2 transfer kinetics of RBC measured with a stopped-flow method are strongly affected by diffusion of O_2 in the medium around the cell. Based on several experimental findings, we have concluded that only O_2 release measurements with a sufficiently high concentration of dithionite can provide a good approximation to the O_2 transfer kinetics which are predominantly determined by the conditions within the RBC and are little influenced by the limitation in the surrounding medium. In the present study, 40 mmol/l dithionite (after mixing) was used, a concentration shown to be high enough to suppress the resistance caused by the medium.

Although considerable differences in the O_2 Hb dissociation velocity constant, k, were found (Table 2), the contribution of chemical reaction between O_2 and Hb to the O_2 transfer resistance of RBC has been shown to be small (Scheid et al., 1986; Piiper et al., 1987; Yamaguchi et al., 1987 a, b). The quantitative role of facilitated diffusion, i.e. diffusion of Hb and HbO_2 within the cell, is uncertain (Scheid et al., 1986; Yamaguchi et al., 1987a). The diffusional resistance imposed by the RBC membrane is generally believed to be very small. Thus, in conditions where the limiting role of the pericellular medium exerts but little influence on O_2 transfer kinetics, the major factor determining O_2 transfer conductance of RBC appears to be the diffusibility of free O_2 inside the RBC, quantified by the Krogh O_2 diffusion constant, Ko_2, which is equal to the product of O_2 solubility, αo_2, and the O_2 diffusion coefficient, Do_2, in the RBC.

Morphological factors determining O_2 transfer conductance

A previous study by our group (Yamaguchi et al., 1987a) has shown that O_2 transfer conductance of single RBC, g_{st}, could be approximately predicted by the Fick diffusion equation for steady state O_2 transfer through a plane sheet with the surface area and effective diffusion path equivalent to those of the real RBC. The ratio of surface area to effective diffusion distance within the cell (S/X) is considered as the decisive geometric factor, but X is not easily estimated with sufficient accuracy.

In order to obtain an approximate value for the effective diffusion path within the cell, O_2 diffusion coefficient, Do_2 (cm^2/sec), and O_2 solubility, αo_2 (mmol/l/Torr), inside the cell of each species were calculated, as influenced by the cytoplasmic Hb concentration, from the data compiled by Kreuzer (1970) and Christoforides and Hedley-White (1969), respectively. The effective diffusion distance, X (μm), was calculated from the Krogh diffusion constant, Ko_2 (= $Do_2 \cdot \alpha o_2$), surface area (S) and g_{st}:

$$X = Ko_2 \cdot S/g_{st} \qquad (6)$$

Although X varied from 0.27 μm (llama or vicuña RBC) to 0.55 μm (turtle RBC), the ratio of X to mean cell thickness was found to range from 0.22 to 0.32, averaging 0.27. This means that the effective diffusion path is about one fourth of the mean cell thickness and nearly independent of the cell shape and size (Table 2).

The surface area/volume ratio, S/V, has been regarded as the appropriate geometric factor determining the diffusive O_2 exchange properties of the RBC (cf. Jones, 1979). According to the data in Tables 1 and 2 the

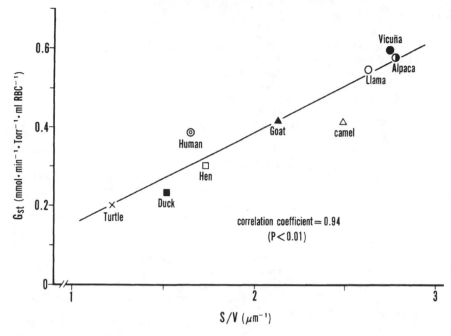

Fig. 3. Effect of S/V on G_{st}. Solid line: linear regression

ratio S/V decreases with increase in cell size, whereas g_{st} increases. This is explainable by our model in which G_{st} is proportional to S/V divided by the effective diffusion path, i.e. to $S/(V \cdot X)$ (Fig. 3).

In conclusion, the analysis of O_2 kinetics of RBC of different size and shape, from various species, shows that size is an important determinant of O_2 transfer conductance of RBC.

SUMMARY

To estimate the significance of the dimensions of RBC on O_2 transfer, the kinetics of O_2 release from RBC into medium containing dithionite (40 mmol/l) was measured, by a stopped-flow technique, for nine different species with varying RBC size (man, llama, vicuna, alpaca, dromedary camel, pygmy goat, domestic hen, muscovy duck and turtle). The observed O_2 transfer kinetics were found to be size-dependent, i.e. the O_2 transfer conductance of the single RBC, g_{st}, was lower, whereas the specific O_2 transer conductance of packed RBC, G_{st}, or of whole blood, θ_{st}, was higher for smaller RBC. The ratio of surface area to effective diffusion path length which was found to be about one fourth of the mean cell thickness irrespective of cell size and cell shape, may be considered as the essential morphological factor determining O_2 transfer efficiency of the single RBC.

REFERENCES

Abdalla, M.A., Maina, J.N., King, A.S., King, K.Z., and Henry, J., 1982, Morphometrics of the avian lung. I. The domestic fow (Gallus gallus variant domesticus), Respir. Physiol. 47: 267.

Canham, P.B., and Burton, A.C., 1968, Distribution of size and shape in populations of normal human red cells, Circ. Res. 22: 405.

Christoforides, C., and Hedley-White, J., 1969, Effect of temperature and hemoglobin concentration on solubility of O_2 in blood, J. Appl. Physiol. 27: 592.

Coin, J.T., and Olson, J.S., 1979, The rate of oxygen uptake by human red blood cells, J. Biol Chem. 254: 1178.

Drabkin, D.L., and Austin, J.H., 1935, Spectrophotometric studies II. Preparations from washed blood cells, nitric oxide hemoglobin and sulfhemoglobin, J. Biol. Chem. 112: 51.

Gaehtgens, P., Schmidt, F., and Will, G., 1981, Comparative rheology of nucleated and non-nucleated red blood cells, Pflügers Arch. 390: 278.

Holland, R.A.B., and Forster, R.E., 1966, The effect of size of red cells on the kinetics of their oxygen uptake, J. gen. Physiol. 49: 727.

Jones, D.A., 1979, The importance of surface area/volume ratio to the rate of oxygen uptake by red cells, J. gen. Physiol. 74: 643.

Kreuzer, F., 1972, Facilitated diffusion of oxygen and its possible significance: a review, Respir. Physiol. 9: 1.

Nguyen-Phu, D., Yamaguchi, K., Scheid, P., and Piiper, J., 1986, Kinetics of oxygen uptake and release by red blood cells of chicken and duck, J. exp. Biol. 125: 15.

Piiper, J., Yamaguchi, K., and Scheid, P., 1987, Effects of temperature on oxygen transfer conductance of human red blood cells, Adv. exp. Med. Biol., in press.

Saint Girons, M.-C., 1970, Morphology of the circulating blood cells, in: "Biology of the Reptilia", Vol. 3, C. Gans, ed., Academic Press, London and New York, pp. 73-91.

Scheid, P., Hook, C., Yamaguchi, K., and Piiper, J., 1986, Factors limiting O_2 transfer of red blood cells: model analysis using results from stopped-flow experiments, Progr. Resp. Res. 21: 11.

Sheeler, P., and Barber, A.A., 1964, Comparative hematology of the turtle, rabbit and rat, Comp. Biochem. Physiol. 11: 139.

Wintrobe, M.M., 1934, Variations in the size and hemoglobin content of erythrocytes in the blood of various vertebrates, Folia Haematol. 51: 32.

Vandegriff, K.D., and Olson, J.S., 1984, The kinetics of O_2 release by human red blood cells in the presence of external sodium dithionite, J. Biol. Chem. 259: 12609.

Yamaguchi, K., Nguyen-Phu, D., Scheid, P., and Piiper, J., 1985, Kinetics of oxygen uptake and release by human red blood cells studied by a stopped-flow technique, J. Appl. Physiol. 58: 1215.

Yamaguchi, K., Jürgens, K.D., Bartels, H., and Piiper, J., 1987a, Oxygen transfer properties and dimensions of red blood cells in high-altitude camelids, dromedary camel and goat, J. Comp. Physiol. B 157: 1.

Yamaguchi, K., Glahn, J., Scheid, P., and Piiper, J., 1987b, Oxygen transfer conductance of human red blood cells at varied pH and temperature, Respir. Physiol. 67: 209.

EMULSIFIED PERFLUOROCHEMICALS FOR OXYGENATION OF MICROBIAL CELL CULTURES?

A.T. King, K.C. Lowe and B.J. Mulligan*

Departments of Zoology and Botany*
University of Nottingham
Nottingham NG7 2RD, U.K.

INTRODUCTION

In industrial fermentation technology the rate of O_2 supply to sub-merged microbial cells has long been recognised as a major limiting factor. This occurs when the O_2 supply rate causes the dissolved O_2 level in the medium to fall below the critical O_2 concentration for the organism. The conventional fermentation method is to bubble air or pure O_2 through the fermenter together with vigorous stirring to improve O_2 transport. However, due to low O_2 solubility in aqueous media (at s.t.p. O_2 is soluble in water to approx. 0.25 mM) and the expense of mechanical stirring, normal systems for oxygenation have no great efficiency.

A number of novel methods have been reported to improve O_2 supply to microbial cultures. One such approach is to generate O_2 in situ, either by adding hydrogen peroxide to the medium and including an organism with high natural catalase activity (Holst et al., 1982), or by co-immobilization of O_2-consuming organisms with O_2 producers (Adlercreutz et al., 1982). An alternative has been to increase O_2 concentration within the medium itself: this has been achieved using organic solvents instead of buffer solutions (Buckland et al., 1975). A more recent method has been to incubate the cell cultures with perfluorochemicals (PFCs) in an attempt to improve O_2 solubility in the medium (Adlercreutz and Mattiasson, 1982; Mattiasson and Adlercreutz, 1984; Damiano and Wang, 1985; Chandler et al., 1987).

PFCs are ring or chain form hydrocarbons with hydrogen atoms replaced by fluorine. PFCs are extremely stable and chemically inert due to the presence of very strong carbon-fluorine bonds and steric protection of the carbon groups by the fluorine atoms. Of much greater importance to the biologist is the fact that PFCs have the ability to dissolve large volumes of O_2 (40%-60% O_2 dissolved per unit volume) and CO_2 (2-3 times greater than O_2) according to Henry's Law (Reiss and Le Blanc, 1982). Emulsions of PFCs dispersed in physiologically-acceptable electrolyte solutions have been widely tested as O_2-carrying resuscitation fluids in several mammalian species (Lowe and Bollands, 1985).

Although great interest has focussed on the use of PFCs as potential oxygen carriers in mammalian systems (Lowe and Bollands, 1985; Lowe, 1986), the effects of PFCs and other emulsion components on growth and structure of microbial cells have not been examined in detail. The purpose of the present experiments, therefore, was to investigate the effects of PFC emulsions and their components on cultures of E. coli and S. cerevisiae.

MATERIALS AND METHODS

Organisms and Media

Saccharomyces cerevisiae (strain X-2180-1B) was grown aerobically at 30°C or 37°C in YP medium (1% yeast extract, 1% Bacto-Peptone, 2% glucose). Escherichia coli (strain HB101 containing plasmid pBR322) was grown at 30°C or 37°C in LB medium containing 50μg/ml ampicillin.

Starter cultures were grown by inoculating 3ml of medium in a 20ml McCartney bottle with a single colony from a plate culture and incubating overnight at 30/37°C in a shaking water bath. Test culture media were set up by adding 1ml of an appropriately diluted sterile stock solution of a test substance [Pluronic F-68 surfactant, perfluorodecalin (FDC) oil, 20% (w/v) FDC emulsion or Fluosol-DA 20% (F-DA) stem emulsion (Green Cross, Japan)] to 4ml of growth medium in a McCartney bottle. Test media were then inoculated with 0.1ml of the appropriate starter culture and incubated at 30/37°C at 180 cycles/min. in a reciprocal shaker. Samples were withdrawn at intervals for measurement of cell growth by absorbance at 600nm. Viable cell counts were also determined by serial dilution of cell samples in growth medium and plating on media solidified with 1.5% agar.

RESULTS

Growth of S. cerevisiae as determined by viable cell counts was unaffected by culture with all concentrations of Pluronic F-68. However, cultures grown in 5-10% Pluronic at 37°C and 10% Pluronic at 30°C showed significantly reduced absorbance at 600nm compared with controls (Figures 1 & 2). In addition, incubation of S. cerevisiae with all concentrations of F-DA, FDC emulsion and FDC oil had no effect on growth rate as measured by viable cell counts.

Absorbance measurements suggested that the growth of E. coli up to 3 hours was unaffected by incubation with Pluronic F-68 but was inhibited by 5-10% Pluronic thereafter (Figure 3). Viable cell counting also indicated possible inhibitory effects at higher concentrations of Pluronic. Preliminary results by viable cell counting indicated that F-DA and FDC emulsion, and FDC oil have no deleterious effects on growth of E. coli.

DISCUSSION

Previous work has shown that O_2-saturated PFC oils and their emulsions can be used to deliver O_2 to microbial cultures (Adlercreutz and Mattiasson, 1982; Mattiasson and Adlercreutz, 1984; Damiano and Wang, 1985). While O_2 enrichment using PFC delivery systems is an attractive proposition, caution is required in the use of such methods since the biological effects of emulsions and their components have not been studied in detail. In this regard, previous work has shown that the Pluronic F-68 surfactant can cause subtle perturbations in mammalian cell structure which may have far reaching physiological consequences (Bucala et al., 1983; Janco et al., 1985; Wake et al., 1985). Therefore careful assessment of the effects of the pluronic surfactant on microbial cells was required before including it in potential O_2-carrying emulsions.

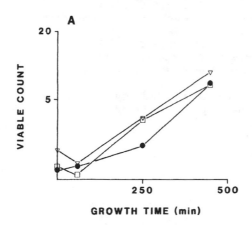

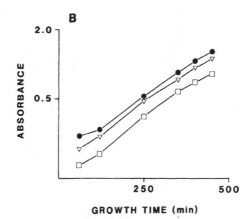

Fig. 1. Changes in A. viable counts (cells x $10^{-6}.ml^{-1}$) or B. absorbance (600nm) of S. cerevisiae cultures grown in 5% ($\triangledown$) or 10% ($\square$) Pluronic F-68 at 37°C compared with controls (●).

The results reported here indicate that pure FDC oil has no adverse effects on S. cerevisiae or E. coli and is, therefore, a biologically inert PFC suitable for O_2-supply to these organisms. FDC oils have a density greater than that of water and are immiscible with aqueous culture media. This property can be exploited to provide a recycling system where PFC is gassed with pure O_2 or air prior to pumping into the culture vessel (Damiano and Wang, 1985).

The situation regarding PFC emulsions is more complex owing to the presence of the Pluronic F-68 emulsifier. The present results show that this surfactant has profound effects on S. cerevisiae which, although not producing a significant decrease in cell viability, are likely to produce changes in cell structure, as reflected by reduced culture absorbance. Light microscopical analysis of S. cerevisiae treated with 10% Pluronic F-68 revealed no obvious change in cell size and thus, this drop in absorbance presumably indicated ultrastructural changes within the cell. The explanation for this effect is obscure but may be related to the observation that the structure of cytoplasmic membranes of mammalian cells is perturbed in the presence of emulsions containing the Pluronic surfactant (Wake et al., 1985). Related work has shown that exposure of yeast cells to 15-30% emulsified FDC produced ultrastructural changes characterised by an increase in cytoplasmic vacuolation, plasmolysis and the appearance of electron dense deposits (Chandler et al., 1987). Any such membrane changes in yeast might be expected to produce wide ranging effects on cell structure and growth, perhaps as a result of osmotic stress but this remains to be determined.

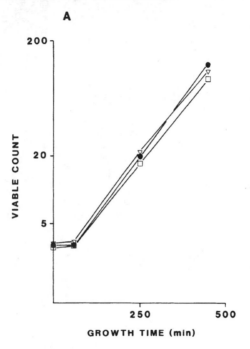

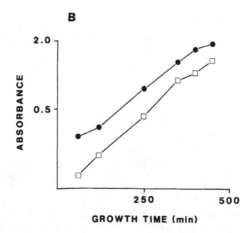

Fig. 2. Changes in A. viable counts (cells x 10^{-6}.ml^{-1}) or B. absorbance (600nm) of <u>S. cerevisiae</u> cultures grown in 5% ($\bigtriangledown$) or 10% ($\square$) Pluronic F-68 at 30°C compared with controls ($\bullet$).

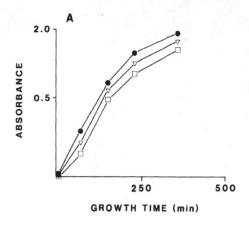

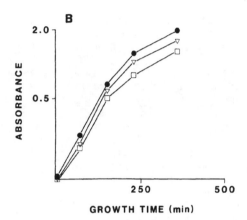

Fig. 3. Changes in absorbance (600nm) of E. coli cultures grown in 5% (▽) or 10% (□) Pluronic F-68 at A. 30°C or B. 37°C compared with controls (●).

In E. coli, the effects of Pluronic F-68 were more subtle: incubation in 5-10% Pluronic had a slight inhibitory effect on cell growth as judged by absorbance, and preliminary data indicates a possible reduction of cell viability. Clearly, these preliminary results suggest differences in the responses of prokaryotic and eucaryotic cells to PFC emulsion constituents but this requires further clarification.

SUMMARY

PFCs and their emulsions may have value for increasing the efficiency of O_2-transport to microbial cultures. Therefore, the effects of emulsion components on growth of S. cerevisiae and E. coli have been examined. Viable cell counts revealed that perfluorodecalin or the commercial emulsion, Fluosol-DA 20%, produced no obvious growth-inhibition over 6h. However, incubation of cells with up to 10% of the Pluronic F-68 surfactant reduced absorbance at 600nm. Further experiments to assess the effects of PFC emulsion components on growth and structure of microbial cells are in progress.

ACKNOWLEDGEMENTS

We are grateful to Drs. T. Suyama and K. Yokoyama of the Green Cross Corporation, Japan, for their gifts of Fluosol-DA. Perfluorodecalin (Flutec PP5) was kindly donated by I.S.C. Chemicals Ltd, Avonmouth, while Pluronic F-68 was generously supplied by the BASF Wyandette Corporation, U.S.A. This work was supported by a grant from the Allied-Lyons Research Fund, University of Nottingham.

REFERENCES

Adlercreutz, P., Holst, O., and Mattiasson, B., 1982, Oxygen supply to immobilized cells: 2. Studies on a co-immobilized algae-bacteria preparation with in situ oxygen generation, Enzyme Microb. Technol., 4:395.

Adlercreutz, P., and Mattiasson, B., 1982, Oxygen supply to immobilized cells: 3. Oxygen supply by hemoglobin or emulsions of perfluorochemicals, Eur. J. Appl. Microbiol. Biotechnol., 16:165.

Bucala, R., Kawakami, M., and Cerami, A., 1983, Cytotoxicity of a perfluorocarbon blood substitute to macrophages in vitro, Science, 220:965.

Buckland, B.C., Dunnill, P., and Lilly, M.D., 1975, The enzymatic transformation of water-soluble reactants in nonaqueous solvents. Conversion of cholesterol to Cholest-4-ene-3-one by a Nocardia sp., Biotechnol. Bioeng., 17:815.

Chandler, D., Davey, M.R., Lowe, K.C., and Mulligan, B.J., 1987, Effects of emulsified perfluorochemicals on growth and ultrastructure of microbial cells in culture, Biotechnol. Letts., 9:185.

Damiano, D., and Wang, S.S., 1985, Novel use of a perfluorocarbon for supplying oxygen to aerobic submerged cultures, Biotechnol. Letts., 7:81.

Holst, O., Enfors, S., and Mattiasson, B., 1982, Oxygenation of immobilized cells using Hydrogen-Peroxide; A model study of Gluconobacter oxydans converting glycerol to dihydroxyacetone, Eur. J. Appl. Microbiol. Biotechnol., 14:64.

Janco, R.L., Virmani, R., Morris, P.J., and Gunter, K., 1985, Perfluorochemical blood substitutes differentially alter human monocyte procoagulant generation and oxidative metabolism, Transfusion, 25:578.

Lowe, K.C., 1986, Blood transfusion or blood substitution?, Vox Sang., 51:257.

Lowe, K.C., and Bollands, A.D., 1985, Physiological effects of perfluorocarbon blood substitutes, Med. Lab. Sci., 42:367.

Mattiasson, B., and Adlercreutz, P., 1984, Use of perfluorochemicals for oxygen supply to immobilized cells, Ann. N.Y. Acad. Sci., 413:545.

Reiss, J.G., and Le Blanc, M., 1982, Solubility and transport phenomena in perfluorochemicals relevant to blood substitution and other biomedical applications, Pure Appl. Chem., 54:2383.

Wake, E.J., Studzinski, G.P., and Bhandal, A., 1985, Changes in human cultured cells exposed to perfluorocarbon emulsion, Transfusion, 25:73.

EFFECT OF HEPATIC BLOOD OXYGENATION ON BILE SECRETION IN RATS

Takakatsu Matsumura, Nobuhiro Sato, Sunao Kawano,
Taizo Hijioka, Hiroshi Eguchi, and Takenobu Kamada

The First Department of Medicine, Osaka University
Medical School Fukushima-ku, Fukushima, Osaka, 553 Japan

SUMMARY

The effect of hepatic hemodynamics and hepatic tissue blood oxygenation
on bile flow was studied in anesthetized rats by reflectance spectrophotometry. The hepatic hemodynamics and blood oxygenation were assessed by
reflectance spetrophotometry. The hepatic ischemia was induced by partial
ligation of portal vein and hepatic hypoxia was induced by inhalation of
low concetration of oxygen.
1. The ischemia decreased hepatic blood volume index and hepatic blood
oxygenation, and diminished bile flow.
2. Respiratory hypoxia suppressed hepatic oxygenation with mininal change
of hepatic blood volume, and it also reduced the bile flow.
3. Bile flow was related hyperbolically with hepatic oxygenation and its
dependency in hepatic ischemia and respiratory hepatic blood hypoxia
identical.
   It is concluded that the hepatic tissue blood oxygenation affects
hepatic energy metabolism, thus affecting the bile secretion.

INTRODUCTION

The energy supply on which liver function depends is produced chiefly
by mitochondrial aerobic metabolism in hepatocytes. The oxygen and
substrates for mitochodrial oxidative phosphorylation are supplied
through the hepatic microcirculation. Therefore, hepatic energy metabolism and cellular function have a relation with hepatic blood flow. We
previously demonstrated that hepatic function in patients with chronic
liver diseases deteriorates as regional hepatic blood flow, volume and
hepatic oxygenation decreased(Sato et al., 1983). The critical level of
hepatic oxygenation for various hepatocellular function, however,
remains undefined.
   The bile secretion is an energy-dependent process involving sodium-
potassium ATPase(Strange, 1984), and is considered to be an index of
hepatic energy status. In this study the relation between the levels of
hepatic tissue oxygenation and bile secretion were examined in anesthetized rats in vivo.

MATERIAL AND METHOD

Sprangue-Dawley rats weighing 200 to 350g were fasted for 24 hours. The abdomen was then incised along the midline under pentobarbital anesthesia (37.5mg/kg intraperitoneally). A fine polyethylene tube was inserted into bile duct, and bile was collected for every 4 min. Bile flow was expressed as microliters per minute. And two series of experiments were performed to obtain various hepatic tissue blood oxygenation levels;(1) In order to diminish the hepatic blood flow, portal vein was constricted to various degree at portal trunk with silk ligature. (2) Rats were ventilated through a tracheal cannula with a mixture of air and nitrogen to suppress oxygen supply to the liver.

Analysis of hepatic hemodynamics and hepatic tissue blood oxygenation by reflectance spectrophotometry

Regional hepatic blood volume index and oxygen saturation of the regional hepatic tissue hemoglobin(Hb-SO2) were determined with reflectance spectrophotomtry as previously reported(Sato et al., 1981). Regional hepatic blood volume index was determined from the difference in absorption between 569nm and 650nm(ΔEr569-650), which had a good corelation with the local tissue hemoglobin concentration. The oxygen saturation of the regional hepatic tissue hemoglobin was estimated from the spectrum of the liver in vivo at wavelengths 569, 577, and 586nm. Regional hepatic oxygen consumption was estimated by the rate of alteration of regional hepatic tissue blood SO2 during pressuring the liver surface with the tip of optic fiber bundle. The spectrophotometric measurement was carried out 3 times during the 4 min of bile collection and mean values were used.

RESULT

Following the partial ligation of portal vein, the regional hepatic blood volume index and oxygen saturation of the hepatic tissue blood oxygenation(Hb-SO2) dropped rapidly to a new steady state. These hemody-

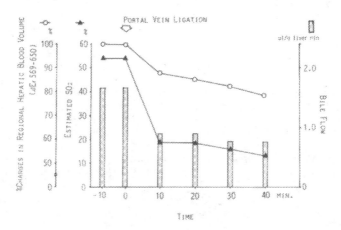

Figure 1. Changes in regional hepatic blood volume, oxygen saturation of regional hepatic tissue hemoblobin, and bile flow of a rat following partial ligation of portal vein.

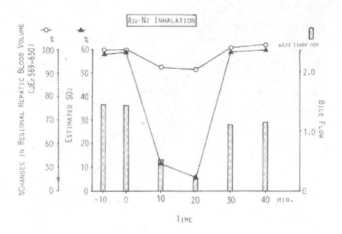

Figure 2. Changes in regional hepatic blood volume, oxygen saturation of
   regional hepatic tissue hemoglobin, and bile flow of a rat following
   air-nitrogen gas inhalation.

namic changes were associated with the decrease in bile flow (Fig. 1).
Various degrees of hepatic tissue hypoxia were induced by the inhalation
of the mixture of air and nitrogen, with minor alteration of hepatic
blood volume index. Bile flow pararelled the changes in hepatic tissue
blood oxygenation(Fig. 2). The relation between hepatic tissue blood
oxygenation and bile flow in cases of ischemia and respiratory hypoxia
were presented in Figure 3. Bile flow was related hyperbolically to the

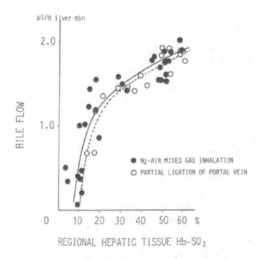

Figure 3. Relationship between bile flow and oxygen saturation of regional hepatic tissue hemoglobin in rats under respiratory hypoxia or
   with partial ligation of portal vein.

hepatic tissue blood oxygenation, and its dependence of bile flow on hepatic tissue blood oxygenation in cases of ischemia was identical to that in cases of hypoxia. Bile flow decreased gradually as hepatic tissue blood oxygenation to 20% and it declined abruptly ar the hepatic tissue blood oxygenation level less than 20%(Fig. 3). Figure 4 showed the relationship between regional hepatic oxygen consumption and bile flow. The bile flow was correlated linearly with regional hepatic oxygen consumption.

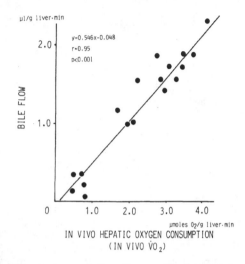

Figure 4. Relationship between bile flow and regional hepatic oxygen consumption in rats under respiratory hypoxia.

DISCUSSION

Bile flow has been shown to fall with the decline in the hepatic energy charge in rats when arterial oxygen tension droped below 60mmHg as a result of inhaling a low concetration of oxygen(Ukikusa et al., 1979). It was also demonstrated that decline of bile secretion corresponded to the fall in hepatic energy charge under hemorrhagic shock in rats without a change in arterial oxygen tension(Ukikusa et al., 1979). From these repors, it is considered that hepatic blood flow and/or hepatic oxygenation may affect the bile secretion. However, no attempt to examine the relationship among hepatic blood flow, hepatic oxygenation and bile secretion was made.

In the present study, we investigated the effect of regional hepatic blood volume index and hepatic tissue blood oxygenation on the bile flow. Hepatic tissue oxygenation was estimated the non-invasive spectrophotometric analysis of hepatic tissue hemoglobin in vivo using reflectance spectrophotometry. Bile flow promptly change when hepatic blood flow and hepatic tissue oxgen levels were altered. The ischmia induced by partial constriction of the portal vein lowered hepatic blood oxygenation and resulted in a decrease in bile flow. Thus, it seem likely that the bile flow depended on the hepatic blood flow. On the other hand, the respiratory hypoxia also diminished the hepatic oxygenation and resulted in suppression of the bile flow. Moreover, the relation of bile flow to the hepatic tissue oxygenation in the case of respiratory hypoxia was identical to that seen in the case of ishemia. Thus, it is likely that bile flow depended on hepatic tissue oxygenation, and the reduction of bile flow in ishemia was mediated by the lowering of hepatic tissue oxygenation.

Bile flow is considered to be a parameter of hepatic cellular energy metabolism because energy-dependent transport process is involved in bile secretion(Strange, 1984). This is consistent with the finding that bile flow was correlated linearly with hepatic oxygen consumption(Fig. 4). Since the bile flow was observed to depend on hepatic blood oxygenation(Fig.3), hepatic energy metabolism is considered to depend on hepatic blood oxygenation. From the dependency of bile flow on hepatic oxygenation(Fig. 3), it seems that hepatic energy metabolism is profoundly depressed when hepatic blood oxygenation level falls to less than 20%. This accords with the observation that few patients with chronic liver disease have Hb-SO2 level below 20%; lower level may be not consistent with survival. Furthermore, bile flow appeared to decrease gradually dependently on hepatic tissue blood oxygenation less than 60%. This evidence suggests the depedency of hepatic energy metabolism on hepatic blood oxygenation even above 20% of hepatic blood HB-SO2. This suggestion may be consistent with our previous demonstrations that the decease in regional blood volume was associated with the lowered hepatic blood oxygenation from 60% to 20% of Hb-SO2, and that the depressed regional blood flow was related to hepatic dysfunction in patients with chronic liver diseases. Therefore, these experimental findings in the present study support the idea that hepatic dydfunction in patients with chronic liver disease is attributable considerably to low hepatic oxygenation.

REFFERENCES

Sato N , Matsumura T, Schichiri M, Kamada T, Abe H ,and Hagihara B, 1981, Hemoperfusion, rate of oxygen consumption and redox levels of mitochondrial cytochrome c(+cl) in liver in situ of anesthetized rats measured by reflectance spectrophotometry. Biochim Biophys Acta 634, 1-10.
Sato N , Hayashi N, Kawano S, Kamada T, and Abe H, 1983, Hepatic hemodynamics in patients with chronic hepatitis or cirrhosis as assessed by organ-reflectance spectrophotometry, Gastroent. 84:611-617.
Strange R.C, 1984, Hepatic bile flow. Physiol. Rev. 64:1055-1102.
Ukikusa M, 1979, The influence of hypoxia and hemorrhage upon adenylate energy charge and bile flow. Surg. Gynecol. and Obst. 149:346-352.

IN VIVO ESTIMATION OF OXYGEN SATURATION OF HEMOGLOBIN IN HEPATIC LOBULES IN RATS

Hiroshi Eguchi, Nobuhiro Sato, Takakatsu Matsumura, Sunao Kawano and Takenobu Kamada

Department of Medicine, Osaka University Medical School Fukushima-ku, Osaka, Japan

SUMMARY

We developed a system for the in vivo estimation of oxygen saturation of hemoglobin in the sinusoidal blood of periportal and pericentral regions of hepatic lobules in rats. There was a marked heterogeneity in blood oxygenation and oxygen gradients within hepatic lobules. The sinusoidal blood oxygenation was reduced in some pericentral regions within hepatic lobules following acute ethanol ingestion, suggesting an occurrence of perivenular hypoxia in the liver after acute ethanol consumption.

INTRODUCTION

Oxygen gradients from periportal(PP) to pericentral(PC) regions within hepatic lobules have long been considered. Ji et al(1982) measured oxygen gradients in perfused livers of rats using NADH fluorescence. However, little is known in in vivo liver. In this study, using in vivo microscopic system(Sato et al., 1986) coupled with a spectrophotometer by optical fiber bundle, we attempted in vivo estimation of oxygen saturation of hemoglobin in the sinusoidal blood localized at PP and PC of hepatic lobules and examined the effect of acute ethanol ingestion on intralobular oxygenation in rats, because the perivenular hypoxia after alcohol consumption may be incriminated in the pathogenesis of alcoholic liver injury.

MATERIAL AND METHOD

Male Sprague-Dawley rats(100-150g) were anesthetized with pento-barbital(35mg/kg, i.p.). Using in vivo microscopic system(Sato et al., 1986), we observed the hepatic microcirculation of a single lobule at the edge of the liver. A diaphragm(Zeiss, d=2.5mm) was set in the camera tube and the plastic fiber bundle of spectrophotometer(UNISOKU, MS-401-01) was attached to the diaphragm(Fig. 1). The microscopic stage was adjusted to center the measuring area(d=87μm) along one sinusoid from PP to PC of hepatic lobules(Fig. 2). We recorded the absorption spectrum of the hemoglobin in the liver tissue by transmitting the light(480-680nm) through the liver. The equation for calculating the oxygen saturation of

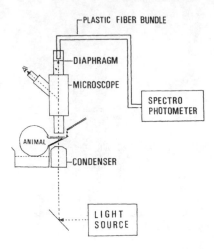

Fig.1. Block diagram of a system for the in vivo estimation of oxygen saturation of hemoglobin in hepatic lobules in rats

hemoglobin(SO_2)(Sato et al., 1981) was

$$SO_2=[\Delta Er_{577}-(9\times\Delta Er_{569}+8\times\Delta Er_{586})/17]/1.485/(\Delta Er_{569}-\Delta Er_{586})$$

Ethanol(1g/kg) was instilled by a gastric tube, and before and 15 min after ethanol ingestion we estimated SO_2 in PP and PC. Statistical analysis were performed with matched-paired Student's t-test.

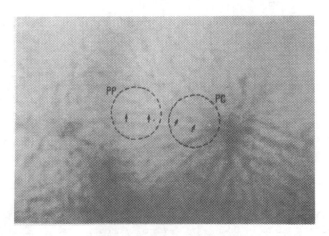

Fig. 2. A representative photograph showing in vivo hepatic lobule in an anesthetized rat,
Circles show the areas(d=87μm) in which transmitted light was guided to the plastic fiber bundle connected with the spectrophotometer. Arrows indicate the sinusoid along which we centered the measuring areas from PP to PC.

RESULTS

Figure 3 shows the representative spectra of one lobule at PP(A) and PC(B). These spectra demonstrated two peaks of oxyhemoglobin(542, 577nm). The intensity of spectrum B was larger than that of spectrum A and the former represented larger hemoglobin concentration than the latter.

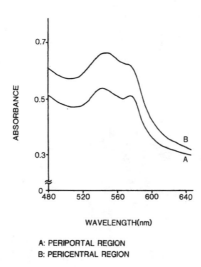

A: PERIPORTAL REGION
B: PERICENTRAL REGION

Fig. 3. Representative spectra of a hepatic lobule at periportal(A) and pericentral(B) regions in an anesthetized rat

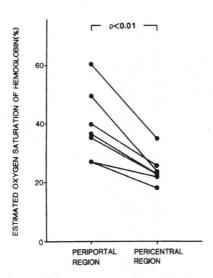

Fig. 4. Comparison of estimated oxygen saturation of hemoglobin between periportal and pericentral regions of hepatic lobules in anesthetized rats

Comparing the SO_2 between PP and PC, the oxygenation in PP were always larger than that in PC, indicating the presence of oxygen gradient from PP to PC. However, the oxygenation level at PP showed a variance(27.1–60.3%). The oxygen gradients within hepatic lobules were also heterogeneous(4.4–25.4%)(Fig. 4).

Following acute ethanol ingestion, the increase of SO_2 was detected in all the measuring areas in PP, while the changes of SO_2 in PC were heterogeneous:4 regions among 6 measured showed an increase while 2 showed a decrease(Fig. 5).

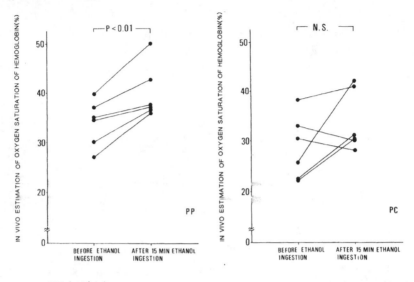

PP:Periportal region
PC:Pericentral region

Fig. 5. Effect of acute ethanol ingestion(1g/kg, p.o.) on estimated oxygen saturation of hemoglobin in periportal and pericentral regions of hepatic lobules in anesthetized rats

DISCUSSION

We developed a system for the in vivo estimation of oxygen saturation of hemoglobin in the sinusoidal blood of hepatic lobules using in vivo microscopic system coupled with a spectrophotometer. The edge of the liver in rats is composed of a single hepatic lobule. Therefore, this region is appropriate for the measurement of tissue hemoglobin oxygenation in PP and PC separately by the means of transmitted light. We set up a diaphragm(d=2.5mm) in the camera tube at the focus plane of the objective image and calculated the measuring area from the formula:

$$d = \frac{D}{M \times OF \times TF}$$

d:diameter of measuring area
D:diameter of inserted diaphragm
M:objective magnification
OF:optovar factor
TF:tube factor

In this study, we selected D=2.5mm, M=23, OF=1.25, TF=1, thus d=87μm. As sinusoidal length was reported to range from 200 to 500μm(Kessler et al., 1973), the diameter of measuring area(87μm) was small enough to examine the oxygenation of hemoglobin in the blood in PP and PC separately(Fig.2). The classical lobules were separated into several hepatic units(Rappaport et al., 1954; Matsumoto et al., 1979) and considerable differences of the hemodynamics among these hepatic units were demonstrated(Rappaport, 1973). Therefore, in order to estimate oxygen gradients within hepatic lobules, we settled the measuring areas along one sinusoid from PP to PC and analyzed the regional hemoglobin absorbance at these two areas. The spectral intensity was larger in PC than PP, and this finding was consistent with the report that the vascular bed of PC was greater than that of PP(Miller et al., 1979).

We found heterogeneous oxygenation in PP and PC, in which many anatomical or physiological factors are presumed to be involved. Ji et al(1982) reported the similar heterogeneity in the hemoglobin-free perfused livers via portal vein without the inflow from the hepatic artery, and they attributed the heterogeneity to the architectural difference of lobular size. The other possible factor is the different contribution of hepatic arterial blood to sinusoidal blood flow(McCuskey, 1966). On the other hand, we reported a marked heterogeneity in sinusoidal blood flow(Sato et al., 1986). Further investigations about hepatic arterial termination and three-dimensional sinusoidal structure are needed.

Heterogeneous oxygen gradients were also reported in the perfused livers of rats(Ji et al., 1982). The heterogeneous sinusoidal blood flows and the difference in sinusoidal length may affect the different oxygen delivery.

Acute ethanol ingestion rose the oxygenation in PP. This rise might be caused by the blood with higher oxygenation. On the other hand, the changes of oxygenation in PC were variable. Previous reports claimed that increased hepatic oxygen consumption due to ethanol metabolism was fully compensated by the increased portal blood flow as whole organ(Shaw et al., 1977; Sato et al., 1980). The decreases of oxygen saturations in some PC, however, suggested the regional imbalance between oxygen supply and demand. We reported heterogeneous responses in sinusoidal blood flow to acute ethanol ingestion(Eguchi et al., 1987). These heterogeneous responses may contribute to the heterogeneous changes of oxygenation in PC, which might induce perivenular hypoxia in some PC of hepatic lobules after acute ethanol consumption.

REFERENCES

   Eguchi, H., Sato, N., Minamiyama, M., Matsumura, T., Kawano, S., and
      Kamada, T., 1987, Heterogeneous response in hepatic sinusoidal
      blood flow to acute ethanol ingestion in rats, Alcohol &
      Alcoholism, Suppl.1:519-521.
   Ji, S., Lemasters, J. J., Christenson, V., and Thurman, R. G., 1982,
      Periportal and pericentral pyridine nucleotide fluorescence
      from the surface of the perfused liver:Evaluation of the
      hypothesis that chronic treatment with ethanol produces
      pericentral hypoxia, Proc. Natl. Acad. Sci., 79:5415-5419.
   Kessler, M., Lang, H., Sinagowitz, E., Rink, R., and Höper, J., 1973,
      Homeostasis of oxygen supply in liver and kidney, in: "Oxygen
      transport to tissue," H. I. Bicher, and D. F. Bruley, eds.,
      Plenum Press, New York.
   Matsumoto, T., Komori, R., Magara, T., Ui, T., Kawakami, M.,
      Tokuda, T., Takasaki, S., Hayashi, H., Jo, K., Hano, H.,
      Fujino, H., and Tanaka, S., 1979, A study on the normal
      structure of the human liver, with special reference to its
      angioarchitecture, Jikeiikai. Med. J., 26:1-40.
   McCuskey, R. S., 1966, A dynamic and static study of hepatic
      arterioles and hepatic sphincters, Am. J. Anat., 119:455-478.
   Miller, D. L., Zanolli, C. S., and Gumucio, J. J., 1979, Quantitative
      morphology of the sinusoids of the hepatic acinus:Quantimet
      analysis of rat liver, Gastroent., 76:965-969.

Rappaport, A. M., Borowy, Z. J., Lougheed, W. M., and Lotto, W. N., 1954, Subdivision of hexagonal liver lobules into a structural and functional unit;role in hepatic physiology and pathology, Anat. Rec., 119:11-34.

Rappaport, A. M., 1973, The microcirculatory hepatic unit, Microvasc. Res., 6:212-228.

Sato, N., Kamada, T., Schichiri, M., Matsumura, T., Abe, H., and Hagihara, B., 1980, Effect of ethanol on hemoperfusion and O_2 sufficiency in livers in vivo, Adv. Exp. Med. Biol., 132:355-362.

Sato, N., Matsumura, T., Schichiri, M., Kamada, T., Abe H., and Hagihara, B., 1981, Hemoperfusion, rate of oxygen consumption and redox levels of mitochondrial cytochrome $c(+c_1)$ in liver in situ of anesthetized rat measured by reflectance spectrophotometry, Biochim. Biophys. Acta., 634:1-10.

Sato, N., Eguchi, H., Inoue, A., Matsumura, T., Kawano, S., and Kamada, T., 1986, Hepatic microcirculation in Zucker fatty rats, in: "Oxygen transport to tissue VIII," I. S. Longmuir, ed., Plenum. Pub. Cor., New York.

Shaw, S., Heller, E., Friedman, H., Baraona, E., and Lieber, C. S., 1977, Increased hepatic oxygenation following ethanol administration in the baboon, Proc. Soc. Exp. Biol. Med., 156:509-513.

MUSCLE PO$_2$ IN THE INITIAL PHASE OF INCREASED LOCAL OXYGEN DEMAND (RHYTHMICAL MUSCLE CONTRACTION) IN RATS WITH PORTA-CAVAL SHUNT (PCA)

M. Günderoth-Palmowski, R. Heinrich[+], P. Palmowski, S. Dette, W. Grauer, and E. H. Egberts

Med. Univ. Klinik, Abtlg. I, 7400 Tübingen, FRG
+ Ruhr Univ. Bochum, Marienhospital 2, 4690 Herne

INTRODUCTION

Portacaval anastomosis (PCA) in rats induces a time-limited hyperdynamic cardiovascular state. This hypercirculatory syndrome is characterized by increased cardiac output, increased systolic ejection rate, and decreased total peripheral resistance (TPR) (Liehr et al. 1976). The development of portahepatic collaterals coincides with normalization of cardiac output, the TPR however remains low (Heinrich 1987).

Mean tissue oxygenation (PO2) of skeletal muscle is increased directly after the PCA-operation and then decreases continuously to preoperative PO2-values (Heinrich 1987). Several studies concerning tissue oxygenation of skeletal muscle after PCA indicate that a time-limited hypercirculation considerably influences the regulating mechanisms of the microcirculation as no normal PO2 distribution could be observed (Boeksteders et al. 1986, Heinrich 1987).

The purpose of this study was to evaluate the influence of PCA on PO2 distribution in skeletal muscle in the initial phase of a rhythmical contraction period.

MATERIAL AND METHODS

Female Sprague-Dawley rats weighing 167.5 + 12.6 g were subjected to PCA by the technique of Bismuth (1963). 56 days after operation tissue PO2 of m. biceps femoris was measured in 8 PCA-rats and 8 age-matched female controls without any surgical trauma. All operations and PO2-measurements were done in the steady-state under continuous ether anaesthesia.

Rhythmical contraction of m. biceps femoris was induced by direct electrical stimulation (2 V, 2 Hz, 10 ms duration). This type of muscle activity was choosen because of its similarity to muscular contraction in physical exercise.

Muscle PO2 was recorded continuously from the start of

the 210 s-long contraction period and up to 140 s after the last contraction. PO2-histograms were taken before and additional at the same time throughout (at 120 s) and after contraction (at 30 s and at 90 s) using a multiwire surface electrode according to Kessler and Lübbers (1966).

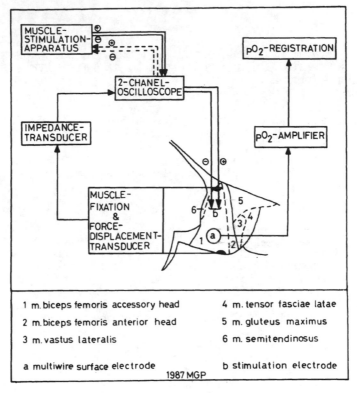

Fig. 1. Experimental design.

RESULTS

Body weight. At the time of PO2-measurement the mean body weight of controls and PCA-rats respectively was 238.3 $\pm$ 15.6 g and 174.3 $\pm$ 22.8 g. All PCA-animals had reached their preoperative body weight by 1o days after operation.

PO2 at rest. Fig. 2 shows the pooled histograms for controls (left) and PCA-rats (right). The mean PO2 of controls was 28.2 mm Hg, in PCA-rats the mean was 29.5 mm Hg without PO2-values in the lower PO2-classes.

Continuously recorded PO2. Fig. 3 shows the continuously recorded PO2 of controls (left) and PCA-rats (right). In both groups mean PO2 increased within the first 30 s of muscle contraction. In controls the PO2 then decreased stepwise to PO2-values which were above the PO2 at rest. In contrast in rats with PCA the PO2 decreased continuously. At the end of contraction in controls muscle PO2 decreased to a lower PO2 (2o mm Hg) than in PCA-animals (25 mm Hg) (p<0.1).

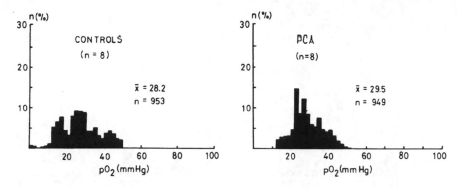

Fig. 2. Pooled histograms of muscle PO2 at rest for controls
(left) and PCA-animals (right).
$\bar{x}$ = mean muscle PO2
n = number of points measured

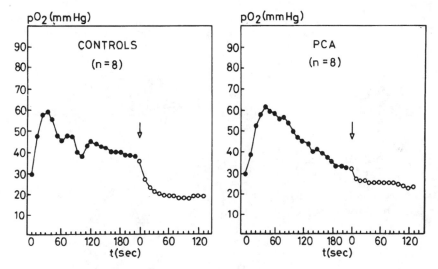

Fig. 3. Continuously recorded muscle PO2 during and after
rhythmical muscle contraction in controls (left) and
PCA-animals (right). The arrows indicate the end of
contraction.

PO2-histograms. Fig. 4 shows the pooled histograms of
controls and rats with PCA during and after the contraction
period. During contraction (at 120 s) mean muscle PO2 of controls was about 43.5 mm Hg and the histogram shows a wide
distribution of PO2-values (between 15 and 80 mm Hg). In PCA-
animals mean PO2 was 42.2 mm Hg and the PO2-values were distributed between 10 and 100 mm Hg. At 30 s and 90 s after the
contraction mean muscle PO2 in controls decreased to 20.0 and
18.9 mm Hg. Both histograms are homogenously distributed. In
rats with PCA the PO2 decreased to 25.4 and 24.3 mm Hg.

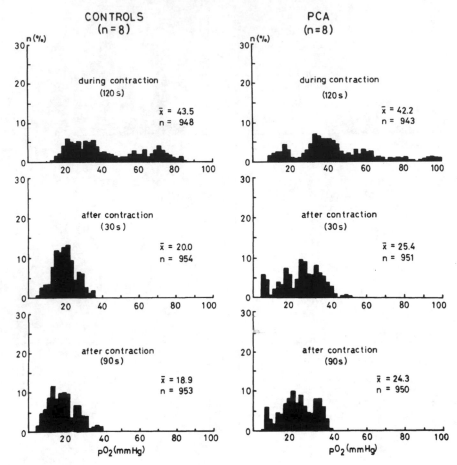

Fig. 4. Pooled histograms of skeletal muscle PO2 during and
after contraction in controls (left) and PCA-animals
(right).
x̄ = mean muscle PO2
n = number of points measured

DISCUSSION

A hypercirculatory syndrome in rats with PCA leads to an
increase of mean muscle PO2 at rest. In the postoperative time
development of portahepatic collaterals coincides with normalization of cardiac output and decreasing values of mean muscle
PO2, whereas PO2-histograms exhibit signs of maldistribution
(Liehr et al. 1976, de Boer et al. 1984, Heinrich 1987).

With the start of muscle contraction mean tissue PO2 increased in controls and in rats with PCA. In both groups muscle
PO2 decreased during contraction. In the initial phase of contraction the mean muscle PO2 in animals with PCA did not differ
from that in controls. In both groups the wide distribution of
PO2-values indicates an increased muscle blood flow during
exercise (Schroeder 1978).

In controls mean tissue PO2-values remained above those measured at rest, whereas muscle PO2 in rats with PCA declined to values comparable to those at rest.

After the end of rhythmical contraction mean muscle PO2 decreased in both groups. This finding is comparable with data from other authors. Schroeder (1978) described a decrease of tissue PO2 after muscle activity and King et al. (1986) found reduced oxygen uptake after contraction. Rosell (1962) published that vasodilatation and increased blood flow due to vasodilator nerve activity induces declined oxygen uptake. Hyman (1959) suggested that vasodilatory nerves serve to enhance non-nutritional blood flow after muscle activity.

Mean tissue PO2 in rats with PCA did not decrease as much as in controls after contraction, furthermore PO2-histograms in rats with PCA were distributed inhomogenously. This different behaviour of tissue PO2 in rats with PCA after muscle contraction might be a further indication of a disturbance of microcirculatory regulation due to the hyperciru-latory syndrome.

SUMMARY

The influence of a hyperdynamic syndrome caused by PCA on PO2 distribution in skeletal muscle of rats during the initial phase of muscle activity was examined. Rhythmical muscle contraction of the m. biceps femoris was induced by direct electrical stimulation. Tissue PO2 of the contracting muscle was recorded continuously from the start of the 210 s-long activity period up to 140 s after the last contraction using a multiwire surface electrode.

In comparison with controls no different behaviour of mean muscle PO2 in the initial phase of contraction was found. After muscle activity mean PO2 decreased to a lower level in rats with PCA than in controls. This might be a further indication of the disturbing influence of a hyperdynamic syndrome on the regulating mechanisms of the microcirculation.

REFERENCES

Bismuth, H., Benhamou, J. P., Lataste, J., 1963, L'anastomose portocave experimentale chez le rat normal. Technique et resultats preliminaires, Presse med., 71:1859.
Boekstegers, P., Heinrich, R., Günderoth-Palmowski, M., Grauer, W., Fleckenstein, W., Schomerus, H., 1986, Tissue-PO2 in isometrically contracting skeletal muscle of rats with portacaval anastomosis (PCA). Adv. exp. med. biol., 200: 429.
Boer, J. E. de, Ostenbroek, R. J., Dongen, J. J. van, Janssen, M. A., Soeters, P. B., 1986, Sequential metabolic characteristics following portacaval shunt in rats, Eur. Surg. Res., 18:96
Heinrich, R., 1987, Messung des Sauerstoffdrucks im Gewebe bei Leberatrophie und Leberzirrhose. Tierexperimentelle Studien und Untersuchungen an Patienten. Univ.-Buchdruckerei Junge und Cohn, Erlangen.
Hyman, S., Rosell, S., Rosen, A., Sonnenschein, R. R., Uvnäs,

B., 1959, Effects of alteration of total muscular blood
flow on local tissue clearance of radio-iodide in the
cat, Acta physiol. scand., 46:358.

Kessler, M., Lübbers, D. W., 1966, Aufbau und Anwendungsmöglichkeiten verschiedener PO2-Elektroden. Pflügers Arch.
ges. Physiol., 291:82.

King, C. E., Dodd, S. L., Stainsby, W. N., Cain, S. M., 1986,
Does contracting muscle restore its energy balance
following severe hypoxia?, Adv. exp. med. biol., 200:539.

Liehr, H., Grün, M., Thiel, H., 1976, Hepatic blood flow and
cardiac output after portacaval anastomosis in the rat,
Acta Hepato.-Gastroenterol., 23:31.

Rosell, S., Uvnäs, B., 1962, Vasomotor nerve activity and
oxygen uptake in skeletal muscle of the anaesthetized
cat, Acta physiol. scand., 54:209.

Schroeder, W., 1978, Die Messung des Sauerstoffdruckes in der
Skelettmuskulatur - Eine quantitative Methode zur Kon-
trolle der Sauerstoffversorgung und Funktion der termina-
len Muskelstrombahn, Herz/Kreislauf, 10:146.

Acknowledgement

Supported by Deutsche Forschungsgemeinschaft (HE 1293/1-2)

CIRCADIAN COURSE OF PO_2-OSCILLATIONS IN SKELETAL MUSCLE OF INTACT AND PORTACAVAL SHUNTED (PCA) RATS

M. Günderoth-Palmowski, R. Heinrich[+], P. Palmowski,
S. Dette, W. Daiss, W. Grauer, and E. H. Egberts

Med. Univ. Klinik, Abtlg. I, 7400 Tübingen, FRG
+ Ruhr Univ. Bochum, Marienhospital 2, 4690 Herne

INTRODUCTION

The measurement of tissue PO_2 is a valid parameter of microcirculation. In healthy subjects variations of mean tissue PO_2 appear to be due to oscillations of heart rate, blood pressure, and peripheral vasomotor changes. An influence of circadian rhythm seems to be possible too (Lübbers 1985).

The deprivation of the liver of portal blood supply by portacaval shunting in healthy rats (PCA) induces time-limited hemodynamic changes (Liehr et al. 1976) and neuro-chemical alterations (Curzon et al. 1975, Cummings et al. 1976, James et al. 1978) as well as changes in electroencephalo-graphic waves (Monmaur et al. 1976, Beaubernard et al. 1977). Campbell (1984) found a loss of circadian locomotor activity.

Mean tissue PO_2 of skeletal muscle is increased directly after operation possibly due to the hyperdynamic syndrome. With normalization of the cardiovascular state mean muscle PO_2 decreases back to preoperative values (Heinrich 1987).

If tissue PO_2 is influenced by circadian rhythm, variations of mean PO_2 of skeletal muscle reflecting a circadian oscillation should be measurable. Therefore the aim of this study was to determine whether muscle PO_2 oscillations are circadian and whether they are altered after time-limited hypercirculation caused by PCA.

MATERIAL AND METHODS

Female Sprague-Dawley rats with a body weight of 142.2 $\pm$ 48.2 g were subjected to portacaval anastomosis by the technique of Bismuth (1963). All operations and PO_2-measurements were carried out during the steady-state under continuous ether anaesthesia. During the whole experiment (from PCA-operation until the last PO_2-measurement) animals with PCA and age-matched female control rats (without any surgical trauma) were kept under constant conditions:

- natural day-night-changes
- air humidity 42 %
- temperature 22-23 °C
- free access to food and water.

Tissue PO2 of m. rectus abdominis of 9 rats with PCA and 10 controls was measured with a multiwire surface electrode according to Kessler and Lübbers (1966) 126 days after operation. In order to get enough values for a circadian course each animal was measured 6 times at intervals of 4 hours. Then curves for each animal and for both groups were calculated. To get a better opinion on the circadian course, each curve was doubled and by this a PO2-course over a 48 hour period was obtained.

RESULTS

Fig. 1 shows the circadian course of mean muscle PO2 of control and PCA-rats. In controls the course was characterized by an amplitude of 14 mm Hg and a period length of approximatly 24 hours, thus the mean PO2-course resembled a sine oscillation. In rats with PCA the course of mean muscle PO2 showed 3 oscillations with different amplitudes, each with a period length of 24 hours.

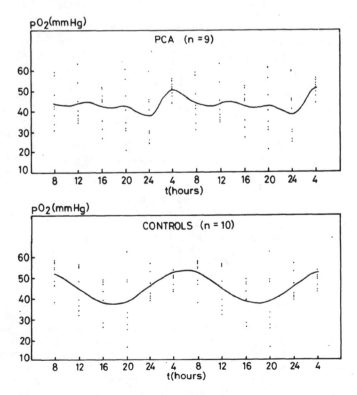

Fig. 1. Circadian course of mean muscle PO2 of controls and rats with PCA 126 days after operation.

Fig. 2 shows the individual PO2 curve of control rat
number 3. This curve is composed of 2 oscillations and was
typical of 90 % of controls, only 1 animal showed a sine
oscillation.

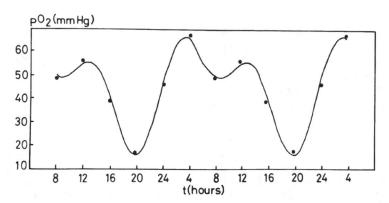

Fig. 2. Circadian course of mean muscle PO2 of control rat
        number 3.

Fig. 3 shows the individual PO2 course of PCA-rat number
173. This curve consists of 2 oscillations, one has an amplitude of 22 mm Hg and the other of 34 mm Hg, both with a
period length of 24 hours. This course too was typical of
90 % of PCA-rats (1 animal showed a sine oscillation). 4 animals showed a PO2-maximum at 20 hr and at 8 hr, 3 at 16 hr
and at 4 hr and 1 at 12 hr and 24 hr. That means that a PO2-
maximum was attained every 12 hours.

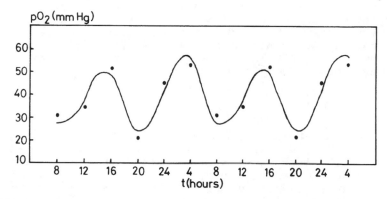

Fig. 3. Circadian course of mean muscle PO2 of PCA-rat
        number 173, 126 days after operation.

DISCUSSION

    In biological systems measurements of all kinds of
circadian oscillations have to be made very carefully. Otherwise the measurement itself triggers (zeitgeber) the examined
oscillation.

In controls the circadian course of mean muscle PO2 resembled a sine oscillation: elevated PO2-values at night and in the morning and low values in the afternoon. This behaviour of mean muscle PO2 coincides with the known circadian rhythm of heart rate, body temperature, and locomotor activity (Meinrath et al. 1978).

The individual values of mean muscle PO2 of controls showed a circadian course, obviously composed of two different components: an endogenous oscillator, driving circadian locomotor activity, and an exogenous one, which is the effect of entrainment of locomotor activity to animal house routine (Dhume et al. 1982). Benessiano et al. (1983) found heart rate and body temperature to increase simultaneously with the beginning of the dark period. In contrast to heart rate and body temperature mean muscle PO2 seems to be influenced by locomotor activity, because PO2 increased very slowly, reaching its first maximum some time after the beginning of the dark period.

The individual values of mean muscle PO2 of portacaval shunted rats were characterized by two oscillations with different amplitudes, each with a period length of 24 hours, whereas the period length between the maxima of PO2-values was 12 hours. In contrast to controls there was no evidence for comparable circadian oscillations of mean muscle PO2 in rats with PCA. This finding coincides with data from the literature. Campbell et al. (1984) described a loss of locomotor activity from the 14th to the 84th postoperative day due to an increase of locomotor activity during the light period. Monmaur et al. (1976) and Beaubernard et al. (1977) found a decrease of slow wave sleep activity and of paradoxical sleep activity during the daylight hours. At a late postoperative stage (98th day) they observed a complete recovery.

It is very difficult to estimate the possible disturbing influence of PO2-measurements on the circadian rhythm of the rat organism. On the other hand since rats with PCA and controls were examined under the same conditions, one can conclude that the demonstrated alterations of circadian oscillations of muscle PO2 in rats with PCA might be due to the former hypercirculatory syndrome.

SUMMARY

In rats several circadian rhythms such as heart rate, body temperature, and locomotor activity are known. Several authors found a loss of day-night-rhythm (locomotor activity, EEG) after portacaval shunting (PCA). The aim of this study was to evaluate whether muscle PO2 oscillations are circadian and whether they are altered after time-limited hypercirculation caused by PCA.

126 days after operation tissue PO2 of m. rectus abdominis of 9 rats with PCA and 10 controls was measured with a multiwire surface electrode. All animals were kept under constant conditions and each animal was measured 6 times at intervals of 4 hours in order to get a circadian PO2 course.

In controls the circadian course of mean muscle PO2 resembled a sine oscillation with high values at night and low values in the afternoon. In PCA-rats the time course of mean muscle PO2 showed 3 oscillations with different amplitudes, each with a period length of 24 hours. Our results indicate that oscillations of muscle PO2 are determined principal by circadian locomotor activity and that time-limited hypercirculation influences the circadian course of mean muscle PO2.

REFERENCES

Beaubernard, C., Salomon, F., Granger, D., Thangrapregassam, M. K., Bismuth, H., 1977, Experimental hepatic encephalopathy. Changes of the level of wakefulness in the rat with portacaval shunt, Biomedicine, 27:69.

Benessiano, J., Levy, B., Samuel, J. L., Leclercq, J. F., Safar, M., Saumont, R., 1983, Circadian changes in heart rate in unanaesthetized normotensive and spontaneously hypertensive rats, Pflügers Arch., 397:70.

Bismuth, H., Benhamou, J. P., Lataste, J., 1963, L'anastomose portocave experimentale chez le rat normal. Technique et resultats preliminaires, Presse med., 71:1859.

Campbell, A., Jeppsson, B., James, J. H., Ziparo, V., Fischer, J. F., 1984, Spontaneous motor activity increase after portacaval anastomosis in rats, Pharm. Biochem. & Behav., 20:875.

Cummings, M. G., Soeters, P. B., James, J. H., Keane, J. M., Fischer, J. E., 1976, Regional brain indoleamine metabolism following chronic portacaval anastomosis in the rats, J. Neurochem., 27:501.

Curzon, G., Kantamaneni, B. D., Fernando, J. C., Woods, M. D., Cavanagh, J. B., 1975, Effects of chronic portacaval anastomosis on brain tryptophan, tyrosine and 5-hydroxytryptamine, J. Neurochem., 24:1065.

Dhume, R. A., Gogate, M. G., 1982, Water as entrainer of circadian running activity in rat, Physiol. & Behav., 28:431.

Heinrich, R., 1987, Messung des Sauerstoffdrucks im Gewebe bei Leberatrophie und Leberzirrhose. Tierexperimentelle Studien und Untersuchungen an Patienten, Univ.-Buchdruckerei Junge und Sohn, Erlangen.

James, J. H., Escourrou, J., Fischer, J. E., 1978, Blood-brain neutral amino acid transport activity is increased after portacaval anastomosis, Science, 200:1395.

Kessler, M., Lübbers, D. W., 1966, Aufbau und Anwendungsmöglichkeiten verschiedener PO2-Elektroden, Pflügers Arch. ges. Physiol., 291:82.

Liehr, H., Grün, M., Thiel, H., 1976, Hepatic blood flow and cardiac output after portacaval anastomosis in the rat, Acta Hepato.-Gastroenterol., 23:31.

Lübbers, D. W., 1985, in: Klinische Sauerstoffdruckmessung. Gewebesauerstoffdruck und transcutaner Sauerstoffdruck bei Erwachsenen, A. M. Ehrly, J. Hauss, R. Huch, eds., Münchner Wissenschaftliche Publikationen, München.

Meinrath, M., D'Amato, M. R., 1979, Interrelationships among heart rate, activity, and body temperature in the rat, Physiol. & Behav., 22:491.

Monmaur, P., Beaubernard, C., Salomon, F., Grange, D., Thangrapregassam, M. J., 1976, Encephalopathie hepatique experi-

mentale. I. Modifications de la duree des differents
etats de sommeil diurne chez le rat avec anastomose portocave, Biol. Gastroenterol., 9:99.

Supported by Deutsche Forschungsgemeinschaft (HE 1293/1-2)

THE EFFECT OF AMINO ACID INFUSION ON PARTIAL PRESSURE OF OXYGEN IN PELVIC URINE OF THE RAT

Gosuke Inoue and Kenji Toba

Department of Medicine, Tokyo Metropolitan Tama Geriatric Hospital, Higashimurayama-shi, Tokyo 189, Japan; and Department of Geriatrics, Faculty of Medicine, University of Tokyo Tokyo 113, Japan

INTRODUCTION

It is well known that glomerular filtration rate (GFR) and urinary concentrating ability are reduced with advancing age. However, the precise mechanism of the aging kidney is unknown. Previous studies in this laboratory have demonstrated that changes in insoluble collagen and acid mucopolysaccharides (glycosaminoglycans) content occur during aging in the renal papilla of the human kidney. This evidence suggests that, in the aging kidney, changes occur in the renal medulla in parallel with the renal cortex. One of the causes of the age-related changes in connective tissue of the renal papilla seems to be local hypoxia (Inoue et al., 1970; Inoue, 1979).

Recently, Brenner et al. (1982) have demonstrated that amino acid infusion results in a marked increase in GFR and have proposed the hypothesis that, with continual high protein intake, sustained hyperfiltration occurs in the renal glomeruli, leading in time to glomerular injury.

Although this hypothesis can explain a decrease in GFR with aging, the effect of amino acid infusion on the renal papilla remains to be determined. The purpose of this study was to examine, by the analysis of the partial pressure of oxygen in the pelvic urine, whether amino acid infusion could modify the local partial pressure of oxygen in the renal papilla. It is known that the partial pressure of oxygen of pelvic urine is in equilibrium with that of the renal medulla (Washington and Holland, 1966).

METHODS

All studies were performed in overnight fasted male Wistar rats weighing 210 to 330 g. Following induction of pentobarbital anesthesia, rats were placed on a temperature-regulated table which maintained body temperature at 36 to 38°C. A tracheostomy was performed and the right femoral vein was cannulated to infuse Ringer's solution and a commercial 3 per cent mixed amino acid solution. The right femoral artery was cannulated for sampling arterial blood. The right ureter was cannulated by a stainless steel pipe catheter to obtain pelvic urine.

After operation, Ringer's solution was infused for 120 minutes initial-

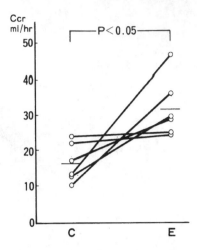

Fig. 1 Effect of amino acid infusion on creatinine clearance (Ccr).
C: control value. E: values after amino acid infusion.

ly at a rate of 9 ml/hour and, then, 20 minute collections of urine were begun to examine urine volume, and sodium and creatinine excretion was measured 2 or 3 times for the calculation of creatinine clearance (Ccr). Next,
a commercial 3 per cent amino acid solution was infused and, after 60 minutes, clearance studies were repeated. The amino acid infusion rate of 270
mg/hour was approximated to that of Meyer et al. (1983). Blood samples for
serum creatinine, sodium and arterial blood gas analysis were obtained at
the mid point of the clearance studies. Ccr was used as a measure of GFR.
Arterial partial pressure of oxygen (PO_2) and urinary partial pressure of
oxygen (UPO_2) were analyzed by a IL system 1303 pH/blood gas analyzer.
Fractional excretion of sodium (FENa, %) was calculated by dividing excreted
sodium by filtered sodium. Filtered sodium was calculated by multiplying
GFR by corresponding plasma sodium.

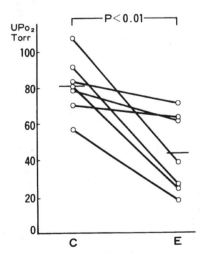

Fig. 2 Effect of amino acid infusion on partial pressure of
oxygen in pelvic urine (UPO_2).

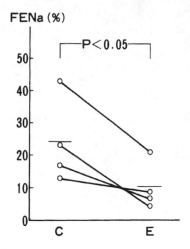

Fig. 3 Changes in fractional excretion of sodium (FENa, %) after
amino acid infusion.

RESULTS

Amino acid infusion resulted in a significant increase in Ccr (p<0.05).
Ccr increased by 45 per cent above the control value of 16 ml/hour (mean),
measured during infusion of Ringer's solution (Fig. 1). Increases of Ccr
with amino acid infusion were associated with significant increases in urine
flow rate from 4.0 ± 0.6 to 7.3 ± 0.8 ml/hour.

On the other hand, UPO_2 decreased by 50 per cent after amino acid infusion compared with the control value (p<0.01; Fig. 2). FENa also decreased significantly with amino acid infusion (p<0.05; Fig. 3). However,
as shown in Fig. 4, PO_2 remained unchanged.

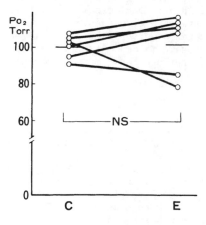

Fig. 4 Effect of amino acid infusion on partial pressure
of oxygen of arterial blood (PO_2).

DISCUSSION

The present study has shown that amino acid infusion elicits local hypoxia in the renal medulla. The decrease in FENa suggests that oxygen consumption occurs owing to sodium reabsorption. The findings may have at least two clinical implications; (1) acute response and (2) chronic effect.

With regard to the acute response, medullary hypoxia occurs in various clinical settings in patients with hyperalimentation. In this connection, it is interesting that amino acid infusion causes a significant increase in metabolic rate and minute ventilation without systemic hypoxia (Weissman et al., 1983). Whether this hyperventilation reaction is an adaptive response to renal medullary hypoxia remains to be defined. Although hypoxia itself does not have a direct deleterious effect on renal tubular function (Cohen, 1979; Gotshall et al., 1983), a recent study of Brezis et al. (1986) indicated that, in the isolated perfused rat kidney, substrate-supportive aerobic metabolic activity for tubular transport may cause hypoxic injury. In this regard, amino acid plays an important role in renal medullary hypoxia.

On the other hand, chronic effects of medullary hypoxia should be taken into consideration in chronic renal diseases including the aging kidney. Since the metabolism of collagen and glycosaminoglycans is modified by local oxygen tension (Hadhazy et al., 1963; Chvapil et al., 1970), age-related changes in connective tissue of the renal papilla may be explained by medullary hypoxia. Thus, high protein intake may affect not only the renal cortex, but also the renal medulla through hypoxia during aging.

SUMMARY

Renal function decreases with age, but the precise mechanism of the decrease is unknown. Recently, Brenner et al. have proposed the hypothesis that high protein intake may injury renal function in various clinical settings including the aging kidney. Since connective tissue in the renal papilla changes with age, high protein intake may have some deleterious effect on the renal papilla. In this study , the effect of amino acid infusion on partial pressure of oxygen of pelvic urine was examined in the rat. After amino acid infusion, glomerular filtration rate significantly increased and partial pressure of oxygen of pelvic urine decreased, while partial pressure of oxygen of arterial blood remained unchanged. The findings imply that high protein intake may affect the renal papilla both acutely and chronically through renal medullary hypoxia.

REFERENCES

Brenner, B.M., Meyer, T.W. and Hostetter, T.H., 1982, Dietary protein and the progressive nature of kidney diseases, N. Eng. J. Med., 307: 652.

Brezis, M., Rosen, S., Spokes, K., Sílva, P. and Epstein, F.H., 1986, Substrates induce hypoxic injury to medullary thick limbs of isolated rat kidneys, Am. J. Physiol., 251: F710.

Chvapil, M., Hurych, J. and Mirejovska, E., 1970, Effect of long-term hypoxia on protein synthesis and in some organs in the rats, Proc. Soc. Exp. Biol. Med., 153: 613.

Gotshall, R.W., Miles, D.S. and Sexson, W.R., 1983, Renal oxygen delivery and consumption during progressive hypoxemia in the anesthetized dog, Proc. Soc. Exp. Biol. Med., 174: 363.

Cohen, J.J., 1979, Is the function of the renal papilla coupled exclusively to an anaerobic pattern of metabolism ?, Am. J. Physiol., 236: F423.

Hadhazy, Cs., Olah, E.H. and Krompecher, St., 1963, Adaptive shift of tissue metabolism in local hypoxia resulting in higher mucopolysaccharide content, Acta biol. Hung., 14: 67.

Inoue, G., Sawada, T., Fukunaga, Y. and Yoshikawa, M., 1970, Levels of acid mucopolysaccharides in aging human kidneys, Gerontologia, 16: 261.

Inoue, G., 1979, Age-related changes in connective tissue of the human kidney and its functional implications, in " Recent advances in gerontology, Proc. XI Internat. Congr. of Gerontol.", H. Orimo, K. Shimada, M. Iriki and D. Maeda, eds, P. 214, Excerpta Medica, Amsterdam.

Meyer, T.W., Ichikawa, I., Zata, R. and Brenner, B.M., 1983, The renal hemodynamic response to amino acid infusion in the rat, Trans. Ass. Amer. Phys., 96: 76.

Washington, J.A.,II and Holland, J.M., 1966, Urine oxygen tension: effects of osmotic and saline diuresis and of ethacrynic acid, Am. J. Physiol., 210: 243.

Weissman, C., Askanazi, J., Rosenbaum, S., Hayman, A.I., Milic-Emili, J. and Kinney, J.M., 1983, Amino acids and respiration, Ann. Intern. Med., 98: 41.

INFLUENCE OF NORMOVOLEMIC HEMODILUTION WITH 10% HES 200[1] ON TISSUE PO_2

IN THE M. TIBIALIS ANT. OF 7 MALE VOLUNTEERS

R. Heinrich*, N. Machac**, M Gunderoth-Palmowski***,
and S. Dette***

*Marienhospital University Clinic, Herne, FRG
**Pfrimmer and Co., Medical Research Dept., Erlangen, FRG
***University Clinic, Tubingen, FRG

INTRODUCTION

The main area of indication for the use of hydroxyethyl starch preparations (HES) is the normovolemic hemodilution in the treatment of peripheral arterial occlusive diseases. There are extensive investigations in patients and healthy subjects available in the literature which demonstrate that the infusion of low molecular HES preparations both changes blood flow properties and, at the same time, improves the oxygen supply to the extremity muscles (Ehrly et al. 1979/1987, Kopp et al. 1982). The studies of the effect of HES infusion or hemodilution on the tissue PO2 in the skeletal muscles were performed either with microneedle electrodes or with multi-wire surface electrodes (Ehrly et al. 1979, survey in Kiesewetter et al. 1986). Both methods can be clinically used to a limited extent only: microneedle probes have the disadvantage of mechanical instability and measurement with the multi-wire surface electrode requires the surgical exposure of the organ. A technically new method for measuring the tissue PO2 in the muscle with fast reacting macro-electrodes[2] (probe diameter 0.3 mm) yields data which are comparable to the results obtained with the microneedle electrode or multi-wire surface electrode (Fleckenstein et al. 1984). The clinical applicability of this PO2 measuring method is guaranteed by the following features: mechanical and electrical stability of the measuring probes, re-usability after sterilization.
The aim of the present study was to examine the changes of tissue PO2 in the M. tibialis anterior of healthy subjects during isovolemic hemodilution with infusion of 500 ml of HES 200.

M e t h o d

Subjects. The study was performed in 7 healthy male subjects, non-smokers, aged between 40 and 57 years (average age: 45.7 years). The study protocol had previously been approved by the Ethics Commission of the Tuebingen University.

---

[1] 10% HES 200/0.5; Pfrimmer & Co. GmbH Erlangen, FRG, in cooperation with Laevosan, Linz, Austria
[2] PO2-Histograph, Eppendorf Geraetebau, Hamburg, FRG

Hemodilution. A sample of 400 ml of venous blood were taken from each subject within 15 min for normovolemic hemodilution. For the determination of laboratory parameters, blood samples of 10 ml were taken at defined times during the course of the study i.e. also during HES infusion (Fig. 1). The total blood volume collected was about 500 ml. It was substituted within 2 hours by infusion of 500 ml of a 10% HES solution (MW 200,000; degree of substitution 0.5).

Clinico-Chemical Parameters. Before, during and after isovolemic hemodilution, the following laboratory parameters were determined together with the PO2 measurements: Hb, Hct, hemogram, serum electrolytes sodium and potassium, serum protein, serum HES concentration, arterial blood gases. Determination of the serum concentration of HES 200 was performed according to Foerster et al. 1981. In addition, closemeshed controls of blood pressure and heart rate were performed.

PO2 Measurement. For measuring the tissue PO2, local anesthesia with 2 ml of Meaverin was strictly limited to the skin. The PO2 needle probes (diameter 0.3 mm, steel-sheated) were gradually inserted into the M. tibialis ant. by a micromanipulator through an Abbocath serving as guide. Further details of this methode have already been published (Fleckenstein 1982, 1984).

Experimental Design. Histograms were recorded at intervals of 30 minutes before, during and after hemodilution (Fig. 1). For each histogram, 200 individual PO2 values were recorded (measuring time 5 minutes). After each histogram an intermediate calibration for calculating the probe drift was performed. The individual histograms of the subjects were pooled to corresponding sum histograms per measuring time, a usual procedure for comparing test results.

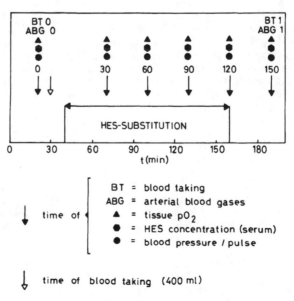

Fig. 1: Experimental design

Results

In none of the 7 cases under observation did infusion of 500 ml of 10% HES 200 per subject result in any allergic reactions. Measurement of tissue PO2 was without complications in all subjects so that all investi-

Tab. 1: Age and laboratory parameters of the test subjects before and after normovolemic hemodilution (* p < 0.0005; n.m. not measured)

Successive no.	1	2	3	4	5	6	7	$\bar{x} \pm s$	Normal value
Age	41	41	57	42	39	48	52	45.7 ± 6.8	
before Hb (g %)	14.8	15.1	14.2	14.0	15.1	15.0	14.7	14.7 ± 0.4	14.0–18.0
after	12.9	13.0	12.3	12.5	11.9	13.0	12.9	12.6 ± 0.4*	
before Hct %	44.5	43.4	40.4	40.9	43.8	46.6	45.1	43.5 ± 2.2	42–52
after	38.8	38.2	38.4	36.2	36.2	39.5	37.6	37.4 ± 1.5*	
before Total protein (g %)	7.4	7.7	6.9	7.2	7.0	7.0	7.5	7.3 ± 0.3	6.5–8.5
after	6.1	6.0	5.5	6.0	5.6	5.9	n.m.	5.9 ± 0.2*	
before Sodium (mEq/l)	142	145	140	138	139	140	142	142 ± 3.8	130–150
after	140	145	134	135	145	146	136	140 ± 5.2	
before Potassium (mEq/l)	3.9	3.6	4.4	4.2	4.0	4.6	4.0	4.1 ± 0.3	3.6–5.2
after	3.9	3.6	4.7	3.9	4.3	4.2	3.7	4.0 ± 0.4	

gations could be performed to the end. Evaluation of blood pressure, heart rate, and measurements of arterial blood gases before, during and after hemodilution yielded values which were all within the normal range; they are therefore not taken into consideration in the representation of results.

Laboratory Parameters. Hemodilution led to an average decrease of the Hb concentration from 14.7 g% ± 0.4 to 12.6 g% ± 0.4 and of the hematocrite (Hct) from 43.5 % ± 2.2 to 37.4 % ± 1.5. Total protein dropped from 7.33 g% ± 0.3 to 5.9 g% ± 0.2. Serum values of sodium (Na+) and potassium (K+) remained unaffected by HES infusion. Table 1 shows the individual values of the subjects.

Figure 2 shows the average course of the serum HES concentration of the subjects during and after infusion. As expected, serum HES values increase during infusion.

Tissue PO2. The mean tissue PO2 in the muscle examined was 16.2 ± 3.5 mmHg before hemodilution (t = 0). As expected, the sum histogram shows an almost bell-shaped configuration. 30 minutes after the start of infusion (t = 30) of HES 200, the mean tissue PO2 rose to 19.1 ± 6.6 mmHg. 60 minutes after the start of infusion (t = 60), a further rise of the tissue PO2 to 27.3 ± 9.3 mmHg was observed. The rise of mean tissue PO2 is paralleled by an increased filling of the PO2 classed above 40 mmHg so that the homogeneous – i.e. bell-shaped – distribution of the PO2 classes within the sum histogram is lost. After 90 minutes (t = 90), the

mean muscle PO2 has slightly dropped to 24.8 ± 7.1 mmHg. After 120 minutes (t = 120), the volume had been restored, but a mean tissue PO2 of 26.4 ± 8.4 mmHg was found, i.e. no further significant change of the mean tissue PO2 as compared to that at time t = 90 was demonstrable.
Thirty minutes after the end of infusion (t = 150), the mean tissue PO2

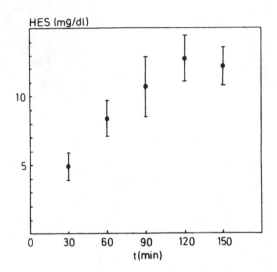

Figure 2: Average course of the serum HES concentration (mg/dl) of the subjects (n = 7) during and after infusion of a 10% HES 200 solution.

remained at 27.1 ± 7.8 mmHg, thus showing no further marked change. The configuration of sum histograms at times t = 90 and t = 120 essentially corresponds to the sum histogram at time t = 60 showing an increased filling of the PO2 classes above 40 mmHg with inhomogeneous histogram configuration. The configuration of the sum histogram at time t = 150 shows an almost bell-shaped distribution as compared with the histograms described above.

Tab. 2: Mean tissue PO2 in the M. tibialis ant. of the subjects before (t = 0), during (t= 30 to t = 120) and after (t = 150) normovolemic hemodilution with 10% HES 200

Successive no.:	1	2	3	4	5	6	7	$\bar{x} \pm s$
t = 0	20.6	17.9	14.7	17.0	19.0	13.4	10.5	16.2 ± 3.5
t = 30	25.9	19.7	22.0	20.4	19.6	6.7	11.3	19.1 ± 6.6
t = 60	42.4	27.7	23.7	33.7	21.4	28.8	13.8	27.3 ± 9.3
t = 90	30.0	29.6	17.1	21.7	17.4	35.6	21.1	24.8 ± 7.1
t = 120	37.3	23.8	26.2	33.8	18.2	31.6	14.2	26.4 ± 8.4
t = 150	27.2	33.0	15.9	19.3	24.3	37.2	32.9	27.1 ± 7.8

The values of the mean muscle PO2 (in mmHg) of the individual subjects at the different measuring times are listed in table 2.

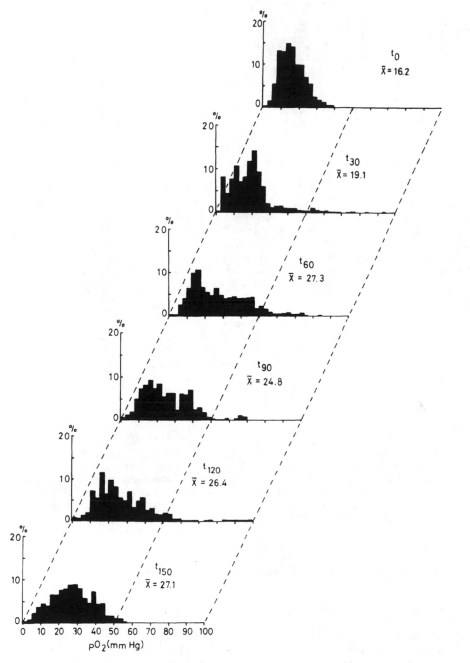

Figure 3: Sum histograms of tissue PO2 of the M. tibialis ant. of 7 healthy subjects before (t = 0), during (t = 30 to t = 120 incl.) and after (t = 150) isovolemic hemodilution with 500 ml of 10% HES 200. (Number of PO2 measuring points per sum histogram: 1400; x = mean value of tissue PO2 (mmHg)).

Discussion

The main clinical field of application of hydroxyethyl starch (HES) with a molecular weight of 200,000 and a degree of substitution of 0.5 is volume deficiency of various origin. In addition, it is administered for therapeutic hemodilution. This principle of reducing the relative volume portion of blood cells - i.e. lowering of the hematocrit value - as a simple therapeutic measure, for example within the scope of stage-adapted therapy of peripheral arterial occlusive disease, is called isovolemic hemodilution. An important therapeutic effect of isovolemic hemodilution with colloidal substances is their positive influence on the flow properties of blood (Ehrly et al. 1979, Landgraf et al. 1981), which leads to an improvement of oxygen supply to the tissues, as direct measurements of tissue PO2 with the multi-wire surface electrode in patients have shown. According to studies performed by Kopp et al. in 1982, isovolemic hemodilution with either low molecular HES (MW 40,000) or Dextran 60 markedly improved muscular oxygen supply in both patient populations, which was documented by an increase of the mean tissue PO2. Since the studies by Rudolfsky et al. (1981, 1982) it is known that hemodilution furthermore leads to an increase of blood flow which, under normovolemic conditions, can be demonstrated over a period of several weeks. This increase of blood flow is explained by a decrease of the peripheral resistance (cardiac afterload) accompanied by an increase of the cardiac output without any increase of the heart rate. The expected reduction of the oxygen transport capacity of the blood can thus be compensated for (Rieger 1982). Sunder-Plassmann was able to demonstrate in animal experiments that not only is collateral perfusion increased by 100%, but also homogenization of the blood flow pattern the capillary system occurs (Sunder-Plassmann et al. 1982). The aim of the present study was to investigate the effect of isovolemic hemodilution with a middle molecular HES solution on oxygen supply and PO2 pressure distribution in the extremity muscles of healthy male subjects. In addition to a significant reduction of the hematocrit (Hct) within the scope of isovolemic hemodilution from about 44% to about 37% and unchanged arterial blood gas values, an increase of the mean muscle PO2 from about 16 mmHg (before hemodilution) to about 27 mmHg (30 minutes after having reached isovolemia) was observed. The increase of the mean tissue PO2 is demonstrable already 30 minutes after the start of infusion (t = 30) of 10% HES 200 (see Table 2). The markedly raised level of the tissue PO2 in the M. tibialis anterior as compared with the initial value is demonstrable both during and after isovolemic hemodilution. In addition to this dilution-related increase of the mean tissue PO2, a marked change of configuration of the PO2 histograms is found. There is an increased filling of the PO2 classes in the pressure ranges above 40 mmHg, which can be interpreted as an increased perfusion of certain areas of the muscular capillary system. In our opinion, the results clearly show that middle molecular HES solutions, too, lead to an improvement of muscular perfusion with an increase of the mean tissue PO2 if used within the scope of isovolemic hemodilution.

Summary

The aim of the present study was to investigate the effect of normovolemic hemodilution with a middle molecular HES solution on oxygen supply and PO2 pressure distribution in the extremity muscles of healthy male subjects. In addition to a significant reduction of the hematocrit (Hct) by hemodilution from about 44% to about 37% and unchanged arterial blood gas values, an increase of the mean muscle PO2 from about 16 mmHg (before hemodilution) to about 27 mmHg (30 minutes after having reached normovolemia) was observed. The increase of the mean tissue PO2 is demonstrable already 30 minutes after the start of infusion (t = 30) (see Table 2). The markedly raised level of the tissue PO2 in the M. tibialis anterior as compared with the initial value is demonstrable both during and after normovolemic hemodilution. In addition to this dilution-related increase of the mean tissue PO2, a marked change of configuration of the PO2 histograms is found. There is an increase filling of the PO2 classes in the pressure ranges above 40 mmHg. The conspicuous increase of the intra-individual dispersion of PO2 mean values during and after infusion of HES 200 cannot be interpreted for the moment. Similar observations were, for example, reported by Landgraf and Ehrly in 1985 when measuring the muscular tissue oxygen pressure in patients with intermittent claudication before, after and during infusion of 0.9% saline solution. This could be interpreted as an increased perfusion of certain areas of the muscular capillary system. In our opinion, the results clearly show that the middle molecular HES solution tested leads to an improvement of muscular perfusion with an increase of the mean tissue PO2 if used within the scope of normovolemic hemodilution.

Acknowledgement

Supported by Deutsche Forschungsgemeinschaft (He 1293/1-2)

References

Ehrly, A.M., Landgraf, H., Saeger-Lorenz, K., Sasse, S., 1979, Verbesserung der Fliesseigenschaften des Blutes nach Infusion von niedermolekularer Hydroxyaethylstaerke (Expafusin) bei gesunden Probanden. Infusionstherapie 6: 331.

Ehrly, A.M., Landgraf, H., Moschner, P.-V., Saeger-Lorenz, K., 1987, Verhalten des Gewebesauerstoffdruckes und der Fliesseigenschaften des Blutes bei Patienten mit Claudicatio intermittens nach Infusion von 500 ml 6%iger Hydroxyaethylstaerkelösung im Vergleich zu physiologischer Kochsalzloesung. VASA 16: 103.

Fleckenstein, W., Weiss, Ch., 1982, Evaluation of PO2-histograms obtained by hypodermic needle electrodes. In: Proceedings of the World Congress on Medical Physics and Biomedical Engineering. Eds.: Bleifeld, Hardess, Leetz, Schaldach. Kuenzel, Goettingen.

Fleckenstein, W., Heinrich, R., Kersting, Th., Schomerus, H., Weiss, Ch., 1984, A new method for the bedside recording of tissue PO2 histograms. Verh. Dtsch. Ges. Inn. Med. 90: 666.

Fleckenstein, W., Weiss, Ch., 1984, A comparison of PO2-histograms from rabbit hindlimb muscle obtained by simultaneous measurements with hypodermic needle electrodes and with surface electrodes. Adv. Exp. Med.Biol. 169: 447.

Foerster, H. Wicarkcyk, C., Dudziak, R., 1981, Bestimmung der Plasmaelimination von Hydroxyaethylstaerke und von Dextran mittels verbesserter analytischer Methodik. Infusiontherapie 2: 88.

Kiesewetter, H., Jung, F., 1986, Rheologische Therapie der peripheren arteriellen Verschlusskrankheit im Stadium IIb. Angio 8: 21.

Kopp, K.H., Sinagowitz, E., Kaeshammer, B., Weidmann, H., 1981, Methodische Probleme bei Sauerstoffpartialdruckmessung im Gewebe von Intensivpatienten. In: Ehrly (ed.): Messung des Gewebesauerstoffdruckes bei Patienten, p.61. Witzstrock-Verlag, Baden-Baden, Koeln, New York.

Kopp, K.H., Kieser, M., Sinagowitz, E., Lund, M.,1982, Der Einfluss von niedermolekularer Hydroxyaethylstaerke und Dextran 60 auf die Muskelsauerstoffversorgung bei Intensivpatienten. Infusionstherapie 9:44-51.

Landgraf, H. Ehrly, A.M., Saeger-Lorenz, K., Vogel, V., 1981, Unter-
    suchung ueber den Einfluss mittelmolekularer HES auf· die Fliess-
    eigenschaften des Blutes gesunder Probanden. Infusionstherapie 8:
    200.
Rieger, H.,1982, Induzierte Blutverduennung (Haemodilution) als neues
    Konzept in der Therapie peripherer Durchblutungsstoerungen. Inter-
    nist 23: 375.
Rudolfsky, G., Strohmenger, H.G., Trexler, S., Brock, F.E., 1981, Iso-
    volaemische Haemodilution bei Patienten mit arterieller Verschluss-
    krankheit im Stadium II. In: Breddin, K. (ed.): Thrombose und
    Atherogenese. G. Witzstrock-Verlag, Baden-Baden, Koeln, New-York.
Rudolfsky, G., Meyer, P., Strohmenger, H.G., 1982, Effects of hemodi-
    lution of resting flow and reactive hyperemia in lower limbs. Bibl.
    Haematol. 47: 157.
Sunder-Plassmann, L., von Hessler, F., Endrich, B., Messmer, K., 1982,
    Improvement of collateral circulation in chronic vascular occlusive
    disease of the lower extremity. Bibl. Haematol. 47: 43.

THE ROLE OF HIGH FLOW CAPILLARY CHANNELS IN THE LOCAL OXYGEN SUPPLY TO SKELETAL MUSCLE

D.K. Harrison, S. Birkenhake, S. Knauf, N. Hagen, I. Beier
and M. Kessler

Institut für Physiologie und Kardiologie der Universität
Erlangen-Nürnberg, Waldstrasse 6, 8520 Erlangen/FRG

INTRODUCTION

Investigations over the years into the local oxygen supply in various
organs have revealed (with the exception of the heart and kidneys) an
interesting discrepancy between the mean tissue and venous pO_2 values. The
latter values were found to be higher than might be expected (Kessler et
al., 1976) indicating a heterogeneity of capillary flow through the organ.
More specifically, studies of the spatial distribution of local pO_2 at the
surface of skeletal muscle have revealed considerable differences in mean
tissue pO_2 values between different capillary supply units (Fig. 1) con-
firming that a large degree of heterogeneity in flow probably exists.

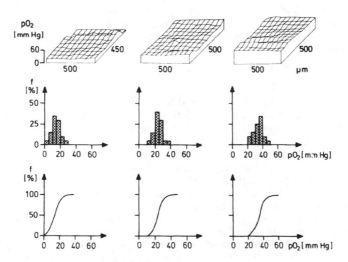

Fig. 1. Topographical representation of the pO_2 distribution in three
neighbouring areas of skeletal muscle.

The investigations we performed into the effects of hypoxia on capillary flow distribution in skeletal muscle (Harrison et al., 1985) showed that large changes in arterial blood flow could occur without any net change in capillary flow as measured at the surface using the H_2 clearance technique (Fig. 2). Preliminary investigations indicated that an opposite reaction to that shown in Fig. 2 could be induced by stimulating the muscle. These results led us to postulate the existence of high flow capillary channels in skeletal muscle which serve a physiological function as a local reserve source of oxygenated blood.

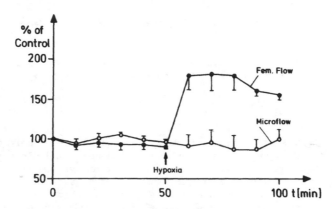

Fig. 2. Mean changes in capillary blood flow in the sartorius muscle and femoral blood flow ($\pm$ SEM) measured in 6 mongrel dogs before and during hypoxaemia (mean P_aO_2 = 33 $\pm$ 3 mmHg).

A summary of our further studies involving the simultaneous measurement of capillary blood flow (microflow), arterial blood flow and local pO_2 in twitching muscle, and the subsequent quantification of changes in distribution of microflow is presented.

METHODS

A Kessler and Lübbers (1966) multiwire surface electrode was used to measure local tissue pO_2. An electrode of the same type, modified for the measurement of hydrogen, was used for recording the capillary H_2 clearances (Kessler et al., 1986).

The needle electrodes (tip diameter 50–70 /um) used for the subsequent measurements of intravascular H_2 clearances were constructed in the following way. 200 /um platinum wire (99.99% pure) was etched electrochemically in potassium cyanide to a tip diameter of about 20 /um and then insulated with glass using a microprocessor-controlled puller designed and built in collaboration with the Institut für Antriebe und Technik, University of Erlangen/Nürnberg. The electrodes were ground to the required diameter, palladinised and dip-coated with a collodion membrane. Each electrode was calibrated before and after every experiment.

In the first series of experiments a small area of the sartorius

muscle was carefully prepared and freed of fascia in anaesthetised, artificially ventilated mongrel dogs. An O-ring holder, which allowed rotation
of the pO_2 or pH_2 electrode about its own axis, was used to keep the
respective sensors in position on the muscle. In this way, pO_2 and microflow could be measured within the same region of tissue. H_2 was administered via the ventilator at an F_iH_2 0.3 for precisely 30 sec and constituted a so-called partial saturation.

Throughout the experiments haemoglobin concentration, haematocrit and
blood gas/acid-base status were measured at frequent intervals, and ECG,
arterial and venous pressure and femoral artery flow were monitored
continuously.

The muscle was made to contract by stimulating the femoral nerve with
a 1.3 volt pulse of 0.5 ms duration. The stimulation frequency was increased stepwise at 25 min intervals with no recovery period in between.
Measurements of all parameters were carried out under control conditions,
at stimulation frequencies of 1, 2, 4, 8 and 20 Hz and after recovery.

HIGH FLOW CAPILLARY CHANNELS

pO_2 histograms measured in the twitching sartorius muscle of six dogs
at the successive stimulation frequencies are presented in Fig. 3. The
arterial pO_2 was kept constant throughout all experiments at 120 $\pm$ 4 mmHg.
Note that only at 8 Hz was there any evidence at all of pO_2 values below
5 mmHg.

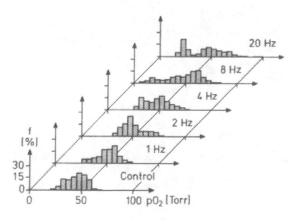

Fig. 3. pO_2 histograms recorded under control conditions and during
stimulation at 1, 2, 4, 8 and 20 Hz. Each histogram represents at
least 624 pO_2 measurements.

The results of the measurements of femoral blood flow and microflow
(Fig. 4) displayed the opposite response during the stimulation experiments to that observed during hypoxia in that here microflow often increased even in the absence of any increase in arterial blood flow. The
six experiments confirmed the first observation that microflow within and
arterial flow to skeletal muscle can react totally independently of each
other depending upon the O_2 supply situation (Fig. 5).

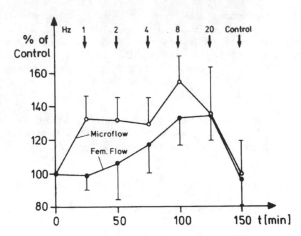

Fig. 4. Mean changes in microflow in the sartorius muscle and arterial
blood flow (± SEM) in 6 dogs during stimulation at increasing
frequencies.

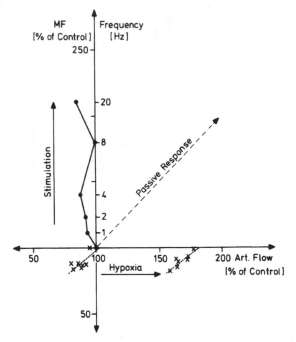

Fig. 5. An example of change in microflow plotted against change in femoral flow measured in the same muscle in one experiment from the
series under control conditions at rest, during hypoxaemia and
during stimulation at increasing frequencies. A clear passive
dependence of microflow upon femoral blood flow can be seen under
resting normoxic and hypoxic conditions. On the other hand, the
hypoxic stimulus produces a horizontal shift to the right whereas
the increased local oxygen demand induces a change in the vertical
direction. The questions which must be posed are: Where does the
blood flow go when a horizontal shift occurs, and where does it
come from in the case of a change in the vertical direction?

This apparent discrepancy between capillary and arterial blood flows led us to postulate the existence of high flow capillary channels which lie too far below the muscle surface to be detected directly using the surface pH_2 electrode (i.e. deeper than 50 /um).

In order to test this hypothesis, we devised a model which involved measuring the venous H_2 clearance function with an intravascular needle electrode. The principle of this model was that it should be possible to identify within the venous washout any components of flow bypassing the catchment volume of the surface electrode (see Fig. 6). Furthermore, it should be possible to quantify any changes in heterogeneity of flow occurring as a result of stimulating the muscle. Needle electrodes were also used to monitor the arterial H_2 input function during the experiments.

In this second series, experiments were carried out on the M vastus medialis of anaesthetised, relaxed, artificially ventilated rabbits under control conditions and during direct electrostimulation of the muscle at a frequency of 2 Hz.

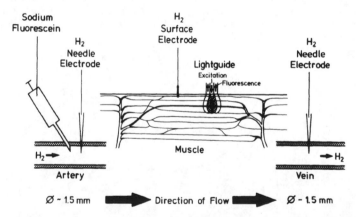

Fig 6. A schematic drawing of the experimental set-up showing the locations of the electrodes and indicating the catchment volume of a single channel of the multiwire surface H_2 electrode. Recently a further method has been introduced, namely the photometric measurement of the washout of sodium fluorescein, and this is indicated by the lightguide shown on the right hand side of the muscle surface together with its catchment volume. The thickly-drawn capillary represents one the postulated high-flow channels which is not detected by the surface electrode, but whose flow would comprise a component of the washout detected by the venous electrode and the lightguide (Harrison et al., 1988 in press).

TREATMENT OF RESULTS

Component analysis of the venous H_2 clearance curves was carried out using the standard curve peeling method (see, for example, Fieschi et al., 1968). Measurements where the electrodes displayed a significant zero drift (greater than 3% per hour) were discarded because an additive

component within an otherwise exponential function can give rise to
serious distortions of the semilogarithmic curves. Intraarterial measurement of the clearance function revealed that the input function resembled
that of a slug injection rather than that of a total saturation. Total
flow was thus calculated using the stoicheiometric method of Zierler
(Zierler, 1965) for the purposes of estimating the total flow and size of
tissue component.

RESULTS AND DISCUSSION

Semilogarithmic plots of a venous H_2 clearance following a 30 sec
partial saturation in the resting and contracting muscle are shown in Fig.
7. Analyses of 8 clearances measured at rest clearly revealed two (but
only two) components of flow. These components represent a flow distribution ($\pm$ S.D.) of 76+13% of the flow to 34+6% of the tissue, and 24+13%
of the flow to 66+6% of the tissue. Stimulation of the muscle always
resulted in an increase in the proportion of the slow and a decrease in
that of the fast compartment. In cases when the muscle was contracting
particularly strongly (5 cases), only one compartment could be
distinguished.

Early results from sodium fluorescein clearances measured simultaneously with tissue H_2 clearances (Harrison et al., 1988 in press) indicate that the high flow channels detected in the venous H_2 washout can
also be seen in the muscle itself by using a sensor with a larger catchment volume (Fig. 6). This indicates that the high flow channels are
present within the muscle itself and not purely in connective tissue as
suggested by Lindbom (1983).

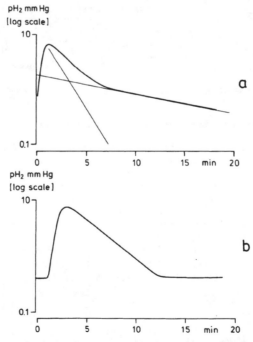

Fig. 7. (a) Semilogarithmic plot, indicating flow components, of a typical
venous H_2 clearance following a 30 sec partial saturation under
control conditions. (b) Semilogarithmic plot of a venous H_2 clearance during local stimulation of the musculus vastus medialis.

The heterogeneity of capillary flow in skeletal muscle has been the object of study and the subject for heated debate for many years. In a recent review Duling and Damon (1987) quite correctly cast doubt upon the merit of the indirect measurement of isolated parameters.

On the other hand, a simple analysis in terms of the Hagen—Poiseuille equation shows that the capillary diameter is the most important factor influencing distribution of flow within a particular capillary supply unit (see Fig. 8) and that even a straightforward Gaussian distribution of capillary diameters would give rise to high and low flow capillary channels.

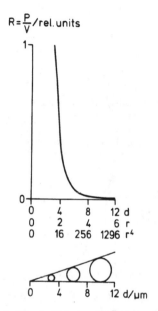

Fig. 8. The influence of capillary diameter upon resistance to flow.

In our studies we adopted a systems approach to the problem and measured, simultaneously as far as possible, those parameters important for the assessment of flow heterogeneity.

The results of our studies show quite clearly that a heterogeneous distribution of capillary flow is to be found in skeletal muscle, but that this heterogeneity provides an oxygen reserve. This reserve, in the form of high flow channels, can be mobilised probably via signals emanating from oxidases within those cells lying at the venous end of the capillary network and which are propagated back along the capillaries to the terminal arterioles (Kessler et al., 1984).

CONCLUSIONS

The heterogeneity of flow found within skeletal muscle seems to be subject to two levels of local regulation. At first reserves to be found between supply units are used up, the effect being a levelling out of the topograms of Fig. 1. The second level of regulation is the diversion of flow from the high flow capillaries in order to meet an increased O_2 demand resulting from, for example, stimulation of the muscle (Figs. 4, 5 and 7).

The capillary signal chain acts convergently in the countercurrent direction resulting first in a very local distribution of flow then secondly in a locally activated diversion of flow from the high flow capillaries. The regulatory mechanism is naturally also subject to the influences upstream of the autonomic nervous system.

ACKNOWLEDGEMENTS

The authors are grateful to Gundi Schuster for her patience and expertise in preparing the figures. We are also most grateful to Günther Schuster and Astrid Rebhan for their valuable technical contributions to this work. Dr. Jens Höper's discussions, comments and criticisms have been invaluable.

REFERENCES

Duling, B.R. and Damon, D.H., 1987, An examination of flow heterogeneity in striated muscle, Circ. Res., 60:1.

Fieschi, C., Isaacs, G. and Kety, S.S., 1968, On the question of heterogeneity of local blood flow in grey matter of the brain, In: "Blood Flow through Organs," Edinburgh: Livingstone.

Harrison, D.K., Knauf, S.K., Vogel, H., Günther H. and Kessler, M., 1985, Redistribution of microcirculation in skeletal muscle during hypoxaemia, Adv. Exp. Med. Biol., 191:387.

Harrison, D.K., Frank, K.H., Birkenhake, S., Hagen, N., Dümmler, W., Appelbaum, K. and Kessler, M., 1988, Direct comparison of fluorescent tracer and hydrogen washout techniques in skeletal muscle, Excerpta Medica, in press.

Kessler, M. and Lübbers, D.W., 1966, Aufbau und Anwendungsmöglichkeit verschiedener PO_2-Elektroden, Pflüg. Arch. ges. Physiol., 291:R82.

Kessler, M., Höper, J. and Krumme, B.A., 1976, Monitoring of tissue perfusion and cellular function, Anesthesiology, 45:184.

Kessler, M., Höper, J., Harrison, D.K., Skolasinska, K., Klövekorn, W.P., Sebening, F., Volkholz, H.J., Beier, I., Kernbach, C., Rettig, V. and Richter, H., 1984, Tissue O_2 supply under normal and pathological conditions, Adv. Exp. Med. Biol., 169:69.

Kessler, M., Harrison, D.K. and Höper, J., 1986, Tissue oxygen measurement techniques, in: "Microcirculatory Technology," C.H. Baker and W.L. Nastuk, eds., New York: Academic Press.

Lindbom, L., 1983, Microvascular blood flow distribution in skeletal muscle. An intravital microscopic study in the rabbit, Acta Physiol. Scand., Suppl. 525.

Zierler, K.L., 1965, Equations for measuring blood flow by external monitoring of radioisotopes, Circ. Res., 16:309.

TISSUE PO$_2$ AND FUNCTIONAL CAPILLARY DENSITY IN CHRONICALLY ISCHEMIC SKELETAL MUSCLE

Michael D. Menger, Frithjof Hammersen*, John Barker**, Gernot Feifel, and Konrad Messmer**

Dept. of General Surgery, University of the Saarland, Homburg, Saar/FRG, *Dept. of Anatomy, Technical University Munich, Munich/FRG, **Dept. of Experimental Surgery, University of Heidelberg, Heidelberg/FRG

INTRODUCTION

Chronic ischemic disease is characterized by diminished perfusion of nutritive capillaries and insufficient tissue oxygenation. Tissue supply becomes inadequate due to a disproportion between supply and demand or as result of changes in the distribution of microvascular blood flow with an increase of the non-nutritional component. Even when ischemia is primarily due to occlusion or stenosis of the larger supplying arteries, nutritional microvascular perfusion deteriorates as result of impaired flow conditions (poststenotic fall of perfusion pressure) and impaired flow properties of the blood (Schmid-Schönbein et al., 1981).

Conservative treatment of chronic ischemic diseases is a problem, which has not yet been resolved. Impaired blood flow in the nutritional microcirculation is recognized today as a key pathophysiologic factor in chronic ischemic diseases.

Current noninvasive diagnostic methods are applicable to the microvasculature of the skin, but not to skeletal muscle (see Mahler et al., 1986). In order to evaluate alterations at the microcirculatory level in chronically ischemic skeletal muscle, further experimental studies are needed to assist in the search for new methods to improve tissue nutrition.

Various studies have been performed to investigate the hemodynamic changes, development of collaterals and the effects of therapeutic measures in chronically ischemic skeletal muscle (Thulesius, 1962; Sanne and Sivertsson, 1968; Conrad et al., 1971; Conrad, 1977; Forst et al., 1984). However, in all these studies assessment of microcirculatory perfusion originated from indirect measurements, not allowing quantification of changes in nutritional tissue supply at the capillary level. Therefore, a new model was developed, permitting quantitative intravital microscopic analysis of the microhemodynamics and determination of tissue PO_2 in skeletal muscle prior to and for a prolonged period of time after induction of ischemia (Menger et al., 1986).

The aim of the present study was to evaluate the changes of nutritional capillary perfusion and of tissue oxygenation in chronically ischemic tissue of skeletal muscle.

METHODS

For our studies we have used the hamster dorsal skin fold chamber, which contains in its observation window skeletal muscle and skin, allowing intravital microscopic observation and determination of tissue PO_2 in the awake animal. In contrast to the hamster cheek pouch, mesenterium or tenuissimus muscle preparation, microcirculatory changes can be analyzed for a prolonged time period, eliminating the effects of anaesthesia and surgical trauma (Endrich and Messmer, 1984). The hamster dorsal skin fold receives its vascular supply from two cranial and two caudal feeding arteries. The double frame chamber is implanted on the skin fold with its observation window being positioned in between the cranial and caudal vascular supply. Ischemia is induced in the cranial part of the skin fold by heat coagulation of the cranial arteries outside of the chambers frame (fig. 1). This technique allows for analysis of the microvascular hemody-

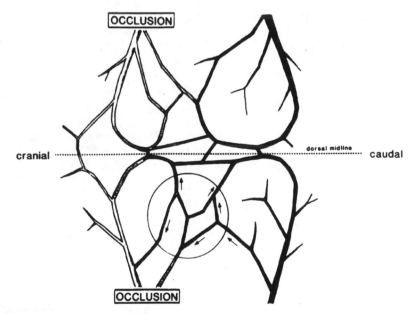

fig. 1: Vascular supply to the dorsal skin fold of the Syrian golden hamster by two cranial and two caudal vessels originating from intercostal arteries. The double frame chamber with its observation window ($\bigcirc$) is positioned between the cranial and caudal feeding vessels. Ischemia is induced by heat coagulation of the cranial arteries outside of the chamber.

namics prior to and during a prolonged period of ischemia in skeletal muscle. The analysis of the microhemodynamics can be performed in the ischemic as well as in the para-ischemic area.

EXPERIMENTAL DESIGN

In 10 Syrian golden hamsters (bw 60-80 g) a dorsal skin fold chamber and two permanent arterial and venous catheters were implanted. The chamber and implantation procedures have been described previously by Endrich et al. (1980). Briefly: Under Nembutal anaesthesia (50mg/kg bw) the animals were fitted with two symetrical teflon-coated aluminium frames, positioned on the dorsal skin fold in such a fashion, that they sandwich the extended double layer of the skin. One layer was completely removed in a circular area ($\emptyset$ 15mm) and the remaining layer, containing skin muscle and subcutaneous tissue was covered with a removable cover glass, incorporated in one of the aluminium frames. Permanent catheters were passed from the dorsal to the ventral side of the neck and placed into the carotid artery and the jugular vein.

A recovery period of 48 hours after the implantation procedure allowed the effects of anesthesia and surgical trauma on the microcirculation to be neglected. Prior to induction of ischemia analyses of microcirculatory hemodynamics were performed by means of intravital microscopy, video techniques and the CAMAS system (Computer Assisted Microcirculation Analysis System) (Zeintl et al., 1986). Ischemia was then induced in the cranial part of the dorsal skin fold, followed by subsequent analyses of microhemodynamics at 1, 4, 7 and 11 days. All observations were performed in the awake animal. The measurements were confined to the skin muscle, excluding microvessels of the subcutaneous and skin tissue. To better visualize the microvasculature during intravital microscopy, contrast enhancement was provided by intravenous application of FITC-Dextran 150.000 (5 mg in 0.1 ml 0.9% NaCl).

In arterioles (30-50 um) and venules (80-120 um) RBC-velocity and diameters were analyzed within the ischemic tissue.

In the capillary bed microvascular changes were analyzed in two regions of interest selected in the ischemic and in the para-ischemic area. Each region, encompassing a surface area of 1.25 mm^2, was divided into 3x3 single windows for determination of functional capillary density (density of red cell perfused capillaries in the area of observation). Additionally, heterogeneity of capillary perfusion, which is the local distribution of capillary densities, was assessed in each animal using the data obtained from the 9 individual windows to calculate the quotient of maximum/minimum difference and mean value of functional capillary density. Capillary RBC-velocity was analyzed in the vessels present in 3 of the 9 windows per region.

In collecting venules (20-60 um diameter) changes of microhemodynamics were investigated in 6 sites of interest in the ischemic and the para-ischemic area. Each site encompasses an area of 0.2 mm^2, containing three collecting venules for determination of diameter and RBC-velocity.

In 11 further animals determination of tissue PO_2 in skeletal muscle was performed prior to and 1, 4, 7 and 11 days after induction of ischemia, using a platinum multiwire electrode as described by Kessler and Grunewald (1969) and Lübbers (1969).

STATISTICS

Data, presenting a normal distribution, was tested by an one-way analysis of variance and paired t-test; results are given as mean $\pm$ SD. Friedman analysis of variances and Wilcoxon test was performed on data, presenting non normal distribution; the results are given as the median (Q1/Q3). The level of significance was defined as $p < 0.05$.

RESULTS

In the arteriolar network collateral flow was observed. The caudal feeding vessels revealed a significantly ($p < 0.001$) increased diameter for the whole observation period following induction of ischemia (11 days); the remaining arteriolar network received blood by collaterals from these feeding vessels. In total, there was an increase in arteriolar diameter of more than 30% compared to initial values.

However, while in the para-ischemic area no changes in functional capillary density and heterogeneity of capillary perfusion were observed, functional capillary density decreased in the ischemic area significantly during the first 7 days. Moreover, there was a more heterogeneous capillary perfusion. The maximum heterogeneity, approximately threefold of control was present at day 4 after induction of ischemia.

Capillary RBC-velocity suffered a reduction throughout the entire ischemia period. After 11 days of ischemia more than 60% of all visible capillaries had velocities below 0.1 mm/sec, mean velocity was reduced from 0.26 (0.20/0.34) mm/sec prior to ischemia to 0.08 (0.05/0.15) at the end of the ischemic observation period.

In collecting venules RBC-velocity was diminished significantly ($p < 0.05$) during the first 4 days of ischemia. The diameters of the collecting venules increased 24 hrs after induction of ischemia ($p < 0.001$), but were found decreased at 4, 7 and 11 days of ischemia. Similar changes were observed in the para-ischemic area.

In the larger venules (80-120um) RBC-velocity was also found reduced after induction of ischemia. After 7 days of ischemia flow velocity was 0.10 (0.08/0.36) mm/sec as compared to the initial value of 0.65 (0.15/1.01) mm/sec ($p < 0.05$). The diameters of these vessels had decreased significantly ($p < 0.01$) from 103.5 (66.0/128.6) um to 64.0 (49.0/102.4) um at the end of the observation period.

In parallel with the decrease of functional capillary density tissue hypoxia was observed after induction of ischemia. The PO_2 histograms were shifted distinctly to lower PO_2 values. The maximum decrease of mean tissue PO_2 (20.5 mmHg, prior to ischemia) was observed at 11 days after induction of ischemia (9.5 mmHg).

CONCLUSION

In the model presented, chronic ischemia in skeletal muscle is characterized by
(a) decreased functional capillary density,
(b) heterogeneous capillary perfusion and
(c) reduced capillary RBC-velocity;
(d) in parallel with the decrease in functional capillary density severe tissue hypoxia was demonstrated.

The increased collateral flow from the caudal vessels was not sufficient to prevent impairment of nutritional capillary perfusion and tissue hypoxia in the cranial part of the preparation.

This model allows one to study the effects of therapeutic measures, such as normovolemic hemodilution or application of vasoactive substances, on the microcirculation in chronically ischemic skeletal muscle in vivo.

SUMMARY

In order to study changes in functional capillary density and tissue PO_2 in chronically ischemic skeletal muscle, a new model, using the Syrian golden hamster was developed. In the hamster dorsal skin fold, which receives its vascular supply from two cranial and two caudal feeding arteries, a double frame chamber was implanted and ischemia was induced in the cranial part by heat coagulation of the cranial arteries outside of the chamber. This technique allows for analysis of microvascular hemodynamics and local tissue PO_2 prior to and during a prolonged period of ischemia in skin muscle.

As result of ischemia the diameters of the arterioles increased ($p < 0.001$) over the whole 11 day observation period. Functional capillary density decreased significantly ($p < 0.01$) during the first 7 days, while capillary RBC-velocity was reduced throughout the 11 days of observation. RBC-velocity in collecting venules was diminished significantly throughout the postischemic observation period. The diameters of the collecting venules first increased upon ischemia ($p < 0.001$) but were found decreased at 4, 7 and 11 days. Measurements of tissue PO_2 demonstrated a marked decrease from a mean PO_2 of 20.5 mmHg prior, to 9.5 mmHg following induction of ischemia.

The model allows for induction of chronic ischemia and is suitable to study the effect of therapeutic measures on the microcirculation in chronically ischemic skeletal muscle in vivo.

REFERENCES

Conrad, M. C., 1977, Effects of therapy on maximal walking time following femoral ligation in the rat, Circulation Res. , 41:775.

Conrad, M. C., Anderson III, J.L., Garrett, Jr., J. B., 1971, Chronic collateral growth after femoral artery occlusion in the dog, J. appl. Physiol. , 31:550.

Endrich, B., Asaishi, K., Götz, A., Messmer, K., 1980, Technical report - a new chamber technique for microvascular studies in unanesthetized hamsters, Res. Exp. Med. , 177:125.

Endrich, B., Messmer, K., 1984, Quantitative analysis of the microcirculation in the awake animal, in : "Handbook of Microsurgery", W. Olszewski, ed. CRC Press, Miami.

Forst, H., Fujita, Y., Racenberg, J., Brückner, U. B., Weiss, Th., Sunder-Plassmann, L., Meßmer, K., 1984, Skelettmuskeldurchblutung bei arterieller Verschschlußkrankheit: Wirkung verschiedener Therapieverfahren, in : "Die Mikrozirkulation der Skelettmuskulatur", F. Hammersen, K. Meßmer, eds., Karger, Basel/New York.

Kessler, M., Grunewald, W. A., 1969, Possibilities of measuring oxygen pressure fields in tissue by multiwire platinum electrodes, Prog. Resp. Res. , 3:147.

Lübbers, D. W., 1969, The meaning of the tissue oxygen distribution curve and its measurement by means of Pt electrodes, Prog. Resp. Res. , 3:112.

Mahler, F., Messmer, K., Hammersen, F. (eds.), 1986, "Techniques in clinical capillary microscopy", Karger Basel/New York.

Menger, M. D., Barker, J., Messmer, K., 1987, A new model for vitalmicroscopic studies of the microcirculation in chronically ischemic tissue, Int. J. Microcirc.: Clin. exp. 6, 81.

Sanne, H., Sivertsson, R., 1968, The effect of exercise on the development of collateral circulation after experimental occlusion of the femoral artery in the cat, Acta physiol. scand. , 73: 257.

Schmid-Schönbein, H., Messmer, K., Rieger, H., 1981, "Hemodilution and flow improvement", Bibl. haematol., vol. 47, Karger, Basel.

Thulesius, O., 1962, Haemodynamic studies on experimental obstruction of the femoral artery in the cat, Acta physiol. scand. , 57: Suppl 199.

Zeintl, H., Tompkins, W. R., Messmer, K., Intaglietta, M., 1986, Static and dynamic microcirculatory video image analysis applied to clinical investigations, in : "Techniques in Clinical Capillary Microscopy", F. Mahler, K. Messmer, F. Hammersen, eds., Karger, Basel/New York.

ELEVATION OF TISSUE PO_2 WITH IMPROVEMENT OF TISSUE PERFUSION BY TOPICALLY APPLIED CO_2

Yoshiaki Komoto, Toshihiko Nakao, Mitsuru Sunakawa, and
Hidenori Yorozu*

Institute for Environment and Diseases, Okayama University
Medical School, Misasa, Tottori and *Research Laboratories
Kao Co., Ltd., Tochigi, Utsunomiya, Japan

INTRODUCTION

Exposure of skin to CO_2 causes local hyperaemia and increases
the number and diameter of functioning cutaneous capillaries (Stein and
Weinstein, 1942; Kowarschik, 1948; Disi, 1958; Witzleb, 1962). Lipid-soluble CO_2 is absorbable into the body through the skin, so that the dissolved
CO_2 acts as a vasodilator of dermal capillaries (McClellan, 1963).
The functional consequences of CO_2 baths on the living body have not
been determined.
In this study, we examined the effects of CO_2 bathing on tissue perfusion in rabbits by gas analysis with medical mass spectrometry.

MATERIAL AND METHODS

a) Experimental protocol (Fig. 1)

A 2 kg rabbit was intubated by tracheotomy under intravenous anaesthesia with 25 mg/kg sodium pentobarbital. Arterial pressure was monitored
by a catheter introduced via the carotid artery under 1 mg/kg general
heparinization. Venous catheterization was performed via the external
jugular vein into the right atrium. The rabbit was then shaved around
the chest, abdomen and legs, and placed in a box from which its neck and
head protruded.
Fresh air was continuously blown at the rabbit so as to prevent CO_2
inhalation. ECG and arterial pressure were monitored on a polygraph, and
anaesthesia was maintained by a constant micro-drip infusion system of 15-
20 drops per minute with 0.3-0.4% sodium pentobarbital.
The ventilation rate was 250 ml/kg/min with assisted respiration of
room air.
Room temperature was maintained at 25±1°C throughout the experiment.
Two calibrated Teflon catheters for monitoring partial pressures of
each tissue gas (PO_2 and PCO_2) were separately placed in the subcutaneous
tissue of the thigh; 30-60 minutes were allowed for partial pressures to
stabilize.

b) Medical mass spectrometry

A medical mass spectrometer, Medspect,© (Chemetron, U.S.A.) was used

637

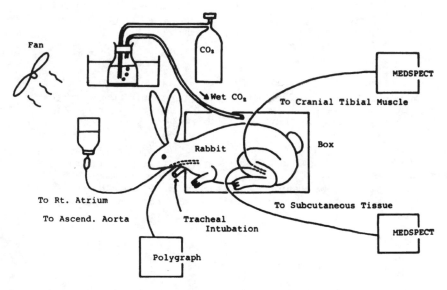

Fan

CO₂

Wet CO₂

MEDSPECT

To Cranial Tibial Muscle

Rabbit

Box

To Rt. Atrium

To Ascend. Aorta

Tracheal
Intubation

To Subcutaneous Tissue

MEDSPECT

Polygraph

Fig. 1.   Diagram of the experimental CO_2 vapour bath.
The rabbit box was placed at a height of 1 meter and covered
with a glass plate. Wet CO_2 was blown into the box, 5 l/min
at 34°C.

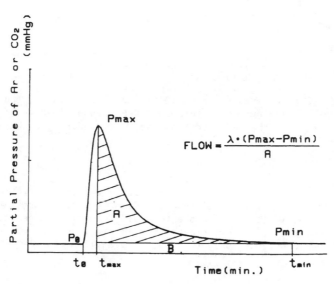

Fig. 2.   Principle for the determination of regional tissue perfusion
volume.
Pmax, Maximum tissue partial pressure of inactive gas (Ar or CO_2)
immediately prior to desaturation; Pmin, Tissue partial pressure
of inactive gas when desaturation is completed; λ, Distribution
coefficient between blood and tissue; λ is presumed to be 1.0
for most tissues except fatty tissue;
A, The area of the desaturation curve lying within the partial
pressure and time axes.
The regional tissue perfusion volume is expressed as ml/100g/min
through an on-line computer system.

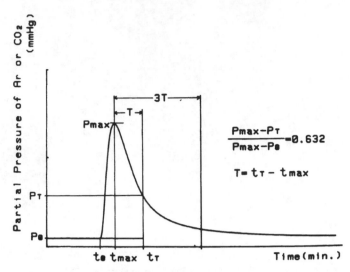

Fig. 3. The time required for a 63.2% decrease from Pmax is the time constant (T). Time measurements are standardized in 3T-time to minimize the deterioration that occurs during an extended period.

for the determination of gas partial pressures in tissue. Probes were inserted into the cranial tibial muscle and the adjacent subcutaneous tissue as atraumatically as possible to avoid formation of hematomata around the sensor area, with a sleeve-clad 14 gauge needle, which was withdrawn, and the catheter slipped through the plastic sleeve, which was then also withdrawn.

c) Determination of tissue perfusion volume

Regional tissue perfusion volume was determined on the basis of a clearance curve for tissue Argon partial pressure. Argon was injected into the cranial tibial muscle close to the sensor area of the catheter during CO_2 bathing and monitored by means of an on-line computer system with mass spectrometry; Kety's clearance theory was applied to Fick's principle for the determination of regional blood flow volume (Figs. 2 and 3) (Kohmoto, 1984).

d) Preparation of CO_2 baths

The rabbit was placed in the bath, and carbon dioxide was bubbled through 20 l of bath water at a constant temperature (36-37°C) in a preliminary study. After stabilization of tissue PO_2 and PCO_2, the water was carbonated to saturation (2210 ppm by Ionalyzer, Model 95-02-00, Orion Research, U.S.A.); this was maintained throughout the experiment.

A 50 g sodium hydrogencarbonate and succinic acid tablet (Kao Co., Ltd., Japan) producing fine CO_2 bubbles in the water was put in the bath water under the same conditions as above (570 ppm).

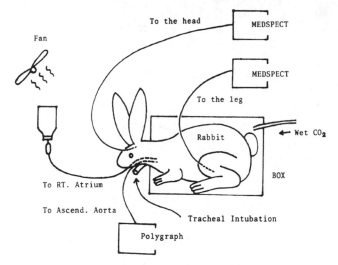

Fig. 4.   Determination of local percutaneous absorption of CO_2.
Calibrated Teflon catheters for monitoring partial pressures of
subcutaneous tissue gases were placed in the leg exposed to CO_2,
and in the head which was kept clear of the CO_2.

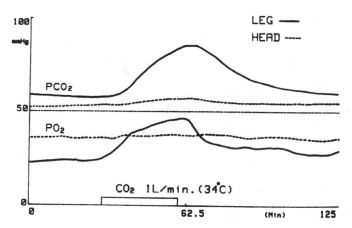

Fig. 5.   Local absorption of CO_2 in the leg of a rabbit is shown by the
change in tissue PCO_2 and PO_2. The curves were obtained by
computer processing of the data to correct for time lag.

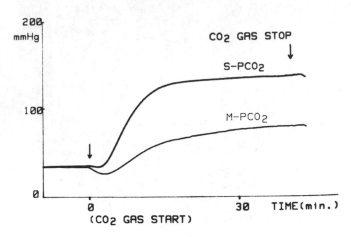

Fig. 6. Changes in CO_2 absorption through skin of a rabbit in a CO_2 bath with time.
CO_2 was blown into the chamber at a rate 1 l/min at 34°C.
S-PCO_2; Subcutaneous tissue partial pressure of CO_2.
M-PCO_2; Muscle tissue partial pressure of CO_2.

RESULTS

The general condition of the rabbit was unchanged during CO_2 bathing in regard to pulse rate and blood pressure.

1) Local absorption of CO_2

To determine the local absorption of CO_2 through skin, subcutaneous tissue partial pressures were compared between sites in the head and leg of the rabbit as illustrated in Fig. 4. The leves of subcutaneous PO_2 and PCO_2 were elevated in the leg exposed to CO_2, whereas subcutaneous PO_2 and PCO_2 at the site in the head, which was not exposed to CO_2, remained unchanged during CO_2 bathing (Fig. 5).
Local absorption of CO_2 in subcutaneous and muscle tissues reached a plateau after about 30 minutes CO_2 bathing (Fig. 6).

2) Changes in tissue partial pressures with CO_2 bathing

Elevation of subcutaneous and muscle PO_2 was followed by elevation of PCO_2 in each tissue as shown in Fig. 7. The raw data shown in Fig. 7 is not corrected for the time lag.
Mean subcutaneous PCO_2 and PO_2 increased by 37.6% and 23.6% respectively, whereas mean muscle PCO_2 and PO_2 increased by 28.8% and 35.1% respectively (n=8).
Levels of PO_2 both in subcutaneous and muscle tissues were maintained for about 2 hours.

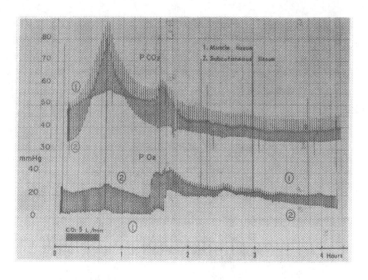

Fig. 7.　Representative changes in tissue partial pressures.
CO_2 vapour bath, 5 l/min at 34°C.

3) Tissue perfusion volume

Tissue perfusion volume ranged from 15.93 to 31.47 ml/100g/min, which was moderately higher (p<0.26) than that observed with tap water bathing (Table 1).

DISCUSSION

There are several ways of increasing O_2 transport into the tissue such as by increasing systemic circulation, inhalation of oxygen and hyperbaric oxygenation. However, we found that percutaneous CO_2 increases O_2 in the tissue.

This improvement of tissue perfusion occurred without any changes in systemic circulation. Tissue perfusion is a dynamic peripheral circulation system involving body fluid including blood, lymph and tissue fluid.

The fluctuation of PO_2 and PCO_2 reflected the dynamic changes of the partial pressure of gases associated with the tissue exposed to the environment.

Oxygen in the tissue following percutaneous CO_2 penetration is derived from the O_2 content of blood in the local circulation. Though we have not clarified the mechanism of facilitating oxygen release by percutaneous CO_2, it is speculated that the increased $[H^+]$ and increased PCO_2 in the red blood cells produces a shift to the right in the hemoglobin dissociation curve, thus facilitating oxygen release at the tissue level (Shapiro et al. 1982).

Increased oxygen release at the tissue level has therapeutic effects on wound healing especially on peripheral arterial damage due to chronic

Table 1.

TISSUE PERFUSION VOLUME WITH BATHS
(ML/100G/MIN, 36-37°C)

No.	TAP WATER	CUM CO_2
1	10.83	20.60
2	11.61	24.63
3	14.54	20.19
4	14.54	29.27
5	17.19	22.09
6	17.37	21.14
7	18.94	27.35
8	21.19	16.96
9	21.54	18.46
10	21.88	18.18
11	22.30	23.59
12	23.29	29.69
13	25.83	19.66
14	25.34	31.47
15	33.53	15.93
16	33.67	34.60
AVG.	20.85	23.36
SD	6.71	5.60
SE	3.56	2.98

(CO_2 BATH; 570 PPM)

occlusive diseases (Zederfeldt and Goldstick, 1969).

The improved tissue perfusion together with the increased tissue oxygen also favourably affects chronic degenerative disorders such as rheumatism, arthrosis deformans and scleroderma, leading to amelioration of chronic pain and restriction of the range of motion (Komoto et al., 1987).

Therefore, daily bathing in a CO_2 bath may facilitate rehabilitation.

SUMMARY

Local absorption of CO_2 was examined, by tissue gas analysis with medical mass spectrometry, in rabbits placed in a bath. The results obtained clearly showed that

(1) Elevation of subcutaneous tissue PO_2 was followed by elevation of PCO_2 in the subcutaneous tissue.
(2) Local absorption of CO_2 through skin was confirmed by elevation of the levels of subcutaneous PO_2 and PCO_2 at the site exposed to CO_2.
(3) Tissue perfusion during the CO_2 bath was moderately higher than that observed with tap water bathing.

The mechanism of oxygen release into the tissue by CO_2 bathing is thought to involve increased $[H^+]$ and PCO_2 in the red blood cells, which shifts the hemoglobin d ssociation curve to the right, thus facilitating oxygen release at the tissue level.
There is, therefore a good possibility, that the improved tissue perfusion together with the increased oxygen in the tissue will favourably affect wound healing and chronic degenerative disorders, thus facilitating rehabilitation.

REFERENCES

1. Disi, A., Local vasodilator action of carbon dioxide on blood vessels of the hand, J. Appl. Physiol., 14: 414 (1958).
2. Kohmoto, T., Local tissue perfusion determined by the clearance curve of tissue partial pressure of CO_2, J. Japn. Coll. Angiol., 24: 1331 (1984), In Japanese with English summary.
3. Komoto, Y., Kohmoto, T., Sunakawa, M., Yagi, N., Yorozu, H., and Matsumoto, Y., Clinical effects of serial artificial CO_2 baths on degenerative disorders in consideration of the improved tissue perfusion (in press, Papers Inst. Environment and Diseases Okayama Univ. Med. School, 1987) in Japanese with English summary.
4. Kowarschik, J., Physikalische Therapie, Die Kohlensäurebäder, Springer-Verlag, Wien, p.120 (1948).
5. McClellan, E. S., Carbon dioxide baths, Medical hydrology, ed., S. Licht, Waverly Press, Baltimore, p. 311 (1963).
6. Shapiro, B. A., Harrison, R. A., and Walton, J. R., Clinical application of blood gases, 3rd Edition, Year Book Medical Publishers, Chicago, p. 79 (1982).
7. Stein, I. D. and Weinstein, I., The value of carbon dioxide baths in the treatment of peripheral vascular disease and allied conditions, Am. Heart J., 23: 349 (1942).
8. Witzleb, E., Kohlensäurewasser, Handbuch der Bäder- u. Klimaheilkunde, ed. W. Amelung und A. Evers, Friedrich-Karl Schattauer-Verlag, Stuttgart, XIII, 8, p. 413 (1962).
9. Zederfeldt, T. K. and Goldstick, T. K., Oxygen and healing, Am. J. Surg., 118: 521 (1969).

RESEARCH ON OXYGEN TRANSPORT IN JAW CYSTS: INVESTIGATION OF CHANGES
OF ERYTHROCYTE MEMBRANE FLUIDITY

Kenya Anezaki

Department of Dentistry and Oral Surgery, School of Medicine
Hirosaki University, Japan

INTRODUCTION

Jaw cysts are follicular lesions within or at the surface of the maxilla or man-
dible. The lumen, cotaining the cyst fluid which resembles serum, is surrounded by a
connective tissue capsule usually lined by epithelium [1,2,8]. It has been established
that cysts contain a few red blood cells; the mean corpuscular volume (MCV) of these
is often increased and the PO_2 values are variable (Fig.1). I have investigated membrane
fluidity and MCV of the cyst erythrocytes, and studied their relation to PO_2. Membrane
fluidity is indicated by the order parameter, S. and an increase in the value of S denotes
a decrease in fluidity [3,4].
Jaw cysts were treated by irrigation of the lumen with antibiotics to remove infection
before extirpation. My investigations were carried out before and after irrigation.

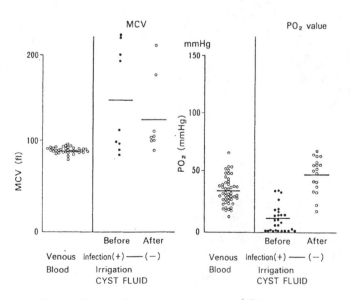

Fig. 1. (a). Change of Mean Corpuscular Volume and PO_2 value before and after
irriqation in overtly infected cases.

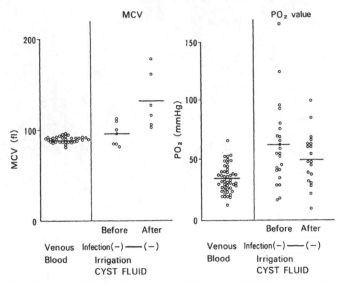

Fig.1.(b). Change of Mean Corpuscular Volume and PO2 value before and after irrigation in symtomless cases.

MATERIALS AND METHODS

1. Measurements of membrane fluidity were made on erythrocytes in 63 samples obtained from jaw cysts in 25 patients who consulted the Department of Oral Surgery in the hospital attached to the School of Medicine, Hirosaki University. As a cotrol 25 venous blood samples were taken from 17 of these patients.

    i) Aspiration and preservation Cyst fluid was preserved in ACD-A fluid (an anticoagulant) with heparin [6,3].

    ii) Highly viscous samples of cyst fluid were treated with hyaluronidase.

    iii) Samples were centrifuged in lymphocyte separation solution(d=1, 119 and 1,077) at approximately 2,800 rpm for 30 min. And they were washed in saline solution, with centrifugation at about 500 rpm for 30 sec.

    iv) Erythrocytes were identified in an auto blood count meter.

    v) Erythrocytes were incubated in 0.006% spin label (5-NS) solution for 2 hr at 37 ℃ according to the BSA exchange method [10,11,12].

    vi) The cells were washed three times in saline solution with centrifugation at about 100 rpm.

    vii) The electron spin resonance (ESR) signal (Fig 2) by Radicsensor JES-3000 (Nihon Denshi) was measured at room temperature in a flat cell. Measurement conditions: field modulation with, 1.7 Gauss; field scanning width, 100 Gauss (maximum); microwave power, 10 mm Watt; scanning velocity, 0.2 Gauss/sec.

2. Calculation of the order parameter, S

The equation for calculating the order parameter, S, is shown below[7]:

$$S = \frac{T_{\parallel}' - T_{\perp}'}{Tzz - \frac{1}{2}\,(Txx + Tyy)} \cdot \frac{a_N}{a_{N'}}$$

in case of 5-Doxylstearic acid spin labelling

$$Tzz - \tfrac{1}{2}\,(Txx + Tyy) = 27.5 \text{ Gauss,}$$
$$a_N = 15.2 \text{ Gauss,} \quad a_{N'} = \tfrac{1}{3}\,(T_{\parallel}' + 2T_{\perp}') \text{ Gauss}$$

$\overline{AD}$, $\overline{BC}$ (Gauss) are values by actual measurement from ESR signal.

    accordingly,

$$S = 1.65 \times \frac{\overline{AD} - \overline{BC} - 1.6}{\overline{AD} + 2\overline{BC} + 3.2} \text{ (no unit)}$$

The order parameter, S, was similarly calculated for erthrocytes in venous blood from the same patients.

$$S = \frac{T'_{\#} - T'_{\perp}}{Tzz - \frac{1}{2}\,(Txx + Tyy)} \cdot \frac{a_N}{a_{N'}} \cong 1.65 \times \frac{\overline{AD} - \overline{BC} - 1.6}{\overline{AD} + 2\,\overline{BC} + 3.2} \quad (\text{ in 5-NS spin labelling})$$

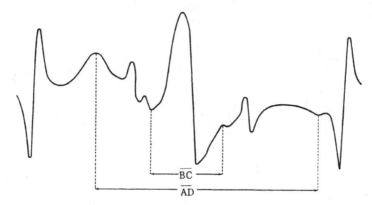

Fig. 2. Electron Spin Resonance (ESR) signal of erythroeyte in jaw cysts.

3. Changes in the lymphocyte and neutrophil segmented percentage were studied in samples of same the cyst fluids.

4. The number of erythrocytes/μl in cyst fluid was measured by an auto blood count meter (Sysmex cc-780, TOA 1·YOU Denshi). Mean values obtained from infected cases and 24 symptomless cases were compared.

Results

1. The mean value (MD) of S for the membrane fluidity of eaythrocytes from cyst fluid is 0.6415 with a range of 0.59 to 0.69. The corresponding values for the venous samples are 0.632, and 0.602 to 0.660. The mean value of S in infected cases is 0.656 and in symptomless cases, 0.627 < 0.656 (see fig. 3)

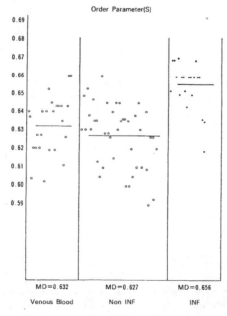

Fig. 3. Comparison of Order parameter(S) for three case: Venous blood, symptomless case, and overtly infected case.

2. In cysts showing symptoms of infection, the PO2 value fell but S, and MCV tended to increase. Antibiotic treatment caused both S and MCV to decrease. In symptomless cases changes in PO2, S and MCV were not correlated unless infection became apparent. (Fig 4,5)

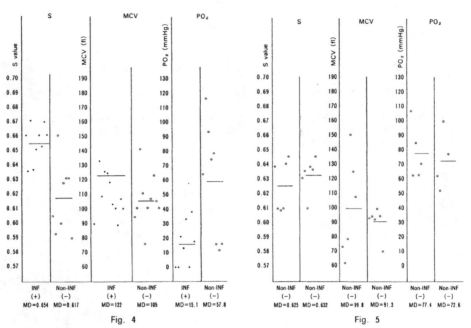

Fig. 4

Fig. 5

Fig.4. Comparison of before (left hand columns) and after (right hand columns) disappearance of infection in overtly infected cases.

Fig.5. Comparison of before (left hand columns)and after (right hand columns) irrigation in symptomless cases.

3. The number of erythrocytes/μl in cases showing symptoms of infection were not necessarily lower than in symptomless cases (Fig.6) but the number of leucocytes/μl was considerably higer in the infected than in the uninfected cysts. (Fig.7)

4. Long-term observation of the white blood cell counts of jaw cysts irrigated with antibiotics showed there was a tendency for the number of lymphocytes to increase and the neutrophils segmented to decrease in both infected and symptomless cases. (Fig.8, Fig.9)

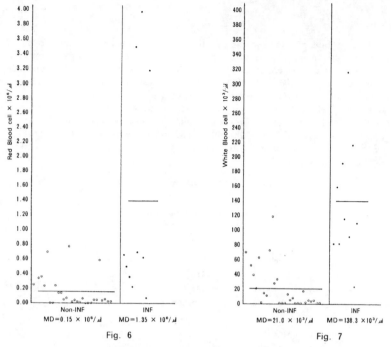

Fig. 6. Red blood cell number in cyst fluid–symptomless cases and overtly infected cases.

Fig. 7. White blood cell number in cyst fluid–Symptomdess cases and overtly infected cases.

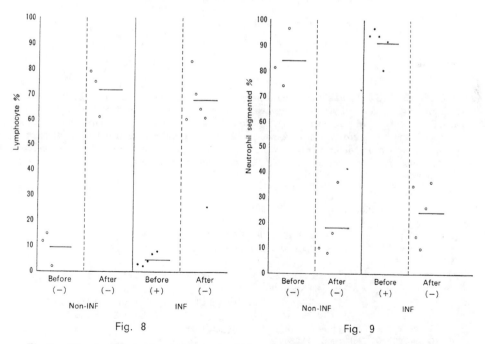

Fig. 8. Change of Lymphocyte count in before and after irrigation of cysts cases observed over a long period.

Fig. 9. Chang of Neutrophil segmented count before and after Irrigation of cysts cases observed over a long period.

Discussion

According to Shear[8] the vascular papillae of the cyst wall are damaged by infection causing the erythrocyte membrane to break down. The results of the present experiment, however, indicate that there is an increase in the number/μl of erythrocytes and leucocytes in overtly infected cases but both decrease in symptomless infected cases. The investigation of membrane fluidity showed this decreased (signalled by an increase in S) in cases exhibiting signs of infection; PO_2 was also reduced in these cases but MCV increased. I suggest that if the leucocyte numbers increase, there is an increased consumption of oxygen by neutrophils[9], and although the number of erythrocytes/μl also rises, the increase in MCV and S reduce the speed at which oxygen is released, hence PO_2 falls. This scheme is shown in Fig. 10.

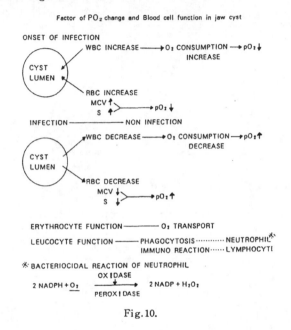

Fig. 10.

Membrane fluidity is usually used as a parameter of the kinetic state of a biological membrane and is defined as the reciprocal of viscosity. When the mobillity of the acyl chain in the phospholipid composing the double membrane is high, membrane viscosity is reduced and the fluidity rises. A decrease in the membrane fluidity of erythrocyts causes a reduction in deformability, which in turn reduces the speed of oxygen release. On the other hand, an increase of Mean Corpuscular volume results in a delay in oxygen emission[10].

MEMBRANE FLUIDITY, MCV and O_2 EMITTING SPEED

1. MEMBRANE FLUIDITY DECREASE
(RESTRICTION OF MEMBRANE PHOSPHOLIPID ACYL CHAIN)

DECREASE OF O_2 DIFFUSION SPEED WITHIN MENBRANE   DECREASE OF DEFORMABILITY

DECREASE OF MOVEMENT OF HEMOGLOBIN WITHIN ERYTHROCYTE

DELAY OF O_2 EMITTING SPEED

2. MCV   INCREASE ⟶ DECREASE OF SURFACE AREA/VOLUME RATIO
⟶ DECREASE OF DEFORMABILITY
⟶ DELAY OF O_2 EMITTING SPEED

This scheme suggests that the PO2 change is related to the changes in S and MCV. The results of the experiments revealed that in overtly infected cysts the drop in PO2 was correlated with in creases in S and MCV, but in symptomless cases this was not necessarily true. This can be partly explained but is not yet established. The white blood cell picture in jaw cysts showed a decreasing tendency in neutrophils and an increasing tendency in lymphocytes during long-term observations. This indicates that the irrigation with antibiotics had produced a clinical cure and eliminated the inflammation, a finding that correlates with the increase in PO2, seen particularly in those cases that originally showed obvious signs of infection. In symptomless cases, although it is true that there was the same tendency for the percentage of leucocytes to change, PO2 did not necessarily increase. As yet the relation of PO2, S and MCV in symptomless cases is obscure as is the relation of the changes in PO2 to the changes in neutrophils.

Summary

I have investigated oxygen transport in the fluid from jaw cysts. Measurements were made of erythrocyte membrane fluidity, mean corpuscular volume, PO2, leucocyte percentage and erythrocyte and leucocyte numbers, both before and after irrigation of the cysts with antibiotics. On the basis of the results it is suggested that the membrane fluidity of erythrocytes in jaw cysts is decreased by infection and the PO2 value falls as a result of this decrease and also because of the increase in mean corpuscular volume and the raised oxygen consumption due to neutrophilia within jaw cysts.

Reference

1) Skang, N.: Soluble proteins in fluid from non keratinizing jaw cysts in man. Int. J. Oral. surg. 62 : 107, 1977.
2) Harris, M. and Toller, P.: The pathogenesis of dental cysts. Br. Med. Bull.. 31 : 159, 1975.
3) Hubbel, W. L. and McConnell, H. M, J. Am. chem. Soc.. 93 : 314, 197.
4) Shiga, T., Maeda, N., Suda, T., Kon, K. and Sekiya, M. : The decreased membrane fluidity of in vitro aged, human erythrocytes. A spin Lahel study. Biophys. Acta.. 553 : 84-95. 1979.
5) Harris, M. et al. : Prostaglandin Production and bone resorption by dental cysts. Nature. 245 : 213, 1973.
6) Harasaki, N. and Harada, M.: Blood preservation and erythrocyte function; A new preservative for blood storage. IGAKUNO AYUMI, 121 : 249, 1982.
7) Ohnishi, S. and McConnel, H. M.. J. Am. Chem. Soc., 87 : 2293, 1965.
8) Shear, M.: Cholesterol in dental cysts. O. S., O. M. & O. P. 16 : 1, 465, 1963.
9) Suzuki, T. and Hoshi, T.. Human Physiology Vol. 1, Blood, NANZANDO, Tokyo, 1980.
10) Maeda, N. and Shiga, T.:Oxygen transport and rheological function of in vitro aged erythrocytes. Medicina Philosophica. 32 : 171, 1984.

EMULSIFIED PERFLUOROCHEMICALS FOR OXYGEN-TRANSPORT TO TISSUES: EFFECTS

ON LYMPHOID SYSTEM AND IMMUNOLOGICAL COMPETENCE

Kenneth C. Lowe

Mammalian Physiology Unit, Department of Zoology
University of Nottingham, Nottingham NG7 2RD, U.K.

INTRODUCTION

     The development of synthetic, physiologically-acceptable oxygen-
transport fluids would be a major innovation for which there would
inevitably be enormous clinical applications. In the last two decades,
considerable interest has focussed on potential oxygen-transport fluids
based on emulsified perfluorochemicals (PFCs). This is a rapidly
expanding field and one which has been the subject of many recent reviews
(see e.g. Lowe and Bollands, 1985; Tremper and Anderson, 1985;
Dellacherie et al., 1986; Lowe, 1986, 1987; Biro and Blais, 1987).

     The present paper describes the effects of emulsified PFCs and their
components on lymphoid tissues and the reticuloendothelial system (RES).
Data from experiments in rodents will be considered and compared, where
possible, with complementary experiments using both animal and human
cells in culture. An objective of this paper is to assess the consequences of in vivo use of PFC emulsions on immune system function and
immunological competence.

Tissue uptake and excretion

     The average particle size of PFC emulsions prepared for biological
uses is generally < 0.25 μm and this allows uptake into lymphoid tissue
macrophages and other RES cells; PFCs are detectable in such cells as
"foamy", membrane-bound vesicles (Nanney et al., 1983). Injection of
high doses of PFCs can lead to enlargement of liver, spleen and other
organs although this is highly variable (Lutz and Metzenauer, 1980; Lowe
and Bollands, 1985) (Table 1). Different PFCs are retained in tissues
for differing periods: for example, perfluorodecalin (FDC) has a body
half-life in rats of approximately 7 days whereas for perfluorotri-
propylamine (FTPA) the corresponding figure is 65 days (Naito and
Yokoyama, 1978; Riess and Le Blanc, 1982). It has been claimed that this
difference is due partly to the presence of an N_2 heteroatom in FTPA
which is generally associated with an increased retention time in
tissues. However, there are also marked species variations in tissue
uptake of the same PFC emulsion (Bollands and Lowe, 1986c, Lowe and
Bollands, 1987).

TABLE 1: Liver and spleen weight changes in rodents injected with emulsified perfluorochemicals

Species	Emulsion	Dose*	Route**	% change in organ weight / [] = day of measurement		Reference
				Liver	Spleen	
Rat	F-DA[†]	10	i.v.	23 [7]	35 [7]	Raven et al. (1987)
Rat	F-DA[††]	25	i.v.	50 [8]	90 [4]	Lutz & Metzenauer (1980)
Rat	F-DA	5 x 10	i.v.	57 [24]	95 [24]	West et al. (1986)
Rat	F-DA[+]	5 x 20	i.v.	118 [5]	400 [5]	Goodman et al. (1984)
Rat	FC-43[+]	10	i.v.	9 [7]	15 [7]	Raven et al. (1987)
Rat	FC-43	4.7 - 10	i.v.	Unc. [1]	Unc. [1]	Caiazza et al. (1984)
Rat	FC-43	40	i.p.	12 [14]	96 [14]	Lowe & Bollands (1985)
Rat	FC-43	100	i.p.	47 [14]	241 [14]	Lowe & Bollands (1985)
Rat	FDC + 1% C-16 HBPO[Δ]	10	i.v.	33 [7]	27 [7]	Raven et al. (1987)
Rat	FMOQ[φ]	T	i.v.	68 [14]	322 [14]	Yokoyama et al. (1984)
Rat	FMOQ	T	i.v.	83 [28]	203 [28]	Yokoyama et al. (1984)
Rat	DMA/NONANE[ψ]	5 x 20	i.v.	142 [5]	416 [5]	Goodman et al. (1984)
Mouse	F-DA	15	i.v.	47 [14]	47 [3]	Mason et al. (1985)
Mouse	F-DA[#]	4 x 15	i.v.	122 [9]	102 [9]	Rockwell et al. (1986)

*   All doses are given as ml/kg body weight (except T = near total exchange transfusion; final haematocrit < 4%).

** i.v. = intravenous injection/infusion; i.p. = intraperitoneal injection.

[†]   F-DA = Fluosol-DA 20% emulsion (Green Cross, Japan).

[††] F-DA stem emulsion only

[+]   FC-43 = Fluosol-43/oxypherol emulsion (Green Cross, Japan).

[Δ]   FDC + 1% C-16 HBPO = Emulsified perfluorodecalin containing 1% of a C-16 higher boiling point oil (HBPO) additive, perfluoroperhydrofluoranthrene (see text for details).

[φ]   FMOQ = Emulsified F-4-methyloctahydroquinolidizine.

[ψ]   DMA/NONANE = Emulsified F-dimethyladamantane/F-trimethylbicyclononane.

[#]   Mice also received 4 x 30 min exposures to carbogen (95% O_2/5% CO_2).

Unc. No. significant change in organ weight.

PFCs are eliminated from the body primarily by expiration and the rate of this excretion depends upon vapour pressure (Naito and Yokoyama, 1978); some PFCs are also transpired through the skin. Excretion of PFCs increases with the intravascular persistence time of the emulsion but is markedly reduced once uptake into tissues has occurred. The elimination of PFCs therefore involves two separate processes occurring at markedly different rates: an initial rapid excretion followed by much slower loss from the tissues. The existence of two such processes can account for the inconsistencies in the literature concerning retention half-times of PFCs in vivo.

There is no convincing evidence that PFCs undergo any form of metabolism in the body and this is often cited in support of claims for the biological inertness of these compounds. Where toxic effects have been observed, these have invariably been attributed to impurities such as fluoroamines (Riess and Le Blanc, 1982) or, when PFCs are injected in emulsified form, to other emulsion constituents, especially in the surfactant components.

Effects of emulsified PFCs on lymphoid tissues and immunological competence

The extent to which PFC emulsion particles accumulate in lymphoid tissues depends upon species used together with dose, route and composition of emulsion administered (Table 1). For example, injection of 5 or 10 ml/kg of the commercial emulsion, Fluosol-DA 20% (F-DA; Green Cross, Japan), either intravenously (i.v.) or intraperitoneally (i.p.) into rats or mice produces differential changes in liver, spleen, thymus and gut mesenteric lymph node (MLN) weights as measured after 8 days (Bollands and Lowe, 1986b; 1987a; Lowe, 1987). Such variability in tissue accumulation and retention of PFC emulsion particles has inevitably introduced complications into studies of the fate of PFCs in the body.

In addition to morphological changes produced by emulsified PFCs in cells of the RES and other tissues (Nanney et al., 1984), interest has also focussed on possible alterations in tissue function induced by these compounds. Lutz and others (Lutz et al., 1982b) observed a transient depression of RES phagocytic function following injection of F-DA stem emulsion in rats while Castro and colleagues reported similar findings in both rats and primates (Castro et al., 1983, 1984). This suppression of RES clearance function by F-DA would help to explain the decreased resistance to bacterial endotoxin seen in rats injected with the emulsion (Lutz et al., 1982a). It is likely that alterations in RES function induced by emulsified PFCs may persist for longer than was initially believed since increases in serum β globulins have been measured in monkeys up to 8 years after exchange-transfusion with either F-DA or Oxypherol/Fluosol-43 (FC-43; Green Cross, Japan) (Rosenblum et al., 1985).

There are inconsistencies in the literature regarding the effects of emulsified PFCs on immune system function and in particular, on the manifestation of humoral responses to immunological challenge. In an initial preliminary report, Shah and colleagues (Shah et al., 1984) noted that pretreatment of Balb/c mice with F-DA or FC-43 i.v. led to a decrease in the in vivo production of antibodies to sheep red blood cells (SRBC). However, subsequent work has produced conflicting results with an increase in the haemagglutination response to SRBC occurring in rats or mice injected i.p. but not i.v. with comparable doses of F-DA (Bollands and Lowe, 1986b; 1987a, Lowe, 1987). This difference may reflect strain variations in response but other factors, notably timing and route administration of emulsion relative to immune challenge, must also be considered. In this regard, Mitsuno and others (Mitsuno et al., 1984) reported that in rats injected with <20 ml/kg F-DA, the emulsion enhanced antibody production when given after immunization whereas it inhibited antibody generation when injected before immunization. However, firm conclusions cannot be drawn from this initial work since neither antigen used nor route of emulsion injection were specified.

More recent studies have revealed that changes in lymphoid tissue weights and plasma antibody titres to SRBC in rodents varied according to the timing and route of a previous or subsequent injection of F-DA (Bollands and Lowe, 1987; Lowe and Bollands, 1987); marked species differences were noted when the responses in rats and mice were compared (Lowe and Bollands, 1987).

It has been proposed that F-DA acts as an adjuvant when injected into the peritoneal cavity of either rats or mice 24 hr prior to SRBC (Bollands and Lowe, 1986b; 1987a). The possibility exists that PFCs may promote release of the intracellular mediator, interleukin-1 (IL-1), from macrophages in a similar manner to that already described for other adjuvants (Allison, 1983). Such IL-1 release would, in turn, potentiate a subsequent immune response. If this is the case, it would help to explain the observed immunopotentiation effects of F-DA and account for some of the variation in humoral immune responses seen in animals receiving the emulsion. While the effects of F-DA or other emulsified PFCs on IL-1 release have not been reported, other particulate material, such as silica, can enhance its release from rabbit macrophages in vitro (Kampschmidt et al., 1986).

The speculation that F-DA can have adjuvant-like effects is supported by the finding it can form complexes with antigens in hormone radioimmunoassay systems (Hammarstrom et al., 1983), a property which has been effectively employed in an immunological agglutination assay (Prather et al., 1986).

Further evidence of potential deleterious effects of emulsified PFCs on immune system function is that near-total transfusion of rats with F-DA inhibited the afferent (induction) phase of a specific immune response but did not alter an ongoing efferent phase (Hodges et al., 1986). This latter effect was attributed to RES blockade by the emulsion as discussed above. Transfusion with F-DA also enhanced the subsequent early specific immune response to SRBC but the active component(s) involved were not identified (Hodges et al., 1986). More recent experiments have shown that transfusion with F-DA inhibited the humoral immune response to type 3 pneumococcal polysaccharide (PPS) when performed 3 days before immunization (Molina et al., 1987). The effects of F-DA transfusion on the immune response to PPS were markedly different to those in which animals were immunized with SRBC (Hodges et al., 1986) and show that the extent of alterations in immune responsiveness produced by emulsified PFCs depends upon whether a T-lymphocyte dependent (SRBC) or T-lymphocyte independent (PPS) system is involved.

Responses to the Pluronic surfactant

There is growing speculation that the polyoxyethylene-polyoxypropyl-amine block to polymer surfacant, Pluronic F-68, which is a component of both F-DA and FC-43 is responsible, at least in part, for some of the adverse effects of emulsified PFCs in vivo (Lowe and Bollands, 1985). In support of this, Lane and Lamkin (1986) reported that Pluronic F-68 was the active agent in F-DA producing impairment of neutrophil migration in mice. This would further explain the earlier findings of decreased resistance to bacterial endotoxin in mice injected with F-DA (Lutz et al., 1982a), since increased susceptibility to infection would therefore be a consequence of treatment with emulsions containing this surfactant. The effects of Pluronic on immune cells in culture is discussed further below.

Pluronic can also contribute to some of the other reported adverse effects of F-DA: for example, it can mimic the complement-activating effects of F-DA both in vivo and in vitro (Vercellotti et al., 1982) and also appears to be responsible for the inhibitory effects of F-DA on both plasma and liver phospholipase A_2 activities (Shakir and Williams, 1982).

McCoy and co-workers have suggested that peroxide derivatives of Pluronic, formed during steam sterilization or long-term storage of the emulsion, could contribute to, but not fully account for, adverse effects of F-DA on arterial endothelial ultrastructure in rats (McCoy et al., 1984). It was noted that alterations in arterial endothelium still occurred in animals transfused with F-DA which had been pre-treated with the anti-oxidant, citric acid, to neutralize any peroxide contaminants. There are no published data on the extent to which such peroxides normally occur in F-DA or on the rate or specific conditions under which they may be generated. However, it is known that peroxide formation can be minimized by high-pressure homogenization under N_2 or CO_2 atmosphere and this emulsification procedure is employed in the manufacture of F-DA (Naito and Yokoyama, 1978).

It is noteworthy that recent work has revealed toxic effects of Pluronic F-68 when injected intravenously in rats in doses of 100-1000 mg/kg: anatomical abnormalities observed included the appearance of pulmonary foam cells together with slight focal degenerative changes in renal proximal tubules (Magnusson et al., 1986). Pluronic also appeared to induce phospholipidosis, comparable to that produced by cationic amphilic drugs. Although the mechanism(s) involved were not determined, one explanation is that Pluronic impaired phospholipid metabolism by inhibiting the degradation of phospholipids by phospholipase. This speculation is supported by the previous findings that Pluronic can inhibit phospholipase A_2 activity in both liver and plasma (Shakir and Williams, 1982).

Novel surfactants

Whether Pluronic or its derivatives are directly responsible for the growing list of reported adverse reactions to commercial PFC emulsions (Lowe and Bollands, 1985) remains to be determined. However, concern about Pluronic has prompted the search for suitable alternative emulsifiers for PFC emulsification: developments in this area include the assessment of perfluorinated surfactants having "fluorophilic" properties. One such surfactant, perfluoroalkylated amine oxide, appears to be tolerated in mice in doses of up to 10 ml/kg with no immediate adverse effects (Clark et al., 1983). A more recent advance has been the development of a novel class of well-defined fluorinated surfactants carefully designed for use in specific PFC systems (Riess and Le Blanc, 1988).

Effects of emulsified PFCs on immune cells in vitro

Evidence that in vivo use of emulsified PFCs can alter immune system function and interfere with normal immunological competence has inevitably prompted studies on the direct effects of such compounds on immune cells in culture. For example, incubation of human blood with either F-DA or FC-43 inhibited phagocytic activity of both neutrophils and monocytes (Virmani et al., 1983, 1984). Exposure to emulsified PFCs also produced increased cytoplasmic vacuolation, decreased chemotaxis and reduced aggregation, adherence and superoxide ion (O_2^-) release in the presence of phorbol myristate acetate (PMA). Similar findings of reduced cell adherence following culture of mouse peritoneal macrophages or non-depleted murine spenocytes with FC-43 or F-DA respectively have also been reported (Bucala et al., 1983; Bollands and Lowe, 1987d).

Current evidence suggests that in vitro cellular responses to emulsified PFCs depends to a great extent upon cell type concerned and species of origin (Lowe and Bollands, 1985; Lowe, 1987). For example, FC-43 is selectively toxic to murine macrophages but not lymphocytes (Bucala et al., 1983). While this difference cannot at present be explained, it has been proposed that this selective cytotoxicity involves either disruption of the cell membrane or adverse effects on the enzyme-regulated oxygen de-toxification systems or a combination of both (Bucala et al., 1983).

In addition to the responses of different cell types, emulsion composition must also be considered: for example, F-DA is more effective than FC-43 in promoting procoagulant generation from human monocytes (Janco et al., 1985). F-DA also impaired stimulated oxidative metabolism (i.e. O_2^- generation) from monocytes whereas FC-43 was unaffective (Janco et al., 1985). F-DA also impaired stimulated oxidative metabolism (i.e. O_2^- generation) from monocytes whereas FC-43 was unaffective (Janco et al., 1985). It is especially noteworthy that some of the adverse cellular effects produced by F-DA can also be reproduced by the Pluronic F-68 emulsifier: Pluronic itself was effective in promoting procoagulant release from monocytes but did not, however, affect stimulated O_2^- production (Janco et al., 1985). Pluronic also appears to be almost entirely responsible for the inhibition of migration produced when human polymorphonuclear cells are exposed to F-DA in vitro (Lane and Lamkin, 1984). It is somewhat paradoxical that Pluronic should be implicated as the active principle responsible for some of the adverse effects of emulsified PFCs on immune cells since this compound has been shown to enhance nutrient uptake into cultured human lymphocytes and thereby increase growth rate (Mizrahi, 1975).

Lymphoid tissue responses to novel PFC emulsions

Recent work has shown that emulsified FDC can be stabilized by the addition of small quantities of polycyclic, perfluorinated, higher boiling point oil (HBPO) additives (Davis et al., 1986; Sharma et al., 1986). The addition of HBPO retards the process of molecular diffusion known as Ostwald ripening which can contribute to the destabilization of PFC emulsions (Davis et al., 1981). Preliminary biocompatibility tests with one such FDC emulsion have been carried out in rats with encouraging results (Davis et al., 1986; Bollands et al., 1987). Work is now in progress to assess the effects of these novel formulations in different biological systems, including cell culture studies.

SUMMARY AND CONCLUSIONS

Data from both in vivo and in vitro studies to assess lymphoid tissue and immune system responses to emulsified PFCs and their components have been discussed. It may be concluded that the extent to which PFC emulsions are retained in lymphoid tissues and their effects on the manifestation of humoral immune responses to immunological "challenges" is variable and depends upon: (1) Emulsion composition; (2) Dose administered; (3) Route of administration; (4) Tissue examined; (5)

Timing of emulsion administration relative to immune challenge; and (6) Species studied.

Such variability in the responses to emulsified PFCs has inevitably introduced difficulties in assessing their effects upon immunological competence. Further work is needed determine the extent to which lymphoid tissue functions may be altered by PFC emulsions and identify the active component(s) and mechanism(s) involved.

Acknowledgements

Some of the original work described in this paper was supported by research grants from I.S.C. Chemicals Ltd., Avonmouth. The assistance of the Green Cross Corporation is gratefully acknowledged.

REFERENCES

Allison, A. C., 1983, Immunological and adjuvants and their mode of action, in: New Approaches to Vaccine Development, Bell, C. and Torrigani, G. (eds) Schwabe, Basel

Biro, G. P. and Blais, P., 1987, Perfluorocarbon blood substitutes, CRC Crit. Rev. Oncol./Hematol., 6:311.

Bollands, A .D. and Lowe, K. C., 1986a, Responses of rat lymphoid tissue to a perfluorocarbon blood substitute, Br. J. Pharmac., 87:118P.

Bollands, A. D. and Lowe, K. C., 1986b, Effects of a perfluorocarbon emulsion, Fluosol-DA, on rat lymphoid tissue and immunological competence, Comp. Biochem. Physiol., 85C:309.

Bollands, A. D. and Lowe, K. C., 1986c, Comparative effects of emulsified perfluorocarbons on lymphoid tissue in rodents, Br. J. Pharmac., 89:664P.

Bollands, A. D. and Lowe, K. C., 1987a, Lymphoid tissue responses to perfluorocarbon emulsion in mice, Comp. Biochem. Physiol., 86C:431.

Bollands, A. D. and Lowe, K. C., 1987b, Lymphoid tissue responses to Fluosol-DA in rats: time course effects relative to immunological challenge, Br. J. Pharmac., 90:180P.

Bollands, A. D. and Lowe, K. C., 1987c, Lymphoid tissue responses to perfluorocarbon emulsion in rat: time course effects relative to immune challenge, Comp. Biochem. Physiol, (in press).

Bollands, A. D. and Lowe, K. C., 1987d, Spleen cell responses to emulsified perfluorocarbons in vitro, Br. J. Pharmac., 91:457P.

Bollands, A. D., Davis, S. S., Lowe, K. C. and Sharma, S. K., 1987, Effects of a novel perfluorocarbon emulsion in rats, Br. J. Pharmac., 90:179P.

Bucala, R., Kawakami, M. and Cerami, A., 1983, Cytotoxicity of perfluorocarbon blood substitute to macrophages in vitro, Science, 220-965.

Caiazza, S., Fanizza, M. and Ferrari, M., 1984, Fluosol 43 particle localization pattern in target organs of rats. An electron microscopical study, Virchows Arch. path. Anat. Physiol., 404:127.

Castro, O., Reindorf, C. A., Socha, W. W. and Rose, A. W., 1983, Perfluorocarbon enhancement of heterologous red cell survival: a reticuloendothelial block effect? Int. Archs. Allergy. appl. Immun., 70:88.

Castro, O., Nesbitt, A. E. and Lyles, D., 1984, Effect of a perfluorocarbon emulsion (Fluosol-DA) on reticuloendothelial system clearance function, Am. J. Hemat., 16:15.

Clark, L. C., Clark, E. W., Moore, R. E., Kinett, D. G. and Inscho, E. I., 1983, Room temperature-stable biocompatible fluorocarbon emulsions. In: Advances in Blood Substitute Research, Progress in Clinical and Biological Research, Vol. 122, Dollu, R. B., Geyer, R. P. and Nemo, G. J., eds. Liss, New York.

Davis, S. S., Round, H. P. and Purewal, T. S., 1981, Ostwald Ripening of emulsion systems. An explanation for the effect of an added third component, J. Coll. Interface Sci., 80:508.

Davis, S. S., Lowe, K. C. and Sharma, S. K., 1986, Novel compositions of emulsified perfluorochemicals for biological applications, Br. J. Pharmac., 89:665P.

Dellacherie, E., Labrude, P., Vigneron, C. and Riess, J. G., 1987, Synthetic carriers of oxygen, CRC Crit. Rev. Therapeutic Drug Carrier Sys. 3:41.

Goodman, R. L., Moore, R. E., Davis, M. E., Stokes, D. and Yuhas, J. M., 1984, Perfluorocarbon emulsions in cancer therapy: preliminary observations on presently available formulations, Int. J. Radiat. Oncol. Biol. Phys., 10:1421.

Hammarstrom, M., Mullins, R. and Sgoutas, D., 1983, Effect of Fluosol-DA on radioimmunoassay results, Clin. Chem. 29:1418.

Hodges, G. R., Worley, S. E., Kemner, J. M., Adbou, N. I. and Clark, G. M., 1986, Effect of exchange-transfusion with Fluosol-DA on splenic distribution and immunocompetence of rats lymphocytes, J. Leukocyte Biol. 39,141.

Janco, R. L., Virmani, R., Morris, P. J. and Gunter, K., 1985, Perfluorochemical blood substitutes differentially alter human monocyte procoagulant generation and oxidative metabolism, Transfusion, 25:578.

Kampschmidt, R. F., Worthington, M. L. and Mesecher, M. I., 1986, Release of interleukin-1 (IL-1) and IL-1-like factors from rabbit macrophages with silica, J. Luekocyte Biol., 39:123.

Lane, T. A. and Lamkin, G. E., 1984, Paralysis of phagocyte migration due to an artificial blood substitute, Blood, 64:400.

Lane, T. A. and Lamkin, G. E., 1986, Increased infection mortality and decreased neutrophil migration due to a component of an artificial blood substitute, Blood, 68:351.

Lowe, K. C., 1986, Blood transfusion or blood substitution? Vox. Sang. 51:257.

Lowe, K. C., 1987, Perfluorocarbons as oxygen-transport fluids, Comp. Biochem. Physiol., (in press).

Lowe, K. C. and Bollands, A. D., 1985, Physiological effects of perfluorocarbon blood substitutes, Med. Lab. Sci., 42:367.

Lowe, K. C. and Bollands, A. D., 1987, Lymphoid tissue responses to emulsified perfluorochemicals: comparative aspects, Biomat., Art. Cells, Art. Org. (in press).

Lutz, J. and Metzenauer, P., 1980, Effects of potential blood substitute (perfluorochemicals) on rat liver and spleen, Pflugers Arch., 387:175.

Lutz, J., Barthel, U. and Metzenauer, P., 1982a, Variation in toxicity of Escherichia coli endotoxin in rats after treatment with perfluorated blood substitutes in mice, Circ. Shock, 9:99.

Lutz, J., Metzenauer, P., Kunz, E. and Heine, W. D., 1982b, Cellular responses after use of perfluorinated blood substitutes, in: Oxygen-Carrying Colloidal Blood Substitutes, Frey, R., Beisbarth, H. and Stosseck, K., eds. Zuckschwerdt, Munich.

Magnusson, G., Olsson, T. and Nyberg, J.- A., 1986, Toxicity of Pluronic F-68, Toxicol. Lett. 30:203.

Mason, K. A., Withers, H. R. and Steckel, R. J., 1985, Acute effects of a perfluorochemical oxygen carrier on normal tissues of the mouse, Radiat. Res., 104:387.

McCoy, L. E., Becker, C. A., Goodin, T. H. and Barnhart, M. I., 1984, Endothelial responses to perfluorochemical emulsion. Scan. electron Micros., 16:311.

Mitsuno, T., Ohyanagi, H. and Yokoyama, K., 1984, Development of a perfluorochemical emulsion as a gas carrier, Artif. Org., 8:25.

Mizrahi, A., 1975, Pluronic polyols in human lymphocyte cell cultures, J. Clin. Microbiol., 2,11.

Molina, N. C., Hodges, G. R., Worley, S. E. and Abdou, N. I., 1987,
Effect of Fluosol-DA 20% on antibody response to type 3 pneumococcal
polysaccharide, Clin. Immun. Immunopath. 42:211.

Naito, R. and Yokoyama, K., 1978, Pefluorochemical blood substitutes,
Technical Information Series No. 5 Green Cross, Osaka.

Nanney, L., Fink, L. M. and Virmani, R., 1984, Perfluorchemicals: morpho-
logic changes in infused liver, spleen, lung and kidney of rabbits,
Arch. Pathol. Lab. Med., 108:631.

Prather, T. L., Grane, J., Keese, C. R. and Giaever, I., 1986, An agglutination assay using emulsified oils, J. Immun. Methods, 87:211.

Raven, P. D., Bollands, A. D., Sharma, S. K. and Lowe, K. C., 1987,
Effects of a novel perfluorochemical emulsion on lymphoid tissues
and immunocompetence in rats, Br. J. Pharmac., (in press).

Riess, J. G. and Le Blanc, M., 1982, Solubility and transport phenomena
in perfluorochemicals relevant to blood substitution and other biomedical applications, Pure Appl. Chem., 54:2383.

Riess, J. G. and Le Blanc, M., 1988, Preparation of fluorocarbon emul-
sions for biomedical applications, Materials and Methods, in: Blood
Substitutes: Physiology and Medical Applications, Lowe, K. C. ed,
Ellis Horwood, Chichester (in press).

Rockwell, S., Irvin, C. G. and Nierenburg, M., 1986, Effect of a
perfluorochemical emulsion on the development of artificial lung
metastases in mice, Clin. Expl. Metastasis, 4:45.

Rosenblum, W. I., Moncure, W. and Behm, G., 1985, Some long-term effects
of exchange transfusion with fluorocarbon emulsions in Macaque
monkeys. Arch. Pathol. Lab. Med., 109:340.

Shah, K. H., Yamamura, Y. and Usuba, A., 1984, Immunopathologic changes
in artificial blood perfluorochemicals, Lab. Invest., 48:62A.

Shakir, K. M. M. and Williams, T. J., 1982, Inhibition of phospholipase
A2 activity by Fluosol, an artificial blood substitute,
Prostaglandins, 23, 919.

Sharma, S. K., Davis, S. S., Johnson, O. L. and Lowe, K. C., 1986,
Physicochemical assessment of novel formulations of emulsified
perfluorocarbons, J. Pharm. Pharmac., 38:5P.

Tremper, K. K. and Anderson, S. T., 1985, Perfluorochemical emulsion
oxygen transport fluids: a clinical review. Ann. Rev. Med. 36:309.

Vercellotti, G. M., Hammerschmidt, D. E., Craddock, P. R. and Jacob, H. S.,
1982, Activation of plasma complement by perfluorocarbon artificial
blood: probable mechanism of adverse pulmonary reactions in patients
and rationale for corticosteroid prophylaxis, Blood 59:1299.

Virmani, R., Warren, D., Rees, D., Fink, L. M. and English, D., 1983,
Effects of perfluorochemical on phagocytic function of leukocytes,
Transfusion, 23:512.

Virmani, R., Fink, L. M., Gunter, K. and English, D., 1984, Effect of
perfluorochemical blood substitutes on human neutrophil function,
Transfusion, 24:343.

West, L, McIntosh, N., Gendler, S., Seymour, C. and Wisdom, C., 1986,
Effects of intravenously infused Fluosol-DA 20% in rats, Int. J.
Radiat. Oncology Biol. Phys., 12:1319.

Yokoyama, K., Suyama, S., Okamoto, H., Watanabe, M., Ohyanagi, H. and
Saitoh, Y., 1984, A perfluorochemical emulsion as an oxygen carrier,
Artif. Org., 8:34.

EFFECTS OF METHIONINE ENKEPHALIN (MET-ENKEPHALIN) ON REGIONAL BLOOD

FLOW AND VASCULAR RESISTANCE: RADIOACTIVE MICROSPHERE TECHNIQUES

H. M. Rhee, P. Eulie and H. Laughlin[*]

Departments of Pharmacology and Physiology[*]
Oral Roberts University School of Medicine
Tulsa, OK 74137 U.S.A.

INTRODUCTION

Ever since Hughes et al. (1975) discovered opioid-like peptides from the brain extract of rat, many advancements in the field of neurophysiology and pharmacology have been made. Methionine enkephalin ([met^5]enkephalin), a simple pentapeptide, has been implicated in many physiological functions such as pain perception, appetite, hormone-release, thermoregulation, gastrointestinal function, and cardiovascular function (Martin, 1983). Although the cardiovascular effects of [met^5]enkephalin depend on many factors, the level of anesthesia (Schaz et al., 1980; Eulie and Rhee, 1984), the sites of peptide injection (Wei et al., 1980; Petty and DeJong, 1983) and the animal species have been shown to be important factors.

Previously we reported that, in anesthetized rabbits, [met^5]enkephalin given i.v. produced bradycardia and hypotension. These effects were naloxone-reversible (Eulie and Rhee, 1984), but was not antagonized by naloxone methobromide, a quartnary analog of naloxone. The hypotensive action of [met^5]enkephalin was also blocked by phentolamine, an α adrenoceptor antagonist. Recently we also showed that [met^5]enkephalin reduced the rate of sympathetic nerve discharge in parallel with the reduction of blood pressure (Rhee et al., 1985b). These data suggest that [met^5]enkephalin might reduce sympathetic tone selectively so that specific vascular beds might be selectively dilated. Therefore, the primary objective of this study was to examine the effect of [met^5]enkephalin on the vascular resistance of anesthetized rabbits.

MATERIALS AND METHODS

Animal Preparation and Recording of Physiological Parameters

Male New Zealand white rabbits weighing between 1.5 and 2.5 kgs were anesthetized using sodium pentobarbital (Abbott-Lab), 30 mg/kg i.v., via the marginal ear vein. A light level of surgical anesthesia was maintained by supplemental doses of pentobarbital administered through a cannulated femoral vein. The rabbits were allowed to respire spontaneously through a tracheostomy, and were maintained at 37°C with a temperature regulator (Gorman Rupp, Bellville, Ohio).

Aortic blood pressure was measured using a Statham P23DB transducer attached to a cannula placed in the descending aorta via the femoral artery. ECG was obtained using needle electrodes in a standard Lead II configuration. Heart rate was obtained electronically from the ECG or arterial pressure curve. All measurements were recorded on a Gould electrostatic recorder (Gould Instrument Co., Cleveland, OH, Model ES1000). A catheter was placed in the left ventricle by cannulating the left common carotid artery with either polyethylene tubing (PE60) or a precalibrated solid state pressure transducer (Millar Model 5F) containing an injection port. Arterial pressure was monitored throughout this procedure to determine if any damage may have occurred to the aortic valve. Whenever damage was confirmed, that animal was eliminated from the study.

Microsphere Techniques

In this study radiolabeled (Sc-46, Sr-85, Sn-113, and Co-57) microspheres of 15.3 ± 0.98 μm diameter were used. This sphere size was chosen to minimize arteriovenous shunting without unfavorably influencing rheological factors (Warren and Ledingham, 1974). Microsphere suspensions consisted of 500,000 spheres in 0.1 ml of 10% dextran, containing less than 0.1% Tween 80. The suspensions were mixed for a minimum of 10 min in a Mettler Electronics ultrasonicator and then on a Vortex vibrator for a minimum of 2 min prior to infusion. This was to assure a uniform distribution of spheres in the suspension medium. This technique has been checked by visual observation (Heymann et. al., 1977; Laughlin et al., 1982) and found to be valid.

Microsphere injections were performed as follows: 1) Withdrawal was started for a reference blood sample, at a rate of 1.02 ml/min, on a Harvard infusion-withdrawal pump. The reference blood sample was drawn from a catheter positioned to rest in the descending aorta. 2) As blood was seen moving into the catheter; 0.1 ml of sphere suspension was slowly infused through the catheter located in the left ventricle followed by 1.0 ml of warm (37°C) saline. 3) Infusion time was between 10 and 15 s. 4) Withdrawal of the reference sample was continued until 2.4 ml of blood was collected or for at least a period of 1 min post-infusion of sphere suspension and saline. This infusion and withdrawal procedure of microspheres was repeated 5 s after either saline or [met[5]]enkephalin (1 mg/kg) infusion. Naloxone (1 mg/kg) was injected 10 min before the infusion of [met[5]]enkephalin to see its antagonism against the effect of [met[5]]enkephalin. There was at least 30 min interval between one treatment to another to insure that the previous treatment did not affect hemodynamic variables of subsequent treatments in the same animal.

After completion of the last microsphere injection, the animal was overdosed with a bolus of pentobarbital. The animal was always sacrificed within one hour of the first injection of spheres. The following 35 tissue samples (2-10 g) were taken: brain, which was divided into right and left cerebrum, cerebellum, diencephalon, and mesencephalon. Lumbar spinal cord was also sampled. Paired right and left samples were taken of eyes, ears, masseter, diaphragm, triceps, pectoral and rect. abdom. muscles, and kidneys. The heart was removed and divided into the following samples: right and left atria and septum, right ventricle, and left ventricular anterior wall with subdivisions of endo, medial, and epicardial tissue, including sino-atrial node and septal marginal trabecular. Also lung, esophageal, duodenal, jejunal and ileal tissue samples were taken. Adrenal, pituitary and pancreatic glands, including testes, were also sampled.

The accuracy of the microsphere technique is dependent upon proper sphere mixing and the number of spheres trapped per tissue sample. Adequacy of microsphere mixing in the left ventricle was monitored in each experiment by comparing blood flow to the right and left kidneys. Heymann et al. (1977) suggested that the number of spheres for injection be calculated to insure all tissue samples would receive an adequate number of spheres. Preliminary observations and calculations indicated that a 500,000 sphere infusion would provide a minimum of 400 spheres per tissue sample, and provide a 95% confidence level I. These calculations were true for all tissues with the exception of skin, ear, and sino-atrial node. Since aortic pressure was reasonably constant during the microsphere infusion, resistance was calculated from mean arterial pressure divided by tissue flow.

Chemicals and Data Analysis

Radiolabeled microspheres were purchased either from New England Nuclear or 3M Company. [Met^5]enkephalin was purchased from Sigma Chemical Co. and naloxone was donated by Endo Research Laboratories. Overall data analysis was performed by a two-way analysis of variance and each datum point was analyzed by unpaired Student's t-test. Significant difference was considered as P is less than 0.05.

RESULTS

Cardiovascular Effect of [Met^5]enkephalin

In pentobarbital anesthetized rabbits, an infusion of 1.2 ml saline over 30 s interval did not produce any measurable changes in cardiovascular parameters such as systolic and diastolic pressure, heart rate and electrocardiogram. An infusion of equivolume of saline containing [met^5]enkephalin (1 mg/kg, body weight) reduced systolic, mean and diastolic pressures (fig. 1). The figure also shows that there was a slight, but actual increase in pulse pressure. The control mean arterial blood pressure was 85.9 ± 4.4 mmHg as summarized in Table 1. [Met^5]enkephalin significantly (P<0.05) reduced blood pressure, which was blocked by the treatment of naloxone (1 mg/kg). Bradycardia induced by [met^5]enkephalin (see fig. 1) was also reversed by naloxone pretreatment, which is consistent with our previous reports (Rhee et al., 1985a and 1985b).

TABLE 1. Effects of [met^5]enkephalin alone or [met^5]enkephalin after naloxone treatment on blood pressure and heart rate in anesthetized rabbits.[a]

Experiments	n	Mean Arterial Pressure (mmHg)	Heart Rate (BPM)
Control	9	85.9 ± 4.4	289 ± 16
[Met^5]enkephalin (1 mg/kg)	9	55.0 ± 7.1[b]	202 ± 11[b]
[Met^5]enkephalin after naloxone (1 mg/kg)	7	83.9 ± 3.8	260 ± 17

[a] All values represent peak effects of drugs and are expressed in mean ± S.E.

[b] Indicate P<0.05, compared to the relevant control.

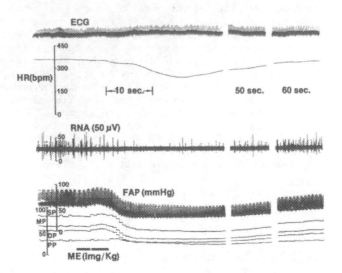

Fig. 1.   Temporal action of [met⁵]enkephalin (ME) on the blood pressure in anesthetized rabbit.   [Met⁵]enkephalin (1 mg/kg
i.v.) was injected in a volume of 0.25 ml at time 0 s and
washed by saline immediately as described under Methods.
Femoral arterial pressure was continuously recorded up to
60 s. SP, MP, DP, and PP stand for systolic, mean, diastolic, and pulse pressure. Numbers on top of the figure indicate time in seconds after an injection of [met⁵]enkepha-
lin. RNA indicates renal sympathetic nerve activity and FAP
stands for femoral arterial blood pressure.

Effects of [Met⁵]enkephalin on Tissue Blood Flow

Blood flows were measured in the tissues listed in the Methods.
[Met⁵]enkephalin-induced hypotension made it difficult to compare
blood flow data since a decrease in vascular resistance combined with
decreased pressure may result in no change in blood flow. Therefore
apparent vascular resistance was calculated as: mean arterial pressure (mmHg)/blood flow (ml/min/100 g). The vascular resistance data
for many of the tissues included in this study are presented in figures 2 and 3. [Met⁵]enkephalin had no statistically significant effect on the vascular resistances in glandular and in cardiac tissues
(data not shown). Vascular resistance was reduced in cerebral hemisphere   after   the   treatment   of   [met⁵]enkephalin   (fig. 2).
[Met⁵]enkephalin had no consistent effect on the blood flow of other
brain tissues.   In the case of pituitary the standard error was high
due to small size of the tissue. Vasoconstriction (increased vascular
resistance) was seen in the intestinal tissues such as duodenum,
jejunum and ileum, while the stomach and spleen showed no change in
vascular resistance (fig. 3). Vascular resistance was also increased
160% in the kidneys from 0.41 ± 0.04 to 0.59 ± 0.08. On the other
hand, vascular resistance was decreased 75% in the triceps muscles,
50% in the pectoral and masseter muscles (Tab. 2). Thus the regional
blood flow data obtained in this study indicate that [met⁵]enkephalin

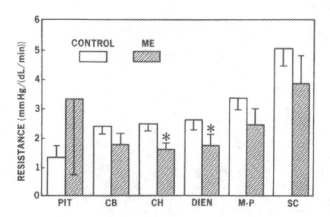

Fig. 2. Effects of [met⁵]enkephalin (ME) on vascular resistance in
brain tissues of anesthetized rabbits. Phentobarbital anesthetized control and ME (1 mg/kg) treated rabbits were prepared as in Methods.   Blood flow was determined after an
injection of microsphere (SC-46, Sr-85, Sn-113, and Co-57) in
saline or saline containing ME as described under Methods.
Based on mean arterial pressure (MAP) resistance (R) was
calculated by the formula, R = MAP/blood flow in ml/min per
100 g of tissues. Vertical bars represent S.E.M., and each
value is the mean of 9 experiments. Pit stands for pituitary
gland; CB for cerebellum; CH for cerebral hemisphere; DIEN
for Diencephalon: M-P for mesencephalon and SC for Spinal
Cord, respectively. * indicates P<0.05.

TABLE 2. Antagonistic effects of naloxone on [met5]enkephalin induced vasodilation in the skeletal muscles of anesthetized rabbits.[a]

Muscle	N[b]	Vascular resistance (mmHg•ml^{-1}•min^{-1}•100g tissue)		
		Control	[Met5]enkephalin alone[c]	[Met5]enkephalin plus naloxone[d]
Right triceps	8	36.79 ± 4.48	9.65 ± 2.35[f]	39.85 ± 5.06
Left triceps	8	36.46 ± 5.61	8.87 ± 1.56[f]	42.42 ± 13.62
Right pectoralis	8	22.29 ± 5.18	9.06 ± 1.98[e]	23.56 ± 5.15
Left pectoralis	8	24.11 ± 6.23	8.77 ± 2.05[e]	21.67 ± 5.24
Right masseter	9	22.19 ± 4.08	13.43 ± 3.11[e]	16.60 ± 3.66
Left masseter	9	20.15 ± 3.14	9.12 ± 2.10[e]	19.22 ± 2.71
Rectus abdomi	9	19.54 ± 2.18	15.79 ± 2.20	20.30 ± 4.93
Diaphragm	9	6.11 ± 1.04	3.96 ± 0.96	6.52 ± 1.28

[a] Vascular resistance was calculated by a formula of (pressure/flow) and all values are mean ± S.E. Mean arterial blood pressures in control, [met5]enkephalin and [met5]enkephalin after naloxone treatment were 85.3, 55.0 and 93.9 mmHg, respectively.

[b] Indicate number of animals used.

[c] [Met5]enkephalin dose was 1 mg/kg body weight.

[d] Naloxone (1 mg/kg) was pretreated 10 min before [met5]enkephalin infusion as indicated in "Methods."

[e,f] *, ** indicate P is less than 0.05 and 0.01 respectively, compared to the relevant control. All other values are not significantly different from the control.

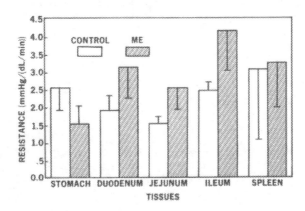

Fig. 3.   Effects of [met5]enkephalin (ME) on vascular resistance in gastrointestinal tissue in anesthetized rabbits. Experiments and calculation of resistance were carried out as in fig. 1. Signs were used in fig. 2. * indicates P<0.05.

induced hypotension is associated with vasoconstriction in intestinal tissue and kidney, and vasodilation of skeletal muscle vascular beds.

Inhibitory Effect of Naloxone on [Met^5]enkephalin Action

The fact that [met^5]enkephalin selectively reduced vascular resistance in skeletal muscle bed lead us to test whether [met^5]enkephalin-induced vasodilation is mediated by the specific opiate receptors. Therefore, to test the possibility, we used naloxone, a specific opiate receptor antagonist. As summarized in table 2, naloxone at the dose of 1 mg/kg reversed the vasodilatory effect of the peptide, which indicates a receptor specific action of [met^5]enkephalin in the skeletal muscle bed in anesthetized rabbits. Naloxone itself, at the dose that we used, did not have any significant effects on heart rate, blood pressures, and renal blood flow including renal nerve activity for up to 1 hr after its treatment. [Met^5]enkephalin had little or no effect on the vascular resistance of other tissues such as aorta, spleen, lung, liver, both ears and eyes (data not shown).

DISCUSSION

The results of this study confirm previous observations from this laboratory (Rhee et al., 1985a and 1985b) and others that i.v. infusions of [met^5]enkephalin in anesthetized rabbits produces decreased cardiac output and hypotension. In addition, the regional flow data indicate that these changes are associated with increased vascular resistance in most visceral tissues and decreased vascular resistance in skeletal muscle tissue. Blood flow measurements with the microsphere technique are based upon the assumption that the labeled spheres are evenly mixed and distributed with the blood. In the experiments included in this study there were no differences in the blood flows determined in the 9 paired tissue samples from the right and left sides of the body. Thus, spheres were adequately mixed upon infusion in the left ventricular chamber (Heymann et al., 1977; Laughlin et al., 1982).

Although cardiovascular effects of [met^5]enkephalin are well known, little is known about the mechanisms of these effects (see the Introduction). Previously our studies (Rhee et al., 1985b; Eulie and Rhee, 1984) indicated that the dominant site of [met^5]enkephalin-induced hypotension is in the central nervous system. The main reasons for this conclusion are: 1) [met^5]enkephalin reduced sympathetic discharge in the renal nerve which subsequently produced hypotension (Rhee et al., 1985b); 2) the hypotension was antagonized by naloxone, but not by the equipotent dose of naloxone methobromide which cannot cross the blood brain barrier; and 3) [met^5]enkephalin was not able to antagonize the hypertension induced by exogenous norepinephrine. This hypothesis is also supported directly or indirectly by others (Feldberg and Wei, 1977 and 1978). We have also observed that [met^5]enkephalin has little or no effect on the dose-response curve of norepinephrine in the helical preparation of rabbit aorta (Rhee et al., 1985a), and has no effect on isolated heart (Eulie and Rhee, 1985). Rending et al. (1980) also reported that [met^5]enkephalin had no effects on papillary muscle preparations, although direct action of opioid peptides on atrial tissue or cultured cells have been presented (Eiden and Ruth, 1982; Laurent et al., 1985).

Total peripheral resistance (TPR) may be decreased by i.v. [met^5]enkephalin. The vascular bed(s) that are involved in the drop in TPR have not been previously known. The purpose of this study was to determine if [met[5]]enkephalin induced hypotension is the result of a generalized withdrawal of sympathetic tone to all tissues or is it limited to one or a few vascular beds. The results of this study clearly show that a generalized sympathetic withdrawal may not be the mechanism. The only vascular bed that had a consistent decrease in resistance during [met[5]]enkephalin treatment was skeletal muscles (Tab. 2), whereas the major visceral tissues (intestine and kidney) had increases in vascular resistance.

The measurements of vascular resistance in the skeletal muscle were made at only one time in the present study. Therefore these data do not provide any information about the time course of the vascular resistance changes. However, the following sequence of reaction seems reasonable at this time: 1) [Met[5]]enkephalin produced a centrally mediated decrease in sympathetic outflow to the heart and skeletal muscle vascular beds (Rhee and Eulie, 1984 and 1985; Rhee et al., 1985a). 2) The decrease in sympathetic tone to the small veins in the skeletal muscle vascular beds causes a decrease in venous return and cardiac output. The decreased cardiac output may also be due to decreased sympathetic influences on the heart. 3) The decreased cardiac output and TPR result in decreased mean aortic pressure. 4) The hypotension produces baroreceptor activation, which reflexly increases in sympathetic tone to the visceral tissues causing increased vascular resistance. Although this sequence is hypothetical, this hypothesis does seem reasonable in view of available information. Bond and Green (1969) demonstrated, by measuring regional blood flows as a function of time following bilateral carotid occlusion, that skeletal muscle vasculature can be the initiating factor in centrally mediated changes in aortic pressure. Complete understanding of the mechanism of [met[5]]enkephalin-induced hypotension will require further studies similar to those of Bond and Green study (1969) on pressure, flows and resistance as functions of time.

The data obtained in this study do not allow the determination of whether the vasoconstriction seen in intestine and kidney, and the vasodilation seen in skeletal muscle are the result of direct effects of [met[5]]enkephalin on vascular smooth muscle or the result of central effects of the drug. As outlined above, we believe that all available information (Eulie and Rhee, 1984; Rhee et al., 1985b) indicates that these effects are primarily the result of central effects. The data in table 2, that the [met[5]]enkephalin effect was antagonized by naloxone, may also support this contention because opiate receptors in the central nervous system specifically interact with central adrenergic receptors, thereby reducing the sympathetic discharge (Rhee et al., 1985b).

A treatment of nitroprusside (15 µg/kg) produced a comparable degree of hypotension to the one induced by [met[5]]enkephalin (Eulie and Rhee, 1984). In the case of nitroprusside there was reflexogenic tachycardia, whereas there was only peptide dose-dependent bradycardia in the hypotension induced by [met[5]]enkephalin (Rhee et al., 1985b). This also suggests that the mechanism of [met[5]]enkephalin action is more complicated than the simple baroreflex response to the vasodilatory action of nitroprusside. In the study naloxone alone did not produce any cardiovascular action at the dose we tested. This indicates that the antagonistic action of naloxone against cardiovascular actions of [met[5]]enkephalin is truly mediated by the interaction between the opioid peptide and its receptors.

In conclusion, the results of this study indicate that i.v. infusion of [met^5]enkephalin in anesthetized rabbits produce decreases in heart rate and mean arterial pressure. During these changes, vascular resistance in most tissues including the coronary circulation and the brain was not changed. However, visceral tissues showed increased vascular resistance whereas skeletal muscle vascular beds had decreased resistance and increased blood flows. These data suggest that the hypotensive effects of [met^5]enkephalin are at least partially due to vasodilation in skeletal muscle vascular beds.

REFERENCES

Bond, R.F., and Green, H.D., 1969, Cardiac output redistribution during bilateral common carotid occlusion, Am. J. Physiol. 216:393-403.

Eiden, L.E., and Ruth, J.A., 1982, Enkephalins modulate the responsiveness of rat atria in vitro to norepinephrine, Peptides 3:475-478.

Eulie, P.J., and Rhee H.M., 1984, Reduction by phentolamine of the hypotensive effect of methionine enkephalin in anesthetized rabbits, Br. J. Pharmacol. 83:783-790.

Eulie, P. and Rhee, H.M., 1985, Mechanism of cardiodepressant actions of enkephalin, New York Acad. of Sci. 435:407-411.

Feldberg, W. and Wei, E., 1977, The central origin and mechanism of cardiovascular effects of morphine as revealed by naloxone in cats, J. Physiol. 272:99-100P.

Feldberg, W. and Wei, E., 1978, Central sites at which morphine acts when producing cardiovascular effects, J. Physiol. 275:57P.

Heymann, M.A., Payne, B.D., Hoffman, J.I.E., and Rudolph, A.M., 1977, Blood Flow Measurements with Radionuclide-labeled Particles, Prog. Cardiovas. Dis. 20:55-79.

Hughes, J., Smith, T.W., Kosterlitz, H.W., Fothergill, L.H., Morgan, B.A., Morris, H.R., 1975, Identification of two related pentapeptides for the brain with potent opiate agonist activity, Nature 258:577-579.

Laughlin, M.H., Armstrong, R.B., White, J., and Rouk, K., 1982, A method of using microspheres to measure muscle blood flow in exercising Rats, J. Appl. Physiol. 52:1629-1635.

Laurent, S., Marsh, J.D., and Smith, T.W., 1985, Enkephalins have a direct positive inotropic effect on cultured cardiac myocytes, Proc. Natl. Acad. Sci. 82:5930-5934.

Martin, W.R., 1983, Pharmacology of opioids, Pharmacal. Rev. 35:283-323.

Petty, M.A. and DeJong, W., 1983, Enkephalins induce a centrally mediated rise in blood pressure in rats, Brain Res. 260:322-325.

Rendig, S.V., Amsterdam, E.A., Henderson, G.L., and Mason, D.T., 1980, Comparative cardiac contractile actions of six narcotic analgesics: Morphine, meperidine, pentazocine, fentanyl, methadone and l-α-acetylmethadol, J. Pharmac. Exp. Ther. 215:259-265.

Rhee, H., Eulie, P., and Tyler, L., 1985a, Characteristics of stress induced by enkephalin stress in intact rabbits, in: "Stress and Heart Diseases," Beamish, et al., ed., Martinus Nijhoff Publ., Boston.

Rhee, H.M., Eulie, P.J., and Peterson, D.F., 1985b, Suppression of renal nerve activity by methionine enkephalin in anesthetized rabbits, J. Pharmacol. Exp. Ther. 234:537-541.

Schaz, K., Stock, G., Simon, W., Schlör, K.-H, Unger, T., Rockhold, R., Ganten, D., 1980, Enkephalin effects on blood pressure, heart rate, and baroreceptor reflex, Hypertension, 2:395-407.

Stickney, J.L. and Eikenburg, D.C., 1981, Peripheral sympatholytic effect of l-α-acetylmethadol, J. Cardiovasc. Pharmac. 3:369-380.

Warren, D.J., and Ledingham, J.G.G., 1974, Measurements of Cardiac Output Distribution Using Microspheres, Some practical and theoretical consideration, Cardiov. Res. 8:570-581.

Wei, E.T., Lee, A., and Chang, J.L., 1980, Cardiovascular effects of peptides related to the enkephalins and β-casomorphin, Life Sci., 1517-1522.

THE EFFECTS OF THE PROLIFERATION OF THE RADIAL ARTERIES OF

THE PLACENTA ON OXYGEN TRANSPORT TO THE FETAL GUINEA PIG

M. Tabata, H. Negishi, T. Yamaguchi, S. Makinoda,
S. Fujimoto, and W. Moll*

Department of Obstetrics and Gynecology, School of Medicine
Hokkaido University, Sapporo, Japan
*Institute fuer Physiologie der Universitaet Regensberg
Federal Republic of Germany

INTRODUCTION

Nutrient and oxygen, which are supplied by the utero-placental blood, are indispensable for the intrauterine growth of fetus. Peeters et al (1) have reported that utero-placental blood flow is well adjusted fetal weight. The regulating mechanism which adjusts maternalplacental blood flow to fetal growth is still unknown. The adjustment of utero-placental blood flow during pregnancy is related to "adaptive changes" in the placental arteries (2), namely, progressive structural changes in the arterial wall associated with arterial widening. In guinea pigs, massive growth has also been shown to be associated with widening of the utero-placental arteries (Fig. 1).

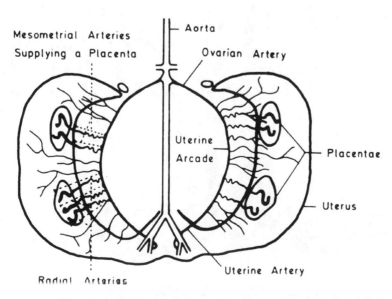

Fig. 1 : Uterine radial artery in guinea pig.

While the terminal portions of radial arteries supplying a placenta increase seven fold in diameter, they increase 30 to 50-fold in weight (Fig. 2), protein content and DNA content (3). Since growth and widening of utero-placental arteries appear to be related to each other, information on the regulation of arterial growth, including regulation of arterial DNA synthesis, may yield a clue as to how arterial widening, and thus placental blood flow, are controlled during pregnancy. For this reason, the rate of ^{3}H-thymidine incorporation into radial arteries during pregnancy as well as during the estrus cycle was studied in guinea pig as a measure of local DNA synthesis.

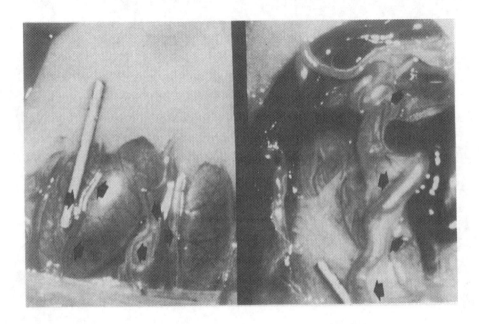

Fig. 2 : Uterine-radial artery in guinea pig (▶)
left; non-pregnant animals, right; pregnant animals

To investigate whether the oxygen transport to fetus is adjusted to fetal growth or not, the amniotic fluid was collected on various days of gestation, and measured oxygen partial pressure.

MATERIALS AND METHODS

Animals

Guinea pigs, of 70 day or more were inspected daily by examination the vaginal membrane. The day on which the vaginal membrane was ruptured was called Day 0 of the estrus cycle. During the estrus cycle, 24 animals were used for the experiment. Animals which had shown at least one normal estrus cycle (average 16 d) were mated and 68 of them were used for the experiments. In some cases, gestational age (GA) was determined from mean crown-to-rump length of the fetus (CRL) using a formula modified after Kaufmann (4).

Preparation of utero-placental arteries

The animals were anesthetized with pentobarbital sodium (Nembutal, Abbott, 30 mg/kg, intraperitoneally). Laparotomy was performed and the distal portions of the uterine radial arteries lying outside the fatty tissue were rapidly separated from surrounding mesometrial tissue. In non-pregnant animals, 4 to 12 arterial segments were examined. All dilated radial arterial segments leading to a placenta were excised in pregnant animals. The average wet weight of the segments were about 200 µg in non-pregnant animals and about 10 mg in mid-gestation.

Incubation

The arteries were incubated for 4 hr in medium 199 contains penicillin (15 IU/ml), streptomycin (15 µg/ml), Vitamin C (0.1 mg/ml) and methyl-^{3}H thymidine (185 k Bq/ml) under 6% CO_2 in air. The pH was adjusted to 7.4 by the addition of sodium bicarbonate.

Measurement of radioactive concentration

The arteries were extracted by shaking for 10 min in 1 ml perchloric acid (10%), dissolved in 200 µl tissue solubilizer (Soluene-350, Packard Instruments) to which 20 µl distilled water was added. The solution was neutralized with acetic acid to avoid chemiluminescence and added to xylene with 2,5-dipenyloxazole and p-bis-(0-methylstyryl)-benzene. A commercial beta-counter was used to determine radioactivity. The efficiency of counting was about 50 per cent.

Oxygen partial pressure of amniotic fluid

The animals were anasthetized with pentobarbital sodium (Nembutal, Abbott, 30 mg/kg intraperitoneally). The uterus was exposed and incised. The amniotic fluid was collected by gestational sac puncture with a syringe, from 16 animals on various days of gestation and measured oxygen partial pressure by Po_2 meter.

RESULTS

The ^{3}H-thymidine incorporation during pregnancy is shown in Fig. 3 with the data of normal estrus cycle before conception.

During the first half of pregnancy, ^{3}H-thymidine was incorporated at the rate of about 1,000 Bq/mg.h (Fig. 4). This value is about 50 times higher than the rate in diestrus animals (20±9 Bq/mg.h) and similar to that at estrus. During the later course of pregnancy the rate of incorporation decreased, being only about 330 Bq/mg.h in the period between Day 40 and Day 50. It was 170 Bq/mg.h between Day 50 and Day 60. The rate near the term decreased to below 100 Bq/mg.h.

The oxygen partial pressure in amniotic fluid was about 110 mmHg on various days of gestation and no significant changes were observed with the course of time (Fig. 5).

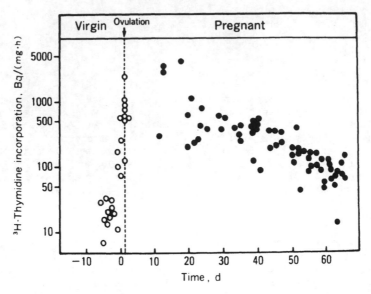

Fig. 3 : ³H-thymidine incorporation to the uterine radial artery
during estrus cycle and pregnancy.

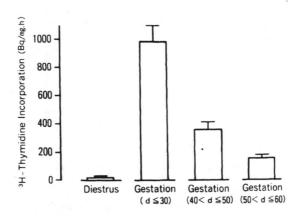

Fig. 4 : ³H-thymidine incorporation to the uterine radial artery
during diestrus and various gestational periods.

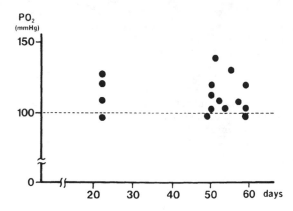

Fig. 5 : The oxygen partial pressure in amniotic fluid on various
days of gestation.

DISCUSSION

In spite of the fetal growth, the oxygen partial pressure in the
amniotic fluid was maintained at about 110 mmHg throughout the pregnancy.
This phenomenon can be brought only by the increase of utero-placental
blood flow. The uterine radial artery which supplies blood flow to
placenta increases 7-fold in diameter, 30-50 fold in weight, protein
content and DNA content (3). Such increases can be achieved by the
proliferation of the arterial wall, which are monitored by ³H-thymidine
incorporation as shown in this study.

In the beginning of pregnancy, the rate of ³H-thymidine incorporation
is increased 50-fold when compared to the rate during diestrus. The
initial rate during pregnancy is similar to the rate observed during
estrus. This observation suggests that the same factors stimulate DNA
synthesis during gestation and estrus. Two possible mechanism may be
proposed to explain the initiation of arterial DNA synthesis at the
biggining of estrus and gestation. Firstly, DNA synthesis may be
initiated by increased blood flow resulting from the development of the
fistula-like placental vascular bed. In this case, arterial growth could
be part of a servomechanism adapting arterial resistance to peripheral
resistance. However, the observation that the rate of ³H-thymidine
incorporation in the initial stage of pregnancy is similar to the rate at
estrus, when the concentration of estradiol is high, suggests a second
possibility, namely that DNA synthesis is controlled by estradiol. It
seems like therefore that estradiol, produced in the vicinity of the
arteries, initiates growth and widening of the radial arteries, which
supply the placenta, thereby increasing placental blood flow. At first,
this concept appears to be inconsistent with the rise of systemic plasma
concentration of estradiol reported by Challis et al (5) and Batra et al
(6), which occurs later in gastation. By contrast, we found that
³H-thymidine incorporation is highest in the first half of pregnancy and

thereafter falls. However, since the maternal plasma concentration of estradiol decreases only slightly when the maternal ovaries are removed (5), it may be assumed that estradiol is also produced in the placenta and/or in the decidua. It is also possible that estrogen is produced inside the arterial wall, perhaps by trophoblastic cells, since these are reported to invade the wall of utero-placental arteries in man (2). In this way the estradiol concentration in the wall of those uterine radial arteries lying close to a placenta may temporarily exceed that in arterial plasma. As we have reported (7), the rate of ^{3}H-thymidine incorporation per vessel length increased with the injected dose of estradiol bensoate (Fig. 6).

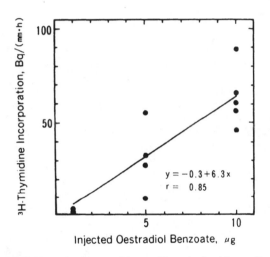

Fig. 6 : ^{3}H-thymidine incorporation after injection of estradiol
benzoate

Thus, the concept of arterial growth control by estradiol during pregnancy appears to be a reasonable working hypothesis. The oxygen partial pressure in amniotic fluid was about 110 mmHg on various day of gestation and no significant changes were observed with time. For these reasons, we can conclude that estrogen is one of the control factors in oxygen transport to fetus, which is consistent with the present state of knowledge but still awaits final proof.

SUMMARY

The adjustment of placental blood flow during pregnancy is related to progressive structual changes in the placental arteries. In guinea pig, massive growth has also been shown to be associated with widening of the uterine radial arteries.

The rate of ³H-thymidine incorporation into uterine radial arteries during pregnancy was studied in guinea pig as a measure of local DNA synthesis. And the oxygen partial pressure was used to investigate how the oxygen transport is done between mother to fetus.

During the first half of pregnancy, ³H-thymidine was incorporated at the rate of about 1,000 Bq/mg.h. The initial rate during pregnancy is similar to the rate observed during estrus. This observation suggest that the same factors stimulate DNA synthesis during gestation and estrus. The rate of ³H-thymidine incorporation per vessel length increased with the injected dose of estradiol benzoate. We suppose that estradiol is one of the control factors in DNA synthesis in uterine radial artery. The oxygen partial pressure in amniotic fluid was about 110 mmHg on various days of gestation and no significant changes were observed with time. For these reasons, estrogen is supposed to be one of the control factors in oxygen transport to fetus.

REFERENCES

1. Peeters, L.L.H., Grutters, G. & Martin, C.B. : "Distribution of cardiac output in the unstressed pregnant guinea pig", American Journal of Obstetrics and Gynecology. 138:1177-1184 (1980).
2. Brosens, I., Robertson, W.B. & Dixon, H.G. : "The physiological response of the vessels of the placental bed to normal pregnancy", Journal of Pathology and Bacteriology. 93:569-579 (1967).
3. Moll, W., Espach, A. & Wrobel, K.-H. : "Growth and dilation of mesometrial arteries in guinea pigs during pregnancy", Placenta. 4:111-124 (1983).
4. Kaufmann, P. : "Die meerschweinchenplacenta und ihre Entwicklung", Zeitschrsft für Anatomie und Entwicklungsge-schichte. 129:83-101 (1969).
5. Challis, J.R.G., Heap, R.B. & Illingworth, D.W. : "Concent-rations of oestrogens in the plasma of nonpregnant, pregnant and lactating guinea-pigs", Journal of Endocrinology. 51:333-345 (1971).
6. Batra, S., Sjoeberg, N.-O. & Thorbert, G. : "Sex steroids in plasma and reproductive tissues of the female guinea pig", Biology of Reproduction. 22:430-437 (1980).
7. Makinoda, S., Shimotomai, K. & Ichinoe, K. : "The effect of estrogen and progesterone on the proliferation of the uterine radial artery in guinea pigs", Acta Obst. Gynaec. Jpn. 36:2356-2357 (1984).

PREFERENTIAL OXYGEN SUPPLY TO THE BRAIN AND UPPER BODY IN THE FETAL PIG

M. Silver, R.J. Barnes, A.L. Fowden and R.S. Comline

Physiological Laboratory, Cambridge
CB2 3EG, England

INTRODUCTION

In most mammals about 50% of the blood returning from the placenta bypasses the liver via the ductus venosus (DV). This oxygenated blood does not undergo substantial mixing in the inferior vena cava and is preferentially distributed through the foramen ovale to the left ventricle and thence to the coronary circulation and the upper body where the fetal brain receives the largest share (Edelstone and Rudolph, 1979). In the Equidae no such shunt exists but the overall level of fetal oxygenation is higher than in most other species (Comline and Silver, 1974). The fetal pig also has no true DV as it is lost early in embryonic life (Dickson, 1956) but a series of channels develops in the fetal liver at a later stage of development (Kaman, 1968). The nature of these channels and the partitioning of flow through them compared with that in the liver itself has been examined using radioactively labelled microspheres (Barnes et al, 1979; Silver et al, 1982). It was shown that 50-60% of the total umbilical blood flow bypassed the liver in the fetal pig near term and that the shunts offered no detectable resistance to spheres of up to 100µ diameter. However, it is not clear from these studies whether the blood velocity within the shunts is sufficient to maintain a separate caval stream so that the head is supplied with better oxygenated blood than the lower body. The present experiments attempt to answer this question.

METHODS

Seven pregnant sows were anaesthetized with 6% (W/V) sodium pentobarbitone after intubation under halothane (Comline et al, 1979). After insertion of arterial and venous catheters into the circumflex iliac vessels of the sow the uterus was exposed and catheters were inserted into the carotid and femoral arteries and the umbilical vein of 1-2 fetuses as described previously (Comline et al, 1979). The fetuses were returned to the uterus, the uterine and abdominal incisions closed and the animal left to recover from the surgery for 30-40 min. Maternal pO_2 was controlled by the inhalation of different concentrations of O_2 in N_2 from a modified Boyles machine attached to the endotracheal tube. Maternal arterial pO_2 was measured at frequent (10-15 min) intervals to establish stable conditions before any fetal

samples were taken. Simultaneous samples were withdrawn from each
fetal vessel and from the maternal artery to determine pO_2, pCO_2, pH,
haemoglobin (Hb), O_2 content ($[O_2]$) and % saturation (SO_2). Not more
than four sets of samples were taken from any one fetus to avoid
depletion of its blood volume and consequent haemodynamic changes.
Blood gas tensions and pH were measured using standard Radiometer
equipment. Oxygen saturation and Hb concentrations were measured using
an oximeter and the O_2 content calculated from these values (assuming
1gHb combines with 0.0598 mM O_2 at STP). All values are expressed as
means ± SE's unless otherwise stated.

RESULTS

     All catheterized fetuses were maintained in good condition
throughout the experiments. Fetal and maternal blood gas and pH values
during normoxemia were within the range of those reported previously
for sows and fetuses in late gestation (Comline et al, 1979; Edelstone,
1980). In all fetuses the level of oxygen in the carotid artery was
higher than that in the femoral artery irrespective of their blood gas
or acid base status. Table 1 summarizes the mean values for pO_2 SO_2
and $[O_2]$ obtained in simultaneous samples from 12 fetuses under basal
conditions. The mean femoral arterial pH was 7.36 ± 0.02 and mean pCO_2
was 49.0 + 1.9 mmHg in these 12 fetuses. The mean differences in pO_2
SO_2 and $[O_2]$ between paired carotid and femoral arterial samples are
also given in Table 1: these differences were all statistically
significant.
     The relationships between carotid or femoral $[O_2]$ and the
corresponding values in the umbilical vein over a wide range of oxygen
concentrations are shown in Fig. 1. Even at the lowest levels of fetal
oxygenation, when the sow was breathing 8% O_2 and umbilical venous pO_2
fell to below 15mmHg (SO_2<20%), the carotid O_2 content was maintained
above that in the femoral artery. There was no significant difference
between the slopes of the repression lines relating umbilical venous
$[O_2]$ to that in the carotid or femoral arteries.

Table 1   Mean pO_2, SO_2 & $[O_2]$ in fetal blood withdrawn

          simultaneously from different vascular sites in 12

          catheterized piglets in utero during maternal normoxaemia

	umbilical vein	femoral artery (FA)	carotid artery (CA)	mean difference CA-FA
pO_2 (mmHg)	37.5 ± 1.41	25.6 ± 0.66	28.6 ± 0.74	3.0 ± 0.38[*]
% SO_2	78.6 ± 2.1	46.4 ± 2.6	54.1 ± 2.6	7.9 ± 0.90[*]
O_2 content (mM)	3.88 ± 0.10	2.49 ± 0.13	2.85 ± 0.11	0.35 ± 0.04[*]

     [*] $p < 0.01$

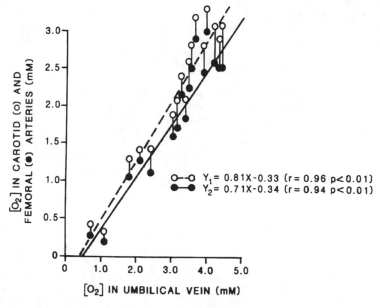

Figure 1 Relationship between umbilical venous O_2 content and that in the carotid (O) or femoral ( ) artery, sampled simultaneously (joined by vertical lines). The regression of Y (carotid (Y_1 ——) or femoral (Y_2 ——) O_2 content) upon X (umbilical venous O_2 content) calculated by the method of least squares. The slopes of the two lines did not differ significantly.

DISCUSSION

The present observations on the differences in the arterial oxygen supply to the upper and lower body of the fetal piglet confirm the hypothesis that the vascular channels in the liver, which link the umbilical vein with the IVC, function in the same way as the true DV of other species. The maintenance of a separate stream of oxygenated blood in the IVC and its preferential distribution through the foramen ovale to the left ventricle, which has been demonstrated in a number of ways in the fetal lamb (Edelstone, 1980), would thus appear to occur in the fetal piglet. Although earlier studies with different sizes of microspheres provided evidence of shunting, and showed that more than 50% of umbilical venous blood bypasses the liver of the piglet, the degree of subsequent mixing in the IVC and in the heart could only be estimated indirectly (Barnes et al, 1979; Silver et al, 1982). The absence of any muscular or nervous tissue round the hepatic shunts and the lack of an endothelial lining suggests that the channels must be purely passive (Silver et al, 1982) so that the rate of flow through them will be dependent upon umbilical venous and IVC pressures.

It has been suggested that the DV in the lamb functions as a passive pressure equalizer, despite its musculature and innervation, by stabilizing the pressure gradient between the umbilical vein and IVC (Edelstone, 1980). In support of this, Edelstone et al (1980) showed that progressive reductions in umbilical blood flow led to increases in flow through the DV thereby maintaining the caval streaming and pressure of the oxygenated blood. In the present studies systematic reductions in umbilical blood flow were not attempted, but preliminary observations have shown that when umbilical blood flow falls the femoral/carotid artery [O_2] differential is maintained or even increased (unpublished observations).

The demonstration of the hepatic shunts in the fetal piglet leaves only the Equidae without any form of hepatic bypass (Barnes et al, 1979; Silver et al, 1982). It seems unlikely that this group are at a disadvantage as a consequence of this anatomical aberration and it may be that the somewhat higher fetal pO_2 and efficient placental exchange mechanisms which exist in this species are sufficient to ensure that all tissues including the brain and head receive adequate oxygen (Comline and Silver, 1974). Alternatively it could be argued that preferential streaming of oxygenated blood through the DV confers little benefit to the fetus because both cerebral and coronary blood flows are already very high (Cohn et al, 1974). In fact the rate of oxygen delivery to the brain remains constant over a wide range of conditions in the fetal lamb, since cerebral blood flow can increase to two to threefold if the fetus becomes hypoxic (Cohn et al, 1974; Peeters et al, 1979); even greater rises in coronary flow occur during hypoxaemia (Cohn et al, 1974). The question of the physiological significance of the more oxygenated blood stream might be resolved by occlusion of the DV or its equivalent for a period covering the major phase of fetal development. Whether any impairment of growth or any specific changes in the normal brain morphology and maturation would be detectable under these conditions remains to be determined.

SUMMARY

In the fetal pig over 50% of the oxgenated umbilical venous blood bypasses the liver via vascular channels of over 100μ diameter which appear to act as a ductus venosus (DV). In the present experiments the oxygen levels (pO_2, SO_2 and O_2 content) in the fetal carotid (CA) and femoral (FA) arteries were measured to determine whether preferential streaming of the oxygenated blood is maintained from the bypass to the foramen ovale and thence through the left heart to the upper body, as in other species. The results were obtained from 12 piglets catheterized and maintained in utero under sodium pentobarbitone anaesthesia. Significant CA-FA differences in pO_2, SO_2 and O_2 content were observed in these fetuses during both normoxaemia and hypoxaemia showing that a higher level of oxygen can be maintained to the fetal head and brain over a wide range of oxygenation. These results confirm that the hepatic channels in the fetal piglet act as a fully functional DV.

Acknowledgements

We thank the technical staff of the Physiological Laboratory for their assistance during these experiments. The work was supported by the Agriculture and Food Research Council.

References

Barnes, R. J., Comline, R. S., Dobson, A., Silver, M., Burton, G. & Steven, D. H., 1979, On the presence of a ductus venosus in the fetal pig in late gestation, J. Devl Physiol., 1:105-110.

Comline, R. S. and Silver, M., 1974, A comparative study of blood gas tensions, oxygen affinity and red cell 2, 3-DPG concentrations in foetal and maternal blood in the mare, cow and sow. J. Physiol., 242:805-826.

Comline, R. S., Fowden, A. L. and Silver, M., 1979, Carbohydrate metabolism in the fetal pig during late gestation. Q. J. Exp. Physiol., 64:277-289.

Cohn, H. E., Sacks, E. J., Heymann, M. A. and Rudolph, A. M., 1974, Cardiovascular responses to hypoxemia and acidemia in fetal lambs. Amer. J. Obstet. & Gyn., 120:817-824.

Dickson, A. D., 1956, The ductus venosus of the pig. J. Anat. 90:143-152.

Edelstone, D. I. and Rudolph, A. M., 1979, Preferential streaming of ductus venosus blood to the brain and heart in fetal lambs. Amer. J. Physiol., 237:H724-H729.

Edelstone, D. I., 1980, Regulation of blood flow through the ductus venosus. J. Devl. Physiol., 2:219-238.

Edelstone, D. I., Rudolph, A. M., Heymann, M. A., 1980, Effects of hypoxemia and decreasing umbilical blood flow on liver and ductus venosus blood flows in fetal lambs. Amer. J. Physiol., 238:H656-H663.

Kaman, J., 1968, Der Umbau des Ductus Venosus des Schweines. 1. Pränatales Stadium. Anat. Anz., 122:252-266.

Peeters, L. L. H., Sheldon, R . E., Jones, M. D., MaKowski, E. L. and Meschia, G., 1979, Blood flow to fetal organs as a function of arterial oxygen content, Amer. J. Obstet. & Gynec., 135:637-646.

Silver, M., Barnes, R. J., Comline, R. S. & Burton, G. J., 1982, Placental blood flow: some fetal and maternal cardiovascular adjustments during gestation. J. Reprod. Fert. Suppl. 31:139-160.

OVARIAN BLOOD FLOW AND OXYGEN TRANSPORT TO THE FOLLICLE

DURING THE PREOVULATORY PERIOD

S. Makinoda, M. Tabata, T. Yamaguchi, K. Nakajin, T. Koyama
and K. Ichinoe

Department of Obstetrics and Gynecology, School of Medicine
Hokkaido University, Sapporo, Japan

INTRODUCTION

During the past few decades a substantial amount of information on the hypothalamo-pituitary-ovarian axis has been obtained and the endocrinologic mechanism of ovulation is almost established. However, the basic physiology underlying the mechanisms responsible for follicle rupture remains obscure.

In 1976, Peppler reported that when the blood flow to the ovary was decreased by hysterectomy or ligation of the uterine artery, the number of ovulations also decreased. This phenomenon suggests the importance of the vascular system in ovulation. Since the vascular system plays an important role in oxygen transport, ovarian blood flow, ovarian vessel morphology and oxygen transport to the follicle during the preovulatory period have been studied to investigate the mechanisms of follicle rupture.

MATERIALS AND METHODS

Animals

Sexually mature Japanese white female rabbits weighing 3.0-4.4 kg were used in all experiments. Animals were caged individually and fed water and dry pellet food ad libitum.

Ovulation was induced by the administration of pregnant mare serum (PMS ; 100 iu, IM) and human chorionic gonadotrophin (hCG ; 100 iu, IV). Ovulation was usually observed between 10 and 13 h after hCG administration.

Measurement of ovarian blood flow

The animals were fixed on a heated board, which was maintained at a temperature of 37°C, and anesthetized with pentobarbital sodium (Nembutal, Abbott, 30 mg/kg, IV). Laparotomy was performed through a dorsal incision and a crossed-thermocouple inserted into the ovary as shown in Fig. 1. Despite the surgical trauma due to the crossed-thermocouple, the number of ovulations was almost the same in these animals as in unoperated ones. The measurement of ovarian blood flow was carried out after the recovery from anesthesia.

The crossed-thermocouple is shown, as a-b-c-d, in Fig. 2. It is made
of 100 μm diameter copper wire (thin line) and constantan wire (thick
line), connected at point f.

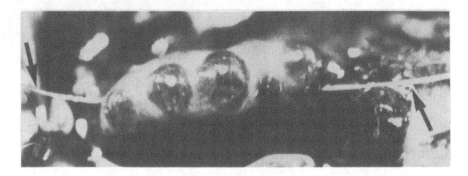

Fig. 1 : The crossed-thermocouple inserted into the ovary (as shown
by the arrows).

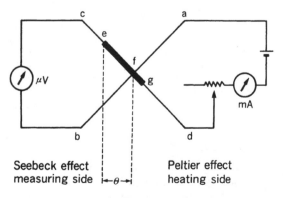

Fig. 2 : Schematic diagram of the crossed-thermocouple method.
A-b-c-d is a crossed-thermocouple. The thin line represents
a copper wire and the thick line, a constantan wire.

The principle of this method is as follows: In the right-hand circuit a-f-g-d, the 6V 450mA current makes point f 3°C warmer than the surrounding tissue as a result of the Peltier effect. Since point f is warmer than the surrounding tissue, there is a temperature difference, θ, between point e and f. If the blood flow is high, point f is cooled by the blood and θ is small. If the blood flow is low, point f is cooled only a little and θ is large. This temperature difference, θ, induces a voltage in the left-hand circuit due to the Seebeck effect. This voltage can be used as a parameter of the blood flow.

The variation of ovarian blood flow was expressed as the percentage ratio based on the post-mortem value (=0%) and the initial value prior to hCG administration (=100%).

Ovarian vessel morphology

Ovaries were excised from a series of 10 rabbits before, and at 2 h intervals after, hCG administration. They were fixed with 10% formalin and sectioned longitudinally. These sections were observed after H and E. staining, with special regard to the diameter of the vessels.

Oxygen transport to follicle

The oxygen partial pressure (P_{O_2}) of the follicle was monitored with a Pt-Ag electrode before and after the inhalation of 100% oxygen. The time to reach the maximal P_{O_2} (Tmax) was compared at 7 h and 12 h after hCG administration.

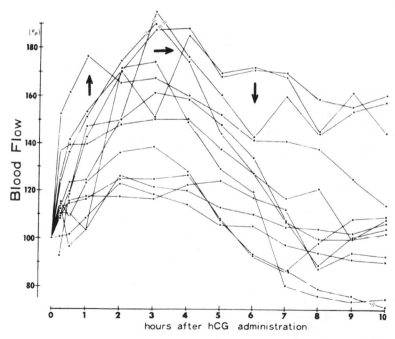

Fig. 3 : The ovarian blood flow after hCG administration in rabbit. The results of all measurements (n=13) are shown. The arrows are indicating direction of change.

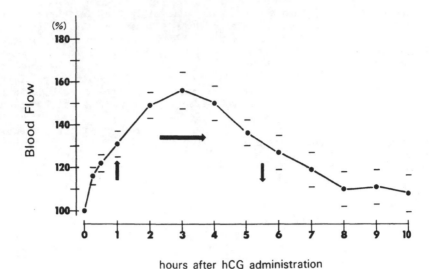

hours after hCG administration

Fig. 4 : The ovarian blood flow after hCG administration in rabbit.
The mean and standard errors (n=13) are shown. The arrows
are indicating direction of change.

RESULTS

Ovarian blood flow

    Fig. 3 shows the results of all 13 experiments on ovarian blood flow.
Although the deviation is large, each line reveals the same character,
that is, an immediate increase of ovarian blood flow after hCG administration, a high percentage ratio for several hours and thereafter a gradual
decrease. The mean values and the standard errors of the ovarian blood
flow are shown in Fig. 4. After hCG administration, the ovarian blood
flow immediately increased to 115.6±4.3% at 15 min, 122.1±5.1% at 30 min
and 131.3±6.2% at 1 h. The high percentage increases were exhibited over
the 2 h to 5 h period, showing a peak value of 155.3±12.7% at 3 h. The
ovarian blood flow decreased gradually from 5 h to 7 h and was maintained
at about 110% after 8 h.

Ovarian vessel morphology

    Fig. 5 and 6 show a similar area of ovarian stromal tissue at the
same magnification. However, the appearance is quite different. In Fig.
5, before hCG administration, the stromal cells are small and the nuclei
pyknotic. The capillaries are not identifiable. After hCG administration, the stromal tissue shows a definite morphological change. The
stromal cells become larger and secrete many granules. The capillaries
dilate with time. At 10 h after hCG administration, as is shown in Fig.
6, the capillaries dilate to 55.5±7.0 μm and many blood cells are found
in the vessels, which suggests congestion. In perifollicular tissues, the
same process is observed. Table 1 shows the diameters of the capillaries
at 2 h intervals after hCG administration.
    At the apex, where the rupture point for the discharge of the oocyte
occurs, the vessels also dilate with time. Just prior to ovulation, some
parts of the apex appear to consist only of the congested capillaries.

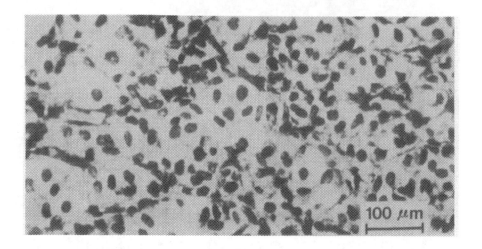

Fig. 5 : The ovarian stromal tissues before hCG administration.
No dilatation of the capillaries is observed.

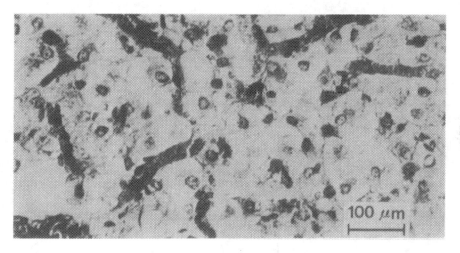

Fig. 6 : The ovarian stromal tissues at 10 h after hCG administration.
The vasodilatation is remarkable and many blood cells are
found in the capillaries.

Oxygen transport to the follicle

The time to reach the maximal Po_2 after the start of 100% O_2 inhalation is shown in Fig. 7. It reveals that oxygen transport to the follicle is reduced just prior to ovulation.

Tab. 1 : The diameter of the ovarian capillaries after hCG administration (n=10, mean+SE).

hours after hCG administration	stromal tissue (μ)	follicle (μ)	
		adjacent to basement membrane	theca interna
0	very fine	very fine	
2	6.5 ± 0.5	5.3 ± 0.3	13.8 ± 0.8
4	12.5 ± 0.9	9.0 ± 1.3	18.0 ± 2.0
6	19.5 ± 1.3	10.0 ± 0.9	22.8 ± 2.2
8	22.5 ± 3.3	15.6 ± 1.4	25.5 ± 1.7
10	55.5 ± 7.0	35.0 ± 3.7	49.0 ± 3.2

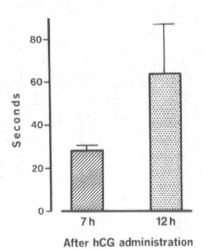

Fig. 7 : The time to reach the maximal follicle Po_2 after the start of 100 O_2 inhalation.

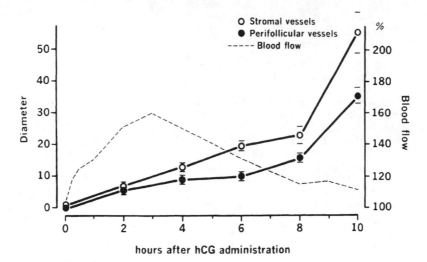

Fig. 8 : The diameter of the ovarian capillaries, and the ovarian
blood flow (dotted line), after hCG administration. The
open circles indicate the stromal capillaries and the
closed circles, the perifollicular capillaries (n=10,
mean+error bars).

DISCUSSION

The rate of vasodilatation is plotted in Fig. 8. Stromal and
perifollicular vasodilatation was detected at 2 h, the vessel diameters
being 6.5±0.5 μm and 5.3±0.3 μm, respectively. Moderate dilatation was
observed during the 4 to 8 h period. At 10 h, just prior to ovulation,
vasodilatation became very marked. The dotted line in Fig. 8 shows the
ovarian blood flow. Up to 3 h after hCG administration, the blood flow
increased and the vessels dilated gradually. But after 3 h, the vasodilatation continued to progress, despite the decrease in blood flow.

Since the blood flow (q) is indicated by

$$q_a = \pi \, r_a^2 \cdot v_a \quad ,$$

where r = radius of the vessel
       v = velocity of the blood flow
       a = time
the velocity of the blood flow is

$$v_a = \frac{q_a}{\pi \cdot r_a^2}$$

In this formula, the following values were obtained as mentioned above.

$$q_4 = 1.505, \quad q_8 = 1.104, \quad r_4 = 4.5, \quad r_8 = 7.8$$

Results were as follows:

$$V_4 : V_8 = \frac{1.505}{\pi \times (4.5)^2} : \frac{1.104}{\pi \times (7.8)^2} = 1.0 : 0.24$$

Hence, the velocity of the blood flow at 8 h is lower than that at 4 h after hCG administration. This proves the existence of congestion in the ovarian vessels. The disturbance of oxygen transport to the follicle also supports the existence of congestion. Similarly, a scanning electron microscopic study performed by Kanzaki et al. (1982) revealed blood congestion, especially in the apical vessels of the follicle.

Fig. 9 shows a follicle just prior to ovulation. Since the oocyte is only 100 μm in diameter, the hole through which the oocyte is discharged could theoretically avoid these vessels. However, ovulation is accompanied by the rupture of the apical vessels in almost all cases. This confirms that these congested vessels play an important role, together with proteolytic enzymes, in the mechanism leading to ovulation.

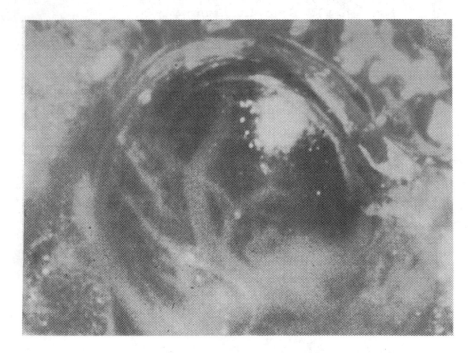

Fig. 9 : The follicle at 10 h after hCG administration.

SUMMARY

The mechanism of ovulation, especially the mechanism of follicle rupture, is still uncertain. Ovarian blood flow, ovarian vessel morphology and oxygen transport to follicle were therefore studied during the preovulatory period.

Japanese white rabbits, weighing 3.0-4.4 kg, were used as the experimental animal. Ovulation was induced by the administration of PMS (100 iu, IM) and hCG (100 iu, IV). The ovulation was observed at 10-13 h after hCG administration. Continuous measurement of ovarian blood flow was facilitated by the crossed-thermocouple inserted into the unilateral ovary. The variation of ovarian blood flow was expressed as the percentage ratio based on the post-mortem value (=0%) and the initial value prior to hCG administration (=100%). Histologic changes of ovarian blood vessels were observed at intervals of every 2 hours after hCG administration. Oxygen transport to follicle was compared at 7 and 12 h after hCG administration.

The ovarian blood flow increased rapidly within 1 h following hCG administration. High percentage increases were demonstrated during 2 h to 5 h, showing a peak value of 155.3±12.7% at 3 h. The ovarian blood flow decreased gradually from 5 h to 8 h and then was maintained at about 110% after 8 h. The perifollicular and stromal vasodilatations were confirmed at 2 h and moderate dilatation was observed during 4 h to 6 h. At 10 h, just prior to ovulation, vasodilatation became most remarkable especially at the apical vessels. These hemodynamic and histologic results, which shows vasodilatation in spite of the decrease of the blood flow after 4 h, suggest a congestion in ovarian blood vessels during the preovulatory period. The oxygen transport to follicle was retarded at 12 h after hCG administration. This shows also congestion in ovarian blood vessels. This blood congestion may relate to a significant role in the mechanism leading to ovulation.

REFERENCES

Kanzaki, H., Okamura, H., Okuda, Y., Takenaka, A., Morimoto, K. & Nishimura, T. (1982). Scanning electron microscopic study of rabbit ovarian follicle microvasculature using resin injection-corrosion casts. J. Anat., 134:697-704.

Peppler, R.D. (1976a). Hysterectomy in the rat. Am. J. Anat., 145:121-125.

Peppler, R.D. (1976b). Effect of uterine artery ligation on ovulation in the rat. Anat. Rec., 184:183-190.

THE ANALYSIS OF PO$_2$ DIFFERENCE BETWEEN AIR SPACE AND ARTERIALIZED BLOOD

IN CHICKEN EGGS WITH RESPECT TO WIDELY ALTERED SHELL CONDUCTANCE

Hiroshi Wakayama and Hiroshi Tazawa

Department of Electronic Engineering
Muroran Institute of Technology
Muroran 050, Japan

INTRODUCTION

In contrast to the convective transport of respiratory gases in the vertebrate lung, the gas exchange of avian eggs takes place by diffusive transport. The diffusion barrier lying between the ambient atmosphere and the capillary blood of the gas exchanger (i.e., the vascularized chorioallantoic membrane) is conveniently divided into two parts by the air space between the fibrous shell membranes. The outer barrier comprises the porous eggshell and outer shell membrane. The resistance to gas diffusion is mainly attributed to the eggshell, thus the conductance of the outer barrier is determined by the porosity of the shell. The air cell which is formed between the outer and inner shell membranes at the blunt end of the egg is assumed to substitute for the air space. The O$_2$ in the air space is transferred through the inner shell membrane and the chorioallantoic membrane to the capillary blood. These two membranes constitute the inner diffusion barrier and its diffusive conductance includes the gas reaction with blood in the chorioallantoic capillaries. While the diffusive conductance of the outer barrier corresponds to the term in the convective transport system of alveolar ventilation multiplied by the capacitance coefficient (Rahn and Paganelli, 1982), the air cell is equivalent to the alveolar space and the conductance of the inner barrier corresponds to the diffusing capacity of the lung. In contrast to the alveolar-arterial Po$_2$ difference in the lung, the difference of Po$_2$ between the air cell and the arterialized blood in the allantoic vein is large in chick embryos (Tazawa et al., 1980). Based upon O$_2$ measurements performed in hypoxia, normoxia and hyperoxia, Piiper et al. (1980) reported that there was a sizable functional arterio-venous shunt amounting to 10-15% of the total chorioallantoic blood flow and that the diffusing capacity of the air cell-blood barrier for O$_2$ was about 7 µl/(min·torr). Because in chicken eggs the diffusive conductance of the outer barrier can be altered widely, we analyzed the Po$_2$ difference between the air space and arterialized blood ($\Delta P_{A_{O_2}} \cdot Pa_{O_2}$) with respect to widely altered diffusive conductance of the shell, the diffusing capacity of the chorioallantoic membrane and the physiological shunt in the allantoic circulation with reference to the results reported by Piiper et al. (1980).

METHODS AND MATERIALS

Fertile chicken eggs were measured for length (L, cm) and maximum breadth diameter (B, cm) to estimate the fresh egg mass (W, g) using the equation of $W = 0.5632 \, B^2 \, L$ (Romanoff and Romanoff, 1949). Eggs were divided into three groups at random; in the first group (conductance-decreased eggs) the shell at the sharp end was covered with epoxy cement for a quarter to half the length of the egg; in the second group (conductance-increased eggs) the shell covering the air cell was removed over a width of 3 to 10 mm; in the third group (control eggs) the eggs remained intact. All eggs were incubated at 38°C and approximately 55% relative humidity for 16 days.

On day 14-15 of incubation, the water vapor conductance of the shell was determined by the 'calibrated egg' technique (Tullett, 1981). The water vapor conductance, G_{H_2O} in mg/(day torr), of the infertile egg which was subsequently used as the 'calibrated egg' was determined first by measuring the weight loss for a given period in a dry atmosphere (buried in silica gel) at 38°C. The G_{H_2O} of an experimental egg was then determined from the ratio of the rate of water loss of the experimental egg to that of the calibrated egg multiplied by the G_{H_2O} of the calibrated egg. For actual measurements, 4 calibrated eggs were used and kept together with the experimental eggs in the same incubator. The G_{H_2O} multiplied by 1.06 (conversion factor from G_{H_2O} in units of mass to G_{O_2} in units of volume) provides the conductance for O_2 (G_{O_2} in ml STP/(day·torr)). For standardization, G_{O_2} was divided by the fresh egg mass to provide mass-specific conductance for O_2 (g_{O_2} in ml STP/(day·torr·g)).

On day 16 of incubation, the O_2 consumption ($\dot{M}_{O_2}$ in ml STPD/day) of individual eggs was determined using a modified Scholander and Edwards respirometer (Scholander and Edwards, 1942) submerged in a 38°C water bath. The respirometer consisted of two equal-sized lucite chambers connected by a U-shaped water-filled glass manometer. As the embryo within an experimental egg, placed in one chamber along with a CO_2 absorber, consumed O_2, the level of water in the manometer was displaced. The displacement was corrected by injecting pure O_2 into the chamber containing the experimental egg. The O_2 consumption was calculated from the volume of O_2 injected in a given time.

After measurement of $\dot{M}_{O_2}$, each egg was subjected to blood sampling from the allantoic vein as described previously (Tazawa, 1971). Immediately after collection of 0.4 ml blood into a Hamilton glass syringe (No.750), the blood was measured for P_{O_2} and pH with an Instrumentation Laboratory blood gas analyzer (type 213 and 326).

Analysis Of The Air Space-Arterialized Blood P_{O_2} Difference

Air space P_{O_2} ($P_{A_{O_2}}$) In diffusive transport in chicken eggs, the $\dot{M}_{O_2}$ is governed by the shell conductance for O_2 (G_{O_2}) multiplied by the P_{O_2} difference between the environment and the air space ($P_{I_{O_2}} - P_{A_{O_2}}$). Since the effective P_{O_2} of the environment ($P_{I_{O_2}}$) is known, the air space P_{O_2} is calculated from the $\dot{M}_{O_2}$ and G_{O_2} measured for each egg as follows,

$$P_{A_{O_2}} = P_{I_{O_2}} - \dot{M}_{O_2}/G_{O_2} \tag{1}$$

Estimation of diffusing capacity (D_{O_2}) and allantoic shunt ($\dot{q}_{sh}$) The $\Delta P_{A_{O_2}} \cdot P_{a_{O_2}}$ is assumed to be dependent upon the diffusing capacity of the inner diffusion barrier (D_{O_2}) and the physiological shunt in the allantoic circulation (referred to as allantoic shunt, $\dot{q}_{sh}$). The D_{O_2} and $\dot{q}_{sh}$ are estimated so that they produce a given $\Delta P_{A_{O_2}} \cdot P_{a_{O_2}}$. The sequence of the iteration process of the Bohr integration procedure using microcomputer is indicated as follows and in Fig. 1,
  (1) D_{O_2} and $\dot{q}_{sh}$ are assigned temporary values by reference to the previous report (Piiper et al., 1980).

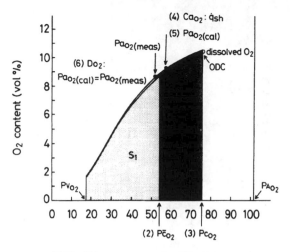

(1) $\dot{M}O_2$ & GO_2 are measured, and DO_2 & $\dot{q}sh$ are given.

Fig. 1. Sequence of estimation for diffusing capacity
(DO_2) and allantoic shunt ($\dot{q}_{sh}$). Numbers (2)-(6)
correspond to those in brackets in the text indicating the analytical sequence.

(2) Mean capillary blood PO_2 ($P\bar{c}_{O_2}$) is calculated from $\dot{M}O_2$ and DO_2 as
follows,

$$P\bar{c}_{O_2} = P_{AO_2} - \dot{M}O_2/DO_2 \tag{2}$$

(3) End-capillary blood PO_2 (Pc_{O_2}) is estimated from the Bohr integration so that the area under the O_2 dissociation curve (ODC), which includes
the dissolved O_2, from PO_2 of the allantoic artery blood (Pv_{O_2}) to $P\bar{c}_{O_2}$ (S_1)
becomes identical with that from $P\bar{c}_{O_2}$ to Pc_{O_2} (S_2). The ODC reported previously for 16-day-old embryos (Tazawa et al., 1976) is,

$$\log PO_2 = 4.872 - 0.453 \, pH + 0.395 \, \log So_2/(100-So_2) \tag{3}$$

where for pH the value determined in each embryo is used. The Pv_{O_2} which is
hyperbolically related to the shell conductance (Nakazawa and Tazawa, 1987)
is estimated from individual values for g_{O_2} as follows,

$$Pv_{O_2} = 20.6 - 0.53/g_{O_2} \tag{4}$$

The O_2 content for a given blood PO_2 is estimated from

$$Co_2 = O_2 \text{ capacity} \cdot So_2/100 + \alpha_{O_2} \cdot Po_2 \tag{5}$$

where the O_2 capacity is taken to be 12 vol% as reported for 16-day-old
embryos (Tazawa and Mochizuki, 1977) and an O_2 solubility coefficient (α_{O_2})
of 0.00311 ml/(100 ml·torr) is used.

(4) The arterialized blood O_2 content in the allantoic vein (Ca_{O_2}) is
estimated by taking $\dot{q}_{sh}$ into consideration as follows,

$$Ca_{O_2} = Cc_{O_2} - (Cc_{O_2} - Cv_{O_2}) \cdot \dot{q}_{sh} \tag{6}$$

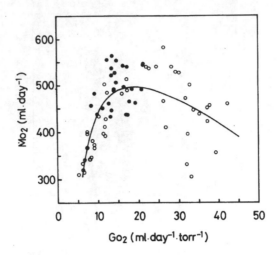

Fig. 2. Oxygen consumption of control (closed circles),
conductance-decreased and conductance-increased
eggs (both indicated by open circles) measured
on day 16 of incubation.

where end-capillary O_2 content (Cc_{O_2}) and mixed arterial O_2 content in the allantoic artery (Cv_{O_2}) are estimated from ODC and dissolved O_2 content for a given Po_2.

(5) Then, Pa_{O_2} is calculated from Ca_{O_2} and ODC.

(6) Procedures (2)-(5) are repeated altering Do_2 until Pa_{O_2} calculated as above equals the measured Pao_2 or the difference between the two values becomes smaller than a given difference. The conductance obtained as above is the Do_2 producing $\Delta P_{AO_2} \cdot Pao_2$ for a given allantoic shunt.

(7) the influence of $\dot{q}_{sh}$ on $\Delta P_{AO_2} \cdot Pao_2$ is examined by repeating the above procedure but altering $\dot{q}_{sh}$ while Do_2 is kept constant.

RESULTS

The O_2 consumption measured for 27 control, 24 conductance-decreased and 23 conductance-increased eggs is plotted for the individual conductances in Fig. 2. The control eggs are shown by closed circles. For widely altered conductances, the $\dot{M}o_2$ increases hyperbolically with increasing Go_2 until the conductance reaches the control range and then decreases. The best-fit regression curve is expressed by the following equation,

$$\dot{M}o_2 = 768 - 2567/Go_2 - 7.17\ Go_2 \qquad (7)$$

where the coefficients of the 2nd and 3rd terms are significant by the Fisher test (F = 68.6 (P<0.01) and 33.0 (P<0.01), respectively). The decrease in $\dot{M}o_2$ at large conductances is attributed to the small size of the embryos, caused by excess water loss (Okuda and Tazawa, 1987).

Fig. 3 represents the Po_2 of arterialized blood collected from the allantoic vein (Pa_{O_2}) and the air space calculated using eq. 1 (P_{AO_2}). The

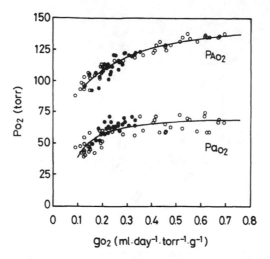

Fig. 3. Oxygen tensions of arterialized blood (Pa_{O_2}) and air space (PA_{O_2}) of 16-day-old embryos with widely altered shell conductances. Control eggs are shown by closed circles.

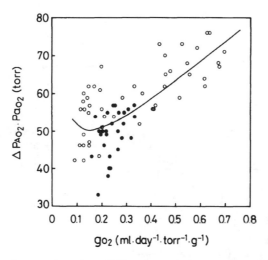

Fig. 4. Oxygen tension differences between air cell gas and arterialized blood of embryos shown in Fig. 3. The solid curve is drawn from the regression equation, $\Delta P_{A_{O_2}} \cdot Pa_{O_2} = 34.3 + 1.18/g_{O_2} + 53.9\ g_{O_2}$.

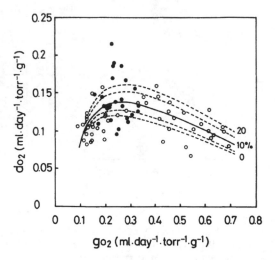

Fig. 5. Diffusing capacity (standardized by egg mass) estimated so that it produces the individual values for $\Delta P_{AO_2} \cdot Pa_{O_2}$ shown in Fig. 4 with a 10% allantoic shunt.

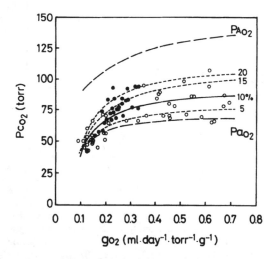

Fig. 6. End-capillary blood Po_2 of individual eggs having the diffusing capacity shown in Fig. 5 and an allantoic shunt of 10%. The solid curve is drawn from the best-fit regression equation.

quadratic regression equations of inverse g_{O_2} expressed by

$$Pa_{O_2} = 72 - 2.86/g_{O_2} - 0.033/(g_{O_2})^2 \qquad\qquad (8)$$
$$\text{and } P_{AO_2} = 150 - 11.09/g_{O_2} + 0.540/(g_{O_2})^2 \qquad\qquad (9)$$

give a good fit for the relation between O_2 tension and conductance. The significance of the regression equation which becomes quadratic is discussed elsewhere (Okuda and Tazawa, 1987). The $\Delta P_{AO_2} \cdot Pa_{O_2}$ increases with increasing g_{O_2} (Fig. 4).

Fig. 5 shows the diffusing capacity and allantoic shunt which produce the $\Delta P_{AO_2} \cdot Pa_{O_2}$ plotted for individual g_{O_2} values in Fig. 4. The diffusing capacity is standardized by fresh egg mass (i.e., mass-specific diffusing capacity, d_{O_2} in ml STP/(day·torr·g)) for comparison with the g_{O_2} of the outer barrier. Each plot indicates d_{O_2} when $\dot{q}_{sh}$ is 10% and the solid curve shows the best-fit regression equation expressed by

$$d_{O_2} = 0.237 - 0.013/g_{O_2} - 0.191\ g_{O_2} \qquad\qquad (10)$$

where the hyperbolic and linear parameters are significant (F=19.6 (P<0.01) and F=23.5 (P<0.01), respectively). For $\dot{q}_{sh}$ of 0, 5, 15 and 20% (N=74, 74, 66, 48, respectively), only the regression curves are shown (dashed lines).

Fig. 6 represents the end-capillary blood Po_2 (Pc_{O_2}) plotted for individual g_{O_2} values when the allantoic shunt is assumed to be 10% and the diffusing capacity has a value shown in Fig. 5. For $\dot{q}_{sh}$ of 5, 15 and 20%, regression curves only are presented (dashed lines). The regression curves for P_{AO_2} and Pa_{O_2} shown in Fig. 3 are also presented (broken lines).

DISCUSSION

In contrast to the alveolar-arterial Po_2 difference in the vertebrate lung, the $\Delta P_{AO_2} \cdot Pa_{O_2}$ of chicken eggs is large (Figs. 3 and 4); the average $\Delta P_{AO_2} \cdot Pa_{O_2}$ of intact control eggs is 49 ±6 (SD) torr (N=27). In the light of the vertebrate lung, it has been suggested that the large O_2 difference could be caused by a combination of three things; a low diffusing capacity, a true arterio-venous shunt in the allantoic circulation, or a nonuniform distribution of 'ventilation' and perfusion in different regions of the chorioallantoic membrane (Wangensteen, 1972). The ventilation multiplied by the capacitance coefficient in the lungs corresponds to the shell diffusive conductance in eggs. It has been shown that there is a gradual reduction in diffusive conductance from the blunt pole of the egg toward the sharp end (Rokitka and Rahn, 1987) and that the respiratory exchange ratio over the air cell region is larger than that over the rest of the egg (Visschedijk, 1968). These findings suggest that the conductance/perfusion ratio of the air cell region is greater than that in the gas space of the rest of the egg. Recently, Paganelli et al. (personal communication) confirmed the presence of regional differences in this ratio and simultaneously suggested that the differences between air cell gas tensions and those of the whole egg are small and therefore air cell gas tensions are reasonably representative of the whole egg. Consequently, the nonuniform distribution of shell conductance and perfusion may not be responsible for the $\Delta P_{AO_2} \cdot Pa_{O_2}$ in control eggs. Instead, inequalities in the diffusing capacity and blood circulation of the chorioallantoic membrane may play a part in the Po_2 difference, which is regarded in the present analysis as the physiological allantoic shunt.

For eggs incubated normally for 16 days, the Do_2 has been reported to be 6.0 μl/(min·torr) (Tazawa and Mochizuki, 1976), 4.9 μl/(min·torr) (Tazawa and Mochizuki, 1977), 7 μl/(min·torr) (Piiper et al., 1980) and 6.8 μl/(min·torr) (Wangensteen and Weibel, 1982), and the diffusing capacity for carbon monox-

ide (Dco), to be 5.1 $\mu l/(min \cdot torr)$ (Bissonnette and Metcalfe, 1978). As to
the allantoic shunt, Piiper et al. (1980) reported it to be 10-15%. In order
to produce the $\Delta P_{AO_2} \cdot Pa_{O_2}$ calculated for the present control eggs (N=27), the
D_{O_2} has to be larger than 4 $\mu l/(min \cdot torr)$ (since the averaged egg mass is
about 60 g, it corresponds to a d_{O_2} of about 0.1 ml/(day$\cdot$torr$\cdot$g)) even if $\dot{q}_{sh}$
is assumed to be zero. Conversely, even if $\dot{q}_{sh}$ reaches 20%, the D_{O_2} has to
be smaller than 10 $\mu l/(min \cdot torr)$ (corresponding to a d_{O_2} of about 0.24 ml/
(day$\cdot$torr$\cdot$g)), because the end-capillary blood P_{O_2} calculated in the present
analysis could not exceed the P_{AO_2}. Similarly, the allantoic shunt could not
exceed 20%. As the allantoic shunt increases, the D_{O_2} should increase to
yield a given $\Delta P_{AO_2} \cdot Pa_{O_2}$ (Fig. 5). As the D_{O_2} is increased, however, the
calculated Pc_{O_2} exceeds P_{AO_2}, limiting the upper $\dot{q}_{sh}$ values in the analysis.
Suppose the allantoic shunt is 10%, the D_{O_2} of control eggs ranges from 4.6
to 8.3 $\mu l/(min \cdot torr)$ (corresponding to a d_{O_2} of about 0.1 to 0.22 ml/(day$\cdot$
torr$\cdot$g)) with an average of 6.1 ±1.0 (SD)(N=27) $\mu l/(min \cdot torr)$. The average
D_{O_2} for $\dot{q}_{sh}$ of 0, 5, 15 and 20% is 5.3 ±0.8 (N=27), 5.6 ±0.9 (N=27), 6.8 ±1.2
(N=22) and 7.4 ±1.1 (N=14) $\mu l/(min \cdot torr)$, respectively.

Alteration of the shell conductance by removing or covering the shell
causes regional differences in the conductance. Because both $\dot{M}o_2$ and Go_2
measured represent average values for the whole egg, the P_{AO_2} calculated from
these variables is representative of the whole egg. In the case of widely
altered conductances, the $\Delta P_{AO_2} \cdot Pa_{O_2}$ is analyzed below for the diffusing
capacity and physiological shunt in the allantoic circulation.

The $\Delta P_{AO_2} \cdot Pa_{O_2}$ becomes large when the shell conductance is increased
from the control range (Fig. 4), and at decreased go_2 the averaged value of
$\Delta P_{AO_2} \cdot Pa_{O_2}$ (54 ±7 (SD) torr, N=24) is significantly larger than the control
(49 ±6 (SD) torr, N=27, P<0.05). Consequently, as the go_2 increases, the
$\Delta P_{AO_2} \cdot Pa_{O_2}$ is first decreased to a minimum value at control go_2 and then increased. As for the diffusing capacity of the inner barrier, on the other
hand, it is first increased with increasing go_2 and then decreased (Fig. 5).
Comparison between changes in the variables shown in Figs. 4 and 5 indicates
that changes in $\Delta P_{AO_2} \cdot Pa_{O_2}$ with go_2 are a mirror image of those in the diffusing capacity. This implies that when the shell conductance is altered by
changing the gas exchange area of the shell, the diffusing capacity of the
inner barrier is decreased, contributing to the large $\Delta P_{AO_2} \cdot Pa_{O_2}$ at low and
high go_2. In addition, regional alteration of the shell conductance may
change the physiological shunt; e.g., the $\dot{q}_{sh}$ may be reduced when the shell
conductance is increased. If this is the case, the d_{O_2} shown in Fig. 5, which
is estimated so as to produce a given $\Delta P_{AO_2} \cdot Pa_{O_2}$ at a given $\dot{q}_{sh}$, will be reduced at increased go_2 at the rate of decrease in $\dot{q}_{sh}$.

If it is assumed that the physiological shunt in the chorioallantoic
circulation is zero, the end-capillary blood P_{O_2} (Pc_{O_2}) is equal to the P_{O_2}
measured for the allantoic vein blood (Pa_{O_2}) and the observed $\Delta P_{AO_2} \cdot Pa_{O_2}$ is
attributable only to a diffusive resistance. Even for a 10% shunt, at re-
duced go_2 about 90% of $\Delta P_{AO_2} \cdot Pa_{O_2}$ is attributable to diffusion limitation
(Fig. 6); i.e., the ratio of $\Delta P_{AO_2} \cdot Pc_{O_2}$ to $\Delta P_{AO_2} \cdot Pa_{O_2}$ averages out to 0.89
±0.06 (SD) (N=24), while at control and increased go_2 the $\Delta P_{AO_2} \cdot Pc_{O_2}/\Delta P_{AO_2} \cdot$
Pa_{O_2} ratio is 0.72 ±0.12 (SD) (N=27) and 0.75 ±0.12 (SD)(N=23), respectively.
The effect of a shunt on blood oxygenation increases with increasing go_2
until go_2 reaches the control range, but the diffusion limitation is still
responsible for 70-75% of $\Delta P_{AO_2} \cdot Pa_{O_2}$ at increased go_2.

The D_{O_2} per unit weight of embryo (on day 16 of incubation) is almost
comparable to the diffusing capacity of human lungs expressed per unit weight
of the body. Nevertheless, in the chorioallantoic gas exchanger, the end-
capillary blood P_{O_2} fails to reach the air space P_{O_2} and the blood has to
leave the chorioallantoic capillary without being fully oxygenated. This may
partially be attributed to the structure of the gas exchanger, and to the

fact that the contact time is short as compared with that in human lungs (Tazawa and Mochizuki, 1976).

For control eggs with a shell diffusive conductance of 0.242 ±0.044 (average ±SD, N=27) ml/(day·torr·g), and supposing the allantoic shunt is 10%, the allantoic diffusing capacity averages out to 0.140 ±0.027 (SD) ml/ (day·torr·g). The ratio of the allantoic diffusing capacity to the shell diffusive conductance is 0.60 ±0.17 (SD). In other words, the resistance of the inner barrier to O_2 diffusion is about 1.7-fold that of the outer barrier (eggshell and outer membrane). A similar ratio has been reported for resistances to CO diffusion in the hen's egg incubated for 16 days (Bissonnette and Metcalfe, 1978). As the shell diffusive conductance is decreased or increased from the control, the resistance of the chorioallantoic membrane to O_2 diffusion becomes more predominant, which may be a main limiting factor for O_2 uptake of chick embryos.

SUMMARY

The gas exchange of chicken eggs takes place by molecular diffusion. The diffusion barrier between ambient atmosphere and erythrocyte hemoglobin of the gas exchanger (the vascularized chorioallantoic membrane) is conveniently divided into two parts by the air space in the fibrous shell membranes; i.e., the outer barrier (mainly the porous eggshell) and the inner barrier (the chorioallantoic membrane and the chemical reaction with hemoglobin). In contrast to the alveolar-arterial Po_2 difference in vertebrate lungs, the difference of Po_2 between the air space and the arterialized blood in the allantoic vein ($\Delta P_{AO_2} \cdot Pa_{O_2}$) is large in chick embryos. The present study analyzed the $\Delta P_{AO_2} \cdot Pa_{O_2}$ in relation to the diffusing capacity of the chorioallantoic membrane (inner barrier) and physiological shunt in the allantoic circulation with respect to widely altered diffusive conductance of the shell (outer barrier).

The shell diffusive conductance (Go_2) was altered of the beginning of incubation, and the O_2 consumption ($\dot{M}o_2$) was measured on day 16. The $\dot{M}o_2$ increased hyperbolically with increasing Go_2, reached a maximum at control values of Go_2 and decreased with further increases in Go_2. From Go_2 and $\dot{M}o_2$, the air space Po_2 was determined. The $\Delta P_{AO_2} \cdot Pa_{O_2}$ was increased in eggs with augmented Go_2 (from about 50 torr in control eggs to 70 torr in conductance-increased eggs). The diffusing capacity and allantoic shunt which produce a given $\Delta P_{AO_2} \cdot Pa_{O_2}$ were estimated employing a microcomputer performing the Bohr integration procedure so that a calculated Pa_{O_2} agreed with the measured Pa_{O_2}. The allantoic shunt is not more than 20%; 10% is likely. The diffusing capacity becomes maximum in intact control eggs and is decreased at both lowered and augmented Go_2. At lowered Go_2, diffusion limitation is responsible for about 90% of $\Delta P_{AO_2} \cdot Pa_{O_2}$ even in the presence of a 10% shunt. The diffusion limitation to blood oxygenation decreases as Go_2 increases, but it is still predominant at augmented Go_2. In control eggs, the resistance of the inner barrier to O_2 diffusion is about 1.7-fold that of the shell (outer barrier) which agrees with the previous reports.

REFERENCES

Bissonnette, J. M. and Metcalfe, J., 1978, Gas exchange of the fertile hen's egg: Components of resistance., Respir. Physiol. 34: 209-218.
Nakazawa, S. and Tazawa, H., 1987, Blood gases and hematological variables of chick embryos with widely altered shell conductance., Comp. Biochem. Physiol. (in press).

Okuda, A. and Tazawa, H., 1987, Oxygen consumption and growth of chick
    embryos with shell conductance altered widely from the beginning of
    incubation., The Physiologist (in press).
Paganelli, C. V., Sotherland, P. R., Olszowka, A. J. and Rahn, H., 1987,
    Regional differences in diffusive conductance/perfusion ratio in the
    shell of the hen's egg., (personal communication).
Piiper, J., Tazawa, H., Ar, A. and Rahn, H., 1980, Analysis of chorioallan-
    toic gas exchange in the chick embryo., Respir. Physiol. 39: 273-284.
Rahn, H. and Paganelli, C. V., 1982, Role of diffusion in gas exchange of the
    avian egg., Fed. Proc. 41: 2134-2136.
Rokitka, M. A. and Rahn, H., 1987, Regional differences in shell conductance
    and pore density of avian eggs., Respir. Physiol. 68: 371-376.
Romanoff, A. L. and Romanoff, A. J., 1949, "The Avian Egg.", John Wiley and
    sons, New York.
Scholander, P. F. and Edwards, G. A., 1942, Micro-respiration apparatus.,
    Rev. Sci. Instrum. 13: 292-295.
Tazawa, H., 1971, Measurement of respiratory parameters in blood of chicken
    embryo., J. Appl. Physiol. 30: 17-20.
Tazawa, H. and Mochizuki, M., 1976, Estimation of contact time and diffusing
    capacity for oxygen in the chorioallantoic vascular plexus.,
    Respir. Physiol. 28: 119-128.
Tazawa, H., Ono, T. and Mochizuki, M., 1976, Oxygen dissociation curve for
    chorioallantoic capillary blood of chicken embryo., J. Appl. Physiol.
    40: 393-398.
Tazawa, H. and Mochizuki, M., 1977, Oxygen analyses of chicken embryo blood.,
    Respir. Physiol. 31: 203-215.
Tazawa, H., Ar, A., Rahn, H. and Piiper, J., 1980, Repetitive and
    simultaneous sampling from the air cell and blood vessels in the chick
    embryo., Respir. Physiol. 39: 265-272.
Tullett, S. G., 1981, Theoretical and practical aspects of eggshell
    porosity., Turkeys 29: 24-28.
Visschedijk, A. H. J., 1968, The air space and embryonic respiration.
    I. The pattern of gaseous exchange in the fertile egg during the
    closing stages of incubation., Brit. Poultry. Sci. 9: 173-184.
Wangensteen, O. D., 1972, Gas exchange by a bird's embryo., Respir. Physiol.
    14: 64-74.
Wangensteen, O. D. and Weibel, E. R., 1982, Morphometric evaluation of
    chorioallantoic oxygen transport in the chick embryo.,
    Respir. Physiol. 47: 1-20.

WIDE VARIATION OF MYOGLOBIN CONTENTS IN GIZZARD

SMOOTH MUSCLES OF VARIOUS AVIAN SPECIES

Y. Enoki, T. Morimoto,* A. Nakatani, S. Sakata, Y. Ohga,
H. Kohzuki, and S. Shimizu

Second Department of Physiology, Nara Medical University
Kashihara, Nara 634, Japan
*Osaka Municipal Tennoji Zoological Garden, Osaka, Osaka 543
Japan

INTRODUCTION

While our knowledge of myoglobin structure has greatly advanced in the
second half of this century, its physiological function in situ is not yet
completely understood. Ample circumstantial evidence such as the extremely high concentration of the protein in the muscles of diving animals
suggest a possible close relation to oxygen transport and utilization in
muscle cells, however, we have a rather poor and controversial literature
on direct experimental evidence (Cole et al, 1978; Cole, 1982; Wittenberg
and Wittenberg, 1975; Jones and Kennedy, 1982).

Myoglobin has generally been considered to exist exclusively in
skeletal and cardiac muscles, but not in smooth muscles. The only probable
exception was originally suggested for chicken gizzard by Kennedy and
Whipple (1928). Later this was further studied by Gröschel-Stewart et al.
(1971) and Blessing and Müller (1974). Very recently we too have attempted
the isolation and characterization of myoglobin from chicken gizzard and
found that a structurally and functionally identical myoglobin was present
in both skeletal and smooth (gizzard) muscles (Enoki et al., 1984).

Absence of teeth in birds, an adaptive change for flying, is considered to endow the gizzard with an important role in food digestion (Duke,
1986). The importance, however, may be different from species to species,
depending upon the feeding habits. Generally speaking, the importance
seems more evident in herbivorous birds which have well developed gizzard
muscles to grind their harder and more fibrous feeds (Duke, 1986). This
higher mechanical performance will necessarily need higher supply of energy
and consequently of oxygen.

In this investigation we determined the gizzard myoglobin content
in a variety of avian species with different feeding habits, and found a
distinctly higher myoglobin content in the herbivores than in the carnivores. The higher myoglobin content was considered to reflect the higher
requirement of mechanical activity and possibly the higher oxygen demand
of the herbivore's gizzard in digestion of food.

MATERIALS AND METHODS

Gizzards and Skeletal Muscles of Birds

The gizzards and breast muscles of 33 species were supplied from fresh autopsy cases from Osaka Municipal Tennoji Zoological Garden. The materials were immediately stored in a deep freeze (-90°C) after trimming off the koilin linings, fascia and fat tissues. Fresh chicken gizzards, leg muscles and blood were obtained from a local poultry farm.

Preparation of Chicken Hemoglobin and Myoglobin

Hemoglobin was prepared from fresh blood by conventional method. Myoglobin was prepared from an aqueous extract of chicken gizzards and from skeletal muscles by the heat denaturation-gel permeation-chromatofocusing procedure as previously reported (Enoki et al., 1984). The proteins were preserved in carboxy form in a tightly stoppered test tube (4°C).

Determination of Myoglobin Contents in Gizzards and Skeletal Muscles

· The spectrophotometric procedure described by Reynafarje (1963) was followed with slight modification. Small pieces of muscles were taken from the specimens, chopped up and well mixed with scissors and a razor blade on an ice-chilled glass plate, divided into tared and capped polypropylene microcentrifuge tubes (Quality: No. 505) and weighed on an electronic balance (AND: ER-180A). Since a gradient of myoglobin content was found across the gizzard wall (Enoki et al., unpublished), the determinations were usually repeated at two loci of different depths, one superficial and the other deeper, and the results were averaged. The samples varied from 10 to 30 mg each depending on the myoglobin content. Ten millimolar K_2HPO_4 solution containing 1 mM Na EDTA (0.8 ml), pre-saturated with CO and chilled, was added and the muscle sample was well homogenized with a handy microhomogenizer (NITI-ON: Physcotron NS-310E) equipped with an NS-4 micro-shaft. The sample tube was held in ice water during homogenization. The homogenate was then spun at 15,000 rpm in a microcentrifuge (Tomy Seiko: MC15A), and the resultant supernate transferred to a new microcentrifuge tube; the air space was flushed with CO after the addition of 0.3 ml of chloroform, and the capped tube with its content was vigorously hand-shaken for a few seconds. The final step was a modification of Reynafarje's original procedure to remove fatty ingredients which gave rise to a cloudiness in the sample which interfered with the optical measurement. The supernatant aqueous layer, after centrifugation at 15,000 rpm, was put in a microcuvette, the air space was flushed with CO, and a few crystals of sodium dithionite were added and the sample was well mixed by several inversions. After the bubbles had disappeared on standing, the sample was scanned for absorbance in the 530 to 580 nm wave length region.

General Analytical Methods

Oxygen dissociation curves were constructed according to a spectrophotometric procedure as conventionally used in our laboratory (Enoki, 1959). Spectrophotometric measurements were performed with a Hitachi spectrophotometer model 124 or a Union high sensitivity spectrophotometer model 401. The concentration of the heme proteins was standardized on Fe basis, which was, in turn, determined by the o-phenanthroline method of Cameron (1964). All the chemicals used were of reagent grade (Nakara Chemicals, Kyoto). All procedures were performed in a room maintained at 20°C.

710

RESULTS AND DISCUSSION

Visible Absorption Spectra of Chicken Carboxy-Hemoglobin and -Myoglobin: Derivation of Equation for Myoglobin Content of Muscle Extract

The absorption spectra are shown in Figure 1, from which the milli-molar absorption coefficients were read for determination of myoglobin content of the muscle extract according to the procedure of Reynafarje (1963): 14.78(CO myoglobin) and 14.00 (CO hemoglobin) at 538 nm, and 11.62 (CO myoglobin) and 14.00 (CO hemoglobin) at 568 nm. No difference was found between the absorptions of the myoglobins from the gizzard and skeletal muscles.

Based on these figures the following equation was derived for the myoglobin content:

$$C = \frac{5.696(A_{538} - A_{568})\ (0.800 + 0.75W)}{W} \tag{1}$$

where C; myoglobin content in mg/g wet muscle, A_{538} and A_{568}; absorbances of the muscle extract at 538 and 568 nm, and W; weight of muscle sample in g. This equation was used in all the following determinations of the various birds.

Oxygen Dissociation Curves of Myoglobins from Chicken Gizzard and Skeletal Muscle

Figure 2 shows the oxygen dissociation curves of the chicken gizzard and skeletal muscle myoglobins as used for determining the absorption spectra(Fig. 1). It is evident that the proteins exhibit functional proper-

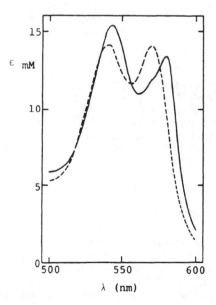

Fig. 1. Visible absorption spectra of chicken carboxy-hemoglobin(----) and -myoglobin(——). Ordinate: millimolar absorption coefficient.

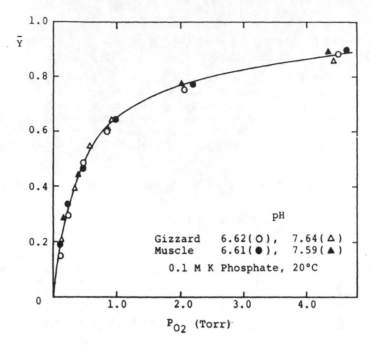

Fig. 2. Oxygen dissociation curves of chicken gizzard and skeletal muscle myoglobins at two different pH's. $\bar{Y}$; fractional saturation of myoglobin with oxygen.

ties characteristic of myoglobin such as the absence of subunit cooperativity, very high oxygen affinity, and apparent lack of the Bohr effect. It should also be noted that identical oxygenation properties were observed independent of the origin of the proteins.

Myoglobin Contents in Avian Gizzards and Skeletal Muscles

The gizzard myoglobin content in 34 bird species so far determined are summarized in Figure 3 and Table 1, together with their breast muscle myoglobin content. No correlation was found between the myoglobin content of the gizzard and of breast muscle in either group. The diving species such as rockhopper penguin (Eudyptes chrysocome) and king penguin (Aptenodytes patagonicus) showed exceptionally higher contents of myoglobin in the skeletal muscle.

It appears that these bird species can be divided into two groups: in one the myoglobin content of the gizzard is high, in the other low. The high myoglobin group, composed of 15 species, showed a mean content ± SD of 7.74 ± 1.81 (mg/g wet muscle) ranging widely from 5.76 in white-throated spine-tailed swift (Hirundapus caudacuta) to 11.94 in European pochard (Aythya ferina). The low myoglobin group with 19 species, exhibited the remarkably low average content of 1.54 ± 0.41 (mg/g wet muscle) with a rather small variation from 0.75 in red-necked phalarope (Phalaropus lobatus) to 2,14 in Rufous-tailed hummingbird (Amazilia tzacatl) and Ural owl (Strix uralensis).

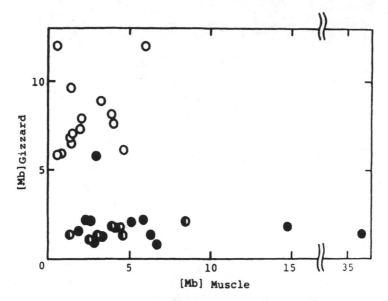

Fig. 3. Myoglobin contents (mg/g muscle) of gizzards (ordinate) and breast
muscles (abscissa) in various avian species. Note the feeding
habits of the birds: herbivorous (O), carnivorous (●), and
omnivorous (◐).

We can see another feature in the results, that the differentiation of
the two groups might be closely correlated with their feeding habits. In
the high myoglobin groups, all the species except one, i.e. white-throated
spine-tailed swift (Hirundapus caudacuta), were herbivores. In the low
myoglobin group, on the contrary, the birds were mostly carnivorous or,
otherwise, omnivorous. The hummingbird (Amazilia tzacatl), albeit omnivorous, feeds primarily on flower honey and sometimes on small insects.
An apparent exception was also found in this group, i.e. rose-ringed
parakeet (Psittacula krameri manilensis) which is described as herbivorous.

As is well documented (Duke, 1986), carnivorous birds such as raptors
employ chemical digestion of foods rather than mechanical and consequently
have poorly developed and thin-walled gizzards. Conversely, herbivorous,
especially granivorous, birds which feed primarily a hard or fibrous
vegetable food require strong mechanical assistance for digestion, and
therefore have two pairs of highly developed smooth muscles in their
gizzards. These differences in the gizzards between the carnivores and
herbivores are clearly reflected in the mechanical activity as measured
experimentally. As previously reported by several workers (see Duke, 1986),
the intraluminal pressure changes during gizzard contractions were quite
different between the two groups of birds with different feeding habits:
8-26 mmHg (buzzard) on one hand and 100-280 mmHg (hen, duck and goose) on
the other. It is evident that the gizzards in the herbivores perform
considerably higher mechanical work and it is highly likely that they
require corespondingly higher energy supply. The higher myoglobin contents
in the herbivores may be considered to be correlated with, and probably to
meet, this energy requirement.

Table 1. Myoglobin Contents of Gizzards and Breast Muscles in Various
Avian Species, with Special Reference to Their Feeding Habits.

Avian Species	Myoglobin Contents (mg/g muscle)		Feeding Habit
	Gizzard[a]	Breast muscle[b]	
1 Chicken	5.85	0.84	H
2 Green Peafowl	6.42	1.40	H
3 Brazilian Teal	8.83	3.21	H
4 Temmink's Tragopan	9.61	1.41	H
5 Australian Shoveller	7.57	3.99	H
6 Green-winged Teal	7.28	1.93	H
7 California Quail	11.96	0.57	H
8 European Pochard	11.94	6.02	H
9 Ringed Teal	6.74	1.32	H
10 Spix's Guan	5.82	0.58	H
11 Keel-billed Toucan	6.09	4.65	H
12 Garganey	7.86	2.01	H
13 Moorhen	7.03	1.54	H
14 Mute Swan	8.13	3.87	H
15 White-throated Spine-tailed Swift	5.76	2.92	C
16 Greater Flamingo	1.31	1.36	C
17 Grey Heron	1.27	6.32	C
18 Little Egret	2.03	5.12	C
19 Brown Hawk Owl	2.03	2.63	C
20 Rufous-tailed Hummingbird	2.14	2.32	O[c]
21 Black-headed Gull	0.87	2.85	C
22 Ural Owl	2.14	5.86	C
23 Woodcock	1.21	3.34	C
24 Gouldian Finch	1.03	2.53	O
25 Rockhopper Penguin	1.77	14.78	C
26 Brown-eared Bulbul	1.29	4.57	O
27 Red-necked Phalarope	0.75	6.66	C
28 Scarlet Ibis	1.51	1.87	C
29 Golden-breasted Starling	1.77	3.95	O
30 Rose-ringed Parakeet	1.71	4.44	H
31 Andean Cock-of-the-Rock	2.03	8.46	O
32 White-crested Laughing Thrush	1.77	4.12	C
33 King Penguin	1.40	35.94	C
34 White-faced Whistling Duck	1.26	3.06	O

H: herbivorous, C: carnivorous, O: omnivorous.
a) mean of 12 determinations (6 determinations each at two different loci
across the gizzard wall); b) mean of 6 determinations; c) feeding primarily
flower honey, and insects.

Another features should be noted here. The gizzard, most typically
in the herbivores, is reported to repeat slow (2-3 times per minute) but
strong tonic contractions. During the contracions it is possible that the
blood and circulatory oxygen supply would be greatly impaired by lateral
compression of the blood vessels in the gizzards, as clearly observed in
the myocardium during the cardiac cycle (Kirk and Honig, 1964). In this
respect, the carnivore's gizzard may be a little different in that it shows
weak and rather peristalsis-like contractions arising in the proventriculus
(glandular stomach) and passing through the gizzard to the duodenum (Duke,
1986). The higher myoglobin content in the herbivore's gizzard may be a
physiological adaptation to this circulatory situation.

In conclusion, we found a distinctly higher level of myoglobin in the gizzard of herbivorous avian species than in carnivorous or omnivorous ones. This higher level may be considered to guarantee a continuous oxygen supply during the strong tonic contractions and to support the higher mechanical activity required in the digestion of hard and fibrous food in herbivores.

Further studies on extended species of birds are now in progress and also an experimental approach has been attempted to elucidate directly the physiological role of myoglobin in muscular function using avian gizzards as an excellent model.

SUMMARY

We determined myoglobin contents of gizzards (muscular stomach) and breast muscles in 34 avian species by a modification of Reynafarje's spectrophotometric procedure. The birds were apparently differentiated into two groups in respect of the gizzard, one with a high myoglobin content (7.74 $\pm$ 1.81 mg/g muscle) and the other with a low (1.54 $\pm$ 0.41 mg/g). In the former group of 15 species all but one were herbivorous, and all but one were carnivorous or else omnivorous in the latter group of 19 species. The myoglobin level was considered to closely correlate with mechanical performance and therefore oxygen demands of the gizzards. It might also be relevant to a circulatory situation during the tonic contractions of this organs.

REFERENCES

Blessing, M. H., and Müller, G., 1974, Myoglobin concentration in the chicken, especially in the gizzard, Comp. Biochem. Physiol.,47A:535.

Cameron, B. F., 1965, Determination of iron in heme compounds. II. Hemoglobin and myoglobin, Anal. Biochem., 11:164.

Cole, R. P., 1982, Myoglobin function in exercising skeletal muscle, Science, 216:523.

Cole, R. P., Wittenberg, B. A., and Caldwell, P. R. B., 1978, Myoglobin function in the isolated fluorocarbon-perfused dog heart, Am. J. Physiol., 234:H567.

Duke, G. E., 1986, Alimentary canal: anatomy, regulation of feeding, and motility, in: "Avian Physiology", P. D. Sturkie, ed., Springer, New York.

Enoki, Y., 1959, Salt effect on hemoglobin-oxygen equilibrium, J. Nara Med. Ass., 10:345.

Enoki, Y., Ohga, Y., Kawase, M., and Nakatani, A., 1984, Identical myoglobin is present in both skeletal and smooth muscles of chicken, Biochim. Biophys. Acta, 789:334.

Gröschel-Stewart, U., Jaroschik, U., and Schwalm, H., 1971, Chicken gizzard, a myoglobin containing smooth muscle, Experientia, 27:512.

Jones, D. P., and Kennedy, F. G., 1982, Intracellular oxygen gradients in cardiac myocytes. Lack of a role for myoglobin in facilitation of intracellular oxygen diffusion, Biochem. Biophys. Res. Commun., 105:419.

Kennedy, R. P., and Whipple, G. H., 1928-1929, The hemoglobin of smooth and striated muscle of the fowl, Am. J. Physiol., 87:192.

Reynafarje, B., 1963, Simplified method for the determination of myoglobin, J. Lab. Clin. Med., 61:138.

Wittenberg, B. A., and Wittenberg, J. B., 1975, Role of myoglobin in the oxygen supply to red skeletal muscle, J. Biol. Chem., 250:9038.

ACKNOWLEDGEMENT

We are grateful to Mr. Yoshihiko Doi, Director, Osaka Municipal Zoological Garden, for his continuous support.

TUMORS

EVALUATION OF OXYGEN DIFFUSION DISTANCES IN HUMAN BREAST
CANCER USING CELL LINE SPECIFIC IN VIVO DATA: ROLE OF VARIOUS
PATHOGENETIC MECHANISMS IN THE DEVELOPMENT OF TUMOR HYPOXIA*

P. Vaupel[1], F. Kallinowski[1], and K. Groebe[2]

[1]Department of Applied Physiology, University of
Mainz, D-6500 Mainz, Fed. Rep. Germany
[2]Department of Mechanical Engineering, University of Rochester, Rochester, NY 14620, USA

INTRODUCTION

Radiobiological hypoxia in malignant tumors has been
shown to originate (i) from spatial and temporal functional
disturbances of tumor microcirculation resulting in a limited
convective O_2 flux in microregions even in tissue areas exhibiting high vascular densities, and (ii) from morphological
abnormalities of the microcirculatory bed leading to a limitation of the diffusive O_2 flux. In addition to these pathogenetic mechanisms, systemic factors (anemia, arterial hypoxia)
can also play a role in the development of tumor hypoxia.

Since tumor tissue oxygenation and the appearance of
tumor hypoxia are of paramount clinical importance, both
experimental and theoretical analyses have been performed.
Oxygen diffusion distances have been calculated for human
(Thomlinson and Gray, 1955) and mouse tumors (Tannock, 1968).
However, in these studies as well as in a recent model described by Degner and Sutherland (1986) data were used for
computation of the critical supply distances which were either
derived from in vitro investigations or were totally unrelated
to the tumor system of interest. Thus, the resulting estimates of the O_2 distribution cannot be regarded as representative descriptions of tissue oxygenation. For this reason, a
re-evaluation of O_2 diffusion distances was performed using
cell line specific in vivo data derived from human breast
cancer xenografts. The main purpose of this paper, however, is
to illustrate qualitatively variations of the oxygenation
within tumor microregions due to different supply conditions.
Although a comprehensive description of the real O_2 supply
situation within tumor tissue is hardly possible due to pronounced intratumor heterogeneities of the microcirculation,
the contribution of different mechanisms to the development of
tumor hypoxia can be estimated. To describe the consequences
following different O_2 supply conditions, the surface defined
by the 1 mm Hg isobar is graphically displayed resulting in

* Supported by DFG grants Va 57/2-4 and Gr 887/1-1

tissue cones surrounding tumor microvessels. Tissue regions beyond that boundary have to be considered as radiobiologically hypoxic (Vaupel et al., 1981).

METHODS

To gain information on the tissue regions in which insufficient O_2 supply leads to radiobiological hypoxia, the three-dimensional differential equation describing convectional and diffusional O_2 transport in respiring tissue has to be solved. This undertaking, however, generates considerable mathematical difficulties and its results are too intricate to facilitate a general understanding of the mechanisms to be discussed here. Therefore, a simplified approach was chosen: The tumor microvessels are conceived of as parallel straight tubes arranged in a hexagonal pattern. On the basis of this arrangement, the PO_2 distribution in a thin cross sectional slab of tissue may be described by a 1-dimensional Krogh model centered at the capillary (axial symmetry assumed, longitudinal diffusion neglected). The Krogh radius R_K, the point of zero radial flux, is determined by either the intercapillary distance (if the capillary PO_2 is large enough so that the whole cross sectional area can be supplied sufficiently) or by the O_2 diffusion distance pertinent to the capillary PO_2 at the actual longitudinal position on the capillary. In the latter case, R_K may be calculated to be that distance at which PO_2 falls to the critical PO_2, below which radioresistance occurs.

In order to consider the effects of the gradual decrease of PO_2 along the microvessel, the whole tissue area is discretized with respect to its longitudinal coordinate to form 200 sequential slabs, each slab receiving its blood from the preceding one. In each increment, the equation for R_K is solved as mentioned above employing the PO_2 of the entering blood for the actual mean capillary PO_2. Then, the O_2 content of the blood is diminished by the amount of oxygen which is consumed inside the slab during the transit of the blood, this new oxygen content being used as input to the next increment. By this, capillary PO_2 and O_2 diffusion distances at varying positions along a microvessel become a function of the overall O_2 consumption in the preceding slabs. Starting at the arterial end of a microvessel and proceeding stepwise down the vessel until the measured venous PO_2 is reached, the O_2 supply area of the entire microvessel can be computed. All calculations are based on tumor cell line specific in vivo data and refer to steady state conditions only (Vaupel et al., 1987).

RESULTS

Based on an O_2 consumption rate of small human breast cancer xenografts ($\dot{V}O_2$ = 30 ul/g/min), Fig. 1 shows the critical O_2 diffusion distance (R_{crit}) beyond that radiobiological hypoxia has to be expected. The interrelationship between R_{crit} and intracapillary PO_2 values is displayed taking into account Krogh's diffusion constant for tumor tissue ($1.9 \cdot 10^{-5}$ ml O_2/cm/min/atm; Grote et al., 1977) and a blood flow rate of 0.3 ml/g/min (mean flow rate of small tumors < 1 g; Vaupel et al., 1987).

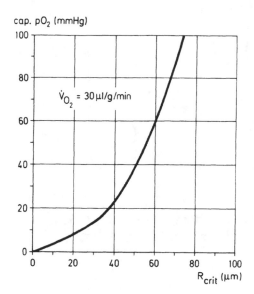

cap. pO_2 (mmHg)

$\dot{V}_{O_2}$ = 30 μl/g/min

R_{crit} (μm)

$\dot{V}_{O_2}$ = 30 μl/g/min
arterial pO_2 = 90 mmHg
venous pO_2 = 45 mmHg
[Hb] = 130 g/l
intercap. distance = 125 μm

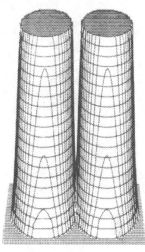

Fig. 1. Interrelationship bet ween oxygen diffusion distances (R_{crit}) in human breast cancer xenografts and capillary PO_2. Calculations are based on speci fic in vivo data (Vaupel et al., 1987)

Fig. 2. Tissue volumes surrounding two straight tumor microvessels beyond which radiobiological hypoxia has to be expected ("radiosensitive" tissue volumes) under the listed "standard" conditions

The respective tissue volumes pertinent to characteristic features of tumor microcirculation or to systemic pathogenetic factors (anemia, arterial hypoxia) are shown in Figs. 3-6.

So far, all calculations have been based on the assump- tion of concurrent flow within adjacent tumor microvessels. As can be seen from Fig. 7 (upper panel), the size of the radio- biologically hypoxic regions (black areas) increases as the blood is deoxygenated (desaturated) along the microvessel. If countercurrent flow is assumed, only minor changes in the size of these hypoxic areas are observed (Fig. 7, lower panel).

Tissue volumes surrounding two straight tumor microves- sels beyond which radiobiologically hypoxia has to be expected ("radiosensitive" tissue volumes) under "standard" conditions are presented in Fig. 2. Parameters describing this "standard" situation are listed above the tissue cone.

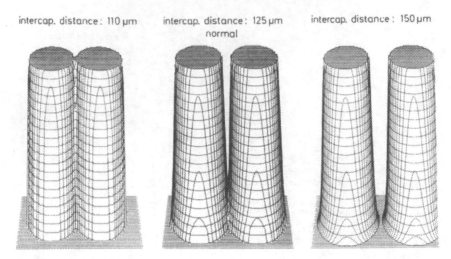

intercap. distance: 110 μm   intercap. distance: 125 μm   intercap. distance: 150 μm
                                        normal

Fig. 3. "Radiosensitive" tissue volumes pertinent to variations of intercapillary distances in breast cancers.

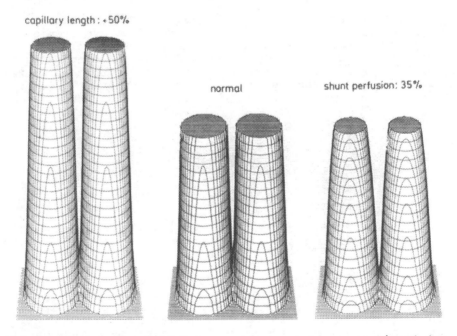

capillary length: +50%

                         normal   shunt perfusion: 35%

Fig. 4. "Radiosensitive" tumor tissue volumes pertinent to
elongation of tumor microvessels (+ 50%) and arterio venous
shunt perfusion (35%)

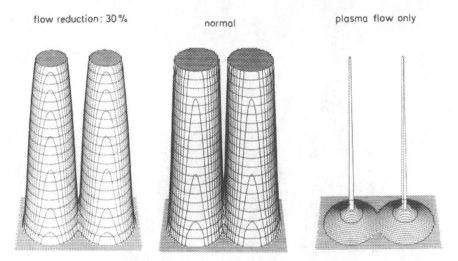

Fig. 5. "Radiosensitive" tumor tissue volumes during blood flow reduction (- 30%) and during plasma flow only

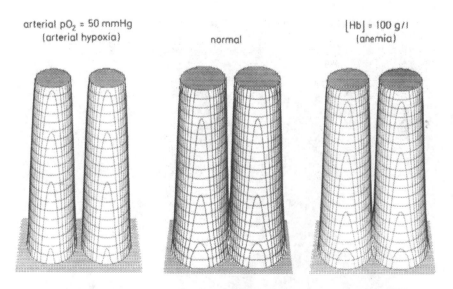

Fig. 6. "Radiosensitive" tumor tissue volumes during arterial hypoxia (PO_2 = 50 mm Hg) and during tumor induced anemia (hemoglobin concentration = 100 g/l)

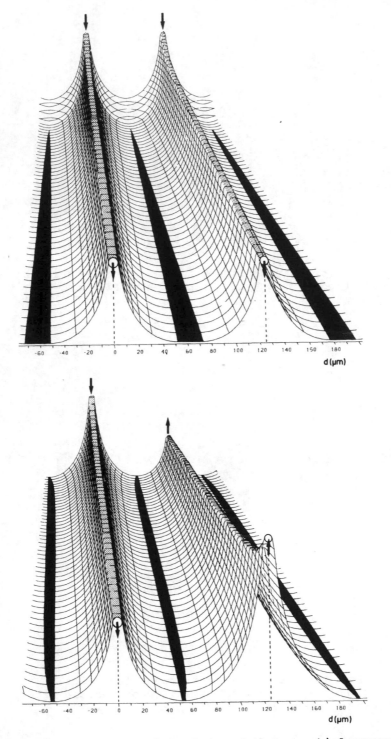

Fig. 7. Relief presentation of computed O_2 partial pressures in tumor microregions surrounding microvessels. The black areas indicate radiobiologically hypoxic regions with PO_2 values < 1 mm Hg (upper panel: concurrent blood flow, lower panel: countercurrent blood flow)

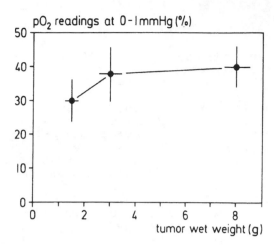

Fig. 8 Percentage of measured PO_2 values < 1 mm Hg in s.c. human breast cancer xenografts as a function of tumor wet weight. Values are means ± SEM (Vaupel et al., 1987)

CONCLUSIONS

1. Considering in vivo data derived from human breast cancer xenografts, there is clear evidence that radiobiological hypoxia is a common feature even for the "standard" situation with intercapillary distances of 125 um. Approximately 20 % of the tissue were found to be radiobiologically hypoxic under these conditions.

2. Based on the material presented, tumor hypoxia may develop at intercapillary distances above 100 - 110 um. If intercapillary distances exceed 140 - 150 um, tumor hypoxia is present at the arterial end of microvessels already.

3. Venous hypoxia evidenced under "standard" conditions is distinctly aggravated by the following tumor inherent changes of the microcirculation: reduced blood flow, increased arterio venous shunt perfusion, reduced red blood cell flux (e.g., during plasma flow only), and elongation of tumor microvessels.

4. Systemic pathogenetic factors which can occur in the tumor patients, anemia and arterial hypoxia, may also be responsible for the expansion of hypoxic tumor areas.

5. As a rule, several of the pathogenetic mechanisms responsible for the development of tumor hypoxia occur simultaneously in malignant tumors. Consequently, significant tissue hypoxia has to be expected, limiting the efficacy of various non-surgical tumor therapies. These theoretical considerations are in agreement with PO_2 measurements in breast cancer xenografts (see Fig. 8).

REFERENCES

Degner, F.L., and Sutherland, R.M., 1986, Theoretical evaluation of expected changes in oxygenation of tumors associated with different hemoglobin levels, Int. J. Radiat. Oncol. Biol. Phys., 12: 1291.

Grote, J., Suesskind, R., and Vaupel, P., 1977, Oxygen diffusivity in tumor tissue (DS-Carcinosarcoma) under temperature conditions within the range of 20-40°C, Pfluegers Arch., 372:37.

Tannock, I. F., 1968, The relation between cell proliferation and the vascular system in a transplanted mouse mammary tumour, Brit. J. Cancer, 22: 258.

Thomlinson, R. H., and Gray, L. H., 1955, The histological structure of some human lung cancers and the possible implications for radiotherapy, Brit. J. Cancer, 9: 539.

Vaupel, P. W., Frinak, S., and Bicher, H. I., 1981, Heterogeneous oxygen partial pressure and pH distribution in C3H mouse mammary adenocarcinoma, Cancer Res., 41: 2008.

Vaupel, P., Fortmeyer, H. P., Runkel, S., and Kallinowski, F., 1987, Blood flow, oxygen consumption, and tissue oxygenation of human breast cancer xenografts in nude rats, Cancer Res., 47: 3496.

PARALLELISM BETWEEN TRANSPORT INHIBITION AND L1210 CELL

GROWTH BY OUABAIN

Hee Min Rhee

Department of Pharmacology
Oral Roberts University School of Medicine
Tulsa, Oklahoma 74137 U.S.A.

INTRODUCTION

It is well documented that cardiac glycosides such as digoxin or digitoxin, have some beneficial influences over the development of tumor in human subjects. For instance, the recurrence rate of tumor after mastectomy among patients who did not receive cardiac glycoside was 9.6 times greater than the risk of recurrence of those who were on digitalis (Stenkvist et al., 1982). The characteristics of tumor cells of the patients with breast cancer who were receiving cardiac glycosides at the time of diagnosis were different from those of patients who did not use the drugs. Tumor cells from the patients who were on digitalis were smaller in size, and more uniform in cellular morphology, density, and structure than those in patients not using digitalis (Stenkvist et al., 1980 and 1979). The biochemical or pharmacological basis of the modifying effect of cardiac glycoside on the biological aggressiveness of breast tumor is yet to be identified.

Cardiac glycosides inhibit sodium and potassium activated adenosinetriphosphatase (Na^+,K^+-ATPase, E.C. 3.6.1.3). Inhibition of Na^+,K^+-ATPase by cardiac glycoside is related to the toxic action of the drug, although the relationship between the therapeutic action of the drug and inhibition of the enzyme is still in controversal (Repke, 1963; Lee and Klaus, 1971; Schwartz et al., 1975; Rhee et al., 1976; Huang et al., 1979). It is not known whether the tumor modifying effect of cardiac glycosides is related to the inhibition of Na^+,K^+-ATPase. Cardiac glycosides inhibit not only fluxes of monovalent cations, but also the movement of sugars and amino acids. In an effort to understand the molecular or cellular mechanism of digitalis action on tumor development, the main objective of this paper was to examine the effect of ouabain on the uptake of thymidine in relation to the growth of L1210 cells.

MATERIALS AND METHODS

Cell Culture and Counting

The lymphocytic mouse leukemia L1210 cells were first described by Law et al. in 1949 and subsequent success in an in-vitro suspension culture of the cell line was reported (Moore et al., 1966). We cultured the cell lines in a medium of 11% fetal bovine serum and

25 μM morpholinopropane sulfonic acid) (MOPS) with Rosewell Park
Memorial Institute (RPMI) 1640 medium (KC Biological, Lenexa, Kansas)
at 37°C in an incubator (Sheldon Manufacturing, Inc., Portland,
Oregon, Model 25). The humidity was maintained 98% under stream of
95% O_2 and 5% CO_2, and fresh medium was added at 48 to 72 hour intervals. During the incubation, cell growth was continually checked by
measuring the number of cells in a volume of cell sample as indicated
under legends of figures. Well suspended cell samples (usually
0.1 ml) were diluted with 9.9 ml of saline and the number of cells
counted in a Coulter counter (Coulter Electronics, Inc., Hialeah,
Florida, Model ZB1). The number of cells was calculated by multiplying the dilution factor from triplicate readings. All chemical
reagents and instruments used were sterile.

Incubation with Cardiac Glycosides

Ouabain·$8H_2O$ (Sigma Chemical Co., St. Louis, MO) was dissolved in
appropriate volume of RPMI 1640 medium under sterile conditions.
Various amounts of ouabain solution were added to the identical number
of cells so that the final ouabain concentration would be $10^{-6}M$,
$10^{-5}M$, 5 x $10^{-5}M$ or $10^{-4}M$ in triplicate tubes. Inhibitory effects of
ouabain on cell growth were checked after 48 hours of incubation as
described above. In some experiments, digoxin or digitoxin was used
instead of ouabain. To study temporal effects of ouabain on cell
growth, cell numbers in the control media or in media containing
various concentrations of ouabain were determined after an incubation
for 4, 8, 12, 24, 48, 72, 96, 120, or 144 hours at 37°C.

Determination of ^3H Ouabain Binding and Uptake

To correlate the inhibitory effect of ouabain on cell growth with
either ouabain binding to the cells or ouabain uptake into the cells,
^3H-ouabain (New England Nuclear Corp., Boston, Mass., S.A. 14Ci/mmol),
was added to the cells in addition to nonlabelled ouabain to make a
final ouabain concentration of $10^{-6}M$, $10^{-5}M$, 5 x $10^{-5}M$ or $10^{-4}M$. Upon
completing incubation, the cells were harvested by centrifugation
after determination of cell number, and an aliquot was counted for
radioactivity in 10 ml of ACS scintillation medium (Amersham Corp.,
Arlington Heights, Illinois) in a liquid scintillation spectrometer
(Parkard Instrument Co., Grove, Illinois, Model 3883). Based on the
number of cells determined by the Coulter counter, ouabain upake
was expressed as picomoles of ouabain per cell. Ouabain binding was
determined by the analysis of bound ouabain after lysis of the cell,
using 20% trichloroacetic acid and subsequent filtration. Based on
the specific activity of ^3H-ouabain used, picomoles of ouabain
associated with a cell was calculated. Effect of ouabain on the
uptake and binding of ^3H-thymidine (New England Nuclear Corp., S.A. 20
Ci/mmol) by the cells was carried out similarly.

RESULTS

Effect of Ouabain on the Rate of L1210 Cell Growth

The initial growth rate for cells incubated in control bottles was
linear up to 72 hour incubation and multiplied at least 10 times of
their initial cell counts in 48 hours. The rate of growth was reduced
significantly after 72 hrs. (fig. 1). Ouabain addition (to a final
concentration of $10^{-6}M$) did not alter significantly the rate of cell
growth. In the presence of ouabain, $10^{-5}M$, cells were able to grow,
although the rate was slightly reduced. Ouabain, 5 x $10^{-5}M$ or higher,

did not allow the cells to divide as evidenced by only small increases in cell number. The concentration of ouabain that inhibited by 50 percent the increase in control cell growth (IC_{50}) was about $1 \times 10^{-5}M$, when analyzed graphically as in fig. 2.

Since ouabain inhibits the division of L1210 cells, it was of interest to determine whether ouabain was rather freely diffusible into the cells or if the cell membrane serves as barrier to the drug. In many cell types, ouabain in the extracellular space is sufficient to demonstrate Na^+,K^+-ATPase inhibition. In unbroken, washed cells the total content of ^3H-ouabain increased linearly with increasing concentration of ouabain in the medium up to $10^{-4}M$. In trichloroacetic acid precipitated broken cells ^3H-ouabain binding increased up to $10^{-5}M$ ouabain, which exerts inhibitory effect on L1210 cell reproduction (fig. 2). As the cells were arrested at high concentration of ouabain there was no further increase of ^3H-ouabain binding to L1210 cells (data not shown).

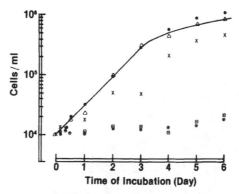

Fig. 1. Incubation time dependent growth of L1210 cell. Logarithmically growing L1210 cells were harvested by centrifugation and 10^4 cells were incubated as described under "Methods" for up to 6 days. In control incubation bottle, ouabain was not added (Δ-Δ). In other bottles, different concentrations of sterile ouabain was added so that the final drug concentrations were $10^{-6}M$ (•-•), $10^{-5}M$ (x-x), $5 \times 10^{-5}M$ (□-□) and $10^{-4}M$ (O-O), respectively. Cell growth was checked in all bottles at 4, 8, 12, 24, 48, 72, 96, 120, and 144 hours after incubation by a Coulter counter without changing the incubation medium. Number of cells was calculated on the basis of dilution factor and expressed in cells per ml of incubation medium. Each point represents mean of at least four independent experiments of triplicate readings. Standard error of the mean was less than 10% of the mean values.

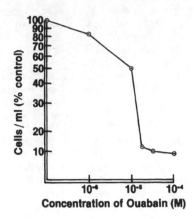

Fig. 2. Effect of ouabain on the growth rate of L1210 cells. Cells
were prepared as in Fig. 1 and incubated without (control) or
in the presence of the indicated final concentration of
ouabain. At the end of 48 hours of incubation, the number of
cells was determined as described in "Method" and expressed
as percentage of control cell number. Each point represents
mean of at least 4 independent experiments of triplicate
readings. Standard error of the mean was under 10%.

Ouabain Uptake and Binding to L1210 Cells

    After incubation for 24 hours ^{3}H-ouabain binding was higher than
its uptake, which indicates the slow diffusion process of ouabain
molecules into the cells (fig. 3). The picture of ouabain uptake and
binding was opposite after an incubation of the cells for 48 hours.
However, the total number of ouabain molecules associated with the
cells was remarkably identical in the early phase of cell growth. As
we continued incubation for 72 hours, the cells had to be in lag phase
without losing the accumulated ouabain inside the cells. Precipita-
tion of protein by 20% trichloroacetic acid caused the cells' release
the bound ouabain (fig. 3).

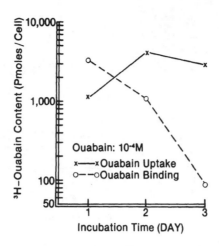

Fig. 3   Time dependent alteration of ^{3}H-ouabain binding and uptake by
L1210 cells.   Cells were prepared as in fig. 1 and incubation
with ^{3}H-ouabain was carried out as in fig. 3 ^{3}H-ouabain
uptake (x-x) and binding (0-0) were determined after an incubation for 24, 48, and 72 hours.   Each point represents mean
of 4 independent experiments with triplicate determinations.

In order to understand the mechanism of ouabain cytotoxic action,
effects of ouabain on the uptake and binding of ^{3}H-thymidine into the
cells was investigated (fig. 4). ^{3}H-thymidine uptake and its subsequent incorporation into cellular DNA were drastically reduced as much
as 95% of control uptake of ^{3}H-thymidine by 10^{-4}M ouabain in 24 hour
incubation. ^{3}H-thymidine binding to the cells was also reduced by
ouabain, 10^{-4}M, which was gradual as the duration of incubation was
prolonged.

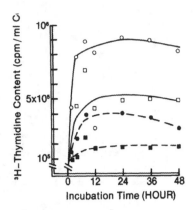

Fig. 4   Effect of 10^{-4}M ouabain on ^{3}H-thymidine uptake by L1210 cells
and on ^{3}H-thymidine binding to L1210 cell DNA.   Cells were
prepared and incubated as in fig. 1.   Sterile ^{3}H-thymidine
(10^7 cpm) was added to incubation bottles without (O and □)
or in the presence (• and ▪) of 10^{-4}M ouabain. Cell growth
was   determined   after   incubation   for   indicated   time
intervals. ^{3}H-thymidine uptake (──) and its binding to DNA
(- - -) were determined as described in "Methods."   Each
point represents mean of three independent experiments with
triplicate determinations.

DISCUSSION

In most mammalian cells ($Na^+ + K^+$)-ATPase exists in the plasma
membrane, which is responsible for maintaining the intracellular Na^+
and K^+ concentrations.   Studies indicate that the purified enzyme
consists of several subunits: two catalytic subunits and a glycoprotein (Hokin et al., 1973).   One of the two catalytic subunits has the
binding site for cardiac steroid as shown by antibodies against the
catalytic subunit (Rhee and Hokin, 1979).   Reconstitution of Na^+ and
K^+ transport system has been demonstrated in phospholipid vesicles
which were incorporated with ($Na^+ + K^+$)-ATPase purified from electric
organ of electric eel or from the rectal gland of shark (Hilden
et al., 1974).   Functional significance of ($Na^+ + K^+$)-ATPase inhibition by cardiac glycosides such as ouabain for its positive inotropic
action has been controversial (Rhee et al., 1976; Huang et al,
1979).

Therefore, it is not unreasonable to associate the cytotoxic action of ouabain with its inhibitory effect on $(Na^+ + K^+)$-ATPase activity. In this study we did not perform an assay of $(Na^+ + K^+)$-ATPase activity or monovalent cation (e.g., $^{86}Rb^+$ uptake) transport in L1210 cells. The main objective was to verify the cytotoxic action of ouabain against L1210 cells, which may explain the beneficial action of cardiac glycosides against certain types of tumors (Stenkvist et al., 1979, 1980, 1982). From this study it is quite clear that ouabain inhibited L1210 growth in drug dose dependency (Fig. 1 and 2) in the logarithmically growing cell population in vitro. The IC_{50} value for ouabain which is about 10^{-5}M is at least 50 to 100 times higher than IC_{50} value obtained from many sensitive species such as dog's in vivo test. Since rodents are notoriously insensitive to many cardiac glycosides (Schwartz et al., 1975), L1210 cells which are originated from mouse lymphocytic leukemia might expect such unusual high IC_{50} values. We also looked into the extent of binding of ouabain to the cells or ouabain taken up by the cell. As summarized in fig. 3, ouabain not only binds to the cell membrane, but also diffuses slowly into the cytoplasmic space. When the concentration of ouabain is 1 nanomole per cell or higher, cell division is slowed or stopped completely and ouabain is released from broken cells.

From the present study, it is not clear exactly how ouabain arrested the L1210 cell growth in vitro, although it does bind to and is accumulated into the cell as above. In order for this cell to grow continuously, ionic milieu of intracellular space of the cell must be maintained, which is essential not only for isotonicity for cellular volume control, but also for intracellular enzymatic activity. By virtue of inhibition of Na^+ and K^+ stoichiometric exchange by ouabain (Lee and Klaus, 1971; Schwartz et al., 1975), this drug also inhibits many Na^+ dependent transport of amino acids and sugar. Thus, we examined the effect of ouabain on the incorporation of thymidine into L1210 cell DNA. Ouabain indeed significantly inhibited (p<0.001) the uptake of ^{3}H-thymidine (fig. 4), which will subsequently interfere with the synthesis of cellular DNA. Mortality and tumor mass of mice with L1210 leukemia have been reduced by a combination treatment of cytotoxic agent with methylxanthine which increases intracellular cyclic AMP levels and results in inhibition of DNA synthesis (Mednieks et al., 1982). It can be concluded that the cytotoxic action of ouabain is at least in part due to its inhibitory effect on DNA synthesis as a result of blockade of the uptake of thymidine, although more work, particularly ouabain effects on ion transport and ATPase activity, should be done in relation to the cytotoxic action of ouabain.

REFERENCES

Hilden, S., Rhee, H. M., and Hokin, L.E., 1974, Sodium transport by phospholipid vesicles containing purified sodium and potassium ion-activated adenosine triphosphatase, J. Biol. Chem. 249:7432-7440.
Hokin, L. E., Dahl, J. L., Deupree, J. D., Dixon, J. F., Hackney, J. F., and Perdue, J.F., 1973, Studies on the characterization of the sodium-potassium transport adenosine triphosphatase. X. Purification of the enzyme from the rectal gland of squalus acanthias, J. Biol. Chem., 248:2593-2605.

Huang, W., Rhee, H. M., Chiu, T. H., and Askari, A., 1979, Re-evaluation of the relationship between the positive inotropic effect of ouabain and its inhibitory effect on (Na$^+$, K$^+$)-dependent adenosinetriphosphatase in rabbit and dog hearts, J. Pharmacol. Exp. Therap., 211:571-582.

Law, E.W., Dunn, T. B,., Boyle, P. J., and Miller, J. H., 1949, Observations of the effect of a folic-acid antagonist on transplantable lympoid leukemias in mice, J. Nat. Cancer Inst., 10:179-192.

Lee, K.S., and Klaus, S., 1971, The subcellular basis for the metabolism of inotropic action of cardiac glycosides, Pharmacol. Rev., 23:193-261.

Mednieks, M.I., Jungmann, R.A., and DeWys, W. D., 1982, Cyclic adenosine 3´:5´-monophosphate-dependent protein phosphorylation and the control of leukemia L1210 cell growth, Cancer Res., 42:2742-2747.

Moore, G.E., Sandberg, A. A., and Ulrich, K., 1966, Suspension cell culture and in vivo and in vitro chromosome constitution of mouse leukemia L1210. J. Nat. Cancer Inst., 36:405-421.

Repke, K., 1963, Metabolism of cardiac glycosides. In Proceedings of the First International Pharmacolgical Meeting, Stockholm, Pergamon Press, Oxford. Vol. 3, pp. 47-73.

Rhee, H.M., Dutta, S., and Marks, B. H., 1976, Cardiac Na,K-ATPase activity during positive inotropic and toxic actions of ouabain, Europ. J. Pharmacol., 37:141-153.

Rhee, H.M., and Hokin, L. E., 1979, Inhibition of ouabain binding to (Na$^+$ K$^+$) ATPase by antibody against the catalytic subunit but not by antibody against the glycoprotein subunit, Biochim. Biophys. Acta, 558:108-112.

Schwartz, A., Lindenmayer, G. E., and Allen, J. C., 1975, The sodium-potassium adenosine triphosphatase: pharmacological, physiological and biochemical aspects. Pharmacol. Rev., 27:3-134.

Stenkvist, B., Bengtsson, E., Eriksson, O., Holmquist, J., Nordin, B., and Westman-Naeser, S., 1979, Cardiac glycosides and breast cancer, Lancet, 1:563.

Stenkvist, B., Bengtsson, E., Eklund, G., Eriksson, O., Holnquist, J., Nordin, B., and Westman-Naeser, S., 1980, Evidence of a modifying influence of heart glycosides on the development of breast cancer, Anal. Quant. Cytol., 2:49-54.

Stenkvist, B., Bentsson, E., Dahlquist, B., Eriksson, O., Jarkrans, T., and Nordin, B., 1982, Cardiac glycosides and breast cancer, revisited, New Eng. J. Med., 306:484.

EFFECTS OF CALCIUM CHANNEL BLOCKERS ON THE RESPIRATION OF HELA CELLS AND
HELA MITOCHONDRIA AND THE GENERATION OF OXIDISING FREE RADICALS IN TISSUES
SUBJECT TO CALCIUM IMBALANCE

T.J. Piva, M. McCabe, and E. McEvoy-Bowe

Department of Chemistry and Biochemistry, James Cook
University of North Queensland, Townsville, Australia, Q4811

INTRODUCTION

The profile of oxygen concentration across a sphere of respiring tissue
is largely determined by the oxygen consumption (QO_2) of the cells in the
tissue; other factors such as diffusional constraints imposed by an extracellular matrix, are trivial in comparison with the overriding effect of the
parameter for oxygen consumption (McCabe et al., 1979).

While a growing solid tumour normally generates a capillary supply, the
extent of angiogenesis often lags behind the optimum oxygen requirements of
the growing spheroid of transformed cells. Under these circumstances there
may be a selection for clones of cells which are resistant to hypoxia. It is
perhaps this tendency which causes radiation treatment to become often a
temporary palliative rather than an effective cure for some cancers since it
is well established that anoxic tissue is increasingly resistant to radiation
damage. Attempts at identifying alternative hypoxic cell radiosensitisers,
i.e. compounds which mimic oxygen in the (in vivo) capacity to generate free
radicals when irradiated, seem to be so far unsuccessful. For this reason it
still seems important to attempt to modulate the oxygen gradient across the
tumour spheroid by selectively limiting what respiration there is and in this
way permitting the elevation of the intra tumour tissue pO_2.

There is evidence that the functioning of calcium channels is responsible
for a significant part of normal tissue respiration, additionally several
drugs known to be calcium channel blockers are already available on the market
for treatment of heart attacks. We have compared two of these drugs to
ascertain which might be the more suitable as a specific inhibitor of respiration of tumour cells. The tumour cell selected was an epithelial type (since
the great majority of spontaneous human tumours are epithelial in origin),
and the drugs investigated were diltiazem and verapamil. Their effects upon
the respiration of whole HeLa cells and upon freshly isolated HeLa cell mitochondria was compared with their effects on beef heart mitochondria.

Since it is known that solid tumours are to a greater or lesser extent
infiltrated by phagocytic cells which are presumably mounting an (albeit sometimes ineffectual) oxidative free radical attack upon the tumour, it is
important to know what consequences might be expected of the intracellular
modulation of calcium to these cell types, and which might be an inadvertant
consequence to treatment of the patient with calcium channel blockers. For

these reasons we have also investigated the role of intracellular calcium on the oxidative free radical generation consequent on the stimulated oxidative burst in phagocytic cells.

MATERIALS AND METHODS

HeLa cells were grown on cytodex I or III microcarrier beads in Media 199. When confluency was obtained (as permicroscopic examination) the beads were used either for respiratory studies or for mitochondrial preparation. HeLa mitochondria were prepared from confluent microcarrier cultures of HeLa cells. Cells were removed from the beads by trypsinisation. Mitochondria from isolated cells were obtained by using the method of Moreadith and Fiskum (1986), except that the digitonin treatment step was omitted.

Beef heart mitochondria were isolated from freshly slaughtered animals as per method 3 of the procedure used by Smith (1967).

The integrity of the mitochondrial preparations was demonstrated by measuring the extent of respiratory control using succinate as the substrate.

Oxygen uptake studies were performed using a standard polarographic oxygen electrode and cell (Rank Bros., Bottisham, Cambridge, UK) modified by the inclusion of a glass sleeve and with the addition of a stirring button constructed from perspex as an oblate spheroid which mimicked the shape of the cell compartment and which almost filled the bottom of the respirometer chamber. In this way volumes of 1 ml or less were adequate to fill the chamber and permitted a significant reduction in the number of HeLa cells needed to prepare the required mitochondrial suspension. The suspending medium was as per Moreadith & Lehninger (1984).

Whole cell and mitochondrial protein was by the method of Lowry et al., (1951).

Preparation of rat thymus and rat peritoneal cells. Rat thymus cells from 7 week female outbred Wistar rats were prepared by sieving isolated minced thymus tissue through a steel mesh. The cells were collected and washed by centrifugation into phosphate buffered saline at pH 7.2 containing 5 mM glucose. The isolated and washed cells were then resuspended in the buffer at concentrations of 2×10^7 cells/ml. Cell viability was assessed by Trypan blue exclusion.

Peritoneal cells were obtained from the peritoneal cavity after peritoneal injection of PBS containing heparin. The abdomen was gently massaged prior to the withdrawal of the fluid, which was centrifuged at 400 g for 7 min. at $4^{\circ}C$. Red cells were lysed by addition of water and the cells were then washed twice in the PBS/glucose.

Measurement of Luminol dependent chemiluminescence to detect O_2^-. essentially the method of Minkenber & Ferber (1984) was enployed, using a scintillation counter set into "out of coincidence" mode. Luminol reacts with oxidising species produced during stimulation to form an electronically excited aminophthalate ion which releases photons on returning to a ground state.

5 ml samples containing 2×10^7 cells/ml were used with the addition and preincubation of luminol (113 μM final concentration). After 5 mins a stimulant of the respiratory burst was added and the resultant chemiluminescent response was measured.

Additions to the system were made during the preincubation stage as follows: (a) Ca^{++} and/or Mg^{++} up to a final concentration of 1 mM; (b) the calcium ionophore A23187 (10 µMolar final concentration); and (c) diltiazem (5 mMolar final concentration).

RESULTS AND DISCUSSION

It was soon realised that mitochondria prepared from HeLa cells present some differences from mitochondria prepared from beef heart. The main manifestation was a diminished degree of respiratory control and an impaired ability to use several substrates for oxidative phosphorylation. Additionally HeLa mitochondria which exhibited a satisfactory respiratory control ratio when freshly prepared tended to deteriorate rather rapidly, even though stored in ice but without freezing. Results of respiration studies were restricted to mitochondrial samples which showed respiratory control ratios with succinate of better than 2. These results are shown in Table 1, from which it can be seen that both diltiazem and verapamil are effective inhibitors of succinate induced mitochondrial respiration, for both HeLa and beef heart. Additionally, it can be seen that for mitochondria from these two cell types at least, diltiazem displays some selectivity for inhibiting the respiration of tumour cell type by a factor of almost 4.

There is clear evidence that regulation by Ca^{2+} is a mechanism whereby the flux of respiratory substrates through the citric acid cycle, is controlled (McCormack & Denton, 1979). Intramitochondrial calcium levels are normally controlled by active transport mechanisms (Lehninger et al., 1978), and regulation by Ca^{2+} is presumably effected via a calcium dependent regulator such as calmodulin. The increased sensitivity of the tumour cell mitochondria to diltiazem may reflect an increased dependency on a partial citric acid cycle containing enzymes known to be regulated by calcium for example, isocitrate dehydrogenase complex (Denton et al., 1978) and/or 2-oxoglutarate dehydrogenase complex (McCormack & Denton, 1979).

The table also shows the effects of diltiazem and verapamil upon the respiration of HeLa spheroids and of isolated (intact) HeLa cells. The spheroids were always completely oxygenated since they were grown to a monolayer confluence over the surface of microspheres.

Table 1. Respiration of HeLa cells and mitochondria from heart muscle and HeLa cells and the effects of diltiazem and verapamil upon succinate stimulated respiration

Verapamil concentration (mMolar)	0	0.5	1.0	1.5	2.0	2.5	3.0	4.0	5.0
Respiration of									
HeLa cell spheroids	100	79	61	47	38	32	28	24	21
HeLa cells individual cell	100	91	83	72	63	50	48	36	30
HeLa cell Mitochondria	100	54	31	17	–				
Beef heart Mitochondria	100	37	19	12	–				

Table 2. Respiration rate of HeLa cells and mitochondria from HeLa
and beef heart, in the presence of diltiazem

Diltiazem concentration (mMolar)	0	0.5	1.0	1.5	2.0	2.5	3.0	4.0	5.0
Respiration of									
HeLa cell spheroids	100	84	69	59	52	47	43	37	32
HeLa cell mitochondria	100	57	38	28	22	18	13	–	–
Beef heart mitochondria	100	80	66	59	51	46	41	35	–

Using a whole cell preparation it was found that addition of succinate
did little to stimulate the respiration rate of the cells, however it did
stabilise the respiration which otherwise tended to slowly decline with
time. Equally the addition of glutamine and glutamate, either alone or both
together did not provoke any change in oxygen uptake by the whole cell
preparation.

While at first sight the data appear to support the suggestion that
addition of calcium channel blockers may ultimately generate oxidative free
radicals in subsequently irradiated tissues (by acting to depress respiration,
and hence enhance the extent of tissue oxygenation), it must be born in mind
that calcium itself is known to play a most significant part in the normal
generation of superoxide or related free radicals in many tissues. For example it is known that Ca^{2+} markedly stimulates the generation of hydrogen
peroxide by heart muscle mitochondria (Boveris, et al., 1972); Boveris & Chance,
1973; Cadenas & Boveris, 1980). Similarly mammalian cell microsomal preparations also seem frequently capable of generating superoxide, and this superoxide generation is markedly stimulated by the addition of Ca^{2+} (Hildebrandt,
et al., 1973).

Additionally tumour tissues would seem generally to be infiltrated with
phagocytic cells, and such cells are certainly capable of generating oxygen
free radicles in response to the presence of calcium ionophores such as Ca
inonophore A23186. Of course the action of this compound is to promote a
sharp Ca influx into the cell – in direct contrast to the action of the calcium channel blockers.

The effects of modulation of the intracellular levels of Ca^{++} in phagocytic cells is shown in Figure 1. from which it can be seen that the availability of calcium as an intracellular signal is a necessary prerequisite for
the (respiratory burst induced) generation of oxidative free radicals.

Of course the oxidative radical generation produced by irradiation
should not necessarily be affected by a lowering of intracellular calcium,
but any assessment of the value of this source of oxidative free radicals
must take into account possible reductions from other sources consequent
on the therapeutic intervention.

Additionally the flavoprotein containing enzyme systems can be potent
generators of oxidative free radicals. Some of these are to be found in the

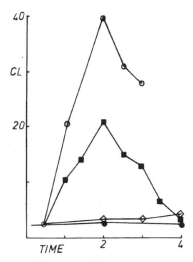

Fig. 1. Changes of the chemiluminescence of rat thymus
cells in response to modulation of intracellular
Ca^{++}.
O - Ca^{++} 1mMolar + Ca ionophore A23187
 - Ca ionophore, no extra Ca^{++}
 - Ca^{++} 1mMolar + Ca channel blocker
 - Control (No Ca^{++}; no Ca ionophor)

mitochondrial (and in some cell types, microsomal) membrane as parts of a
cytochrome P_{450} complex. Some, like the xanthine/hypoxanthine oxidase
system may be activated by low levels of oxygen in the tissue, such that
subsequent transient elevations of tissue oxygen levels will generate a
burst of oxidative free radical formation. Thus any attempt at chemo-
therapeutic intervention into the superoxide generating capacity of a
tissue, should take full account of all sources and potential sources of
oxidative free radical production within the tissue.

REFERENCES

Boveris, A. and Chance, B., 1973, Mitochondrial generation of hydrogen
        peroxide, Biochem. J., 134:707-716.
Boveris, A., Ochino, N., and Chance, B., 1972, Cellular production of
        hydrogen peroxide, Biochem. J.,, 128:617-630.
Cadenas, E. and Boveris, A., 1980, Enhancement of hydrogen peroxide for-
        mation by protophores and ionophores in antimycin supplemented
        mitochondria, Biochem. J., 188:31-37.
Denton, R.M., Richards, D.A., and Chin, J.S., 1978, Calcium ions and the
        regulation of NAD limited isocitrate dehydrogenase from the mito-
        chondria of rat heart and other tissues, Biochem. J., 176:899-906.
Hilderbrandt, A.G., Speck, M., and Roots, I., 1973, Possible control of
        hydrogen peroxide production and degradation in microsomes during
        mixed function oxidation reaction, Biochem. Biophys. Res. Comm.,
        54:968-976.
Lehninger, A.L., Reynafarje, B., Vercesi, A., and Tew, W.P., 1978, Cha-
        racteristics of energy dependent calcium influx and efflux systems
        of mitochondria, Ann. N.Y. Acad. Sci., 307:160-174.
Lowry, O.H., Rosebrough, N.J., Farr, A.L., and Randall, R.V., 1951, Protein
        measurement with the folin phenol reagent, J. Biol. Chem., 193:
        265-275.

McCabe, M., Adam, K., and Maguire, D., 1979, The effect of temperature upon the combined diffusional and kinetic parameters of tissue respiration, J. Theoret. Biol., 78:51–59.

McCormack, J.G. and Denton, R.M., 1979, The effects of calcium ions and adenine nucleotides on the activity of pig heart 2-oxoglutarate dehydrogenase complex, Biochem. J., 180:533–544.

Minkenberg, I. and Ferber, E., 1984, Lucigenin dependent chemiluminescence as a new assay for NADPH-oxidase activity in particulate fractions of human polymorphonuclear leucocytes, J. Immunol. Methods, 71:61–67.

Moreadith, R.W. and Fiskum, G.M., 1986, Isolation of mitochondria from ascites tumour cells permeabilised with digitonin, Anal. Biochem., 37:360–367.

Moreadith, R.W. and Lehninger, A.L., 1984, The pathways of glutamate and glutamine oxidation by tumour cell mitochindria, J. Biol. Chem, 259:6215–6221.

Smith, A.L., 1967, Preparation, properties and conditions for assay of mitochondria slaughterhouse material small scale, Methods in Enzymol, 10:81–86.

THE EFFECTS OF HYPO- AND HYPERTHERMIA ON THE OXYGEN PROFILE OF A TUMOUR

SPHEROID

D. Maguire, M. McCabe[*], and T. Piva[*]

School of Biological Sciences, Griffith University, St. Lucia
Australia, Q4067
[*]Department of Biochemistry, James Cook University of North
Queensland, Townsville, Australia, Q4811

INTRODUCTION

The extent of oxygenation of a growing tumour spheroid is a matter of
some importance since several different therapies are all significantly
affected by the oxygen status of the tumour cells. Additionally hyperthermia
has become an additional option as a treatment or adjunct for treatment of
tumours.

In a previous publication (McCabe et al., 1979) we have considered the
role of temperature on both the diffusional and kinetic parameters of tissue
respiration for a sheet of tissue. Here we consider the effects of temperature variations upon the oxygen profile of a respiring sphere of tissue. A
modified form of the Warburg equation (Warburg, 1923) is presented, which
allows a prediction to be made of the change in oxygen gradient profile (within a respiring tissue spheroid) as a function of temperature.

Energies of activation for the respiratory complex of both a tumour cell
type (HeLa cell) and for a normal human epithelial cell type (sheets of fresh
human epidermis) are measured. The results enable a prediction of the consequence of hypo- and hyperthermia on the oxygen status of a respiring tissue,
supplied with oxygen by diffusion from a capillary bed completely external to
the tissue, and when the tissue has a simple geometry such as a sheet or
sphere.

MATERIALS AND METHODS

HeLa spheroids were grown on Cytodex microcarrier beads (Pharmacia). Culture
was continued until confluency, in Medium 199 supplemented with 10% foetal
calf serum.

Preparation of sheets of human epidermis. Normal epidermis was raised as a
blister by gentle sustained suction over a period of several hours, essentially by the method of Mustakallio et al., (1964). The discs used contained
either 5 mm or 2 mm holes and collected 5 to 19 blisters respectively. The
blisters were raised on the inner surface of the arm using a vacuum of 150
mmHg for 2 to 3 hours. The blisters were then removed with a scalpel.

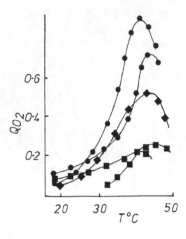
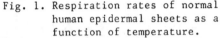
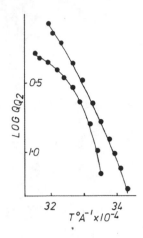

Fig. 1. Respiration rates of normal
human epidermal sheets as a
function of temperature.

Fig. 2. Arrhenius plot of respiration
rate of human epidermis.

Protein content of the samples was measured using the material salvaged from
the respirometer cell at the termination of the experiment. The total protein was estimated using Folin-Cioclteu's phenol reagent as described in
Miller's (1959) modification of the method of Lowry et al. (1951).

Respiration rates were measured with a Rank polarographic oxygen electrode
system. The temperature of the apparatus could be maintained within 0.1°C
with an outside water bath and circulation pump. In order to obtain several
measurements of respiration rate at different temperatures but within a short
space of time (in order to minimise time dependent or irreversible alteration of rate), two water baths were connected to the jacket of the respirometer by a two way tap. As a rate was being measured using one water bath
the second could be readjusted to the next temperature. In this way numerous
readings of respiration rate as a function of temperature were obtained for
each sample within a short space of time. As an additional check that the
observed changes of respiration rate were exclusively a consequence of temperature shifts, the samples were periodically returned to 30°C to ensure
that respiration at this temperature was unchanged.

RESULTS

    Respiration rates were measured for HeLa spheroids and for fresh sheets
of normal human epithelial cells (epidermis). Energies of activation were
calculated from appropriate parts of the derived Arrhenius plots and are
shown in Table 1.

Table 1. Measured values for μ (apparent energy of activation of the respiratory complex) for whole HeLa cells and human epidermal cells, as a function of the temperature range.

Temperature	μ, Human Epidermis	μ, HeLa Cells
25–30	16	3.6
30–35	14	3.6
35–40	12	3.6
40–45	8	−5.0
45–50	−12	

DISCUSSION

The temperature modified form of the Warburg equation

The usual form of the Warburg equation for the respiration of a sheet of tissue is:

$$\mu = C - a/_{2DK} (Hx - x^2)$$

where μ is the oxygen partial pressure in ats at a point x cm inside the surface of the respiring splice of total thickness H cm. C is the external oxygen partial pressure, also in ats, and a is the respiration rate defined as ml oxygen consumed (reduced to STP) per ml tissue per min. and D_K is the Kroch diffusion coefficient, expressed as ml oxygen (reduced to STP) passing through 1 cm^2 of surface per min. in response to an oxygen gradient of lat/cm tissue.

When μ = 0 at x = H/2, the tissue is just oxygenated to the centre of the slice, and under these conditions

$$H = (8CD_K/a)^{\frac{1}{2}}$$

Substitutions for D_K and a as functions which explicity include temperature can be made (McCabe et al., 1979) and the above equation then has the form:

$$H = B^* \sqrt{C} \ \exp \ (f/2RT)$$

where $B^* = 80\alpha R/\pi r_o$

and $f = \mu - (\Delta H + B)$

where α is the Bunsen adsorption coefficient of oxygen in the tissue and R is the gas constant in cals mole^{-1} deg^{-1} and r_o is a constant (the hydrated radius of the oxygen molecule). μ is a complex temperature coefficient which is a function of the energies of activation of the enzymes of the respiratory complex, and ΔH is the differential heat of solution of one mole of the gas in a saturated solution at a temp. T^oK. B is an experimentally derived constant relating the viscous behaviour of water as a function of temperature.

A comparable series of substitution into the Warburg equation as expressed to describe the respiration of a sphere of tissue rather than a sheet, gives a substituted equation of the following final form:

$$R = 0.866 \ B^* \sqrt{C} \ \exp \ (f/2RT)$$

Table 2. Variation of the calculated value of (ΔH + B) as a
function of temperature

Temperature $^{\circ}C$	ΔH cal mol^{-1}	B cal mol^{-1}	(ΔH + B) cal mol^{-1}
0.0	−4000	5100	1100
5.0	−3800	4800	1000
10.0	−3600	4500	900
15.0	3400	4300	900
20.0	−3200	4200	1000
25.0	−2900	4100	1200
30.0	−2600	4000	1400
35.0	−2400	3900	1500
40.0	−2200	3700	1500
45.0	−2100	3500	1400
50.0	−2000	3200	1200

Both of the above functions 1 and 2 are dominated by the values of the
exponential part of the equation, and thus whether H or R (the thickness of
the oxygenated rim of a sheet or sphere of tissue) either increases or
decreases with changing temperature, depends upon the sign of the function f.
Now it is possible to assign values to this function since it is dependent
upon the energy of activation of the respiratory complex (μ), and the changing values of ΔH and B with temperature. Table 2 shows the calculated values
for (ΔH + B) obtained from the data of Battino & Cleaver (1966) and from the
International Critical Tables (1929).

Table 3.

Temperature $^{\circ}C$	Human Epidermis		HeLa cells	
25–30	f/2RT	12.1	f/2RT	1.92
	H	1033	H	1.39
	R	888	R	1.19
30–35	f/2RT	10.1	f/2RT	1.76
	H	141	H	1.18
	R	121	R	1.0
35–40	f/2RT	8.4	f/2RT	1.59
	H	24	H	1.0
	R	20.6	R	0.86
40–45	f/2RT	5.2	f/2RT	−2.79
	H	1.0	H	3.26
	R	0.86	R	2.80
45–50	f/2RT	−8.4		
	H	24		
	R	20.6		

Table 3 shows the values for the function f/2RT for different temperature ranges, and for the appropriate energies of activation of the respiratory complexes. H is the corresponding thickness of the oxygenated rim of tissue from a parallel (slice) of tissue and R is the appropriate thickness of the oxygenated rim of a sphere of tissue.

It can be seen from Table 3, that hyperthermia tends to diminish the thickness of the oxygenated rim of tissue both for the cancer cell type and for the epidermal layer. Epidermis does in fact normally suffer large swings of temperature during its normal functioning, and so one might expect some selection for a respiratory complex which can help the tissue to accommodate to these large temperature fluctuations without serious dislocation of the extent of oxygenation of the tissue. In general a respiratory complex which is indifferent to temperature shifts would seem to be desirable from this point of view. The above measurements do however, show rather large energies of activation suggesting that the opening or closing of the subdermal capillary bed and the consequent warming or cooling of the blood is capable of compensating for the temperature induced fluctuations of QO_2.

While cancer cells have not of necessity been selected to accommodate to temperature shifts, they are probably selected to survive in a hypoxic environment. The measured low energy of activation for the respiratory complex of the tumour cell type suggests that if diffusion from outside the body of the tumour mass is the only means of supply, and if the adjacent capillary bed is not subject to any temperature variation, then an increase in the temperature of the tumour will result in a rather minimal increase in deoxygenation. Of course no matter how carefully the heat is localised for delivery to the tumour, there must inevitably be some warming of adjacent tissue, if only as a consequence of the thermal conductance of the tissue. Equally adjacent capillaries will inevitably be warmed to a greater or lesser extent. Warming the blood will of course promote a large increase in the oxygen gradient which may well compensate for the shift towards deoxygenation by the tumour particularly since the tumour appears to have a low activation energy for the respiratory complex. As a general rule it might be that the more broadly beamed the thermal radiance, the greater will be the tendency for a high oxygen status within the tumour, while the more narrowly beamed, the less will be the consequent oxygenation of the tumour core.

REFERENCES

Battino, R. and Clever, H.L., 1966, Solubility of gases in liquids, Chem. Rev., 66:395-463.
International Critical Tables of Numerical Data, Physics, Chemistry and Technology, 1929, Compiled by C.J. West, The Maple Press Co., York, Pa, USA.
Lowry, O.H., Rosebrough, N.J., Farr, A.L., and Randau, R.J., 1951, Protein measurements with the folin phanol reagent, J. Biol. Chem., 193: 265-275.
McCabe, M., Adam, K., and Maguire, D., 1979, The effect of temperature upon the combined diffusional and kinetic parameters of tissue respiration, J. Theor. Biol., 78:51-59.
Miller, G.L., 1959, Protein determination for large numbers of samples, Analyt. Chem., 31:p.964.
Mustakallio, K.K., Kustala, U., Nieminen, E., and Leikola, E., 1964, Lipids of the human epidermis: separation of the epidermis in vivo by a section technique, Arch. Biochem. Cosmeto., 7:11-16.
Warburg, O., 1923, Experiments on surviving carcinous tissue, Methods Biochem. Z., 142:317-333.

A COMPUTER SIMULATION OF SIMULTANEOUS HEAT AND OXYGEN TRANSPORT

DURING HETEROGENEOUS THREE DIMENSIONAL TUMOR HYPERTHERMIA

Kyung A. Kang, Duane F. Bruley, and Haim Bicher

Chemical Engr. Dept., U.C. Davis, Davis, CA 95616
Cal Poly State Univ., San Luis Obispo, CA 93401
and Valley Cancer Institute, Panorama City, CA 91409

INTRODUCTION

The ultimate goal of mathematical modelling and simulation of hyperthermia is the real time prediction of heat application and oxygen supply during therapy to maximize the damage to cancer cells with minimum damage to normal cells. Many mathematical models used for biomedical research are based on the assumption of lumped parameters for normal tissue, tumor, and capillaries (e.g. convection in the commonly used bioheat equation - Strohbehn, 1984). Using hyperthermia alone or with other modelities, such as radiation therapy with hypertensive oxygen (Boerema et al., 1964; Wada and Iwa, 1970) or chemotherapy, it is necessary to use different physical or chemical properties between tumor, normal tissue, and vasculature of the system. Therefore, in order for the mathematical model to be useful for hyperthermia, it is essential to consider the heterogeneities of the biological system.

Mathematical models of mass (oxygen) transfer and heat transfer for this simulated system include diffusion (conduction), convection, and reaction (oxygen consumption) with different parameter for tumor, normal tissue, and capillaries.

SYSTEM SIMULATED AND PARAMETERS USED

The geometry simulated is a cube of tissue with different numbers of capillaries parallel to the z axis with tumor cells at the center of this tissue (Fig. 1). This geometry has been chosen to demonstrate the importance of the heterogeneities.

(1) Volume of System Simulated: 1120 x 1120 x 1120 μm cubic tissue.

(2) Volume of Tumor: 400 x 400 x 400 μm cube located center of the
tissue.

(3) Volume of a Capillary: 80 x 80 x 1120 μm

Tumor cells have very poor vasculature compared to normal cells. Therefore, for this simulated system, capillaries are located only in the normal tissue.

Simulation (a): Six capillaries (Capillaries 7, 8, 9, 10, 11, and 12) are located right and left hand side of tumor.

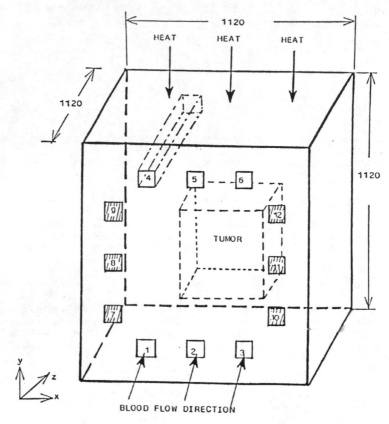

Fig 1. System Simulated. Tissue Cube with Different Capillary
Configurations and Tumor at the Center of the tissue.
(Unit used: μm).

Simulation (b): Six capillaries (Capillaries 1, 2, 3, 4, 5, and 6) are
located above and below of the tumor.

Simulation (c): Twelve capillaries (Capillaries 1, 2, 3, 4, 5, 6, 7, 8,
9, 10, 11, and 12) are around the tumor.

(4) Blood Velocity: Experimental data suggests that blood flow markedly
increases as large as 20 folds during hyperthermia (Song et al.,
1984). For the simulation, the initial blood velocity is at 1000
μm/sec increasing up to ten times the normal value after one hour of
heat treatment. This relation may be expressed as

$$V_b = 1000 \ (1.0 + 10 \ t/3600) \quad \mu m/sec \tag{1}$$

where V_b is the capillary blood velocity and t is the time in sec.

(5) Simulation Time: 3 sec.

(6) Density of Blood, Tissue and Tumor: 1 g/ml.

(7) Specific heat of Tissue and Blood: 3.65×10^3 J/Kg-°C (Busch, 1982).

(8) Specific Heat of Tumor: 3.15×10^3 J/Kg-°C (Busch, 1982).

(9) Oxygen Diffusion Coefficient in Tissue: 1120 μm^2/sec (Reneau, 1967).

(10) Oxygen Diffusion Coefficient in Tumor: Experimental values are between 15-20 μm^2/sec (Grote et al., 1977; Boag, 1970). For the simulation, the value, 20 μm^2/sec is used.

(11) Oxygen Consumption Rate of Normal Tissue: 0.000834 cm^3 O_2/cm^3 tissue/sec.

(12) Oxygen Consumption Rate of Tumor: Oxygen consumption rate of tumor is known to be much lower than that of normal tissue and zeroth order (Evans and Naylor, 1962). In the simulation, this value is chosen to be twenty times less than normal tissue, 0.0000417 cm^3 O_2 /cm^3 tissue/sec.

(13) Solubility of Oxygen in Tissue: 0.000029 cm^3 O_2/cm^3 tissue mmHg.

(14) Solubility of Oxygen in Blood: 0.0000342 cm^3 O_2/cm^3 tissue mmHg.

ILLUSTRATIVE EQUATIONS AND INITIAL AND BOUNDARY CONDITIONS

When conventional numerical solution techniques are employed, the following set of deterministic partial differential equations would characterize the system. The Williford-Bruley method does not solve these equations directly but relies on the definition of the Green's Function with probability distribution functions at each nodal point in the solution domain (Williford, 1974; Kang, 1984).

A. Oxygen Transfer

(1) Equations

In the Tissue and Tumor, oxygen is transfered by diffusion and reaction only.

$$\partial^2 C/\partial x^2 + \partial^2 C/\partial y^2 + \partial^2 C/\partial z^2 - Rx = (1/D)(\partial C/\partial t) \tag{2}$$

where C is partial pressure, Rx is the oxygen consumption, and D is oxygen diffusion coefficient of tissue or tumor.

In the capillaries, oxygen is dissociated nonlinearly from the red blood cell. For the simulation, Hill's Equation is used.

$$\lambda = (K\ C^n) / (1.0 + k\ C^n) \tag{3}$$

where,

λ= fractional saturation of oxygen chemically combined with hemoglobin

k, n= constants. For the human blood n is approximately 2.2 and k is 0.001.

Then, the oxygen transfer equation becomes

$$\partial^2 C/\partial x^2 + \partial^2 C/\partial y^2 + \partial^2 C/\partial z^2 - Vb\ \partial C/\partial z = (OXDIS/Db)(\partial C/\partial t) \tag{4}$$

where Vb is the blood velocity in the capillary, Db is the diffusion coefficient in blood, and OXDIS is as follows,

$$OXDIS = 1.0 + (N\ n\ k\ C^n) / \{Sb\ (1.0 + k\ C^n)^2\} \tag{5}$$

where Sb is the oxygen solubility in blood and N is the oxygen capacity in Blood (Kang, 1984).

(2) Initial Condition

$$Co(x,y,z,t=0.0) = 0.0 \quad mm\ Hg \tag{6}$$

$$Cf(z=L,\ t=0.0) = 100.0 \quad mm\ Hg \tag{7}$$

$$Cb = Ci - (Ci-Cf)(z/L) \tag{8}$$

where Cb is the oxygen partial pressure in blood, Ci is the inlet blood partial pressure, Cf is the initial outlet blood partial pressure, z is the position at the point of interest, and L is the total system length in z direction, 1120 μm.

(3) Boundary Conditions

$$Cb(z=0) = 150.0 \quad mm\ Hg \tag{9}$$

At the boundaries,

$$\partial C/\partial x = \partial C/\partial y = \partial C/\partial z = 0.0. \tag{10}$$

B. Heat Transfer

(1) Equation

$$\partial^2 T/\partial x^2 + \partial^2 T/\partial y^2 + \partial^2 T/\partial z^2 - Vb\ \partial T/\partial z = (1/a)\ (\partial T/\partial t) \tag{11}$$

where T is temperature of Tissue, Blood, or Tumor, Vb is blood Velocity (in tissue or tumor, this value is 0.0) and a is thermal diffusivity.

(2) Initial Condition

$$To(x,y,z,t=0.0) = 36.0 \quad {}^\circ C \tag{12}$$

(3) Boundary Conditions

$$T(y=L) = 45.0\ {}^\circ C, \tag{13}$$
where L is 1120 μm.

At the boundaries,

$$\partial T/\partial x = \partial T/\partial y = \partial T/\partial z = 0.0. \tag{14}$$

TECHNIQUE AND COMPUTER USED FOR THE SIMULATION

The technique used for obtaining the solution is a deterministic probabilistic numerical technique, the Williford-Bruley (W-B) method. This technique calculates the solution by using transient density function on the three dimensional space divided by grids, which saves computation time as compared with actual random walks using the Monte Carlo technique directly (Williford, 1972 and Kang, 1984). This method also has advantages in the respect that the computation is rather simple such that three dimensional time dependent conduction, convection, and reaction problem can be solved by it readily.

The computer used for this simulation was the VAX 11/785 at University of California at Davis, Davis, California.

RESULTS AND DISCUSSION

Oxygen and temperature change for three cases with different capillary arrangement were simulated. Profiles for oxygen and temperature on the plane z=560 μm, which is half of the total length in z direction, at the time, t=3 seconds are presented in figures 2 to 7.

A. Oxygen Transfer

For oxygen transfer, at time t=0.0, the oxygen partial pressure of the entire system is assumed to be 0.0 except in the capillary region. The initial oxygen partial pressure of capillaries are 150.0 mm Hg at the inlet of the capillaries and 100.0 mm Hg at the outlet and between the inlet and outlet, is assumed to be linear. These assumptions are because the steady state oxygen concentration profiles are not known, however, it is desired to know the effect of the different capillary arrangements on final oxygen distribution

Values of the oxygen profile on figures are nondimensionalized by dividing the oxygen partial pressure, C, by Cin=35 mm Hg, which is known to be the average oxygen partial pressure in normal tissue.

$$Cn = C/Cin, \qquad (15)$$

where Cn is nondimensionalized oxygen partial pressure

For example, Cn=1.0 in figures has 35.0 mm Hg of oxygen partial pressure value.

Fig. 2. shows the isobaric lines when oxygen is supplied by six capillaries both side of the tumor. Notice that the oxygen concentration in most of the tumor is less than 0.1, especially, the upper and lower part of the tumor remain with its oxygen concentration 0.0, and the upper and lower parts of the normal tissue oxygen concentration is very low, also.

Fig. 4. shows the oxygen concentration profiles when three capillaries above and three capillaries below supply oxygen the tumor. This gives the same results as the previous case except x axis and y axis are changed to y and x axis.

Fig. 6. shows the profiles when twelve capillaries are surround the tumor. As can be seen in the figures, the oxygen concentration in the tumor is much higher than two previous cases and the oxygen is supplied well and almost the entire region has higher concentration than 0.1

From these three oxygen concenntration profiles, it is demonstrate that the system with fewer and localized capillaries needs more time or higher pressure oxygen supply than the system with dense capillaries of well distributed arrangement in order to obtain same effect.

B. Heat Transfer

Figs 3, 5, and 7 show the isothermal lines in the system during the hyperthermia. When the time, t = 0.0, the temperature of the entire system is To=36.0 °C except heated plane, the plane y=1120.0 μm. The temperature of heated plane is Tfin = 45.0 °C. The temperature values are also nondimensionalized as follows.

$$Tn = (T-To)/(Tfin-To), \qquad (16)$$

where Tn is normalized temperature and T is temperature of interest.

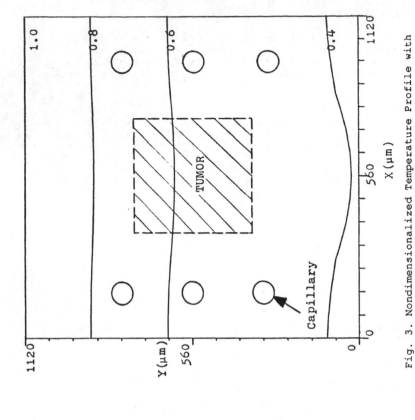

Fig. 2. Nondimensionalized Oxygen Partial pressure
with Three Capillaries each side of the Tumor
during Hyperthermia. Nondimensionalized
Partial Pressure, $Cn=C/Cin$, where C is Tissue
Oxygen Partial Pressure and Cin is 35 mmHg.

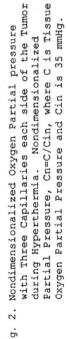

Fig. 3. Nondimensionalized Temperature Profile with
Three Capillaries each side of the Tumor
during Hyperthermia. Nondimensionalized
Temperature, $Tn=(T-Tin)/(Tfin-Tin)$, where T
is Tissue Temperature, Tin is 36, and Tfin
is 45 °C.

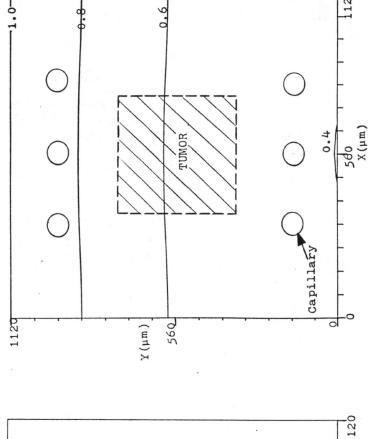

Fig. 4. Nondimensionalized Oxygen Partial pressure with Three Capillaries Above and Three Below the Tumor during Hyperthermia. Nondimensionalized Partial Pressure, $C_n = C/C_{in}$, where C is Tissue Oxygen Partial Pressure and C_{in} is 35 mm Hg.

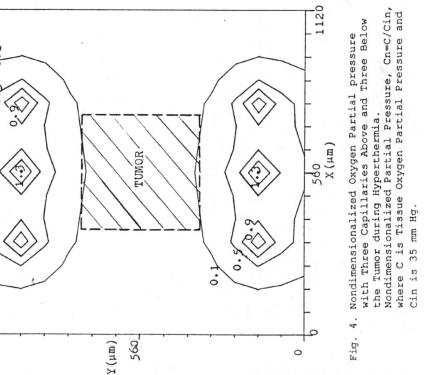

Fig. 5. Nondimensionalized Temperature Profile with Three Capillaries Above and Three Below the Tumor during Hyperthermia. Nondimensionalized Temperature, $T_n = (T-T_{in})/(T_{fin}-T_{in})$, where T is Tissue Temperature, T_{in} is 36, and T_{fin} is 45 °C.

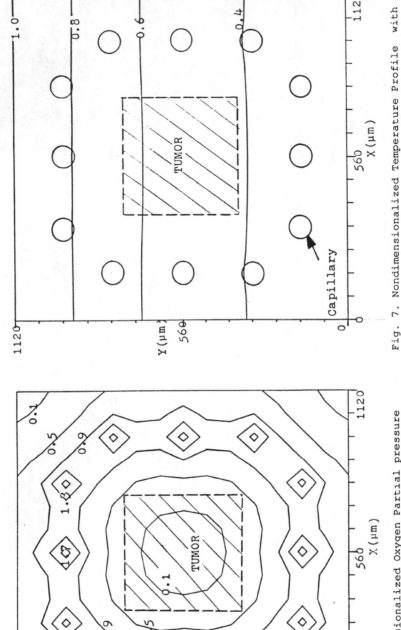

Fig. 6. Nondimensionalized Oxygen Partial pressure with Twelve Capillaries around the Tumor during Hyperthermia. Nondimensionalized Partial Pressure, $Cn=C/Cin$, where C is Tissue Oxygen Partial Pressure and Cin is 35 mmHg.

Fig. 7. Nondimensionalized Temperature Profile with Twelve Capillaries around the Tumor during Hyperthermia. Nondimensionalized Temperature, $Tn=(T-Tin)/(Tfin-Tin)$, where T is Tissue Temperature, Tin is 36, and Tfin is 45 ^{o}C.

For example, if Tn is 1.0, the temperature is 45 °C and if Tn is
0.0, then the temperature is 36.0 °C.

When there is six capillaries both sides of the tumor the isothermal
lines become slightly convex because of the heat dissipation by the
capillaries at both sides and the lowest isothermal line in the system is
Tn = 0.4 between y =40 and 100 μm (Fig. 3). However, with same number
of capillaries, when the capillaries are arranged above and below the
tumor (Fig. 5), entire system temperature is higher than the capillary
arrangement both sides of the tumor.

For the case with twelve capillaries around the tumor (Fig. 7), the
temperature if the system is much lower than two previous case.
Temperature, Tn, at y position below 320 μm is lower than 0.4, which is
much lower than two previous cases.

Therefore, when there are dense capillary arrangement in the system
to be treated, the time for the therapy should be set longer than the
system with fewer blood vessels

CONCLUSION

It can be concluded from the computer simulation that,

(1) It is important to use heterogeneous models for the simulation of
    oxygen and heat transfer in tumor hyperthermia treatment.

(2) Williford-Bruley computational strategy is an efficient and effective
    method for solving three dimensional, time dependent biological
    transfer problems.

(3) A meaningful extension of this work would be the
    comparisonexperimental results with theoretical prediction.

ABSTRACT

Hyperthermia is a developing modelity for the treatment of cancer.
This therapy is occasionally used by itself, however, usually it is used
as an adjuvate with chemo or radiation therapy. The mechanism for this
treatment is based on the fact that cancer cells are heated
preferentially by heat application due to lower vascularity in the tumor
tissue as compared with the surrounding normal tissue and that, when used
with radiation therapy or chemo therapy, higher oxygen partial pressure
in the tumor results in increased tumor cell damage.

Appropriate mathematical models and their real time prediction of
oxygen and temperature profiles could be very helpful in achieving
optimal results via hyperthermia and to avoid possible danger which might
occur during the treatment. Because of the complexity and the
heterogeneous nature of physiological system, it is necessary to include
heterogeneous properties in the mathematical models for them to be useful
for biomedical calculations. Of course, it is much more difficult to
solve mathematically the heterogeneous system than the homogeneous one.

In this paper, the importance of the implementation of
heterogeneities in the heat and mass transport for biological system
mathematical modelling is discussed. Results of a three dimensional
computer simulation of mass and heat transfer in tumor tissue with
different capillary geometries during hyperthermia are demonstrated. The
method used for the computer simulation is a deterministic/ probabilistic
technique, Williford-Bruley calculational strategy.

REFFERENCES

Boag, J., 1970, Correspondence, Cell Respiration as a Function of Oxygen
        Tension", Int. J. Radiat. Biol., 18, 475-478.
Boerema, I., Brummelkamp, W., and Meijne, G., 1964, "Clinical Application
        of Hyperbaric Oxygen", Elsevier Publishing Co, New York.
Busch, N., Bruley, D., and Bicher, H.,1982, Computer Modeling of Tumor
        Hypertermia (A Dynamic Lumped Parameter Model), in:
        "Hyperthermia", Adv. Exp. Med. Biol.", 157, 185-189, Plenum Press,
        New York.
Busch, N., Bruley, D., and Bicher, H., 1982, Identification of Variable
        Regions in 'vivo' Spherical Tumors: A Mathematical Investigation",
        in: "Hyperthermia", Adv. Exp. Med. Biol., 157, 1-7, Plenum Press,
        New York.
Dickinson, R., 1984, An Ultrasound for Local Hyperthermia Using Scanned
        Transducer IEEE Trans.Biomed. Engr., BME-31, 120-125.
Evans, N. and Naylor, P, 1977, The Effect of Oxygen Breathing and
        Radiotherapy upon the Grote, J., Susskind, R., and Vaupel, P.,
        Oxygen Diffusivity in Tumor Tissue (DS-Carcinoma) under
        Temperature Conditions within the Range of 20-40 º C, Pflug.
        Arch., 372, 37-42
Hahn, G., 1984 Hyperthermia for the Engineer: A Short Biological
        Premier", IEEE Trans. Biomed Engr., BME-31, 3-8.
Kang, K., 1984, An Analysis of the Williford Bruley Technique, A
        Simulation of Oxygen Transport in Brain Tissue, MS Thesis,
        Biomedical Engineering Department, Louisiana Tech Univ.
Ozisic, N., 1980, "Heat Conduction", John Wiley and Sons, New York.
Reneau, Jr., D., Bruley, D., and Knisely, M, 1967, A Mathematical
        Simulation of Oxygen Release, Diffusion and Consumption in the
        Capillaries and Tissue of the human Brain, in: "Chemical
        Engineering in Medicine and Biology", Plenum Press, New York, 135-
        241.
Song, C., Lokshina, A., Rhee,J., Patten,M., and Levitt, S., 1984,
        Implication of Blood Flow in Hyperthermic Treatment of Tumors,
        IEEE Trans. Biomed. Engr., BME-31, 9-16.
Strohbehn, J. and Roemer R., 1984, A Survey of Computer Simulations of
        Hyperthermia Treatments", IEEE Trans. Biomed. Engr., BME-31, 136-
        149.
Tannock, B, 1972, Oxygen Diffusion and the Distribution of Cellular
        Radiosensitivity in Tumours", Brit. J. Radiol., 45, 515-524.
Wada, J. and Iwa, T., 1970, "Hyperbaric Medicine", Igadu Shoin, Tokyo.
Williford, Jr., C., Bruley, D., and Artigue, R., 1974, Probabilistic
        Modelling of Oxygen Transport in Brain Tissue, NeuroResearch 2,
        153-170.

*The page numbers given are the first pages of the papers in which these subjects are covered.